CALCULUS

For Business, Economics, and the Social and Life Sciences

BRIEF NINTH EDITION

CALCULUS

For Business, Economics, and the Social and Life Sciences

LAURENCE D. HOFFMANN
Salomon Smith Barney

GERALD L. BRADLEY
Claremont McKenna College

Boston   Burr Ridge, IL   Dubuque, IA   Madison, WI   New York   San Francisco   St. Louis
Bangkok   Bogotá   Caracas   Kuala Lumpur   Lisbon   London   Madrid   Mexico City
Milan   Montreal   New Delhi   Santiago   Seoul   Singapore   Sydney   Taipei   Toronto

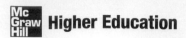
CALCULUS FOR BUSINESS, ECONOMICS, AND THE SOCIAL AND LIFE SCIENCES, BRIEF
NINTH EDITION

Published by McGraw-Hill, a business unit of The McGraw-Hill Companies, Inc., 1221 Avenue of the Americas, New York, NY 10020. Copyright © 2007 by The McGraw-Hill Companies, Inc. All rights reserved. No part of this publication may be reproduced or distributed in any form or by any means, or stored in a database or retrieval system, without the prior written consent of The McGraw-Hill Companies, Inc., including, but not limited to, in any network or other electronic storage or transmission, or broadcast for distance learning.

Some ancillaries, including electronic and print components, may not be available to customers outside the United States.

This book is printed on acid-free paper.

3 4 5 6 7 8 9 0 VNH/VNH 0 9 8 7

ISBN-13   978–0–07–305191–8
ISBN-10   0–07–305191–8

Publisher: *Elizabeth J. Haefele*
Senior Sponsoring Editor: *Elizabeth Covello*
Director of Development: *David Dietz*
Developmental Editor: *Dan Seibert*
Senior Marketing Manager: *Nancy Anselment Bradshaw*
Senior Project Manager: *Vicki Krug*
Senior Production Supervisor: *Kara Kudronowicz*
Senior Media Project Manager: *Sandra M. Schnee*
Lead Media Producer: *Jeff Huettman*
Designer: *Rick D. Noel*
Cover Designer: *Maureen McCutcheon*
(USE) Cover Image: *©Getty Images,* ae8374.001, Gas pump
Senior Photo Research Coordinator: *Lori Hancock*
Supplement Producer: *Melissa M. Leick*
Compositor: *The GTS Companies*
Typeface: *10/12 Times*
Printer: *Von Hoffmann Corporation*

Photo Credits: Opener 1, p. 93: © Ed Eckstein/Corbis; Opener 2, p. 185(top): © Corbis Royalty Free; p. 185(left): © Nigel Cattlin/Photo Researchers, Inc.; p. 185(right): © Runk/Schoenberger/Grant Heilman; Opener 3, p. 282: © Vol. 23/PhotoDisc/Getty Images; Opener 4, p. 357(top): © The McGraw-Hill Companies, Inc./Jill Braaten, photographer; p. 357(bottom): © J.A. Jackman/Texas A&M University; Opener 5, p. 460(top): © Richard Klune/ Corbis; p. 460(bottom): © Gage/Custom Medical Stock Photos; 5.50: (c) Vol. 4/PhotoDisc/Getty Images; Opener 6, p. 505(top): © AFP/Getty Images; p. 505(bottom): © ImagingBody; Opener 7(left), p. 604(top): Maps a la carte, Inc.; Opener 7(right): US Geological Survey; Appendix: © Eric Meola/Getty Images.

Library of Congress Cataloging-in-Publication Data

Hoffmann, Laurence D., 1943–
    Calculus for business, economics, and the social and life sciences. — 9th ed. / Laurence D. Hoffmann, Gerald L. Bradley, with Kenneth H. Rosen.
        p.   cm.
    Includes index.
    ISBN 978–0-07–305191–8 — ISBN 0–07–305191–8 (acid-free paper)
    1. Calculus—Textbooks. I. Title: Calculus. II. Bradley, Gerald L., 1940–. III. Rosen, Kenneth H. IV. Title.

QA303.2.H64   2007
515—dc22
                                                  2005025832
                                                  CIP

www.mhhe.com

In memory of our parents
Doris and Banesh Hoffmann
and
Mildred and Gordon Bradley

CONTENTS

CHAPTER   4   Exponential and Logarithmic Functions

CHAPTER   5   Integration

CHAPTER   6   Additional Topics in Integration

PREFACE

Calculus for Business, Economics, and the Social and Life Sciences, Brief Edition, is designed for a one- or two-semester course in calculus taken by students who have had the equivalent of two years of high school algebra. Our applied introduction to calculus in real-world contexts provides a sound, intuitive understanding of the basic concepts students need as they pursue careers in business; economics; finance and investment; the life, health, and environmental sciences; the physical sciences, and the social sciences. We strive to teach techniques of differential and integral calculus without sacrificing mathematical accuracy. We carefully and completely state the main definitions and theorems, motivating and explaining them intuitively or geometrically when possible. We endeavor to make every concept and idea easy to understand. And we illuminate the many practical and important applications of calculus that students will find in their careers and continued studies.

The Eighth Edition of *Calculus for Business, Economics, and the Social and Life Sciences* was extremely successful. In the past five years more than 100,000 students worldwide have studied from this text. When asked, "What makes our text more teachable for your course?" users point to our applied orientation to concepts, our problem-solving approach, our straightforward and concise writing style, and our strong and comprehensive exercise sets. Our goal has always been to assist instructors by organizing material so that each section in the text corresponds to a single lecture and each chapter contains the right amount of material for an examination. Students need to know how calculus applies to their particular area of interest. To aid instructors in matching applications to a particular audience, we have organized applications sections according to their usefulness in business and economics or in the social and life sciences. Our applied exercises are labeled for easy identification, and each of the over 400 individual applications in the text are conveniently indexed behind the front and back covers.

Although the Eighth Edition has been an extremely successful tool for learning, many reviewers and long-time users provided suggestions for improvement to help us build on our strengths. These individuals gave us crucial guidance in preparing this Ninth Edition. They helped us focus on the most important topics and applications that make this book well suited for students in business, economics, and the life sciences. They also highlighted the need for additional and improved exercises and applications, streamlined coverage and clearer explanations of certain topics, and updated data to keep our real-world applications fresh and relevant. And finally they gave us valuable feedback about the role of technology in applied calculus to help us build a new and substantially expanded MathZone website accompanying this text. With great effort and attention to detail we have implemented their suggestions, resulting in a more effective Ninth Edition for both students and instructors.

**Improvements to
This Edition**

There have been a number of significant and important improvements within this Ninth Edition, many of which have been made in response to user and reviewer recommendations.

Enhanced Topic Coverage

Every section in the text underwent careful analysis and extensive review to ensure the most beneficial and clear presentation. Additional steps and definition boxes were added when necessary for greater clarity and precision, and discussions and introductions were rewritten as needed to improve presentation. A more detailed list of chapter-by-chapter changes can be found later in this preface.

Improved Exercise Sets

Two hundred new routine and applied problems have been added to the already extensive exercise sets. Based on feedback from users, routine problems have been added where needed to ensure that students have more than enough practice to master basic skills. Furthermore, a wealth of new applied problems has been added to help demonstrate the practicality of the material. These new problems come from many fields of study, including economics, business, biology, medicine, and other areas of life science. Great care has been taken to update applied examples and exercises whenever possible.

Just-In-Time Reviews

Just-In-Time Reviews are used to quickly remind students of important concepts from college algebra or precalculus as they are being used in examples and discussions. Each such review is placed in a box adjacent to the location in the text where the reviewed topic material is used, and these quick reminders alert students to key facts without distracting from the material under discussion. This feature proved hugely popular after its debut in the Eighth Edition, so we have added more Just-In-Time Reviews at key points in the Ninth Edition.

Mathematical Modeling

The role of mathematical modeling appears as a recurrent theme throughout the text in applications involving both managerial and life science. The modeling introduction in Chapter 1 has been rewritten and expanded and is reinforced by many new applied examples and exercises throughout the text. New discussions on modeling have been added in the Ninth Edition to coverage of integration in Chapter 5 and differential equations in Chapter 6.

Technology

The Explore! boxes in the Ninth Edition have been updated to improve compatibility with current calculators such as the TI-84 Plus and have been rewritten for clarity. The chapter-ending Explore! Update sections have also been improved to provide clearer hints and solutions to selected Explore! boxes.

Graphing Calculator Introduction

The introduction to graphing calculators appearing on pages xix through xxviii has been updated for the Ninth Edition for compatibility with current calculators such as the TI-84 Plus. The material includes instructions regarding common calculator keystrokes, calculator terminology, and introductions to more advanced calculator applications that are developed in more detail at appropriate locations in the text. The Calculator Introduction can serve as a primer for students unfamiliar with the use of graphing calculators or as a guide to improved usage for others with some calculator experience.

Chapter-by-Chapter Changes

Users of the previous edition of this book will find the following detailed list of changes useful. This list highlights the many improvements in this edition made possible by reviewer and user feedback.

Chapter 1: Functions, Graphs, and Limits

- Many new routine problems and new applied problems covering applications for air pollution, architecture, cost management, earnings, production, real estate, and more have been added.
- The mathematical modeling introduction in Section 1.4 has been rewritten and expanded with a step-by-step guide.
- The Market Equilibrium subsection in Section 1.4 has been rewritten for clarity.

Chapter 2: Differentiation: Basic Concepts

- Many new routine problems and new applied problems covering applications for cardiology, consumer expenditures, cost of production, depreciation, manufacturing output, quality of life, and more have been added.
- The introduction to the derivative in Section 2.1 has been rewritten and expanded for clarity and better detail.
- An example previewing horizontal tangent lines has been added to Section 2.3.

Chapter 3: Additional Applications of the Derivative

- Many new routine problems and new applied problems covering applications for account management, advertising, average cost, botany, government spending, housing starts, inventory, mortgage refinancing, population distribution, spread of a disease, voting patterns, and more have been added.
- The Inflection Points subsection in Section 3.2 has been rewritten for clarity, with a new definition box and rewritten procedure box.

Chapter 4: Exponential and Logarithmic Functions

- Many new routine problems and new applied problems covering applications for concentration of a drug and tripling time have been added.
- The Properties of Logarithms subsection in Section 4.2 has been rewritten for clarity and better detail, and a new table has been added showing corresponding properties of exponential and logarithmic functions.

Chapter 5: Integration

- Many new routine problems and new applied problems covering applications for average production, conservation, horticulture, lottery payout, present value of an investment, temperature, and more have been added.
- Coverage on the average value of a function in Section 5.4 has been considerably enhanced, with rewritten examples, expanded discussion, and new definition boxes.
- Section 5.6 has been rewritten and extensively expanded. New subsections on Population Density and The Volume of a Solid of Revolution have been added along with new examples, figures, and definition boxes, and the discussion of survival and renewal has been expanded and generalized.

Chapter 6: Additional Topics in Integration

- Many new routine problems and new applied problems covering applications for consumer's surplus, future value of an investment, nuclear waste, population density, worker efficiency, and more have been added.
- The differential equations introduction in Section 6.2 has been rewritten and expanded with a discussion on modeling. A new subsection on Logistic Growth and several new examples and figures have also been added to enhance this section.
- A new business example on interpreting data with numerical integration has been added to Section 6.4.

Chapter 7: Calculus of Several Variables

- Many new routine problems and new applied problems covering applications for allocation of labor, amortization of debt, construction, consumer demand, constant returns to scale, landscaping, packaging, retail sales, and more have been added.
- Section 7.2 features a new subsection on the The Chain Rule for Derivatives containing new discussion, new definition and procedure boxes, new computational and applied examples, and a corresponding new block of problems.
- Coverage of the method of Lagrange multipliers in Section 7.5 has been rewritten for clarity, with an updated procedure box.
- Section 7.6 features new subsections on Double Integrals over Nonrectangular Regions and Applications of Double Integrals.
- Joint probability density functions are now covered in Chapter 10 of the Expanded Ninth Edition of this text.

KEY FEATURES OF THIS TEXT

SECTION 4.2 Logarithmic Functions **311**

Doubling Time In the introductory paragraph at the beginning of this section, you were asked how long it would take for a particular investment to double in value. This question is answered in Example 4.2.10.

EXAMPLE | 4.2.10

If $1,000 is invested at 8% annual interest, compounded continuously, how long will it take for the investment to double? Would the doubling time change if the principal were something other than $1,000?

Solution

With a principal of $1,000, the balance after t years is $B(t) = 1,000e^{0.08t}$, so the investment doubles when $B(t) = $2,000$; that is, when

$$2,000 = 1,000e^{0.08t}$$

Dividing by 1,000 and taking the natural logarithm on each side of the equation, we get

$$2 = e^{0.08t}$$
$$\ln 2 = 0.08t$$
$$t = \frac{\ln 2}{0.08} \approx 8.66 \text{ years}$$

If the principal had been P_0 dollars instead of $1,000, the doubling time would satisfy

$$2P_0 = P_0e^{0.08t}$$
$$2 = e^{0.08t}$$

which is exactly the same equation we had with $P_0 = $1,000$, so once again, the doubling time is 8.66 years.

> **A Procedure for Solving Related Rates Problems**
> 1. Draw a figure (if appropriate) and assign variables.
> 2. Find a formula relating the variables.
> 3. Use implicit differentiation to find how the rates are related.
> 4. Substitute any given numerical information into the equation in step 3 to find the desired rate of change.

Here are four applied problems involving related rates.

EXAMPLE | 2.6.5

The manager of a company determines that when q hundred units of a particular commodity are produced, the total cost of production is C thousand dollars, where $C^2 - 3q^3 = 4,275$. When 1,500 units are being produced, the level of production is increasing at the rate of 20 units per week. What is the total cost at this time and at what rate is it changing?

Solution

We want to find $\frac{dC}{dt}$ when $q = 15$ (1,500 units) and $\frac{dq}{dt} = 0.2$ (20 units per week with q measured in hundreds of units). Diffe[...]

4,275 with respect to time, we get

$$2C \frac{dC}{dt} \ldots$$

Definitions

Definitions and key concepts are set off in shaded boxes to provide easy referencing for the student.

Applications

Throughout the text great effort is made to ensure that topics are applied to practical problems soon after their introduction, providing methods for dealing with both routine computations and applied problems. These problem-solving methods and strategies are introduced in applied examples and practiced throughout in the exercise sets. Many new applied examples were added to this Ninth Edition from a wide variety of sources with careful attention paid to updating obsolete or outdated data. A complete listing of all applications used in the text can be found in the Index of Selected Applications behind the front and back covers.

Procedural Examples and Boxes

This text endeavors to approach each new topic with careful clarity by providing step-by-step problem-solving techniques. These techniques are demonstrated in the numerous procedural examples working through these problems, as well as in the frequent procedural summary boxes highlighting the techniques demonstrated.

The First Derivative Test for Relative Extrema ■ Let c be a critical number for $f(x)$ [that is, $f(c)$ is defined and either $f'(c) = 0$ or $f'(c)$ does not exist]. Then the critical point $P(c, f(c))$ is

A **relative maximum** if $f'(x) > 0$ to the left of c and $f'(x) < 0$ to the right of c

A **relative minimum** if $f'(x) < 0$ to the left of c and $f'(x) > 0$ to the right of c

Not a relative extremum if $f'(x)$ has the same sign on both sides of c

EXAMPLE | 3.1.3

Find all critical numbers of the function

$$f(x) = 2x^4 - 4x^2 + 3$$

and classify each critical point as a relative maximum, a relative minimum, or neither.

Just-In-Time Review

When the fundamental theorem of calculus

$$\int_a^b f(x)\,dx = F(b) - F(a)$$

is used to evaluate a definite integral, remember to compute *both* $F(b)$ and $F(a)$, even when $a = 0$.

Solution

An antiderivative of $f(x) = e^{-x} + \sqrt{x}$ is $F(x) = -e^{-x} + \frac{2}{3}x^{3/2}$, so the definite integral is

$$\int_0^1 (e^{-x} + \sqrt{x})\,dx = \left(-e^{-x} + \frac{2}{3}x^{3/2} \right)\Big|_0^1$$

$$= \left[-e^{-1} + \frac{2}{3}(1)^{3/2} \right] - \left[-e^0 + \frac{2}{3}(0) \right]$$

$$= \frac{5}{3} - \frac{1}{e} \approx 1.299$$

Our definition of the definite integral was motivated by computing area, which is a nonnegative quantity. However, since the definition does not require $f(x) \geq 0$, it is quite possible for a definite integral to be negative, as illustrated in Example 5.3.5.

Just-In-Time Reviews

These references, located in the margins, are used to quickly remind students of important concepts from college algebra or precalculus as they are being used in examples and discussions. More detailed discussion can still be found in Appendix A: Algebra Review.

3-15

In Problems 49 through 52, the graph of a function f is given. In each case, sketch a possible graph for f'.

49.

50.

51.

52.

Exercise Sets

Always a strong feature of past editions, the Ninth Edition offers 200 new problems to increase the effectiveness of these exercise sets. Routine problems have been added where needed to ensure students have enough practice to master basic skills, and a wealth of new applied problems have been added to help demonstrate the practicality of the material. Answers for the odd-numbered end-of-section and Chapter Review problems can be found at the back of the text, with fully worked-out solutions provided in the *Student's Solutions Manual*.

---

a paragraph on whether you think such methods are valid. You may wish to begin with the reference cited in this problem.

44. **NATIONAL CONSUMPTION** Assume that total national consumption is given by a function $C(x)$, where x is the total national income. The derivative $C'(x)$ is called the **marginal propensity to consume.** Then $S = x - C$ represents total national savings, and $S'(x)$ is called **marginal propensity to save.** Suppose the consumption function is $C(x) = 8 - 0.8x - 0.8\sqrt{x}$. Find the marginal propensity to consume, and determine the value of x that results in the smallest total savings.

45. **SENSITIVITY TO DRUGS** Body reaction to drugs is often modeled* by an equation of the form

$$R(D) = D^2\left(\frac{C}{2} - \frac{D}{3} \right)$$

where D is the dosage and C (a constant) is the maximum dosage that can be given. The rate of change of $R(D)$ with respect to D is called the **sensitivity.**
   a. Find the value of D for which the sensitivity is the greatest. What is the greatest sensitivity? (Express your answer in terms of C.)
   b. What is the reaction (in terms of C) when the dosage resulting in greatest sensitivity is used?

46. **AERODYNAMICS** In designing airplanes, an important feature is the so-called drag factor; that is, the retarding force exerted on the plane by the air. One model measures drag as a function of the form

$$F(v) = Av^2 + \frac{B}{v^2}$$

where A and B are positive constants. It is found experimentally that drag is minimized when $v = 160$ mph. Use this information to find the ratio $\frac{B}{A}$.

47. **ELECTRICITY** When a resistor of R ohms is connected across a battery with electromotive force E volts and internal resistance r ohms, a current of I amperes will flow, generating P watts of power, where

$$I = \frac{E}{r + R} \quad \text{and} \quad P = I^2 R$$

Assuming r is constant, what choice of R results in maximum power?

48. **SURVIVAL OF AQUATIC LIFE** It is known that a quantity of water that occupies 1 liter at 0°C will occupy

$$V(T) = \left(\frac{-6.8}{10^8} \right)T^3 + \left(\frac{8.5}{10^6} \right)T^2 - \left(\frac{6.4}{10^5} \right)T + 1$$

liters when the temperature is T°C, for $0 \leq T \leq 30$.
   a. Use a graphing utility to graph $V(T)$ for $0 \leq T \leq 10$. The density of water is maximized when $V(T)$ is minimized. At what temperature does this occur?
   b. Does the answer to part (a) surprise you? It should. Water is the only common liquid whose maxi-mum density occurs *above* its freezing point (0°C for water). Read an article on the survival of aquatic life during the winter and then write a paragraph on how the property of water examined in this problem is related to such survival.

49. **BLOOD PRODUCTION** A useful model for the production $p(x)$ of blood cells involves a function of the form

$$p(x) = \frac{Ax}{B + x^m}$$

where x is the number of cells present, and A, B, and m are positive constants.[†]
   a. Find the rate of blood production $R(x) = p'(x)$ and determine where $R(x) = 0$.
   b. Find the rate at which $R(x)$ is changing with respect to x and determine where $R'(x) = 0$.
   c. If $m > 1$, does the nonzero critical number you found in part (b) correspond to a relative maximum or a relative minimum? Explain.

50. **AMPLITUDE OF OSCILLATION** In physics, it can be shown that a particle forced to oscillate in a resisting medium has amplitude $A(r)$ given by

$$A(r) = \frac{1}{(1 - r^2)^2 + kr^2}$$

where r is the ratio of the forcing frequency to the natural frequency of oscillation and k is a positive constant that measures the damping effect of the resisting medium. Show that $A(r)$ has exactly one positive critical number. Does it correspond to a relative maximum or a relative minimum? Can anything be said about the *absolute* extrema of $A(r)$?

*R. M. Thrall et al., *Some Mathematical Models in Biology*, U. of Michigan, 1967.

[†]M. C. Mackey and L. Glass, "Oscillations and Chaos in Physiological Control Systems," *Science*, Vol. 197, pp 287–289.

l cost of producing x s $C(x)$ thousand dol-

$9x + 242$

$C(x)/x$ is the average

) increasing? For g?

on x is average cost inimum average cost? The total cost of pro-mmodity is given by the cost curve and arginal cost increase oduction?

Let $p = (10 - 3x)^2$ which x hundred units e sold, and let $R(x) =$ from the sale of the nue $R'(x)$ and sketch

the revenue and marginal revenue curves on the same graph. For what level of production is revenue maximized?

56. **PROFIT UNDER A MONOPOLY** To produce x units of a particular commodity, a monopolist has a total cost of

$$C(x) = 2x^2 + 3x + 5$$

and total revenue of $R(x) = xp(x)$, where $p(x) = 5 - 2x$ is the price at which the x units will be sold. Find the profit function $P(x) = R(x) - C(x)$ and sketch its graph. For what level of production is profit maximized?

57. **MEDICINE** The concentration of a drug t hours after being injected into the arm of a patient is given by

$$C(t) = \frac{0.15t}{t^2 + 0.81}$$

Sketch the graph of the concentration function. When does the maximum concentration occur?

Writing Exercises

Every exercise set includes writing problems that are related to issues raised in the examples and exercises, designated by the writing icon. These problems challenge a student's critical thinking skills and invite students to research topics on their own, exploring the language of mathematics.

Calculator Exercises

Each section includes numerous problems that can be completed only through the use of a graphing utility. These exercises are clearly labeled by the  calculator icon.

Important Terms, Symbols, and Formulas

Secant line and tangent line (96 and 98)

Average and instantaneous rates of change (97–98)

Difference quotient

$$\frac{f(x+h)-f(x)}{h} \quad (99)$$

Derivative of $f(x)$: $f'(x) = \lim_{h\to 0}\frac{f(x+h)-f(x)}{h}$ (99)

Differentiation (96)

Differentiable function (99)

The derivative $f'(x_0)$ as slope of the tangent line to
$y = f(x)$ at $(x_0, f(x_0))$ (99)

The derivative $f'(x_0)$ as the rate of change of $f(x)$ with
respect to x when $x = x_0$ (99)

Derivative notation for $f(x)$: $f'(x)$ and $\dfrac{df}{dx}$ (103)

A differentiable function is continuous (104)

The constant rule: $\dfrac{d}{dx}[c] = 0$ (109)

The power rule: $\dfrac{d}{dx}[x^n] = nx^{n-1}$ (110)

The constant multiple rule: $\dfrac{d}{dx}[cf(x)] = c\,\dfrac{df}{dx}$ (112)

Relative rate of change of $Q(x)$: $\dfrac{Q'(x)}{Q(x)}$ (112)

Percentage rate of change of $Q(x)$: $\dfrac{100Q'(x)}{Q(x)}$ (112)

Rectilinear motion: position $s(t)$
velocity $v(t) = s'(t)$
acceleration $a(t) = v'(t)$ (115)

The motion of a projectile:

height $H(t) = -\dfrac{1}{2}gt^2 + V_0 t + H_0$ (116)

The product rule:

$$\frac{d}{dx}[f(x)g(x)] = f(x)\frac{dg}{dx} + g(x)\frac{df}{dx} \quad (122)$$

The quotient rule: For $g(x) \neq 0$

$$\frac{d}{dx}\left[\frac{f(x)}{g(x)}\right] = \frac{g(x)\dfrac{df}{dx} - f(x)\dfrac{dg}{dx}}{g^2(x)} \quad (124)$$

The second derivative of $f(x)$: $f''(x) = [f'(x)]' = \dfrac{d^2f}{dx^2}$

gives the rate of change of $f'(x)$ (127)

Notation for the nth derivative $\dfrac{d^n f}{}$

Chapter Review

Essential to a student's understanding, the Chapter Review material aids in synthesizing the important concepts discussed within the chapter. The Important Terms, Symbols, and Formulas section provides a master list of all key technical terms and mathematics discussed throughout the chapter.

Checkup for Chapter 2

1. In each case, find the derivative $\dfrac{dy}{dx}$.

   a. $y = 3x^4 - 4\sqrt{x} + \dfrac{5}{x^2} - 7$

   b. $y = (3x^3 - x + 1)(4 - x^2)$

   c. $y = \dfrac{5x^2 - 3x + 2}{1 - 2x}$

   d. $y = (3 - 4x + 3x^2)^{3/2}$

2. Find the second derivative of the function
   $f(t) = t(2t+1)^2$.

3. Find an equation for the tangent line to the curve
   $y = x^2 - 2x + 1$ at the point where $x = -1$.

4. Find the rate of change of the function
   $f(x) = \dfrac{x+1}{1-3x}$ with respect to x when $x = 1$.

5. b. When is the object stationary? When is it advancing? Retreating?

   c. What is the total distance traveled by the object for $0 \le t \le 2$?

7. **PRODUCTION COST** Suppose the cost of producing x units of a particular commodity is $C(x) = 0.04x^2 + 5x + 73$ hundred dollars.

   a. Use marginal cost to estimate the cost of producing the sixth unit.

   b. What is the actual cost of producing the sixth unit?

8. **INDUSTRIAL OUTPUT** At a certain factory, the daily output is $Q = 500L^{3/4}$ units, where L denotes the size of the labor force in worker-hours. Currently, 2,401 worker-hours of labor are used each day. Use calculus (increments) to estimate the effect on output of increasing the size of the labor force by 200 worker-hours from its current level.

9. **PEDIATRIC MEASUREMENT** Pediatricians use the formula $S = 0.2029w^{0.425}$ to estimate the surface area S (m^2) of a child 1 meter tall who weighs w kilograms (kg). A particular child weighs 30 kg and is gaining weight at the rate of 0.13 kg per week while remaining 1 meter tall. At what rate is this child's surface area changing?

10. **GROWTH OF A TUMOR** A cancerous tumor is roughly spherical in shape. Estimate the percentage error that can be allowed in the measurement of the radius r to ensure there will be no more than an 8% error in the calculation of volume using the formula $V = \dfrac{4}{3}\pi r^3$.

Chapter Checkup

The Chapter Checkup provides a quick quiz for students to test their understanding of the concepts introduced in the chapter. All answers to this quiz are provided in the Answer Key at the back of the text, with fully worked-out solutions provided in the *Student's Solutions Manual*.

Review Problems

In Problems 1 and 2, use the definition of the derivative to find $f'(x)$.

1. $f(x) = x^2 - 3x + 1$

2. $f(x) = \dfrac{1}{x-2}$

In Problems 3 through 13, find the derivative of the given function.

3. $f(x) = 6x^4 - 7x^3 + 2x + \sqrt{2}$

4. $f(x) = x^3 - \dfrac{1}{3x^3} + 2\sqrt{x} - \dfrac{3}{x} + \dfrac{1-2x}{x^3}$

5. $y = \dfrac{2 - x^2}{3x^2 + 1}$

6. $y = (x^3 + 2x - 7)(3 + x - x^2)$

7. $f(x) = (5x^4 - 3x^2 + 2x + 1)^{10}$

8. $f(x) = \sqrt{x^2 + 1}$

9. $y = \left(x + \dfrac{1}{x}\right)^2 - \dfrac{5}{\sqrt{3x}}$

10. $y = \left(\dfrac{x+1}{1-x}\right)^2$

11. $f(x) = (3x + 1)\sqrt{6x + 5}$

12. $f(x) = \dfrac{(3x+1)^3}{(1-3x)^4}$

13. $y = \sqrt{\dfrac{1-2x}{3x+2}}$

In Problems 14 through 17, find an equation for the tangent line to the graph of the given function at the specified point.

14. $f(x) = x^2 - 3x + 2$; $x = 1$

15. $f(x) = \dfrac{4}{x-3}$; $x = 1$

16. $f(x) = \dfrac{x}{x^2 + 1}$; $x = 0$

17. $f(x) = \sqrt{x^2 + 5}$; $x = -2$

18. In each of these cases, find the rate of change of $f(t)$ with respect to t at the given value of t.

   a. $f(t) = t^3 - 4t^2 + 5t\sqrt{t} - 5$ at $t = 4$

   b. $f(t) = \dfrac{2t^2 - 5}{1 - 3t}$ at $t = 1$

   c. $f(t) = t^3(t^2 - 1)$ at $t = 0$

   d. $f(t) = (t^2 - 3t + 6)^{1/2}$ at $t = 1$

19. In each of these cases, find the percentage rate of change of the function $f(t)$ with respect to t at the given value of t.

   a. $f(t) = t^2 - 3t + \sqrt{t}$ at $t = 4$

   b. $f(t) = t^3(3 - 2t)^3$ at $t = 1$

   c. $f(t) = \dfrac{1}{t+1}$ at $t = 0$

20. Use the chain rule to find $\dfrac{dy}{dx}$.

   a. $y = 5u^2 + u - 1$; $u = 3x + 1$

   b. $y = -\dfrac{1}{u^2}$; $u = 2x + 3$

21. Use the chain rule to find $\dfrac{dy}{dx}$ for the given value of x.

   a. $y = u^3 - 4u^2 + 5u + 2$; $u = x^2 + 1$; for $x = 1$

   b. $y = \sqrt{u}$, $u = x^2 + 2x - 4$; for $x = 2$

   c. $y = \left(\dfrac{u-1}{u+1}\right)^{1/2}$, $u = \sqrt{x-1}$; for $x = \dfrac{34}{9}$

22. Find the second derivative of each of these functions:

   a. $f(x) = 6x^5 - 4x^3 + 5x^2 - 2x + \dfrac{1}{x}$

   b. $z = \dfrac{2}{1 + x^2}$

   c. $y = (3x^2 + 2)^4$

   d. $f(x) = 2x(x + 4)^3$

   e. $f(x) = \dfrac{x-1}{(x+1)^2}$

23. Find $\dfrac{dy}{dx}$ by implicit differentiation.

   a. $5x + 3y = 12$

   b. $x^2 y = 1$

   c. $(2x + 3y)^5 = x + 1$

   d. $(1 - 2xy^3)^5 = x + 4y$

24. Use implicit differentiation to find the slope of the line that is tangent to the given curve at the specified point.

   a. $xy^3 = 8$; $(1, 2)$

   b. $x^2 y - 2xy^3 + 6 = 2x + 2y$; $(0, 3)$

Review Problems

A wealth of additional routine and applied problems are provided within the end-of-chapter exercise sets, offering further opportunities for practice.

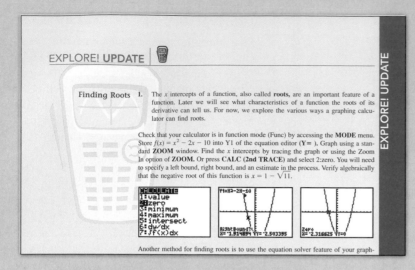

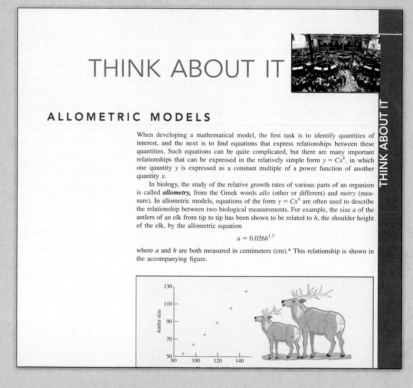

Explore! Technology

For those choosing to include a graphing focus in their course, the Explore! boxes provide an optional path through the material. These explorations, tied to specific examples, continue to guide students in the use of graphing calculators and challenge their understanding of the topics presented. Each chapter concludes with an Explore! Update section that provides solutions and hints to selected boxes throughout the chapter.

Think About It

The modeling-based Think About It essays show students how material introduced in the chapter can be used to construct useful mathematical models while explaining the modeling process. The exercise sets following each essay are an excellent starting point for student projects and can be covered in the classroom or serve as the basis for group discussions. Instructors can access complete worked-out solutions to these problems from the Hoffmann/Bradley MathZone website.

Also available . . .

Applied Calculus for Business, Economics, and the Social and Life Sciences, Expanded Ninth Edition

The Expanded Ninth Edition of *Applied Calculus for Business, Economics, and the Social and Life Sciences* contains all the material present in the Brief Ninth Edition of *Calculus for Business, Economics, and the Social and Life Sciences,* plus four additional chapters:

- Chapter 8: Differential Equations
- Chapter 9: Infinite Series and Taylor Series Approximations
- Chapter 10: Probability and Calculus
- Chapter 11: Trigonometric Functions

The Expanded Edition meets the needs of instructors who cover topics in one or more of these four chapters together with material from the initial seven chapters in their one- or two-term course. The additional chapters feature more of the real-world applications, strong exercise sets, and easy-to-understand presentation for which *Calculus for Business, Economics, and the Social and Life Sciences* is renowned. For more information about the Expanded Ninth Edition, please contact your McGraw-Hill representative. ISBN–13: 978-0-07-305192-5 (ISBN–10: 0-07-305192-6)

Supplements

Student's Solutions Manual

The *Student's Solutions Manual* accompanying this Brief Ninth Edition contains comprehensive, worked-out solutions for all odd-numbered problems in the text, including the Checkup (all problems) and Review Problems (odd-numbered problems) at the end of each chapter. Detailed calculator instructions and keystrokes are provided with the solutions to problems marked in the text by the calculator icon, and these instructions have been fully revised and updated for compatibility with the latest calculator models. ISBN-13: 978-0-07-325729-X (ISBN-10: 0-07-325729-X)

Instructor's Solutions Manual

The *Instructor's Solutions Manual* accompanying this Brief Ninth Edition contains comprehensive, worked-out solutions for all even-numbered problems in the text, including the Review Problems at the end of each chapter. ISBN-13: 978-0-07-325728-0 (ISBN–10: 0-07-325728-1)

Instructor's Testing and Resource CD-ROM

Brownstone Diploma testing software, available on cross-platform CD-ROM, offers instructors a quick and easy way to create customized exams and view student results. The software utilizes an electronic test bank of well over 1,000 short-answer, multiple-choice, and true/false questions tied directly to the text, with many new questions added for the Ninth Edition. Instructors can sort questions by section, difficulty level, and type; edit existing questions or add new ones; create multiple versions of questions using algorithmically randomized variables; export tests to word processors or the Web; and create and manage course grade books. ISBN-13: 978-0-07-325731-0 (ISBN–10: 0-07-325731-1)

MathZone 3.0 www.mathzone.com

Free, Easy, Has it all . . .

McGraw-Hill's MathZone 3.0 is a complete, online tutorial and course management system for mathematics and statistics, designed for greater ease of use than any other system available. Free upon adoption, MathZone allows instructors to create and share courses and assignments with colleagues and adjuncts in a matter of a few clicks of the mouse. All assignments, questions, and online tutoring are tied to text-specific materials in *Calculus for Business, Economics, and the Social and Life Sciences*. MathZone courses are customized to your textbook, but you can edit questions and algorithms, import your own content, create your own slides and presentations using an image bank of artwork from the text, and create announcements and due dates for assignments. MathZone has automatic grading and reporting of easy-to-assign algorithmically generated homework, quizzing, and testing. All student activity within MathZone is automatically graded and accessible to instructors through a fully integrated, exportable grade book.

MathZone also provides a wide variety of interactive student tutorials, including practice problems; e-Professors, a collection of over 100 step-by-step animated instructions for solving exercises from the text; video lectures; interactive applets; and NetTutor, a live personalized online tutoring service.

Acknowledgments

As in past editions, we have enlisted the feedback of professors teaching from our text as well as those using other texts to point out possible areas for improvement. Our reviewers provided a wealth of detailed information on both our content and the changing needs of their course, and many changes we have made were a direct result of consensus among these review panels. This text owes its considerable success to their valuable contributions, and we thank every individual involved in this process.

Faiz Al-Rubaee, *University of North Florida*
George Anastassiou, *University of Memphis*
Dan Anderson, *University of Iowa*
Don Bensy, *Suffolk County Community College*
Neal Brand, *University of North Texas*
Randall Brian, *Vincennes University*
Paul W. Britt, *Louisiana State University—Baton Rouge*
Albert Bronstein, *Purdue University*
James F. Brooks, *Eastern Kentucky University*
Beverly Broomell, *SUNY—Suffolk*
Laura Cameron, *University of New Mexico*
Rick Carey, *University of Kentucky*
Steven Castillo, *Los Angeles Valley College*
Deanna Caveny, *College of Charleston*
Gerald R. Chachere, *Howard University*
Terry Cheng, *Irvine Valley College*
William Chin, *DePaul University*
Lynn Cleaveland, *University of Arkansas*
Dominic Clemence, *North Carolina A&T State University*
Charles C. Clever, *South Dakota State University*
Peter Colwell, *Iowa State University*
Cecil Coone, *Southwest Tennessee Community College*
Charles Brian Crane, *Emory University*
Raul Curto, *University of Iowa*
Jean F. Davis, *Texas State University—San Marcos*
John Davis, *Baylor University*
Karahi Dints, *Northern Illinois University*
Ken Dodaro, *Florida State University*
Eugene Don, *Queens College*
Dora Douglas, *Wright State University*
Peter Dragnev, *Indiana University—Purdue University, Fort Wayne*
Bruce Edwards, *University of Florida*

Margaret Ehrlich, *Georgia State University*
Maurice Ekwo, *Texas Southern University*
George Evanovich, *St. Peters' College*
Haitao Fan, *Georgetown University*
Klaus Fischer, *George Mason University*
Michael Freeze, *University of North Carolina—Wilmington*
Constantine Georgakis, *DePaul University*
Sudhir Goel, *Valdosta State University*
Ronnie Goolsby, *Winthrop College*
Lauren Gordon, *Bucknell University*
Angela Grant, *University of Memphis*
John Gresser, *Bowling Green State University*
Murli Gupta, *George Washington University*
Doug Hardin, *Vanderbilt University*
Jonathan Hatch, *University of Delaware*
Celeste Hernandez, *Richland College*
William Hintzman, *San Diego State University*
Matthew Hudock, *St. Philips College*
Joel W. Irish, *University of Southern Maine*
Zonair Issac, *Vanderbilt University*
Erica Jen, *University of Southern California*
Shafiu Jibrin, *Northern Arizona University*
Victor Kaftal, *University of Cincinnati*
Sheldon Kamienny, *University of Southern California*
Georgia Katsis, *DePaul University*
Fritz Keinert, *Iowa State University*
Melvin Kiernan, *St. Peter's College*
Donna Krichiver, *Johnson County Community College*
Harvey Lambert, *University of Nevada*
Donald R. LaTorre, *Clemson University*
Melvin Lax, *California State University at Long Beach*
Robert Lewis, *El Camino College*
W. Conway Link, *Louisiana State University—Shreveport*
James Liu, *James Madison University*
Yingjie Liu, *University of Illinois at Chicago*
Jeanette Martin, *Washington State University*
James E. McClure, *University of Kentucky*
Mark McCombs, *University of North Carolina*
Ann B. Megaw, *University of Texas at Austin*
Fabio Milner, *Purdue University*
Kailash Misra, *North Carolina State University*
Mohammad Moazzam, *Salisbury State University*
Rebecca Muller, *Southeastern Louisiana University*
Karla Neal, *Louisiana State University*
Cornelius Nelan, *Quinnipiac University*
Devi Nichols, *Purdue University—West Lafayette*
Jaynes Osterberg, *University of Cincinnati*
Ray Otto, *Wright State University*
Hiram Paley, *University of Illinois*
Virginia Parks, *Georgia Perimeter College*
Shahla Peterman, *University of Missouri—St. Louis*

Murray Peterson, *College of Marin*
Lefkios Petevis, *Kirkwood Community College*
Cyril Petras, *Lord Fairfax Community College*
Natalie Priebe, *Rensselaer Polytechnic Institute*
Georgia Pyrros, *University of Delaware*
Richard Randell, *University of Iowa*
Mohsen Razzaghi, *Mississippi State University*
Nathan P. Ritchey, *Youngstown State University*
Arthur Rosenthal, *Salem State College*
Judith Ross, *San Diego State University*
Robert Sacker, *University of Southern California*
Katherine Safford, *St. Peter's College*
Mansour Samimi, *Winston-Salem State University*
Dolores Schaffner, *University of South Dakota*
Thomas J. Sharp, *West Georgia College*
Robert E. Sharpton, *Miami-Dade Community College*
Anthony Shershin, *Florida International University*
Minna Shore, *University of Florida International University*
Ken Shores, *Arkansas Tech University*
Jane E. Sieberth, *Franklin University*
Marlene Sims, *Kennesaw State University*
Brian Smith, *Parkland College*
Nancy Smith, *Kent State University*
Joseph F. Stokes, *Western Kentucky University*
Hugo Sun, *California State University—Fresno*
Keith Stroyan, *University of Iowa*
Martin Tangora, *University of Illinois at Chicago*
Tuong Ton-That, *University of Iowa*
Lee Topham, *North Harris Community College*
George Trowbridge, *University of New Mexico*
Dinh Van Huynh, *Ohio University*
Maria Elena Verona, *University of Southern California*
Kimberly Vincent, *Washington State University*
Karen Vorwerk, *Westfield State College*
Charles C. Votaw, *Fort Hays State University*
Hiroko Warshauer, *Southwest Texas State University*
Pam Warton, *Bowling Green State University*
Jonathan Weston-Dawkes, *University of North Carolina*
Henry Wyzinski, *Indiana University—Northwest*
Paul Yun, *El Camino College*
Xiao-Dong Zhang, *Florida Atlantic University*
Jay Zimmerman, *Towson University*

Special thanks go to those instrumental in checking each problem and page for accuracy, including Devilyna Nichols, Cindy Trimble, Brad Davis, and Jaqui Bradley. Reginald Luke and Cindy Trimble were instrumental in creating the Explore! material for this edition. Special thanks as well go to Cornelius Nelan and Charles Brian Crane for providing a wealth of specific, detailed suggestions for improvement that were particularly helpful in preparing this Ninth Edition. Finally, we wish to thank our McGraw-Hill team, Liz Covello, Nancy Anselment, David Dietz, Dan Seibert, and Vicki Krug for their patience, dedication, and sustaining support.

CALCULATOR INTRODUCTION

The following discussion is intended as a brief introduction to standard graphing utility features found on most graphing calculators. A typical commercially available graphing calculator is the TI-84 Plus, produced by Texas Instruments, Inc. This introduction and subsequent Explore! Updates and solutions use this calculator with OS* 2.30 as its basis. The TI-84 Plus (Silver Edition) keyboard incorporates the standard numerical, editing, and arrow cursor keys, as well as those keys accessing mathematical and scientific operations and graphical and programming features. Consult with the manual accompanying your calculator for more specific instructions.

FIGURE 1

FIGURE 2

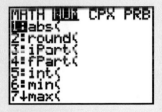

FIGURE 3

General Access

Pressing the **MATH** key, for instance, yields a pull-down screen with multiple menu options (Figures 1 and 2), which can be selected by number or by moving the arrow down cursor key to the desired selection and pressing the **ENTER** key. Note that there are additional selections obtained through the down arrow, such as options **8:nDeriv(** and **0:Solver.** Pressing the right arrow key gains use of additional screens (Figure 3) where the absolute value function is the first item available under the numerical **NUM** menu. Complex numbers (**CPX**) and Probability (**PRB**) round off the rest of the menu screens.

Two colored keys on the keyboard provide entry into additional layers of calculator functions: the green **ALPHA** key provides access to the alphabet and other symbols, while the blue **2nd** function key gains entry into blue-colored commands. For example, pressing the **2nd** key and then **MODE (QUIT)** allows access to a "homescreen" on which text or calculations can be written. Try writing the screen shown in Figure 4. To obtain the blank space, use the key sequence **ALPHA,** then **0.** You can lock the **ALPHA** key by pressing **2nd ALPHA.** The equal sign is available as the first item of the **TEST (2nd MATH)** menu, as in Figure 5. We can use the memory storage capacity of the alpha variables to execute numerical calculations. For example, use the **STO→** key (just above the **ON** key) to create the screen shown in Figure 6, in which the length of the hypotenuse of a right triangle is obtained.

---

*OS 2.30 is available for download from the Texas Instruments website, http://education.ti.com/.

FIGURE 4 **FIGURE 5** **FIGURE 6**

Equation Solver A more effective method of solving equations is accessible through the **EQUATION SOLVER** of the **MATH** menu, where we have written the Pythagorean equation shown in Figure 7, with all terms on the right side of the equation. Pressing **ENTER** yields Figure 8, with inputted values of 5 for A and 13 for C, in an attempt to solve for the value of B. Place a reasonable starter value there and then execute the green-colored command **SOLVE (ALPHA ENTER)** to obtain the anticipated solution (Figure 9). Notice that we have the ability to solve for any variable, by giving desired or required values for the other variables.

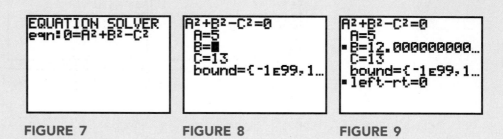

FIGURE 7 **FIGURE 8** **FIGURE 9**

Mode The predominant feature of a graphing calculator is its ability to present, visually display, and analyze the many functions that occur in calculus. Suppose you were interested in the relationship between the two functions $f(x) = x^2 - 1$ and $g(x) = 5 - x$. Press the **MODE** key, and observe the default settings shown in Figure 10. From row 4 we notice that we are in function (**FUNC**) mode, as opposed to parametric (**PAR**), polar (**POL**), or sequence (**SEQ**). The setting of **CONNECTED** from row 5 refers to the fact that the graphs will be generated in connected fashion rather than with intermittent dots. The setting of **SEQUENTIAL** from row 6 refers to the order in which the graphs will be generated, one after the previous rather than all at once (simultaneously). Now press the **Y=** key, called the "equation editor," located in the upper left of the keyboard, and place $f(x)$ into Y1 and $g(x)$ into Y2, as shown in Figure 11.

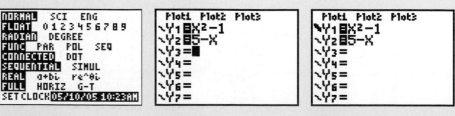

FIGURE 10 **FIGURE 11** **FIGURE 12**

Graphing Style

We can select different graphing styles for each function, by moving the cursor to the left of the Y function symbol, and pressing the **ENTER** key in succession. The graphing style choices that appear in order are bold, shading above the function, shading below, ball with trailing line, ball by itself, disconnected line, and then back to the initial standard graphing style. Pressing the **ENTER** key once while the cursor is to the left of Y1 will yield a bold graphing style for $f(x) = x^2 - 1$, as shown in Figure 12.

Viewing Windows

Most graphing calculators have several optional viewing windows set to prescribed dimensions. In the TI-84 Plus, these viewing windows are accessed through the **ZOOM** key (located in the middle of the top row in the keyboard). Pressing this key will give you Figure 13. A standard window, **6:ZStandard,** generates the graph shown in Figure 14. The dimensions of this "standard window" are $-10 \le X \le 10$ and $-10 \le Y \le 10$, with the axes scaled in unit tick marks, designated by $X_{scl} = 1$ and $Y_{scl} = 1$, respectively. Press **WINDOW** and see the screen shown in Figure 15.

FIGURE 13

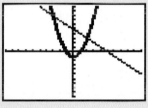

FIGURE 14

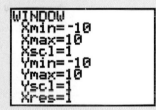

FIGURE 15

Decimal Window

Another useful type of viewing rectangle is the "decimal window," accessed through the **ZOOM** menu, **4:ZDecimal.** The decimal window dimensions are denoted $[-4.7, 4.7]1$ by $[-3.1, 3.1]1$ to designate the X and Y range of values, respectively, where the 1 values after the enclosure marks indicate that both $X_{scl} = 1$ and $Y_{scl} = 1$ (see Figure 16). The special dimensions of the decimal window permit the cursor to trace over decimal values of the X variable, in increments of 0.1. Modifying the Y dimensions of this window allows for a better view of the graphs of both functions, as shown in Figure 17, where $-2 \le Y \le 10$. When window dimensions are modified, the **GRAPH** key (located at the top right of the keyboard) must be used in order to obtain the desired graph, shown in Figure 18.

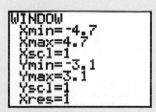

FIGURE 16

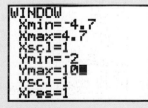

FIGURE 17

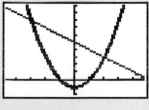

FIGURE 18

Trace   Movement along the graph of the function is handled using the **TRACE** key (top row of the keyboard), maneuvering right or left on a graph using the appropriate right or left arrow keys, as demonstrated in Figure 19. Note that the x values of the tracing point are decimal ones, since the graph was generated using a modified decimal window. Use the up or down arrow keys to skip from this function to the other one, in this case the function $g(x) = 5 - x$, as displayed in Figure 20. We could continue tracing on both graphs to find the points of intersection, one of which appears to be the point $(2, 3)$, as shown in Figure 21.

 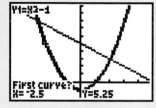 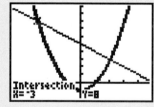

FIGURE 19   **FIGURE 20**   **FIGURE 21**

Intersection-Finding Feature   To demonstrate one of the capabilities of graphing calculators, we review the intersection-finding feature of the TI-84 Plus, found through the **CALC** key **(2nd TRACE),** as shown in Figure 22. Note that we can also find functional values, roots, minimum and maximum values, as well as some calculus operations. Let us find the other point of intersection of $f(x)$ and $g(x)$. Select **5:intersect** and identify the two curves and an estimated point of intersection, pressing **ENTER** successively to attain the next step in the process (Figure 23). Then the calculator will attempt to seek the exact intersection point, as shown in Figure 24.

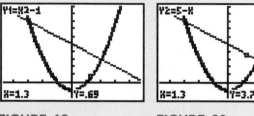

FIGURE 22   **FIGURE 23**   **FIGURE 24**

Symbolic Representation   Another useful feature of a graphing calculator is its ability to represent functions symbolically. We have already placed functions into Y1 and Y2 of the equation editor, **Y=.** There are many occasions when we need to use these symbols in equations and expressions. For example, suppose we wanted to write Y1(5) to evaluate $f(x) = x^2 - 1$ at $x = 5$. How is this done? The symbol **Y1** (or **Y2**) can be found via the **VARS** key, arrowing right to **Y-VARS,** selecting **1:Function,** and then choosing **Y1** (or **Y2**). These steps are illustrated in Figures 25 through 27.

FIGURE 25　　　　　**FIGURE 26**　　　　　**FIGURE 27**

The Y1 and Y2 symbols can be manipulated in functional combinations, just as one would $f(x)$ and $g(x)$. For example, sums, products, and composites as well as translations and transformations are easily constructed. With Y1 and Y2 as defined, write Y3 = Y1(X + 2). First, however, deselect (turn off) Y1 and Y2, so that their graphs do not appear on the screen. This is done by moving the cursor to the equal sign and pressing **ENTER** to deselect the desired function, as in Figure 28, where Y1 and Y2 have been deselected. The resulting graph, using a **ZOOM** standard window and with the **TRACE** key activated, appears in Figure 29. What equation does this function have? Finally, deselect Y3 and write Y4 = Y1(Y2), the composite $f(g(x))$. Its graph appears in Figure 30. Again, what function is this?

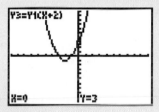
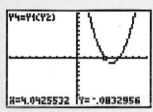

FIGURE 28　　　　　**FIGURE 29**　　　　　**FIGURE 30**

Selecting Windows

One of the key challenges in presenting a function pictorially is to create the best window to view the significant portion of its graph. The standard or decimal window available through the **ZOOM** key may not display the graph appropriately. Viewing a table of values may be the best first step to getting a fix on functional behavior. As an example, let us explore how to present the best window dimensions for the graph of the function $f(x) = 15 + 30x/(x^2 + 1)$, which we place into Y1. First, create a table setting to be able to see the behavior of the function near the origin. Press **2nd WINDOW (TBL SET)** and change the table settings to obtain a starting x value of $x = -3$ in unit incremental values; that is, set TblStart $= -3$ and ΔTbl $= 1$, as shown in Figure 31. Now press the **TABLE** key (**2nd GRAPH**) to obtain the table of values shown in Figure 32.

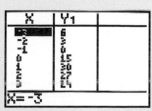

FIGURE 31　　　　　**FIGURE 32**

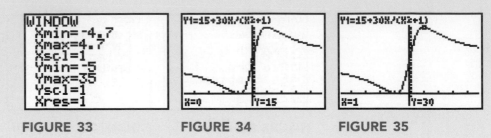

FIGURE 33   **FIGURE 34**   **FIGURE 35**

Observing the table of values gives a good idea of the range of functional values and a reasonable domain with which to work. It appears that we should set the range of y values a little larger than $[0, 30]$, say, $-5 \leq Y \leq 35$. Since the scope of x values should be beyond $x = -3$ to 3, we use a modified decimal window, namely $-4.7 \leq X \leq 4.7$. Press the following keys: **ZOOM, 4:ZDecimal,** and then **WINDOW,** and change the settings as given in Figure 33. These dimensions would be expressed as $[-4.7, 4.7]1$ by $[-5, 35]1$. Finally, pressing **GRAPH** creates the graph shown in Figure 34, where the **TRACE** key identifies the function and specific points on the graph.

The graph shown in Figure 34 has one apparent maximum point, located near $x = 1$. The **TRACE** key can be very helpful here, especially, since the modified decimal window allows tracing over decimal values, as shown in Figure 35. Most graphing calculators have a maximum–finding feature to assist in locating this important point. Press the **CALC** key (**2nd TRACE), 4:maximum,** as shown in Figure 36. You will need to specify a reasonable left bound (Figure 37), right bound (Figure 38), and estimate (Figure 39), pressing the **ENTER** key as required, before the program homes in on the maximal point, which is identified as $x = 0.99999881$, in Figure 40. Note that this point is technically $x = 1$ within the tolerance limits of the calculator, but this can be verified analytically through calculus methods as exactly $x = 1$, as will be demonstrated in Chapter 3.

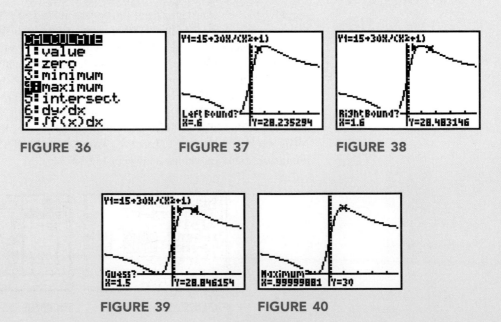

FIGURE 36   **FIGURE 37**   **FIGURE 38**

FIGURE 39   **FIGURE 40**

Modeling

Another important feature of a graphing calculator is its capacity to model statistical data, that is, finding and displaying the best function to represent a set of data obtained experimentally or theoretically. Consider the following table of data, giving the population of the United States (in millions) from census surveys for the past several decades, as cited in the *Britannica Almanac 2003*.

Year	1940	1950	1960	1970	1980	1990	2000
U.S. Pop.	131.67	151.33	179.32	203.30	226.54	248.72	281.42

The first step is to store these population data in the graphing calculator lists Press the **STAT** key, located in the middle of the keyboard, and select the first option **1:Edit** (Figure 41), showing up to three of the possible six lists, L1 to L6, as well as an unnamed list. You may have to select option **5:SetUpEditor,** if some of these lists are hidden or missing. Now enter the data as shown in Figure 42. If there are any lists of unwanted data, move the cursor to the top of the specific list, press the **CLEAR** key, and then press the down arrow, clearing out all the old data and getting ready for the first new data entry, as we have done for List L3 (Figure 43).

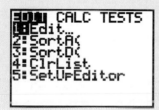

FIGURE 41   **FIGURE 42**   **FIGURE 43**

The second phase of the data presentation process is to set up the statistical plots. Press the **STAT PLOT** key (**2nd Y=**) to reveal a screen with three plot selections, as shown in Figure 44. Choose the first option and select the settings, line by line for the 5 rows, to coincide with the display shown in Figure 45. In Line 2 we specify the data type, by arrowing right and pressing the **ENTER** key to select in order one of the following choices: scatterplot, line graph, histogram, box plot with outliers identified, regular boxplot, normal plot. We have chosen the first selection, scatterplot, for our bivariate data. The variable Year in List 1 is our **Xlist,** while U.S. Population in List 2 is **Ylist.** The graphing mark is a small square symbol.

FIGURE 44   **FIGURE 45**

Now that the plot settings have been selected, we use the **ZOOM** feature of the TI-84 Plus, selecting **9:ZoomStat,** as shown in Figure 46, to produce an instant graph, shown in Figure 47, for which the **TRACE** key can be pressed to identify the values of each data point. The shape of the graph appears linear, so we attempt to find the best straight line of the form $Y = aX + b$ that fits the population data. This process is called linear regression, based on the Gaussian least-squares method, which produces the regression coefficients a and b for the linear function that best approximates the data points. Prepare a cleared homescreen, press the **STAT** key, arrow right once to **CALC** and select option **4:LinReg(ax+b),** shown in Figure 48. Press this selection, and then write the three symbols: L1, L2, and Y1, separated by commas, as shown in Figure 49. Recall that Y1 is obtained through the sequence: **VARS, Y-VARS, 1:Function, 1:Y1.** Finally, press **ENTER** to obtain the desired results in Figure 50.

FIGURE 46 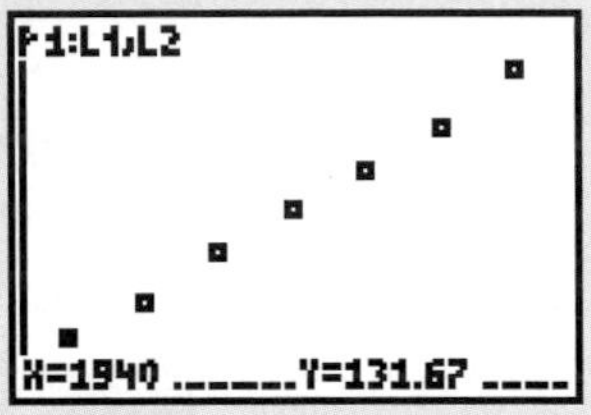   FIGURE 47

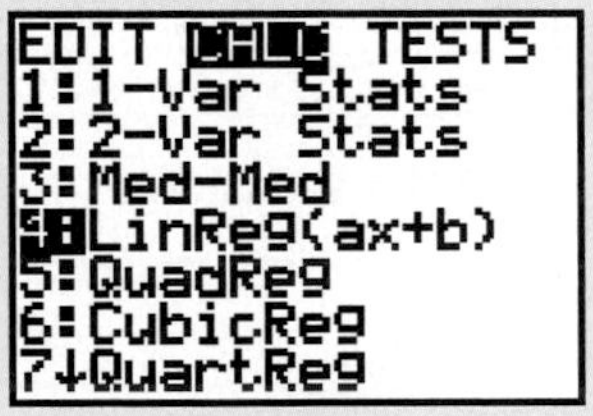

FIGURE 48   FIGURE 49 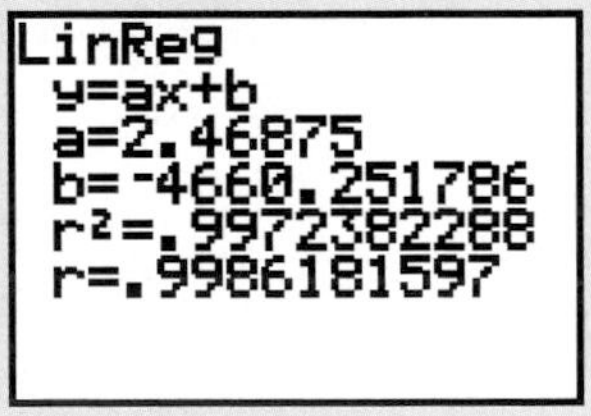   FIGURE 50

The regression coefficient a represents the slope of the function Y1, and its value, $a \approx 2.47$, indicates that the United States had an average yearly increase of 2.47 million people over the time period. The y intercept, $b \approx -4,660$, has no relevant interpretation, being the alleged U.S. population in the year zero, much beyond the scope of the data. The r value is the correlation coefficient, a value between -1 and 1, which indicates how well the linear equation fits the data. Since $r \approx 0.9986$, this is a very good fit, as we shall next display.

Press the **ZOOM, 9:ZoomStat** keys and observe the scatterplot and linear regression function in Figure 51. The **TRACE** key shows that, as an example, when X = 1970, the actual population was Y = 203.30 million, as shown in Figure 51, while the predicted population, based on the linear equation, is 203.19 million people, as shown in Figure 52. When in **TRACE** mode, use the up/down arrows to switch between the scatterplot and the regression line. With the cursor on the line, we can also type any value for X within the data range, such as X = 2003, and the predicted y value will appear, in this case, 284.65 million people (Figure 53).

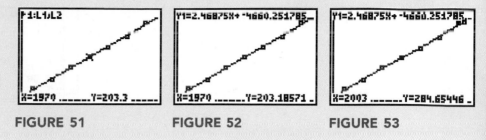

FIGURE 51 **FIGURE 52** **FIGURE 53**

Demographers expect a population to grow exponentially, since the rate of growth of a particular population is usually proportional to the population itself. Let us attempt to fit an exponential curve to the data. Press **STAT,** arrow right to **CALC,** and then arrow up to option **0:ExpReg,** as in Figure 54. Add L1, L2, Y2 after the ExpReg command, so that the exponential curve will be placed into the Y2 location of the equation editor. The display in Figure 55 will result. In Figure 56, we have deselected Y1 and changed Y2 to a bold graphing style to show what the predicted 2003 population would be, based on this exponential curve.

FIGURE 54 **FIGURE 55** **FIGURE 56**

Polynomial Regression

We conclude this data analysis example by fitting a polynomial function to the population data. There are sufficient bends in the data pattern to attempt a curve of the form, $Y3 = aX^4 + bX^3 + cX^2 + dX + e$, a fourth-degree polynomial called a quartic regression form. Press **STAT,** arrow right to **CALC,** and choose selection **7:QuartReg,** as shown in Figure 48. Set up the screen displayed in Figure 57, which yields Figures 58 and 59.

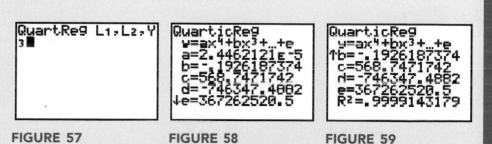

FIGURE 57 **FIGURE 58** **FIGURE 59**

Finally, press **Y=** and deselect Y1 and Y2 and only show Y3, as seen in Figure 60. Tracing to the year 2003 yields the predicted value of 295.36 million people, based on the quartic regression equation, as seen in Figure 61.

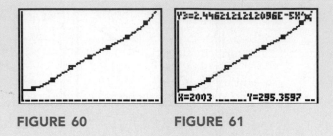

FIGURE 60   **FIGURE 61**

The U.S. Census Bureau uses sophisticated modeling techniques to predict further U.S. populations [*Projections of the Population of the U.S. by Age, Sex, and Location: 1995–2050* (2002), U.S. Bureau of the Census]. For the year 2010, the U.S. projection is 299.86 million. How do the three regression models compare with this projection? The table here provides the comparisons:

Model Predicting Pop. in millions	2010	R^2
Linear regression	301.94	0.998618
Exponential regression	325.10	0.991521
Quartic regression	339.71	0.999914
U.S. Census Bureau projection	299.86	

Based on the 7 data points (U.S. Population for Decades, 1940–2000), the three regression models overestimate the U.S. Census Bureau projection, with the linear estimate the closest. The quartic regression fits the points the best, where R^2 serves as a measure of model fit. Some caution must be noted, however, since a polynomial of higher degree $(n - 1)$ can be made to fit a finite set of n data points perfectly.

CALCULUS

For Business, Economics, and the Social and Life Sciences

CHAPTER 1

Supply and demand determine the price of stock and other commodities.

FUNCTIONS, GRAPHS, AND LIMITS

SECTION 1.1 | Functions

In many practical situations, the value of one quantity may depend on the value of a second. For example, the consumer demand for beef may depend on the current market price; the amount of air pollution in a metropolitan area may depend on the number of cars on the road; or the value of a rare coin may depend on its age. Such relationships can often be represented mathematically as **functions.**

Loosely speaking, a function consists of two sets and a rule that associates elements in one set with elements in the other. For instance, suppose you want to determine the effect of price on the number of units of a particular commodity that will be sold at that price. To study this relationship, you need to know the set of admissible prices, the set of possible sales levels, and a rule for associating each price with a particular sales level. Here is the definition of function we shall use.

> **Function** ■ A **function** is a rule that assigns to each object in a set A exactly one object in a set B. The set A is called the **domain** of the function, and the set of assigned objects in B is called the **range.**

For most functions in this book, the domain and range will be collections of real numbers and the function itself will be denoted by a letter such as f. The value that the function f assigns to the number x in the domain is then denoted by $f(x)$ (read as "f of x"), which is often given by a formula, such as $f(x) = x^2 + 4$.

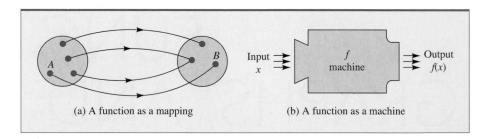

(a) A function as a mapping   (b) A function as a machine

FIGURE 1.1 Interpretations of the function $f(x)$.

It may help to think of such a function as a "mapping" from numbers in A to numbers in B (Figure 1.1a), or as a "machine" that takes a given number from A and converts it into a number in B through a process indicated by the functional rule (Figure 1.1b). For instance, the function $f(x) = x^2 + 4$ can be thought of as an "f machine" that accepts an input x, then squares it and adds 4 to produce an output $y = x^2 + 4$. No matter how you choose to think of a functional relationship, it is important to remember that it assigns one and only one number in the range (output) to each number in the domain (input). Here is an example.

EXPLORE!

Store $f(x) = x^2 + 4$ into your graphing utility. Evaluate at $x = -3, -1, 0, 1,$ and 3. Make a table of values. Repeat using $g(x) = x^2 - 1$. Explain how the values of $f(x)$ and $g(x)$ differ for each x value.

EXAMPLE | 1.1.1

Find $f(3)$ if $f(x) = x^2 + 4$.

Solution

$$f(3) = 3^2 + 4 = 13$$

Observe the convenience and simplicity of the functional notation. In Example 1.1.1, the compact formula $f(x) = x^2 + 4$ completely defines the function, and you can indicate that 13 is the number the function assigns to 3 by simply writing $f(3) = 13$.

It is often convenient to represent a functional relationship by an equation $y = f(x)$, and in this context, x and y are called **variables.** In particular, since the numerical value of y is determined by that of x, we refer to y as the **dependent variable** and to x as the **independent variable.** Note that there is nothing sacred about the symbols x and y. For example, the function $y = x^2 + 4$ can just as easily be represented by $s = t^2 + 4$ or by $w = u^2 + 4$.

Functional notation can also be used to describe tabular data. For instance, Table 1.1 lists the average tuition and fees for private 4-year colleges at 5-year intervals from 1973 to 2003.

TABLE 1.1   Average Tuition and Fees for 4-Year Private Colleges

Academic Year Ending in	Period n	Tuition and Fees
1973	1	$1,898
1978	2	$2,700
1983	3	$4,639
1988	4	$7,048
1993	5	$10,448
1998	6	$13,785
2003	7	$18,273

SOURCE: *Annual Survey of Colleges,* The College Board, New York.

We can describe this data as a function f defined by the rule

$$f(n) = \begin{bmatrix} \text{average tuition and fees at the} \\ \text{beginning of the } n\text{th 5-year period} \end{bmatrix}$$

Thus, $f(1) = 1,898, f(2) = 2,700, \ldots, f(7) = 18,273$. Note that the domain of f is the set of integers $A = \{1, 2, \ldots, 7\}$.

The use of functional notation is illustrated further in Examples 1.1.2 and 1.1.3. In Example 1.1.2, notice that letters other than f and x are used to denote the function and its independent variable.

Just-In-Time **Review**

Recall that $x^{a/b} = \sqrt[b]{x^a}$ whenever a and b are positive integers. Example 1.1.2 uses the case when $a = 1$ and $b = 2$; $x^{1/2}$ is another way of expressing $\sqrt{x}$.

EXAMPLE | 1.1.2

If $g(t) = (t - 2)^{1/2}$, find (if possible) $g(27)$, $g(5)$, $g(2)$, and $g(1)$.

Solution

Rewrite the function as $g(t) = \sqrt{t - 2}$. (If you need to brush up on fractional powers, consult the discussion of exponential notation in Appendix A. Then

$$g(27) = \sqrt{27 - 2} = \sqrt{25} = 5$$
$$g(5) = \sqrt{5 - 2} = \sqrt{3} \approx 1.7321$$

and
$$g(2) = \sqrt{2 - 2} = \sqrt{0} = 0$$

However, $g(1)$ is undefined since

$$g(1) = \sqrt{1-2} = \sqrt{-1}$$

and negative numbers do not have real square roots.

Functions are often defined using more than one formula, where each individual formula describes the function on a subset of the domain. A function defined in this way is sometimes called a **piecewise-defined function.** Here is an example of such a function.

EXAMPLE 1.1.3

Find $f\left(-\dfrac{1}{2}\right)$, $f(1)$, and $f(2)$ if

$$f(x) = \begin{cases} \dfrac{1}{x-1} & \text{if } x < 1 \\ 3x^2 + 1 & \text{if } x \geq 1 \end{cases}$$

Solution

Since $x = -\dfrac{1}{2}$ satisfies $x < 1$, use the top part of the formula to find

$$f\left(-\frac{1}{2}\right) = \frac{1}{-1/2 - 1} = \frac{1}{-3/2} = -\frac{2}{3}$$

However, $x = 1$ and $x = 2$ satisfy $x \geq 1$, so $f(1)$ and $f(2)$ are both found by using the bottom part of the formula:

$$f(1) = 3(1)^2 + 1 = 4 \qquad \text{and} \qquad f(2) = 3(2)^2 + 1 = 13$$

Domain Convention ▪ Unless otherwise specified, if a formula (or several formulas, as in Example 1.1.3) is used to define a function f, then we assume the domain of f to be the set of all numbers for which $f(x)$ is defined (as a real number). We refer to this as the **natural domain** of f. Determining the natural domain of a function often amounts to excluding all numbers x that result in dividing by 0 or taking the square root of a negative number. This procedure is illustrated in Example 1.1.4.

EXAMPLE 1.1.4

Find the domain and range of each of these functions

a. $f(x) = \dfrac{1}{x-3}$ 　　　　　　　　**b.** $g(t) = \sqrt{t-2}$

Solution

a. Since division by any number other than 0 is possible, the domain of f is the set of all numbers $x \neq 3$. The range of f is the set of all numbers y except 0, since for any $y \neq 0$, there is an x such that $y = \dfrac{1}{x-3}$; in particular, $x = 3 + \dfrac{1}{y}$.

b. Since negative numbers do not have real square roots, $g(t)$ can be evaluated only when $t - 2 \geq 0$, so the domain of g is the set of all numbers t such that $t \geq 2$. The range of g is the set of all nonnegative numbers, for if $y \geq 0$ is any such number, there is a t such that $y = \sqrt{t - 2}$; namely, $t = y^2 + 2$.

Functions Used in Economics

There are several basic functions that occur frequently in applications involving economics. A **demand function** $p = D(x)$ is a function that relates the unit price p for a particular commodity to the number of units x demanded by consumers at that price. The **total revenue** is given by the product

$$R(x) = (\text{number of items sold})(\text{price per item})$$
$$= xp = xD(x)$$

If $C(x)$ is the **total cost** of producing the x units, then the **profit** derived from their sale at the unit price p is given by the function

$$P(x) = R(x) - C(x) = xD(x) - C(x)$$

These terms are used in Example 1.1.5.

EXAMPLE 1.1.5

Market research indicates that consumers will buy x thousand units of a particular kind of coffee maker when the unit price is

$$p = -0.27x + 51$$

dollars. The cost of producing the x thousand units is

$$C(x) - 2.23x^2 + 3.5x + 85$$

thousand dollars.

a. What are the demand, revenue, and profit functions, $D(x)$, $R(x)$, and $P(x)$, for this production process?

b. For what values of x is production of the coffee makers profitable?

Solution

a. The demand function is $D(x) = -0.27x + 51$, so the revenue is

$$R(x) = xD(x) = -0.27x^2 + 51x$$

thousand dollars, and the profit is

$$\begin{aligned} P(x) &= R(x) - C(x) \\ &= -0.27x^2 + 51x - (2.23x^2 + 3.5x + 85) \\ &= -2.5x^2 + 47.5x - 85 \end{aligned}$$

thousand dollars.

b. Production is profitable when $P(x) > 0$. We find that

$$\begin{aligned} P(x) &= -2.5x^2 + 47.5x - 85 \\ &= -2.5(x^2 - 19x + 34) \\ &= -2.5(x - 2)(x - 17) \end{aligned}$$

Just-In-Time **Review**

The product of two numbers is positive if they have the same sign and is negative if they have different signs. That is, $ab > 0$ if $a > 0$ and $b > 0$ and also if $a < 0$ and $b < 0$. On the other hand, $ab < 0$ if $a < 0$ and $b > 0$ or if $a > 0$ and $b < 0$.

EXPLORE!

Refer to Example 1.1.6, and store the cost function $C(q)$ into Y1 as
    X³ − 30X² + 500X + 200
Construct a **TABLE** of values for $C(q)$ using your calculator, setting TblStart at X = 5 with an increment ΔTbl = 1 unit. On the table of values observe the cost of manufacturing the 10th unit.

Since the coefficient -2.5 is negative, it follows that $P(x) > 0$ only if the terms $(x - 2)$ and $(x - 17)$ have different signs; that is, when $x - 2 > 0$ and $x - 17 < 0$. Thus, production is profitable for $2 < x < 17$.

Example 1.1.6 illustrates how functional notation is used in a practical situation. Notice that to make the algebraic formula easier to interpret, letters suggesting the relevant practical quantities are used for the function and its independent variable. (In this example, the letter C stands for "cost" and q stands for "quantity" manufactured.)

EXAMPLE | 1.1.6

Suppose the total cost in dollars of manufacturing q units of a certain commodity is given by the function $C(q) = q^3 - 30q^2 + 500q + 200$.

a.   Compute the cost of manufacturing 10 units of the commodity.

b.   Compute the cost of manufacturing the 10th unit of the commodity.

Solution

a.   The cost of manufacturing 10 units is the value of the total cost function when $q = 10$. That is,

$$\begin{aligned} \text{Cost of 10 units} &= C(10) \\ &= (10)^3 - 30(10)^2 + 500(10) + 200 \\ &= \$3{,}200 \end{aligned}$$

b.   The cost of manufacturing the 10th unit is the difference between the cost of manufacturing 10 units and the cost of manufacturing 9 units. That is,

$$\text{Cost of 10th unit} = C(10) - C(9) = 3{,}200 - 2{,}999 = \$201$$

Composition of Functions

There are many situations in which a quantity is given as a function of one variable that, in turn, can be written as a function of a second variable. By combining the functions in an appropriate way, you can express the original quantity as a function of the second variable. This process is called **composition of functions** or **functional composition.**

For instance, suppose environmentalists estimate that when p thousand people live in a certain city, the average daily level of carbon monoxide in the air will be $c(p)$ parts per million, and that separate demographic studies indicate the population in t years will be $p(t)$ thousand. What level of pollution should be expected in t years? You would answer this question by substituting $p(t)$ into the pollution formula $c(p)$ to express c as a composite function of t.

We shall return to the pollution problem in Example 1.1.11 with specific formulas for $c(p)$ and $p(t)$, but first you need to see a few examples of how composite functions are formed and evaluated. Here is a definition of functional composition.

> **Composition of Functions** ■ Given functions $f(u)$ and $g(x)$, the composition $f(g(x))$ is the function of x formed by substituting $u = g(x)$ for u in the formula for $f(u)$.

Note that the composite function $f(g(x))$ "makes sense" only if the domain of f contains the range of g. In Figure 1.2, the definition of composite function is illustrated as an "assembly line" in which "raw" input x is first converted into a transitional product $g(x)$ that acts as input the f machine uses to produce $f(g(x))$.

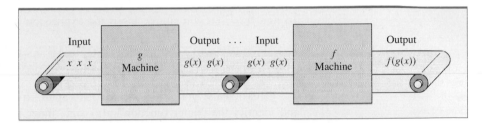

FIGURE 1.2   The composition $f(g(x))$ as an assembly line.

EXAMPLE 1.1.7

Find the composite function $f(g(x))$, where $f(u) = u^2 + 3u + 1$ and $g(x) = x + 1$.

Solution

Replace u by $x + 1$ in the formula for $f(u)$ to get

$$f(g(x)) = (x + 1)^2 + 3(x + 1) + 1$$
$$= (x^2 + 2x + 1) + (3x + 3) + 1$$
$$= x^2 + 5x + 5$$

EXPLORE!

Store the functions $f(x) = x^2$ and $g(x) = x + 3$ into Y1 and Y2, respectively, of the function editor. Deselect (turn off) Y1 and Y2. Set Y3 = Y1(Y2) and Y4 = Y2(Y1). Show graphically (using **ZOOM** Standard) and analytically (by table values) that $f(g(x))$ represented by Y3 and $g(f(x))$ represented by Y4 are not the same functions. What are the explicit equations for both of these composites?

NOTE   By reversing the roles of f and g in the definition of composite function, you can define the composition $g(f(x))$, and, in general, $f(g(x))$ and $g(f(x))$ will *not* be the same. For instance, with the functions in Example 1.1.7, you first write

$$g(w) = w + 1 \qquad \text{and} \qquad f(x) = x^2 + 3x + 1$$

and then replace w by $x^2 + 3x + 1$ to get

$$g(f(x)) = (x^2 + 3x + 1) + 1$$
$$= x^2 + 3x + 2$$

which is equal to $f(g(x)) = x^2 + 5x + 5$ only when $x = -3/2$ (you should verify this). ∎

Example 1.1.7 could have been worded more compactly as follows: Find the composite function $f(x + 1)$ where $f(u) = u^2 + 3u + 1$. The use of this compact notation is illustrated further in Example 1.1.8.

EXPLORE!

Refer to Example 1.1.8. Store $f(x) = 3x^2 + 1/x + 5$ into Y1. Write Y2 = Y1(X − 1). Construct a table of values for Y1 and Y2 for 0, 1, . . . , 6. What do you notice about the values for Y1 and Y2?

EXAMPLE 1.1.8

Find $f(x - 1)$ if $f(x) = 3x^2 + \dfrac{1}{x} + 5$.

Solution

At first glance, this problem may look confusing because the letter x appears both as the independent variable in the formula defining f and as part of the expression

$x - 1$. Because of this, you may find it helpful to begin by writing the formula for f in more neutral terms, say as

$$f(\square) = 3(\square)^2 + \frac{1}{\square} + 5$$

To find $f(x - 1)$, you simply insert the expression $x - 1$ inside each box, getting

$$f(x - 1) = 3(x - 1)^2 + \frac{1}{x - 1} + 5$$

Occasionally, you will have to "take apart" a given composite function $g(h(x))$ and identify the "outer function" $g(u)$ and "inner function" $h(x)$ from which it was formed. The procedure is demonstrated in Example 1.1.9.

EXAMPLE 1.1.9

If $f(x) = \dfrac{5}{x - 2} + 4(x - 2)^3$, find functions $g(u)$ and $h(x)$ such that $f(x) = g(h(x))$.

Solution

The form of the given function is

$$f(x) = \frac{5}{\square} + 4(\square)^3$$

where each box contains the expression $x - 2$. Thus, $f(x) = g(h(x))$, where

$$\underbrace{g(u) = \frac{5}{u} + 4u^3}_{outer\ function} \qquad \text{and} \qquad \underbrace{h(x) = x - 2}_{inner\ function}$$

Actually, in Example 1.1.9, there are infinitely many pairs of functions $g(u)$ and $h(x)$ that combine to give $g(h(x)) = f(x)$. [For example, $g(u) = \dfrac{5}{u + 1} + 4(u + 1)^3$ and $h(x) = x - 3$.] The particular pair selected in the solution to this example is the most natural one and reflects most clearly the structure of the original function $f(x)$.

EXAMPLE 1.1.10

A **difference quotient** is an expression of the general form

$$\frac{f(x + h) - f(x)}{h}$$

where f is a given function of x and h is a number. Difference quotients will be used in Chapter 2 to define the *derivative*, one of the fundamental concepts of calculus. Find the difference quotient for $f(x) = x^2 - 3x$.

Solution

You find that

$$\frac{f(x + h) - f(x)}{h} = \frac{[(x + h)^2 - 3(x + h)] - [x^2 - 3x]}{h}$$

$$= \frac{[x^2 + 2xh + h^2 - 3x - 3h] - [x^2 - 3x]}{h} \qquad \textit{expand the numerator}$$

$$= \frac{2xh + h^2 - 3h}{h} \qquad \textit{combine terms in the numerator}$$

$$= 2x + h - 3 \qquad \textit{divide by } h$$

Example 1.1.11 illustrates how a composite function may arise in an applied problem.

EXAMPLE | 1.1.11

An environmental study of a certain community suggests that the average daily level of carbon monoxide in the air will be $c(p) = 0.5p + 1$ parts per million when the population is p thousand. It is estimated that t years from now the population of the community will be $p(t) = 10 + 0.1t^2$ thousand.

a. Express the level of carbon monoxide in the air as a function of time.

b. When will the carbon monoxide level reach 6.8 parts per million?

Solution

a. Since the level of carbon monoxide is related to the variable p by the equation

$$c(p) = 0.5p + 1$$

and the variable p is related to the variable t by the equation

$$p(t) = 10 + 0.1t^2$$

it follows that the composite function

$$c(p(t)) = c(10 + 0.1t^2) = 0.5(10 + 0.1t^2) + 1 = 6 + 0.05t^2$$

expresses the level of carbon monoxide in the air as a function of the variable t.

b. Set $c(p(t))$ equal to 6.8 and solve for t to get

$$6 + 0.05t^2 = 6.8$$
$$0.05t^2 = 0.8$$
$$t^2 = \frac{0.8}{0.05} = 16$$
$$t = \sqrt{16} = 4 \qquad \textit{discard } t = -4$$

That is, 4 years from now the level of carbon monoxide will be 6.8 parts per million.

P R O B L E M S | **1.1**

In Problems 1 through 11, compute the indicated values of the given function.

1. $f(x) = 3x^2 + 5x - 2$; $f(0), f(-2), f(1)$

2. $h(t) = (2t + 1)^3$; $h(-1), h(0), h(1)$

3. $g(x) = x + \dfrac{1}{x}$; $g(-1), g(1), g(2)$

4. $f(x) = \dfrac{x}{x^2 + 1}$; $f(2), f(0), f(-1)$

5. $h(t) = \sqrt{t^2 + 2t + 4}$; $h(2), h(0), h(-4)$

6. $g(u) = (u + 1)^{3/2}$; $g(0), g(-1), g(8)$

7. $f(t) = (2t - 1)^{-3/2}; f(1), f(5), f(13)$

8. $g(x) = 4 + |x|; g(-2), g(0), g(2)$

9. $f(x) = x - |x - 2|; f(1), f(2), f(3)$

10. $h(x) = \begin{cases} -2x + 4 & \text{if } x \le 1 \\ x^2 + 1 & \text{if } x > 1 \end{cases}; h(3), h(1), h(0), h(-3)$

11. $f(t) = \begin{cases} 3 & \text{if } t < -5 \\ t + 1 & \text{if } -5 \le t \le 5; f(-6), f(-5), f(16) \\ \sqrt{t} & \text{if } t > 5 \end{cases}$

In Problems 12 through 15, determine whether the given function has the set of all real numbers as its domain.

12. $f(x) = \dfrac{x + 1}{x^2 - 1}$

13. $g(x) = \dfrac{x}{1 + x^2}$

14. $h(t) = \sqrt{t^2 + 1}$

15. $f(t) = \sqrt{1 - t}$

In Problems 16 through 21, determine the domain of the given function.

16. $f(x) = x^3 - 3x^2 + 2x + 5$

17. $g(x) = \dfrac{x^2 + 5}{x + 2}$

18. $f(t) = \dfrac{t + 1}{t^2 - t - 2}$

19. $f(x) = \sqrt{2x + 6}$

20. $h(s) = \sqrt{s^2 - 4}$

21. $f(t) = \dfrac{t + 2}{\sqrt{9 - t^2}}$

In Problems 22 through 29, find the composite function $f(g(x))$.

22. $f(u) = u^2 + 4, g(x) = x - 1$

23. $f(u) = 3u^2 + 2u - 6, g(x) = x + 2$

24. $f(u) = (2u + 10)^2, g(x) = x - 5$

25. $f(u) = (u - 1)^3 + 2u^2, g(x) = x + 1$

26. $f(u) = \dfrac{1}{u}, g(x) = x^2 + x - 2$

27. $f(u) = \dfrac{1}{u^2}, g(x) = x - 1$

28. $f(u) = u^2, g(x) = \dfrac{1}{x - 1}$

29. $f(u) = \sqrt{u + 1}, g(x) = x^2 - 1$

In Problems 30 through 33, find the difference quotient of f; namely, $\dfrac{f(x + h) - f(x)}{h}$.

30. $f(x) = 2x + 3$

31. $f(x) = x^2$

32. $f(x) = 4x - x^2$

33. $f(x) = \dfrac{1}{x}$

In Problems 34 through 37, obtain the composite functions $f(g(x))$ and $g(f(x))$, and find all (if any) values of x such that $f(g(x)) = g(f(x))$.

34. $f(x) = x^2 + 1, g(x) = 1 - x$

35. $f(x) = \sqrt{x}, g(x) = 1 - 3x$

36. $f(x) = \dfrac{1}{x}, g(x) = \dfrac{4 - x}{2 + x}$

37. $f(x) = \dfrac{2x + 3}{x - 1}, g(x) = \dfrac{x + 3}{x - 2}$

In Problems 38 through 45, find the indicated composite function.

38. $f(x + 1)$ where $f(x) = x^2 + 5$

39. $f(x - 2)$ where $f(x) = 2x^2 - 3x + 1$

40. $f(x + 3)$ where $f(x) = (2x - 6)^2$

41. $f(x - 1)$ where $f(x) = (x + 1)^5 - 3x^2$

42. $f\left(\dfrac{1}{x}\right)$ where $f(x) = 3x + \dfrac{2}{x}$

43. $f(x^2 + 3x - 1)$ where $f(x) = \sqrt{x}$

44. $f(x^2 - 2x + 9)$ where $f(x) = 2x - 20$

45. $f(x + 1)$ where $f(x) = \dfrac{x - 1}{x}$

In Problems 46 through 51, find functions h(x) and g(u) such that f(x) = g(h(x)).

46. $f(x) = (x^5 - 3x^2 + 12)^3$

47. $f(x) = (x - 1)^2 + 2(x - 1) + 3$

48. $f(x) = \sqrt{3x - 5}$

49. $f(x) = \dfrac{1}{x^2 + 1}$

50. $f(x) = \sqrt{x + 4} - \dfrac{1}{(x + 4)^3}$

51. $f(x) = \sqrt[3]{2 - x} + \dfrac{4}{2 - x}$

52. MANUFACTURING COST   Suppose the total cost in dollars of manufacturing q units of a certain commodity is given by the function
$$C(q) = q^3 - 30q^2 + 400q + 500.$$
  a. Compute the cost of manufacturing 20 units.
  b. Compute the cost of manufacturing the 20th unit.

53. WORKER EFFICIENCY   An efficiency study of the morning shift at a certain factory indicates that an average worker who arrives on the job at 8:00 A.M. will have assembled
$$f(x) = -x^3 + 6x^2 + 15x$$
television sets x hours later.
  a. How many sets will such a worker have assembled by 10:00 A.M.? [*Hint:* At 10:00 A.M., $x = 2$.]
  b. How many sets will such a worker assemble between 9:00 and 10:00 A.M.?

54. TEMPERATURE CHANGE   Suppose that t hours past midnight, the temperature in Miami was $C(t) = -\dfrac{1}{6}t^2 + 4t + 10$ degrees Celsius.
  a. What was the temperature at 2:00 A.M.?
  b. By how much did the temperature increase or decrease between 6:00 and 9:00 P.M.?

55. POPULATION GROWTH   It is estimated that t years from now, the population of a certain suburban community will be $P(t) = 20 - \dfrac{6}{t + 1}$ thousand.
  a. What will be the population of the community 9 years from now?
  b. By how much will the population increase during the 9th year?
  c. What happens to $P(t)$ as t gets larger and larger? Interpret your result.

56. EXPERIMENTAL PSYCHOLOGY   To study the rate at which animals learn, a psychology student performed an experiment in which a rat was sent repeatedly through a laboratory maze. Suppose that the time required for the rat to traverse the maze on the nth trial was approximately
$$T(n) = 3 + \dfrac{12}{n}$$
minutes.
  a. What is the domain of the function T?
  b. For what values of n does $T(n)$ have meaning in the context of the psychology experiment?
  c. How long did it take the rat to traverse the maze on the 3rd trial?
  d. On which trial did the rat first traverse the maze in 4 minutes or less?
  e. According to the function T, what will happen to the time required for the rat to traverse the maze as the number of trials increases? Will the rat ever be able to traverse the maze in less than 3 minutes?

57. BLOOD FLOW   Biologists have found that the speed of blood in an artery is a function of the distance of the blood from the artery's central axis. According to **Poiseuille's law,**[*] the speed (in centimeters per second) of blood that is r centimeters from the central axis of an artery is given by the function $S(r) = C(R^2 - r^2)$, where C is a constant and R is the radius of the artery. Suppose that for a certain artery, $C = 1.76 \times 10^5$ and $R = 1.2 \times 10^{-2}$ centimeters.
  a. Compute the speed of the blood at the central axis of this artery.

---

[*]Edward Batschelet, *Introduction to Mathematics for Life Scientists,* 3rd ed., New York: Springer-Verlag, 1979, pp. 101–103.

b. Compute the speed of the blood midway between the artery's wall and central axis.

58. **DISTRIBUTION COST**   Suppose that the number of worker-hours required to distribute new telephone books to $x\%$ of the households in a certain rural community is given by the function

$$W(x) = \frac{600x}{300 - x}.$$

   a. What is the domain of the function W?
   b. For what values of x does $W(x)$ have a practical interpretation in this context?
   c. How many worker-hours were required to distribute new telephone books to the first 50% of the households?
   d. How many worker-hours were required to distribute new telephone books to the entire community?
   e. What percentage of the households in the community had received new telephone books by the time 150 worker-hours had been expended?

59. **ISLAND ECOLOGY**   Observations show that on an island of area A square miles, the average number of animal species is approximately equal to $s(A) = 2.9\sqrt[3]{A}$.
   a. On average, how many animal species would you expect to find on an island of area 8 square miles?
   b. If s_1 is the average number of species on an island of area A and s_2 is the average number of species on an island of area $2A$, what is the relationship between s_1 and s_2?
   c. How big must an island be to have an average of 100 animal species?

60. **POPULATION DENSITY**   Observations suggest that for herbivorous mammals, the number of animals N per square kilometer can be estimated by the formula $N = \dfrac{91.2}{m^{0.73}}$, where m is the average mass of the animal in kilograms.
   a. Assuming that the average elk on a particular reserve has mass 300 kg, approximately how many elk would you expect to find per square kilometer in the reserve?
   b. Using this formula, it is estimated that there is less than one animal of a certain species per square kilometer. How large can the average animal of this species be?
   c. One species of large mammal has twice the average mass as a second species. If a particular reserve contains 100 animals of the larger

species, how many animals of the smaller species would you expect to find there?

61. **IMMUNIZATION**   Suppose that during a nationwide program to immunize the population against a certain form of influenza, public health officials found that the cost of inoculating $x\%$ of the population was approximately $C(x) = \dfrac{150x}{200 - x}$ million dollars.
   a. What is the domain of the function C?
   b. For what values of x does $C(x)$ have a practical interpretation in this context?
   c. What was the cost of inoculating the first 50% of the population?
   d. What was the cost of inoculating the second 50% of the population?
   e. What percentage of the population had been inoculated by the time 37.5 million dollars had been spent?

62. **POSITION OF A MOVING OBJECT**   A ball has been dropped from the top of a building. Its height (in feet) after t seconds is given by the function $H(t) = -16t^2 + 256$.
   a. What is the height of the ball after 2 seconds?
   b. How far will the ball travel during the third second?
   c. How tall is the building?
   d. When will the ball hit the ground?

63. **AIR POLLUTION**   An environmental study of a certain suburban community suggests that the average daily level of carbon monoxide in the air will be $c(p) = 0.4p + 1$ parts per million when the population is p thousand. It is estimated that t years from now the population of the community will be $p(t) = 8 + 0.2t^2$ thousand.
   a. Express the level of carbon monoxide in the air as a function of time.
   b. What will the carbon monoxide level be 2 years from now?
   c. When will the carbon monoxide level reach 6.2 parts per million?

64. **MANUFACTURING COST**   At a certain factory, the total cost of manufacturing q units during the daily production run is $C(q) = q^2 + q + 900$ dollars. On a typical workday, $q(t) = 25t$ units are manufactured during the first t hours of a production run.
   a. Express the total manufacturing cost as a function of t.

b. How much will have been spent on production by the end of the third hour?

c. When will the total manufacturing cost reach $11,000?

In Problems 65 through 68, the demand function D(x) and the total cost function C(x) for a particular commodity are given in terms of the level of production x. In each case, find:
(a) The revenue R(x) and profit P(x).
(b) All values of x for which production of the commodity is profitable.

65. $D(x) = -0.02x + 29$
$C(x) = 1.43x^2 + 18.3x + 15.6$

66. $D(x) = -0.37x + 47$
$C(x) = 1.38x^2 + 15.15x + 115.5$

67. $D(x) = -0.5x + 39$
$C(x) = 1.5x^2 + 9.2x + 67$

68. $D(x) = -0.09x + 51$
$C(x) = 1.32x^2 + 11.7x + 101.4$

69. CONSUMER DEMAND An importer of Brazilian coffee estimates that local consumers will buy approximately $Q(p) = \dfrac{4,374}{p^2}$ kilograms of the coffee per week when the price is p dollars per kilogram. It is estimated that t weeks from now the price of this coffee will be

$$p(t) = 0.04t^2 + 0.2t + 12$$

dollars per kilogram

a. Express the weekly consumer demand for the coffee as a function of t.

b. How many kilograms of the coffee will consumers be buying from the importer 10 weeks from now?

c. When will the demand for the coffee be 30.375 kilograms?

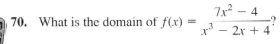 **70.** What is the domain of $f(x) = \dfrac{7x^2 - 4}{x^3 - 2x + 4}$?

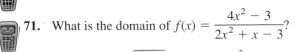 **71.** What is the domain of $f(x) = \dfrac{4x^2 - 3}{2x^2 + x - 3}$?

72. For $f(x) = 2\sqrt{x - 1}$ and $g(x) = x^3 - 1.2$, find $g(f(4.8))$. Use two decimal places.

73. For $f(x) = 2\sqrt{x - 1}$ and $g(x) = x^3 - 1.2$, find $f(g(2.3))$. Use two decimal places.

74. COST OF EDUCATION The accompanying Table 1.2 gives the average annual total fixed costs (tuition, fees, room and board) for undergraduates by institution type in constant (inflation-adjusted) 2002 dollars for the academic years 1987–1988 to 2002–2003. Define the **cost of education index** (CEI) for a particular academic year to be the ratio of the total fixed cost for that year to the total fixed cost for the base academic year ending in 1990. For example, for 4-year public institutions in the academic year ending in 2000, the cost of education index is

$$CEI(2000) = \frac{8,311}{6,476} = 1.28$$

a. Compute the CEI for your particular type of institution for each of the 16 academic years shown in the table. What is the average annual increase in CEI over the 16-year period for your type of institution?

b. Compute the CEI for all four institution types for the academic year ending in 2003 and interpret your results.

 c. Write a paragraph on the cost of education index. Can it continue to rise as it has? What do you think will happen eventually?

75. VALUE OF EDUCATION The accompanying Table 1.3 gives the average income in constant (inflation-adjusted) 2002 dollars by educational attainment for persons at least 18 years old for the decade 1991–2000. Define the **value of education index** (VEI) for a particular level of education in a given year to be the ratio of average income earned in that year to the average income earned by the lowest level of education (no high school diploma) for the same year. For example, for a person with a bachelor's degree in 1995, the value of education index is

$$VEI(1995) = \frac{43,450}{16,465} = 2.64$$

a. Compute the VEI for each year in the decade 1991–2000 for the level of education you hope to attain.

b. Compare the VEI for the year 2000 for the four different educational levels requiring at least a high school diploma. Interpret your results.

TABLE 1.2   Average Annual Total Fixed Costs of Education (Tuition, Fees, Room and Board) by Institutional Type in Constant (Inflation-Adjusted) 2002 Dollars

Sector/Year	87–88	88–89	89–90	90–91	91–92	92–93	93–94	94–95	95–96	96–97	97–98	98–99	99–00	00–01	01–02	02–03
2-yr public	1,112	1,190	1,203	1,283	1,476	1,395	1,478	1,517	1,631	1,673	1,701	1,699	1,707	1,752	1,767	1,914
2-yr private	10,640	11,159	10,929	11,012	11,039	11,480	12,130	12,137	12,267	12,328	12,853	13,052	13,088	13,213	13,375	14,202
4-yr public	6,382	6,417	6,476	6,547	6,925	7,150	7,382	7,535	7,680	7,784	8,033	8,214	8,311	8,266	8,630	9,135
4-yr private	13,888	14,852	14,838	15,330	15,747	16,364	16,765	17,216	17,560	17,999	18,577	18,998	19,368	19,636	20,783	21,678

All data are **unweighted averages,** intended to reflect the average prices set by institutions.   SOURCE: *Annual Survey of Colleges.* The College Board, New York, NY.

TABLE 1.3   Average Annual Income by Educational Attainment in Constant 2002 Dollars

Level of Education/Year	1991	1992	1993	1994	1995	1996	1997	1998	1999	2000
No high school diploma	16,582	16,344	15,889	16,545	16,465	17,135	17,985	17,647	17,346	18,727
High school diploma	24,007	23,908	24,072	24,458	25,180	25,289	25,537	25,937	26,439	27,097
Some college	27,017	26,626	26,696	26,847	28,037	28,744	29,263	30,304	30,561	31,212
Bachelor's degree	41,178	41,634	43,529	44,963	43,450	43,505	45,150	48,131	49,149	51,653
Advanced degree	60,525	62,080	69,145	67,770	66,581	69,993	70,527	69,777	72,841	72,175

SOURCE: U.S. Census Bureau website (www.census.gov/hhes/income/histinc/p28).

SECTION   1.2 | The Graph of a Function

Graphs have visual impact. They also reveal information that may not be evident from verbal or algebraic descriptions. Two graphs depicting practical relationships are shown in Figure 1.3.

The graph in Figure 1.3a describes the variation in total industrial production in a certain country over a 4-year period of time. Notice that the highest point on the graph occurs near the end of the third year, indicating that production was greatest at that time.

The graph in Figure 1.3b represents population growth when environmental factors impose an upper bound on the possible size of the population. It indicates that the *rate* of population growth increases at first and then decreases as the size of the population gets closer and closer to the upper bound.

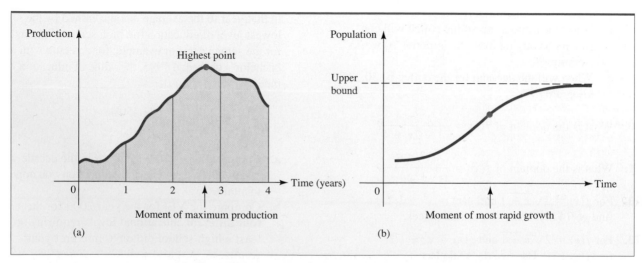

FIGURE 1.3   (a) A production function. (b) Bounded population growth.

To represent a function $y = f(x)$ geometrically as a graph, it is common practice to use a rectangular coordinate system on which units for the independent variable x are marked on the horizontal axis and those for the dependent variable y are marked on the vertical axis.

The Graph of a Function ■ The graph of a function f consists of all points (x, y) where x is in the domain of f and $y = f(x)$; that is, all points of the form $(x, f(x))$.

In Chapter 3, you will see efficient techniques involving calculus that can be used to draw accurate graphs of functions. For many functions, however, you can make a fairly good sketch by plotting a few points, as illustrated in Example 1.2.1.

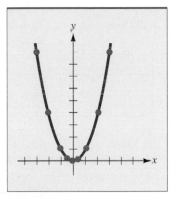

FIGURE 1.4 The graph of $y = x^2$.

EXAMPLE | 1.2.1

Graph the function $f(x) = x^2$.

Solution

Begin by constructing the table

x	-3	-2	-1	$-\dfrac{1}{2}$	0	$\dfrac{1}{2}$	1	2	3
$y = x^2$	9	4	1	$\dfrac{1}{4}$	0	$\dfrac{1}{4}$	1	4	9

Then plot the points (x, y) and connect them with the smooth curve shown in Figure 1.4.

NOTE Many different curves pass through the points in Example 1.2.1. Several of these curves are shown in Figure 1.5. There is no way to guarantee that the curve we pass through the plotted points is the actual graph of f. However, in general, the more points that are plotted, the more likely the graph is to be reasonably accurate. ■

EXPLORE!

Store $f(x) = x^2$ into Y1 of the equation editor, using a bold graphing style. Represent $g(x) = x^2 + 2$ by Y2 = Y1 + 2 and $h(x) = x^2 - 3$ by Y3 = Y1 − 3. Use **ZOOM** decimal graphing to show how the graphs of $g(x)$ and $h(x)$ relate to that of $f(x)$. Now deselect Y2 and Y3 and write Y4 = Y1(X + 2) and Y5 = Y1(X − 3). Explain how the graphs of Y1, Y4, and Y5 relate.

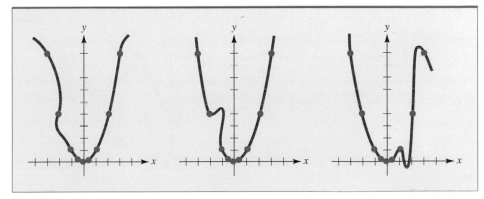

FIGURE 1.5 Other graphs through the points in Example 1.2.1.

Example 1.2.2 illustrates how to sketch the graph of a function defined by more than one formula.

EXPLORE!

Certain functions that are defined piecewise can be entered into a graphing calculator using indicator functions in sections. For example, the absolute value function,

$$f(x) = |x| = \begin{cases} x & \text{if } x \geq 0 \\ -x & \text{if } x < 0 \end{cases}$$

can be represented by Y1 = X(X ≥ 0) + (−X)(X < 0). Now represent the function in Example 1.2.2, using indicator functions, and graph it with an appropriate viewing window. [*Hint:* You will need to represent the interval, 0 < X < 1, by the boolean expression, (0 < X)(X < 1).]

EXAMPLE 1.2.2

Graph the function

$$f(x) = \begin{cases} 2x & \text{if } 0 \leq x < 1 \\ \dfrac{2}{x} & \text{if } 1 \leq x < 4 \\ 3 & \text{if } x \geq 4 \end{cases}$$

Solution

When making a table of values for this function, remember to use the formula that is appropriate for each particular value of x. Using the formula $f(x) = 2x$ when $0 \leq x < 1$, the formula $f(x) = \dfrac{2}{x}$ when $1 \leq x < 4$, and the formula $f(x) = 3$ when $x \geq 4$, you can compile this table:

x	0	$\dfrac{1}{2}$	1	2	3	4	5	6
$f(x)$	0	1	2	1	$\dfrac{2}{3}$	3	3	3

Now plot the corresponding points $(x, f(x))$ and draw the graph as in Figure 1.6. Notice that the pieces for $0 \leq x < 1$ and $1 \leq x < 4$ are connected to one another at $(1, 2)$ but that the piece for $x \geq 4$ is separated from the rest of the graph. [The "open dot" at $\left(4, \dfrac{1}{2}\right)$ indicates that the graph approaches this point but that the point is not actually on the graph.]

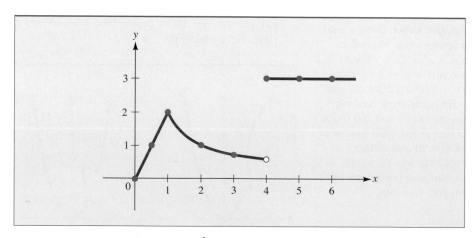

FIGURE 1.6 The graph of $f(x) = \begin{cases} 2x & 0 \leq x < 1 \\ \dfrac{2}{x} & 1 \leq x < 4 \\ 3 & x \geq 4 \end{cases}$

Intercepts

The points (if any) where a graph crosses the x axis are called **x intercepts,** and similarly, a **y intercept** is a point where the graph crosses the y axis. Intercepts are key features of a graph and can be determined using algebra or technology in conjunction with these criteria.

EXPLORE!

Using your graphing utility, locate the x intercepts of $f(x) = -x^2 + x + 2$. These intercepts can be located by first using the **ZOOM** button and then confirmed by using the root finding feature of the graphing utility. Do the same for $g(x) = x^2 + x - 4$. What radical form do these roots have?

How to Find the x and y Intercepts

To find any x intercept of a graph, set $y = 0$ and solve for x. To find any y intercept, set $x = 0$ and solve for y. For a function f, the only y intercept is $y_0 = f(0)$, but finding x intercepts may be difficult.

EXAMPLE | 1.2.3

Graph the function $f(x) = -x^2 + x + 2$. Include all x and y intercepts.

Solution

The y intercept is $f(0) = 2$. To find the x intercepts, solve the equation $f(x) = 0$. Factoring, we find that

$$-x^2 + x + 2 = 0$$
$$-(x + 1)(x - 2) = 0 \quad \text{factor}$$
$$x = -1, \; x = 2 \quad \text{uv = 0 if and only if u = 0 or}$$
$$\text{v = 0}$$

Thus, the x intercepts are $(-1, 0)$ and $(2, 0)$.

Next, make a table of values and plot the corresponding points $(x, f(x))$.

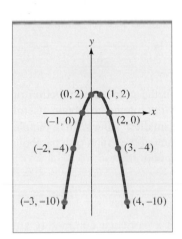

FIGURE 1.7 The graph of $f(x) = -x^2 + x + 2$.

x	-3	-2	-1	0	1	2	3	4
$f(x)$	-10	-4	0	2	2	0	-4	-10

The graph of f is shown in Figure 1.7.

> **NOTE** The factoring in Example 1.2.3 is fairly straightforward, but in other problems, you may need to review the factoring procedure provided in Appendix A2.

Graphing Parabolas

The graphs in Figures 1.4 and 1.7 are called **parabolas.** In general, the graph of $y = Ax^2 + Bx + C$ is a parabola as long as $A \neq 0$. All parabolas have a "U shape," and the parabola $y = Ax^2 + Bx + C$ opens up if $A > 0$ and down if $A < 0$. The "peak" or "valley" of the parabola is called its **vertex,** and it always occurs where $x = \dfrac{-B}{2A}$ (Figure 1.8; see also Problem 58). These features of the parabola are easily obtained by the methods of calculus developed in Chapter 3. Note that to get a reasonable sketch of the parabola $y = Ax^2 + Bx + C$, you need only determine three key features:

1. The location of the vertex $\left(\text{where } x = \dfrac{-B}{2A} \right)$

2. Whether the parabola opens up ($A > 0$) or down ($A < 0$)

3. Any intercepts

For instance, in Example 1.2.3, the parabola $y = -x^2 + x + 2$ opens downward (since $A = -1$ is negative) and has its vertex (high point) at $x = \dfrac{-B}{2A} = \dfrac{-1}{2(-1)} = \dfrac{1}{2}$.

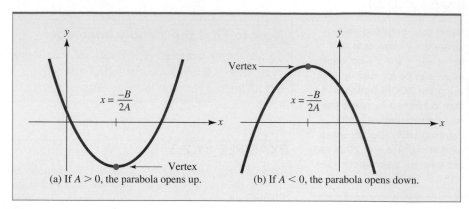

(a) If $A > 0$, the parabola opens up.   (b) If $A < 0$, the parabola opens down.

FIGURE 1.8   The graph of the parabola $y = Ax^2 + Bx + C$.

In Chapter 3, we will develop a procedure in which the graph of a function of practical interest is first obtained by calculus and then interpreted to obtain useful information about the function, such as its largest and smallest values. In Example 1.2.4 we preview this procedure by using what we know about the graph of a parabola to determine the maximum revenue obtained in a production process.

EXAMPLE | 1.2.4

A manufacturer determines that when x hundred units of a particular commodity are produced, they can all be sold for a unit price given by the demand function $p = 60 - x$ dollars. At what level of production is revenue maximized? What is the maximum revenue?

Solution

The revenue derived from producing x hundred units and selling them all at $60 - x$ dollars is $R(x) = x(60 - x)$ hundred dollars. Note that $R(x) \geq 0$ only for $0 \leq x \leq 60$. The graph of the revenue function

$$R(x) = x(60 - x) = -x^2 + 60x$$

is a parabola that opens downward (since $A = -1 < 0$) and has its high point (vertex) at

$$x = \frac{-B}{2A} = \frac{-60}{2(-1)} = 30$$

as shown in Figure 1.9. Thus, revenue is maximized when $x = 30$ hundred units are produced, and the corresponding maximum revenue is

$$R(30) = 30(60 - 30) = 900$$

hundred dollars. The manufacturer should produce 3,000 units and at that level of production should expect maximum revenue of \$90,000.

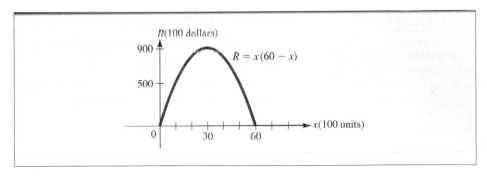

FIGURE 1.9   A revenue function.

Note that we can also find the largest value of $R(x) = -x^2 + 60x$ by completing the square:

$$R(x) = -x^2 + 60x = -(x^2 - 60x) \qquad \textit{factor out } -1, \textit{ the coefficient of } x$$
$$= -(x^2 - 60x + 900) + 900 \qquad \textit{complete the square inside}$$
$$\underbrace{\qquad}_{-900} \; \underbrace{\qquad}_{+900} \qquad \textit{parentheses by adding}$$
$$\qquad (-60/2)^2 - 900$$
$$= -(x - 30)^2 + 900$$

Thus, $R(30) = 0 + 900 = 900$ and if c is any number other than 30, then

$$R(c) = -(c - 30)^2 + 900 < 900 \qquad \textit{since } -(c - 30)^2 < 0$$

so the maximum revenue is 900 when $x = 30$.

Intersections of Graphs

Sometimes it is necessary to determine when two functions are equal. For instance, an economist may wish to compute the market price at which the consumer demand for a commodity will be equal to supply. Or a political analyst may wish to predict how long it will take for the popularity of a certain challenger to reach that of the incumbent. We shall examine some of these applications in Section 1.4.

In geometric terms, the values of x for which two functions $f(x)$ and $g(x)$ are equal are the x coordinates of the points where their graphs intersect. In Figure 1.10, the graph of $y = f(x)$ intersects that of $y = g(x)$ at two points, labeled P and Q. To find the points of intersection algebraically, set $f(x)$ equal to $g(x)$ and solve for x. This procedure is illustrated in Example 1.2.5.

Just-In-Time **Review**

The **quadratic formula** is used in Example 1.2.5. Recall that this result says that the equation $Ax^2 + Bx + C = 0$ has real solutions if and only if $B^2 - 4AC \geq 0$, in which case, the solutions are

$$r_1 = \frac{-B + \sqrt{B^2 - 4AC}}{2A}$$

and

$$r_2 = \frac{-B - \sqrt{B^2 - 4AC}}{2A}$$

A review of the quadratic formula may be found in Appendix A2.

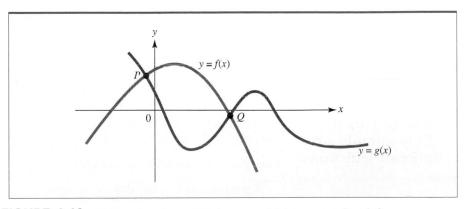

FIGURE 1.10   The graphs of $y = f(x)$ and $y = g(x)$ intersect at P and Q.

EXAMPLE 1.2.5

Find all points of intersection of the graphs of $f(x) = 3x + 2$ and $g(x) = x^2$.

Solution

You must solve the equation $x^2 = 3x + 2$. Rewrite the equation as $x^2 - 3x - 2 = 0$ and apply the quadratic formula to obtain

$$x = \frac{-(-3) \pm \sqrt{(-3)^2 - 4(1)(-2)}}{2(1)} = \frac{3 \pm \sqrt{17}}{2}$$

The solutions are

$$x = \frac{3 + \sqrt{17}}{2} \approx 3.56 \qquad \text{and} \qquad x = \frac{3 - \sqrt{17}}{2} \approx -0.56$$

(The computations were done on a calculator, with results rounded off to two decimal places.)

Computing the corresponding y coordinates from the equation $y = x^2$, you find that the points of intersection are approximately $(3.56, 12.67)$ and $(-0.56, 0.31)$. [As a result of round-off errors, you will get slightly different values for the y coordinates if you substitute into the equation $y = 3x + 2$.] The graphs and the intersection points are shown in Figure 1.11.

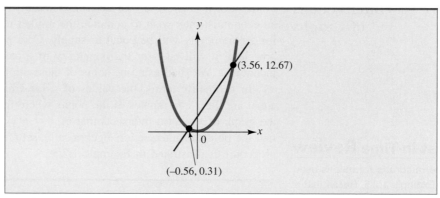

FIGURE 1.11 The intersection of the graphs of $f(x) = 3x + 2$ and $g(x) = x^2$.

Power Functions, Polynomials, and Rational Functions

A **power function** is a function of the form $f(x) = x^n$, where n is a real number. For example, $f(x) = x^2$, $f(x) = x^{-3}$, and $f(x) = x^{1/2}$ are all power functions. So are $f(x) = \frac{1}{x^2}$ and $f(x) = \sqrt[3]{x}$ since they can be rewritten as $f(x) = x^{-2}$ and $f(x) = x^{1/3}$, respectively.

A **polynomial** is a function of the form

$$p(x) = a_n x^n + a_{n-1}x^{n-1} + \cdots + a_1 x + a_0$$

EXPLORE!

Use your calculator to graph the third-degree polynomial $f(x) = x^3 - x^2 - 6x + 3$. Conjecture the values of the x intercepts and confirm them using the root finding feature of your calculator.

where n is a nonnegative integer and $a_0, a_1, \ldots, a_n$ are constants. If $a_n \neq 0$, the integer n is called the **degree** of the polynomial. For example, $f(x) = 3x^5 - 6x^2 + 7$ is a polynomial of degree 5. It can be shown that the graph of a polynomial of degree n is an unbroken curve that crosses the x axis no more than n times. To illustrate some of the possibilities, the graphs of three polynomials of degree 3 are shown in Figure 1.12.

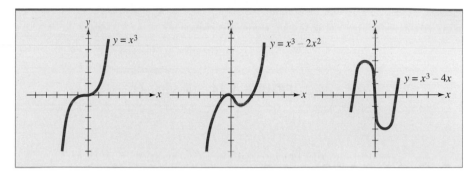

FIGURE 1.12   Three polynomials of degree 3.

A quotient $\dfrac{p(x)}{q(x)}$ of two polynomials $p(x)$ and $q(x)$ is called a **rational function.** Such functions appear throughout this text in examples and exercises. Graphs of three rational functions are shown in Figure 1.13. You will learn how to sketch such graphs in Section 3.3 of Chapter 3.

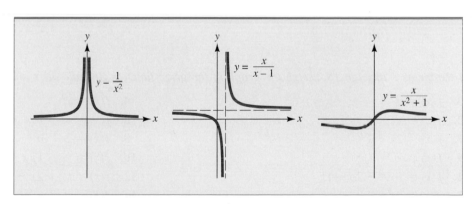

FIGURE 1.13   Graphs of three rational functions.

The Vertical Line Test

It is important to realize that not every curve is the graph of a function (Figure 1.14). For instance, suppose the circle $x^2 + y^2 = 5$ were the graph of some function $y = f(x)$. Then, since the points $(1, 2)$ and $(1, -2)$ both lie on the circle, we would have $f(1) = 2$ and $f(1) = -2$, contrary to the requirement that a function assigns one and *only* one value to each number in its domain. Here is a geometric rule for determining whether a curve is the graph of a function.

The Vertical Line Test ■ A curve is the graph of a function if and only if no vertical line intersects the curve more than once.

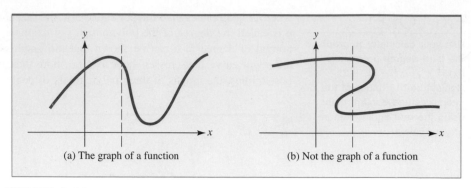

FIGURE 1.14 The vertical line test.

PROBLEMS 1.2

In Problems 1 and 2, classify each function as a polynomial, a power function, or a rational function. If the function is not one of these types, classify it as "different."

1. a. $f(x) = x^{1.4}$
    b. $f(x) = -2x^3 - 3x^2 + 8$
    c. $f(x) = (3x - 5)(4 - x)^2$
    d. $f(x) = \dfrac{3x^2 - x + 1}{4x + 7}$

2. a. $f(x) = -2 + 3x^2 + 5x^4$
    b. $f(x) = \sqrt{x} + 3x$
    c. $f(x) = \dfrac{(x - 3)(x + 7)}{-5x^3 - 2x^2 + 3}$
    d. $f(x) = \left(\dfrac{2x + 9}{x^2 - 3}\right)^3$

In Problems 3 through 18, sketch the graph of the given function. Include all x and y intercepts.

3. $f(x) = x$

4. $f(x) = x^2$

5. $f(x) = x^3$

6. $f(x) = x^4$

7. $f(x) = 2x - 1$

8. $f(x) = 2 - 3x$

9. $f(x) = x(2x + 5)$

10. $f(x) = (x - 1)(x + 2)$

11. $f(x) = -x^2 - 2x + 15$

12. $f(x) = x^2 + 2x - 8$

13. $f(x) = \sqrt{x}$

14. $f(x) = -x^3 + 1$

15. $f(x) = \begin{cases} x - 1 & \text{if } x \le 0 \\ x + 1 & \text{if } x > 0 \end{cases}$

16. $f(x) = \begin{cases} 2x - 1 & \text{if } x < 2 \\ 3 & \text{if } x \ge 2 \end{cases}$

17. $f(x) = \begin{cases} x^2 + x - 3 & \text{if } x < 1 \\ 1 - 2x & \text{if } x \ge 1 \end{cases}$

18. $f(x) = \begin{cases} 9 - x & \text{if } x \le 2 \\ x^2 + x - 2 & \text{if } x > 2 \end{cases}$

In Problems 19 through 24, find the points of intersection (if any) of the given pair of curves and draw the graphs.

19. $y = 3x + 5$ and $y = -x + 3$

20. $y = 3x + 8$ and $y = 3x - 2$

21. $y = x^2$ and $y = 3x - 2$

22. $y = x^2 - x$ and $y = x - 1$

23. $3y - 2x = 5$ and $y + 3x = 9$

24. $2x - 3y = -8$ and $3x - 5y = -13$

In each of Problems 25 through 28, the graph of a function $f(x)$ is given. In each case find:
  (a) The y intercept.
  (b) All x intercepts.
  (c) The largest value of $f(x)$ and the value(s) of x for which it occurs.
  (d) The smallest value of $f(x)$ and the value(s) of x for which it occurs.

25.

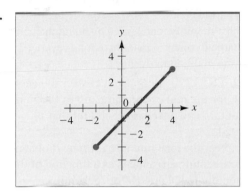

26.

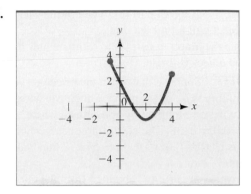

27.

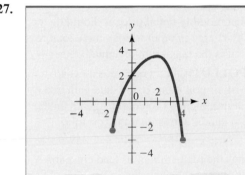

28.

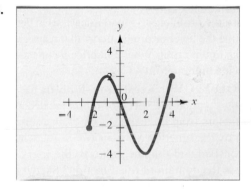

29. **MANUFACTURING COST** A manufacturer can produce cassette tape recorders at a cost of $40 apiece. It is estimated that if the tape recorders are sold for p dollars apiece, consumers will buy $120 - p$ of them a month. Express the manufacturer's monthly profit as a function of price, graph this function, and use the graph to estimate the optimal selling price.

30. **RETAIL SALES** A bookstore can obtain an atlas from the publisher at a cost of $10 per copy and estimates that if it sells the atlas for x dollars per copy, approximately $20(22 - x)$ copies will be sold each month. Express the bookstore's monthly profit from the sale of the atlas as a function of price, graph this function, and use the graph to estimate the optimal selling price.

31. **CONSUMER EXPENDITURE** Suppose $x = -200p + 12,000$ units of a particular commodity are sold each month when the market price is p dollars per unit. The total monthly consumer expenditure E is the total amount of money spent by consumers during each month.
   a. Express total monthly consumer expenditure E as a function of the unit price p, and sketch the graph of $E(p)$.
   b. Discuss the economic significance of the p intercepts of the expenditure function $E(p)$.
   c. Use the graph in part (a) to determine the market price that generates the greatest total monthly consumer expenditure.

32. MOTION OF A PROJECTILE If an object is thrown vertically upward from the ground with an initial speed of 160 feet per second, its height (in feet) t seconds later is given by the function $H(t) = -16t^2 + 160t$.
  a. Graph the function $H(t)$.
  b. Use the graph in part (a) to determine when the object will hit the ground.
  c. Use the graph in part (a) to estimate how high the object will rise.

33. PROFIT Suppose that when the price of a certain commodity is p dollars per unit, then x hundred units will be purchased by consumers, where $p = -0.05x + 38$. The cost of producing x hundred units is $C(x) = 0.02x^2 + 3x + 574.77$ hundred dollars.
  a. Express the profit P obtained from the sale of x hundred units as a function of x. Sketch the graph of the profit function.
  b. Use the profit curve found in part (a) to determine the level of production x that results in maximum profit. What unit price p corresponds to maximum profit?

34. BLOOD FLOW Recall from Problem 57, Section 1.1, that the speed of blood located r centimeters from the central axis of an artery is given by the function $S(r) = C(R^2 - r^2)$, where C is a constant and R is the radius of the artery.* What is the domain of this function? Sketch the graph of $S(r)$.

35. ROAD SAFETY When an automobile is being driven at v miles per hour, the average driver requires D feet of visibility to stop safely, where $D = 0.065v^2 + 0.148v$. Sketch the graph of $D(v)$.

36. POSTAGE RATES Effective June 30, 2002, the cost of mailing a first-class letter, flat, or parcel was 37 cents for the first ounce and 23 cents for each additional ounce or fraction of an ounce. Let $P(w)$ be the postage required for mailing a parcel weighing w ounces, for $0 \leq w \leq 10$.
  a. Describe $P(w)$ as a piecewise-defined function.
  b. Sketch the graph of P.

37. REAL ESTATE A real estate company manages 150 apartments in Irvine, California. All the apartments can be rented at a price of $1,200 per month,

but for each $100 increase in the monthly rent, there will be five additional vacancies.
  a. Express the total monthly revenue R obtained from renting apartments as a function of the monthly rental price p for each unit.
  b. Sketch the graph of the revenue function found in part (a).
  c. What monthly rental price p should the company charge in order to maximize total revenue? What is the maximum revenue?

38. REAL ESTATE Suppose it costs $500 each month for the real estate company in Problem 37 to maintain and advertise each unit that is unrented.
  a. Express the total monthly revenue R obtained from renting apartments as a function of the monthly rental price p for each unit.
  b. Sketch the graph of the revenue function found in part (a).
  c. What monthly rental price p should the company charge in order to maximize total revenue? What is the maximum revenue?

39. AIR POLLUTION Lead emissions are a major source of air pollution. Using data gathered by the U.S. Environmental Protection Agency in the 1990s, it can be shown that the formula
$$N(t) = -35t^2 + 299t + 3,347$$
estimates the total amount of lead emission N (in thousands of tons) occurring in the United States t years after the base year 1990.
  a. Sketch the graph of the pollution function $N(t)$.
  b. Approximately how much lead emission did the formula predict for the year 1995? (The actual amount was about 3,924 thousand tons.)
  c. Based on this formula, when during the decade 1990–2000 would you expect the maximum lead emission to have occurred?
  d. Can this formula be used to predict the current level of lead emission? Explain.

40. ARCHITECTURE An arch over a road has a parabolic shape. It is 6 meters wide at the base and is just tall enough to allow a truck 5 meters high and 4 meters wide to pass.
  a. Assuming that the arch has an equation of the form $y = ax^2 + b$, use the given information to find a and b. Explain why this assumption is reasonable.
  b. Sketch the graph of the arch equation you found in part (a).

---

*Edward Batschelet, *Introduction to Mathematics for Life Scientists*, 3rd ed., New York: Springer-Verlag, 1979, pp. 101–103.

In Problems 41 through 44, use the vertical line test to determine whether or not the given curve is the graph of a function

41.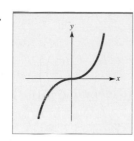

42.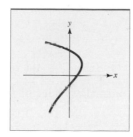

43.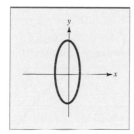

44.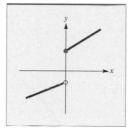

45. What viewing rectangle should be used to get an adequate graph for the quadratic function
$$f(x) = -9x^2 + 3,600x - 358,200?$$

46. What viewing rectangle should be used to get an adequate graph for the quadratic function
$$f(x) = 4x^2 - 2,400x + 355,000?$$

47. **a.** Graph the functions $y = x^2$ and $y = x^2 + 3$. How are the graphs related?

b. Without further computation, graph the function $y = x^2 - 5$.

c. Suppose $g(x) = f(x) + c$, where c is a constant. How are the graphs of f and g related? Explain.

48. **a.** Graph the functions $y = x^2$ and $y = -x^2$. How are the graphs related?

b. Suppose $g(x) = -f(x)$. How are the graphs of f and g related? Explain.

49. **a.** Graph the functions $y = x^2$ and $y = (x - 2)^2$. How are the graphs related?

b. Without further computation, graph the function $y = (x + 1)^2$.

c. Suppose $g(x) = f(x - c)$, where c is a constant. How are the graphs of f and g related? Explain.

50. It costs $90 to rent a piece of equipment plus $21 for every day of use.

a. Make a table showing the number of days the equipment is rented and the cost of renting for 2 days, 5 days, 7 days, and 10 days.

b. Write an algebraic expression representing the cost y as a function of the number of days x.

c. Graph the expression in part (b).

51. A company that manufactures lawn mowers has determined that a new employee can assemble N mowers per day after t days of training, where
$$N(t) = \frac{45t^2}{5t^2 + t + 8}$$

a. Make a table showing the numbers of mowers assembled for training periods of lengths $t = 2$ days, 3 days, 5 days, 10 days, 50 days.

b. Based on the table in part (a), what do you think happens to $N(t)$ for very long training periods?

c. Use your calculator to graph $N(t)$.

52. Use your graphing utility to graph $y = x^4$, $y = x^4 - x$, $y = x^4 - 2x$, and $y = x^4 - 3x$ on the same coordinate axes, using $[-2, 2]1$ by $[-2, 5]1$. What effect does the added term involving x have on the shape of the graph? Repeat using $y = x^4$, $y = x^4 - x^3$, $y = x^4 - 2x^3$, and $y = x^4 - 3x^3$. Adjust the viewing rectangle appropriately.

53. Graph $f(x) = \frac{-9x^2 - 3x - 4}{4x^2 + x - 1}$. Determine the values of x for which the function is defined.

54. Graph $f(x) = \frac{8x^2 + 9x + 3}{x^2 + x - 1}$. Determine the values of x for which the function is defined.

55. Graph $g(x) = -3x^3 + 7x + 4$ and find the x intercepts.

56. Show that the distance d between the two points (x_1, y_1) and (x_2, y_2) is given by the formula
$$d = \sqrt{(x_2 - x_1)^2 + (y_2 - y_1)^2}$$

[*Hint:* Apply the Pythagorean theorem to the right triangle whose hypotenuse is the line segment joining the two given points.] Then use the distance formula to find the distance between these points:

a. $(5, -1)$ and $(2, 3)$

b. $(2, 6)$ and $(2, -1)$

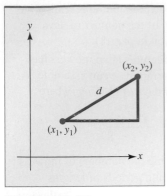

PROBLEM 56

57. Use the distance formula in Problem 56 to show that the circle with center (a, b) and radius R has the equation
$$(x - a)^2 + (y - b)^2 = R^2$$

58. Show that the vertex of the parabola $y = Ax^2 + Bx + C$ $(A \neq 0)$ occurs at the point where $x = \dfrac{-B}{2A}$.

[*Hint:* First verify that $Ax^2 + Bx + C = A\left[\left(x + \dfrac{B}{2A}\right)^2 + \left(\dfrac{C}{A} - \dfrac{B^2}{4A^2}\right)\right]$

Then note that the largest or smallest value of $f(x) = Ax^2 + Bx + C$ must occur where $x = \dfrac{-B}{2A}$.]

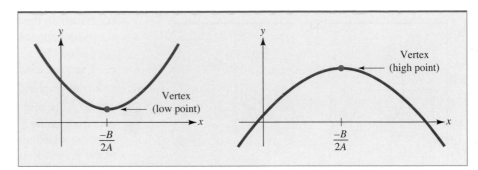

PROBLEM 58

SECTION 1.3 | Linear Functions

In many practical situations, the rate at which one quantity changes with respect to another is constant. Here is a simple example from economics.

EXAMPLE 1.3.1

A manufacturer's total cost consists of a fixed overhead of $200 plus production costs of $50 per unit. Express the total cost as a function of the number of units produced and draw the graph.

Solution

Let x denote the number of units produced and $C(x)$ the corresponding total cost. Then,

$$\text{Total cost} = (\text{cost per unit})(\text{number of units}) + \text{overhead}$$

where

$$\text{Cost per unit} = 50$$
$$\text{Number of units} = x$$
$$\text{Overhead} = 200$$

Hence,

$$C(x) = 50x + 200$$

The graph of this cost function is sketched in Figure 1.15.

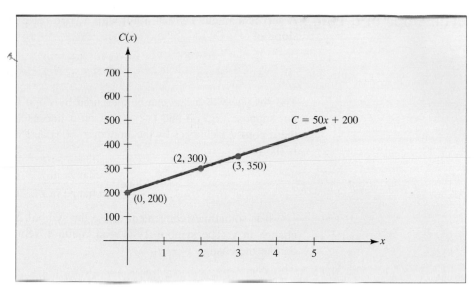

FIGURE 1.15 The cost function, $C(x) = 50x + 200$.

The total cost in Example 1.3.1 increases at a constant rate of $50 per unit. As a result, its graph in Figure 1.15 is a straight line that increases in height by 50 units for each 1-unit increase in x.

In general, a function whose value changes at a constant rate with respect to its independent variable is said to be a **linear function.** This is because the graph of such a function is a straight line. In algebraic terms, a linear function is a function of the form

$$f(x) = a_1 x + a_0$$

where a_0 and a_1 are constants. For example, the functions $f(x) = \dfrac{3}{2} + 2x, f(x) = -5x$, and $f(x) = 12$ are all linear. Linear functions are traditionally written in the form

$$y = mx + b$$

where m and b are constants. This standard notation will be used in the discussion that follows.

Linear Functions ■ A linear function is a function that changes at a constant rate with respect to its independent variable.

The graph of a linear function is a straight line.

The equation of a linear function can be written in the form

$$y = mx + b$$

where m and b are constants.

The Slope of a Line   A surveyor might say that a hill with a "rise" of 2 feet for every foot of "run" has a **slope** of

$$m = \frac{\text{rise}}{\text{run}} = \frac{2}{1} = 2$$

The steepness of a line can be measured by slope in much the same way. In particular, suppose (x_1, y_1) and (x_2, y_2) lie on a line as indicated in Figure 1.16. Between these points, x changes by the amount $x_2 - x_1$ and y by the amount $y_2 - y_1$. The slope is the ratio

$$\text{Slope} = \frac{\text{change in } y}{\text{change in } x} = \frac{y_2 - y_1}{x_2 - x_1}$$

It is sometimes convenient to use the symbol Δy instead of $y_2 - y_1$ to denote the change in y. The symbol Δy is read "delta y." Similarly, the symbol Δx is used to denote the change $x_2 - x_1$.

The Slope of a Line ■ The slope of the nonvertical line passing through the points (x_1, y_1) and (x_2, y_2) is given by the formula

$$\text{Slope} = \frac{\Delta y}{\Delta x} = \frac{y_2 - y_1}{x_2 - x_1}$$

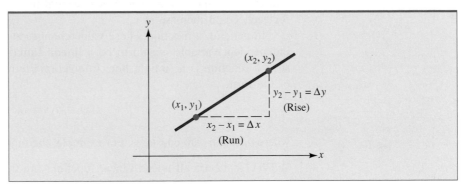

FIGURE 1.16 Slope $= \dfrac{y_2 - y_1}{x_2 - x_1} = \dfrac{\Delta y}{\Delta x}$.

The use of this formula is illustrated in Example 1.3.2.

EXAMPLE | 1.3.2

Find the slope of the line joining the points $(-2, 5)$ and $(3, -1)$.

Solution

$$\text{Slope} = \frac{\Delta y}{\Delta x} = \frac{-1 - 5}{3 - (-2)} = \frac{-6}{5}$$

The line is shown in Figure 1.17.

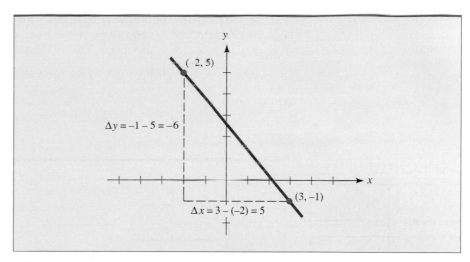

FIGURE 1.17   The line joining $(-2, 5)$ and $(3, -1)$

    The sign and magnitude of the slope of a line indicate the line's direction and steepness, respectively. The slope is positive if the height of the line increases as x increases and is negative if the height decreases as x increases. The absolute value of the slope is large if the slant of the line is severe and small if the slant of the line is gradual. The situation is illustrated in Figure 1.18.

EXPLORE!

Store the varying slope values $\{2, 1, 0.5, -0.5, -1, -2\}$ into List 1, using the **STAT** menu and the **EDIT** option. Display a family of straight lines through the origin, similar to Figure 1.18, by placing $Y1 = L1 * X$ into your calculator's equation editor. Graph using a **ZOOM** Decimal Window and **TRACE** the values for the different lines at X = 1.

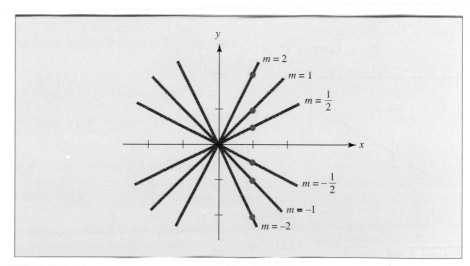

FIGURE 1.18   The direction and steepness of a line.

Horizontal and Vertical Lines

Horizontal and vertical lines (Figures 1.19a and 1.19b) have particularly simple equations. The y coordinates of all of the points on a horizontal line are the same. Hence, a horizontal line is the graph of a linear function of the form $y = b$, where b is a constant. The slope of a horizontal line is zero, since changes in x produce no changes in y.

The x coordinates of all the points on a vertical line are equal. Hence, vertical lines are characterized by equations of the form $x = c$, where c is a constant. The slope of a vertical line is undefined. This is because only the y coordinates of points on a vertical line can change, and so the denominator of the quotient $\dfrac{\text{change in } y}{\text{change in } x}$ is zero.

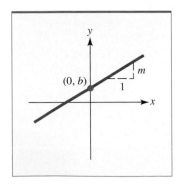

Determine what three slope values must be placed into List 1 so that Y1 = L1∗X + 1 creates the screen pictured here.

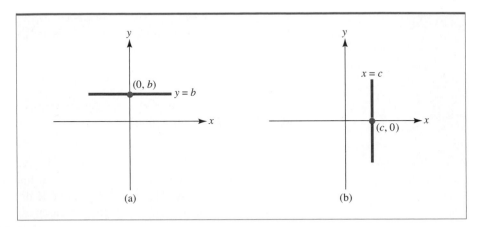

FIGURE 1.19   Horizontal and vertical lines.

The Slope-Intercept Form of the Equation of a Line

The constants m and b in the equation $y = mx + b$ of a nonvertical line have geometric interpretations. The coefficient m is the slope of the line. To see this, suppose that (x_1, y_1) and (x_2, y_2) are two points on the line $y = mx + b$. Then, $y_1 = mx_1 + b$ and $y_2 = mx_2 + b$, and so

$$\text{Slope} = \frac{y_2 - y_1}{x_2 - x_1} = \frac{(mx_2 + b) - (mx_1 + b)}{x_2 - x_1}$$

$$= \frac{mx_2 - mx_1}{x_2 - x_1} = \frac{m(x_2 - x_1)}{x_2 - x_1} = m$$

The constant b in the equation $y = mx + b$ is the value of y corresponding to $x = 0$. Hence, b is the height at which the line $y = mx + b$ crosses the y axis, and the corresponding point $(0, b)$ is the y intercept of the line. The situation is illustrated in Figure 1.20.

Because the constants m and b in the equation $y = mx + b$ correspond to the slope and y intercept, respectively, this form of the equation of a line is known as the **slope-intercept form.**

FIGURE 1.20   The slope and y intercept of the line $y = mx + b$.

The Slope-Intercept Form of the Equation of a Line ■ The equation

$$y = mx + b$$

is the equation of the line whose slope is m and whose y intercept is $(0, b)$.

The slope-intercept form of the equation of a line is particularly useful when geometric information about a line (such as its slope or y intercept) is to be determined from the line's algebraic representation. Here is a typical example.

EXAMPLE | 1.3.3

Find the slope and y intercept of the line $3y + 2x = 6$ and draw the graph.

Solution

First put the equation $3y + 2x = 6$ in slope-intercept form $y = mx + b$. To do this, solve for y to get

$$3y - -2x + 6 \qquad \text{or} \qquad y = -\frac{2}{3}x + 2$$

It follows that the slope is $-\dfrac{2}{3}$ and the y intercept is $(0, 2)$.

To graph a linear function, plot two of its points and draw a straight line through them. In this case, you already know one point, the y intercept $(0, 2)$. A convenient choice for the x coordinate of the second point is $x = 3$, since the corresponding y coordinate is $y = -\dfrac{2}{3}(3) + 2 = 0$. Draw a line through the points $(0, 2)$ and $(3, 0)$ to obtain the graph shown in Figure 1.21.

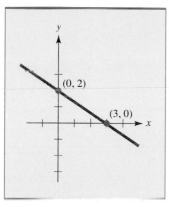

FIGURE 1.21   The line $3y + 2x = 6$.

The Point-Slope Form of the Equation of a Line

Geometric information about a line can be obtained readily from the slope-intercept formula $y = mx + b$. There is another form of the equation of a line, however, that is usually more efficient for problems in which the geometric properties of a line are known and the goal is to find the equation of the line.

The Point-Slope Form of the Equation of a Line ▪ The equation

$$y - y_0 = m(x - x_0)$$

is an equation of the line that passes through the point (x_0, y_0) and that has slope equal to m.

The point-slope form of the equation of a line is simply the formula for slope in disguise. To see this, suppose the point (x, y) lies on the line that passes through a given point (x_0, y_0) and that has slope m. Using the points (x, y) and (x_0, y_0) to compute the slope, you get

$$\frac{y - y_0}{x - x_0} = m$$

which you can put in point-slope form

$$y - y_0 = m(x - x_0)$$

by simply multiplying both sides by $x - x_0$.

The use of the point-slope form of the equation of a line is illustrated in Examples 1.3.4 and 1.3.5.

 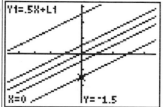

EXPLORE!

Determine the y intercept values needed in List L1 so that the function $Y1 = 0.5X + L1$ creates the screen shown here.

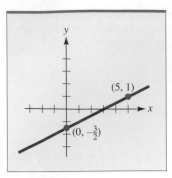

FIGURE 1.22 The line $y = \frac{1}{2}x - \frac{3}{2}$.

EXAMPLE 1.3.4

Find the equation of the line that passes through the point (5, 1) and whose slope is equal to $\frac{1}{2}$.

Solution

Use the formula $y - y_0 = m(x - x_0)$ with $(x_0, y_0) = (5, 1)$ and $m = \frac{1}{2}$ to get

$$y - 1 = \frac{1}{2}(x - 5)$$

which you can rewrite as

$$y = \frac{1}{2}x - \frac{3}{2}$$

The graph is shown in Figure 1.22.

For practice, solve the problem in Example 1.3.4 using the slope-intercept formula. Notice that the solution based on the point-slope formula is more efficient.

In Chapter 2, the point-slope formula will be used extensively for finding the equation of the tangent line to the graph of a function at a given point. Example 1.3.5 illustrates how the point-slope formula can be used to find the equation of a line through two given points.

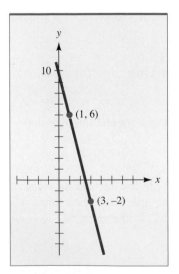

FIGURE 1.23 The line $y = -4x + 10$.

EXAMPLE 1.3.5

Find the equation of the line that passes through the points (3, −2) and (1, 6).

Solution

First compute the slope

$$m = \frac{6 - (-2)}{1 - 3} = \frac{8}{-2} = -4$$

Then use the point-slope formula with (1, 6) as the given point (x_0, y_0) to get

$$y - 6 = -4(x - 1) \qquad \text{or} \qquad y = -4x + 10$$

Convince yourself that the resulting equation would have been the same if you had chosen (3, −2) to be the given point (x_0, y_0). The graph is shown in Figure 1.23.

NOTE The general form for the equation of a line is $Ax + By + C = 0$, where A, B, C are constants, with A and B not both equal to 0. If $B = 0$, the line is vertical, and when $B \neq 0$, the equation $Ax + By + C = 0$ can be rewritten as

$$y = \left(\frac{-A}{B}\right)x + \left(\frac{-C}{B}\right)$$

Comparing this equation with the slope-intercept form $y = mx + b$, we see that the slope of the line is given by $m = -A/B$ and the y intercept by $b = -C/B$. The line is horizontal (slope 0) when $A = 0$.   ■

Practical
Applications

If the rate of change of one quantity with respect to a second quantity is constant, the function relating the quantities must be linear. The constant rate of change is the slope of the corresponding line. Examples 1.3.6 and 1.3.7 illustrate techniques you can use to find the appropriate linear functions in such situations.

EXAMPLE | 1.3.6

Since the beginning of the year, the price of whole wheat bread at a local discount supermarket has been rising at a constant rate of 2 cents per month. By November first, the price had reached $1.56 per loaf. Express the price of the bread as a function of time and determine the price at the beginning of the year.

Solution

Let x denote the number of months that have elapsed since the first of the year and y the price of a loaf of bread (in cents). Since y changes at a constant rate with respect to x, the function relating y to x must be linear, and its graph is a straight line. Since the price y increases by 2 each time x increases by 1, the slope of the line must be 2. The fact that the price was 156 cents ($1.56) on November first, 10 months after the first of the year, implies that the line passes through the point (10, 156). To write an equation defining y as a function of x, use the point-slope formula

$$y - y_0 = m(x - x_0)$$

with

$$m = 2, x_0 = 10, y_0 = 156$$

to get $\qquad y - 156 = 2(x - 10) \qquad$ or $\qquad y = 2x + 136$

The corresponding line is shown in Figure 1.24. Notice that the y intercept is (0, 136), which implies that the price of bread at the beginning of the year was $1.36 per loaf.

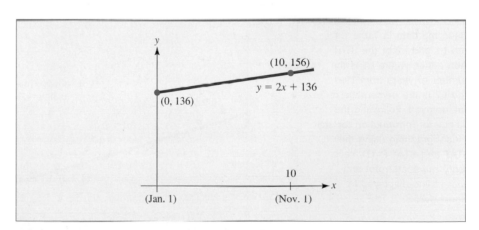

FIGURE 1.24 The rising price of bread: $y = 2x + 136$.

Sometimes it is hard to tell how two quantities, x and y, in a data set are related by simply examining the data. In such cases, it may be useful to graph the data to see if the points (x, y) follow a clear pattern (say, lie along a line). Here is an example of this procedure.

TABLE 1.4 Percentage of Civilian Unemployment 1991–2000

Year	Number of Years from 1991	Percentage of Unemployed
1991	0	6.8
1992	1	7.5
1993	2	6.9
1994	3	6.1
1995	4	5.6
1996	5	5.4
1997	6	4.9
1998	7	4.5
1999	8	4.2
2000	9	4.0

SOURCE: U.S. Bureau of Labor Statistics, Bulletin 2307; and *Employment and Earnings,* monthly.

EXAMPLE | 1.3.7

Table 1.4 lists the percentage of the labor force that was unemployed during the decade 1991–2000. Plot a graph with time (years after 1991) on the x axis and percentage of unemployment on the y axis. Do the points follow a clear pattern? Based on these data, what would you expect the percentage of unemployment to be in the year 2005?

Solution

The graph is shown in Figure 1.25. Note that except for the initial point (0, 6.8), the pattern is roughly linear. There is not enough evidence to infer that unemployment is linearly related to time, but the pattern does suggest that we may be able to get useful information by finding a line that "best fits" the data in some meaningful way. One such procedure, called "least-squares approximation," requires the approximating line to be positioned so that the sum of squares of vertical distances from the data points to the line is minimized. The least-squares procedure, which will be developed in Section 7.4 of Chapter 7, can be carried out on your calculator. When this procedure is applied to the unemployment data in this example, it produces the "best-fitting line" $y = -0.389x + 7.338$, as displayed in Figure 1.25. We can then use this formula to attempt a prediction of the unemployment rate in the year 2005 (when $x = 14$):

$$y(14) = -0.389(14) + 7.338 = 1.892$$

Thus, based on least-squares extrapolation from the given data, our best guess is that there will be roughly a 1.9% unemployment rate in 2005.

EXPLORE!

Place the data in Table 1.4 into L1 and L2 of the **STAT** data editor, where L1 is the number of years from 1991 and L2 is the percentage of unemployed. Following the Calculator Introduction for statistical graphing using the **STAT** and **STAT PLOT** keys, verify the scatterplot and best-fit line displayed in Figure 1.25.

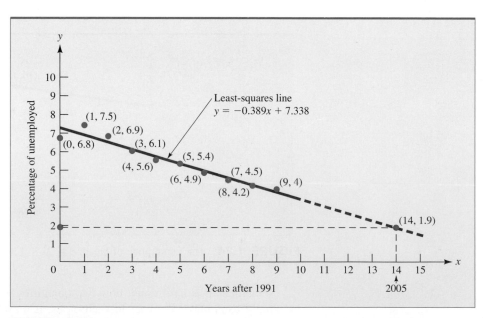

FIGURE 1.25 Percentage of unemployed in the United States for 1991–2000.

NOTE Care must be taken when making predictions by extrapolating from known data, especially when the data set is as small as the one in Example 1.3.7. In particular, the economy began to weaken after the year 2000, but the least-squares line in Figure 1.23 predicts a steadily decreasing unemployment rate. Is this reasonable? In Problem 59, you are asked to explore this question by first using the Internet to obtain unemployment data for years subsequent to 2000 and then comparing this new data with the values predicted by the least-squares line. ■

Parallel and Perpendicular Lines

In applications, it is sometimes necessary or useful to know whether two given lines are parallel or perpendicular. A vertical line is parallel only to other vertical lines and is perpendicular to any horizontal line. Cases involving nonvertical lines can be handled by these criteria.

Parallel and Perpendicular Lines ■ Let m_1 and m_2 be the slopes of the nonvertical lines L_1 and L_2. Then

L_1 and L_2 are **parallel** if and only if $m_1 = m_2$.

L_1 and L_2 are **perpendicular** if and only if $m_2 = \dfrac{-1}{m_1}$.

These criteria are illustrated in Figure 1.26. Geometric proofs are outlined in Exercises 60 and 61. We close this section with an example illustrating one way the criteria can be used.

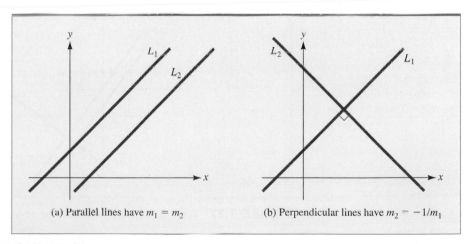

(a) Parallel lines have $m_1 = m_2$ (b) Perpendicular lines have $m_2 = -1/m_1$

FIGURE 1.26

EXAMPLE | 1.3.8

Let L be the line $4x + 3y = 3$.

a. Find the equation of a line L_1 parallel to L through $P(-1, 4)$.

b. Find the equation of a line L_2 perpendicular to L through $Q(2, -3)$.

Solution

By rewriting the equation $4x + 3y = 3$ in the slope-intercept form $y = -\dfrac{4}{3}x + 1$, we see that L has slope $m_L = -\dfrac{4}{3}$.

a. Any line parallel to L must also have slope $m = -\dfrac{4}{3}$. The required line L_1 contains $P(-1, 4)$, so

$$y - 4 = -\frac{4}{3}(x + 1)$$

$$y = -\frac{4}{3}x + \frac{8}{3}$$

b. A line perpendicular to L must have slope $m = -\dfrac{1}{m_L} = \dfrac{3}{4}$. Since the required line L_2 contains $Q(2, -3)$, we have

$$y + 3 = \frac{3}{4}(x - 2)$$

$$y = \frac{3}{4}x - \frac{9}{2}$$

The given line L and the required lines L_1 and L_2 are shown in Figure 1.27.

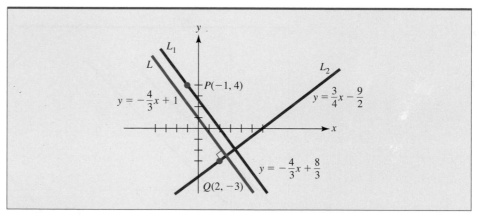

FIGURE 1.27   Lines parallel and perpendicular to a given line L.

PROBLEMS   **1.3**

In Problems 1 through 6, find the slope (if possible) of the line that passes through the given pair of points.

1. $(2, -3)$ and $(0, 4)$

2. $(-1, 2)$ and $(2, 5)$

3. $(2, 0)$ and $(0, 2)$

4. $(5, -1)$ and $(-2, -1)$

5. $(2, 6)$ and $(2, -4)$

6. $\left(\dfrac{2}{3}, -\dfrac{1}{5}\right)$ and $\left(-\dfrac{1}{7}, \dfrac{1}{8}\right)$

In Problems 7 through 10, find the slope and intercepts of the line shown. Then find an equation for the line.

7.

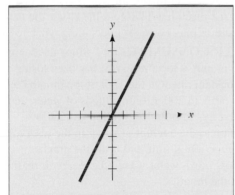

8.

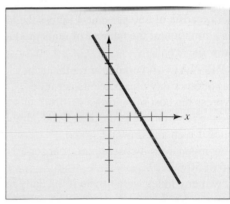

9.

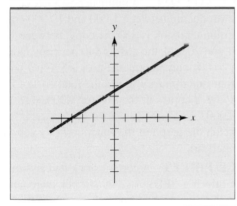

10.

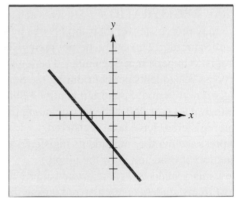

In Problems 11 through 18, find the slope and intercepts of the line whose equation is given and sketch the graph of the line.

11. $x = 3$

12. $y = 5$

13. $y - 3x$

14. $y - 3x - 6$

15. $3x + 2y = 6$

16. $5y - 3x = 4$

17. $\dfrac{x}{2} + \dfrac{y}{5} = 1$

18. $\dfrac{x + 3}{-5} + \dfrac{y - 1}{2} = 1$

In Problems 19 through 34, write an equation for the line with the given properties.

19. Through $(2, 0)$ with slope 1

20. Through $(-1, 2)$ with slope $\dfrac{2}{3}$

21. Through $(5, -2)$ with slope $-\dfrac{1}{2}$

22. Through $(0, 0)$ with slope 5

23. Through $(2, 5)$ and parallel to the x axis

24. Through $(2, 5)$ and parallel to the y axis

25. Through $(1, 0)$ and $(0, 1)$

26. Through $(2, 5)$ and $(1, -2)$

27. Through $\left(-\dfrac{1}{5}, 1\right)$ and $\left(\dfrac{2}{3}, \dfrac{1}{4}\right)$

28. Through $(-2, 3)$ and $(0, 5)$

29. Through $(1, 5)$ and $(3, 5)$

30. Through $(1, 5)$ and $(1, -4)$

31. Through $(4, 1)$ and parallel to the line $2x + y = 3$

32. Through $(-2, 3)$ and parallel to the line $x + 3y = 5$

33. Through $(3, 5)$ and perpendicular to the line $x + y = 4$

34. Through $\left(-\dfrac{1}{2}, 1\right)$ and perpendicular to the line $2x + 5y = 3$

35. MANUFACTURING COST A manufacturer's total cost consists of a fixed overhead of $5,000 plus production costs of $60 per unit. Express the total cost as a function of the number of units produced and draw the graph.

36. CAR RENTAL A certain car rental agency charges $35 per day plus 55 cents per mile.
   a. Express the cost of renting a car from this agency for 1 day as a function of the number of miles driven and draw the graph.
   b. How much does it cost to rent a car for a 1-day trip of 50 miles?
   c. How many miles were driven if the daily rental cost was $72?

37. COURSE REGISTRATION Students at a state college may preregister for their fall classes by mail during the summer. Those who do not preregister must register in person in September. The registrar can process 35 students per hour during the September registration period. Suppose that after 4 hours in September, a total of 360 students (including those who preregistered) have been registered.
   a. Express the number of students registered as a function of time and draw the graph.
   b. How many students were registered after 3 hours?
   c. How many students preregistered during the summer?

38. MEMBERSHIP FEES Membership in a swimming club costs $250 for the 12-week summer season. If a member joins after the start of the season, the fee is prorated; that is, it is reduced linearly.
   a. Express the membership fee as a function of the number of weeks that have elapsed by the time the membership is purchased and draw the graph.
   b. Compute the cost of a membership that is purchased 5 weeks after the start of the season.

39. LINEAR DEPRECIATION A doctor owns $1,500 worth of medical books which, for tax purposes, are assumed to depreciate linearly to zero over a 10-year period. That is, the value of the books decreases at a constant rate so that it is equal to zero at the end of 10 years. Express the value of the books as a function of time and draw the graph.

40. LINEAR DEPRECIATION A manufacturer buys $20,000 worth of machinery that depreciates linearly so that its trade-in value after 10 years will be $1,000.
   a. Express the value of the machinery as a function of its age and draw the graph.
   b. Compute the value of the machinery after 4 years.

 c. When does the machinery become worthless? The manufacturer might not wait this long to dispose of the machinery. Discuss the issues the manufacturer may consider in deciding when to sell.

41. WATER CONSUMPTION Since the beginning of the month, a local reservoir has been losing water at a constant rate. On the 12th of the month the reservoir held 200 million gallons of water, and on the 21st it held only 164 million gallons.
   a. Express the amount of water in the reservoir as a function of time and draw the graph.
   b. How much water was in the reservoir on the 8th of the month?

42. PRINTING COST A publisher estimates that the cost of producing between 1,000 and 10,000 copies of a certain textbook is $50 per copy; between 10,001 and 20,000, the cost is $40 per copy; and between 20,001 and 50,000, the cost is $35 per copy.
   a. What function $F(N)$ gives the total cost of producing N copies of the text for $1,000 \leq N \leq 50,000$?
   b. Sketch the graph of the function $F(N)$ you found in part (a).

43. STOCK PRICES A certain stock had an initial public offering (IPO) price of $10 per share, and is traded 24 hours a day. Sketch the graph of the share price over a 2-year period for each of the following cases:
   a. The price increases steadily to $50 a share over the first 18 months and then decreases steadily to $25 per share over the next 6 months.
   b. The price takes just 2 months to rise at a constant rate to $15 a share, then slowly drops to $8 over the next 9 months before steadily rising to $20.
   c. The price steadily rises to $60 a share during the first year, at which time, an accounting scandal is uncovered. The price gaps down to $25 a share, then steadily decreases to $5 over the next 3 months before rising at a constant rate to close at $12 at the end of the 2-year period.

44. AN ANCIENT FABLE In Aesop's fable about the race between the tortoise and the hare, the tortoise trudges along at a slow, constant rate from start to finish. The hare starts out running steadily at a much more rapid pace, but halfway to the finish line, stops to take a nap. Finally, the hare wakes up, sees the tortoise near the finish line, desperately resumes his old rapid pace, and is nosed out by a hair. On the same coordinate plane, graph the respective distances of the tortoise and the hare from the starting line regarded as functions of time.

45. **GROWTH OF A CHILD** The average height H in centimeters of a child of age A years can be estimated by the linear function $H = 6.5A + 50$. Use this formula to answer these questions.
   a. What is the average height of a 7-year-old child?
   b. How old must a child be to have an average height of 150 cm?
   c. What is the average height of a newborn baby? Does this answer seem reasonable?
   d. What is the average height of a 20-year-old? Does this answer seem reasonable?

46. **CAR POOLING** To encourage motorists to form car pools, the transit authority in a certain metropolitan area has been offering a special reduced rate at toll bridges for vehicles containing four or more persons. When the program began 30 days ago, 157 vehicles qualified for the reduced rate during the morning rush hour. Since then, the number of vehicles qualifying has been increasing at a constant rate, and today 247 vehicles qualified.
   a. Express the number of vehicles qualifying each morning for the reduced rate as a function of time and draw the graph.
   b. If the trend continues, how many vehicles will qualify during the morning rush hour 14 days from now?

47. **TEMPERATURE CONVERSION**
   a. Temperature measured in degrees Fahrenheit is a linear function of temperature measured in degrees Celsius. Use the fact that 0° Celsius is equal to 32° Fahrenheit and 100° Celsius is equal to 212° Fahrenheit to write an equation for this linear function.
   b. Use the function you obtained in part (a) to convert 15° Celsius to Fahrenheit.
   c. Convert 68° Fahrenheit to Celsius.
   d. What temperature is the same in both the Celsius and Fahrenheit scales?

48. **ENTOMOLOGY** It has been observed that the number of chirps made by a cricket each minute depends on the temperature. Crickets won't chirp if the temperature is 38°F or less, and observations yield the following data:

Number of chirps (C)	0	5	10	20	60
Temperature T (°F)	38	39	40	42	50

   a. Express T as a linear function of C.
   b. How many chirps would you expect to hear if the temperature is 75°F? If you hear 37 chirps in a 30-second period of time, what is the approximate temperature?

49. **APPRECIATION OF ASSETS** The value of a certain rare book doubles every 10 years. In 1900, the book was worth $100.
   a. How much was it worth in 1930? In 1990? What about the year 2000?
   b. Is the value of the book a linear function of its age? Answer this question by interpreting an appropriate graph.

50. **AIR POLLUTION** In certain parts of the world, the number of deaths N per week have been observed to be linearly related to the average concentration x of sulfur dioxide in the air. Suppose there are 97 deaths when $x = 100$ mg/m^3 and 110 deaths when $x = 500$ mg/m^3.
   a. What is the functional relationship between N and x?
   b. Use the function in part (a) to find the number of deaths per week when $x = 300$ mg/m^3. What concentration of sulfur dioxide corresponds to 100 deaths per week?
    c. Research data on how air pollution affects the death rate in a population.* Summarize your results in a one-paragraph essay.

51. **COLLEGE ADMISSIONS** The average scores of incoming students at an eastern liberal arts college in the SAT mathematics examination have been declining at a constant rate in recent years. In 1995, the average SAT score was 575, while in 2000 it was 545.
   a. Express the average SAT score as a function of time.
   b. If the trend continues, what will the average SAT score of incoming students be in 2005?
   c. If the trend continues, when will the average SAT score be 527?

---

*You may find the following articles helpful: D. W. Dockery, J. Schwartz, and J. D. Spengler, "Air Pollution and Daily Mortality: Associations with Particulates and Acid Aerosols," *Environ. Res.*, Vol. 59, 1992, pp. 362–373; Y. S. Kim, "Air Pollution, Climate, Socioeconomics Status and Total Mortality in the United States," *Sci. Total Environ.*, Vol. 42, 1985, pp. 245–256.

52.   **NUTRITION**   Each ounce of Food I contains 3 g of carbohydrate and 2 g of protein, and each ounce of Food II contains 5 g of carbohydrate and 3 g of protein. Suppose x ounces of Food I are mixed with y ounces of Food II. The foods are combined to produce a blend that contains exactly 73 g of carbohydrate and 46 g of protein.
   a.   Explain why there are $3x + 5y$ g of carbohydrate in the blend and why we must have $3x + 5y = 73$. Find a similar equation for protein. Sketch the graphs of both equations.
   b.   Where do the two graphs in part (a) intersect? Interpret the significance of this point of intersection.

53.   **ACCOUNTING**   For tax purposes, the book value of certain assets is determined by depreciating the original value of the asset linearly over a fixed period of time. Suppose an asset originally worth V dollars is linearly depreciated over a period of N years, at the end of which it has a scrap value of S dollars.
   a.   Express the book value B of the asset t years into the N-year depreciation period as a linear function of t. [*Hint:* Note that $B = V$ when $t = 0$ and $B = S$ when $t = N$.]
   b.   Suppose a $50,000 piece of office equipment is depreciated linearly over a 5-year period, with a scrap value of $18,000. What is the book value of the equipment after three years?

54.   **ALCOHOL ABUSE CONTROL**   Ethyl alcohol is metabolized by the human body at a constant rate (independent of concentration). Suppose the rate is 10 milliliters per hour.
   a.   How much time is required to eliminate the effects of a liter of beer containing 3% ethyl alcohol?
   b.   Express the time T required to metabolize the effects of drinking ethyl alcohol as a function of the amount A of ethyl alcohol consumed.
      c.   Discuss how the function in part (b) can be used to determine a reasonable "cutoff" value for the amount of ethyl alcohol A that each individual may be served at a party.

55.   Graph $y = \dfrac{25}{7}x + \dfrac{13}{2}$ and $y = \dfrac{144}{45}x + \dfrac{630}{229}$ on the same set of coordinate axes using $[-10, 10]1$ by $[-10, 10]1$. Are the two lines parallel?

56.   Graph $y = \dfrac{54}{270}x - \dfrac{63}{19}$ and $y = \dfrac{139}{695}x - \dfrac{346}{14}$ on the same set of coordinate axes using $[-10, 10]1$ by $[-10, 10]1$ for a starting range. Adjust the range settings until both lines are displayed. Are the two lines parallel?

57.   **EQUIPMENT RENTAL**   A rental company rents a piece of equipment for a $60.00 flat fee plus an hourly fee of $5.00 per hour.
   a.   Make a chart showing the number of hours the equipment is rented and the cost for renting the equipment for 2 hours, 5 hours, 10 hours, and t hours of time.
   b.   Write an algebraic expression representing the cost y as a function of the number of hours t. Assume t can be measured to any decimal portion of an hour. (In other words, assume t is any nonnegative real number.)
   c.   Graph the expression from part (b).
   d.   Use the graph to approximate, to two decimal places, the number of hours the equipment was rented if the bill is $216.25 (before taxes).

58.   **ASTRONOMY**   The following table gives the length of year L (in earth years) of each planet in the solar system along with the mean (average) distance D of the planet from the sun, in astronomical units (1 astronomical unit is the mean distance of the earth from the sun).

Planet	Mean Distance from Sun, D	Length of Year, L
Mercury	0.388	0.241
Venus	0.722	0.615
Earth	1.000	1.000
Mars	1.523	1.881
Jupiter	5.203	11.862
Saturn	9.545	29.457
Uranus	19.189	84.013
Neptune	30.079	164.783
Pluto	39.463	248.420

SOURCE: Kendrick Frazier, *The Solar System*, Alexandria, VA: Time/Life Books, 1985, p. 37.

   a.   Plot the point (D, L) for each planet on a coordinate plane. Do these quantities appear to be linearly related?

b. For each planet, compute the ratio $\dfrac{L^2}{D^3}$. Interpret what you find by expressing L as a function of D.

c. What you have discovered in part (b) is one of Kepler's laws, named for the German astronomer Johannes Kepler (1571–1630). Read an article about Kepler and describe his place in the history of science.

59. UNEMPLOYMENT RATE   In the solution to Example 1.3.7, we observed that the line that best fits the data in that example in the sense of least-squares approximation has the equation $y = -0.389x + 7.338$. Interpret the slope of this line in terms of the rate of unemployment. The data in the example stop at the year 2000. Use the Internet to find unemployment data for the years since 2000. Does the least-squares approximating line do a good job of predicting this new data? Explain.

60. PARALLEL LINES   Show that two nonvertical lines are parallel if and only if they have the same slope.

61. PERPENDICULAR LINES   Show that if a non-vertical line L_1 with slope m_1 is perpendicular to a line L_2 with slope m_2, then $m_2 = -1/m_1$. [*Hint:* Find expressions for the slopes of the lines L_1 and L_2 in the accompanying figure. Then apply the Pythagorean theorem along with the distance formula from Problem 56, Section 1.2, to the right triangle OAB to obtain the required relationship between m_1 and m_2.]

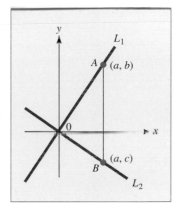

PROBLEM 61

SECTION   1.4   Functional Models

Practical problems in business, economics, and the physical and life sciences are often too complicated to be precisely described by simple formulas, and one of our basic goals is to develop mathematical methods for dealing with such problems. Toward this end, we shall use a procedure called **mathematical modeling,** which may be described in terms of four stages:

Stage 1 (Formulation): Given a real-world situation (for example, the U.S. trade deficit, the AIDS epidemic, global weather patterns), we make enough simplifying assumptions to allow a mathematical formulation. This may require gathering and analyzing data and using knowledge from a variety of different areas to identify key variables and establish equations relating those variables. This formulation is called a **mathematical model**.

Stage 2 (Analysis of the Model): We use mathematical methods to analyze or "solve" the mathematical model. Calculus will be the primary tool of analysis in this text, but in practice, a variety of tools, such as statistics, numerical analysis, and computer science, may be brought to bear on a particular model.

Stage 3 (Interpretation): After the mathematical model has been analyzed, any conclusions that may be drawn from the analysis are applied to the original real-world problem, both to gauge the accuracy of the model and to make predictions. For instance, analysis of a model of a particular business may predict that profit will be maximized by producing 200 units of a certain commodity.

Stage 4 (Testing and Adjustment): In this final stage, the model is tested by gathering new data to check the accuracy of any predictions inferred from the analysis. If the predictions are not confirmed by the new evidence, the assumptions of the model are adjusted and the modeling process is repeated. Referring to the business example described in stage 3, it may be found that profit begins to wane at a production level significantly less than 200 units, which would indicate that the model requires modification.

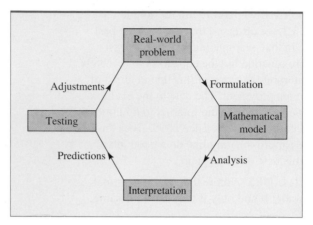

FIGURE 1.28 A diagram illustrating the mathematical modeling procedure.

The four stages of mathematical modeling are displayed in Figure 1.28. In a good model, the real-world problem is idealized just enough to allow mathematical analysis but not so much that the essence of the underlying situation has been compromised. For instance, if we assume that weather is strictly periodic, with rain occurring every 10 days, the resulting model would be relatively easy to analyze but would clearly be a poor representation of reality.

In preceding sections, you have seen models representing quantities such as manufacturing cost, price and demand, air pollution levels, and population size, and you will encounter many more in subsequent chapters. Some of these models, especially those analyzed in the Think About It essays at the end of each chapter, are more detailed and illustrate how decisions are made about assumptions and predictions. Constructing and analyzing mathematical models is one of the most important skills you will learn in calculus, and the process begins with learning how to set up and solve word problems. Examples 1.4.1 through 1.4.8 illustrate a variety of techniques for addressing such problems.

Elimination of Variables

In Examples 1.4.1 and 1.4.2 the quantity you are seeking is expressed most naturally in terms of two variables. You will have to eliminate one of these variables before you can write the quantity as a function of a single variable.

EXAMPLE | 1.4.1

The highway department is planning to build a picnic area for motorists along a major highway. It is to be rectangular with an area of 5,000 square yards and is to be fenced

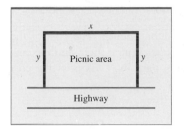

FIGURE 1.29   Rectangular picnic area.

off on the three sides not adjacent to the highway. Express the number of yards of fencing required as a function of the length of the unfenced side.

Solution

It is natural to start by introducing two variables, say x and y, to denote the lengths of the sides of the picnic area (Figure 1.29). Expressing the number of yards F of required fencing in terms of these two variables, we get

$$F = x + 2y$$

Since the goal is to express the number of yards of fencing as a function of x alone, you must find a way to express y in terms of x. To do this, use the fact that the area is to be 5,000 square yards and write

$$\underbrace{xy}_{\text{area}} = 5{,}000$$

Solve this equation for y

$$y = \frac{5{,}000}{x}$$

and substitute the resulting expression for y into the formula for F to get

$$F(x) = x + 2\left(\frac{5{,}000}{x}\right) = x + \frac{10{,}000}{x}$$

A graph of the relevant portion of this rational function is sketched in Figure 1.30. Notice that there is some length x for which the amount of required fencing is minimal. In Chapter 3, you will compute this optimal value of x using calculus.

EXPLORE!

Store $f(x) = x + \dfrac{10{,}000}{x}$ into the equation editor of your graphing calculator. Use the table feature (**TBLSET**) to explore where a minimal value of $f(x)$ might occur. For example, alternatively try setting the initial value of x (**TblStart**) at zero, increasing at increments (**ΔTbl**) of 100, 50, and 10. Finally, use what you have discovered to establish a proper viewing window to view $f(x)$ and to answer the question as to the dimensions of the picnic area with minimal fencing.

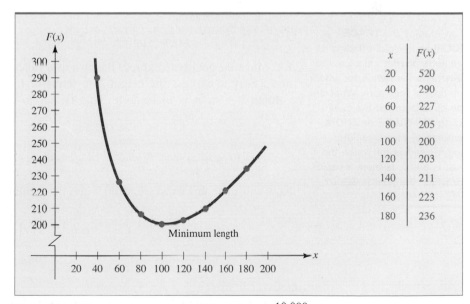

FIGURE 1.30   The length of fencing: $F(x) = x + \dfrac{10{,}000}{x}$.

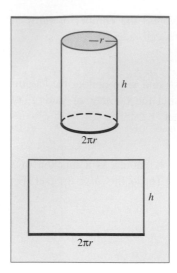

FIGURE 1.31 Cylindrical can for Example 1.4.2.

EXPLORE!

Use the table feature **(TBLSET)** of your graphing calculator to discover an appropriate viewing window for graphing

$$C(r) = 6\pi r^2 + \frac{96\pi}{r}$$

Now employ the **TRACE**, **ZOOM** or minimum-finding methods of your calculator to determine the radius for which the cost is minimum. What are the dimensions of the can? Using the **TRACE** or **ZOOM** features of your calculator, find the radius for which the cost is $3.00. Is there a radius for which the cost is $2.00?

EXAMPLE | 1.4.2

A cylindrical can is to have capacity (volume) of 24π cubic inches. The cost of the material used for the top and bottom of the can is 3 cents per square inch, and the cost of the material used for the curved side is 2 cents per square inch. Express the cost of constructing the can as a function of its radius.

Solution

Let r denote the radius of the circular top and bottom, h the height of the can, and C the cost (in cents) of constructing the can. Then,

$$C = \text{cost of top} + \text{cost of bottom} + \text{cost of side}$$

where, for each component of cost,

$$\text{Cost} = (\text{cost per square inch})(\text{number of square inches})$$
$$= (\text{cost per square inch})(\text{area})$$

The area of the circular top (or bottom) is πr^2, and the cost per square inch of the top (or bottom) is 3 cents. Hence,

$$\text{Cost of top} = 3\pi r^2 \quad \text{and} \quad \text{Cost of bottom} = 3\pi r^2$$

To find the area of the curved side, imagine the top and bottom of the can removed and the side cut and spread out to form a rectangle, as shown in Figure 1.31. The height of the rectangle is the height h of the can. The length of the rectangle is the circumference $2\pi r$ of the circular top (or bottom) of the can. Hence, the area of the rectangle (or curved side) is $2\pi rh$ square inches. Since the cost of the side is 2 cents per square inch, it follows that

$$\text{Cost of side} = 2(2\pi rh) = 4\pi rh$$

Putting it all together,

$$C = 3\pi r^2 + 3\pi r^2 + 4\pi rh = 6\pi r^2 + 4\pi rh$$

Since the goal is to express the cost as a function of the radius alone, you must find a way to express the height h in terms of r. To do this, use the fact that the volume $V = \pi r^2 h$ is to be 24π. That is, set $\pi r^2 h$ equal to 24π and solve for h to get

$$\pi r^2 h = 24\pi \quad \text{or} \quad h = \frac{24}{r^2}$$

Now substitute this expression for h into the formula for C:

$$C(r) = 6\pi r^2 + 4\pi r\left(\frac{24}{r^2}\right)$$

or

$$C(r) = 6\pi r^2 + \frac{96\pi}{r}$$

A graph of the relevant portion of this cost function is sketched in Figure 1.32. Notice that there is some radius r for which the cost is minimal. In Chapter 3, you will learn how to find this optimal radius using calculus.

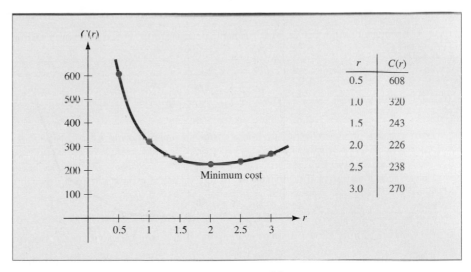

FIGURE 1.32 The cost function: $C(r) = 6\pi r^2 + \dfrac{96\pi}{r}$.

r	$C(r)$
0.5	608
1.0	320
1.5	243
2.0	226
2.5	238
3.0	270

EXAMPLE | 1.4.3

During a drought, residents of Marin County, California, were faced with a severe water shortage. To discourage excessive use of water, the county water district initiated drastic rate increases. The monthly rate for a family of four was $1.22 per 100 cubic feet of water for the first 1,200 cubic feet, $10 per 100 cubic feet for the next 1,200 cubic feet, and $50 per 100 cubic feet thereafter. Express the monthly water bill for a family of four as a function of the amount of water used.

Solution

Let x denote the number of hundred-cubic-feet units of water used by the family during the month and $C(x)$ the corresponding cost in dollars. If $0 \leq x \leq 12$, the cost is simply the cost per unit times the number of units used:

$$C(x) = 1.22x$$

If $12 < x \leq 24$, each of the first 12 units costs $1.22, and so the total cost of these 12 units is $1.22(12) = 14.64$ dollars. Each of the remaining $x - 12$ units costs $10, and hence the total cost of these units is $10(x - 12)$ dollars. The cost of all x units is the sum

$$C(x) = 14.64 + 10(x - 12) = 10x - 105.36$$

If $x > 24$, the cost of the first 12 units is $1.22(12) = 14.64$ dollars, the cost of the next 12 units is $10(12) = 120$ dollars, and that of the remaining $x - 24$ units is $50(x - 24)$ dollars. The cost of all x units is the sum

$$C(x) = 14.64 + 120 + 50(x - 24) = 50x - 1,065.36$$

Combining these three formulas, we can express the total cost as the piecewise-defined function

$$C(x) = \begin{cases} 1.22x & \text{if } 0 \leq x \leq 12 \\ 10x - 105.36 & \text{if } 12 < x \leq 24 \\ 50x - 1,065.36 & \text{if } x > 24 \end{cases}$$

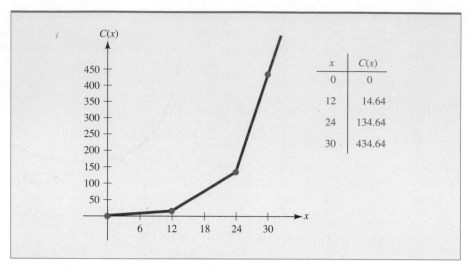

x	$C(x)$
0	0
12	14.64
24	134.64
30	434.64

FIGURE 1.33 The cost of water in Marin County.

The graph of this function is shown in Figure 1.33. Notice that the graph consists of three line segments, each one steeper than the preceding one. What aspect of the practical situation is reflected by the increasing steepness of the lines?

Proportionality

In constructing mathematical models, it is often important to consider proportionality relationships. Three of the most important kinds of proportionality are defined as follows:

> **Proportionality** ■ The quantity Q is said to be:
>
> *directly proportional* to x if $Q = kx$ for some constant k
>
> *inversely proportional* to x if $Q = k/x$ for some constant k
>
> *jointly proportional* to x and y if $Q = kxy$ for some constant k

Here is an example from biology.

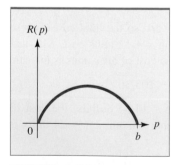

FIGURE 1.34 The rate of bounded population growth: $R(p) = kp(b - p)$.

EXAMPLE | 1.4.4

When environmental factors impose an upper bound on its size, population grows at a rate that is jointly proportional to its current size and the difference between its current size and the upper bound. Express the rate of population growth as a function of the size of the population.

Solution

Let p denote the size of the population, $R(p)$ the corresponding rate of population growth, and b the upper bound placed on the population by the environment. Then

$$\text{Difference between population and bound} = b - p$$

and so

$$R(p) = kp(b - p)$$

where k is the constant of proportionality.

A graph of this factored polynomial is sketched in Figure 1.34. In Chapter 3, you will use calculus to compute the population size for which the rate of population growth is greatest.

Modeling in Business and Economics

Business and economic models often involve issues such as pricing, cost control, and optimization of profit. In Chapter 3, we shall examine a variety of such models. Here is an example in which profit is expressed as a function of the selling price of a particular product.

EXAMPLE | 1.4.5

A manufacturer can produce blank premium-quality videotapes at a cost of $2 per cassette. The cassettes have been selling for $5 apiece, and at that price, consumers have been buying 4,000 cassettes a month. The manufacturer is planning to raise the price of the cassettes and estimates that for each $1 increase in the price, 400 fewer cassettes will be sold each month.

a. Express the manufacturer's monthly profit as a function of the price at which the cassettes are sold.

b. Sketch the graph of the profit function. What price corresponds to maximum profit? What is the maximum profit?

Solution

a. Begin by stating the desired relationship in words.

$$\text{Profit} = (\text{number of cassettes sold})(\text{profit per cassette})$$

Since the goal is to express profit as a function of price, the independent variable is price and the dependent variable is profit. Let p denote the price at which each cassette will be sold and let $P(p)$ be the corresponding monthly profit.

Next, express the number of cassettes sold in terms of the variable p. You know that 4,000 cassettes are sold each month when the price is $5 and that 400 fewer will be sold each month for each $1 increase in price. Since the number of $1 increases is the difference $p - 5$ between the new and old selling prices, you must have

$$
\begin{aligned}
\text{Number of cassettes sold} &= 4,000 - 400(\text{number of } \$1 \text{ increases}) \\
&= 4,000 - 400(p - 5) \\
&= 6,000 - 400p
\end{aligned}
$$

The profit per cassette is simply the difference between the selling price p and the cost $2. Thus,

$$\text{Profit per cassette} = p - 2$$

and the total profit is

$$
\begin{aligned}
P(p) &= (\text{number of cassettes sold})(\text{profit per cassette}) \\
&= (6,000 - 400p)(p - 2) \\
&= -400p^2 + 6,800p - 12,000
\end{aligned}
$$

b. The graph of $P(p)$ is the downward opening parabola shown in Figure 1.35. Maximum profit will occur at the value of p that corresponds to the highest point on the profit graph. This is the vertex of the parabola, which we know occurs where

$$p = \frac{-B}{2A} = \frac{-(6,800)}{2(-400)} = 8.5$$

EXPLORE!

Store the function Y1 = $-400X^2 + 6,800X - 12,000$ into the equation editor of your graphing calculator. Use the **TBLSET** feature to set the initial value of x at 5 in TblStart with unit (1) increment for ΔTbl. Then, construct an appropriate viewing window for the graph of this profit function. Now employ the **TRACE**, **ZOOM** or minimum finding method on your calculator to confirm the optimal cost and profit as depicted in Figure 1.35.

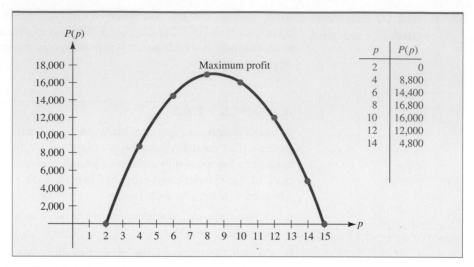

FIGURE 1.35   The profit function $P(p) = (6,000 - 400p)(p - 2)$.

Thus, profit is maximized when the manufacturer charges $8.50 for each cassette, and the maximum monthly profit is

$$P_{max} = P(8.5) = -400(8.5)^2 + 6,800(8.5) - 12,000$$
$$= \$16,900$$

Market Equilibrium   Recall from Section 1.1 that the **demand function** $D(x)$ for a commodity relates the number of units x that are produced to the unit price $p = D(x)$ at which all x units are demanded (sold) in the marketplace. Similarly, the **supply function** $S(x)$ gives the corresponding price $p = S(x)$ at which producers are willing to supply x units to the marketplace. Usually, as the price of a commodity increases, more units of the commodity will be supplied and fewer will be demanded. Likewise, as the production level x increases, the supply price $p = S(x)$ also increases but the demand price $p = D(x)$ decreases. This means that a typical supply curve is rising, while a typical demand curve is falling, as indicated in Figure 1.36.

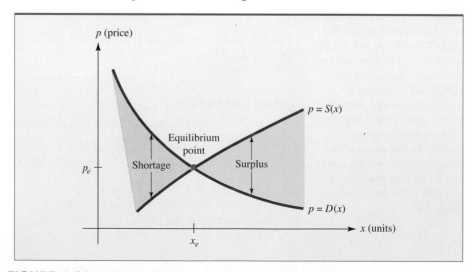

FIGURE 1.36   Market equilibrium occurs when supply equals demand.

The **law of supply and demand** says that in a competitive market environment, supply tends to equal demand, and when this occurs, the market is said to be in **equilibrium.** Thus, market equilibrium occurs precisely at the production level x_e, where $S(x_e) = D(x_e)$. The corresponding unit price p_e is called the **equilibrium price;** that is,

$$p_e = D(x_e) = S(x_e)$$

When the market is not in equilibrium, it has a **shortage** when demand exceeds supply $[D(x) > S(x)]$ and a **surplus** when supply exceeds demand $[S(x) > D(x)]$. This terminology is illustrated in Figure 1.36 and in Example 1.4.6.

EXPLORE!

Following Example 1.4.6, store $S(x) = x^2 + 14$ into Y1 and $D(x) - 174 - 6x$ into Y2. Use a viewing window [5, 35]5 by [0, 200]50 to observe the shortage and surplus sectors. Check if your calculator can shade these sectors by a command such as **SHADE (Y2, Y1).** What sector is this?

EXAMPLE | 1.4.6

Market research indicates that manufacturers will supply x units of a particular commodity to the marketplace when the price is $p = S(x)$ dollars per unit and that the same number of units will be demanded (bought) by consumers when the price is $p = D(x)$ dollars per unit, where the supply and demand functions are given by

$$S(x) = x^2 + 14 \qquad \text{and} \qquad D(x) = 174 - 6x$$

a. At what level of production x and unit price p is market equilibrium achieved?

b. Sketch the supply and demand curves, $p = S(x)$ and $p = D(x)$, on the same graph and interpret.

Solution

a. Market equilibrium occurs when

$$\begin{aligned} S(x) &= D(x) \\ x^2 + 14 &= 174 - 6x \\ x^2 + 6x - 160 &= 0 \qquad \text{subtract } 174 - 6x \text{ from both sides} \\ (x - 10)(x + 16) &= 0 \qquad \text{factor} \\ x = 10 \quad &\text{or} \quad x = -16 \end{aligned}$$

Since only positive values of the production level x are meaningful, we reject $x = -16$ and conclude that equilibrium occurs when $x_e - 10$. The corresponding equilibrium price can be obtained by substituting $x = 10$ into either the supply function or the demand function. Thus,

$$p_e - D(10) - 174 - 6(10) = 114$$

b. The supply curve is a parabola and the demand curve is a line, as shown in Figure 1.37. Notice that no units are supplied to the market until the price reaches $14 per unit and that 29 units are demanded when the price is 0. For $0 \le x < 10$, there is a market shortage since the supply curve is below the demand curve. The supply curve crosses the demand curve at the equilibrium point (10, 114), and for $10 < x \le 29$, there is a market surplus.

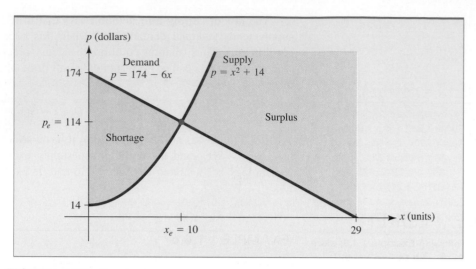

FIGURE 1.37 Supply, demand, and equilibrium point for Example 1.4.6.

Break-Even Analysis

Intersections of graphs arise in business in the context of **break-even analysis.** In a typical situation, a manufacturer wishes to determine how many units of a certain commodity have to be sold for total revenue to equal total cost. Suppose x denotes the number of units manufactured and sold, and let $C(x)$ and $R(x)$ be the corresponding total cost and total revenue, respectively. A pair of cost and revenue curves is sketched in Figure 1.38.

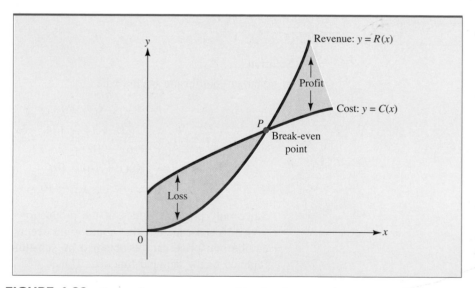

FIGURE 1.38 Cost and revenue curves, with a break-even point at P.

Because of fixed overhead costs, the total cost curve is initially higher than the total revenue curve. Hence, at low levels of production, the manufacturer suffers a loss. At higher levels of production, however, the total revenue curve is the higher one and the manufacturer realizes a profit. The point at which the two curves cross is called the **break-even point,** because when total revenue equals total cost, the manufacturer breaks even, experiencing neither a profit nor a loss. Here is an example.

EXAMPLE | 1.4.7

A manufacturer can sell a certain product for $110 per unit. Total cost consists of a fixed overhead of $7,500 plus production costs of $ 60 per unit.

a.   How many units must the manufacturer sell to break even?

b.   What is the manufacturer's profit or loss if 100 units are sold?

c.   How many units must be sold for the manufacturer to realize a profit of $1,250?

Solution

If x is the number of units manufactured and sold, the total revenue is given by $R(x) = 110x$ and the total cost by $C(x) = 7,500 + 60x$.

a.   To find the break-even point, set $R(x)$ equal to $C(x)$ and solve:

$$110x = 7,500 + 60x$$
$$50x = 7,500$$

so that $$x = 150$$

It follows that the manufacturer will have to sell 150 units to break even (see Figure 1.39).

EXPLORE!

Following Example 1.4.7, put $C(x) = 7,500 + 60x$ into Y1 and $R(x) = 110x$ into Y2. Use the viewing window [0, 250]50 by [−1,000, 20,000]5,000 with **TRACE** and **ZOOM** or the intersection-finding features of your graphing calculator to confirm the break-even point.

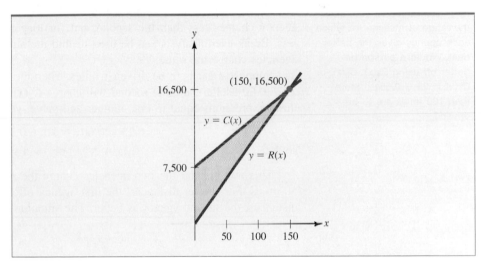

FIGURE 1.39   Revenue $R(x) = 110x$ and cost $C(x) = 7,500 + 60x$.

b.   The profit $P(x)$ is revenue minus cost. Hence,

$$P(x) = R(x) - C(x) = 110x - (7,500 + 60x) = 50x - 7,500$$

The profit from the sale of 100 units is

$$P(100) = 50(100) - 7,500$$
$$= -2,500$$

The minus sign indicates a negative profit (that is, a loss), which was expected since 100 units is less than the break-even level of 150 units. It follows that the manufacturer will lose $2,500 if 100 units are sold.

c.   To determine the number of units that must be sold to generate a profit of $1,250, set the formula for profit $P(x)$ equal to 1,250 and solve for x. You get

$$P(x) = 1,250$$
$$50x - 7,500 = 1,250$$
$$50x = 8,750$$
$$x = \frac{8,750}{50} = 175$$

from which you can conclude that 175 units must be sold to generate the desired profit.

Example 1.4.8 illustrates how break-even analysis can be used as a tool for decision making.

EXPLORE!

Refer to Example 1.4.8. Place $C_1(x) = 25 + 0.6x$ into Y1 and $C_2(x) = 30 + 0.5x$ into Y2 of the equation editor of your graphing calculator. Use the viewing window [−25, 250]25 by [−10, 125]50 to determine the range of mileage for which each agency gives the better deal. Would a person be better off using $C_1(x)$, $C_2(x)$, or $C_3(x) = 23 + 0.55x$ if more than 100 miles are to be driven?

EXAMPLE | 1.4.8

A certain car rental agency charges $25 plus 60 cents per mile. A second agency charges $30 plus 50 cents per mile. Which agency offers the better deal?

Solution

The answer depends on the number of miles the car is driven. For short trips, the first agency charges less than the second, but for long trips, the second agency charges less. Break-even analysis can be used to find the number of miles for which the two agencies charge the same.

Suppose a car is to be driven x miles. Then the first agency will charge $C_1(x) = 25 + 0.60x$ dollars and the second will charge $C_2(x) = 30 + 0.50x$ dollars. If you set these expressions equal to one another and solve, you get

$$25 + 0.60x = 30 + 0.50x$$

so that $\qquad\qquad 0.1x = 5 \qquad$ or $\qquad x = 50$

This implies that the two agencies charge the same amount if the car is driven 50 miles. For shorter distances, the first agency offers the better deal, and for longer distances, the second agency is better. The situation is illustrated in Figure 1.40.

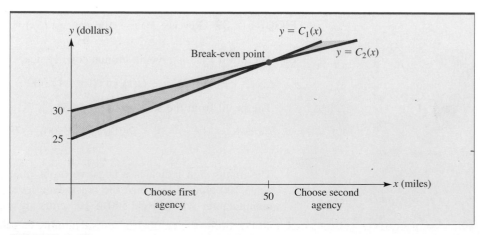

FIGURE 1.40   Car rental costs at competing agencies.

PROBLEMS 1.4

1. **FENCING**   A farmer wishes to fence off a rectangular field with 1,000 feet of fencing. If the long side of the field is along a stream (and does not require fencing), express the area of the field as a function of its width.

2. **LANDSCAPING**   A landscaper wishes to make a rectangular flower garden that is twice as long as it is wide. Express the area of the garden as a function of its width.

3. The sum of two numbers is 18. Express the product of the numbers as a function of the smaller number.

4. The product of two numbers is 318. Express the sum of the numbers as a function of the smaller number.

5. **SALES REVENUE**   Each unit of a certain commodity costs $p = 35x + 15$ cents when x units of the commodity are produced. If all x units are sold at this price, express the revenue derived from the sales as a function of x.

6. **FENCING**   A city recreation department plans to build a rectangular playground 3,600 square meters in area. The playground is to be surrounded by a fence. Express the length of the fencing as a function of the length of one of the sides of the playground, draw the graph, and estimate the dimensions of the playground requiring the least amount of fencing.

7. **AREA**   Express the area of a rectangular field whose perimeter is 320 meters as a function of the length of one of its sides. Draw the graph and estimate the dimensions of the field of maximum area.

8. **PACKAGING**   A closed box with a square base is to have a volume of 1,500 cubic inches. Express its surface area as a function of the length of its base.

9. **PACKAGING**   A closed box with a square base has a surface area of 4,000 square centimeters. Express its volume as a function of the length of its base.

In Problems 10 through 14, you need to know that a cylinder of radius r and height h has volume $V = \pi r^2 h$ and lateral (side) surface area $S = 2\pi rh$. A circular disk of radius r has area $A = \pi r^2$.

10. **PACKAGING**   A soda can holds 12 fluid ounces (approximately 6.89π cubic inches).

Express the surface area of the can as a function of its radius.

11. **PACKAGING**   A closed cylindrical can has surface area 120π square inches. Express the volume of the can as a function of its radius.

12. **PACKAGING**   A closed cylindrical can has a radius r and height h.
    a. If the surface area S of the can is a constant, express the volume V of the can in terms of S and r.
    b. If the volume V of the can is a constant, express the surface area S in terms of V and r.

13. **PACKAGING**   A cylindrical can is to hold 4π cubic inches of frozen orange juice. The cost per square inch of constructing the metal top and bottom is twice the cost per square inch of constructing the cardboard side. Express the cost of constructing the can as a function of its radius if the cost of the side is 0.02 cent per square inch.

14. **PACKAGING**   A cylindrical can with no top has been made from 27π square inches of metal. Express the volume of the can as a function of its radius.

15. **POPULATION GROWTH**   In the absence of environmental constraints, population grows at a rate proportional to its size. Express the rate of population growth as a function of the size of the population.

16. **RADIOACTIVE DECAY**   A sample of radium decays at a rate proportional to the amount of radium remaining. Express the rate of decay of the sample as a function of the amount remaining.

17. **TEMPERATURE CHANGE**   The rate at which the temperature of an object changes is proportional to the difference between its own temperature and the temperature of the surrounding medium. Express this rate as a function of the temperature of the object.

18. **THE SPREAD OF AN EPIDEMIC**   The rate at which an epidemic spreads through a community is jointly proportional to the number of people who have caught the disease and the number who have not. Express this rate as a function of the number of people who have caught the disease.

19. **POLITICAL CORRUPTION** The rate at which people are implicated in a government scandal is jointly proportional to the number of people already implicated and the number of people involved who have not yet been implicated. Express this rate as a function of the number of people who have been implicated.

20. **PRODUCTION COST** At a certain factory, setup cost is directly proportional to the number of machines used and operating cost is inversely proportional to the number of machines used. Express the total cost as a function of the number of machines used.

21. **TRANSPORTATION COST** A truck is hired to transport goods from a factory to a warehouse. The driver's wages are figured by the hour and so are inversely proportional to the speed at which the truck is driven. The cost of gasoline is directly proportional to the speed. Express the total cost of operating the truck as a function of the speed at which it is driven.

PEDIATRIC DRUG DOSAGE *Several different formulas have been proposed for determining the appropriate dose of a drug for a child in terms of the adult dosage. Suppose that A milligrams (mg) is the adult dose of a certain drug and C is the appropriate dosage for a child of age N years. Then* Cowling's rule *says that*

$$C = \left(\frac{N + 1}{24}\right)A$$

while Friend's rule *says that*

$$C = \frac{2NA}{25}$$

Problems 22 through 24 require these formulas.

22. If an adult dose of ibuprofen is 300 mg, what dose does Cowling's rule suggest for an 11-year-old child? What dose does Friend's rule suggest for the same child?

23. Assume an adult dose of $A = 300$ mg, so that Cowling's rule and Friend's rule become functions of the child's age N. Sketch the graphs of these two functions.

24. For what child's age is the dosage suggested by Cowling's rule the same as that predicted by Friend's rule? For what ages does Cowling's rule suggest a larger dosage than Friend's rule? For what ages does Friend's rule suggest the larger dosage?

25. **PEDIATRIC DRUG DOSAGE** As an alternative to Friend's rule and Cowling's rule, pediatricians sometimes use the formula

$$C = \frac{SA}{1.7}$$

to estimate an appropriate drug dosage for a child whose surface area is S square meters, when the adult dosage of the drug is A milligrams (mg). In turn, the surface area of the child's body is estimated by the formula

$$S = 0.0072W^{0.425}H^{0.725}$$

where W and H are, respectively, the child's weight in kilograms (kg) and height in centimeters (cm).

   a. The adult dosage for a certain drug is 250 mg. How much of the drug should be given to a child who is 91 cm tall and weighs 18 kg?

   b. A drug is prescribed for two children, one of whom is twice as tall and twice as heavy as the other. Show that the larger child should receive approximately 2.22 times as much of the drug as the smaller child.

26. **AUCTION BUYER'S PREMIUM** Usually, when you purchase a lot in an auction, you pay not only your winning bid price but also a buyer's premium. At one auction house, the buyer's premium is 17.5% of the winning bid price for purchases up to $50,000. For larger purchases, the buyer's premium is 17.5% of the first $50,000 plus 10% of the purchase price above $50,000.

   a. Find the total price a buyer pays (bid price plus buyer's premium) at this auction house for purchases of $1,000, $25,000, and $100,000.

   b. Express the total purchase price of a lot at this auction house as a function of the final (winning) bid price. Sketch the graph of this function.

27. **TRANSPORTATION COST** A bus company has adopted the following pricing policy for groups that wish to charter its buses. Groups containing no more than 40 people will be charged a fixed amount of $2,400 (40 times $60). In groups containing between 40 and 80 people, everyone will pay $60 minus 50 cents for each person in excess of 40. The company's lowest fare of $40 per person will be offered to groups that have 80 members or more. Express the bus company's revenue as a function of the size of the group. Draw the graph.

28. INCOME TAX The accompanying table represents the 2004 federal income tax rate schedule for single taxpayers.

   a. Express an individual's income tax as a function of the taxable income x for $0 \leq x \leq 146{,}750$ and draw the graph.

   b. The graph in part (a) should consist of four line segments. Compute the slope of each segment. What happens to these slopes as the taxable income increases? Explain the behavior of the slopes in practical terms

If the Taxable Income Is		The Income Tax Is	
Over	But Not Over		Of the Amount Over
0	$7,150	10%	0
$7,150	$29,050	$715 + 15%	$7,150
$29,050	$70,350	$4,000 + 25%	$29,050
$70,350	$146,750	$14,325 + 28%	$70,350

SOURCE: Bankrate.com.

29. ADMISSION FEES A local natural history museum charges admission to groups according to the following policy. Groups of fewer than 50 people are charged a rate of $3.50 per person, while groups of 50 people or more are charged a reduced rate of $3 per person.

   a. Express the amount a group will be charged for admission as a function of its size and draw the graph.

   b. How much money will a group of 49 people save in admission costs if it can recruit 1 additional member?

30. CONSTRUCTION COST A closed box with a square base is to have a volume of 250 cubic meters. The material for the top and bottom of the box costs $2 per square meter, and the material for the sides costs $1 per square meter. Express the construction cost of the box as a function of the length of its base.

31. CONSTRUCTION COST An open box with a square base is to be built for $48. The sides of the box will cost $3 per square meter, and the base will cost $4 per square meter. Express the volume of the box as a function of the length of its base.

32. CONSTRUCTION COST An open box is to be made from a square piece of cardboard, 18 inches by 18 inches, by removing a small square from each corner and folding up the flaps to form the sides. Express the volume of the resulting box as a function of the length x of a side of the removed squares. Draw the graph and estimate the value of x for which the volume of the resulting box is greatest.

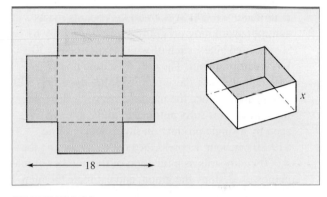

PROBLEM 32

33. POSTER DESIGN A rectangular poster contains 25 square centimeters of print surrounded by margins of 2 centimeters on each side and 4 centimeters on the top and bottom. Express the total area of the poster (printing plus margins) as a function of the width of the printed portion.

34. RETAIL SALES A bookstore can obtain a certain gift book from the publisher at a cost of $3 per book. The bookstore has been offering the book at the price of $15 per copy, and at this price, has been selling 200 copies a month. The bookstore is planning to lower its price to stimulate sales and estimates that for each $1 reduction in the price, 20 more books will be sold each month. Express the bookstore's monthly profit from the sale of this book as a function of the selling price, draw the graph, and estimate the optimal selling price.

35. RETAIL SALES A manufacturer has been selling lamps at the price of $30 per lamp, and at this price consumers have been buying 3,000 lamps a month. The manufacturer wishes to raise the price and estimates that for each $1 increase in the price, 1,000 fewer lamps will be sold each month. The manufacturer can produce the lamps at a cost of $18 per lamp. Express the manufacturer's monthly profit as a function of the price that the lamps are sold, draw the graph, and estimate the optimal selling price.

36. DISTANCE A truck is 300 miles due east of a car and is traveling west at the constant speed of 30 miles per hour. Meanwhile, the car is going north at the constant speed of 60 miles per hour. Express the distance between the car and truck as a function of time.

37. PRODUCTION COST A company has received an order from the city recreation department to manufacture 8,000 Styrofoam kickboards for its summer swimming program. The company owns several machines, each of which can produce 30 kickboards an hour. The cost of setting up the machines to produce these particular kickboards is $20 per machine. Once the machines have been set up, the operation is fully automated and can be overseen by a single production supervisor earning $19.20 per hour. Express the cost of producing the 8,000 kickboards as a function of the number of machines used, draw the graph, and estimate the number of machines the company should use to minimize cost.

38. AGRICULTURAL YIELD A Florida citrus grower estimates that if 60 orange trees are planted, the average yield per tree will be 400 oranges. The average yield will decrease by 4 oranges per tree for each additional tree planted on the same acreage. Express the grower's total yield as a function of the number of additional trees planted, draw the graph, and estimate the total number of trees the grower should plant to maximize yield.

39. HARVESTING Farmers can get $3 per bushel for their potatoes on July first, and after that, the price drops by 2 cents per bushel per day. On July first, a farmer has 140 bushels of potatoes in the field and estimates that the crop is increasing at the rate of 1 bushel per day. Express the farmer's revenue from the sale of the potatoes as a function of the time at which the crop is harvested, draw the graph, and estimate when the farmer should harvest the potatoes to maximize revenue.

MARKET EQUILIBRIUM *In Problems 40–43, supply and demand functions, S(x) and D(x) are given for a particular commodity in terms of the level of production x. In each case:*
(a) Find the value of x_e for which equilibrium occurs and the corresponding equilibrium price p_e.
(b) Sketch the graphs of the supply and demand curves, p = S(x) and p = D(x), on the same graph.
(c) For what values of x is there a market shortage? A market surplus?

40. $S(x) = 4x + 200$ and $D(x) = -3x + 480$

41. $S(x) = 3x + 150$ and $D(x) = -2x + 275$

42. $S(x) = x^2 + x + 3$ and $D(x) = 21 - 3x^2$

43. $S(x) = 2x + 7.43$ and $D(x) = -0.21x^2 - 0.84x + 50$

44. SUPPLY AND DEMAND When electric blenders are sold for p dollars apiece, manufacturers will supply $\dfrac{p^2}{10}$ blenders to local retailers, while the local demand will be $60 - p$ blenders. At what market price will the manufacturers' supply of electric blenders be equal to the consumers' demand for the blenders? How many blenders will be sold at this price?

45. SUPPLY AND DEMAND Producers will supply x units of a certain commodity to the market when the price is $p = S(x)$ dollars per unit, and consumers will demand (buy) x units when the price is $p = D(x)$ dollars per unit, where

$$S(x) = 2x + 15 \qquad \text{and} \qquad D(x) = \frac{385}{x + 1}$$

a. Find the equilibrium production level x_e and the equilibrium price p_e.

b. Draw the supply and demand curves on the same graph.

c. Where does the supply curve cross the y axis? Describe the economic significance of this point.

46. SPY STORY The hero of a popular spy story has escaped from the headquarters of an international diamond smuggling ring in the tiny Mediterranean country of Azusa. Our hero, driving a stolen milk truck at 72 kilometers per hour, has a 40-minute head start on his pursuers, who are chasing him in a Ferrari going 168 kilometers per hour. The distance from the smugglers' headquarters to the border, and freedom, is 83.8 kilometers. Will he make it?

47. AIR TRAVEL Two jets bound for Los Angeles leave New York 30 minutes apart. The first travels 550 miles per hour, while the second goes 650 miles per hour. At what time will the second plane pass the first?

48. BREAK-EVEN ANALYSIS A furniture manufacturer can sell dining room tables for $70 apiece. The manufacturer's total cost consists of a fixed overhead of $8,000 plus production costs of $30 per table.

a. How many tables must the manufacturer sell to break even?

b. How many tables must the manufacturer sell to make a profit of $6,000?

c. What will be the manufacturer's profit or loss if 150 tables are sold?

d. On the same set of axes, graph the manufacturer's total revenue and total cost functions. Explain how the overhead can be read from the graph.

49. **PUBLISHING DECISION** An author must decide between two publishers who are vying for his new book. Publisher A offers royalties of 1% of net proceeds on the first 30,000 copies and 3.5% on all copies in excess of that figure, and expects to net $2 on each copy sold. Publisher B will pay no royalties on the first 4,000 copies sold but will pay 2% on the net proceeds of all copies sold in excess of 4,000 copies, and expects to net $3 on each copy sold. Suppose the author expects to sell N copies. State a simple criterion based on N for deciding how to choose between the publishers.

50. **CHECKING ACCOUNT** The charge for maintaining a checking account at a certain bank is $12 per month plus 10 cents for each check that is written. A competing bank charges $10 per month plus 14 cents per check. Find a criterion for deciding which bank offers the better deal.

51. **PHYSIOLOGY** The pupil of the human eye is roughly circular. If the intensity of light I entering the eye is proportional to the area of the pupil, express I as a function of the radius r of the pupil.

52. **RECYCLING** To raise money, a service club has been collecting used bottles that it plans to deliver to a local glass company for recycling. Since the project began 80 days ago, the club has collected 24,000 pounds of glass for which the glass company currently offers 1 cent per pound. However, since bottles are accumulating faster than they can be recycled, the company plans to reduce by 1 cent each day the price it will pay for 100 pounds of used glass. Assuming that the club can continue to collect bottles at the same rate and that transportation costs make more than one trip to the glass company infeasible, express the club's revenue from its recycling project as a function of the number of additional days the project runs. Draw the graph and estimate when the club should conclude the project and deliver the bottles to maximize its revenue.

53. **BIOCHEMISTRY** In biochemistry, the rate R of an enzymatic reaction is found to be given by the equation

$$R = \frac{R_m[S]}{K_m + [S]}$$

where K_m is a constant (called the **Michaelis constant**), R_m is the maximum possible rate, and $[S]$ is the substrate concentration.* Rewrite this equation so that $y = \dfrac{1}{R}$ is expressed as a function of $x = \dfrac{1}{[S]}$, and sketch the graph of this function.

(This graph is called the **Lineweaver-Burk double-reciprocal plot**.)

54. **SUPPLY AND DEMAND** Producers will supply q units of a certain commodity to the market when the price is $p = S(q)$ dollars per unit, and consumers will demand (buy) q units when the price is $p = D(q)$ dollars per unit, where

$$S(q) = aq + b \qquad \text{and} \qquad D(q) = cq + d$$

where a, b, c, and d are all constants.

a. What can you say about the signs of the constants a, b, c, and d if the supply and demand curves are as shown in the accompanying figure?

b. Express the equilibrium production level q_e and the equilibrium price p_e in terms of the coefficients a, b, c, and d.

c. Use your answer in part (b) to determine what happens to the equilibrium production level q_e as a increases. What happens to q_e as d increases?

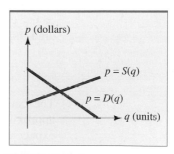

PROBLEM 54

*Mary K. Campbell, *Biochemistry*, Philadelphia: Saunders College Publishing, 1991, pp. 221–226.

55. PUBLISHING PROFIT It costs a book publisher $74,200 to prepare a book for publication (typesetting, illustrating, editing, and so on); printing and binding costs are $5.50 per book. The book is sold to bookstores for $19.50 per copy.

a. Make a table showing the cost of producing 2,000, 4,000, 6,000, and 8,000 books. Use four significant digits.

b. Make a table showing the revenue from selling 2,000, 4,000, 6,000, and 8,000 books. Use four significant digits.

c. Write an algebraic expression representing the cost y as a function of the number of books x that are produced.

d. Write an algebraic expression representing the revenue y as a function of the number of books x sold.

e. Graph both functions on the same coordinate axes.

f. Use **TRACE** and **ZOOM** to find where cost equals revenue.

g. Use the graph to determine how many books need to be made to produce revenue of at least $85,000. How much profit is made for this number of books?

SECTION 1.5 | Limits

As you will see in subsequent chapters, calculus is an enormously powerful branch of mathematics with a wide range of applications, including curve sketching, optimization of functions, analysis of rates of change, and computation of area and probability. What gives calculus its power and distinguishes it from algebra is the concept of limit, and the purpose of this section is to provide an introduction to this important concept. Our approach will be intuitive rather than formal. The ideas outlined here form the basis for a more rigorous development of the laws and procedures of calculus and lie at the heart of much of modern mathematics.

Intuitive Introduction to the Limit

Roughly speaking, the limit process involves examining the behavior of a function $f(x)$ as x approaches a number c that may or may not be in the domain of f. Limiting behavior occurs in a variety of practical situations. For instance, absolute zero, the temperature T_c at which all molecular activity ceases, can be approached but never actually attained in practice. Similarly, economists who speak of profit under ideal conditions or engineers profiling the ideal specifications of a new engine are really dealing with limiting behavior.

To illustrate the limit process, consider a manager who determines that when $x \%$ of her company's plant capacity is being used, the total cost of operation is C hundred thousand dollars, where

$$C(x) = \frac{8x^2 - 636x - 320}{x^2 - 68x - 960}$$

The company has a policy of rotating maintenance in an attempt to ensure that approximately 80% of capacity is always in use. What cost should the manager expect when the plant is operating at this ideal capacity?

It may seem that we can answer this question by simply evaluating $C(80)$, but attempting this evaluation results in the meaningless fraction $\frac{0}{0}$. However, it is still possible to evaluate $C(x)$ for values of x that approach 80 from the right ($x > 80$, when capacity is temporarily overutilized) and from the left ($x < 80$, when capacity is underutilized). A few such calculations are summarized in the following table.

	x approaches 80 from the left ⟩				⟨ *x* approaches 80 from the right		
x	79.8	79.99	79.999	80	80.0001	80.001	80.04
C(*x*)	6.99782	6.99989	6.99999	✕	7.000001	7.00001	7.00043

The values of $C(x)$ displayed on the lower line of this table suggest that $C(x)$ approaches the number 7 as x gets closer and closer to 80. Thus, it is reasonable for the manager to expect a cost of \$700,000 when 80% of plant capacity is utilized.

The functional behavior in this example can be described by saying "$C(x)$ has the limiting value 7 as x approaches 80" or, equivalently, by writing

$$\lim_{x \to 80} C(x) = 7$$

More generally, the limit of $f(x)$ as x approaches the number c can be defined informally as follows.

Limit ■ If $f(x)$ gets closer and closer to a number L as x gets closer and closer to c from both sides, then L is the limit of $f(x)$ as x approaches c. The behavior is expressed by writing

$$\lim_{x \to c} f(x) = L$$

Geometrically, the limit statement $\lim_{x \to c} f(x) = L$ means that the height of the graph $y = f(x)$ approaches L as x approaches c, as shown in Figure 1.41. This interpretation is illustrated along with the tabular approach to computing limits in Example 1.5.1.

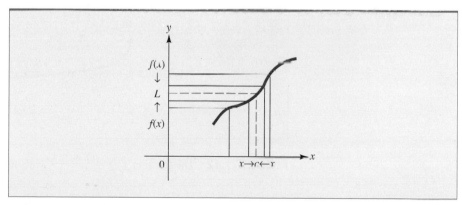

FIGURE 1.41 If $\lim_{x \to c} f(x) = L$, the height of the graph of f approaches L as x approaches c.

EXAMPLE │ **1.5.1**

Use a table to estimate the limit

$$\lim_{x \to 1} \frac{\sqrt{x} - 1}{x - 1}$$

EXPLORE!

Graph $f(x) = \dfrac{\sqrt{x} - 1}{x - 1}$, using the modified decimal viewing window [0, 4.7]1 by [−1.1, 2.1]1. Trace values near $x = 1$. Also construct a table of values, using an initial value of 0.97 for x with an incremental change of 0.01. Describe what you observe. Now use an initial value of 0.997 for x with an incremental change of 0.001. Specifically what happens as x approaches 1 from either side? What would be the most appropriate value for $f(x)$ at $x = 1$ to fill up the hole in the graph?

Solution

Let

$$f(x) = \frac{\sqrt{x} - 1}{x - 1}$$

and compute $f(x)$ for a succession of values of x approaching 1 from the left and from the right:

				$x \to 1 \leftarrow x$			
x	0.99	0.999	0.9999	1	1.00001	1.0001	1.001
$f(x)$	0.50126	0.50013	0.50001	✕	0.499999	0.49999	0.49988

The numbers on the bottom line of the table suggest that $f(x)$ approaches 0.5 as x approaches 1; that is,

$$\lim_{x \to 1} \frac{\sqrt{x} - 1}{x - 1} = 0.5$$

The graph of $f(x)$ is shown in Figure 1.42. The limit computation says that the height of the graph of $y = f(x)$ approaches $L = 0.5$ as x approaches 1. This corresponds to the "hole" in the graph of $f(x)$ at $(1, 0.5)$. We will compute this same limit using an algebraic procedure in Example 1.5.6.

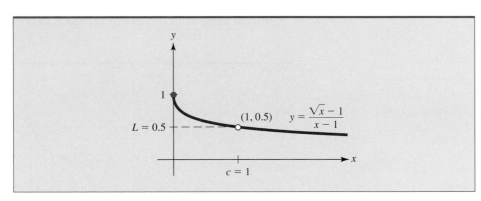

FIGURE 1.42 The function $f(x) = \dfrac{\sqrt{x} - 1}{x - 1}$ tends toward $L = 0.5$ as x approaches $c = 1$.

It is important to remember that limits describe the behavior of a function *near* a particular point, not necessarily *at* the point itself. This is illustrated in Figure 1.43. For all three functions graphed, the limit of $f(x)$ as x approaches 3 is equal to 4. Yet the functions behave quite differently at $x = 3$ itself. In Figure 1.43a, $f(c)$ is equal to the limit 4; in Figure 1.43b, $f(c)$ is different from 4; and in Figure 1.43c, $f(c)$ is not defined at all.

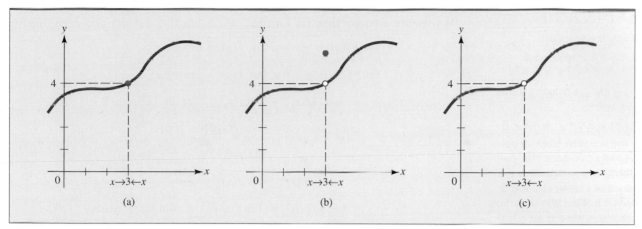

FIGURE 1.43   Three functions for which $\lim\limits_{x \to 3} f(x) = 4$.

Figure 1.44 shows the graph of two functions that do not have a limit as x approaches 2. The limit does not exist in Figure 1.44a because $f(x)$ tends toward 5 as x approaches 2 from the right and tends toward a different value, 3, as x approaches 2 from the left. The function in Figure 1.44b has no finite limit as x approaches 2 because the values of $f(x)$ increase without bound as x tends toward 2 and hence tend to no finite number L. Such so-called *infinite limits* will be discussed later in this section.

EXPLORE!

Graph $f(x) = \dfrac{2}{(x-2)^2}$ using the window [0, 4]1 by [−5, 40]5. Trace the graph on both sides of $x = 2$ to view the behavior of $f(x)$ about $x = 2$. Also display the table value of the function with the incremental change of x set to 0.01 and the initial value $x = 1.97$. What happens to the values of $f(x)$ as x approaches 2?

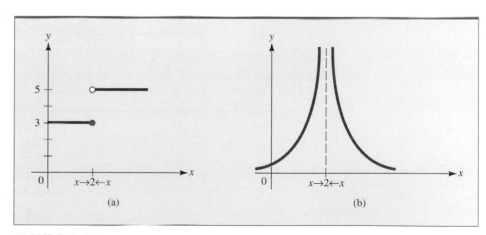

FIGURE 1.44   Two functions for which $\lim\limits_{x \to 2} f(x)$ does not exist.

Properties of Limits   Limits obey certain algebraic rules that can be used in computations. These rules, which should seem plausible on the basis of our informal definition of limit, are proved formally in more theoretical courses. They are important because they simplify the calculation of limits of algebraic functions.

Algebraic Properties of Limits ■ If $\lim\limits_{x \to c} f(x)$ and $\lim\limits_{x \to c} g(x)$ exist, then

$$\lim_{x \to c} [f(x) + g(x)] = \lim_{x \to c} f(x) + \lim_{x \to c} g(x)$$

$$\lim_{x \to c} [f(x) - g(x)] = \lim_{x \to c} f(x) - \lim_{x \to c} g(x)$$

$$\lim_{x \to c} [kf(x)] = k \lim_{x \to c} f(x) \quad \text{for any constant } k$$

$$\lim_{x \to c} [f(x)g(x)] = [\lim_{x \to c} f(x)][\lim_{x \to c} g(x)]$$

$$\lim_{x \to c} \frac{f(x)}{g(x)} = \frac{\lim\limits_{x \to c} f(x)}{\lim\limits_{x \to c} g(x)} \quad \text{if } \lim_{x \to c} g(x) \neq 0$$

$$\lim_{x \to c} [f(x)]^p = [\lim_{x \to c} f(x)]^p \quad \text{if } [\lim_{x \to c} f(x)]^p \text{ exists}$$

That is, the limit of a sum, a difference, a multiple, a product, a quotient, or a power exists and is the sum, difference, multiple, product, quotient, or power of the individual limits, as long as all expressions involved are defined.

Here are two elementary limits that we will use along with the limit rules to compute limits involving more complex expressions.

Limits of Two Linear Functions ■ For any constant k,

$$\lim_{x \to c} k = k \qquad \text{and} \qquad \lim_{x \to c} x = c$$

That is, the limit of a constant is the constant itself, and the limit of $f(x) = x$ as x approaches c is c.

In geometric terms, the limit statement $\lim\limits_{x \to c} k = k$ says that the height of the graph of the constant function $f(x) = k$ approaches k as x approaches c. Similarly, $\lim\limits_{x \to c} x = c$ says that the height of the linear function $f(x) = x$ approaches c as x approaches c. These statements are illustrated in Figure 1.45.

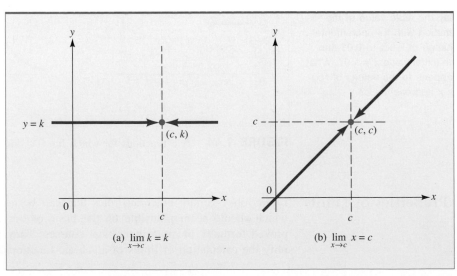

FIGURE 1.45 Limits of two linear functions.

Computation of Limits

Examples 1.5.2 through 1.5.6 illustrate how the properties of limits can be used to calculate limits of algebraic functions. In Example 1.5.2, you will see how to find the limit of a polynomial.

EXAMPLE | 1.5.2

Find $\lim\limits_{x \to -1} (3x^3 - 4x + 8)$.

Solution

Apply the properties of limits to obtain

$$\lim_{x \to -1} (3x^3 - 4x + 8) = 3\left(\lim_{x \to -1} x\right)^3 - 4\left(\lim_{x \to -1} x\right) + \lim_{x \to -1} 8$$
$$= 3(-1)^3 - 4(-1) + 8 = 9$$

In Example 1.5.3, you will see how to find the limit of a rational function whose denominator does not approach zero.

EXAMPLE | 1.5.3

Find $\lim\limits_{x \to 1} \dfrac{3x^3 - 8}{x - 2}$.

Solution

Since $\lim\limits_{x \to 1} (x - 2) \neq 0$, you can use the quotient rule for limits to get

$$\lim_{x \to 1} \frac{3x^3 - 8}{x - 2} = \frac{\lim\limits_{x \to 1} (3x^3 - 8)}{\lim\limits_{x \to 1} (x - 2)} = \frac{3\lim\limits_{x \to 1} x^3 - \lim\limits_{x \to 1} 8}{\lim\limits_{x \to 1} x - \lim\limits_{x \to 1} 2} = \frac{3 - 8}{1 - 2} = 5$$

In general, you can use the properties of limits to obtain these formulas, which can be used to evaluate many limits that occur in practical problems.

Limits of Polynomials and Rational Functions ■ If $p(x)$ and $q(x)$ are polynomials, then

$$\lim_{x \to c} p(x) = p(c)$$

and

$$\lim_{x \to c} \frac{p(x)}{q(x)} = \frac{p(c)}{q(c)} \quad \text{if } q(c) \neq 0$$

In Example 1.5.4, the denominator of the given rational function approaches zero, while the numerator does not. When this happens, you can conclude that the limit does not exist. The absolute value of such a quotient increases without bound and hence does not approach any finite number.

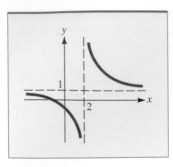

FIGURE 1.46 The graph of $f(x) = \dfrac{x + 1}{x - 2}$.

EXPLORE!

Graph $y = \dfrac{x + 1}{x - 2}$ using an enlarged decimal window [−9.4, 9.4]1 by [−6.2, 6.2]1. Use the **TRACE** key to approach $x = 2$ from the left side and the right side. Also create a table of values, using an initial value of 1.97 for x and increasing in increments of 0.01. Describe what you observe.

EXAMPLE 1.5.4

Find $\displaystyle \lim_{x \to 2} \frac{x + 1}{x - 2}$.

Solution

The quotient rule for limits does not apply in this case since the limit of the denominator is

$$\lim_{x \to 2} (x - 2) = 0$$

Since the limit of the numerator is $\displaystyle \lim_{x \to 2} (x + 1) = 3$, which is not equal to zero, you can conclude that the limit of the quotient does not exist.

The graph of the function $f(x) = \dfrac{x + 1}{x - 2}$ in Figure 1.46 gives you a better idea of what is actually happening in this example. Note that $f(x)$ increases without bound as x approaches 2 from the right and decreases without bound as x approaches 2 from the left.

In Example 1.5.5, the numerator and the denominator of the given rational function *both* approach zero. When this happens, you should try to simplify the function algebraically to find the desired limit.

EXAMPLE 1.5.5

Find $\displaystyle \lim_{x \to 1} \frac{x^2 - 1}{x^2 - 3x + 2}$.

Solution

As x approaches 1, both the numerator and the denominator approach zero, and you can draw no conclusion about the size of the quotient. To proceed, observe that the given function is not defined when $x = 1$ but that for all other values of x, you can divide the numerator and denominator by $x - 1$ to obtain

$$\frac{x^2 - 1}{x^2 - 3x + 2} = \frac{(x - 1)(x + 1)}{(x - 1)(x - 2)} = \frac{x + 1}{x - 2} \qquad x \neq 1$$

(Since $x \neq 1$, you are not dividing by zero.) Now take the limit as x approaches (but is not equal to) 1 to get

$$\lim_{x \to 1} \frac{x^2 - 1}{x^2 - 3x + 2} = \frac{\displaystyle \lim_{x \to 1} (x + 1)}{\displaystyle \lim_{x \to 1} (x - 2)} = \frac{2}{-1} = -2$$

The graph of the function $f(x) = \dfrac{x^2 - 1}{x^2 - 3x + 2}$ is shown in Figure 1.47. Note that it is like the graph in Figure 1.46 with a hole at the point $(1, -2)$.

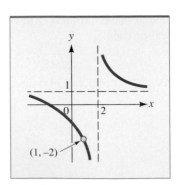

FIGURE 1.47 The graph of $f(x) = \dfrac{x^2 - 1}{x^2 - 3x + 2}$.

In general, when both the numerator and denominator of a quotient approach zero as x approaches c, your strategy will be to simplify the quotient algebraically (as in Example 1.5.5 by canceling $x - 1$). In most cases, the simplified form of the quotient will be valid for all values of x except $x = c$. Since you are interested in the behavior of the quotient *near* $x = c$ and not *at* $x = c$, you may use the simplified form of the quotient to calculate the limit. In Example 1.5.6, we use this technique to obtain the limit we estimated using a table in Example 1.5.1.

EXAMPLE | 1.5.6

Find $\lim\limits_{x \to 1} \dfrac{\sqrt{x} - 1}{x - 1}$.

Solution

Both the numerator and denominator approach 0 as x approaches 1. To simplify the quotient, we rationalize the numerator (that is, multiply numerator and denominator by $\sqrt{x} + 1$) to get

$$\frac{\sqrt{x} - 1}{x - 1} = \frac{(\sqrt{x} - 1)(\sqrt{x} + 1)}{(x - 1)(\sqrt{x} + 1)} = \frac{x - 1}{(x - 1)(\sqrt{x} + 1)} = \frac{1}{\sqrt{x} + 1} \qquad x \neq 1$$

and then take the limit to obtain

$$\lim_{x \to 1} \frac{\sqrt{x} - 1}{x - 1} = \lim_{x \to 1} \frac{1}{\sqrt{x} + 1} = \frac{1}{2}$$

Just-In-Time Review

Recall that

$(a - b)(a + b) = a^2 - b^2$

In Example 1.5.6, we use this identity with $a = \sqrt{x}$ and $b = 1$.

Limits Involving Infinity

"Long-term" behavior is often a matter of interest in business and economics or the physical and life sciences. For example, a biologist may wish to know the population of a bacterial colony or a population of fruit flies after an indefinite period of time, or a business manager may wish to know how the average cost of producing a particular commodity is affected as the level of production increases indefinitely.

In mathematics, the infinity symbol ∞ is used to represent either unbounded growth or the result of such growth. Here are definitions of limits involving infinity we will use to study "long-term behavior."

Limits at Infinity ■ If the values of the function $f(x)$ approach the number L as x increases without bound, we write

$$\lim_{x \to +\infty} f(x) = L$$

Similarly, we write

$$\lim_{x \to -\infty} f(x) = M$$

when the functional values $f(x)$ approach the number M as x decreases without bound.

Geometrically, the limit statement $\lim\limits_{x \to +\infty} f(x) = L$ means that as x increases without bound, the graph of $f(x)$ approaches the horizontal line $y = L$, while $\lim\limits_{x \to -\infty} f(x) = M$ means that the graph of $f(x)$ approaches the line $y = M$ as x decreases without bound. The lines $y = L$ and $y = M$ that appear in this context are called **horizontal asymptotes** of the graph of $f(x)$. There are many different ways for a graph to have horizontal asymptotes, one of which is shown in Figure 1.48. We will have more to say about asymptotes in Chapter 3 as part of a general discussion of graphing with calculus.

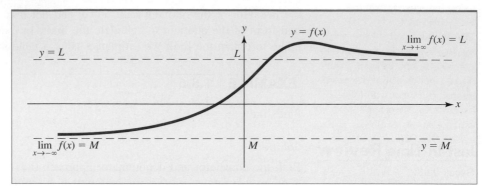

FIGURE 1.48 A graph illustrating limits at infinity and horizontal asymptotes.

The algebraic properties of limits listed earlier in this section also apply to limits at infinity. In addition, since any reciprocal power $1/x^k$ for $k > 0$ becomes smaller and smaller in absolute value as x either increases or decreases without bound, we have these useful rules:

Reciprocal Power Rules ▪ If A and k are constants with $k > 0$ and x^k is defined for all x, then

$$\lim_{x \to +\infty} \frac{A}{x^k} = 0 \qquad \text{and} \qquad \lim_{x \to -\infty} \frac{A}{x^k} = 0$$

The use of these rules is illustrated in Example 1.5.7.

EXAMPLE | 1.5.7

Find $\displaystyle \lim_{x \to +\infty} \frac{x^2}{1 + x + 2x^2}$

Solution

To get a feeling for what happens with this limit, we evaluate the function

$$f(x) = \frac{x^2}{1 + x + 2x^2}$$

at $x = 100$, $1{,}000$, $10{,}000$, and $100{,}000$ and display the results in the table:

				$x \to +\infty$
x	100	1,000	10,000	100,000
$f(x)$	0.49749	0.49975	0.49997	0.49999

The functional values on the bottom line in the table suggest that $f(x)$ tends toward 0.5 as x grows larger and larger. To confirm this observation analytically, we divide each term in $f(x)$ by the highest power that appears in the denominator $1 + x + 2x^2$; namely,

EXPLORE!

Graph $t(x) = \dfrac{x^2}{1 + x + 2x^2}$
using the viewing window
$[-20, 20]5$ by $[0, 1]1$. Now
TRACE the graph to the right
for large values of x, past
$x = 30, 40$, and beyond. What
do you notice about the corresponding y values and the
behavior of the graph? What
would you conjecture as the
value of $f(x)$ as $x \to \infty$?

by x^2. This enables us to find $\lim\limits_{x \to +\infty} f(x)$ by applying reciprocal power rules as follows.

$$\lim_{x \to +\infty} \frac{x^2}{1 + x + 2x^2} = \lim_{x \to +\infty} \frac{x^2/x^2}{1/x^2 + x/x^2 + 2x^2/x^2}$$

$$= \frac{\lim\limits_{x \to +\infty} 1}{\lim\limits_{x \to +\infty} 1/x^2 + \lim\limits_{x \to +\infty} 1/x + \lim\limits_{x \to +\infty} 2} \qquad \textit{several algebraic properties of limits}$$

$$= \frac{1}{0 + 0 + 2} = 0.5 \qquad \textit{reciprocal power rule}$$

The graph of $f(x)$ is shown in Figure 1.49. For practice, verify that $\lim\limits_{x \to -\infty} f(x) = 0.5$ also.

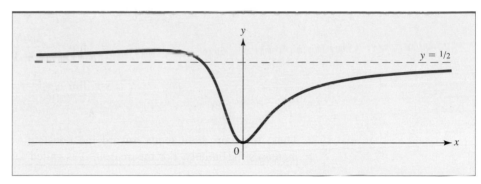

FIGURE 1.49   The graph of $f(x) = \dfrac{x^2}{1 + x + 2x^2}$.

Here is a general description of the procedure for evaluating a limit of a rational function at infinity.

Procedure for Evaluating a Limit at Infinity of $f(x) = p(x)/q(x)$

Step 1. Divide each term in $f(x)$ by the highest power x^k that appears in the denominator polynomial $q(x)$.

Step 2. Compute $\lim\limits_{x \to +\infty} f(x)$ or $\lim\limits_{x \to -\infty} f(x)$ using algebraic properties of limits and the reciprocal power rules.

EXAMPLE | 1.5.8

Find $\lim\limits_{x \to +\infty} \dfrac{2x^2 + 3x + 1}{3x^2 - 5x + 2}$.

Solution

The highest power in the denominator is x^2. Divide the numerator and denominator by x^2 to get

$$\lim_{x \to +\infty} \frac{2x^2 + 3x + 1}{3x^2 - 5x + 2} = \lim_{x \to +\infty} \frac{2 + 3/x + 1/x^2}{3 - 5/x + 2/x^2} = \frac{2 + 0 + 0}{3 - 0 + 0} = \frac{2}{3}$$

EXAMPLE | 1.5.9

If a crop is planted in soil where the nitrogen level is N, then the crop yield Y can be modeled by the *Michaelis-Menten* function

$$Y(N) = \frac{AN}{B + N} \qquad N \geq 0$$

where A and B are positive constants. What happens to crop yield as the nitrogen level is increased indefinitely?

Solution

We wish to compute

$$\lim_{N \to +\infty} Y(N) = \lim_{N \to +\infty} \frac{AN}{B + N}$$

$$= \lim_{N \to +\infty} \frac{AN/N}{B/N + N/N}$$

$$= \lim_{N \to +\infty} \frac{A}{B/N + 1} = \frac{A}{0 + 1}$$

$$= A$$

Thus, the crop yield tends toward the constant value A as the nitrogen level N increases indefinitely. For this reason, A is called the *maximum attainable yield*.

Infinite Limits   We say that $\lim_{x \to c} f(x)$ is an **infinite limit** if $f(x)$ increases or decreases without bound as $x \to c$. Technically, such a limit does not exist, but more information can be given about the behavior of the function by writing

$$\lim_{x \to c} f(x) = +\infty$$

if $f(x)$ increases without bound as $x \to c$ or

$$\lim_{x \to c} f(x) = -\infty$$

if $f(x)$ decreases without bound as $x \to c$. This notation is illustrated in Example 1.5.10 for the case where $x \to +\infty$.

EXAMPLE | 1.5.10

Find $\lim\limits_{x \to +\infty} \dfrac{-x^3 + 2x + 1}{x - 3}$.

Solution

The highest power in the denominator is x. Divide numerator and denominator by x to get

$$\lim_{x \to +\infty} \frac{-x^3 + 2x + 1}{x - 3} = \lim_{x \to +\infty} \frac{-x^2 + 2 + 1/x}{1 - 3/x}$$

Since

$$\lim_{x \to +\infty}\left(-x^2 + 2 + \frac{1}{x}\right) = -\infty \qquad \text{and} \qquad \lim_{x \to +\infty}\left(1 - \frac{3}{x}\right) = 1$$

it follows that

$$\lim_{x \to +\infty} \frac{-x^3 + 2x + 1}{x - 3} = -\infty$$

PROBLEMS | 1.5

In Problems 1 through 6, find $\lim_{x \to a} f(x)$ if it exists.

1.   **2.**   **3.**

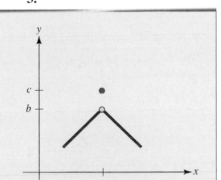

4.   **5.**   **6.**

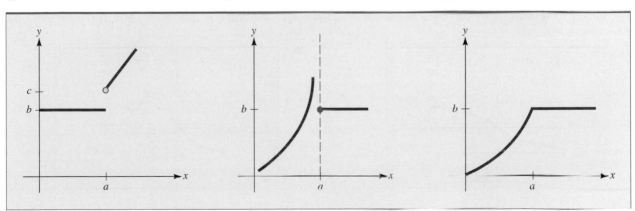

In Problems 7 through 26, find the indicated limit if it exists.

7. $\lim_{x \to 2} (3x^2 - 5x + 2)$

8. $\lim_{x \to -1} (x^3 - 2x^2 + x - 3)$

9. $\lim_{x \to 0} (x^5 - 6x^4 + 7)$

10. $\lim_{x \to -1/2} (1 - 5x^3)$

11. $\lim_{x \to 3} (x - 1)^2(x + 1)$

12. $\lim_{x \to -1} (x^2 + 1)(1 - 2x)^2$

13. $\lim\limits_{x \to 1/3} \dfrac{x + 1}{x + 2}$

14. $\lim\limits_{x \to 1} \dfrac{2x + 3}{x + 1}$

15. $\lim\limits_{x \to 5} \dfrac{x + 3}{5 - x}$

16. $\lim\limits_{x \to 3} \dfrac{2x + 3}{x - 3}$

17. $\lim\limits_{x \to 1} \dfrac{x^2 - 1}{x - 1}$

18. $\lim\limits_{x \to 3} \dfrac{9 - x^2}{x - 3}$

19. $\lim\limits_{x \to 5} \dfrac{x^2 - 3x - 10}{x - 5}$

20. $\lim\limits_{x \to 2} \dfrac{x^2 + x - 6}{x - 2}$

21. $\lim\limits_{x \to 4} \dfrac{(x + 1)(x - 4)}{(x - 1)(x - 4)}$

22. $\lim\limits_{x \to 0} \dfrac{x(x^2 - 1)}{x^2}$

23. $\lim\limits_{x \to -2} \dfrac{x^2 - x - 6}{x^2 + 3x + 2}$

24. $\lim\limits_{x \to 1} \dfrac{x^2 + 4x - 5}{x^2 - 1}$

25. $\lim\limits_{x \to 4} \dfrac{\sqrt{x} - 2}{x - 4}$

26. $\lim\limits_{x \to 9} \dfrac{\sqrt{x} - 3}{x - 9}$

For Problems 27 through 36, find $\lim\limits_{x \to +\infty} f(x)$ *and* $\lim\limits_{x \to -\infty} f(x)$. *If the limiting value is infinite, indicate whether it is* $+\infty$ *or* $-\infty$.

27. $f(x) = x^3 - 4x^2 - 4$

28. $f(x) = 1 - x + 2x^2 - 3x^3$

29. $f(x) = (1 - 2x)(x + 5)$

30. $f(x) = (1 + x^2)^3$

31. $f(x) = \dfrac{x^2 - 2x + 3}{2x^2 + 5x + 1}$

32. $f(x) = \dfrac{1 - 3x^3}{2x^3 - 6x + 2}$

33. $f(x) = \dfrac{2x + 1}{3x^2 + 2x - 7}$

34. $f(x) = \dfrac{x^2 + x - 5}{1 - 2x - x^3}$

35. $f(x) = \dfrac{3x^2 - 6x + 2}{2x - 9}$

36. $f(x) = \dfrac{1 - 2x^3}{x + 1}$

In Problems 37 and 38, the graph of a function f(x) is given. Use the graph to determine $\lim\limits_{x \to +\infty} f(x)$ *and* $\lim\limits_{x \to -\infty} f(x)$.

37.

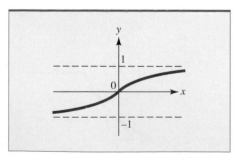

38.

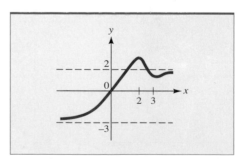

In Problems 39 through 42, complete the table by evaluating f(x) at the specified values of x. Then use the table to estimate the indicated limit or show it does not exist.

39. $f(x) = x^2 - x$; $\lim\limits_{x \to 2} f(x)$

x	1.9	1.99	1.999	2	2.001	2.01	2.1
$f(x)$							

40. $f(x) = x - \dfrac{1}{x}$; $\lim\limits_{x \to 0} f(x)$

x	−0.09	−0.009	0	0.0009	0.009	0.09
$f(x)$						

41. $f(x) = \dfrac{x^3 + 1}{x - 1}; \lim\limits_{x \to 1} f(x)$

x	0.9	0.99	0.999	1	1.001	1.01	1.1
$f(x)$							

42. $f(x) = \dfrac{x^3 + 1}{x + 1}; \lim\limits_{x \to -1} f(x)$

x	−1.1	−1.01	−1.001	−1	−0.999	−0.99	−0.9
$f(x)$							

In Problems 43 through 50, find the indicated limit or show that it does not exist using the following facts about limits involving the functions $f(x)$ and $g(x)$:

$$\lim_{x \to c} f(x) = 5 \qquad \text{and} \qquad \lim_{x \to \infty} f(x) = -3$$

$$\lim_{x \to c} g(x) = -2 \qquad \text{and} \qquad \lim_{x \to \infty} g(x) = 4$$

43. $\lim\limits_{x \to c} [2f(x) - 3g(x)]$

44. $\lim\limits_{x \to c} f(x)\, g(x)$

45. $\lim\limits_{x \to c} \dfrac{f(x)}{g(x)}$

46. $\lim\limits_{x \to c} \dfrac{2f(x) - g(x)}{5g(x) + 2f(x)}$

47. $\lim\limits_{x \to \infty} \sqrt{g(x)}$

48. $\lim\limits_{x \to \infty} \dfrac{2f(x) + g(x)}{x + f(x)}$

49. A wire is stretched horizontally, as shown in the accompanying figure. An experiment is conducted in which different weights are attached at the center and the corresponding vertical displacements are measured. When too much weight is added, the wire snaps. Based on the data in the following table, what do you think is the maximum possible displacement for this kind of wire?

Weight W (lb)	15	16	17	18	17.5	17.9	17.99
Displacement y (in.)	1.7	1.75	1.78	Snaps	1.79	1.795	Snaps

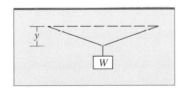

PROBLEM 49

50. PRODUCTION A business manager determines that t months after production begins on a new product, the number of units produced will be P thousand, where

$$P(t) = \frac{6t^2 + 5t}{(t + 1)^2}$$

What happens to production in the long run (as $t \to \infty$)?

51. PER CAPITA EARNINGS Studies indicate that t years from now, the population of a certain country will be $p = 0.2t + 1,500$ thousand people, and that the gross earnings of the country will be E million dollars, where

$$E(t) = \sqrt{9t^2 + 0.5t + 179}$$

a. Express the per capita earnings of the country $P = E/p$ as a function of time t. (Take care with the units.)

b. What happens to the per capita earnings in the long run (as $t \to \infty$)?

52. BACTERIAL GROWTH The accompanying graph shows how the growth rate $R(T)$ of a bacterial colony changes with temperature T.*

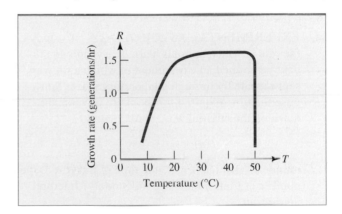

PROBLEM 52

a. Over what range of values of T does the growth rate $R(T)$ double?

b. What can be said about the growth rate for $25 < T < 45$?

c. What happens when the temperature reaches roughly 45°C? Does it make sense to compute $\lim\limits_{T \to 50} R(T)$?

d. Write a paragraph describing how temperature affects the growth rate of a species.

*Source: Michael D. La Grega, Phillip L. Buckingham, and Jeffrey C. Evans, *Hazardous Waste Management*. New York: McGraw-Hill, 1994, pp. 565–566. Reprinted by permission.

53. **ANIMAL BEHAVIOR**   In some animal species, the intake of food is affected by the amount of vigilance maintained by the animal while feeding. In essence, it is hard to eat hardy while watching for predators that may eat you. In one model,* if the animal is foraging on plants that offer a bite of size S, the intake rate of food, $I(S)$, is given by a function of the form

$$I(S) = \frac{aS}{S + c}$$

where a and c are positive constants.
   a. What happens to the intake $I(S)$ as bite size S increases indefinitely? Interpret your result.

   b. Read an article on various ways that the food intake rate may be affected by scanning for predators. Then write a paragraph on how mathematical models may be used to study such behavior in zoology. The reference cited in this problem offers a good starting point.

54. **EXPERIMENTAL PSYCHOLOGY**   To study the rate at which animals learn, a psychology student performed an experiment in which a rat was sent repeatedly through a laboratory maze. Suppose the time required for the rat to traverse the maze on the nth trial was approximately

$$T(n) = \frac{5n + 17}{n}$$

minutes. What happens to the time of traverse as the number of trials n increases indefinitely? Interpret your result.

55. **AVERAGE COST**   A business manager determines that the total cost of producing x units of a particular commodity may be modeled by the function

$$C(x) = 7.5x + 120,000$$

(dollars). The average cost is $A(x) = \dfrac{C(x)}{x}$. Find $\lim\limits_{x \to +\infty} A(x)$ and interpret your result.

56. Solve Problems 17 through 26 by using the **TRACE** feature of your calculator to make a table of x and $f(x)$ values near the number x is approaching.

57. The accompanying graph represents a function $f(x)$ that oscillates between 1 and -1 more and more frequently as x approaches 0 from either the right or the left. Does $\lim\limits_{x \to 0} f(x)$ exist? If so, what is its value? [*Note:* For students with experience in trigonometry, the function $f(x) = \sin\left(\dfrac{1}{x}\right)$ behaves in this way.]

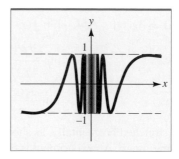

PROBLEM 57

58. If \$1,000 is invested at 5% compounded n times per year, the balance after 1 year will be $1,000(1 + 0.05x)^{1/x}$, where $x = \dfrac{1}{n}$ is the length of the compounding period. For example, if $n = 4$ the compounding period is $\dfrac{1}{4}$ year long. For what is called *continuous compounding* of interest, the balance after 1 year is given by the limit

$$B = \lim_{x \to 0^+} 1,000(1 + 0.05x)^{1/x}$$

Estimate the value of this limit by filling in the second line of the following table:

x	1	0.1	0.01	0.001	0.0001
$1,000(1 + 0.05x)^{1/x}$					

---

*A. W. Willius and C. Fitzgibbon, "Costs of Vigilance in Foraging Ungulates," *Animal Behavior,* Vol. 47, Pt. 2 (Feb. 1994).

59. Evaluate the limit

$$\lim_{x \to +\infty} \frac{a_n x^n + a_{n-1} x^{n-1} + \cdots + a_1 x + a_0}{b_m x^m + b_{m-1} x^{m-1} + \cdots + b_1 x + b_0}$$

for constants $a_0, a_1, \ldots, a_n$ and $b_0, b_1, \ldots, b_m$ in each of the following cases:

a. $n < m$

b. $n = m$

c. $n > m$

[*Note:* There are two possible answers, depending on the signs of a_n and b_m.]

SECTION 1.6 One-Sided Limits and Continuity

The dictionary defines "continuity" as an "unbroken or uninterrupted succession." Continuous behavior is certainly an important part of our lives. For instance, the growth of a tree is continuous, as are the motion of a rocket and the volume of water flowing into a bathtub. In this section, we shall discuss what it means for a function to be continuous, and shall examine a few important properties of such functions.

One-Sided Limits

Informally, a continuous function is one whose graph can be drawn without the "pen" leaving the paper (Figure 1.50a). Not all functions have this property, but those that do play a special role in calculus. A function is *not* continuous where its graph has a "hole or gap" (Figure 1.50b), but what do we really mean by "holes and gaps" in a graph? To describe such features mathematically, we require the concept of a **one-sided limit** of a function; that is, a limit in which the approach is either from the right or from the left, rather than from both sides as required for the "two-sided" limit introduced in Section 1.5.

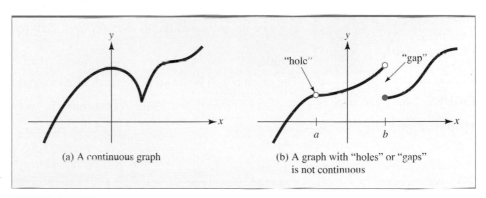

(a) A continuous graph

(b) A graph with "holes" or "gaps" is not continuous

FIGURE 1.50

For instance, Figure 1.51 shows the graph of inventory I as a function of time t for a company that immediately restocks to level L_1 whenever the inventory falls to a certain minimum level L_2 (this is called *just-in-time inventory*). Suppose the first restocking time occurs at $t = t_1$. Then as t tends toward t_1 from the left, the limiting value of $I(t)$ is L_2, while if the approach is from the right, the limiting value is L_1.

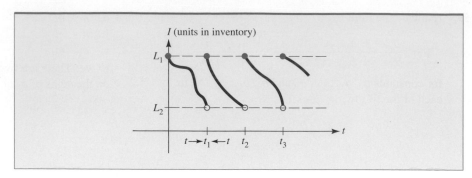

FIGURE 1.51   One-sided limits in a just-in-time inventory example.

Here is the notation we will use to describe one-sided limiting behavior.

> **One-Sided Limits**   ■   If $f(x)$ approaches L as x tends toward c from the left $(x < c)$, we write $\lim\limits_{x \to c^-} f(x) = L$. Likewise, if $f(x)$ approaches M as x tends toward c from the right $(c < x)$, then $\lim\limits_{x \to c^+} f(x) = M$.

If this notation is used in our inventory example, we would write

$$\lim_{t \to t_1^-} I(t) = L_2 \qquad \text{and} \qquad \lim_{t \to t_1^+} I(t) = L_1$$

Here are two more examples involving one-sided limits.

EXAMPLE | 1.6.1

For the function

$$f(x) = \begin{cases} 1 - x^2 & \text{if } 0 \le x < 2 \\ 2x + 1 & \text{if } x \ge 2 \end{cases}$$

evaluate the one-sided limits $\lim\limits_{x \to 2^-} f(x)$ and $\lim\limits_{x \to 2^+} f(x)$.

Solution

The graph of $f(x)$ is shown in Figure 1.52. Since $f(x) = 1 - x^2$ for $0 \le x < 2$, we have

$$\lim_{x \to 2^-} f(x) = \lim_{x \to 2^-} (1 - x^2) = -3$$

Similarly, $f(x) = 2x + 1$ if $x \ge 2$, so

$$\lim_{x \to 2^+} f(x) = \lim_{x \to 2^+} (2x + 1) = 5$$

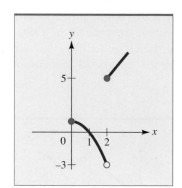

FIGURE 1.52   The graph of $f(x) = \begin{cases} 1 - x^2 & \text{if } 0 \le x < 2 \\ 2x + 1 & \text{if } x \ge 2 \end{cases}$

EXPLORE!

Refer to Example 1.6.2. Graph $f(x) = \dfrac{x - 2}{x - 4}$ using the window [0, 9.4]1 by [−4, 4]1 to verify the limit results as x approaches 4 from the left and the right. Now trace $f(x)$ for large positive or negative values of x. What do you observe?

EXAMPLE | 1.6.2

Find $\lim \dfrac{x - 2}{x - 4}$ as x approaches 4 from the left and from the right.

Solution

First, note that for $2 < x < 4$ the quantity

$$f(x) = \frac{x - 2}{x - 4}$$

is negative, so as x approaches 4 from the left, $f(x)$ *decreases* without bound. We denote this fact by writing

$$\lim_{x \to 4^-} \frac{x - 2}{x - 4} = -\infty$$

Likewise, as x approaches 4 from the right (with $x > 4$), $f(x)$ increases without bound and we write

$$\lim_{x \to 4^+} \frac{x - 2}{x - 4} = +\infty$$

The graph of f is shown in Figure 1.53.

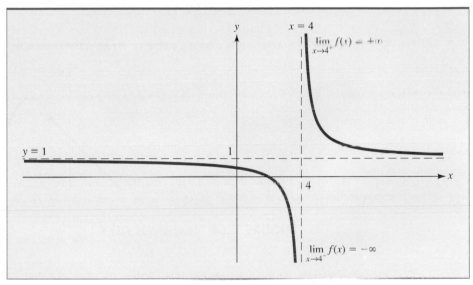

FIGURE 1.53 The graph of $f(x) = \dfrac{x - 2}{x - 4}$.

Notice that the two-sided limit $\lim_{x \to 4} f(x)$ does *not* exist for the function in Example 1.6.2 since the functional values $f(x)$ do not approach a single value L as x tends toward 4 from each side. In general, we have the following useful criterion for the existence of a limit.

Existence of a Limit ■ The two-sided limit $\lim_{x \to c} f(x)$ exists if and only if the two one-sided limits $\lim_{x \to c^-} f(x)$ and $\lim_{x \to c^+} f(x)$ both exist and are equal, and then

$$\lim_{x \to c} f(x) = \lim_{x \to c^-} f(x) = \lim_{x \to c^+} f(x)$$

EXPLORE!

Re-create the piecewise linear function $f(x)$ defined in the Explore! box on page 62. Verify graphically that $\lim_{x \to 2^-} f(x) = 3$ and $\lim_{x \to 2^+} f(x) = 5$.

EXAMPLE | 1.6.3

Determine whether $\lim_{x \to 1} f(x)$ exists, where

$$f(x) = \begin{cases} x + 1 & \text{if } x < 1 \\ -x^2 + 4x - 1 & \text{if } x \geq 1 \end{cases}$$

Solution

Computing the one-sided limits at $x = 1$, we find

$$\lim_{x \to 1^-} f(x) = \lim_{x \to 1^-} (x + 1) = (1) + 1 = 2 \quad \textit{since } f(x) = x + 1 \textit{ when } x < 1$$

and

$$\lim_{x \to 1^+} f(x) = \lim_{x \to 1^+} (-x^2 + 4x - 1) \quad \textit{since } f(x) = -x^2 + 4x - 1 \textit{ when } x \geq 1$$
$$= -(1)^2 + 4(1) - 1 = 2$$

Since the two one-sided limits are equal, it follows that the two-sided limit of $f(x)$ at $x = 1$ exists, and we have

$$\lim_{x \to 1} f(x) = \lim_{x \to 1^-} f(x) = \lim_{x \to 1^+} f(x) = 2$$

The graph of $f(x)$ is shown in Figure 1.54.

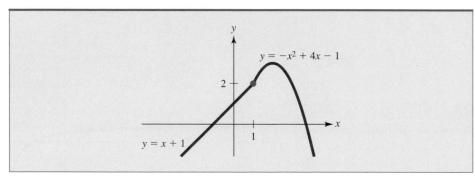

FIGURE 1.54   The graph of $f(x) = \begin{cases} x + 1 & \text{if } x < 1 \\ -x^2 + 4x - 1 & \text{if } x \geq 1 \end{cases}$.

Continuity   At the beginning of this section, we observed that a continuous function is one whose graph has no "holes or gaps." A "hole" at $x = c$ can arise in several ways, three of which are shown in Figure 1.55.

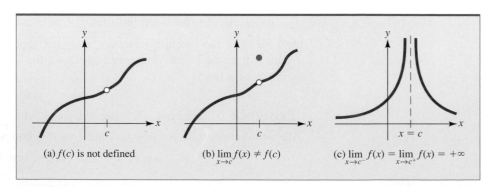

FIGURE 1.55   Three ways the graph of a function can have a "hole" at $x = c$.

The graph of $f(x)$ will have a "gap" at $x = c$ if the one-sided limits $\lim_{x \to c^-} f(x)$ and $\lim_{x \to c^+} f(x)$ are not equal. Three ways this can happen are shown in Figure 1.56.

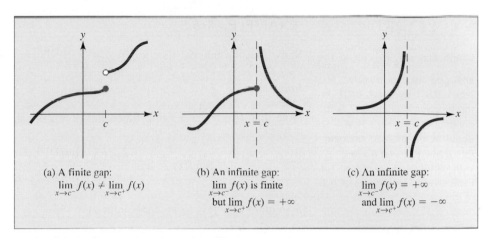

FIGURE 1.56 Three ways for the graph of a function to have a "gap" at $x = c$.

So what properties will guarantee that $f(x)$ does not have a "hole or gap" at $x = c$? The answer is surprisingly simple. The function must be defined at $x = c$, it must have a finite, two-sided limit at $x = c$, and $\lim\limits_{x \to c} f(x)$ must equal $f(c)$. To summarize:

> **Continuity** ▪ A function f is continuous at c if all three of these conditions are satisfied:
>
> **a.** $f(c)$ is defined
>
> **b.** $\lim\limits_{x \to c} f(x)$ exists
>
> **c.** $\lim\limits_{x \to c} f(x) = f(c)$
>
> If $f(x)$ is not continuous at c, it is said to have a **discontinuity** there.

Continuity of Polynomials and Rational Functions

Recall that if $p(x)$ and $q(x)$ are polynomials, then

$$\lim_{x \to c} p(x) = p(c)$$

and

$$\lim_{x \to c} \frac{p(x)}{q(x)} = \frac{p(c)}{q(c)} \quad \text{if } q(c) \neq 0$$

These limit formulas can be interpreted as saying that **a polynomial or a rational function is continuous wherever it is defined.** This is illustrated in Examples 1.6.4 through 1.6.7.

EXAMPLE | 1.6.4

Show that the polynomial $p(x) = 3x^3 - x + 5$ is continuous at $x = 1$.

Solution

Verify that the three criteria for continuity are satisfied. Clearly $p(1)$ is defined; in fact, $p(1) = 7$. Moreover, $\lim\limits_{x \to 1} p(x)$ exists and $\lim\limits_{x \to 1} p(x) = 7$. Thus,

$$\lim_{x \to 1} p(x) = 7 = p(1)$$

as required for $p(x)$ to be continuous at $x = 1$.

EXPLORE!

Graph $f(x) = \dfrac{x+1}{x-2}$ using the enlarged decimal window [−9.4, 9.4]1 by [−6.2, 6.2]1. Is the function continuous? Is it continuous at $x = 2$? How about at $x = 3$? Also examine this function using a table with an initial value of x at 1.8, increasing in increments of 0.2.

EXPLORE!

Store $h(x)$ of Example 1.6.6 (c) into the equation editor as Y1 = (X + 1)(X < 1) + (2 − X)(X ≥ 1). Use a decimal window with a dot graphing style. Is this function continuous at $x = 1$? Use the **TRACE** key to display the value of the function at $x = 1$ and to find the limiting values of y as x approaches 1 from the left side and from the right side.

EXAMPLE 1.6.5

Show that the rational function $f(x) = \dfrac{x+1}{x-2}$ is continuous at $x = 3$.

Solution

Note that $f(3) = \dfrac{3+1}{3-2} = 4$. Since $\lim\limits_{x \to 3} (x-2) \neq 0$, you find that

$$\lim_{x \to 3} f(x) = \lim_{x \to 3} \frac{x+1}{x-2} = \frac{\lim\limits_{x \to 3}(x+1)}{\lim\limits_{x \to 3}(x-2)} = \frac{4}{1} = 4 = f(3)$$

as required for $f(x)$ to be continuous at $x = 3$.

EXAMPLE 1.6.6

Discuss the continuity of each of the following functions:

a. $f(x) = \dfrac{1}{x}$ **b.** $g(x) = \dfrac{x^2-1}{x+1}$ **c.** $h(x) = \begin{cases} x+1 & \text{if } x < 1 \\ 2-x & \text{if } x \geq 1 \end{cases}$

Solution

The functions in parts (a) and (b) are rational and are therefore continuous wherever they are defined (that is, wherever their denominators are not zero).

a. $f(x) = \dfrac{1}{x}$ is defined everywhere except $x = 0$, and thus it is continuous for all $x \neq 0$ (Figure 1.57a).

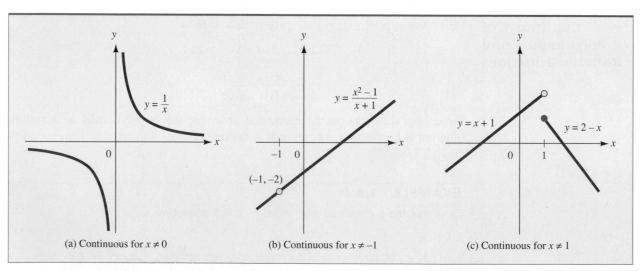

FIGURE 1.57   Functions for Example 1.6.6.

b. Since $x = -1$ is the only value of x for which $g(x)$ is undefined, $g(x)$ is continuous except at $x = -1$ (Figure 1.57b).

EXPLORE!

Graph $f(x) = \dfrac{x^3 - 8}{x - 2}$, using a standard window. Does this graph appear continuous? Now use a modified decimal window [−4.7, 4.7]1 by [0, 14.4]1, and describe what you observe. Which case in Example 1.6.6 does this resemble?

c. This function is defined in two pieces. First check for continuity at $x = 1$, the value of x that separates the two pieces. You find that $\lim\limits_{x \to 1} h(x)$ does not exist, since $h(x)$ approaches 2 from the left and 1 from the right. Thus, $h(x)$ is not continuous at 1 (Figure 1.57c). However, since the polynomials $x + 1$ and $2 - x$ are each continuous for every value of x, it follows that $h(x)$ is continuous at every number x other than 1.

EXAMPLE | 1.6.7

For what value of the constant A is the following function continuous for all real x?

$$f(x) = \begin{cases} Ax + 5 & \text{if } x < 1 \\ x^2 - 3x + 4 & \text{if } x \geq 1 \end{cases}$$

Solution

Since $Ax + 5$ and $x^2 - 3x + 4$ are both polynomials, it follows that $f(x)$ will be continuous everywhere except possibly at $x = 1$. Moreover, $f(x)$ approaches $A + 5$ as x approaches 1 from the left and approaches 2 as x approaches 1 from the right. Thus, for $\lim\limits_{x \to 1} f(x)$ to exist, we must have $A + 5 = 2$ or $A = -3$, in which case

$$\lim_{x \to 1} f(x) = 2 = f(1)$$

This means that f is continuous for all x only when $A = -3$.

Continuity on an Interval

For many applications of calculus, it is useful to have definitions of continuity on open and closed intervals.

> **Continuity on an Interval** ■ A function $f(x)$ is said to be continuous on an open interval $a < x < b$ if it is continuous at each point $x = c$ in that interval.
>
> Moreover, f is continuous on the closed interval $a \leq x \leq b$ if it is continuous on the open interval $a < x < b$ and
>
> $$\lim_{x \to a^+} f(x) = f(a) \qquad \text{and} \qquad \lim_{x \to b^-} f(x) = f(b)$$

In other words, continuity on an interval means that the graph of f is "one piece" throughout the interval.

EXAMPLE | 1.6.8

Discuss the continuity of the function

$$f(x) = \frac{x + 2}{x - 3}$$

on the open interval $-2 < x < 3$ and on the closed interval $-2 \leq x \leq 3$.

Solution

The rational function $f(x)$ is continuous for all x except $x = 3$. Therefore, it is continuous on the open interval $-2 < x < 3$ but not on the closed interval $-2 \leq x \leq 3$, since it is discontinuous at the endpoint 3 (where its denominator is zero). The graph of f is shown in Figure 1.58.

FIGURE 1.58 The graph of $f(x) = \dfrac{x + 2}{x - 3}$.

The Intermediate Value Property

An important feature of continuous functions is the **intermediate value property,** which says that if $f(x)$ is continuous on the interval $a \le x \le b$ and L is a number between $f(a)$ and $f(b)$, then $f(c) = L$ for some number c between a and b (see Figure 1.59). In other words, *a continuous function attains all values between any two of its values.* For instance, a girl who weighs 5 lb at birth and 100 lb at age 12 must have weighed exactly 50 lb at some time during her 12 years of life, since her weight is a continuous function of time.

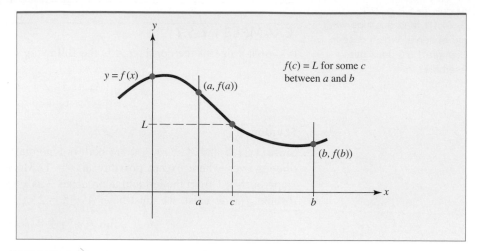

FIGURE 1.59 The intermediate value property.

The intermediate value property has a variety of applications. In Example 1.6.9, we show how it can be used to estimate a solution of a given equation.

EXAMPLE | 1.6.9

Show that the equation $x^2 - x - 1 = \dfrac{1}{x + 1}$ has a solution for $1 < x < 2$.

Solution

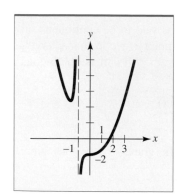

FIGURE 1.60 The graph of $y = x^2 - x - 1 - \dfrac{1}{x + 1}$.

Let $f(x) = x^2 - x - 1 - \dfrac{1}{x + 1}$. Then $f(1) = -\dfrac{3}{2}$ and $f(2) = \dfrac{2}{3}$. Since $f(x)$ is continuous for $1 \le x \le 2$ and the graph of f is below the x axis at $x = 1$ and above the x axis at $x = 2$, it follows from the intermediate value property that the graph must cross the x axis somewhere between $x = 1$ and $x = 2$ (see Figure 1.60). In other words, there is a number c such that $1 < c < 2$ and $f(c) = 0$, so

$$c^2 - c - 1 = \frac{1}{c + 1}$$

NOTE The root-location procedure described in Example 1.6.9 can be continued to estimate the root c to any desired degree of accuracy. For instance, the midpoint of the interval $1 \le x \le 2$ is $d = 1.5$ and $f(1.5) = -0.65$, so the root c must lie in the interval $1.5 < x < 2$ (since $f(2) > 0$), and so on.

"That's nice," you say, "but I can use the solve utility on my calculator to find a much more accurate estimate for c with much less effort." You are right, of course, but how do you think your calculator makes its estimation? Perhaps not by the method just described, but certainly by some similar algorithmic procedure. It is important to understand such procedures as you use the technology that utilizes them. ■

PROBLEMS 1.6

In Problems 1–4, find the one-sided limits $\lim\limits_{x\to 2^-} f(x)$ *and* $\lim\limits_{x\to 2^+} f(x)$ *from the given graph of f and determine whether* $\lim\limits_{x\to 2} f(x)$ *exists.*

1.

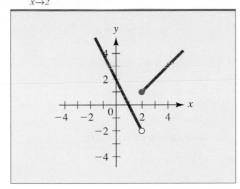

2.

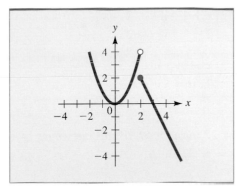

3.

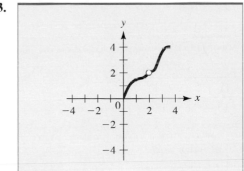

4.

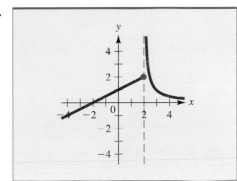

In Problems 5–14, find the indicated one-sided limit. If the limiting value is infinite, indicate whether it is $+\infty$ *or* $-\infty$.

5. $\lim\limits_{x\to 4^+} (3x^2 - 9)$

6. $\lim\limits_{x\to 2^-} \dfrac{x+3}{x+2}$

7. $\lim\limits_{x\to 3^+} \sqrt{3x - 9}$

8. $\lim\limits_{x\to 2^+} \dfrac{x^2 + 4}{x - 2}$

9. $\lim\limits_{x\to 0^+} (x - \sqrt{x})$

10. $\lim\limits_{x\to 1^-} \dfrac{x - \sqrt{x}}{x - 1}$

11. $\lim\limits_{x\to 3^+} \dfrac{\sqrt{x+1} - 2}{x - 3}$

12. $\lim\limits_{x\to 5^+} \dfrac{\sqrt{2x - 1} - 3}{x - 5}$

13. $\lim\limits_{x\to 3^-} f(x)$ and $\lim\limits_{x\to 3^+} f(x)$,

where $f(x) = \begin{cases} 2x^2 - x & \text{if } x < 3 \\ 3 - x & \text{if } x \geq 3 \end{cases}$

14. $\lim\limits_{x\to -1^-} f(x)$ and $\lim\limits_{x\to -1^+} f(x)$,

where $f(x) = \begin{cases} \dfrac{1}{x - 1} & \text{if } x < -1 \\ x^2 + 2x & \text{if } x \geq -1 \end{cases}$

In Problems 15–26, decide if the given function is continuous at the specified value of x.

15. $f(x) = 5x^2 - 6x + 1$ at $x = 2$

16. $f(x) = x^3 - 2x^2 + x - 5$ at $x = 0$

17. $f(x) = \dfrac{x + 2}{x + 1}$ at $x = 1$

18. $f(x) = \dfrac{2x - 4}{3x - 2}$ at $x = 2$

19. $f(x) = \dfrac{x + 1}{x - 1}$ at $x = 1$

20. $f(x) = \dfrac{2x + 1}{3x - 6}$ at $x = 2$

21. $f(x) = \dfrac{\sqrt{x} - 2}{x - 4}$ at $x = 4$

22. $f(x) = \dfrac{\sqrt{x} - 2}{x - 4}$ at $x = 2$

23. $f(x) = \begin{cases} x + 1 & \text{if } x \le 2 \\ 2 & \text{if } x > 2 \end{cases}$ at $x = 2$

24. $f(x) = \begin{cases} x + 1 & \text{if } x < 0 \\ x - 1 & \text{if } x \ge 0 \end{cases}$ at $x = 0$

25. $f(x) = \begin{cases} x^2 + 1 & \text{if } x \le 3 \\ 2x + 4 & \text{if } x > 3 \end{cases}$ at $x = 3$

26. $f(x) = \begin{cases} \dfrac{x^2 - 1}{x + 1} & \text{if } x < -1 \\ x^2 - 3 & \text{if } x \ge -1 \end{cases}$ at $x = -1$

In Problems 27 through 40, list all the values of x for which the given function is not continuous.

27. $f(x) = 3x^2 - 6x + 9$

28. $f(x) = x^5 - x^3$

29. $f(x) = \dfrac{x + 1}{x - 2}$

30. $f(x) = \dfrac{3x - 1}{2x - 6}$

31. $f(x) = \dfrac{3x + 3}{x + 1}$

32. $f(x) = \dfrac{x^2 - 1}{x + 1}$

33. $f(x) = \dfrac{3x - 2}{(x + 3)(x - 6)}$

34. $f(x) = \dfrac{x}{(x + 5)(x - 1)}$

35. $f(x) = \dfrac{x}{x^2 - x}$

36. $f(x) = \dfrac{x^2 - 2x + 1}{x^2 - x - 2}$

37. $f(x) = \begin{cases} 2x + 3 & \text{if } x \le 1 \\ 6x - 1 & \text{if } x > 1 \end{cases}$

38. $f(x) = \begin{cases} x^2 & \text{if } x \le 2 \\ 9 & \text{if } x > 2 \end{cases}$

39. $f(x) = \begin{cases} 3x - 2 & \text{if } x < 0 \\ x^2 + x & \text{if } x \ge 0 \end{cases}$

40. $f(x) = \begin{cases} 2 - 3x & \text{if } x \le -1 \\ x^2 - x + 3 & \text{if } x > -1 \end{cases}$

41.   WEATHER   Suppose air temperature on a certain day is 30°F. Then the equivalent windchill temperature (in °F) produced by a wind with speed v miles per hour (mph) is given by*

$$W(v) = \begin{cases} 30 & \text{if } 0 \le v \le 4 \\ 1.25v - 18.67\sqrt{v} + 62.3 & \text{if } 4 < v < 45 \\ -7 & \text{if } v \ge 45 \end{cases}$$

  a. What is the windchill temperature when $v = 20$ mph? When $v = 50$ mph?
  b. What wind speed produces a windchill temperature of 0°F?
  c. Is the windchill function $W(v)$ continuous at $v = 4$? What about at $v = 45$?

42.   ELECTRIC FIELD INTENSITY   If a hollow sphere of radius R is charged with one unit of static electricity, then the field intensity $E(x)$ at a point P located x units from the center of the sphere satisfies:

$$E(x) = \begin{cases} 0 & \text{if } 0 < x < R \\ \dfrac{1}{2x^2} & \text{if } x = R \\ \dfrac{1}{x^2} & \text{if } x > R \end{cases}$$

Sketch the graph of $E(x)$. Is $E(x)$ continuous for $x > 0$?

---

*Adapted from *UMAP Module No. 658,* "Windchill," by W. Bosch and L. G. Cobb, 1984, pp. 244–247.

43. POSTAGE The "postage function" $p(x)$ can be described as follows:

$$p(x) = \begin{cases} 37 & \text{if } 0 < x \le 1 \\ 60 & \text{if } 1 < x \le 2 \\ 83 & \text{if } 2 < x \le 3 \\ \vdots \\ 290 & \text{if } 11 < x \le 12 \end{cases}$$

where x is the weight of a letter in ounces and $p(x)$ is the corresponding postage in cents. Sketch the graph of $p(x)$ for $0 < x \le 6$. For what values of x is $p(x)$ discontinuous for $0 < x \le 6$?

44. WATER POLLUTION A ruptured pipe in a North Sea oil rig produces a circular oil slick that is y meters thick at a distance x meters from the rupture. Turbulence makes it difficult to directly measure the thickness of the slick at the source (where $x = 0$), but for $x > 0$, it is found that

$$y = \frac{0.5(x^2 + 3x)}{x^3 + x^2 + 4x}$$

Assuming the oil slick is continuously distributed, how thick would you expect it to be at the source?

45. ENERGY CONSUMPTION The accompanying graph shows the amount of gasoline in the tank of Sue's car over a 30-day period. When is the graph discontinuous? What do you think happens at these times?

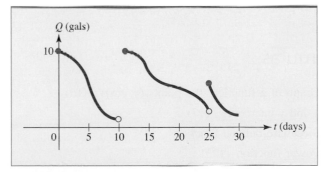

PROBLEM 45

46. INVENTORY The accompanying graph shows the number of units in inventory at a certain business over a 2-year period. When is the graph discontinuous? What do you think is happening at those times?

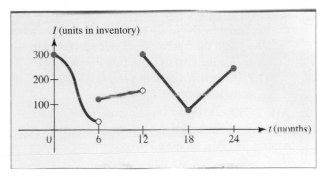

PROBLEM 46

47. COST-BENEFIT ANALYSIS In certain situations, it is necessary to weigh the benefit of pursuing a certain goal against the cost of achieving that goal. For instance, suppose that in order to remove $x\%$ of the pollution from an oil spill, it costs C thousands of dollars, where

$$C(x) = \frac{12x}{100 - x}$$

a. How much does it cost to remove 25% of the pollution? 50%?
b. Sketch the graph of the cost function.
c. What happens as $x \to 100^-$? Is it possible to remove all the pollution?

48. EARNINGS On January 1, 2005, Sam started working for Acme Corporation with an annual salary of $48,000, paid each month on the last day of that month. On July 1, he received a commission of $2,000 for his work, and on September 1, his base salary was raised to $54,000 per year. Finally, on December 21, he received a Christmas bonus of 1% of his base salary.

a. Sketch the graph of Sam's cumulative earnings E as a function of time t (days) during the year 2005.
b. For what values of t is the graph of $E(t)$ discountinuous?

49. COST MANAGEMENT A business manager determines that when $x\%$ of her company's plant capacity is being used, the total cost of operation is C hundred thousand dollars, where

$$C(x) = \frac{8x^2 - 636x - 320}{x^2 - 68x - 960}$$

a. Find $C(0)$ and $C(100)$.

b. Explain why the result of part (a) cannot be used along with the intermediate value property to show that the cost of operation is exactly $700,000 when a certain percentage of plant capacity is being used.

50. AIR POLLUTION It is estimated that t years from now the population of a certain suburban community will be p thousand people, where

$$p(t) = 20 - \frac{7}{t + 2}$$

An environmental study indicates that the average level of carbon monoxide in the air will be c parts per million when the population is p thousand, where

$$c(p) = 0.4\sqrt{p^2 + p + 21}$$

What happens to the level of pollution c in the long run (as $t \to \infty$)?

In Problems 51 and 52 find the values of the constant A such that the function $f(x)$ will be continuous for all x.

51. $f(x) = \begin{cases} Ax - 3 & \text{if } x < 2 \\ 3 - x + 2x^2 & \text{if } x \geq 2 \end{cases}$

52. $f(x) = \begin{cases} 1 - 3x & \text{if } x < 4 \\ Ax^2 + 2x - 3 & \text{if } x \geq 4 \end{cases}$

53. Discuss the continuity of the function

$$f(x) = x\left(1 + \frac{1}{x}\right)$$ on the open interval $0 < x < 1$

and on the closed interval $0 \leq x \leq 1$.

54. Discuss the continuity of the function

$$f(x) = \begin{cases} x^2 - 3x & \text{if } x < 2 \\ 4 + 2x & \text{if } x \geq 2 \end{cases}$$

on the open interval $0 < x < 2$ and on the closed interval $0 \leq x \leq 2$.

55. Show that the equation $\sqrt[3]{x - 8} + 9x^{2/3} = 29$ has at least one solution for the interval $0 \leq x \leq 8$.

56. Show that the equation $\sqrt[3]{x} = x^2 + 2x - 1$ must have at least one solution on the interval $0 \leq x \leq 1$.

57. Investigate the behavior of $f(x) = \dfrac{2x^2 - 5x + 2}{x^2 - 4}$ when x is near to (a) 2 and (b) -2. Does the limit exist at these values of x? Is the function continuous at these values of x?

58. Explain why there must have been some time in your life when your weight in pounds was the same as your height in inches.

59. Explain why there is a time every hour when the hour hand and minute hand of a clock coincide.

60. At age 15, Nan is twice as tall as her 5-year-old brother Dan, but on Dan's 21st birthday, they find that he is 6 inches taller. Explain why there must have been a time when they were exactly the same height.

CHAPTER SUMMARY

Important Terms, Symbols, and Formulas

Function   (2)

Functional notation: $f(x)$ (2)

Domain and range of a function   (2)

Domain convention   (4)

Independent and dependent variables   (3)

Functions used in economics:

Demand   (5)

Supply   (48)

Cost   (5)

Revenue   (5)

Profit   (5)

Composition of functions: $g(h(x))$ (6)

Graph of a function: the points $(x, f(x))$ (15)

x and y intercepts   (17)

Piecewise-defined functions   (4)

Power function   (20)

Polynomial   (20)

Rational function   (21)

Vertical line test   (21)

Linear function; constant rate of change   (27)

Slope: $m = \dfrac{\Delta y}{\Delta x} = \dfrac{y_2 - y_1}{x_2 - x_1}$ (28)

Slope-intercept formula: $y = mx + b$ (30)

Point-slope formula: $y - y_0 = m(x - x_0)$ (31)

Criteria for lines to be parallel or perpendicular (35)

Mathematical modeling (41)

Direct proportionality: $Q = kx$ (46)

Inverse proportionality: $Q = \dfrac{k}{x}$ (46)

Joint proportionality: $Q = kxy$ (46)

Market equilibrium; law of supply and demand (49)

Shortage and surplus (49)

Break-even analysis (50)

Limit of a function: $\lim\limits_{x \to c} f(x)$ (59)

Limits at infinity:

$$\lim_{x \to +\infty} f(x) \text{ and } \lim_{x \to -\infty} f(x) \quad (65)$$

Reciprocal power rules:

$$\lim_{x \to +\infty} \frac{A}{x^k} = 0 \text{ and } \lim_{x \to -\infty} \frac{A}{x^k} = 0 \quad k > 0 \quad (66)$$

Limits at infinity of a rational function $f(x) = \dfrac{p(x)}{q(x)}$:

Divide all terms in $f(x)$ by the highest power x^k in the denominator $q(x)$ and use the reciprocal power rules. (67)

Horizontal asymptote (65)

Infinite limit: $\lim\limits_{x \to c} f(x) = +\infty$ or $\lim\limits_{x \to c} f(x) = -\infty$ (68)

One-sided limits:

$$\lim_{x \to c^-} f(x) \text{ and } \lim_{x \to c^+} f(x) \quad (74)$$

Existence of a limit: $\lim\limits_{x \to c} f(x)$ exists if and only if $\lim\limits_{x \to c^-} f(x)$ and $\lim\limits_{x \to c^+} f(x)$ exist and are equal. (75)

Continuity of $f(x)$ at $x = c$:

$$\lim_{x \to c} f(x) = f(c) \quad (77)$$

Continuity on an interval (79)

Discontinuity (77)

Continuity of polynomials and rational functions (77)

Intermediate value property (80)

Checkup for Chapter 1

1. Specify the domain of the function

$$f(x) = \frac{2x - 1}{\sqrt{4 - x^2}}$$

2. Find the composite function $g(h(x))$, where

$$g(u) = \frac{1}{2u + 1} \quad \text{and} \quad h(x) = \frac{x + 2}{2x + 1}$$

3. Find an equation for each of these lines:
   a. Through the point $(-1, 2)$ with slope $-\dfrac{1}{2}$
   b. With slope 2 and y intercept -3

4. Sketch the graph of each of these functions. Be sure to show all intercepts and any high or low points.
   a. $f(x) = 3x - 5$ b. $f(x) = -x^2 + 3x + 4$

5. Find each of these limits. If the limit is infinite, indicate whether it is $+\infty$ or $-\infty$.
   a. $\lim\limits_{x \to -1} \dfrac{x^2 + 2x - 3}{x - 1}$
   b. $\lim\limits_{x \to 1} \dfrac{x^2 + 2x - 3}{x - 1}$
   c. $\lim\limits_{x \to 1} \dfrac{x^2 - x - 1}{x - 2}$
   d. $\lim\limits_{x \to +\infty} \dfrac{2x^3 + 3x - 5}{-x^2 + 2x + 7}$

6. Determine whether this function $f(x)$ is continuous at $x = 1$:

$$f(x) = \begin{cases} 2x + 1 & \text{if } x \le 1 \\ \dfrac{x^2 + 2x - 3}{x - 1} & \text{if } x > 1 \end{cases}$$

7. **PRICE OF GASOLINE** Since the beginning of the year, the price of unleaded gasoline has been increasing at a constant rate of 2 cents per gallon per month. By June first, the price had reached $1.80 per gallon.
   a. Express the price of unleaded gasoline as a function of time and draw the graph.
   b. What was the price at the beginning of the year?
   c. What will be the price on October first?

8. **SUPPLY AND DEMAND** Suppose it is known that producers will supply x units of a certain commodity to the market when the price is $p = S(x)$ dollars per unit and that the same number of units will be demanded (bought) by consumers when the price is $p = D(x)$ dollars per unit, where

$$S(x) = x^2 + A \quad \text{and} \quad D(x) = Bx + 59$$

for constants A and B. It is also known that no units will be supplied until the unit price is at least \$3 and that market equilibrium occurs when $x = 7$ units.

a. Use this information to find A and B and the equilibrium unit price.

b. Sketch the supply and demand curves on the same graph.

c. What is the difference between the supply price and the demand price when 5 units are produced? When 10 units are produced?

9. **BACTERIAL POPULATION** The population (in thousands) of a colony of bacteria t minutes after the introduction of a toxin is given by the function

$$f(t) = \begin{cases} t^2 + 7 & \text{if } 0 \le t < 5 \\ -8t + 72 & \text{if } t \ge 5 \end{cases}$$

a. When does the colony die out?

b. Explain why the population must be 10,000 sometime between $t = 1$ and $t = 7$.

10. **MUTATION** In a study of mutation in fruit flies, researchers radiate flies with X-rays and determine that the mutation percentage M increases linearly with the X-ray dosage D, measured in kilo-Roentgens (kR). When a dose of $D = 3$ kR is used, the percentage of mutations is 7.7%, while a dose of 5 kR results in a 12.7% mutation percentage. Express M as a function of D. What percentage of the flies will mutate even if no radiation is used?

Review Problems

1. Specify the domain of each of these functions:

a. $f(x) = x^2 - 2x + 6$

b. $f(x) = \dfrac{x - 3}{x^2 + x - 2}$

c. $f(x) = \sqrt{x^2 - 9}$

2. As advances in technology result in the production of increasingly powerful and compact calculators, the price of calculators currently on the market drops. Suppose that x months from now, the price of a certain model will be $P(x) = 40 + \dfrac{30}{x + 1}$ dollars.

a. What will be the price 5 months from now?

b. By how much will the price drop during the fifth month?

c. When will the price be \$43?

d. What happens to the price in the "long run" (as x becomes very large)?

3. Find the composite function $g(h(x))$.

a. $g(u) = u^2 + 2u + 1, h(x) = 1 - x$

b. $g(u) = \dfrac{1}{2u + 1}, h(x) = x + 2$

c. $g(u) = \sqrt{1 - u}, h(x) = 2x + 4$

4. a. Find $f(x - 2)$ if $f(x) = x^2 - x + 4$.

b. Find $f(x^2 + 1)$ if $f(x) = \sqrt{x} + \dfrac{2}{x - 1}$.

c. Find $f(x + 1) - f(x)$ if $f(x) = x^2$.

5. Find functions $h(x)$ and $g(u)$ such that $f(x) = g(h(x))$.

a. $f(x) = (x^2 + 3x + 4)^5$

b. $f(x) = (3x + 1)^2 + \dfrac{5}{2(3x + 2)^3}$

6. **ENVIRONMENTAL ANALYSIS** An environmental study of a certain community suggests that the average daily level of smog in the air will be $Q(p) = \sqrt{0.5p + 19.4}$ units when the population is p thousand. It is estimated that t years from now, the population will be $p(t) = 8 + 0.2t^2$ thousand.

a. Express the level of smog in the air as a function of time.

b. What will the smog level be 3 years from now?

c. When will the smog level reach 5 units?

7. Find c so that the curve $y = 3x^2 - 2x + c$ passes through the point $(2, 4)$.

8. Graph these functions.

a. $f(x) = x^2 + 2x - 8$

b. $f(x) = 3 + 4x - 2x^2$

9. Find the slope and y intercept of the given line and draw the graph.
   a. $y = 3x + 2$
   b. $5x - 4y = 20$
   c. $2y + 3x = 0$
   d. $\dfrac{x}{3} + \dfrac{y}{2} = 4$

10. Find equations for these lines:
   a. Slope 5 and y intercept $(0, -4)$
   b. Slope -2 and contains $(1, 3)$
   c. x intercept $(3, 0)$ and y intercept $\left(0, -\dfrac{2}{3}\right)$
   d. Contains $(5, 4)$ and is parallel to $2x + y = 3$
   e. Contains $(-1, 3)$ and is perpendicular to $5x - 3y = 7$

11. **EDUCATIONAL FUNDING**   A private college in the southwest has launched a fund-raising campaign. Suppose that college officials estimate that it will take $f(x) = \dfrac{10x}{150 - x}$ weeks to reach $x\%$ of their goal.
   a. Sketch the relevant portion of the graph of this function.
   b. How long will it take to reach 50% of the campaign's goal?
   c. How long will it take to reach 100% of the goal?

12. **CONSUMER EXPENDITURE**   The demand for a certain commodity is given by $D(x) = -50x + 800$; that is, x units of the commodity will be demanded by consumers when the price is $p = D(x)$ dollars per unit. Total consumer expenditure $E(x)$ is the amount of money consumers pay to buy x units of the commodity.
   a. Express consumer expenditure as a function of x, and sketch the graph of $E(x)$.
   b. Use the graph in part (a) to determine the level of production x at which consumer expenditure is largest. What price p corresponds to maximum consumer expenditure?

13. **MICROBIOLOGY**   A spherical cell of radius r has volume $V = \dfrac{4}{3}\pi r^3$ and surface area $S = 4\pi r^2$. Express V as a function of S. If S is doubled, what happens to V?

14. **NEWSPAPER CIRCULATION**   The circulation of a newspaper is increasing at a constant rate. Three months ago the circulation was 3,200. Today it is 4,400.
   a. Express the circulation as a function of time and draw the graph.
   b. What will be the circulation 2 months from now?

15. Find the points of intersection (if any) of the given pair of curves and draw the graphs.
   a. $y = -3x + 5$ and $y = 2x - 10$
   b. $y = x + 7$ and $y = -2 + x$
   c. $y = x^2 - 1$ and $y = 1 - x^2$
   d. $y = x^2$ and $y = 15 - 2x$

16. **OPTIMAL SELLING PRICE**   A manufacturer can produce bookcases at a cost of \$80 apiece. Sales figures indicate that if the bookcases are sold for x dollars apiece, approximately $150 - x$ will be sold each month. Express the manufacturer's monthly profit as a function of the selling price x, draw the graph, and estimate the optimal selling price.

17. **OPTIMAL SELLING PRICE**   A retailer can obtain cameras from the manufacturer at a cost of \$150 apiece. The retailer has been selling the cameras at the price of \$340 apiece, and at this price, consumers have been buying 40 cameras a month. The retailer is planning to lower the price to stimulate sales and estimates that for each \$5 reduction in the price, 10 more cameras will be sold each month. Express the retailer's monthly profit from the sale of the cameras as a function of the selling price. Draw the graph and estimate the optimal selling price.

18. **STRUCTURAL DESIGN**   A cylindrical can with no top is to be constructed for 80 cents. The cost of the material used for the bottom is 3 cents per square inch, and the cost of the material used for the curved side is 2 cents per square inch. Express the volume of the can as a function of its radius.

19. **MANUFACTURING EFFICIENCY**   A manufacturing firm has received an order to make 400,000 souvenir silver medals commemorating the 35th anniversary of the landing of Apollo 11 on the moon. The firm owns several machines, each of which can produce 200 medals per hour. The cost of setting up the machines to produce the medals is \$80 per machine, and the total operating cost is \$5.76 per hour. Express the cost of producing the 400,000 medals as a function of the number of machines used. Draw the graph and estimate the number of machines the firm should use to minimize cost.

20. **INVENTORY ANALYSIS**   A businessman maintains inventory over a particular 30-day month as follows:

days 1–9   30 units

days 10–15   17 units

days 16–23   12 units

days 24–30   steadily decreasing from 12 units
                to 0 units

Sketch the graph of the inventory as a function of time t (days). At what times is the graph discontinuous?

21. **BREAK-EVEN ANALYSIS**   A manufacturer can sell a certain product for $80 per unit. Total cost consists of a fixed overhead of $4,500 plus production costs of $50 per unit.

  a. How many units must the manufacturer sell to break even?

  b. What is the manufacturer's profit or loss if 200 units are sold?

  c. How many units must the manufacturer sell to realize a profit of $900?

22. **PRODUCTION MANAGEMENT**   During the summer, a group of students builds kayaks in a converted garage. The rental for the garage is $1,500 for the summer, and the materials needed to build a kayak cost $125. The kayaks can be sold for $275 apiece.

  a. How many kayaks must the students sell to break even?

  b. How many kayaks must the students sell to make a profit of at least $1,000?

23. **LEARNING**   Some psychologists believe that when a person is asked to recall a set of facts, the rate at which the facts are recalled is proportional to the number of relevant facts in the subject's memory that have not yet been recalled. Express the recall rate as a function of the number of facts that have been recalled.

24. **COST EFFICIENT DESIGN**   A cable is to be run from a power plant on one side of a river 900 meters wide to a factory on the other side, 3,000 meters downstream. The cable will be run in a straight line from the power plant to some point P on the opposite bank and then along the bank to the factory. The cost of running the cable across the water is $5 per meter, while the cost over land is $4 per meter. Let x be the distance from P to the point directly across the river

from the power plant. Express the cost of installing the cable as a function of x.

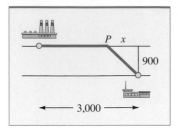

PROBLEM 24

25. **CONSTRUCTION COST**   A window with a 20-foot perimeter (frame) is to be comprised of a semicircular stained glass pane above a rectangular clear pane, as shown in the accompanying figure. Clear glass costs $3 per square foot and stained glass costs $10 per square foot. Express the cost of the window as a function of the radius of the stained glass pane.

PROBLEM 25

26. **MANUFACTURING OVERHEAD**   A furniture manufacturer can sell end tables for $70 apiece. It costs the manufacturer $30 to produce each table, and it is estimated that revenue will equal cost when 200 tables are sold. What is the overhead associated with the production of the tables? [*Note:* Overhead is the cost when 0 units are produced.]

27. **MANUFACTURING COST**   A manufacturer is capable of producing 5,000 units per day. There is a fixed (overhead) cost of $1,500 per day and a variable cost of $2 per unit produced. Express the daily cost C as a function of the number of units produced, and sketch the graph of $C(x)$. Is $C(x)$ continuous? If not, where do its discontinuities occur?

28. At what time between 3 P.M. and 4 P.M. will the minute hand coincide with the hour hand? [*Hint:* The hour hand moves $\dfrac{1}{12}$ as fast as the minute hand.]

29. **PROPERTY TAX** A homeowner is trying to decide between two competing property tax propositions. With Proposition A, the homeowner will pay $100 plus 8% of the assessed value of her home, while Proposition B requires a payment of $1,900 plus 2% of the assessed value. Assuming the homeowner's only consideration is to minimize her tax payment, develop a criterion based on the assessed value V of her home for deciding between the propositions.

In Problems 30 through 43, either find the given limit or show it does not exist. If the limit is infinite, indicate whether it is $+\infty$ or $-\infty$.

30. $\displaystyle\lim_{x\to 2}\frac{x^2 - 3x}{x + 1}$

31. $\displaystyle\lim_{x\to 1}\frac{x^2 + x - 2}{x^2 - 1}$

32. $\displaystyle\lim_{x\to 1}\left(\frac{1}{x^2} - \frac{1}{x}\right)$

33. $\displaystyle\lim_{x\to 2}\frac{x^3 - 8}{2 - x}$

34. $\displaystyle\lim_{x\to -\infty}\left(2 + \frac{1}{x^2}\right)$

35. $\displaystyle\lim_{x\to 0}\left(2 - \frac{1}{x^3}\right)$

36. $\displaystyle\lim_{x\to 0^-}\left(x^3 - \frac{1}{x^2}\right)$

37. $\displaystyle\lim_{x\to -\infty}\frac{x}{x^2 + 5}$

38. $\displaystyle\lim_{x\to -\infty}\frac{x^3 - 3x + 5}{2x + 3}$

39. $\displaystyle\lim_{x\to -\infty}\frac{x^4 + 3x^2 - 2x + 7}{x^3 + x + 1}$

40. $\displaystyle\lim_{x\to -\infty}\frac{x(x - 3)}{7 - x^2}$

41. $\displaystyle\lim_{x\to -\infty}\frac{1 + \dfrac{1}{x} + \dfrac{1}{x^2}}{x^2 + 3x - 1}$

42. $\displaystyle\lim_{x\to 0^+}\sqrt{x\left(1 + \frac{1}{x^2}\right)}$

43. $\displaystyle\lim_{x\to 0^-}x\sqrt{1 - \frac{1}{x}}$

In Problems 44 through 47, list all values of x for which the given function is not continuous.

44. $f(x) = 5x^3 - 3x + \sqrt{x}$

45. $f(x) = \dfrac{x^2 - 1}{x + 3}$

46. $g(x) = \dfrac{x^3 + 5x}{(x - 2)(2x + 3)}$

47. $h(x) = \begin{cases} x^3 + 2x - 33 & \text{if } x \le 3 \\ \dfrac{x^2 - 6x + 9}{x - 3} & \text{if } x > 3 \end{cases}$

48. In each of these cases, find the value of the constant A that makes the given function $f(x)$ continuous for all x.

   a. $f(x) = \begin{cases} 2x + 3 & \text{if } x < 1 \\ Ax - 1 & \text{if } x \ge 1 \end{cases}$

   b. $f(x) = \begin{cases} \dfrac{x^2 - 1}{x + 1} & \text{if } x < -1 \\ Ax^2 + x - 3 & \text{if } x \ge -1 \end{cases}$

49. The radius of the earth is roughly 4,000 miles, and an object located x miles from the center of the earth weighs $w(x)$ lb, where

$$w(x) = \begin{cases} Ax & \text{if } x \le 4{,}000 \\ \dfrac{B}{x^2} & \text{if } x > 4{,}000 \end{cases}$$

and A and B are positive constants. Assuming that $w(x)$ is continuous for all x, what must be true about A and B? Sketch the graph of $w(x)$.

50. Graph $f(x) = \dfrac{3x^2 - 6x + 9}{x^2 + x - 2}$. Determine the values of x where the function is undefined.

51. Graph $y = \dfrac{21}{9}x - \dfrac{84}{35}$ and $y = \dfrac{654}{279}x - \dfrac{54}{10}$ on the same set of coordinate axes using $[-10, 10]1$ by $[-10, 10]1$. Are the two lines parallel?

52. For $f(x) = \sqrt{x + 3}$ and $g(x) = 5x^2 + 4$, find (a) $f(g(-1.28))$ and (b) $g(f(\sqrt{2}))$. Use three decimal place accuracy.

53. Graph $y = \begin{cases} x^2 + 1 & \text{if } x \le 1 \\ x^2 - 1 & \text{if } x > 1 \end{cases}$. Find the points of discontinuity.

54. Graph the function $f(x) = \dfrac{x^2 - 3x - 10}{1 - x} - 2$. Find the x and y intercepts. For what values of x is this function defined?

55. The accompanying graph represents a function $g(x)$ that oscillates more and more frequently as x approaches 0 from either the right or the left but with decreasing magnitude. Does $\lim\limits_{x \to 0} g(x)$ exist? If so, what is its value? [*Note:* For students with experience in trigonometry, the function $g(x) = |x| \sin(1/x)$ behaves in this way.]

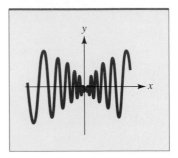

PROBLEM 55

Explore! Updates are included at the end of each chapter of this textbook. These updates provide additional instruction and exercises for exploring calculus using a graphing calculator or giving hints and solutions for selected Explore! boxes found in each chapter. An attempt has been made to use function keys that are standard on most handheld graphing utilities. The exact names of the function keys on your particular calculator may vary, depending on the brand. Please consult your calculator user manual for more specific details.

Complete solutions for all Explore! boxes throughout the text can be accessed via the book specific website www.mhhe.com/hoffmann.

Solution for Explore! on Page 2

Write $f(x) = x^2 + 4$ in the function editor (**Y=** key) of your graphing calculator. While on a cleared homescreen (**2nd MODE CLEAR**), locate the symbol for Y1 through the **VARS** key, arrowing right to **Y-VARS** and selecting **1:Function** and **1:Y1.** (Also see the Calculator Introduction at the front of the text.) Y1({−3, −1, 0, 1, 3}) yields the desired functional values, all at once. Or you can do each value individually, such as Y1(−3).

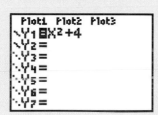

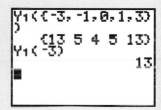

NOTE An Explore! tip is that it is easier to view a table of values, especially for several functions. Set up the desired table parameters through **TBLSET (2nd WINDOW).** Now enter $g(x) = x^2 − 1$ into **Y2** of the equation editor (**Y=**). Observing the values of Y1 and Y2, we notice that they differ by a fixed constant of −5, since the two functions are simply vertical translations of $f(x) = x^2$.

Solution for Explore! on Page 4 (middle)

The graph of the piecewise-defined function Y1 = 2(X < 1) + (−1)(X ≥ 1) is shown in the following figure, and the table to its right shows the functional values at X = −2, 0, 1, and 3. Recall that the inequality symbols can be accessed through the **TEST (2nd MATH)** key.

EXPLORE! UPDATE

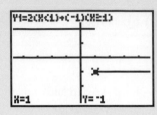
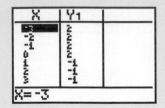

Solution for Explore! on Page 4 (bottom)

The function $f(x) = 1/(x - 3)$ appears to have a break at $x = 3$ and not be defined there. So the domain consists of all real numbers $x \neq 3$. The function $g(x) = \sqrt{(x - 2)}$ is not defined for $x < 2$, as indicated by the following graph in the middle screen and the table on the right.

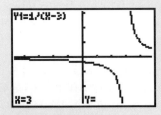
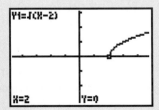
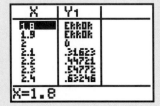

Solution for Explore! on Page 44

Use **TBLSET** with **TblStart = 1** and **ΔTbl = 1** to obtain the following table on the left which indicates a minimal cost in the low hundreds occurring within the interval [1, 4]. Then, graph using **WINDOW** dimensions of [1, 4]1 by [100, 400]50 to obtain the following middle screen, where we have utilized the minimum-finding option in **CALC** (see Calculator Introduction at the front of the text) to locate an apparent minimal value of about 226 cents at a radius of 2 inches. Thus, no radius will create a cost of $2.00, which is less than the minimal cost. Graphing Y2 = 300 and utilizing the intersection-finding feature shows that the cost is $3.00 when the radius is 1.09 in. or 3.33 in.

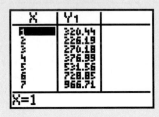
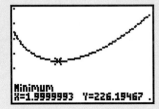
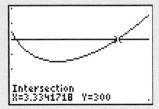

Solution for Explore! on Page 79

The graph of $f(x) = (x^3 - 8)/(x - 2)$ appears continuous based on the window, $[-6, 6]1$ by $[-2, 10]1$. However, examination of this graph using a modified decimal window, $[-4.7, 4.7]1$ by $[0, 14.4]1$ shows an exaggerated hole at $x = 2$.

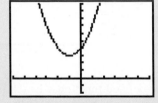
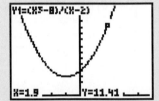
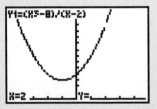

The function $f(x)$ is not continuous, specifically at $x = 2$, where it is not defined. The situation is similar to Figure 1.57(b) on page 78. What value of y would fill up the hole in the graph here? Answer: $\lim_{x \to 2} f(x) = 12$.

THINK ABOUT IT

ALLOMETRIC MODELS

When developing a mathematical model, the first task is to identify quantities of interest, and the next is to find equations that express relationships between these quantities. Such equations can be quite complicated, but there are many important relationships that can be expressed in the relatively simple form $y = Cx^k$, in which one quantity y is expressed as a constant multiple of a power function of another quantity x.

In biology, the study of the relative growth rates of various parts of an organism is called **allometry,** from the Greek words *allo* (other or different) and *metry* (measure). In allometric models, equations of the form $y = Cx^k$ are often used to describe the relationship between two biological measurements. For example, the size a of the antlers of an elk from tip to tip has been shown to be related to h, the shoulder height of the elk, by the allometric equation

$$a = 0.026h^{1.7}$$

where a and h are both measured in centimeters (cm).* This relationship is shown in the accompanying figure.

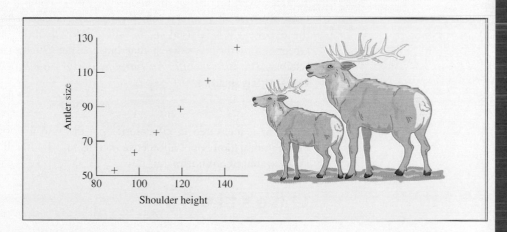

Whenever possible, allometric models are developed using basic assumptions from biological (or other) principles. For example, it is reasonable to assume that the body volume and hence the weight of most animals is proportional to the cube of the linear dimension of the body, such as height for animals that stand up or length for four-legged animals. Thus, it is reasonable to expect the weight of a snake to be proportional to the cube of its length, and indeed, observations of the hognose snake of

---

*Frederick R. Adler, *Modeling the Dynamics of Life,* Pacific Grove, CA: Brooks-Cole Publishing, 1998, p. 61.

Kansas indicate that the weight w (grams) and length L (meters) of a typical such snake are related by the equation*

$$w = 440L^3$$

At times, observed data may be quite different from the results predicted by the model. In such a case, we seek a better model. For the western hognose snake, it turns out that the equations $w = 446L^{2.99}$ and $w = 429L^{2.90}$ provide better approximations to the weight of male and female western hognose snakes, respectively. However, there is no underlying biological reason why we should use exponents less than three. It just turns out these equations are slightly better fits.

The *basal metabolic rate M* of an animal is the rate of heat produced by its body when it is resting and fasting. Basal metabolic rates have been studied since the 1830s. Allometric equations of the form $M = cw^r$, for constants c and r, have long been used to build models relating the basal metabolic rate to the body weight w of an animal. The development of such a model is based on the assumption that basal metabolic rate M is proportional to S, the surface area of the body, so that $M = aS$ where a is a constant. To set up an equation relating M and w, we need to relate the weight w of an animal to its surface area S. Assuming that all animals are shaped like spheres or cubes, and that the weight of an animal is proportional to its volume, we can show (see Exercises 1 and 2) that the surface area is proportional to $w^{2/3}$, so that $S = bw^{2/3}$ where b is a constant. Putting the equations $M = aS$ and $S = bw^{2/3}$ together, we obtain the allometric equation

$$M = abw^{2/3} = kw^{2/3}$$

where $k = ab$.

However, this is not the end of the story. When more refined modeling assumptions are used, it is found that the basal metabolic rate M is better approximated if the exponent 3/4 is used in the allometric equation rather than the exponent 2/3. Observations further suggest that the constant 70 be used in this equation (see M. Kleiber, *The Fire of Life, An Introduction to Animal Energetics*, Wiley, 1961). This gives us the equation

$$M = 70w^{3/4}$$

where M is measured in kilocalories per day and w is measured in kilograms. Additional information about allometric models may be found in our Web Resources Guide at www.mhhe.com/hoffmann.

Questions

1. What weight does the allometric equation $w = 440L^3$ predict for a western hognose snake 0.7 meters long? If this snake is male, what does the equation $w = 446L^{2.99}$ predict for its weight? If it is female, what does the equation $w = 429L^{2.90}$ predict for its weight?

2. What basal metabolic rates are predicted by the equation $M = 70w^{3/4}$ for animals with weights of 50 kg, 100 kg, and 350 kg?

---

*Edward Batschelet, *Introduction to Mathematics for Life Scientists*, 3rd ed., New York: Springer-Verlag, 1979, p. 178.

3. Observations show that the brain weight h, measured in grams, of adult female primates is given by the allometric equation $h = 0.064w^{0.822}$, where w is the weight in grams (g) of the primate. What are the predicted brain weights of female primates weighing 5 kg, 10 kg, 25 kg, 50 kg, and 100 kg? (Recall that 1 kg = 1,000 g).

4. If $y = Cx^k$ where C and k are constants, then y and x are said to be related by a positive allometry if $k > 1$ and a negative allometry if $k < 1$. The *weight-specific metabolic rate* of an animal is defined to be the basal metabolic rate M of the animal divided by its weight w; that is, M/w. Show that if the basal metabolic rate is the positive allometry of the weight in Exercise 2, then the weight-specific metabolic rate is a negative allometry equation of the weight.

5. Show that if we assume all animal bodies are shaped like cubes, then the surface area S of an animal body is proportional to $V^{2/3}$, where V is the volume of the body. By combining this fact with the assumption that the weight w of an animal is proportional to its volume, show that S is proportional to $w^{2/3}$.

6. Show that if we assume all animal bodies are shaped like spheres, then the surface area S of an animal body is proportional to $V^{2/3}$, where V is the volume of the body. By combining this fact with the assumption that the weight w of an animal is proportional to its volume, show that S is proportional to $w^{2/3}$. [*Hint:* Recall that a sphere of radius r has surface area $4\pi r^2$ and volume $\dfrac{4}{3}\pi r^3$.]

CHAPTER | 2

The acceleration of a moving object is found by differentiating its velocity.

DIFFERENTIATION: BASIC CONCEPTS

SECTION 2.1 | The Derivative

Calculus is the mathematics of change, and the primary tool for studying change is a procedure called **differentiation.** In this section, we shall introduce this procedure and examine some of its uses, especially in computing rates of change. Here, and later in this chapter, we shall encounter rates such as velocity, acceleration, production rates with respect to labor level or capital expenditure, the rate of growth of a population, the infection rate of a susceptible population during an epidemic, and many others.

Calculus was developed in the seventeenth century by Isaac Newton (1642–1727) and G. W. Leibniz (1646–1716) and others at least in part in an attempt to deal with two geometric problems:

Tangent problem: Find a tangent line at a particular point on a given curve.

Area problem: Find the area of the region under a given curve.

The area problem involves a procedure called **integration** in which quantities such as area, average value, present value of an income stream, and blood flow rate are computed as a special kind of limit of a sum. We shall study this procedure in Chapters 5 and 6. The tangent problem is closely related to the study of rates of change, and we will begin our study of calculus by examining this connection.

Rates of Change and Slope

Recall from Section 1.3 that a linear function $L(x) = mx + b$ changes at the constant rate m with respect to the independent variable x. That is, the rate of change of $L(x)$ is given by the slope or steepness of its graph, the line $y = mx + b$ (Figure 2.1a). For a function $f(x)$ that is not linear, the rate of change is not constant but varies with x. In particular, when $x = c$ the rate is given by the steepness of the graph of $f(x)$ at the point $P(c, f(c))$, which can be measured by the slope of the *tangent line* to the graph at P (Figure 2.1b). The relationship between rate of change and slope is illustrated in Example 2.1.1.

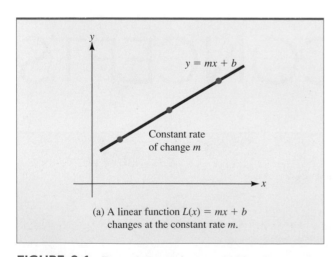

(a) A linear function $L(x) = mx + b$ changes at the constant rate m.

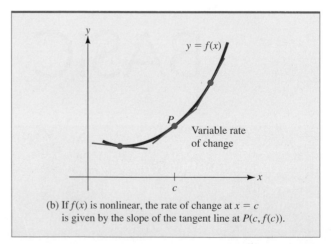

(b) If $f(x)$ is nonlinear, the rate of change at $x = c$ is given by the slope of the tangent line at $P(c, f(c))$.

FIGURE 2.1 Rate of change is measured by slope.

EXAMPLE 2.1.1

The graph shown in Figure 2.2 gives the relationship between the percentage of unemployment U and the corresponding percentage of inflation I. Use the graph to estimate the rate at which I changes with respect to U when the level of unemployment is 3% and again when it is 10%.

Solution

From the figure, we estimate the slope of the tangent line at the point (3, 14), corresponding to $U = 3$, to be approximately -14. That is, when unemployment is 3%, inflation I is *decreasing* at the rate of 14 percentage points for each percentage point increase in unemployment U.

At the point (10, -5), the slope of the tangent line is approximately -0.4, which means that when there is 10% unemployment, inflation is decreasing at the rate of only 0.4 percentage point for each percentage point increase in unemployment.

In Example 2.1.2, we show how slope and a rate of change can be computed analytically using a limit.

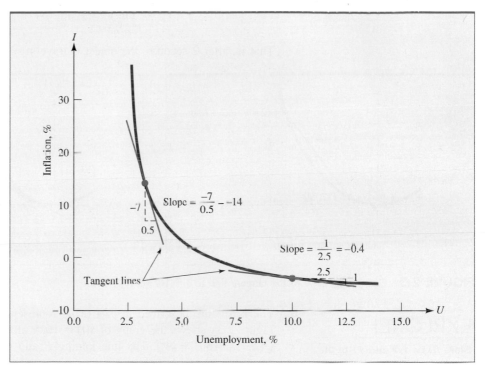

FIGURE 2.2 Inflation as a function of unemployment.

Source: Adapted from Robert Eisner, *The Misunderstood Economy: What Counts and How to Count It*, Boston, MA: Harvard Business School Press, 1994, p. 173.

EXAMPLE | 2.1.2

If air resistance is neglected, an object dropped from a great height will fall $s(t) = 16t^2$ feet in t seconds.

a. What is the object's velocity after $t = 2$ seconds?

b. How is the velocity found in part (a) related to the graph of $s(t)$?

Solution

a. The velocity after 2 seconds is the *instantaneous* rate of change of $s(t)$ at $t = 2$. Unless the falling object has a speedometer, it is hard to simply "read" its velocity. However, we can measure the distance it falls as time t changes by a small amount h from $t = 2$ to $t = 2 + h$ and then compute the *average* rate of change of $s(t)$ over the time period $[2, 2 + h]$ by the ratio

$$v_{ave} = \frac{\text{distance traveled}}{\text{elapsed time}} = \frac{s(2 + h) - s(2)}{(2 + h) - 2}$$

$$= \frac{16(2 + h)^2 - 16(2)^2}{h} = \frac{16(4 + 4h + h^2) - 16(4)}{h}$$

$$= \frac{64h + 16h^2}{h} = 64 + 16h$$

Since the elapsed time h is small, we would expect the average velocity v_{ave} to closely approximate the instantaneous ("speedometer") velocity v_{ins} at time $t = 2$. Thus, it is reasonable to compute the instantaneous velocity by the limit

$$v_{ins} = \lim_{h \to 0} v_{ave} = \lim_{h \to 0} (64 + 16h) = 64$$

That is, after 2 seconds, the object is traveling at the rate of 64 feet per second.

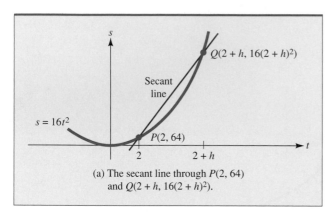

(a) The secant line through $P(2, 64)$ and $Q(2 + h, 16(2 + h)^2)$.

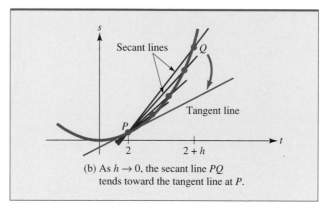

(b) As $h \to 0$, the secant line PQ tends toward the tangent line at P.

FIGURE 2.3 Computing slope of the tangent line to $s = 16t^2$ at $P(2, 64)$.

EXPLORE!

Store $f(x) = 1/x$ into Y1 of the equation editor of your graphing calculator. Graph Y1 using the modified decimal window [0, 4.7]1 by [0, 3.1]1. If this curve were replaced by a polygonal sequence of line segments, how would the slope values for these line segments behave? To see these values, write

Y2 = (Y1(X + H) − Y1(X))/H

in the equation editor, and store 0.5 into H (that is, 0.5 → H). Adjust **TBLSET** with TblStart = 1 and ΔTbl = H. In examining the **TABLE** of values, what do the values of Y2 represent? What is happening when we change H to 0.1 (in ΔTbl as well)?

b. The procedure described in part (a) is represented geometrically in Figure 2.3. Figure 2.3a shows the graph of $s(t) = 16t^2$, along with the points $P(2, 64)$ and $Q(2 + h, 16(2 + h)^2)$. The line joining P and Q is called a *secant line* of the graph of $s(t)$ and has slope

$$m_{sec} = \frac{16(2 + h)^2 - 64}{(2 + h) - 2} = 64 + 16h$$

As indicated in Figure 2.3b, when we take smaller and smaller values of h, the corresponding secant lines PQ tend toward the position of what we intuitively think of as the tangent line to the graph of $s(t)$ at P. This suggests that we can compute the slope m_{tan} of the tangent line by finding the limiting value of m_{sec} as h tends toward 0; that is,

$$m_{tan} = \lim_{h \to 0} m_{sec} = \lim_{h \to 0} (64 + 16h) = 64$$

Thus, the slope of the tangent line to the graph of $s(t) = 16t^2$ at the point where $t = 2$ is the same as the instantaneous rate of change of $s(t)$ with respect to t when $t = 2$.

The procedure used in Example 2.1.2 to find the velocity of a falling body can be used to find other rates of change. Suppose we wish to find the rate at which the function $f(x)$ is changing with respect to x when $x = c$. We begin by finding the **average rate of change** of $f(x)$ as x varies from $x = c$ to $x = c + h$, which is given by the ratio

$$\text{Rate}_{ave} = \frac{\text{change in } f(x)}{\text{change in } x} = \frac{f(c + h) - f(c)}{(c + h) - c}$$

$$= \frac{f(c + h) - f(c)}{h}$$

This ratio can be interpreted geometrically as the slope of the secant line from the point $P(c, f(c))$ to the point $Q(c + h, f(c + h))$, as shown in Figure 2.4a.

EXPLORE!

A graphing calculator can simulate secant lines approaching a tangent line. Store $f(x) = (x - 2)^2 + 1$ into Y1 of the equation editor, selecting a **BOLD** graphing style. Store the values {−2.0, −1.6, −1.1, −0.6, −0.2} into L1 (list 1) using the **STAT** edit menu. Store

$$L1*(x - 2) + 1$$

into Y2 of the equation editor. Graph using a modified decimal window [0, 4.7]1 by [0, 3.1]1. Describe what you observe. What is the equation of the limiting tangent line?

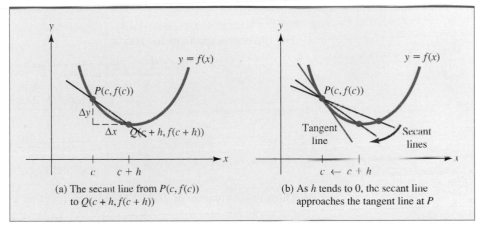

FIGURE 2.4   Secant lines approximating a tangent line.

We then compute the instantaneous rate of change of $f(x)$ at $x = c$ by finding the limiting value of the average rate as h tends to 0; that is,

$$\text{Rate}_{ins} = \lim_{h \to 0} \text{rate}_{ave} = \lim_{h \to 0} \frac{f(c + h) - f(c)}{h}$$

This limit also gives the slope of the tangent line to the curve $y = f(x)$ at the point $P(c, f(c))$, as indicated in Figure 2.4b.

The Derivative   The expression

$$\frac{f(x + h) - f(x)}{h}$$

is called a **difference quotient** for the function $f(x)$, and we have just seen that rates of change and slope can both be computed by finding the limit of an appropriate difference quotient as h tends to 0. To unify the study of these and other applications that involve taking the limit of a difference quotient, we introduce this terminology and notation.

> **The Derivative of a Function** ■ The **derivative** of the function $f(x)$ with respect to x is the function $f'(x)$ given by
>
> $$f'(x) = \lim_{h \to 0} \frac{f(x + h) - f(x)}{h}$$
>
> [read $f'(x)$ as "f prime of x"]. The process of computing the derivative is called **differentiation,** and we say that $f(x)$ is **differentiable** at $x = c$ if $f'(c)$ exists; that is, if the limit that defines $f'(x)$ exists when $x = c$.

NOTE We use "h" to increment the independent variable in difference quotients to simplify algebraic computations. However, when it is important to emphasize that, say, the variable x is being incremented, we will denote the increment by Δx (read as "delta x"). Similarly, Δt and Δs denote small (incremental) changes in the variables t and s, respectively. This notation is used extensively in Section 2.5. ∎

EXAMPLE 2.1.3

Find the derivative of the function $f(x) = 16x^2$.

Solution

The difference quotient for $f(x)$ is

$$\frac{f(x + h) - f(x)}{h} = \frac{16(x + h)^2 - 16x^2}{h}$$

$$= \frac{16(x^2 + 2hx + h^2) - 16x^2}{h}$$

$$= \frac{32hx + 16h^2}{h} \qquad \textit{combine terms}$$

$$= 32x + 16h \qquad \textit{cancel common h terms}$$

Thus, the derivative of $f(x) = 16x^2$ is the function

$$f'(x) = \lim_{h \to 0} \frac{f(x + h) - f(x)}{h} = \lim_{h \to 0} (32x + 16h)$$

$$= 32x$$

Once we have computed the derivative of a function $f(x)$, it can be used to streamline any computation involving the limit of the difference quotient of $f(x)$ as h tends to 0. For instance, notice that the function in Example 2.1.3 is essentially the same as the distance function $s = 16t^2$ that appears in the falling body problem in Example 2.1.1. Using the result of Example 2.1.3, the velocity of the falling body at time $t = 2$ can now be found by simply substituting $t = 2$ into the formula for the derivative $s'(t)$:

$$\text{Velocity} = s'(2) = 32(2) = 64$$

Likewise, the slope of the tangent line to the graph of $s(t)$ at the point $P(2, 64)$ is given by

$$\text{Slope} = s'(2) = 64$$

For future reference, our observations about rates of change and slope may be summarized as follows in terms of the derivative notation.

Slope as a Derivative ■ The slope of the tangent line to the curve $y = f(x)$ at the point $(c, f(c))$ is $m_{\text{tan}} = f'(c)$.

Instantaneous Rate of Change as a Derivative ■ The rate of change of $f(x)$ with respect to x when $x = c$ is given by $f'(c)$.

Just-In-Time Review

Recall that $(a + b)^3 = a^3 + 3a^2b + 3ab^2 + b^3$. This is a special case of the binomial theorem for the exponent 3 and is used to expand the numerator of the difference quotient in Example 2.1.4.

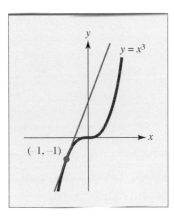

FIGURE 2.5 The graph of $y = x^3$.

In Example 2.1.4, we find the equation of a tangent line. Then in Example 2.1.5, we consider a business application involving a rate of change.

EXAMPLE | 2.1.4

First compute the derivative of $f(x) = x^3$, and then use it to find the slope of the tangent line to the curve $y = x^3$ at the point where $x = -1$. What is the equation of the tangent line at this point?

Solution

According to the definition of the derivative

$$f'(x) = \lim_{h \to 0} \frac{f(x + h) - f(x)}{h} = \lim_{h \to 0} \frac{(x + h)^3 - x^3}{h}$$

$$= \lim_{h \to 0} \frac{(x^3 + 3x^2h + 3xh^2 + h^3) - x^3}{h} = \lim_{h \to 0} (3x^2 + 3xh + h^2)$$

$$= 3x^2$$

Thus, the slope of the tangent line to the curve $y = x^3$ at the point where $x = -1$ is $f'(-1) = 3(-1)^2 = 3$ (Figure 2.5). To find an equation for the tangent line, we also need the y coordinate of the point of tangency; namely, $y = (-1)^3 = -1$. Therefore, the tangent line passes through the point $(-1, -1)$ with slope 3. By applying the point-slope formula, we obtain the equation

$$y - (-1) = 3[x - (-1)]$$

or

$$y = 3x + 2$$

EXPLORE!

Store $f(x) = x^3$ into Y1 of the equation editor and graph using a decimal window. Access the tangent line option of the **DRAW** key (**2nd PRGM**) and, using the left arrow, move the cursor to $(-1, -1)$ on your graph. Press **ENTER** and observe what happens. Does the equation exactly fit the result in Example 2.1.4? Explain.

EXAMPLE | 2.1.5

A manufacturer determines that when x thousand units of a particular commodity are produced, the profit generated will be

$$P(x) = -400x^2 + 6,800x - 12,000$$

dollars. At what rate is profit changing with respect to the level of production x when 9,000 units are produced?

Solution

We find that

$$P'(x) = \lim_{h \to 0} \frac{P(x + h) - P(x)}{h}$$

$$= \lim_{h \to 0} \frac{[-400(x + h)^2 + 6,800(x + h) - 12,000] - (-400x^2 + 6,800x - 12,000)}{h}$$

$$= \lim_{h \to 0} \frac{-400h^2 - 800hx + 6,800h}{h} \quad expand \ (x + h)^2 \ and \ combine \ terms$$

$$= \lim_{h \to 0} (-400h - 800x + 6,800) \quad cancel \ common \ h \ terms$$

$$= -800x + 6,800$$

Thus, when the level of production is $x = 9$ (9,000 units), the profit is changing at the rate of

$$P'(9) = -800(9) + 6,800 = -400$$

dollars per thousand units.

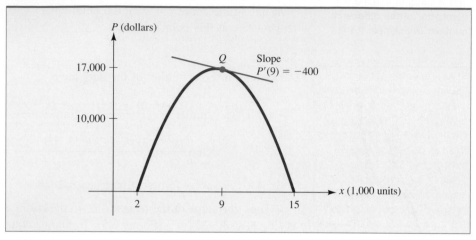

FIGURE 2.6   The graph of the profit function $P(x) = -400x^2 + 6,800x - 12,000$.

In Example 2.1.5, we found that $P'(9) = -400$, which means that the tangent line to the profit curve $y = P(x)$ is sloped *downward* at the point Q where $x = 9$, as shown in Figure 2.6. Since the tangent line at Q is sloped downward, the profit curve must be falling at Q, so profit must be *decreasing* when 9,000 units are being produced.

The significance of the sign of the derivative is summarized in the following box and in Figure 2.7. We will have much more to say about the relationship between the shape of a curve and the sign of derivatives in Chapter 3, where we develop a general procedure for curve sketching.

Significance of the Sign of the Derivative $f'(x)$. ■ If the function f is differentiable at $x = c$, then

$$f \text{ is increasing at } x = c \text{ if } f'(c) > 0$$

and

$$f \text{ is decreasing at } x = c \text{ if } f'(c) < 0$$

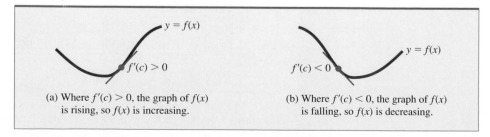

(a) Where $f'(c) > 0$, the graph of $f(x)$ is rising, so $f(x)$ is increasing.

(b) Where $f'(c) < 0$, the graph of $f(x)$ is falling, so $f(x)$ is decreasing.

FIGURE 2.7   The significance of the sign of the derivative $f'(c)$.

Derivative Notation

The derivative $f'(x)$ of $y = f(x)$ is sometimes written as $\dfrac{dy}{dx}$ or $\dfrac{df}{dx}$ (read as "dee y, dee x" or "dee f, dee x"). In this notation, the value of the derivative at $x = c$ [that is, $f'(c)$] is written as

$$\left.\frac{dy}{dx}\right|_{x=c} \qquad \text{or} \qquad \left.\frac{df}{dx}\right|_{x=c}$$

For example, if $y = x^2$, then

$$\frac{dy}{dx} = 2x$$

and the value of this derivative at $x = -3$ is

$$\left.\frac{dy}{dx}\right|_{x=-3} = \left.2x\right|_{x=-3} = 2(-3) = -6$$

The $\dfrac{dy}{dx}$ notation for derivative suggests slope, $\dfrac{\Delta y}{\Delta x}$, and can also be thought of as "the rate of change of y with respect to x." Sometimes it is convenient to condense a statement such as

$$\text{"when } y = x^2, \text{ then } \frac{dy}{dx} = 2x\text{"}$$

by writing simply

$$\frac{d}{dx}(x^2) = 2x$$

which reads, "the derivative of x^2 with respect to x is $2x$."

     Example 2.1.6 illustrates how the different notational forms for the derivative can be used.

EXPLORE!

Many graphing calculators have a special utility for computing derivatives numerically, called the *numerical derivative* (nDeriv). It can be accessed via the **MATH** key. This derivative can also be accessed through the **CALC (2nd TRACE)** key, especially if a graphical presentation is desired. For instance, store $f(x) = \sqrt{x}$ into Y1 of the equation editor and display its graph using a decimal window. Use the $\dfrac{dy}{dx}$ option of the **CALC** key and observe the value of the numerical derivative at $x = 1$.

EXAMPLE | 2.1.6

First compute the derivative of $f(x) = \sqrt{x}$, then use it to:

a. Find the equation of the tangent line to the curve $y = \sqrt{x}$ at the point where $x = 4$.

b. Find the rate at which $y = \sqrt{x}$ is changing with respect to x when $x = 1$.

Solution

The derivative of $y = \sqrt{x}$ with respect to x is given by

$$
\begin{aligned}
\frac{d}{dx}(\sqrt{x}) &= \lim_{h\to 0}\frac{f(x+h)-f(x)}{h} = \lim_{h\to 0}\frac{\sqrt{x+h}-\sqrt{x}}{h}\\[2mm]
&= \lim_{h\to 0}\frac{(\sqrt{x+h}-\sqrt{x})(\sqrt{x+h}+\sqrt{x})}{h(\sqrt{x+h}+\sqrt{x})}\\[2mm]
&= \lim_{h\to 0}\frac{x+h-x}{h(\sqrt{x+h}+\sqrt{x})} = \lim_{h\to 0}\frac{h}{h(\sqrt{x+h}+\sqrt{x})}\\[2mm]
&= \lim_{h\to 0}\frac{1}{\sqrt{x+h}+\sqrt{x}} = \frac{1}{2\sqrt{x}} \qquad \text{if } x > 0
\end{aligned}
$$

a. When $x = 4$, the corresponding y coordinate on the graph of $f(x) = \sqrt{x}$ is $y = \sqrt{4} = 2$, so the point of tangency is $P(4, 2)$. Since $f'(x) = \dfrac{1}{2\sqrt{x}}$, the slope of the tangent line to the graph of $f(x)$ at the point $P(4, 2)$ is given by

$$f'(4) = \frac{1}{2\sqrt{4}} = \frac{1}{4}$$

and by substituting into the point-slope formula, we find that the equation of the tangent line at P is

$$y - 2 = \frac{1}{4}(x - 4)$$

or

$$y = \frac{1}{4}x + 1$$

b. The rate of change of $y = \sqrt{x}$ when $x = 1$ is

$$\left. \frac{dy}{dx} \right|_{x=1} = \frac{1}{2\sqrt{1}} = \frac{1}{2}$$

NOTE Notice that the function $f(x) = \sqrt{x}$ in Example 2.1.6 is defined at $x = 0$ but its derivative $f'(x) = \dfrac{1}{2\sqrt{x}}$ is not. This example shows that a function and its derivative do not always have the same domain. ∎

Differentiability and Continuity

If a function $f(x)$ is differentiable where $x = c$, then the graph of $y = f(x)$ has a non-vertical tangent line at the point $P(c, f(c))$ and at all points on the graph that are "near" P. We would expect such a function to be continuous at $x = c$ since a graph with a tangent line at the point P certainly cannot have a "hole" or "gap" at P. To summarize:

Continuity of a Differentiable Function ▪ If the function $f(x)$ is differentiable at $x = c$, then it is also continuous at $x = c$.

Verification of this observation is outlined in Problem 52. Notice that we are *not* claiming that a continuous function must be differentiable. Indeed it can be shown that a continuous function $f(x)$ will not be differentiable at $x = c$ if $f'(x)$ becomes infinite at $x = c$ or if the graph of $f(x)$ has a "sharp" point at $P(c, f(c))$; that is, a point where the curve makes an abrupt change in direction. If $f(x)$ is continuous at $x = c$ but $f'(c)$ is infinite, the graph of f may have a "vertical tangent" at the point $P(c, f(c))$ (Figure 2.8a) or a "cusp" at P (Figure 2.8b). The absolute value function $f(x) = |x|$ is continuous for all x but has a "sharp point" at the origin $(0, 0)$ (see Figure 2.8c and Problem 51). Another graph with a "sharp point" is shown in Figure 2.8d.

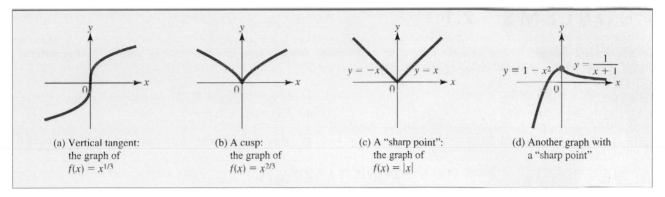

(a) Vertical tangent: the graph of $f(x) = x^{1/3}$

(b) A cusp: the graph of $f(x) = x^{2/3}$

(c) A "sharp point": the graph of $f(x) = |x|$

(d) Another graph with a "sharp point"

FIGURE 2.8   Graphs of four continuous functions that are not differentiable at $x = 0$.

EXPLORE!

Store $f(x) = $ abs(X) into Y1 of the equation editor. The absolute value function can be obtained through the **MATH** key by accessing the **NUM** menu. Use a decimal window and compute the numerical derivative $\dfrac{dy}{dx}$ at $x = 0$. What do you observe and how does this answer reconcile with Figure 2.8c? The curve is too sharp at the point (0, 0) to possess a well-defined tangent line there. Hence, the derivative of $f(x) = $ abs(x) does not exist at $x = 0$. Note that the numerical derivative must be used with caution at cusps and unusual points. Try computing the numerical derivative of $y = \dfrac{1}{x}$ at $x = 0$, and explain how such a result could occur numerically.

In general, the functions you encounter in this text will be differentiable at almost all points. In particular, polynomials are everywhere differentiable and rational functions are differentiable wherever they are defined.

An example of a practical situation involving a continuous function that is not always differentiable is provided by the circulation of blood in the human body.* It is tempting to think of blood coursing smoothly through the veins and arteries in a constant flow, but in actuality, blood is ejected by the heart into the arteries in discrete bursts. This results in an *arterial pulse,* which can be used to measure the heart rate by applying pressure to an accessible artery, as at the wrist. A blood pressure curve showing a pulse is sketched in Figure 2.9. Notice how the curve quickly changes direction at the minimum (*diastolic*) pressure, as a burst of blood is injected by the heart into the arteries, causing the blood pressure to rise rapidly until the maximum (*systolic*) pressure is reached, after which the pressure gradually declines as blood is distributed to tissue by the arteries. The blood pressure function is continuous but is not differentiable at $t = 0.75$ seconds where the infusion of blood occurs.

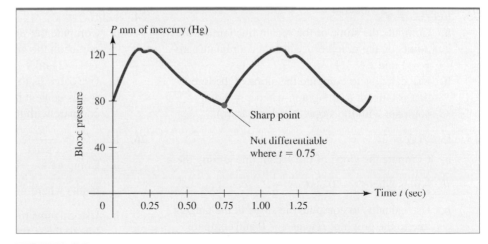

FIGURE 2.9   Graph of blood pressure showing an arterial pulse.

*This example is adapted from F. C. Hoppensteadt and C. S. Peskin, *Mathematics in Medicine and the Life Sciences,* New York: Springer-Verlag, 1992, p. 131.

PROBLEMS 2.1

In Problems 1 through 8, compute the derivative of the given function and find the slope of the line that is tangent to its graph for the specified value of the independent variable.

1. $f(x) = 5x - 3$; $x = 2$

2. $f(x) = x^2 - 1$; $x = -1$

3. $f(x) = 2x^2 - 3x - 5$; $x = 0$

4. $f(x) = x^3 - 1$; $x = 2$

5. $g(t) = \dfrac{2}{t}$; $t = \dfrac{1}{2}$

6. $f(x) = \dfrac{1}{x^2}$; $x = 2$

7. $f(x) = \sqrt{x}$; $x = 9$

8. $h(u) = \dfrac{1}{\sqrt{u}}$; $u = 4$

In Problems 9 through 16, compute the derivative of the given function and find the equation of the line that is tangent to its graph for the specified value $x = c$.

9. $f(x) = x^2$; $c = 1$

10. $f(x) = 3$; $c = -4$

11. $f(x) = 7 - 2x$; $c = 5$

12. $f(x) = 2 - 3x^2$; $c = 1$

13. $f(x) = \dfrac{-2}{x}$; $c = -1$

14. $f(x) = x^3 - 1$; $c = 1$

15. $f(x) = 2\sqrt{x}$; $c = 4$

16. $f(x) = \dfrac{3}{x^2}$; $c = \dfrac{1}{2}$

In Problems 17 through 22, find the rate of change $\dfrac{dy}{dx}$ where $x = x_0$.

17. $y = 3$; $x_0 = 2$

18. $y = 6 - 2x$; $x_0 = 3$

19. $y = x(1 - x)$; $x_0 = -1$

20. $y = \dfrac{1}{2 - x}$; $x_0 = -3$

21. $y = x^2 - 2x$; $x_0 = 1$

22. $y = x - \dfrac{1}{x}$; $x_0 = 1$

23. Let $f(x) = x^3$.
  a. Compute the slope of the secant line joining the points on the graph of f whose x coordinates are $x = 1$ and $x = 1.1$.
  b. Use calculus to compute the slope of the line that is tangent to the graph when $x = 1$ and compare with the slope found in part (a).

24. Let $f(x) = x^2$.
  a. Compute the slope of the secant line joining the points on the graph of f whose x coordinates are $x = -2$ and $x = -1.9$.
  b. Use calculus to compute the slope of the line that is tangent to the graph when $x = -2$ and compare with the slope found in part (a).

25. Let $f(x) = 2x - x^2$.
  a. Compute the slope of the secant line joining the points where $x = 0$ and $x = \dfrac{1}{2}$.
  b. Use calculus to compute the slope of the tangent line to the graph of $f(x)$ at $x = 0$ and compare with the slope found in part (a).

26. Let $f(x) = \dfrac{x}{x - 1}$.
  a. Compute the slope of the secant line joining the points where $x = -1$ and $x = -\dfrac{1}{2}$.
  b. Use calculus to compute the slope of the tangent line to the graph of $f(x)$ at $x = -1$ and compare with the slope found in part (a).

27. Let $f(x) = 3x^2 - x$.

   a. Find the average rate of change of $f(x)$ with respect to x as x changes from $x = 0$ to $x = \dfrac{1}{16}$.

   b. Use calculus to find the instantaneous rate of change of $f(x)$ at $x = 0$ and compare with the average rate found in part (a).

28. Let $s(t) = \sqrt{t}$.

   a. Find the average rate of change of $s(t)$ with respect to t as t changes from $t = 1$ to $t = \dfrac{1}{4}$.

   b. Use calculus to find the instantaneous rate of change of $s(t)$ at $t = 1$ and compare with the average rate found in part (a).

29. Let $s(t) = \dfrac{t - 1}{t + 1}$.

   a. Find the average rate of change of $s(t)$ with respect to t as t changes from $t = -\dfrac{1}{2}$ to $t = 0$.

   b. Use calculus to find the instantaneous rate of change of $s(t)$ to t $-\dfrac{1}{2}$ and compare with the average rate found in part (a).

30. Let $f(x) = x(1 - 2x)$.

   a. Find the average rate of change of $f(x)$ with respect to x as x changes from $x = 0$ to $x = \dfrac{1}{2}$.

   b. Use calculus to find the instantaneous rate of change of $f(x)$ at $x = 0$ and compare with the average rate found in part (a).

31. Fill in the missing interpretations in the table here using the example as a guide.

	If $f(t)$ represents . . .	then $\dfrac{f(t_0 + h) - f(t_0)}{h}$ represents . . . and	$\lim\limits_{h \to 0} \dfrac{f(t_0 + h) - f(t_0)}{h}$ represents . . .
Example	The number of bacteria in a colony at time t	The average rate of change of the bacteria population during the time interval $[t_0, t_0 + h]$	The instantaneous rate of change of the bacteria population at time $t = t_0$
	a. The temperature in San Francisco t hours after midnight on a certain day		
	b. The blood alcohol level t hours after a person consumes an ounce of alcohol		
	c. The 30-year fixed mortgage rate t years after 2000		

32. Fill in the missing interpretations in the table here using the example as a guide.

	If $f(x)$ represents . . .	then $\dfrac{f(x_0 + h) - f(x_0)}{h}$ represents . . . and	$\lim\limits_{h \to 0} \dfrac{f(x_0 + h) - f(x_0)}{h}$ represents . . .
Example	The cost of producing x units of a particular commodity	The average rate of change of cost as production changes from x_0 to $x_0 + h$ units	The instantaneous rate of change of cost with respect to production level when x_0 units are produced
	a. The revenue obtained when x units of a particular commodity are produced		
	b. The amount of unexpended fuel (lb) left in a rocket when it is x feet above ground		
	c. The volume (cm^3) of a cancerous growth 6 months after injection of x mg of an experimental drug		

33. RENEWABLE RESOURCES   The accompanying graph shows how the volume of lumber *V* in a tree varies with time *t* (the age of the tree). Use the graph to estimate the rate at which *V* is changing with respect to time when *t* = 30 years. What seems to be happening to the rate of change of *V* as *t* increases without bound (that is, in the "long run")?

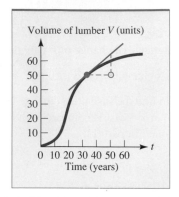

PROBLEM 33   Graph showing how the volume of lumber *V* in a tree varies with time *t*.

Source: Adapted from Robert H. Frank, *Microeconomics and Behavior,* 2nd ed., New York: McGraw-Hill, 1994, p. 623.

34. POPULATION GROWTH   The accompanying graph shows how a population *P* of fruit flies (*Drosophila*) changes with time *t* (days) during an experiment. Use the graph to estimate the rate at which the population is growing after 20 days and also after 36 days. At what time is the population growing at the greatest rate?

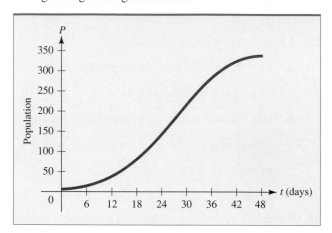

PROBLEM 34   Growth curve for a population of fruit flies.

Source: Adapted from E. Batschelet, *Introduction to Mathematics for Life Scientists,* 3rd ed., New York: Springer-Verlag, 1979, p. 355.

35. THERMAL INVERSION   Air temperature usually decreases with increasing altitude. However, during the winter, thanks to a phenomenon called *thermal inversion,* the temperature of air warmed by the sun in mountains above a fog may rise above the freezing point, while the temperature at lower elevations remains near or below 0°C. Use the accompanying graph to estimate the rate at which temperature *T* is changing with respect to altitude *h* at an altitude of 1,000 meters and also at 2,000 meters.

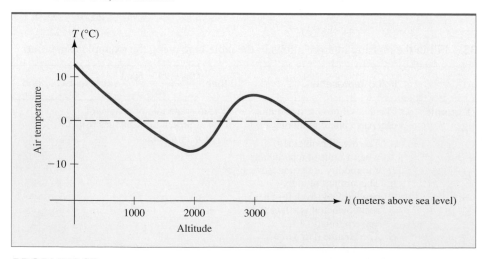

PROBLEM 35

Source: E. Batschelet, *Introduction to Mathematics for Life Scientists,* 3rd ed., New York: Springer-Verlag, 1979, p. 150.

36. VELOCITY   A toy rocket rises vertically in such a way that t seconds after lift off, it is

$$h(t) = -\frac{1}{2}t^2 + 30t \text{ feet above ground.}$$

a. How high is the rocket after 40 seconds?

b. What is the average velocity of the rocket over the first 40 seconds of flight (between $t = 0$ and $t = 40$)?

c. What is the (instantaneous) velocity of the rocket at lift-off ($t = 0$)? What is its velocity after 40 seconds?

37. PROFIT   A manufacturer can produce tape recorders at a cost of $20 apiece. It is estimated that if the tape recorders are sold for p dollars apiece, consumers will buy $q = 120 - p$ recorders each month.

a. Express the manufacturer's profit P as a function of q.

b. What is the average rate of profit obtained as the level of production increases from $q = 0$ to $q = 20$?

c. At what rate is profit changing when $q = 20$ recorders are produced? Is the profit increasing or decreasing at this level of production?

38. ANIMAL BEHAVIOR   Experiments indicate that when a flea jumps, its height (in meters) after t seconds is given by the function

$$H(t) = 4.4t - 4.9t^2$$

a. Find $H'(t)$. At what rate is $H(t)$ changing after 1 second? Is it increasing or decreasing?

b. At what time is $H'(t) = 0$? What is the significance of this time?

c. When does the flea "land" (return to its initial height)? At what rate is $H(t)$ changing at this time? Is it increasing or decreasing?

39. PROFIT   A manufacturer of videocassettes determines that when x hundred units are produced, the profit will be $P(x) = 4,000(15 - x)(x - 2)$ dollars.

a. Find $P'(x)$.

b. Determine where $P'(x) = 0$. What is the significance of the level of production x_m where this occurs?

40. MANUFACTURING OUTPUT   At a certain factory, it is determined that an output of Q units is to be expected when L worker-hours of labor are employed, where

$$Q(L) = 3,100\sqrt{L}$$

a. Find the average rate of change of output as the labor employment changes from $L = 3,025$ worker-hours to 3,100 worker-hours.

b. Use calculus to find the instantaneous rate of change of output with respect to labor level when $L = 3025$.

41. COST OF PRODUCTION   A business manager determines that the cost of producing x units of a particular commodity is C thousands of dollars, where

$$C(x) = 0.04x^2 + 5.1x + 40$$

a. Find the average cost as the level of production changes from $x = 10$ to $x = 11$ units.

b. Use calculus to find the instantaneous rate of change of cost with respect to production level when $x = 10$ and compare with the average cost found in part (a). Is the cost increasing or decreasing when 10 units are being produced?

42. CONSUMER EXPENDITURE   The demand for a particular commodity is given by $D(x) = -35x + 200$; that is, x units will be sold (demanded) at a price of $p = D(x)$ dollars per unit.

a. Consumer expenditure $E(x)$ is the total amount of money consumers pay to buy x units. Express consumer expenditure E as a function of x.

b. Find the average change in consumer expenditure as x changes from $x = 4$ to $x = 5$.

c. Use calculus to find the instantaneous rate of change of expenditure with respect to x when $x = 4$. Is expenditure increasing or decreasing when $x = 4$?

43. UNEMPLOYMENT   In economics, the graph in Figure 2.2 is called the **Phillips curve,** after A. W. Phillips, a New Zealander associated with the London School of Economics. Until Phillips published his ideas in the 1950s, many economists believed that unemployment and inflation were linearly related. Read an article on the Phillips curve (the source cited with Figure 2.2 would be a good place to start) and write a paragraph on the nature of unemployment in the U.S. economy.

44. BLOOD PRESSURE   Refer to the graph of blood pressure as a function of time shown in Figure 2.9.

a. Estimate the average rate of change in blood pressure over the time periods [0.7, 0.75] and [0.75, 0.8]. Interpret your results.

 b. Write a paragraph on the dynamics of the arterial pulse. Pages 131–136 in the reference given with Figure 2.9 is a good place to start, and there is an excellent list of annotated references to related topics on pp. 137–138.

45. CARDIOLOGY A study conducted on a patient undergoing cardiac catheterization indicated that the diameter of the aorta was approximately D millimeters (mm) when the aortic pressure was p (mm of mercury), where

$$D(p) = -0.0009p^2 + 0.13p + 17.81$$

for $50 \leq p \leq 120$.

a. Find the average rate of change of the aortic diameter D as p changes from $p = 60$ to $p = 61$.

b. Use calculus to find the instantaneous rate of change of diameter D with respect to aortic pressure p when $p = 60$. Is the pressure increasing or decreasing when $p = 60$?

c. For what value of p is the instantaneous rate of change of D with respect to p equal to 0? What is the significance of this pressure?

46. a. Find the derivative of the linear function $f(x) = 3x - 2$.

b. Find the equation of the tangent line to the graph of this function at the point where $x = -1$.

c. Explain how the answers to parts (a) and (b) could have been obtained from geometric considerations with no calculation whatsoever.

47. a. Find the derivatives of the functions $y = x^2$ and $y = x^2 - 3$ and account geometrically for their similarity.

b. Without further computation, find the derivative of the function $y = x^2 + 5$.

48. a. Find the derivative of the function $y = x^2 + 3x$.

b. Find the derivatives of the functions $y = x^2$ and $y = 3x$ separately.

c. How is the derivative in part (a) related to those in part (b)?

d. In general, if $f(x) = g(x) + h(x)$, what would you guess is the relationship between the derivative of f and those of g and h?

49. a. Compute the derivatives of the functions $y = x^2$ and $y = x^3$.

b. Examine your answers in part (a). Can you detect a pattern? What do you think is the derivative of $y = x^4$? How about the derivative of $y = x^{27}$?

50. Use calculus to prove that if $y = mx + b$, the rate of change of y with respect to x is constant.

51. Let f be the absolute value function; that is,

$$f(x) = \begin{cases} x & \text{if } x \geq 0 \\ -x & \text{if } x < 0 \end{cases}$$

Show that

$$f'(x) = \begin{cases} 1 & \text{if } x > 0 \\ -1 & \text{if } x < 0 \end{cases}$$

and explain why f is not differentiable at $x = 0$.

52. Let f be a function that is differentiable at $x = c$.

a. Explain why

$$f'(c) = \lim_{x \to c} \frac{f(x) - f(c)}{x - c}$$

b. Use the result of part (a) together with the fact that

$$f(x) - f(c) = \left[\frac{f(x) - f(c)}{x - c} \right] (x - c)$$

to show that

$$\lim_{x \to c} [f(x) - f(c)] = 0$$

c. Explain why the result obtained in part (b) shows that f is continuous at $x = c$.

53. Show that $f(x) = \dfrac{|x^2 - 1|}{x - 1}$ is not differentiable at $x = 1$.

54. Find the x values at which the peaks and valleys of the graph of $y = 2x^3 - 0.8x^2 + 4$ occur. Use four decimal places.

55. Find the slope of the line that is tangent to the graph of the function $f(x) = \sqrt{x^2 + 2x} - \sqrt{3x}$ at the point where $x = 3.85$ by filling in the following chart. Record all calculations using five decimal places.

h	−0.02	−0.01	−0.001	—0—	0.001	0.01	0.02
$x + h$							
$f(x)$							
$f(x + h)$							
$f(x + h) - f(x)$							
$\dfrac{}{h}$							

SECTION 2.2 Techniques of Differentiation

If we had to use the limit definition every time we wanted to compute a derivative, it would be both tedious and difficult to use calculus in applications. Fortunately, this is not necessary, and in this section and the next, we develop techniques that greatly simplify the process of differentiation. We begin with a rule for the derivative of a constant.

The Constant Rule ■ For any constant c,

$$\frac{d}{dx}[c] = 0$$

That is, the derivative of a constant is zero.

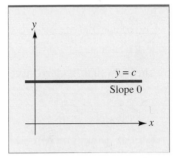

FIGURE 2.10 The graph of $f(x) = c$.

You can see this by considering the graph of a constant function $f(x) = c$, which is a horizontal line (see Figure 2.10). Since the slope of such a line is 0 at all its points, it follows that $f'(x) = 0$. Here is a proof using the limit definition:

$$f'(x) = \lim_{h \to 0} \frac{f(x + h) - f(x)}{h}$$

$$= \lim_{h \to 0} \frac{c - c}{h} = 0 \qquad \text{since } f(c + h) = c \text{ for all } x$$

EXPLORE!

A function and its derivative can be graphed simultaneously on the same screen. As an example, put $x^2 + 2x$ into Y1, using a bold graphing style. Put nDeriv(Y1, X, X) into Y2, where nDeriv is an option on the **MATH** key menu. Graph Y1 and Y2 using the enlarged decimal window $[-4.7, 4.7]1$ by $[-3.1, 9.1]1$. Now use the *dy/dx* function of the **CALC(2nd TRACE)** key to find the slope of Y1 at the *x* values X = −2, −1, 0, 1, and 2. How do these values for *dy/dx* compare with the corresponding values of Y2 for the respective X values? What do you conclude? As an extra challenge, find the explicit equation for Y2.

EXAMPLE 2.2.1

$$\frac{d}{dx}[-15] = 0$$

The next rule is one of the most useful because it tells you how to take the derivative of any power function $f(x) = x^n$. Note that the rule applies not only to functions like $f(x) = x^5$ but also to those such as $g(x) = \sqrt[5]{x^4} = x^{4/5}$ and $h(x) = \frac{1}{x^3} = x^{-3}$. Functions such as $x^{\sqrt{2}}$ can also be differentiated using the power rule, but we do not define what is meant by such functions until Chapter 4.

The Power Rule ■ For any real number n,

$$\frac{d}{dx}[x^n] = nx^{n-1}$$

In words, to find the derivative of x^n, reduce the exponent n of x by 1 and multiply your new power of x by the original exponent.

According to the power rule, the derivative of $y = x^3$ is $\frac{d}{dx}(x^3) = 3x^2$, which agrees with the result found directly in Example 2.1.4 of Section 2.1. You can use the power rule to differentiate radicals and reciprocals by first converting them to power functions with fractional and negative exponents, respectively. (You can find a review of exponential notation in Appendix A1 at the back of the book.) For example, recall that $\sqrt{x} = x^{1/2}$, so the derivative of $y = \sqrt{x}$ is

$$\frac{d}{dx}(\sqrt{x}) = \frac{d}{dx}(x^{1/2}) = \frac{1}{2}x^{-1/2}$$

which agrees with the result of Example 2.1.6 of Section 2.1. In Example 2.2.2, we verify the power rule for a reciprocal power function.

EXAMPLE | 2.2.2

Verify the power rule for the function $F(x) = \dfrac{1}{x^2} = x^{-2}$ by showing that its derivative is $F'(x) = -2x^{-3}$.

Just-In-Time Review

Here is the rule for simplifying a complex fraction:

$$\frac{A/B}{C/D} = \frac{AD}{BC}$$

Solution

The derivative of $F(x)$ is given by

$$F'(x) = \lim_{h \to 0} \frac{F(x+h) - F(x)}{h} = \lim_{h \to 0} \frac{\dfrac{1}{(x+h)^2} - \dfrac{1}{x^2}}{h}$$

$$= \lim_{h \to 0} \frac{\dfrac{x^2 - (x+h)^2}{x^2(x+h)^2}}{h} \qquad \textit{put the numerator over a common denominator}$$

$$= \lim_{h \to 0} \frac{[x^2 - (x^2 + 2hx + h^2)]}{x^2 h(x+h)^2} \qquad \textit{simplify complex fraction}$$

$$= \lim_{h \to 0} \frac{-2xh - h^2}{x^2 h(x+h)^2} \qquad \textit{combine terms in the numerator}$$

$$= \lim_{h \to 0} \frac{-2x - h}{x^2(x+h)^2} \qquad \textit{cancel common h terms}$$

$$= \frac{-2x}{x^2(x^2)}$$

$$= \frac{-2}{x^3} = -2x^{-3}$$

Just-In-Time Review

Recall that $x^{-n} = 1/x^n$ when n is a positive integer, and that $x^{a/b} = \sqrt[b]{x^a}$ whenever a and b are positive integers.

as claimed by the power rule.

Here are several additional applications of the power rule:

$$\frac{d}{dx}(x^7) = 7x^{7-1} = 7x^6$$

$$\frac{d}{dx}(\sqrt[3]{x^2}) = \frac{d}{dx}(x^{2/3}) = \frac{2}{3}x^{2/3-1} = \frac{2}{3}x^{-1/3}$$

$$\frac{d}{dx}\left(\frac{1}{x^5}\right) = \frac{d}{dx}(x^{-5}) = -5x^{-5-1} = -5x^{-6}$$

$$\frac{d}{dx}(x^{1.3}) = 1.3x^{1.3-1} = 1.3x^{0.3}$$

A general proof for the power rule in the case where n is a positive integer is lined in Problem 74. The case where n is a negative integer and the case where n a rational number ($n = r/s$ for integers r and s with $s \neq 0$) are handled in Section 2.5 and Section 2.6, respectively.

The constant rule and power rule provide simple formulas for finding derivatives of a class of important functions, but to differentiate more complicated expressions, we need to know how to manipulate derivatives algebraically. The next two rules tell us that derivatives of multiples and sums of functions are multiples and sums of the corresponding derivatives

The Constant Multiple Rule ▪ If c is a constant and $f(x)$ is differentiable, then so is $cf(x)$ and

$$\frac{d}{dx}[cf(x)] = c\frac{d}{dx}[f(x)]$$

That is, the derivative of a multiple is the multiple of the derivative.

EXAMPLE 2.2.3

$$\frac{d}{dx}(3x^4) = 3\frac{d}{dx}(x^4) = 3(4x^3) = 12x^3$$

$$\frac{d}{dx}\left(\frac{-7}{\sqrt{x}}\right) = \frac{d}{dx}(-7x^{-1/2}) = -7\left(\frac{-1}{2}x^{-3/2}\right) = \frac{7}{2}x^{-3/2}$$

The Sum Rule ▪ If $f(x)$ and $g(x)$ are differentiable, then so is the sum $S(x) = f(x) + g(x)$ and $S'(x) = f'(x) + g'(x)$; that is,

$$\frac{d}{dx}[f(x) + g(x)] = \frac{d}{dx}[f(x)] + \frac{d}{dx}[g(x)]$$

In words, the derivative of a sum is the sum of the separate derivatives.

EXAMPLE 2.2.4

$$\frac{d}{dx}[x^{-2} + 7] = \frac{d}{dx}[x^{-2}] + \frac{d}{dx}[7] = -2x^{-3} + 0 = 2x^{-3}$$

$$\frac{d}{dx}[2x^5 - 3x^{-7}] = 2\frac{d}{dx}[x^5] - 3\frac{d}{dx}[x^{-7}] = 2(5x^4) - 3(-7x^{-8})$$

$$= 10x^4 + 21x^{-8}$$

By combining the power rule, the constant multiple rule, and the sum rule, you can differentiate any polynomial. Here is an example.

EXAMPLE 2.2.5

Differentiate the polynomial $y = 5x^3 - 4x^2 + 12x - 8$.

Solution

Differentiate this sum term by term to get

$$\frac{dy}{dx} = \frac{d}{dx}[5x^3] + \frac{d}{dx}[-4x^2] + \frac{d}{dx}[12x] + \frac{d}{dx}[-8]$$
$$= 15x^2 - 8x^1 + 12x^0 + 0 \qquad \textit{recall } x^0 = 1$$
$$= 15x^2 - 8x + 12$$

EXAMPLE 2.2.6

It is estimated that x months from now, the population of a certain community will be $P(x) = x^2 + 20x + 8,000$.

a. At what rate will the population be changing with respect to time 15 months from now?

b. By how much will the population actually change during the 16th month?

Solution

a. The rate of change of the population with respect to time is the derivative of the population function. That is,

$$\text{Rate of change} = P'(x) = 2x + 20$$

The rate of change of the population 15 months from now will be

$$P'(15) = 2(15) + 20 = 50 \text{ people per month}$$

b. The actual change in the population during the 16th month is the difference between the population at the end of 16 months and the population at the end of 15 months. That is,

$$\text{Change in population} = P(16) - P(15) = 8,576 - 8,525$$
$$= 51 \text{ people}$$

NOTE In Example 2.2.6, the actual change in population during the 16th month in part (b) differs from the monthly rate of change at the beginning of the month in part (a) because the rate varies during the month. The instantaneous rate in part (a) can be thought of as the change in population that would occur during the 16th month if the rate of change of population were to remain constant. ∎

Relative and Percentage Rates of Change

In many practical situations, the rate of change of a quantity Q is not as significant as its *relative* rate of change, which is defined as the ratio

$$\text{Relative change} = \frac{\text{change in } Q}{\text{size of } Q}$$

For example, a yearly rate of change of 500 individuals in a city whose total population is 5 million yields a negligible relative (per capita) rate of change of

$$\frac{500}{5,000,000} = 0.0001$$

or 0.01%, while the same rate of change in a town of 2,000 would yield a relative rate of change of

$$\frac{500}{2,000} = 0.25$$

or 25%, which would have enormous impact on the town.

Since the rate of change of a quantity $Q(x)$ is measured by the derivative $Q'(x)$, we can express relative rate of change and the associated percentage rate of change in the following forms.

Relative and Percentage Rates of Change ■ The **relative rate of change** of a quantity $Q(x)$ with respect to x is given by the ratio

$$\begin{bmatrix} \text{Relative rate of} \\ \text{change of } Q(x) \end{bmatrix} = \frac{Q'(x)}{Q(x)} = \frac{dQ/dx}{Q}$$

The corresponding **percentage rate of change** of $Q(x)$ with respect to x is

$$\begin{bmatrix} \text{Percentage rate of} \\ \text{change of } Q(x) \end{bmatrix} = \frac{100\,Q'(x)}{Q(x)}$$

EXPLORE!

Refer to Example 2.2.7. Compare the rate of change of GDP at time $t = 10$ for $N(t)$ with the rate of a new GDP given by

$$N_1(t) = 2t^2 + 2t + 100$$

Graph both functions using x as the independent variable and the graphing window

$$[3, 12.4]1 \text{ by } [90, 350]0$$

where a 0 scale corresponds to a display of no markings for the y axis. How do the percentage rates of change of GDP compare for 2005?

EXAMPLE | 2.2.7

The gross domestic product (GDP) of a certain country was $N(t) = t^2 + 5t + 106$ billion dollars t years after 1995.

a. At what rate was the GDP changing with respect to time in 2005?

b. At what percentage rate was the GDP changing with respect to time in 2005?

Solution

a. The rate of change of the GDP is the derivative $N'(t) = 2t + 5$. The rate of change in 2005 was $N'(10) = 2(10) + 5 = 25$ billion dollars per year.

b. The percentage rate of change of the GDP in 2005 was

$$100\,\frac{N'(10)}{N(10)} = 100\,\frac{25}{256} \approx 9.77\% \text{ per year}$$

EXAMPLE | 2.2.8

Experiments indicate that the biomass $Q(t)$ of a fish species in a given area of the ocean changes at the rate

$$\frac{dQ}{dt} = rQ\left(1 - \frac{Q}{a}\right)$$

where r is the natural growth rate of the species and a is a constant.* Find the percentage rate of growth of the species. What if $Q(t) > a$?

---

*Adapted from W. R. Derrick and S. I. Grossman, *Introduction to Differential Equations,* 3rd ed., St. Paul, MN: West Publishing, 1987, p. 52, problem 20. The authors note that the problem was originally one of many models described in the text *Mathematical Bioeconomics,* by C. W. Clark (Wiley-Interscience, 1976).

Solution

The percentage rate of change of $Q(t)$ is

$$\frac{100\, Q'(t)}{Q(t)} = \frac{100\, rQ\left(1 - \dfrac{Q}{a}\right)}{Q} = 100r\left(1 - \frac{Q}{a}\right)$$

Notice that the percentage rate decreases as Q increases and becomes zero when $Q = a$. If $Q > a$, the percentage rate is negative, which means the biomass is actually decreasing.

Rectilinear Motion   Motion of an object along a line is called *rectilinear motion*. For example, the motion of a rocket early in its flight can be regarded as rectilinear. When studying rectilinear motion, we will assume that the object involved is moving along a coordinate axis. If the function $s(t)$ gives the *position* of the object at time t, then the rate of change of $s(t)$ with respect to t is its *velocity* $v(t)$ and the time derivative of velocity is its *acceleration* $a(t)$. That is, $v(t) = s'(t)$ and $a(t) = v'(t)$.

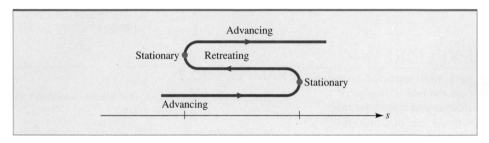

FIGURE 2.11   A diagram for rectilinear motion.

The object is said to be *advancing* (moving forward) when $v(t) > 0$, and *retreating* (moving backward) when $v(t) < 0$. When $v(t) = 0$, the object is neither advancing nor retreating and is said to be *stationary*. See Figure 2.11. Finally, the object is *accelerating* when $a(t) > 0$ and *decelerating* when $a(t) < 0$. To summarize:

> **Rectilinear Motion** ■ If the **position** at time t of an object moving along a straight line is given by $s(t)$, then the object has
>
> $$\textbf{velocity} \quad v(t) = s'(t) = \frac{ds}{dt}$$
>
> and
>
> $$\textbf{acceleration} \quad a(t) = v'(t) = \frac{dv}{dt}$$
>
> The object is **advancing** when $v(t) > 0$, **retreating** when $v(t) < 0$, and **stationary** when $v(t) = 0$. It is **accelerating** when $a(t) > 0$ and **decelerating** when $a(t) < 0$.

If position is measured in meters and time in seconds, velocity is measured in meters per second (m/sec) and acceleration in meters per second per second (written as m/sec^2). Similarly, if position is measured in feet, velocity is measured in feet per second (ft/sec) and acceleration in feet per second per second (ft/sec^2).

EXPLORE!

Try this. Set your graphing utility in parametric, dotted, and simultaneous modes. Store X1T = T, Y1T = 0.5, X2T = T, and Y2T = T^3 – 6T^2 + 9T + 5. Graph using a viewing rectangle of [0, 5]1 by [0, 10]1, and 0 ≤ t ≤ 4 with a step of 0.2. Describe what you observe. What does the vertical axis represent? What does the horizontal line of dots mean?

EXAMPLE | 2.2.9

The position at time t of an object moving along a line is given by $s(t) = t^3 - 6t^2 + 9t + 5$.

a. Find the velocity of the object and discuss its motion between times $t = 0$ and $t = 4$.

b. Find the total distance traveled by the object between times $t = 0$ and $t = 4$.

c. Find the acceleration of the object and determine when the object is accelerating and decelerating between times $t = 0$ and $t = 4$.

Solution

a. The velocity is $v(t) = \dfrac{ds}{dt} = 3t^2 - 12t + 9$. The object will be stationary when

$$v(t) = 3t^2 - 12t + 9 = 3(t - 1)(t - 3) = 0$$

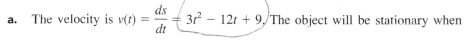

that is, at times $t = 1$ and $t = 3$. Otherwise, the object is either advancing or retreating, as described in the following table.

Interval	Sign of $v(t)$	Description of the Motion
$0 < t < 1$	$+$	Advances from $s(0) - 5$ to $s(1) = 9$
$1 < t < 3$	$-$	Retreats from $s(1) = 9$ to $s(3) = 5$
$3 < t < 4$	$+$	Advances from $s(3) = 5$ to $s(4) = 9$

The motion of the object is summarized in the diagram in Figure 2.12.

b. The object travels from $s(0) = 5$ to $s(1) = 9$, then back to $s(3) = 5$, and finally to $s(4) = 9$. Thus, the total distance traveled by the object is

$$D = \underbrace{|9 - 5|}_{0 < t < 1} + \underbrace{|5 - 9|}_{1 < t < 3} + \underbrace{|9 - 5|}_{3 < t < 4} = 12$$

c. The acceleration of the object is

$$a(t) = \frac{dv}{dt} = 6t - 12 = 6(t - 2) \quad (t = 2)$$

The object will be accelerating $[a(t) > 0]$ when $2 < t < 4$ and decelerating $[a(t) < 0]$ when $0 < t < 2$.

FIGURE 2.12 The motion of an object:
$s(t) = t^3 - 6t^2 + 9t + 5$.

The Motion of a Projectile

An important example of rectilinear motion is the motion of a projectile. Suppose an object is projected (e.g., thrown, fired, or dropped) vertically in such a way that the only acceleration acting on the object is the constant downward acceleration g due to gravity. Near sea level, g is approximately 32 ft/sec^2 (or 9.8 m/sec^2). It can be shown that at time t, the height of the object is given by the formula

$$H(t) = \frac{-1}{2}gt^2 + V_0t + H_0$$

where H_0 and V_0 are the initial height and velocity of the object, respectively. Here is an example using this formula.

EXAMPLE | 2.2.10

Suppose a person standing at the top of a building 112 feet high throws a ball vertically upward with an initial velocity of 96 ft/sec (see Figure 2.13).

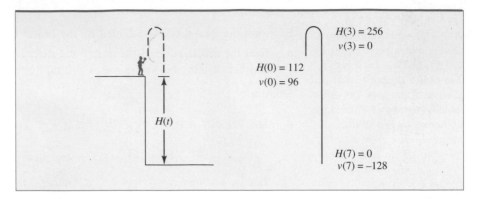

$H(3) = 256$
$v(3) = 0$

$H(0) = 112$
$v(0) = 96$

$H(t)$

$H(7) = 0$
$v(7) = -128$

FIGURE 2.13 The motion of a ball thrown upward from the top of a building.

a. Find the ball's height and velocity at time t.

b. When does the ball hit the ground and what is its impact velocity?

c. When is the velocity 0? What is the significance of this time?

d. How far does the ball travel during its flight?

Solution

a. Since $g = 32$, $V_0 = 96$, and $H_0 = 112$, the height of the ball above the ground at time t is

$$H(t) = -16t^2 + 96t + 112 \text{ feet}$$

The velocity at time t is

$$v(t) = \frac{dH}{dt} = -32t + 96 \text{ ft/sec}$$

b. The ball hits the ground when $H = 0$. Solve the equation $-16t^2 + 96t + 112 = 0$ to find that this occurs when $t = 7$ and $t = -1$ (verify). Disregarding the negative time $t = -1$, which is not meaningful in this context, conclude that impact occurs when $t = 7$ seconds and that the impact velocity is

$$v(7) = -32(7) + 96 = -128 \text{ ft/sec}$$

(The negative sign means the ball is coming down at the moment of impact.)

c. The velocity is zero when $v(t) = -32t + 96 = 0$, which occurs when $t = 3$ seconds. For $t < 3$, the velocity is positive and the ball is rising, and for $t > 3$, the ball is falling (see Figure 2.8). Thus, the ball is at its highest point when $t = 3$ seconds.

d. The ball starts at $H(0) = 112$ feet and rises to a maximum height of $H(3) = 256$ feet before falling to the ground. Thus,

$$\text{Total distance traveled} = \underbrace{(256 - 112)}_{up} + \underbrace{256}_{down} = 400 \text{ feet}$$

PROBLEMS 2.2

In Problems 1 through 25, differentiate the given function. Simplify your answers.

1. $y = x^{-4}$

2. $y = x^{7/3}$

3. $y = -2$

4. $y = x^{3.7}$

5. $y = \pi r^2$

6. $y = \frac{4}{3}\pi r^3$

7. $y = \sqrt{2x}$

8. $y = 2\sqrt[4]{x^3}$

9. $y = \frac{9}{\sqrt{t}}$

10. $y = \frac{3}{2t^2}$

11. $y = x^2 + 2x + 3$

12. $y = 3x^5 - 4x^3 + 9x - 6$

13. $f(x) = x^9 - 5x^8 + x + 12$

14. $f(x) = \frac{1}{4}x^8 - \frac{1}{2}x^6 - x + 2$

15. $f(x) = -0.02x^3 + 0.3x$

16. $f(u) = 0.07u^4 - 1.21u^3 + 3u - 5.2$

17. $y = \frac{1}{t} + \frac{1}{t^2} - \frac{1}{\sqrt{t}}$

18. $y = \frac{3}{x} - \frac{2}{x^2} + \frac{2}{3x^3}$

19. $f(x) = \sqrt{x^3} + \frac{1}{\sqrt{x^3}}$

20. $f(t) = 2\sqrt{t^3} + \frac{4}{\sqrt{t}} - \sqrt{2}$

21. $y = -\frac{x^2}{16} + \frac{2}{x} - x^{3/2} + \frac{1}{3x^2} + \frac{x}{3}$

22. $y = \frac{-7}{x^{1.2}} + \frac{5}{x^{-2.1}}$

23. $y = -\frac{2}{x^2} + x^{2/3} + \frac{1}{2\sqrt{x}} + \frac{x^2}{4} + \sqrt{5} + \frac{x+2}{3}$

24. $y = x^2(x^3 - 6x + 7)$ [*Hint:* Multiply first.]

25. $y = \frac{x^5 - 4x^2}{x^3}$ [*Hint:* Divide first.]

In Problems 26 through 31 find the equation of the line that is tangent to the graph of the given function at the specified point.

26. $y = x^5 - 3x^3 - 5x + 2;\ (1, -5)$

27. $y = -x^3 - 5x^2 + 3x - 1;\ (-1, -8)$

28. $y = \sqrt{x^3} - x^2 + \frac{16}{x^2};\ (4, -7)$

29. $y = 1 - \frac{1}{x} + \frac{2}{\sqrt{x}};\ \left(4, \frac{7}{4}\right)$

30. $y = 2x^4 - \sqrt{x} + \frac{3}{x};\ (1, 4)$

31. $y = (x^2 - x)(3 + 2x);\ (-1, 2)$

In Problems 32 through 37 find the equation of the line that is tangent to the graph of the given function at the point $(c, f(c))$ for the specified value of $x = c$.

32. $f(x) = x^4 - 3x^3 + 2x^2 - 6;\ x = 2$

33. $f(x) = -2x^3 + \frac{1}{x^2};\ x = -1$

34. $f(x) = x^3 + \sqrt{x};\ x = 4$

35. $f(x) = x - \frac{1}{x^2};\ x = 1$

36. $f(x) = x(\sqrt{x} - 1);\ x = 4$

37. $f(x) = -\frac{1}{3}x^3 + \sqrt{8x};\ x = 2$

In Problems 38 through 43 find the rate of change of the given function f(x) with respect to x for the prescribed value x = c.

38. $f(x) = x^3 - 3x + 5; x = 2$

39. $f(x) = 2x^4 + 3x + 1; x = -1$

40. $f(x) = \sqrt{x} + 5x; x = 4$

41. $f(x) = x - \sqrt{x} + \dfrac{1}{x^2}; x = 1$

42. $f(x) = \dfrac{2}{x} - x\sqrt{x}; x = 1$

43. $f(x) = \dfrac{x + \sqrt{x}}{\sqrt{x}}; x = 1$

44. NEWSPAPER CIRCULATION It is estimated that t years from now, the circulation of a local newspaper will be $C(t) = 100t^2 + 400t + 5,000$.
   a. Derive an expression for the rate at which the circulation will be changing with respect to time t years from now.
   b. At what rate will the circulation be changing with respect to time 5 years from now? Will the circulation be increasing or decreasing at that time?
   c. By how much will the circulation actually change during the sixth year?

45. AIR POLLUTION An environmental study of a certain suburban community suggests that t years from now, the average level of carbon monoxide in the air will be $Q(t) = 0.05t^2 + 0.1t + 3.4$ parts per million.
   a. At what rate will the carbon monoxide level be changing with respect to time 1 year from now?
   b. By how much will the carbon monoxide level change this year?
   c. By how much will the carbon monoxide level change over the next 2 years?

46. WORKER EFFICIENCY An efficiency study of the morning shift at a certain factory indicates that an average worker who arrives on the job at 8:00 A.M. will have assembled $f(x) = -x^3 + 6x^2 + 15x$ transistor radios x hours later.
   a. Derive a formula for the rate at which the worker will be assembling radios after x hours.
   b. At what rate will the worker be assembling radios at 9:00 A.M.?
   c. How many radios will the worker actually assemble between 9:00 A.M. and 10:00 A.M.?

47. EDUCATIONAL TESTING It is estimated that x years from now, the average SAT mathematics score of the incoming students at an eastern liberal arts college will be $f(x) = -6x + 582$.
   a. Derive an expression for the rate at which the average SAT score will be changing with respect to time x years from now.

   b. What is the significance of the fact that the expression in part (a) is a constant? What is the significance of the fact that the constant in part (a) is negative?

48. PUBLIC TRANSPORTATION After x weeks, the number of people using a new rapid transit system was approximately $N(x) = 6x^3 + 500x + 8,000$.
   a. At what rate was the use of the system changing with respect to time after 8 weeks?
   b. By how much did the use of the system change during the eighth week?

49. PROPERTY TAX Records indicate that x years after 2000, the average property tax on a three-bedroom home in a certain community was $T(x) = 20x^2 + 40x + 600$ dollars.
   a. At what rate was the property tax increasing with respect to time in 2000?
   b. By how much did the tax change between the years 2000 and 2004?

50. GROWTH OF A TUMOR A cancerous tumor is modeled as a sphere of radius R cm. At what rate is the volume $V = \dfrac{4}{3}\pi R^3$ changing with respect to R when $R = 0.75$ cm?

51. SPREAD OF AN EPIDEMIC A medical research team determines that t days after an epidemic begins, $N(t) = 10t^3 + 5t + \sqrt{t}$ people will be infected, for $0 \le t \le 20$. At what rate is the infected population increasing on the ninth day?

52. ADVERTISING A manufacturer of motorcycles estimates that if x thousand dollars are spent on advertising, then

$$M(x) = 2,300 + \frac{125}{x} - \frac{517}{x^2} \qquad 3 \le x \le 18$$

cycles will be sold. At what rate will sales be changing when \$9,000 is spent on advertising? Are sales increasing or decreasing for this level of advertising expenditure?

53. **COST MANAGEMENT** A company uses a truck to deliver its products. To estimate costs, the manager models gas consumption by the function

$$G(x) = \frac{1}{250}\left(\frac{1,200}{x} + x\right)$$

gal/mile, assuming that the truck is driven at a constant speed of x miles per hour, for $x \geq 5$. The driver is paid $20 per hour to drive the truck 250 miles, and gasoline costs $1.90 per gallon.
   a. Find an expression for the total cost $C(x)$ of the trip.
   b. At what rate is the cost $C(x)$ changing with respect to x when the truck is driven at 40 miles per hour? Is the cost increasing or decreasing at that speed?

54. **PHYSICAL CHEMISTRY** According to Debye's formula in physical chemistry, the orientation polarization P of a gas satisfies

$$P = \frac{4}{3}\pi N\left(\frac{\mu^2}{3kT}\right)$$

where μ, k, and N are positive constants and T is the temperature of the gas. Find the rate of change of P with respect to T.

In Problems 55 and 56 find the relative rate of change of $f(x)$ with respect to x when $x = 1$.

55. $f(x) = 2x^3 - 5x^2 + 4$

56. $f(x) = x + \dfrac{1}{x}$

57. **ANNUAL EARNINGS** The gross annual earnings of a certain company were $A(t) = 0.1t^2 + 10t + 20$ thousand dollars t years after its formation in 2000.
   a. At what rate were the gross annual earnings of the company growing with respect to time in 2004?
   b. At what percentage rate were the gross annual earnings growing with respect to time in 2004?

58. **POPULATION GROWTH** It is estimated that t years from now, the population of a certain town will be $P(t) = t^2 + 200t + 10,000$.
   a. Express the percentage rate of change of the population as a function of t, simplify this function algebraically, and draw its graph.
   b. What will happen to the percentage rate of change of the population in the long run (that is, as t grows very large)?

59. **SALARY INCREASES** Your starting salary will be $45,000, and you will get a raise of $2,000 each year.
   a. Express the percentage rate of change of your salary as a function of time and draw the graph.
   b. At what percentage rate will your salary be increasing after 1 year?
   c. What will happen to the percentage rate of change of your salary in the long run?

60. **GROSS DOMESTIC PRODUCT** The gross domestic product of a certain country is growing at a constant rate. In 1995 the GDP was 125 billion dollars, and in 2003 it was 155 billion dollars. If this trend continues, at what percentage rate will the GDP be growing in 2010?

61. **POPULATION GROWTH** It is projected that x months from now, the population of a certain town will be $P(x) = 2x + 4x^{3/2} + 5,000$.
   a. At what rate will the population be changing with respect to time 9 months from now?
   b. At what percentage rate will the population be changing with respect to time 9 months from now?

62. **SPREAD OF AN EPIDEMIC** A disease is spreading in such a way that after t weeks, the number of people infected is
   $$N(t) = 5,175 - t^3(t - 8) \qquad 0 \leq t \leq 8$$
   a. At what rate is the epidemic spreading after 3 weeks?
   b. Suppose health officials declare the disease to have reached epidemic proportions when the percentage rate of change of N is at least 25%. Over what time period is this epidemic criterion satisfied?
   c. Read an article on epidemiology and write a paragraph on how public health policy is related to the spread of an epidemic.

63. **ORNITHOLOGY** An ornithologist determines that the body temperature of a certain species of bird fluctuates over roughly a 17-hour period according to the cubic formula
   $$T(t) = -68.07t^3 + 30.98t^2 + 12.52t + 37.1$$
   for $0 \leq t \leq 0.713$, where T is the temperature in degrees Celsius measured t days from the beginning of a period.

a. Compute and interpret the derivative $T'(t)$.
b. At what rate is the temperature changing at the beginning of the period ($t = 0$) and at the end of the period ($t = 0.713$)? Is the temperature increasing or decreasing at each of these times?
c. At what time is the temperature not changing (neither increasing nor decreasing)? What is the bird's temperature at this time? Interpret your result.

RECTILINEAR MOTION *In Problems 64 through 67, s(t) is the position of a particle moving along a straight line at time t.*

(a) *Find the velocity and acceleration of the particle.*
(b) *Find all times in the given interval when the particle is stationary.*

64. $s(t) = t^2 - 2t + 6$ for $0 \leq t \leq 2$
65. $s(t) = 3t^2 + 2t - 5$ for $0 \leq t \leq 1$
66. $s(t) = t^3 - 9t^2 + 15t + 25$ for $0 \leq t \leq 6$
67. $s(t) = t^4 - 4t^3 + 8t$ for $0 \leq t \leq 4$

68. MOTION OF A PROJECTILE A stone is dropped from a height of 144 feet.
a. When will the stone hit the ground?
b. With what velocity does it hit the ground?

69. MOTION OF A PROJECTILE You are standing on the top of a building and throw a ball vertically upward. After 2 seconds, the ball passes you on the way down, and 2 seconds after that, it hits the ground below.
a. What is the initial velocity of the ball?
b. How high is the building?
c. What is the velocity of the ball when it passes you on the way down?
d. What is the velocity of the ball as it hits the ground?

70. SPY STORY Our friend, the spy who escaped from the diamond smugglers in Chapter 1 (Problem 46 of Section 1.4), is on a secret mission in space. An encounter with an enemy agent leaves him with a mild concussion and temporary amnesia. Fortunately, he has a book that gives the formula for the motion of a projectile and the values of g for various heavenly bodies (32 ft/sec² on earth, 5.5 ft/sec² on the moon, 12 ft/sec² on Mars, and 28 ft/sec² on Venus). To deduce his whereabouts, he throws a rock vertically upward (from ground level) and notes that it reaches a maximum height of 37.5 ft and hits the ground 5 seconds after it leaves his hand. Where is he?

71. Find numbers a, b, and c such that the graph of the function $f(x) = ax^2 + bx + c$ will have x intercepts at $(0, 0)$ and $(5, 0)$, and a tangent with slope 1 when $x = 2$.

72. Find the equations of all of the tangents to the graph of the function

$$f(x) = x^2 - 4x + 25$$

that pass through the origin $(0, 0)$.

73. Prove the sum rule for derivatives. [*Hint:* Note that the difference quotient for $f + g$ can be written as

$$\frac{(f + g)(x + h) - (f + g)(x)}{h} = \frac{[f(x + h) + g(x + h)] - [f(x) + g(x)]}{h}$$

74. a. If $f(x) = x^4$, show that $\dfrac{f(x + h) - f(x)}{h} = 4x^3 + 6x^2h + 4xh^2 + h^3$

b. If $f(x) = x^n$ for positive integer n, show that

$$\frac{f(x + h) - f(x)}{h} = nx^{n-1} + \frac{n(n - 1)}{2}x^{n-2}h + \cdots + nxh^{n-2} + h^{n-1}$$

c. Use the result in part (b) in the definition of the derivative to prove the power rule:

$$\frac{d}{dx}[x^n] = nx^{n-1}$$

75. POLLUTION CONTROL   It has been suggested that one way to reduce worldwide carbon dioxide (CO_2) emissions is to impose a single tax that would apply to all nations. The accompanying graph shows the relationship between different levels of the carbon tax and the percentage of reduction in CO_2 emissions.

   a. What tax rate would have to be imposed to achieve a worldwide reduction of 50% in CO_2 emissions?

   b. Use the graph to estimate the rate of change of the percentage reduction in CO_2 emissions when the tax rate is $200 per ton.

   c. Read an article on CO_2 emissions and write a paragraph on how public policy can be used to control air pollution.*

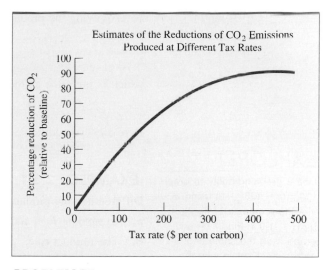

PROBLEM 75   *Source:* Barry C. Field, *Environmental Economics: An Introduction,* New York: McGraw-Hill, 1994, p. 441.

SECTION   2.3 | Product and Quotient Rules; Higher-Order Derivatives

Based on your experience with the multiple and sum rules in Section 2.2, you may think that the derivative of a product of functions is the product of separate derivatives, but it is easy to see that this conjecture is false. For instance, if $f(x) = x^2$ and $g(x) = x^3$, then $f'(x) = 2x$ and $g'(x) = 3x^2$, so

$$f'(x)g'(x) = (2x)(3x^2) = 6x^3$$

while $f(x)g(x) = x^2x^3 = x^5$ and

$$[f(x)g(x)]' = (x^5)' = 5x^4$$

The correct formula for differentiating a product can be stated as follows.

The Product Rule   ■   If $f(x)$ and $g(x)$ are differentiable at x, then so is their product $P(x) = f(x)g(x)$ and

$$\frac{d}{dx}[f(x)g(x)] = f(x)\frac{d}{dx}[g(x)] + g(x)\frac{d}{dx}[f(x)]$$

or equivalently,

$$(fg)' = fg' + gf'$$

In words, the derivative of the product fg is f times the derivative of g plus g times the derivative of f.

---

*You may wish to begin your research by reading Chapter 12, "Incentive-Based Strategies: Emission Taxes and Subsidies," and Chapter 15, "Federal Air Pollution-Control Policy," of Barry C. Field, *Environmental Economics: An Introduction,* New York: McGraw-Hill, 1994.

Applying the product rule to our introductory example, we find that

$$(x^2x^3)' = x^2(x^3)' + (x^3)(x^2)'$$
$$= (x^2)(3x^2) + (x^3)(2x) = 3x^4 + 2x^4 = 5x^4$$

which is the same as the result obtained by direct computation:

$$(x^2x^3)' = (x^5)' = 5x^4$$

Here are two additional examples.

EXAMPLE │ 2.3.1

Differentiate the product $P(x) = (x - 1)(3x - 2)$ by

a. Expanding $P(x)$ and using the polynomial rule.

b. The product rule.

Solution

a. We have $P(x) = 3x^2 - 5x + 2$, so $P'(x) = 6x - 5$.

b. By the product rule

$$P'(x) = (x - 1)\frac{d}{dx}[3x - 2] + (3x - 2)\frac{d}{dx}[x - 1]$$
$$= (x - 1)(3) + (3x - 2)(1) = 6x - 5$$

EXAMPLE │ 2.3.2

For the curve $y = (2x + 1)(2x^2 - x - 1)$:

a. Find y'.

b. Find an equation for the tangent line to the curve at the point where $x = 1$.

c. Find all points on the curve where the tangent line is horizontal.

Solution

a. Using the product rule, we get

$$y' = (2x + 1)[2x^2 - x - 1]' + [2x + 1]'(2x^2 - x - 1)$$
$$= (2x + 1)(4x - 1) + (2)(2x^2 - x - 1)$$

b. When $x = 1$, the corresponding y value is

$$y(1) = [2(1) + 1][2(1)^2 - 1 - 1] = 0$$

so the point of tangency is (1, 0). The slope at $x = 1$ is

$$y'(1) = [2(1) + 1][4(1) - 1] + 2[2(1)^2 - 1 - 1] = 9$$

Substituting into the point-slope formula, we find that the tangent line at (1, 0) has the equation

$$y - 0 = 9(x - 1)$$

or $$y = 9x - 9$$

c. Horizontal tangents occur where the slope is zero; that is, where $y' = 0$. Expanding the expression for the derivative and combining terms, we obtain

$$y' = (2x + 1)(4x - 1) + (2)(2x^2 - x - 1) = 12x^2 - 3$$

Solving $y' = 0$, we find that

$$y' = 12x^2 - 3 = 0$$

$$x^2 = \frac{3}{12} = \frac{1}{4} \qquad \text{add 3 to each side and}$$
$$\text{divide by 12}$$

$$x = \frac{1}{2} \quad \text{and} \quad x = -\frac{1}{2}$$

Substituting $x = \dfrac{1}{2}$ and $x = -\dfrac{1}{2}$ into the formula for y, we get $y\left(\dfrac{1}{2}\right) = -2$ and $y\left(-\dfrac{1}{2}\right) = 0$, so horizontal tangents occur at the points $\left(\dfrac{1}{2}, -2\right)$ and $\left(-\dfrac{1}{2}, 0\right)$ on the curve.

EXAMPLE | 2.3.3

A manufacturer determines that t months after a new product is introduced to the market, $x(t) = t^2 + 3t$ hundred units can be produced and then sold at a price of $p(t) = -2t^{3/2} + 30$ dollars per unit.

a. Express the revenue $R(t)$ for this product as a function of time.

b. At what rate is revenue changing with respect to time after 4 months? Is revenue increasing or decreasing at this time?

Solution

a. The revenue is given by

$$R(t) = x(t)p(t) = (t^2 + 3t)(-2t^{3/2} + 30)$$

hundred dollars.

b. The rate of change of revenue $R(t)$ with respect to time is given by the derivative $R'(t)$, which we find using the product rule:

$$R'(t) = (t^2 + 3t)\frac{d}{dt}[-2t^{3/2} + 30] + (-2t^{3/2} + 30)\frac{d}{dt}[t^2 + 3t]$$

$$= (t^2 + 3t)\left[-2\left(\frac{3}{2}t^{1/2}\right)\right] + (-2t^{3/2} + 30)[2t + 3]$$

At time $t = 4$, the revenue is changing at the rate

$$R'(4) = [(4)^2 + 3(4)][-3(4)^{1/2}] + [-2(4)^{3/2} + 30][2(4) + 3]$$
$$= -14$$

Thus, after 4 months, the revenue is changing at the rate of 14 hundred dollars ($1,400) per month. It is *decreasing* at that time since $R'(4)$ is negative.

A proof of the product rule is given at the end of this section. It is also important to be able to differentiate quotients of functions, and for this purpose, we have the following rule, a proof of which is outlined in Problem 71.

CAUTION: A common error is to assume that $\left(\dfrac{f}{g}\right)' = \dfrac{f'}{g'}$.

The Quotient Rule ▪ If $f(x)$ and $g(x)$ are differentiable functions, then so is the quotient $Q(x) = f(x)/g(x)$ and

$$\frac{d}{dx}\left[\frac{f(x)}{g(x)}\right] = \frac{g(x)\dfrac{d}{dx}[f(x)] - f(x)\dfrac{d}{dx}[g(x)]}{g^2(x)} \quad \text{if } g(x) \neq 0$$

or equivalently,

$$\left(\frac{f}{g}\right)' = \frac{gf' - fg'}{g^2}$$

NOTE The quotient rule is probably the most complicated formula you have had to learn so far in this book. It may help to note that the quotient rule resembles the product rule except it contains a minus sign, which makes the order of terms in the numerator very important. Begin by squaring the denominator g, and then, while still thinking of g, copy it into the numerator. This gets you started with the proper order in the numerator, and you can easily write down the rest while thinking of the product rule. Don't forget to insert the minus sign, without which the quotient rule would not have been so hard to remember in the first place. This whimsical version of the quotient rule may help you remember its form:

$$d\left[\frac{\text{hi}}{\text{ho}}\right] = \frac{\text{ho } d(\text{hi}) - \text{hi } d(\text{ho})}{\text{ho ho}}$$

EXAMPLE 2.3.4

Differentiate the quotient $Q(x) = \dfrac{x^2 - 5x + 7}{2x}$ by

a. Dividing through first.

b. Using the quotient rule.

Just-In-Time Review

Recall that

$$\frac{A + B}{C} = \frac{A}{C} + \frac{B}{C}$$

but that

$$\frac{A}{B + C} \neq \frac{A}{B} + \frac{A}{C}$$

Solution

a. Dividing by the denominator $2x$, we get

$$Q(x) = \frac{1}{2}x - \frac{5}{2} + \frac{7}{2}x^{-1}$$

so

$$Q'(x) = \frac{1}{2} - 0 + \frac{7}{2}(-x^{-2}) = \frac{1}{2} - \frac{7}{2x^2}$$

b. By the quotient rule

$$Q'(x) = \frac{(2x)\dfrac{d}{dx}[x^2 - 5x + 7] - (x^2 - 5x + 7)\dfrac{d}{dx}[2x]}{(2x)^2}$$

$$= \frac{(2x)(2x - 5) - (x^2 - 5x + 7)(2)}{4x^2} = \frac{2x^2 - 14}{4x^2} = \frac{1}{2} - \frac{7}{2x^2}$$

EXAMPLE | 2.3.5

A biologist models the effect of introducing a toxin to a bacterial colony by the function

$$P(t) = \frac{t + 1}{t^2 + t + 4}$$

where P is the population of the colony (in millions) t hours after the toxin is introduced.

a. At what rate is the population changing when the toxin is introduced? Is the population increasing or decreasing at this time?

b. At what time does the population begin to decrease? By how much does the population increase before it begins to decline?

Solution

a. The rate of change of the population with respect to time is given by the derivative $P'(t)$, which we compute using the quotient rule:

$$P'(t) = \frac{(t^2 + t + 4)\dfrac{d}{dt}[t + 1] - (t + 1)\dfrac{d}{dt}[t^2 + t + 4]}{(t^2 + t + 4)^2}$$

$$= \frac{(t^2 + t + 4)(1) - (t + 1)(2t + 1)}{(t^2 + t + 4)^2}$$

$$= \frac{-t^2 - 2t + 3}{(t^2 + t + 4)^2}$$

The toxin is introduced when $t = 0$, and at that time the population is changing at the rate

$$P'(0) = \frac{(0 + 0 + 3)}{(0 + 0 + 4)^2} = \frac{3}{16} = 0.1875$$

That is, the population is initially changing at the rate of 0.1875 million (187,500) bacteria per hour, and it is increasing since $P'(0) > 0$.

b. The population is decreasing when $P'(t) < 0$. Since the numerator of $P'(t)$ can be factored as

$$-t^2 - 2t + 3 = -(t^2 + 2t - 3) = -(t - 1)(t + 3)$$

we can write

$$P'(t) = \frac{-(t - 1)(t + 3)}{(t^2 + t + 4)^2}$$

The denominator $(t^2 + t + 4)^2$ and the factor $t + 3$ are both positive for all $t \geq 0$, which means that

$$\text{for } 0 \leq t < 1 \quad P'(t) > 0 \text{ and } P(t) \text{ is increasing}$$
$$\text{for } t > 1 \quad P'(t) < 0 \text{ and } P(t) \text{ is decreasing}$$

Thus, the population begins to decline after 1 hour.
The initial population of the colony is

$$P(0) = \frac{0 + 1}{0 + 0 + 4} = \frac{1}{4}$$

million, and after 1 hour, the population is

$$P(1) = \frac{1+1}{1+1+4} = \frac{1}{3}$$

million. Therefore, before the population begins to decline, it increases by

$$P(1) - P(0) = \frac{1}{3} - \frac{1}{4} = \frac{1}{12}$$

million; that is, by approximately 83,333 bacteria.

A Word of Advice The quotient rule is somewhat cumbersome, so don't use it unnecessarily. Consider Example 2.3.6.

EXAMPLE | 2.3.6

Differentiate the function $y = \frac{2}{3x^2} - \frac{x}{3} + \frac{4}{5} + \frac{x+1}{x}$.

Solution

Don't use the quotient rule! Instead, rewrite the function as

$$y = \frac{2}{3}x^{-2} - \frac{1}{3}x + \frac{4}{5} + 1 + x^{-1}$$

and then apply the power rule term by term to get

$$\frac{dy}{dx} = \frac{2}{3}(-2x^{-3}) - \frac{1}{3} + 0 + 0 + (-1)x^{-2}$$

$$= -\frac{4}{3}x^{-3} - \frac{1}{3} - x^{-2}$$

$$= -\frac{4}{3x^3} - \frac{1}{3} - \frac{1}{x^2}$$

The Second Derivative In applications, it may be necessary to compute the rate of change of a function that is itself a rate of change. For example, the acceleration of a car is the rate of change with respect to time of its velocity, which in turn is the rate of change with respect to time of its position. If the position is measured in miles and time in hours, the velocity (rate of change of position) is measured in miles per hour, and the acceleration (rate of change of velocity) is measured in miles per hour, per hour.

Statements about the rate of change of a rate of change are used frequently in economics. In inflationary times, for example, you may hear a government economist assure the nation that although inflation is increasing, it is doing so at a decreasing rate. That is, prices are still going up, but not as quickly as they were before.

The rate of change of the function $f(x)$ with respect to x is the derivative $f'(x)$, and likewise, the rate of change of the function $f'(x)$ with respect to x is *its* derivative $(f'(x))'$. This notation is awkward, so we write the derivative of the derivative of $f(x)$ as $(f'(x))' = f''(x)$ and refer to it as the *second derivative* of $f(x)$ (read $f''(x)$ as "f double prime of x"). If $y = f(x)$, then the second derivative of y with respect to x is written as y'' or as $\frac{d^2y}{dx^2}$. Here is a summary of the terminology and notation used for second derivatives.

The Second Derivative ■ The second derivative of a function is the derivative of its derivative. If $y = f(x)$, the second derivative is denoted by

$$\frac{d^2y}{dx^2} \quad \text{or} \quad f''(x)$$

The second derivative gives the rate of change of the rate of change of the original function.

> **NOTE** The derivative $f'(x)$ is sometimes called the **first derivative** to distinguish it from the **second derivative** $f''(x)$. ■

You don't have to use any new rules to find the second derivative of a function. Just find the first derivative and then differentiate again.

EXPLORE!

A graphing calculator can be used to create and display the graph of a higher-order derivative. Put into Y1 the function

$f(x) = 5x^4 - 3x^2 - 3x + 7$

Then write

Y2 = nDeriv(Y1, X, X)

and

Y3 = nDeriv(Y2, X, X)

changing the graphing style of Y3 to bold. Graph these three functions using the window $[-1, 1]1$ by $[-10, 10]1$.

EXAMPLE   2.3.7

Find the second derivative of the function $f(x) = 5x^4 - 3x^2 - 3x + 7$.

Solution

Compute the first derivative

$$f'(x) = 20x^3 - 6x - 3$$

and then differentiate again to get

$$f''(x) = 60x^2 - 6$$

EXAMPLE   2.3.8

Find the second derivative of $y = x^2(3x + 1)$.

Solution

According to the product rule,

$$\frac{d}{dx}[x^2(3x + 1)] = x^2 \frac{d}{dx}[3x + 1] + (3x + 1)\frac{d}{dx}[x^2]$$
$$= x^2(3) + (3x + 1)(2x)$$
$$= 9x^2 + 2x$$

Therefore, the second derivative is

$$\frac{d^2y}{dx^2} = \frac{d}{dx}[9x^2 + 2x]$$
$$= 18x + 2$$

> **NOTE** Before computing the second derivative of a function, always take time to simplify the first derivative as much as possible. The more complicated the form of the first derivative, the more tedious the computation of the second derivative.

The second derivative will be used in Section 3.2 to obtain information about the shape of a graph, and in Sections 3.4 and 3.5 in optimization problems. Here is a more elementary application illustrating the interpretation of the second derivative as the rate of change of a rate of change.

EXAMPLE | 2.3.9

An efficiency study of the morning shift at a certain factory indicates that an average worker who arrives on the job at 8:00 A.M. will have produced

$$Q(t) = -t^3 + 6t^2 + 24t$$

units t hours later.

a. Compute the worker's rate of production at 11:00 A.M.

b. At what rate is the worker's rate of production changing with respect to time at 11:00 A.M.?

Solution

a. The worker's rate of production is the first derivative

$$R(t) = Q'(t) = -3t^2 + 12t + 24$$

of the output $Q(t)$. The rate of production at 11:00 A.M. ($t = 3$) is

$$R(3) = Q'(3) = -3(3)^2 + 12(3) + 24 = 33$$
$$= 33 \text{ units per hour}$$

b. The rate of change of the rate of production is the second derivative

$$R'(t) = Q''(t) = -6t + 12$$

of the output function. At 11:00 A.M., this rate is

$$R'(3) = Q''(3) = -6(3) + 12$$
$$= -6 \text{ units per hour per hour}$$

The minus sign indicates that the worker's rate of production is decreasing; that is, the worker is slowing down. The rate of this decrease in efficiency at 11:00 A.M. is 6 units per hour per hour.

Recall from Section 2.2 that the **acceleration** $a(t)$ of an object moving along a straight line is the derivative of the velocity $v(t)$, which in turn is the derivative of the position function $s(t)$. Thus, the acceleration may be thought of as the second derivative of position; that is,

$$a(t) = \frac{d^2s}{dt^2}$$

This notation is used in Example 2.3.10.

EXPLORE!

Change t to x in $s(t)$, $v(t)$, and $a(t)$ in Example 2.3.10. Use a graphing utility to graph $v(x)$ and $a(x)$ on the same coordinate axes using a viewing rectangle of [0, 2]0.1 by [−5, 5]0.5. Explain what is happening to $v(x)$ when $a(x)$ is zero. Then use your calculator to see what effect changing $s(t)$ to $s_1(t) = 2t^3 - 3t^2 + 4t$ has on $v(t)$ and $a(t)$.

EXAMPLE | 2.3.10

If the position of an object moving along a straight line is given by $s(t) = t^3 - 3t^2 + 4t$ at time t, find its velocity and acceleration.

Solution

The velocity of the object is

$$v(t) = \frac{ds}{dt} = 3t^2 - 6t + 4$$

and its acceleration is

$$a(t) = \frac{dv}{dt} = \frac{d^2s}{dt^2} = 6t - 6$$

Higher-Order Derivatives

If you differentiate the second derivative $f''(x)$ of a function $f(x)$ one more time, you get the third derivative $f'''(x)$. Differentiate again and you get the fourth derivative, which is denoted by $f^{(4)}(x)$ since the prime notation $f''''(x)$ is cumbersome. In general, the derivative obtained from $f(x)$ after n successive differentiations is called the **nth derivative** or **derivative of order n** and is denoted by $f^{(n)}(x)$.

The nth Derivative ■ For any positive integer n, the nth derivative of a function is obtained from the function by differentiating successively n times. If the original function is $y = f(x)$, the nth derivative is denoted by

$$\frac{d^n y}{dx^n} \quad \text{or} \quad f^{(n)}(x)$$

EXAMPLE 2.3.11

Find the fifth derivative of each of these functions:

a. $f(x) = 4x^3 + 5x^2 + 6x - 1$

b. $y = \dfrac{1}{x}$

Solution

a. $f'(x) = 12x^2 + 10x + 6$

   $f''(x) = 24x + 10$

   $f'''(x) = 24$

   $f^{(4)}(x) = 0$

   $f^{(5)}(x) = 0$

b. $\dfrac{dy}{dx} = \dfrac{d}{dx}(x^{-1}) = -x^{-2} = -\dfrac{1}{x^2}$

   $\dfrac{d^2 y}{dx^2} = \dfrac{d}{dx}(-x^{-2}) = 2x^{-3} = \dfrac{2}{x^3}$

   $\dfrac{d^3 y}{dx^3} = \dfrac{d}{dx}(2x^{-3}) = -6x^{-4} = -\dfrac{6}{x^4}$

   $\dfrac{d^4 y}{dx^4} = \dfrac{d}{dx}(-6x^{-4}) = 24x^{-5} = \dfrac{24}{x^5}$

   $\dfrac{d^5 y}{dx^5} = \dfrac{d}{dx}(24x^{-5}) = -120x^{-6} = -\dfrac{120}{x^6}$

Derivation of the Product Rule

The product and quotient rules are not easy to prove. In both cases, the key is to express the difference quotient of the given expression (the product fg or quotient f/g) in terms of difference quotients of f and g. Here is a proof of the product rule. The proof for the quotient rule is outlined in Problem 71.

To show that $\dfrac{d}{dx}(fg) = f\dfrac{dg}{dx} + g\dfrac{df}{dx}$, begin with the appropriate difference quotient and rewrite the numerator by subtracting and adding the quantity $f(x + h)g(x)$ as follows:

$$\frac{d}{dx}(fg) = \lim_{h \to 0} \frac{f(x + h)g(x + h) - f(x)g(x)}{h}$$

$$= \lim_{h \to 0} \left[\frac{f(x + h)g(x + h) - f(x + h)g(x)}{h} + \frac{f(x + h)g(x) - f(x)g(x)}{h} \right]$$

$$= \lim_{h \to 0} \left(f(x + h)\left[\frac{g(x + h) - g(x)}{h} \right] + g(x)\left[\frac{f(x + h) - f(x)}{h} \right] \right)$$

Now let h approach zero. Since

$$\lim_{h \to 0} \frac{f(x + h) - f(x)}{h} = \frac{df}{dx}$$

$$\lim_{h \to 0} \frac{g(x + h) - g(x)}{h} = \frac{dg}{dx}$$

and

$$\lim_{h \to 0} f(x + h) = f(x) \qquad \textit{continuity of } f(x)$$

it follows that

$$\frac{d}{dx}(fg) = f\frac{dg}{dx} + g\frac{df}{dx}$$

PROBLEMS 2.3

In Problems 1 through 20, differentiate the given function.

1. $f(x) = (2x + 1)(3x - 2)$

2. $f(x) = (x - 5)(1 - 2x)$

3. $y = 10(3u + 1)(1 - 5u)$

4. $y = 400(15 - x^2)(3x - 2)$

5. $f(x) = \dfrac{1}{3}(x^5 - 2x^3 + 1)\left(x - \dfrac{1}{x}\right)$

6. $f(x) = -3(5x^3 - 2x + 5)(\sqrt{x} + 2x)$

7. $y = \dfrac{x + 1}{x - 2}$

8. $y = \dfrac{2x - 3}{5x + 4}$

9. $f(t) = \dfrac{t}{t^2 - 2}$

10. $f(x) = \dfrac{1}{x - 2}$

11. $y = \dfrac{3}{x + 5}$

12. $y = \dfrac{t^2 + 1}{1 - t^2}$

13. $f(x) = \dfrac{x^2 - 3x + 2}{2x^2 + 5x - 1}$

14. $f(t) = \dfrac{t^2 + 2t + 1}{t^2 + 3t - 1}$

15. $f(x) = \dfrac{(2x - 1)(x + 3)}{x + 1}$

16. $g(x) = \dfrac{(x^2 + x + 1)(4 - x)}{2x - 1}$

17. $f(x) = (2 + 5x)^2$

18. $f(x) = \left(x + \dfrac{1}{x}\right)^2$

19. $g(t) = \dfrac{t^2 + \sqrt{t}}{2t + 5}$

20. $h(x) = \dfrac{x}{x^2 - 1} + \dfrac{4 - x}{x^2 + 1}$

In Problems 21 through 25, find an equation for the tangent line to the given curve at the point where $x = x_0$.

21. $y = (5x - 1)(4 + 3x);\ x_0 = 0$

22. $y = (x^2 + 3x - 1)(2 - x);\ x_0 = 1$

23. $y = \dfrac{x}{2x + 3};\ x_0 = -1$

24. $y = \dfrac{x + 7}{5 - 2x};\ x_0 = 0$

25. $y = (3\sqrt{x} + x)(2 - x^2),\ x_0 = 1$

26. $y = \dfrac{2x - 1}{1 - x^3};\ x_0 = 0$

In Problems 27 through 31, find all points on the graph of the given function where the tangent line is horizontal.

27. $f(x) = (x + 1)(x^2 - x - 2)$

28. $f(x) = (x - 1)(x^2 - 8x + 7)$

29. $f(x) = \dfrac{x + 1}{x^2 + x + 1}$

30. $f(x) = \dfrac{x^2 + x - 1}{x^2 - x + 1}$

31. $f(x) = x^3(x - 5)^2$

In Problems 32 through 35, find the rate of change $\dfrac{dy}{dx}$ for the prescribed value of x_0.

32. $y = (x^2 + 2)(x + \sqrt{x});\ x_0 = 4$

33. $y = (x^2 + 3)(5 - 2x^3);\ x_0 = 1$

34. $y = \dfrac{2x - 1}{3x + 5};\ x_0 = 1$

35. $y = x + \dfrac{3}{2 - 4x};\ x_0 = 0$

The normal line to the curve $y = f(x)$ at the point $P(x_0, f(x_0))$ is the line perpendicular to the tangent line at P. In Problems 36 through 39, find an equation for the normal line to the given curve at the prescribed point.

36. $y = x^2 + 3x - 5;\ (0, -5)$

37. $y = \dfrac{2}{x} - \sqrt{x};\ (1, 1)$

38. $y = (x + 3)(1 - \sqrt{x});\ (1, 0)$

39. $y = \dfrac{5x + 7}{2 - 3x};\ (1, -12)$

40. a. Differentiate the function $y = 2x^2 - 5x - 3$.
   b. Now factor the function in part (a) as $y = (2x + 1)(x - 3)$ and differentiate using the product rule. Show that the two answers are the same.

41. a. Use the quotient rule to differentiate the function $y = \dfrac{2x - 3}{x^3}$.
   b. Rewrite the function as $y = x^{-3}(2x - 3)$ and differentiate using the product rule.
   c. Rewrite the function as $y = 2x^{-2} - 3x^{-3}$ and differentiate.
   d. Show that your answers to parts (a), (b), and (c) are the same.

In Problems 42 through 47, find the second derivative of the given function. In each case, use the appropriate notation for the second derivative and simplify your answer. (Don't forget to simplify the first derivative as much as possible before computing the second derivative.)

42. $f(x) = 5x^{10} - 6x^5 - 27x + 4$

43. $f(x) = \dfrac{2}{5}x^5 - 4x^3 + 9x^2 - 6x - 2$

44. $y = 5\sqrt{x} + \dfrac{3}{x^2} + \dfrac{1}{3\sqrt{x}} + \dfrac{1}{2}$

45. $y = \dfrac{2}{3x} - \sqrt{2x} + \sqrt{2}x - \dfrac{1}{6\sqrt{x}}$

46. $y = (x^2 - x)\left(2x - \dfrac{1}{x}\right)$

47. $y = (x^3 + 2x - 1)(3x + 5)$

48. DEMAND AND REVENUE The manager of a company that produces graphing calculators determines that when x thousand calculators are produced, they will all be sold when the price is

$$p(x) = \frac{1,000}{0.3x^2 + 8}$$

dollars per calculator.
a. At what rate is demand $p(x)$ changing with respect to the level of production x when 3,000 ($x = 3$) calculators are produced?
b. The revenue derived from the sale of x thousand calculators is $R(x) = xp(x)$ thousand dollars. At what rate is revenue changing when 3,000 calculators are produced? Is revenue increasing or decreasing at this level of production?

49. SALES The manager of the Many Facets jewelry store models total sales by the function

$$S(t) = \frac{2,000t}{4 + 0.3t}$$

where t is the time (years) since the year 2000 and S is measured in thousands of dollars.
a. At what rate are sales changing in the year 2002?
b. What happens to sales in the "long run" (that is, as $t \to +\infty$)?

50. PROFIT Bea Johnson, the owner of the Bea Nice boutique, estimates that when a particular kind of perfume is priced at p dollars per bottle, she will sell

$$B(p) = \frac{500}{p + 3} \qquad p \geq 5$$

bottles per month at a total cost of

$$C(p) = 0.2p^2 + 3p + 200 \qquad \text{dollars}$$

a. Express Bea's profit $P(p)$ as a function of the price p per bottle.
b. At what rate is the profit changing with respect to p when the price is $12 per bottle? Is profit increasing or decreasing at that price?

51. ADVERTISING A company manufactures a "thin" DVD burner kit that can be plugged into personal computers. The marketing manager determines that t weeks after an advertising campaign begins, $P(t)$ percent of the potential market is aware of the burners, where

$$P(t) = 100\left[\frac{t^2 + 5t + 5}{t^2 + 10t + 30}\right]$$

a. At what rate is the market percentage $P(t)$ changing with respect to time after 5 weeks? Is the percentage increasing or decreasing at this time?
b. What happens to the percentage $P(t)$ in the "long run"; that is, as $t \to +\infty$? What happens to the rate of change of $P(t)$ as $t \to +\infty$?

52. BACTERIAL POPULATION A bacterial colony is estimated to have a population of

$$P(t) = \frac{24t + 10}{t^2 + 1}$$

million t hours after the introduction of a toxin.
a. At what rate is the population changing 1 hour after the toxin is introduced ($t = 1$)? Is the population increasing or decreasing at this time?
b. At what time does the population begin to decline?

53. POLLUTION CONTROL A large city commissions a study that indicates that spending money on pollution control is effective up to a point but eventually becomes wasteful. Suppose it is known that when x million dollars is spent on controlling pollution, the percentage of pollution removed is given by

$$P(x) = \frac{100\sqrt{x}}{0.03x^2 + 9}$$

a. At what rate is the percentage of pollution removal $P(x)$ changing when 16 million dollars are spent? Is the percentage increasing or decreasing at this level of expenditure?
b. For what values of x is $P(x)$ increasing? For what values of x is $P(x)$ decreasing?

54. PHARMACOLOGY An oral painkiller is administered to a patient, and t hours later, the concentration of drug in the patient's bloodstream is given by

$$C(t) = \frac{2t}{3t^2 + 16}$$

a. At what rate $R(t)$ is the concentration of drug in the patient's bloodstream changing t hours after being administered? At what rate is $R(t)$ changing at time t?
b. At what rate is the concentration of drug changing after 1 hour? Is the concentration changing at an increasing or decreasing rate at this time?

c. When does the concentration of the drug begin to decline?

d. Over what time period is the concentration changing at a declining rate?

55. WORKER EFFICIENCY An efficiency study of the morning shift at a certain factory indicates that an average worker arriving on the job at 8:00 A.M. will have produced $Q(t) = -t^3 + 8t^2 + 15t$ units t hours later.

a. Compute the worker's rate of production $R(t) = Q'(t)$.

b. At what rate is the worker's rate of production changing with respect to time at 9:00 A.M.?

56. POPULATION GROWTH It is estimated that t years from now, the population of a certain suburban community will be $P(t) = 20 - \dfrac{6}{t+1}$ thousand.

a. Derive a formula for the rate at which the population will be changing with respect to time t years from now.

b. At what rate will the population be growing 1 year from now?

c. By how much will the population actually increase during the second year?

d. At what rate will the population be growing 9 years from now?

e. What will happen to the rate of population growth in the long run?

In Problems 57 through 60, the position s(t) of an object moving along a straight line is given. In each case:

(a) Find the object's velocity v(t) and acceleration a(t).

(b) Find all times t when the acceleration is 0.

57. $s(t) = 3t^5 - 5t^3 - 7$

58. $s(t) = 2t^4 - 5t^3 + t - 3$

59. $s(t) = -t^3 + 7t^2 + t + 2$

60. $s(t) = 4t^{5/2} - 15t^2 + t - 3$

61. VELOCITY An object moves along a straight line so that after t minutes, its distance from its starting point is $D(t) = 10t + \dfrac{5}{t+1} - 5$ meters.

a. At what velocity is the object moving at the end of 4 minutes?

b. How far does the object actually travel during the fifth minute?

62. ACCELERATION After t hours of an 8-hour trip, a car has gone $D(t) = 64t + \dfrac{10}{3}t^2 - \dfrac{2}{9}t^3$ kilometers.

a. Derive a formula expressing the acceleration of the car as a function of time.

b. At what rate is the velocity of the car changing with respect to time at the end of 6 hours? Is the velocity increasing or decreasing at this time?

c. By how much does the velocity of the car actually change during the seventh hour?

63. DRUG DOSAGE One biological model* suggests that the human body's reaction to a dose of medicine can be measured by a function of the form

$$F = \frac{1}{3}(KM^2 - M^3)$$

where K is a positive constant and M is the amount of medicine absorbed in the blood. The derivative $S = \dfrac{dF}{dM}$ can be thought of as a measure of the sensitivity of the body to the medicine.

a. Find the sensitivity S.

b. Find $\dfrac{dS}{dM} = \dfrac{d^2F}{dM^2}$ and give an interpretation of the second derivative.

64. BLOOD CELL PRODUCTION A biological model† measures the production of a certain type of white blood cell (*granulocytes*) by the function

$$p(x) = \frac{Ax}{B + x^m}$$

where A and B are positive constants, the exponent m is positive, and x is the number of cells present.

a. Find the rate of production $p'(x)$.

b. Find $p''(x)$ and determine all values of x for which $p''(x) = 0$ (your answer will involve m).

---

*Thrall et al., *Some Mathematical Models in Biology*, U.S. Dept. of Commerce, 1967.

†M. C. Mackey and L. Glass, "Oscillations and Chaos in Physiological Control Systems," *Science*, Vol. 197, pp. 287–289.

c. Read an article on blood cell production and write a paragraph on how mathematical methods can be used to model such production. A good place to start is with the article, "Blood Cell Population Model, Dynamical Diseases, and Chaos" by W. B. Gearhart and M. Martelli, UMAP Module 1990, Arlington, MA: Consortium for Mathematics and Its Applications, Inc., 1991.

65. **ACCELERATION** If an object is dropped or thrown vertically, its height (in feet) after t seconds is $H(t) = -16t^2 + S_0 t + H_0$, where S_0 is the initial speed of the object and H_0 its initial height.
   a. Derive an expression for the acceleration of the object.
   b. How does the acceleration vary with time?
   c. What is the significance of the fact that the answer to part (a) is negative?

66. Find $f^{(4)}(x)$ if $f(x) = x^5 - 2x^4 + x^3 - 3x^2 + 5x - 6$.

67. Find $\dfrac{d^3 y}{dx^3}$ if $y = \sqrt{x} - \dfrac{1}{2x} + \dfrac{x}{\sqrt{2}}$.

68. **a.** Show that
$$\frac{d}{dx}[fgh] = fg\frac{dh}{dx} + fh\frac{dg}{dx} + gh\frac{df}{dx}$$
   [*Hint:* Apply the product rule twice.]
   b. Find $\dfrac{dy}{dx}$ where $y = (2x + 1)(x - 3)(1 - 4x)$.

69. **a.** By combining the product rule and the quotient rule, find an expression for $\dfrac{d}{dx}\left[\dfrac{fg}{h}\right]$.
   b. Find $\dfrac{dy}{dx}$, where $y = \dfrac{(2x + 7)(x^2 + 3)}{3x + 5}$.

70. The product rule tells you how to differentiate the product of any two functions, while the constant multiple rule tells you how to differentiate products in which one of the factors is constant. Show that the two rules are consistent. In particular, use the product rule to show that $\dfrac{d}{dx}[cf] = c\dfrac{df}{dx}$ if c is a constant.

71. Derive the quotient rule. [*Hint:* Show that the difference quotient for f/g is
$$\frac{1}{h}\left[\frac{f(x+h)}{g(x+h)} - \frac{f(x)}{g(x)}\right] = \frac{g(x)f(x+h) - f(x)g(x+h)}{g(x+h)g(x)h}$$
Before letting h approach zero, rewrite this quotient using the trick of subtracting and adding $g(x)f(x)$ in the numerator.]

72. Prove the power rule $\dfrac{d}{dx}[x^n] = nx^{n-1}$ for the case where $n = -p$ is a negative integer. [*Hint:* Apply the quotient rule to $y = x^{-p} = \dfrac{1}{x^p}$.]

73. Use a graphing utility to sketch the curve $f(x) = x^2(x - 1)$, and on the same set of coordinate axes, draw the tangent line to the graph of $f(x)$ at $x = 1$. Use **TRACE** and **ZOOM** to find where $f'(x) = 0$.

74. Use a graphing utility to sketch the curve $f(x) = \dfrac{3x^2 - 4x + 1}{x + 1}$, and on the same set of coordinate axes, draw the tangent lines to the graph of $f(x)$ at $x = -2$ and at $x = 0$. Use **TRACE** and **ZOOM** to find where $f'(x) = 0$.

75. Graph $f(x) = x^4 + 2x^3 - x + 1$ using a graphing utility with a viewing rectangle of $[-5, 5]1$ by $[0, 2]0.5$. Use **TRACE** and **ZOOM,** or other graphing utility methods, to find the minima and maxima of this function. Find the derivative function $f'(x)$ algebraically and graph $f(x)$ and $f'(x)$ on the same axes using a viewing rectangle of $[-5, 5]1$ by $[-2, 2]0.5$. Use **TRACE** and **ZOOM** to find the x intercepts of $f'(x)$. Explain why the maximum or minimum of $f(x)$ occurs at the x intercepts of $f'(x)$.

76. Repeat Problem 75 for the product function $f(x) = x^3(x - 3)^2$.

SECTION 2.4 | The Chain Rule

In many practical situations, the rate at which one quantity is changing can be expressed as the product of other rates. For example, suppose a car is traveling at 50 miles/hour at a particular time when gasoline is being consumed at the rate of 0.1 gal/mile. Then, to find out how much gasoline is being used each hour, you would multiply the rates:

$$(0.1 \text{ gal/mile})(50 \text{ miles/hour}) = 5 \text{ gal/hour}$$

Or, suppose the total manufacturing cost at a certain factory is a function of the number of units produced, which in turn is a function of the number of hours the factory has been operating. If C, q, and t denote the cost, units produced, and time, respectively, then

$$\frac{dC}{dq} = \left[\begin{array}{c} \text{rate of change of cost} \\ \text{with respect to output} \end{array} \right] \quad \text{(dollars per unit)}$$

and

$$\frac{dq}{dt} = \left[\begin{array}{c} \text{rate of change of output} \\ \text{with respect to time} \end{array} \right] \quad \text{(units per hour)}$$

The product of these two rates is the rate of change of cost with respect to time; that is,

$$\frac{dC}{dt} = \frac{dC}{dq}\frac{dq}{dt} \quad \text{(dollars per hour)}$$

This formula is a special case of an important result in calculus called the **chain rule.**

The Chain Rule ■ If $y = f(u)$ is a differentiable function of u and $u = g(x)$ is in turn a differentiable function of x, then the composite function $y = f(g(x))$ is a differentiable function of x whose derivative is given by the product

$$\frac{dy}{dx} = \frac{dy}{du}\frac{du}{dx}$$

or, equivalently, by

$$\frac{dy}{dx} = f'(g(x))g'(x)$$

NOTE One way to remember the chain rule is to pretend the derivatives $\frac{dy}{du}$ and $\frac{du}{dx}$ are quotients and to "cancel" du; that is,

$$\frac{dy}{dx} = \frac{dy}{d\!\!\!/u}\frac{d\!\!\!/u}{dx} \quad ■$$

To illustrate the use of the chain rule, suppose you wish to differentiate the function $y = (3x + 1)^2$. Your first instinct may be to "guess" that the derivative is

$$\frac{dy}{dx} = \frac{d}{dx}[(3x + 1)^2] = 2(3x + 1)$$

$$= 6x + 2$$

But this guess cannot be correct since expanding $(3x + 1)^2$ and differentiating each term yields

$$\frac{dy}{dx} = \frac{d}{dx}[(3x + 1)^2] = \frac{d}{dx}[9x^2 + 6x + 1] = 18x + 6$$

which is 3 times our "guess" of $6x + 2$. However, if you write $y = (3x + 1)^2$ as $y = u^2$ where $u = 3x + 1$, then

$$\frac{dy}{du} = \frac{d}{du}[u^2] = 2u \qquad \text{and} \qquad \frac{du}{dx} = \frac{d}{dx}[3x + 1] = 3$$

and the chain rule tells you that

$$\frac{dy}{dx} = \frac{dy}{du}\frac{du}{dx} = (2u)(3)$$
$$= 6(3x + 1) = 18x + 6$$

which coincides with the correct answer found earlier by expanding $(3x + 1)^2$. Examples 2.4.1 and 2.4.2 illustrate various ways of using the chain rule.

EXAMPLE | 2.4.1

Find $\dfrac{dy}{dx}$ if $y = (x^2 + 2)^3 - 3(x^2 + 2)^2 + 1$.

Solution

Note that $y = u^3 - 3u^2 + 1$, where $u = x^2 + 2$. Thus,

$$\frac{dy}{du} = 3u^2 - 6u \qquad \text{and} \qquad \frac{du}{dx} = 2x$$

and according to the chain rule,

$$\frac{dy}{dx} = \frac{dy}{du}\frac{du}{dx} = (3u^2 - 6u)(2x)$$

In Sections 2.1 through 2.3, you have seen several applications (e.g., slope, rates of change) that require the evaluation of a derivative at a particular value of the independent variable. There are two basic ways of doing this when the derivative is computed using the chain rule.

For instance, suppose in Example 2.4.1 we wish to evaluate $\dfrac{dy}{dx}$ when $x = -1$. One way to proceed would be to first express the derivative in terms of x alone by substituting $x^2 + 2$ for u as follows:

$$\frac{dy}{dx} = (3u^2 - 6u)(2x) = [3(x^2 + 2)^2 - 6(x^2 + 2)](2x) \quad \textit{replace u with } x^2 + 2$$
$$= 6x(x^2 + 2)[(x^2 + 2) - 2] \qquad \textit{factor out } 6x(x^2 + 2)$$
$$= 6x(x^2 + 2)(x^2) \qquad \qquad \textit{combine terms in the brackets}$$
$$= 6x^3(x^2 + 2)$$

Then, substituting $x = -1$ into this expression, we would get

$$\left.\frac{dy}{dx}\right|_{x=-1} = 6(-1)^3[(-1)^2 + 2] = -18$$

Alternatively, we could compute $u(-1) = (-1)^2 + 2 = 3$ and then substitute directly into the formula $\dfrac{dy}{dx} = (3u^2 - 6u)(2x)$ to obtain

$$\left.\frac{dy}{dx}\right|_{x=-1} = (3u^2 - 6u)(2x)\Big|_{\substack{x=-1 \\ u=3}}$$

$$= [3(3)^2 - 6(3)][2(-1)] = (9)(-2) = -18$$

Both methods yield the correct result, but since it is easier to substitute numbers than algebraic expressions, the second (numerical) method is often preferable, unless for some reason you need to have the derivative function $\dfrac{dy}{dx}$ expressed in terms of x alone. In Example 2.4.2, the numerical method for evaluating a derivative computed with the chain rule is used to find the slope of a tangent line.

EXAMPLE | 2.4.2

Consider the function $y = \dfrac{u}{u+1}$, where $u = 3x^2 - 1$.

a. Use the chain rule to find $\dfrac{dy}{dx}$.

b. Find an equation for the tangent line to the graph of $y(x)$ at the point where $x = 1$.

Solution

a. We find that

$$\frac{dy}{du} = \frac{(u+1)(1) - u(1)}{(u+1)^2} = \frac{1}{(u+1)^2} \qquad \textit{quotient rule}$$

and

$$\frac{du}{dx} = 6x$$

According to the chain rule, it follows that

$$\frac{dy}{dx} = \frac{dy}{du}\frac{du}{dx} = \left[\frac{1}{(u+1)^2}\right](6x) = \frac{6x}{(u+1)^2}$$

b. To find an equation for the tangent line to the graph of $y(x)$ at $x = 1$, we need to know the value of y and the slope at the point of tangency. Since

$$u(1) = 3(1)^2 - 1 = 2$$

the value of y when $x = 1$ is

$$y(1) = \frac{(2)}{(2)+1} = \frac{2}{3} \qquad \textit{substitute } u(1) = 2 \textit{ for } u$$

and the slope is

$$\left.\frac{dy}{dx}\right|_{\substack{x=1 \\ u=2}} = \frac{6(1)}{(2+1)^2} = \frac{6}{9} = \frac{2}{3} \qquad \textit{substitute } x = 1 \textit{ and } u(1) = 2 \textit{ for } u$$

Therefore, by applying the point-slope formula for the equation of a line, we find that the tangent line to the graph of $y(x)$ at the point where $x = 1$ has the equation

$$\frac{y - 2/3}{x - 1} = \frac{2}{3}$$

or, equivalently, $y = \frac{2}{3}x$.

In many practical problems, a quantity is given as a function of one variable, which, in turn, can be written as a function of a second variable, and the goal is to find the rate of change of the original quantity with respect to the second variable. Such problems can be solved by means of the chain rule. Here is an example.

EXAMPLE 2.4.3

The cost of producing x units of a particular commodity is $C(x) = \frac{1}{3}x^2 + 4x + 53$ dollars, and the production level t hours into a particular production run is $x(t) = 0.2t^2 + 0.03t$ units. At what rate is cost changing with respect to time after 4 hours?

Solution

We find that

$$\frac{dC}{dx} = \frac{2}{3}x + 4 \qquad \text{and} \qquad \frac{dx}{dt} = 0.4t + 0.03$$

so according to the chain rule,

$$\frac{dC}{dt} = \frac{dC}{dx}\frac{dx}{dt} = \left(\frac{2}{3}x + 4\right)(0.4t + 0.03)$$

When $t = 4$, the level of production is

$$x(4) = 0.2(4)^2 + 0.03(4) = 3.32 \text{ units}$$

and by substituting $t = 4$ and $x = 3.32$ into the formula for $\frac{dC}{dt}$, we get

$$\frac{dC}{dt}\bigg|_{t=4} = \left[\frac{2}{3}(3.32) + 4\right][0.4(4) + 0.03] = 10.1277$$

Thus, after 4 hours, cost is increasing at the rate of approximately \$10.13 per hour.

Sometimes when dealing with a composite function $y = f(g(x))$ it may help to think of f as the "outer" function and g as the "inner" function, as indicated here:

$$\textit{"outer" function} \longrightarrow$$
$$y = f(g(x))$$
$$\longleftarrow \textit{"inner" function}$$

Then the chain rule

$$\frac{dy}{dx} = f'(g(x))g'(x)$$

says that *the derivative of $y = f(g(x))$ with respect to x is given by the derivative of the outer function evaluated at the inner function times the derivative of the inner function.* In Example 2.4.4, we emphasize this interpretation by using a "box" ($\square$) to indicate the location and role of the inner function in computing a derivative with the chain rule.

EXAMPLE | 2.4.4

Differentiate the function $f(x) = \sqrt{x^2 + 3x + 2}$.

Solution

The form of the function is

$$f(x) = (\boxed{})^{1/2}$$

where the box $\square$ contains the expression $x^2 + 3x + 2$. Then

$$(\square)' = (x^2 + 3x + 2)' = 2x + 3$$

and according to the chain rule, the derivative of the composite function $f(x)$ is

$$f'(x) = \frac{1}{2}(\square)^{-1/2}(\square)'$$

$$= \frac{1}{2}(\square)^{-1/2}(2x + 3)$$

$$= \frac{1}{2}(x^2 + 3x + 2)^{-1/2}(2x + 3) = \frac{2x + 3}{2\sqrt{x^2 + 3x + 2}}$$

The General Power Rule

In Section 2.2, you learned the rule

$$\frac{d}{dx}[x^n] = nx^{n-1}$$

for differentiating power functions. By combining this rule with the chain rule, you obtain the following rule for differentiating functions of the general form $[h(x)]^n$.

The General Power Rule ■ For any real number n and differentiable function h,

$$\frac{d}{dx}[h(x)]^n = n[h(x)]^{n-1}\frac{d}{dx}[h(x)]$$

To derive the general power rule, think of $[h(x)]^n$ as the composite function

$$[h(x)]^n = g[h(x)] \quad \text{where } g(u) = u^n$$

Then,

$$g'(u) = nu^{n-1} \quad \text{and} \quad h'(x) = \frac{d}{dx}[h(x)]$$

and, by the chain rule,

$$\frac{d}{dx}[h(x)]^n = \frac{d}{dx}g[h(x)] = g'[h(x)]h'(x) = n[h(x)]^{n-1}\frac{d}{dx}[h(x)]$$

The use of the general power rule is illustrated in Examples 2.4.5 through 2.4.7.

EXAMPLE | 2.4.5

Differentiate the function $f(x) = (2x^4 - x)^3$.

Solution

One way to solve this problem is to expand the function and rewrite it as

$$f(x) = 8x^{12} - 12x^9 + 6x^6 - x^3$$

and then differentiate this polynomial term by term to get

$$f'(x) = 96x^{11} - 108x^8 + 36x^5 - 3x^2$$

But see how much easier it is to use the general power rule. According to this rule,

$$f'(x) = \left[3(2x^4 - x)^2\right]\frac{d}{dx}[2x^4 - x] = 3(2x^4 - x)^2(8x^3 - 1)$$

Not only is this method easier, but the answer even comes out in factored form!

In Example 2.4.6, the solution to Example 2.4.4 is written more compactly with the aid of the general power rule.

EXAMPLE | 2.4.6

Differentiate the function $f(x) = \sqrt{x^2 + 3x + 2}$.

Solution

Rewrite the function as $f(x) = (x^2 + 3x + 2)^{1/2}$ and apply the general power rule:

$$f'(x) = \left[\frac{1}{2}(x^2 + 3x + 2)^{-1/2}\right]\frac{d}{dx}[x^2 + 3x + 2]$$

$$= \frac{1}{2}(x^2 + 3x + 2)^{-1/2}(2x + 3)$$

$$= \frac{2x + 3}{2\sqrt{x^2 + 3x + 2}}$$

EXAMPLE | 2.4.7

Differentiate the function $f(x) = \dfrac{1}{(2x + 3)^5}$.

Solution

Although you can use the quotient rule, it is easier to rewrite the function as

$$f(x) = (2x + 3)^{-5}$$

and apply the general power rule to get

$$f'(x) = \left[-5(2x + 3)^{-6}\right]\frac{d}{dx}[2x + 3] = -5(2x + 3)^{-6}(2) = -\frac{10}{(2x + 3)^6}$$

The chain rule is often used in combination with the other rules you learned in Sections 2.2 and 2.3. Example 2.4.8 involves the product rule.

EXAMPLE 2.4.8

Differentiate the function $f(x) = (3x + 1)^4(2x - 1)^5$ and simplify your answer. Then find all values of $x = c$ for which the tangent line to the graph of $f(x)$ at $(c, f(c))$ is horizontal.

Just-In-Time Review

If $m > n$ and $q > p$, then

$rA^m B^p + sA^n D^q$
$= A^n B^p(rA^{m-n} + sB^{q-p})$

for any constants r and s.

Solution

First apply the product rule to get

$$f'(x) = (3x + 1)^4 \frac{d}{dx}[(2x - 1)^5] + (2x - 1)^5 \frac{d}{dx}[(3x + 1)^4]$$

Continue by applying the general power rule to each term:

$$f'(x) = (3x + 1)^4[5(2x - 1)^4(2)] + (2x - 1)^5[4(3x + 1)^3(3)]$$
$$= 10(3x + 1)^4(2x - 1)^4 + 12(2x - 1)^5(3x + 1)^3$$

Finally, simplify your answer by factoring:

$$f'(x) = 2(3x + 1)^3(2x - 1)^4[5(3x + 1) + 6(2x - 1)]$$
$$= 2(3x + 1)^3(2x - 1)^4[15x + 5 + 12x - 6]$$
$$= 2(3x + 1)^3(2x - 1)^4(27x - 1)$$

The tangent line to the graph of $f(x)$ is horizontal at points $(c, f(c))$ where $f'(c) = 0$. By solving

$$f'(x) = 2(3x + 1)^3 (2x - 1)^4(27x - 1) = 0$$

we see that $f'(c) = 0$ where

$$3c + 1 = 0 \quad \text{and} \quad 2c - 1 = 0 \quad \text{and} \quad 27c - 1 = 0$$

that is, at $c = -\dfrac{1}{3}$, $c = \dfrac{1}{2}$, and $c = \dfrac{1}{27}$.

EXAMPLE 2.4.9

Find the second derivative of the function $f(x) = \dfrac{3x - 2}{(x - 1)^2}$.

Solution

Using the quotient rule, along with the extended power rule (applied to $(x - 1)^2$), we get

$$f'(x) = \frac{(x - 1)^2(3) - (3x - 2)[2(x - 1)(1)]}{(x - 1)^4}$$
$$= \frac{(x - 1)[3(x - 1) - 2(3x - 2)]}{(x - 1)^4}$$
$$= \frac{3x - 3 - 6x + 4}{(x - 1)^3}$$
$$= \frac{1 - 3x}{(x - 1)^3}$$

Using the quotient rule again, this time applying the extended power rule to $(x - 1)^3$, we find that

$$f''(x) = \frac{(x - 1)^3(-3) - (1 - 3x)[3(x - 1)^2(1)]}{(x - 1)^6}$$

$$= \frac{-3(x - 1)^2[(x - 1) + (1 - 3x)]}{(x - 1)^6}$$

$$= \frac{-3(-2x)}{(x - 1)^4} = \frac{6x}{(x - 1)^4}$$

EXPLORE!

Most graphing calculators can compute the value of a derivative at a point. To illustrate, store the function $C(x) = \sqrt{[0.5(3.1 + 0.1x^2)^2 + 17]}$ into **Y1** of the equation editor and graph using a standard window (**ZOOM 6**). Find the dy/dx command through the **CALC (2nd TRACE)** key. Then press 3 and **ENTER** to get the derivative value at X = 3. You can draw the tangent line to the curve at X = 3, through the **DRAW (2nd PRGM)** key, pressing 3 then **ENTER**, to obtain the equation of the tangent line with the proper slope value.

EXAMPLE | 2.4.10

An environmental study of a certain suburban community suggests that the average daily level of carbon monoxide in the air will be $c(p) = \sqrt{0.5p^2 + 17}$ parts per million when the population is p thousand. It is estimated that t years from now, the population of the community will be $p(t) = 3.1 + 0.1t^2$ thousand. At what rate will the carbon monoxide level be changing with respect to time 3 years from now?

Solution

The goal is to find $\dfrac{dc}{dt}$ when $t = 3$. Since

$$\frac{dc}{dp} = \frac{1}{2}(0.5p^2 + 17)^{-1/2}[0.5(2p)] = \frac{1}{2}p(0.5p^2 + 17)^{-1/2}$$

and

$$\frac{dp}{dt} = 0.2t$$

it follows from the chain rule that

$$\frac{dc}{dt} = \frac{dc}{dp}\frac{dp}{dt} = \frac{1}{2}p(0.5p^2 + 17)^{-1/2}(0.2t) = \frac{0.1pt}{\sqrt{0.5p^2 + 17}}$$

When $t = 3$,

$$p(3) = 3.1 + 0.1(3)^2 = 4$$

and by substituting $t = 3$ and $p = 4$ into the formula for $\dfrac{dc}{dt}$, we get

$$\frac{dc}{dt} = \frac{0.1(4)(3)}{\sqrt{0.5(4)^2 + 17}}$$

$$= \frac{1.2}{\sqrt{25}} = \frac{1.2}{5} = 0.24 \text{ parts per million per year}$$

EXAMPLE | 2.4.11

The manager of an appliance manufacturing firm determines that when blenders are priced at p dollars apiece, the number sold each month can be modeled by

$$D(p) = \frac{8,000}{p}$$

The manager estimates that t months from now, the unit price of the blenders will be $p(t) = 0.06t^{3/2} + 22.5$ dollars. At what rate will the monthly demand for blenders $D(p)$ be changing 25 months from now? Will it be increasing or decreasing at this time?

Solution

We want to find $\dfrac{dD}{dt}$ when $t = 25$. We have

$$\frac{dD}{dp} = \frac{d}{dp}\left[\frac{8,000}{p}\right] = \frac{-8,000}{p^2}$$

and

$$\frac{dp}{dt} = \frac{d}{dt}[0.06t^{3/2} + 22.5] = 0.06\left(\frac{3}{2}t^{1/2}\right) = 0.09t^{1/2}$$

so it follows from the chain rule that

$$\frac{dD}{dt} = \frac{dD}{dp}\frac{dp}{dt} = \left[\frac{-8,000}{p^2}\right](0.09t^{1/2})$$

When $t = 25$, the unit price is

$$p(25) = 0.06(25)^{3/2} + 22.5 = 30 \text{ dollars}$$

and we have

$$\frac{dD}{dt}\bigg|_{\substack{t=25 \\ p=30}} = \left[\frac{-8,000}{30^2}\right][0.09(25)^{1/2}] = -4$$

That is, 25 months from now, the demand for blenders will be changing at the rate of 4 units per month and will be decreasing since $\dfrac{dD}{dt}$ is negative.

PROBLEMS 2.4

In Problems 1 through 10, use the chain rule to compute the derivative $\dfrac{dy}{dx}$ and simplify your answer.

1. $y = u^2 + 1;\ u = 3x - 2$

2. $y = 2u^2 - u + 5;\ u = 1 - x^2$

3. $y = \sqrt{u};\ u = x^2 + 2x - 3$

4. $y = u^2 + 2u - 3;\ u = \sqrt{x}$

5. $y = \dfrac{1}{u^2};\ u = x^2 + 1$

6. $y = \dfrac{1}{u};\ u = 3x^2 + 5$

7. $y = \dfrac{1}{\sqrt{u}};\ u = x^2 - 9$

8. $y = u^2 + u - 2;\ u = \dfrac{1}{x}$

9. $y = \dfrac{1}{u - 1};\ u = x^2$

10. $y = u^2;\ u = \dfrac{1}{x - 1}$

In Problems 11 through 16, use the chain rule to compute the derivative $\dfrac{dy}{dx}$ for the given value of x.

11. $y = 3u^4 - 4u + 5;\ u = x^3 - 2x - 5$ for $x = 2$

12. $y = u^5 - 3u^2 + 6u - 5;\ u = x^2 - 1$ for $x = 1$

13. $y = \sqrt{u}$; $u = x^2 - 2x + 6$ for $x = 3$

14. $y = 3u^2 - 6u + 2$; $u = \dfrac{1}{x^2}$ for $x = \dfrac{1}{3}$

15. $y = \dfrac{1}{u}$; $u = 3 - \dfrac{1}{x^2}$ for $x = \dfrac{1}{2}$

16. $y = \dfrac{1}{u+1}$; $u = x^3 - 2x + 5$ for $x = 0$

In Problems 17 through 36, differentiate the given function and simplify your answer.

17. $f(x) = (2x + 1)^4$

18. $f(x) = \sqrt{5x^6 - 12}$

19. $f(x) = (x^5 - 4x^3 - 7)^8$

20. $f(t) = (3t^4 - 7t^2 + 9)^5$

21. $f(t) = \dfrac{1}{5t^2 - 6t + 2}$

22. $f(x) = \dfrac{2}{(6x^2 + 5x + 1)^2}$

23. $g(x) = \dfrac{1}{\sqrt{4x^2 + 1}}$

24. $f(s) = \dfrac{1}{\sqrt{5s^3 + 2}}$

25. $f(x) = \dfrac{3}{(1 - x^2)^4}$

26. $f(x) = \dfrac{2}{3(5x^4 + 1)^2}$

27. $h(s) = (1 + \sqrt{3s})^5$

28. $g(x) = \sqrt{1 + \dfrac{1}{3x}}$

29. $f(x) = (x + 2)^3(2x - 1)^5$

30. $f(x) = 2(3x + 1)^4(5x - 3)^2$

31. $G(x) = \sqrt{\dfrac{3x + 1}{2x - 1}}$

32. $f(y) = \left(\dfrac{y + 2}{2 - y}\right)^3$

33. $f(x) = \dfrac{(x + 1)^5}{(1 - x)^4}$

34. $F(x) = \dfrac{(1 - 2x)^2}{(3x + 1)^3}$

35. $f(y) = \dfrac{3y + 1}{\sqrt{1 - 4y}}$

36. $f(x) = \dfrac{1 - 5x^2}{\sqrt[3]{3 + 2x}}$

In Problems 37 through 42, find an equation of the line that is tangent to the graph of f for the given value of x.

37. $f(x) = (3x^2 + 1)^2$; $x = -1$

38. $f(x) = (x^2 - 3)^5(2x - 1)^3$; $x = 2$

39. $f(x) = \dfrac{1}{(2x - 1)^6}$; $x = 1$

40. $f(x) = \left(\dfrac{x + 1}{x - 1}\right)^3$; $x = 3$

41. $f(x) = \sqrt[3]{\dfrac{x}{x + 2}}$; $x = -1$

42. $f(x) = x^2\sqrt{2x + 3}$; $x = -1$

In Problems 43 through 48, find all values of $x = c$ so that the tangent line to the graph of $f(x)$ at $(c, f(c))$ will be horizontal.

43. $f(x) = (x^2 + x)^2$

44. $f(x) = x^3(2x^2 + x - 3)^2$

45. $f(x) = \dfrac{x}{(3x - 2)^2}$

46. $f(x) = \dfrac{2x + 5}{(1 - 2x)^3}$

47. $f(x) = \sqrt{x^2 - 4x + 5}$

48. $f(x) = (x - 1)^2(2x + 3)^3$

In Problems 49 and 50 differentiate the given function $f(x)$ by two different methods, first by using the general power rule and then by using the product rule. Show that the two answers are the same.

49. $f(x) = (3x + 5)^2$

50. $f(x) = (7 - 4x)^2$

In Problems 51 through 56, find the second derivative of the given function.

51. $f(x) = (3x + 1)^5$

52. $f(t) = \dfrac{2}{5t + 1}$

53. $h(t) = (t^2 + 5)^8$

54. $y = (1 - 2x^3)^4$

55. $f(x) = \sqrt{1 + x^2}$

56. $f(u) = \dfrac{1}{(3u^2 - 1)^2}$

57. ANNUAL EARNINGS   The gross annual earnings of a certain company are $f(t) = \sqrt{10t^2 + t + 229}$ thousand dollars t years after its formation in January 2005.
  a. At what rate will the gross annual earnings of the company be growing in January 2010?
  b. At what percentage rate will the gross annual earnings be growing in January 2010?

58. MANUFACTURING COST   At a certain factory, the total cost of manufacturing q units is $C(q) = 0.2q^2 + q + 900$ dollars. It has been determined that approximately $q(t) = t^2 + 100t$ units are manufactured during the first t hours of a production run. Compute the rate at which the total manufacturing cost is changing with respect to time 1 hour after production commences.

59. CONSUMER DEMAND   An importer of Brazilian coffee estimates that local consumers will buy approximately $D(p) = \dfrac{4,374}{p^2}$ pounds of the coffee per week when the price is p dollars per pound. It is also estimated that t weeks from now, the price of Brazilian coffee will be $p(t) - 0.02t^2 + 0.1t + 6$ dollars per pound.
  a. At what rate will the demand for coffee be changing with respect to price when the price is $9?
  b. At what rate will the demand for coffee be changing with respect to time 10 weeks from now? Will the demand be increasing or decreasing at this time?

60. AIR POLLUTION   It is estimated that t years from now, the population of a certain suburban community will be $p(t) = 20 - \dfrac{6}{t + 1}$ thousand. An environmental study indicates that the average daily level of carbon monoxide in the air will be $c(p) = 0.5\sqrt{p^2 + p + 58}$ parts per million when the population is p thousand.

  a. At what rate will the level of carbon monoxide be changing with respect to population when the population is 18 thousand people?
  b. At what rate will the carbon monoxide level be changing with respect to time 2 years from now? Will the level be increasing or decreasing at this time?

61. CONSUMER DEMAND   When a certain commodity is sold for p dollars per unit, consumers will buy $D(p) = \dfrac{40,000}{p}$ units per month. It is estimated that t months from now, the price of the commodity will be $p(t) = 0.4t^{3/2} + 6.8$ dollars per unit. At what percentage rate will the monthly demand for the commodity be changing with respect to time 4 months from now?

62. ANIMAL BEHAVIOR   In a research paper,[*] V. A. Tucker and K. Schmidt-Koenig demonstrated that a species of Australian parakeet (the Budgerigar) expends energy (calories per gram of mass per kilometer) according to the formula

$$E = \frac{1}{v}[0.074(v - 35)^2 + 22]$$

where v is the bird's velocity (in km/hr). Find a formula for the rate of change of E with respect to velocity v.

63. MAMMALIAN GROWTH   Observations show that the length L in millimeters (mm) from nose to tip of tail of a Siberian tiger can be estimated using the function $L = 0.25w^{2.6}$, where w is the weight of the tiger in kilograms (kg). Furthermore, when a tiger is less than 6 months old, its weight (kg) can be estimated in terms of its age A in days by the function $w = 3 + 0.21A$.

---

*V. A. Tucker and K. Schmidt-Koenig, "Flight Speeds of Birds in Relation to Energetics and Wind Directions," *The Auk*, Vol. 88, 1971, pp. 97–107.

a. At what rate is the length of a Siberian tiger increasing with respect to its weight when it weighs 60 kg?

b. How long is a Siberian tiger when it is 100 days old? At what rate is its length increasing with respect to time at this age?

64. **PRODUCTION**   The number of units Q of a particular commodity that will be produced when L worker-hours of labor are employed is modeled by

$$Q(L) = 300L^{1/3}$$

Suppose that the labor level varies with time in such a way that t months from now, $L(t)$ worker-hours will be employed, where

$$L(t) = \sqrt{739 + 3t - t^2}$$

for $0 \le t \le 12$.

a. How many worker-hours will be employed in producing the commodity 5 months from now? How many units will be produced at this time?

b. At what rate will production be changing with respect to time 5 months from now? Will production be increasing or decreasing at this time?

65. **PRODUCTION**   The number of units Q of a particular commodity that will be produced with K thousand dollars of capital expenditure is modeled by

$$Q(K) = 500K^{2/3}$$

Suppose that capital expenditure varies with time in such a way that t months from now there will be $K(t)$ thousand dollars of capital expenditure, where

$$K(t) = \frac{2t^4 + 3t + 149}{t + 2}$$

a. What will be the capital expenditure 3 months from now? How many units will be produced at this time?

b. At what rate will production be changing with respect to time 5 months from now? Will production be increasing or decreasing at this time?

66. **QUALITY OF LIFE**   A demographic study models the population p (in thousands) of a community by the function

$$p(Q) = 3Q^2 + 4Q + 200$$

where Q is a quality-of-life index that ranges from $Q = 0$ (extremely poor quality) to $Q = 10$

(excellent quality). Suppose the index varies with time in such a way that t years from now,

$$Q(t) = \frac{t^2 + 2t + 3}{2t + 1}$$

for $0 \le t \le 10$.

a. What value of the quality-of-life index should be expected 4 years from now? What will be the corresponding population at this time?

b. At what rate is the population changing with respect to time 4 years from now? Is the population increasing or decreasing at this time?

67. **DEPRECIATION**   The value V (in thousands of dollars) of an industrial machine is modeled by

$$V(N) = \left(\frac{3N + 430}{N + 1} \right)^{2/3}$$

where N is the number of hours the machine is used each day. Suppose further that usage varies with time in such a way that

$$N(t) = \sqrt{t^2 - 10t + 45}$$

where t is the number of months the machine has been in operation.

a. How many hours per day will the machine be used 9 months from now? What will be the value of the machine at this time?

b. At what rate is the value of the machine changing with respect to time 9 months from now? Will the value be increasing or decreasing at this time?

68. **WATER POLLUTION**   When organic matter is introduced into a body of water, the oxygen content of the water is temporarily reduced by oxidation. Suppose that t days after untreated sewage is dumped into a particular lake, the proportion of the usual oxygen content in the water of the lake that remains is given by the function

$$P(t) = 1 - \frac{12}{t + 12} + \frac{144}{(t + 12)^2}$$

a. At what rate is the oxygen proportion $P(t)$ changing after 10 days? Is the proportion increasing or decreasing at this time?

b. Is the oxygen proportion increasing or decreasing after 15 days?

c. If there is no new dumping, what would you expect to eventually happen to the proportion of oxygen? Use a limit to verify your conjecture.

69. INSECT GROWTH The growth of certain insects varies with temperature. Suppose a particular species of insect grows in such a way that the volume of an individual is

$$V(T) = 0.41(-0.01T^2 + 0.4T + 3.52) \quad \text{cm}^3$$

when the temperature is $T°C$, and that its mass is m grams, where

$$m(V) = \frac{0.39V}{1 + 0.09V}$$

a. Find the rate of change of the insect's volume with respect to temperature.

b. Find the rate of change of the insect's mass with respect to volume.

c. When $T = 10°C$, what is the insect's volume? At what rate is the insect's mass changing with respect to temperature when $T = 10°C$?

70. COMPOUND INTEREST If $10,000 is invested at an annual rate r (expressed as a decimal) compounded weekly, the total amount (principal P and interest) accumulated after 10 years is given by the formula

$$A = 10,000\left(1 + \frac{r}{52}\right)^{520}$$

a. Find the rate of change of A with respect to r.

b. Find the percentage rate of change of A with respect to r when $r = 0.05$ (that is, 5%).

71. LEARNING When you first begin to study a topic or practice a skill, you may not be very good at it, but in time, you will approach the limits of your ability. One model for describing this behavior involves the function

$$T = aL\sqrt{L - b}$$

where T is the time required for a particular person to learn the items on a list of L items and a and b are positive constants.

a. Find the derivative $\dfrac{dT}{dL}$ and interpret it in terms of the learning model.

b. Read and discuss in one paragraph an article on how learning curves can be used to study worker productivity.*

*You may wish to begin your research by consulting Philip E. Hicks, *Industrial Engineering and Management: A New Perspective*, 2nd ed., Chapter 6, New York: McGraw-Hill, 1994, pp. 267–293.

72. An object moves along a straight line with velocity

$$v(t) = (2t + 9)^2(8 - t)^3 \quad \text{for } 0 \le t \le 5$$

a. Find the acceleration $a(t)$ of the object at time t.

b. When is the object stationary for $0 \le t < 5$? Find the acceleration at each such time.

c. When is the acceleration zero for $0 \le t \le 5$? Find the velocity at each such time.

d. Use the graphing utility of your calculator to draw the graphs of the velocity $v(t)$ and acceleration $a(t)$ on the same screen.

e. The object is said to be *speeding up* when $v(t)$ and $a(t)$ have the same sign (both positive or both negative). Use your calculator to determine when (if ever) this occurs for $0 \le t \le 5$.

73. An object moves along a straight line in such a way that its position at time t is given by

$$s(t) = (3 + t - t^2)^{3/2} \quad \text{for } 0 \le t \le 2$$

a. What are the object's velocity $v(t)$ and acceleration $a(t)$ at time t?

b. When is the object stationary for $0 \le t \le 2$? Where is the object and what is its acceleration at each such time?

c. When is the acceleration zero for $0 \le t \le 2$? What are the object's position and velocity at each such time?

d. Use the graphing utility of your calculator to draw the graphs of the object's position $s(t)$, velocity $v(t)$, and acceleration $a(t)$ on the same screen for $0 \le t \le 2$.

e. The object is said to be *slowing down* when $v(t)$ and $a(t)$ have opposite signs (one positive, the other negative). Use your calculator to determine when (if ever) this occurs for $0 \le t \le 2$.

74. Suppose $L(x)$ is a function with the property that $L'(x) = \dfrac{1}{x}$. Use the chain rule to find the derivatives of the following functions and simplify your answers.

a. $f(x) = L(x^2)$

b. $f(x) = L\left(\dfrac{1}{x}\right)$

c. $f(x) = L\left(\dfrac{2}{3\sqrt{x}}\right)$

d. $f(x) = L\left(\dfrac{2x + 1}{1 - x}\right)$

75. Prove the general power rule for $n = 2$ by using the product rule to compute $\dfrac{dy}{dx}$ if $y = [h(x)]^2$.

76. Prove the general power rule for $n = 3$ by using the product rule and the result of Problem 75 to compute $\dfrac{dy}{dx}$ if $y = [h(x)]^3$. [*Hint:* Begin by writing y as $h(x)[h(x)]^2$.]

77. Store the function $f(x) = \sqrt[3]{3.1x^2 + 19.4}$ in your graphing utility. Use the numeric differentiation feature of your utility to calculate $f'(1)$ and $f'(-3)$. Explore the graph of $f(x)$. How many horizontal tangents does the graph have?

78. Store the function $f(x) = (2.7x^3 - 3\sqrt{x} + 5)^{2/3}$ in your graphing utility. Use the numeric differentiation feature of the utility to calculate $f'(0)$ and $f'(4.3)$. Explore the graph of $f(x)$. How many horizontal tangents does it have?

SECTION   2.5 Marginal Analysis and Approximations Using Increments

Calculus is an important tool in economics. We briefly discussed sales and production in Chapter 1, where we introduced economic quantities such as cost, revenue, profit and supply, demand, and market equilibrium. In this section, we will use the derivative to explore rates of change involving economic quantities.

Marginal Analysis

In economics,* the use of the derivative to approximate the change in a quantity that results from a 1-unit increase in production is called **marginal analysis.** For instance, suppose $C(x)$ is the total cost of producing x units of a particular commodity. If x_0 units are currently being produced, then the derivative

$$C'(x_0) = \lim_{h \to 0} \frac{C(x_0 + h) - C(x_0)}{h}$$

is called the **marginal cost** of producing x_0 units. The limiting value that defines this derivative is approximately equal to the difference quotient of $C(x)$ when $h = 1$; that is,

$$C'(x_0) \approx \frac{C(x_0 + 1) - C(x_0)}{1} = C(x_0 + 1) - C(x_0)$$

where the symbol $\approx$ is used to indicate that this is an approximation, not an equality. But $C(x_0 + 1) - C(x_0)$ is just the cost of increasing the level of production by one unit, from x_0 to $x_0 + 1$. To summarize:

> **Marginal Cost** ■ If $C(x)$ is the total cost of producing x units of a commodity, then the **marginal cost** of producing x_0 units is the derivative $C'(x_0)$, which approximates the additional cost $C(x_0 + 1) - C(x_0)$ incurred when the level of production is increased by one unit, from x_0 to $x_0 + 1$.

---

*Economists and business managers view the topics we are about to discuss from a slightly different perspective. A good source for the economist's view is the text by J. M. Henderson and R. E. Quandt, *Microeconomic Theory,* New York: McGraw-Hill, 1986. The viewpoint of business management may be found in D. Salvatore, *Management Economics,* New York: McGraw-Hill, 1989, which is an excellent source of practical applications and case studies.

The geometric relationship between the marginal cost $C'(x_0)$ and the additional cost $C(x_0 + 1) - C(x_0)$ is shown in Figure 2.14.

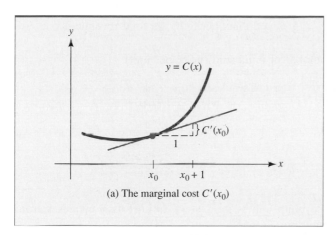

(a) The marginal cost $C'(x_0)$

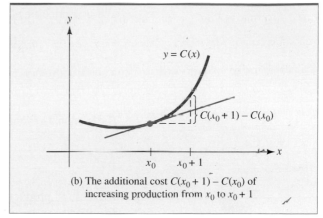

(b) The additional cost $C(x_0 + 1) - C(x_0)$ of increasing production from x_0 to $x_0 + 1$

FIGURE 2.14 Marginal cost $C'(x_0)$ approximates $C(x_0 + 1) - C(x_0)$.

The preceding discussion applies not only to cost, but also to other economic quantities. Here is a summary of what is meant by marginal revenue and marginal profit and how these marginal quantities can be used to estimate 1-unit changes in revenue and profit.

Marginal Revenue and Marginal Profit ■ Suppose $R(x)$ is the revenue generated when x units of a particular commodity are produced, and $P(x)$ is the corresponding profit. When $x = x_0$ units are being produced, then:

The **marginal revenue** is $R'(x_0)$. It approximates $R(x_0 + 1) - R(x_0)$, the additional revenue generated by producing one more unit.

The **marginal profit** is $P'(x_0)$. It approximates $P(x_0 + 1) - P(x_0)$, the additional profit obtained by producing one more unit.

Marginal analysis is illustrated in Example 2.5.1.

EXPLORE!

Refer to Example 2.5.1. Graph the cost function $C(x)$ and revenue function $R(x)$ together using the window [0, 20]10 by [0, 400]50. Graph the tangent line to $C(x)$ at the value $X = 8$. Notice why the marginal cost, represented by the tangent line, is a good approximation to $C(x)$ at $X = 8$. Now draw the tangent line to $R(x)$ at $X = 8$, also demonstrating that the marginal revenue is a good approximation to $R(x)$ near $X = 8$.

EXAMPLE 2.5.1

A manufacturer estimates that when x units of a particular commodity are produced, the total cost will be $C(x) = \frac{1}{8}x^2 + 3x + 98$ dollars, and furthermore, that all x units will be sold when the price is $p(x) = \frac{1}{3}(75 - x)$ dollars per unit.

a. Find the marginal cost and the marginal revenue.

b. Use marginal cost to estimate the cost of producing the ninth unit.

c. What is the actual cost of producing the ninth unit?

d. Use marginal revenue to estimate the revenue derived from the sale of the ninth unit.

e. What is the actual revenue derived from the sale of the ninth unit?

Solution

a. The marginal cost is $C'(x) = \frac{1}{4}x + 3$. Since x units of the commodity are

sold at a price of $p(x) = \frac{1}{3}(75 - x)$ dollars per unit, the total revenue is

$$R(x) = \text{(number of units sold)(price per unit)}$$

$$= xp(x) = x\left[\frac{1}{3}(75 - x)\right] = 25x - \frac{1}{3}x^2$$

The marginal revenue is

$$R'(x) = 25 - \frac{2}{3}x$$

b. The cost of producing the ninth unit is the change in cost as x increases from 8 to 9 and can be estimated by the marginal cost

$$C'(8) = \frac{1}{4}(8) + 3 = \$5$$

c. The actual cost of producing the ninth unit is

$$C(9) - C(8) = \$5.13$$

which is reasonably well approximated by the marginal cost $C'(8) = \$5$.

d. The revenue obtained from the sale of the ninth unit is approximated by the marginal revenue

$$R'(8) = 25 - \frac{2}{3}(8) = \$19.67$$

e. The actual revenue obtained from the sale of the ninth unit is

$$R(9) - R(8) = \$19.33$$

In Example 2.5.2, a marginal economic quantity is used to analyze a production process.

EXAMPLE | 2.5.2

A manufacturer of digital cameras estimates that when x hundred cameras are produced, the total profit will be

$$P(x) = -0.0035x^3 + 0.07x^2 + 25x - 200$$

thousand dollars.

a. Find the marginal profit function.

b. What is the marginal profit when the level of production is $x = 10$, $x = 50$, and $x = 80$?

c. Interpret these results.

Solution

a. The marginal profit is given by the derivative

$$P'(x) = -0.0035(3x^2) + 0.07(2x) + 25$$
$$= -0.0105x^2 + 0.14x + 25$$

b. The marginal profit at $x = 10$, 50, and 80 is

$$P'(10) = -0.0105(10)^2 + 0.14(10) + 25 = 25.35$$
$$P'(50) = -0.0105(50)^2 + 0.14(50) + 25 = 5.75$$
$$P'(80) = -0.0105(80)^2 + 0.14(80) + 25 = -31$$

c. The fact that $P'(10) = 25.35$ means that a 1-unit increase in production from 10 to 11 hundred cameras increases profit by approximately 25.35 thousand dollars ($25,350), so the manager may be inclined to increase production at this level. However, since $P'(50) = 5.75$, increasing the level of production from 50 units to 51 increases the profit by only about $5,750, thus providing relatively little incentive for the manager to make a change. Finally, since $P'(80) = -31$ is negative, the profit will actually *decrease* by approximately $31,000 if the production level is raised from 80 to 81 units. The manager may wish to consider decreasing the level of production in this case.

Approximation by Increments

Marginal analysis is an important example of a general approximation procedure based on the fact that since

$$f'(x_0) = \lim_{h \to 0} \frac{f(x_0 + h) - f(x_0)}{h}$$

then for small h, the derivative $f'(x_0)$ is approximately equal to the difference quotient

$$\frac{f(x_0 + h) - f(x_0)}{h}$$

We indicate this approximation by writing

$$f'(x_0) \approx \frac{f(x_0 + h) - f(x_0)}{h}$$

or, equivalently,

$$f(x_0 + h) - f(x_0) \approx f'(x_0)h$$

To emphasize that the incremental change is in the variable x, we write $h = \Delta x$ (read Δx as "delta x") and summarize the incremental approximation formula as follows.

> **Approximation by Increments** ■ If $f(x)$ is differentiable at $x = x_0$ and Δx is a small change in x, then
>
> $$f(x_0 + \Delta x) \approx f(x_0) + f'(x_0)\Delta x$$
>
> or, equivalently, if $\Delta f = f(x_0 + \Delta x) - f(x_0)$, then
>
> $$\Delta f \approx f'(x_0)\Delta x$$

Here is an example of how this approximation formula can be used in economics.

EXAMPLE 2.5.3

Suppose the total cost in dollars of manufacturing q units of a certain commodity is $C(q) = 3q^2 + 5q + 10$. If the current level of production is 40 units, estimate how the total cost will change if 40.5 units are produced.

Solution

In this problem, the current value of production is $q = 40$ and the change in production is $\Delta q = 0.5$. By the approximation formula, the corresponding change in cost is

$$\Delta C = C(40.5) - C(40) \approx C'(40)\Delta q = C'(40)(0.5)$$

Since

$$C'(q) = 6q + 5 \quad \text{and} \quad C'(40) = 6(40) + 5 = 245$$

it follows that

$$\Delta C \approx [C'(40)](0.5) = 245(0.5) = \$122.50$$

For practice, compute the actual change in cost caused by the increase in the level of production from 40 to 40.5 and compare your answer with the approximation. Is the approximation a good one?

Suppose you wish to compute a quantity Q using a formula $Q(x)$. If the value of x used in the computation is not precise, then its inaccuracy is passed on or *propagated* to the computed value of Q. Example 2.5.4 shows how such **propagated error** can be estimated.

EXPLORE!

Refer to Example 2.5.4. Store the volume $V = \dfrac{1}{6}\pi x^3$ into Y1, where x measures the diameter of the spherical tumor. Write

$$Y2 = Y1(X + 0.05) - Y1(X)$$

to compute the incremental change in volume, and

$$Y3 = nDeriv(Y1, X, X){\cdot}(0.05)$$

for the differential change in volume. Set **TblStart = 2.4** and ΔTbl = 0.05 in **TBLSET (2nd WINDOW)**. Now examine the **TABLE** of values, particularly for Y2 and Y3. Observe the results for X = 2.5. Now perform similar calculations for the accuracy of the volume V if the diameter x is measured accurately within 1%.

EXAMPLE 2.5.4

During a medical procedure, the size of a roughly spherical tumor is estimated by measuring its diameter and using the formula $V = \dfrac{4}{3}\pi R^3$ to compute its volume. If the diameter is measured as 2.5 cm with a maximum error of 2%, how accurate is the volume measurement?

Solution

A sphere of radius R and diameter $x = 2R$ has volume

$$V = \frac{4}{3}\pi R^3 = \frac{4}{3}\pi\left(\frac{x}{2}\right)^3 = \frac{1}{6}\pi x^3$$

so, the volume using the estimated diameter $x = 2.5$ cm is

$$V = \frac{1}{6}\pi(2.5)^3 \approx 8.181 \text{ cm}^3$$

The error made in computing this volume using the diameter 2.5 when the actual diameter is $2.5 + \Delta x$ is

$$V = V(2.5 + \Delta x) - V(2.5) \approx V'(2.5)\Delta x$$

The measurement of the diameter can be off by as much as 2%; that is, by as much as $0.02(2.5) = 0.05$ cm in either direction. Hence, the maximum error in the measurement of the diameter is $\Delta x = \pm 0.05$, and the corresponding maximum error in the calculation of volume is

$$\text{Maximum error in volume} = \Delta V \approx [V'(2.5)](\pm 0.05)$$

Since

$$V'(x) = \frac{1}{6}\pi(3x^2) = \frac{1}{2}\pi x^2 \qquad \text{and} \qquad V'(2.5) = \frac{1}{2}\pi(2.5)^2 \approx 9.817$$

it follows that

$$\text{Maximum error in volume} = (9.817)(\pm 0.05) \approx \pm 0.491$$

Thus, at worst, the calculation of the volume as 8.181 cm^3 is off by 0.491 cm^3, so the actual volume V must satisfy

$$7.690 \leq V \leq 8.672$$

In Example 2.5.5, the desired change in the function is given, and the goal is to estimate the necessary corresponding change in the variable.

EXAMPLE | 2.5.5

The daily output at a certain factory is $Q(L) = 900L^{1/3}$ units, where L denotes the size of the labor force measured in worker-hours. Currently, 1,000 worker-hours of labor are used each day. Use calculus to estimate the number of additional worker-hours of labor that will be needed to increase daily output by 15 units.

Solution

Solve for ΔL using the approximation formula

$$\Delta Q \approx Q'(L)\Delta L$$

with $\qquad\qquad \Delta Q = 15 \qquad L = 1,000 \qquad \text{and} \qquad Q'(L) = 300L^{-2/3}$

to get $\qquad\qquad\qquad\qquad\qquad 15 \approx 300(1,000)^{-2/3}\,\Delta L$

or $\qquad\qquad\qquad \Delta L \approx \frac{15}{300}(1,000)^{2/3} = \frac{15}{300}(10)^2 = 5 \text{ worker-hours}$

Approximation of Percentage Change

In Section 2.2, we defined the **percentage rate of change** of a quantity as the change in the quantity as a percentage of its size prior to the change; that is,

$$\text{Percentage change} = 100 \, \frac{\text{change in quantity}}{\text{size of quantity}}$$

This formula can be combined with the approximation formula and written in functional notation as follows.

> **Approximation Formula for Percentage Change** ■ If Δx is a (small) change in x, the corresponding percentage change in the function $f(x)$ is
>
> $$\text{Percentage change in } f = 100\frac{\Delta f}{f(x)} \approx 100\frac{f'(x)\Delta x}{f(x)}$$

EXAMPLE | 2.5.6

The GDP of a certain country was $N(t) = t^2 + 5t + 200$ billion dollars t years after 1997. Use calculus to estimate the percentage change in the GDP during the first quarter of 2005.

Solution

Use the formula

$$\text{Percentage change in } N \approx 100\frac{N'(t)\Delta t}{N(t)}$$

with $t = 8$ $\Delta t = 0.25$ and $N'(t) = 2t + 5$

to get

$$\text{Percentage change in } N \approx 100\frac{N'(8)0.25}{N(8)}$$

$$= 100\frac{[2(8) + 5](0.25)}{(8)^2 + 5(8) + 200}$$

$$\approx 1.73\%$$

Differentials Sometimes the increment Δx is referred to as the *differential of x* and is denoted by dx, and then our approximation formula can be written as $df \approx f'(x)\,dx$. If $y = f(x)$, the *differential of y* is defined to be $dy = f'(x)\,dx$. To summarize:

> **Differentials** ■ The **differential of x** is $dx = \Delta x$, and if $y = f(x)$ is a differentiable function of x, then $dy = f'(x)\,dx$ is the **differential of y.**

EXAMPLE | 2.5.7

In each case, find the differential of $y = f(x)$.
a. $f(x) = x^3 - 7x^2 + 2$
b. $f(x) = (x^2 + 5)(3 - x - 2x^2)$

Solution

a. $dy = f'(x)\,dx = [3x^2 - 7(2x)]\,dx = (3x^2 - 14x)\,dx$
b. By the product rule,

$$dy = f'(x)\,dx = [(x^2 + 5)(-1 - 4x) + (2x)(3 - x - 2x^2)]\,dx$$

A geometric interpretation of the approximation of Δy by the differential dy is shown in Figure 2.15. Note that since the slope of the tangent line at $P(x, f(x))$ is $f'(x)$, the differential $dy = f'(x)\,dx$ is the change in the height of the tangent that corresponds to a change from x to $x + \Delta x$. On the other hand, Δy is the change in the height of the curve corresponding to this change in x. Hence, approximating Δy by the differential dy is the same as approximating the change in the height of a curve by the change in height of the tangent line. If Δx is small, it is reasonable to expect this to be a good approximation.

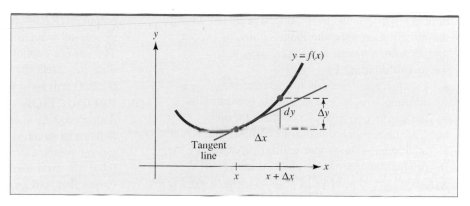

FIGURE 2.15   Approximation of Δy by the differential dy.

PROBLEMS 2.5

MARGINAL ANALYSIS   *In Problems 1 through 6, $C(x)$ is the total cost of producing x units of a particular commodity and $p(x)$ is the price at which all x units will be sold. Assume $p(x)$ and $C(x)$ are in dollars.*
   (a) Find the marginal cost and the marginal revenue.
   (b) Use marginal cost to estimate the cost of producing the fourth unit.
   (c) Find the actual cost of producing the fourth unit.
   (d) Use marginal revenue to estimate the revenue derived from the sale of the fourth unit.
   (e) Find the actual revenue derived from the sale of the fourth unit.

1. $C(x) = \dfrac{1}{5}x^2 + 4x + 57$; $p(x) = \dfrac{1}{4}(36 - x)$

2. $C(x) = \dfrac{1}{4}x^2 + 3x + 67$; $p(x) = \dfrac{1}{5}(45 - x)$

3. $C(x) = \dfrac{1}{3}x^2 + 2x + 39$; $p(x) = -x^2 - 4x + 80$

4. $C(x) = \dfrac{5}{9}x^2 + 5x + 73$; $p(x) = -x^2 - 2x + 33$

5. $C(x) = \dfrac{1}{4}x^2 + 43$; $p(x) = \dfrac{3 + 2x}{1 + x}$

6. $C(x) = \dfrac{2}{7}x^2 + 65$; $p(x) = \dfrac{12 + 2x}{3 + x}$

In Problems 7 through 32, use increments to make the required estimate.

7. Estimate how much the function $f(x) = x^2 - 3x + 5$ will change as x increases from 5 to 5.3.

8. Estimate how much the function $f(x) = \dfrac{x}{x + 1} - 3$ will change as x decreases from 4 to 3.8.

9. Estimate the percentage change in the function $f(x) = x^2 + 2x - 9$ as x increases from 4 to 4.3.

10. Estimate the percentage change in the function $f(x) = 3x + \dfrac{2}{x}$ as x decreases from 5 to 4.6.

11. **MARGINAL ANALYSIS** A manufacturer's total cost is $C(q) = 0.1q^3 - 0.5q^2 + 500q + 200$ dollars, where q is the number of units produced.
    a. Use marginal analysis to estimate the cost of manufacturing the fourth unit.
    b. Compute the actual cost of manufacturing the fourth unit.

12. **MARGINAL ANALYSIS** A manufacturer's total monthly revenue is $R(q) = 240q - 0.05q^2$ dollars when q units are produced and sold during the month. Currently, the manufacturer is producing 80 units a month and is planning to increase the monthly output by 1 unit.
    a. Use marginal analysis to estimate the additional revenue that will be generated by the production and sale of the 81st unit.
    b. Use the revenue function to compute the actual additional revenue that will be generated by the production and sale of the 81st unit.

13. **MARGINAL ANALYSIS** Suppose the total cost in dollars of manufacturing q units is $C(q) = 3q^2 + q + 500$.
    a. Use marginal analysis to estimate the cost of manufacturing the 41st unit.
    b. Compute the actual cost of manufacturing the 41st unit.

14. **AIR POLLUTION** An environmental study of a certain community suggests that t years from now, the average level of carbon monoxide in the air will be $Q(t) = 0.05t^2 + 0.1t + 3.4$ parts per million. By approximately how much will the carbon monoxide level change during the coming 6 months?

15. **NEWSPAPER CIRCULATION** It is projected that t years from now, the circulation of a local newspaper will be $C(t) = 100t^2 + 400t + 5,000$. Estimate the amount by which the circulation will increase during the next 6 months. [*Hint:* The current value of the variable is $t = 0$.]

16. **MANUFACTURING** A manufacturer's total cost is $C(q) = 0.1q^3 - 0.5q^2 + 500q + 200$ dollars when the level of production is q units. The current level of production is 4 units, and the manufacturer is planning to increase this to 4.1 units. Estimate how the total cost will change as a result.

17. **MANUFACTURING** A manufacturer's total monthly revenue is $R(q) = 240q - 0.05q^2$ dollars when q units are produced during the month. Currently, the manufacturer is producing 80 units a month and is planning to decrease the monthly

output by 0.65 unit. Estimate how the total monthly revenue will change as a result.

18. **EFFICIENCY** An efficiency study of the morning shift at a certain factory indicates that an average worker arriving on the job at 8:00 A.M. will have assembled $f(x) = -x^3 + 6x^2 + 15x$ transistor radios x hours later. Approximately how many radios will the worker assemble between 9:00 and 9:15 A.M.?

19. **PRODUCTION** At a certain factory, the daily output is $Q(K) = 600K^{1/2}$ units, where K denotes the capital investment measured in units of $1,000. The current capital investment is $900,000. Estimate the effect that an additional capital investment of $800 will have on the daily output.

20. **PRODUCTION** At a certain factory, the daily output is $Q(L) = 60,000L^{1/3}$ units, where L denotes the size of the labor force measured in worker-hours. Currently 1,000 worker-hours of labor are used each day. Estimate the effect on output that will be produced if the labor force is cut to 940 worker-hours.

21. **PROPERTY TAX** A projection made in January of 2000 determined that x years later, the average property tax on a three-bedroom home in a certain community will be $T(x) = 60x^{3/2} + 40x + 1,200$ dollars. Estimate the percentage change by which the property tax will increase during the first half of the year 2008.

22. **POPULATION GROWTH** A 5-year projection of population trends suggests that t years from now, the population of a certain community will be $P(t) = -t^3 + 9t^2 + 48t + 200$ thousand.
    a. Find the rate of change of population $R(t) = P'(t)$ with respect to time t.
    b. At what rate does the population growth rate $R(t)$ change with respect to time?
    c. Use increments to estimate how much $R(t)$ changes during the first month of the fourth year. What is the actual change in $R(t)$ during this time period?

23. **PRODUCTION** At a certain factory, the daily output is $Q = 3,000K^{1/2}L^{1/3}$ units, where K denotes the firm's capital investment measured in units of $1,000 and L denotes the size of the labor force measured in worker-hours. Suppose that the current capital investment is $400,000 and that 1,331 worker-hours of labor are used each day. Use marginal analysis to estimate the effect that an additional capital investment of $1,000 will

have on the daily output if the size of the labor force is not changed.

24. **PRODUCTION** The daily output at a certain factory is $Q(L) = 300L^{2/3}$ units, where L denotes the size of the labor force measured in worker-hours. Currently, 512 worker-hours of labor are used each day. Estimate the number of additional worker-hours of labor that will be needed to increase daily output by 12.5 units.

25. **MANUFACTURING** A manufacturer's total cost is $C(q) = \dfrac{1}{6}q^3 + 642q + 400$ dollars when q units are produced. The current level of production is 4 units. Estimate the amount by which the manufacturer should decrease production to reduce the total cost by $130.

26. **GROWTH OF A CELL** A certain cell has the shape of a sphere. The formulas $S = 4\pi r^2$ and $V = \dfrac{4}{3}\pi r^3$ are used to compute the surface area and volume of the cell, respectively. Estimate the effect on S and V produced by a 1% increase in the radius r.

27. **CARDIAC OUTPUT** *Cardiac output* is the volume (cubic centimeters) of blood pumped by a person's heart each minute. One way of measuring cardiac output C is by Fick's formula

$$C = \frac{a}{x - b}$$

where x is the concentration of carbon dioxide in the blood entering the lungs from the right heart and a and b are positive constants. If x is measured as $x = c$ with a maximum error of 3%, what is the maximum percentage error that can be incurred by measuring cardiac output with Fick's formula? (Your answer will be in terms of a, b, and c.)

28. **MEDICINE** A tiny spherical balloon is inserted into a clogged artery. If the balloon has an inner diameter of 0.01 millimeter (mm) and is made from material 0.0005 mm thick, approximately how much material is inserted into the artery? [*Hint:* Think of the amount of material as a change in volume ΔV, where $V = \dfrac{4}{3}\pi r^3$.]

29. **ARTERIOSCLEROSIS** In *arteriosclerosis*, fatty material called plaque gradually builds up on the walls of arteries, impeding the flow of blood,

which, in turn, can lead to stroke and heart attacks. Consider a model in which the carotid artery is represented as a circular cylinder with cross-sectional radius $R = 0.3$ cm and length L. Suppose it is discovered that plaque 0.07 cm thick is distributed uniformly over the inner wall of the carotid artery of a particular patient. Use increments to estimate the percentage of the total volume of the artery that is blocked by plaque. [*Hint:* The volume of a cylinder of radius R and length L is $V = \pi R^2 L$. Does it matter that we have not specified the length L of the artery?]

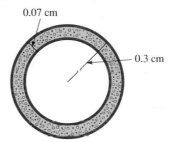

0.07 cm

0.3 cm

PROBLEM 29

30. **BLOOD CIRCULATION** In Problem 57, Section 1.1, we introduced an important law attributed to the French physician, Jean Poiseuille. Another law discovered by Poiseuille says that the volume of the fluid flowing through a small tube in unit time under fixed pressure is given by the formula $V = kR^4$, where k is a positive constant and R is the radius of the tube. This formula is used in medicine to determine how wide a clogged artery must be opened to restore a healthy flow of blood.

     a. Suppose the radius of a certain artery is increased by 5%. Approximately what effect does this have on the volume of blood flowing through the artery?

     b. Read an article on the cardiovascular system and write a paragraph on the flow of blood.*

31. **EXPANSION OF MATERIAL** The (linear) **thermal expansion coefficient** of an object is defined to be

$$\sigma = \frac{L'(T)}{L(T)}$$

---

*You may wish to begin your research by consulting such textbooks as Elaine N. Marieb, *Human Anatomy and Physiology,* 2nd ed., Redwood City, CA: The Benjamin/Cummings Publishing Co., 1992, and Kent M. Van De Graaf and Stuart Ira Fox, *Concepts of Human Anatomy and Physiology,* 3rd ed., Dubuque, IA: Wm. C. Brown Publishers, 1992.

where $L(T)$ is the length of the object when the temperature is T. Suppose a 50-meter span of a bridge is built of steel with $\sigma = 1.4 \times 10^{-5}$ per degree Celsius. Approximately how much will the length change during a year when the temperature varies from $-20°C$ (winter) to $35°C$ (summer)?

32. RADIATION   Stefan's law in physics states that a body emits radiant energy according to the formula $R(T) = kT^4$, where R is the amount of energy emitted from a surface whose temperature is T (in degrees kelvin) and k is a positive constant. Estimate the percentage change in R that results from a 2% increase in T.

Newton's Method ■ Tangent line approximations can be used in a variety of ways. *Newton's method* for approximating the roots of an equation $f(x) = 0$ is based on the idea that if we start with a "guess" x_0 that is close to an actual root c, we can often obtain an improved guess by finding the x intercept x_1 of the tangent line to the curve $y = f(x)$ at $x = x_0$ (see the figure). The process can then be repeated until a desired degree of accuracy is attained. In practice, it is usually easier and faster to use the graphing utility, **ZOOM,** and **TRACE** features of your calculator to find roots, but the ideas behind Newton's method are still important. Problems 33 through 37 involve Newton's method.

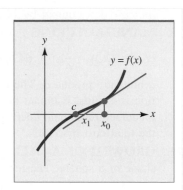

33. Show that when Newton's method is applied repeatedly, the nth approximation is obtained from the $(n-1)$st approximation by the formula

$$x_n = x_{n-1} - \frac{f(x_{n-1})}{f'(x_{n-1})} \quad n = 1, 2, 3, \ldots$$

[*Hint:* First find x_1 as the x intercept of the tangent line to $y = f(x)$ at $x = x_0$. Then use x_1 to find x_2 in the same way.]

EXPLORE!

Store into Y1 the function $f(x) = x^3 - x^2 - 1$ and graph using a decimal window to observe that there is a root between X = 1 and 2. Now write Y2 = nDeriv(Y1, X, X). On a clear homescreen, store the starting value 1 into X, 1 → X, and write X − Y1(X)/Y2(X) → X. Now press **ENTER** successively and observe the sequence of resulting values. Note how many iterations were needed so that two consecutive approximations agree to four decimal places. Repeat this process using X = −2. Note whether a stable result occurs, and how many iterations are required.

34. Let $f(x) = x^3 - x^2 - 1$.
   a. Use your graphing utility to graph $f(x)$. Note that there is only one root located between 1 and 2. Use **ZOOM** and **TRACE** or other utility features to find the root.
   b. Using $x_0 = 1$, estimate the root by applying Newton's method until two consecutive approximations agree to four decimal places.
   c. Take the root you found graphically in part (a) and the root you found by Newton's method in part (b) and substitute each into the equation $f(x) = 0$. Which is more accurate?

35. Let $f(x) = x^4 - 4x^3 + 10$. Use your graphing utility to graph $f(x)$. Note that there are two roots of the equation $f(x) = 0$. Estimate each root using Newton's method and then check your results using **ZOOM** and **TRACE** or other utility features.

36. The ancient Babylonians (circa 1700 B.C.) approximated $\sqrt{N}$ by applying the formula

$$x_{n+1} = \frac{1}{2}\left(x_n + \frac{N}{x_n}\right) \quad \text{for } n = 1, 2, 3, \ldots$$

   a. Show that this formula can be derived from the formula for Newton's method in Problem 33, and then use it to estimate $\sqrt{1,265}$. Repeat the formula until two consecutive approximations agree to four decimal places. Use your calculator to check your result.

 b. The spy (Problem 70, Section 2.2) wakes up one morning in Babylonia and finds that his calculator has been stolen. Create a spy story problem based on using the ancient formula to compute a square root.

37. Sometimes Newton's method fails no matter what initial value x_0 is chosen (unless we are lucky enough to choose the root itself). Let $f(x) = \sqrt[3]{x}$ and choose x_0 arbitrarily ($x_0 \neq 0$).

    a. Show that $x_{n+1} = -2x_n$ for $n = 0, 1, 2, \ldots$ so that the successive "guesses" generated by Newton's method are $x_0, -2x_0, 4x_0, \ldots$.

    b. Use your graphing utility to graph $f(x)$ and use an appropriate utility to draw the tangent lines to the graph of $y = f(x)$ at the points that correspond to $x_0, -2x_0, 4x_0, \ldots$. Why do these numbers fail to estimate a root of $f(x) = 0$?

SECTION 2.6 | Implicit Differentiation and Related Rates

The functions you have worked with so far have all been given by equations of the form $y = f(x)$ in which the dependent variable y on the left is given explicitly by an expression on the right involving the independent variable x. A function in this form is said to be in **explicit form.** For example, the functions

$$y = x^2 + 3x + 1 \qquad y = \frac{x^3 + 1}{2x - 3} \qquad \text{and} \qquad y = \sqrt{1 - x^2}$$

are all functions in explicit form.

Sometimes practical problems will lead to equations in which the function y is not written explicitly in terms of the independent variable x; for example, equations such as

$$x^2 y^3 - 6 = 5y^3 + x \qquad \text{and} \qquad x^2 y + 2y^3 = 3x + 2y$$

Since it has not been solved for y, such an equation is said to **define** y *implicitly as a function of* x and the function y is said to be in **implicit form.**

Differentiation of Functions in Implicit Form

Suppose you have an equation that defines y implicitly as a function of x and you want to find the derivative $\dfrac{dy}{dx}$. For instance, you may be interested in the slope of a line that is tangent to the graph of the equation at a particular point. One approach might be to solve the equation for y explicitly and then differentiate using the techniques you already know. Unfortunately, it is not always possible to find y explicitly. For example, there is no obvious way to solve for y in the equation $x^2 y + 2y^3 = 3x + 2y$.

Moreover, even when you can solve for y explicitly, the resulting formula is often complicated and unpleasant to differentiate. For example, the equation $x^2 y^3 - 6 = 5y^3 + x$ can be solved for y to give

$$x^2 y^3 - 5y^3 = x + 6$$
$$y^3(x^2 - 5) = x + 6$$
$$y = \left(\frac{x + 6}{x^2 - 5}\right)^{1/3}$$

The computation of $\dfrac{dy}{dx}$ for this function in explicit form would be tedious, involving both the chain rule and the quotient rule. Fortunately, there is a simple technique based on the chain rule that you can use to find $\dfrac{dy}{dx}$ without first solving for y explicitly. This technique, known as **implicit differentiation,** consists of differentiating both sides of the given (defining) equation with respect to x and then solving algebraically for $\dfrac{dy}{dx}$. Here is an example illustrating the technique.

EXPLORE!

Store $\dfrac{dy}{dx}$ from Example 2.6.1 in your graphing utility. Substitute $x = 2$ in the given expression of the implicit function and solve for y. Store this value as Y. Store 2 as X. Find the slope of the tangent line to $y = f(x)$ by evaluating $\dfrac{dy}{dx}$ at these values. Finally, find an equation for the tangent line to the function in **Example 2.6.1** at $x = 2$.

EXAMPLE 2.6.1

Find $\dfrac{dy}{dx}$ if $x^2 y + y^2 = x^3$.

Solution

You are going to differentiate both sides of the given equation with respect to x. So that you won't forget that y is actually a function of x, temporarily replace y by $f(x)$ and begin by rewriting the equation as

$$x^2 f(x) + (f(x))^2 = x^3$$

Now differentiate both sides of this equation term by term with respect to x:

$$\frac{d}{dx}[x^2 f(x) + (f(x))^2] = \frac{d}{dx}[x^3]$$

$$\underbrace{\left[x^2 \frac{df}{dx} + f(x)\frac{d}{dx}(x^2)\right]}_{\frac{d}{dx}[x^2 f(x)]} + \underbrace{2 f(x)\frac{df}{dx}}_{\frac{d}{dx}[(f(x))^2]} = \underbrace{3x^2}_{\frac{d}{dx}(x^3)}$$

Thus, we have

$$x^2 \frac{df}{dx} + f(x)(2x) + 2 f(x)\frac{df}{dx} = 3x^2 \qquad \textit{gather all } \frac{df}{dx} \textit{ terms}$$

$$x^2 \frac{df}{dx} + 2 f(x)\frac{df}{dx} = 3x^2 - 2x f(x) \qquad \textit{on one side of the equation}$$

$$[x^2 + 2 f(x)]\frac{df}{dx} = 3x^2 - 2x f(x) \qquad \textit{combine terms}$$

$$\frac{df}{dx} = \frac{3x^2 - 2x f(x)}{x^2 + 2 f(x)} \qquad \textit{solve for } \frac{df}{dx}$$

Finally, replace $f(x)$ by y to get

$$\frac{dy}{dx} = \frac{3x^2 - 2xy}{x^2 + 2y}$$

NOTE Temporarily replacing y by $f(x)$ as in Example 2.6.1 is a useful device for illustrating the implicit differentiation process, but as soon as you feel comfortable with the technique, try to leave out this unnecessary step and differentiate the given equation directly. Just keep in mind that y is really a function of x and remember to use the chain rule when it is appropriate. ■

Here is an outline of the procedure.

Implicit Differentiation ■ Suppose an equation defines y implicitly as a differentiable function of x. To find $\dfrac{dy}{dx}$:

1. Differentiate both sides of the equation with respect to x. Remember that y is really a function of x and use the chain rule when differentiating terms containing y.

2. Solve the differentiated equation algebraically for $\dfrac{dy}{dx}$.

Computing the Slope of a Tangent Line by Implicit Differentiation

In Examples 2.6.2 and 2.6.3, you will see how to use implicit differentiation to find the slope of a tangent line.

EXAMPLE | 2.6.2

Find the slope of the tangent line to the circle $x^2 + y^2 = 25$ at the point $(3, 4)$. What is the slope at the point $(3, -4)$?

Solution

Differentiating both sides of the equation $x^2 + y^2 = 25$ with respect to x, you get

$$2x + 2y \frac{dy}{dx} = 0$$

$$\frac{dy}{dx} = \frac{-x}{y}$$

The slope at $(3, 4)$ is the value of the derivative when $x = 3$ and $y = 4$:

$$\frac{dy}{dx}\bigg|_{(3,4)} = \frac{-x}{y}\bigg|_{\substack{x=3 \\ y=4}} = \frac{-3}{4}$$

Similarly, the slope at $(3, -4)$ is the value of $\dfrac{dy}{dx}$ when $x = 3$ and $y = -4$:

$$\frac{dy}{dx}\bigg|_{(3,-4)} = \frac{-x}{y}\bigg|_{\substack{x=3 \\ y=-4}} = \frac{-3}{(-4)} = \frac{3}{4}$$

The graph of the circle is shown in Figure 2.16 together with the tangent lines at $(3, 4)$ and $(3, -4)$.

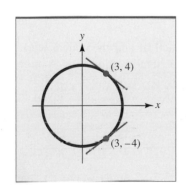

FIGURE 2.16 The graph of the circle $x^2 + y^2 = 25$.

EXAMPLE | 2.6.3

Find all points on the graph of the equation $x^2 - y^2 = 2x + 4y$ where the tangent line is horizontal. Does the graph have any vertical tangents?

Solution

Differentiate both sides of the given equation with respect to x to get

$$2x - 2y\frac{dy}{dx} = 2 + 4\frac{dy}{dx}$$

$$\frac{dy}{dx} = \frac{2x - 2}{4 + 2y}$$

There will be a horizontal tangent at each point on the graph where the slope is zero; that is, where the *numerator* $2x - 2$ of $\dfrac{dy}{dx}$ is zero:

$$2x - 2 = 0$$

$$x = 1$$

To find the corresponding value of y, substitute $x = 1$ into the given equation and solve using the quadratic formula (or your calculator):

$$1 - y^2 = 2(1) + 4y$$

$$y^2 + 4y + 1 = 0$$

$$y = -0.27, -3.73$$

Thus, the given graph has horizontal tangents at the points $(1, -0.27)$ and $(1, -3.73)$.

Since the slope of a vertical line is undefined, the given graph can have a vertical tangent only where the *denominator* $4 + 2y$ of $\dfrac{dy}{dx}$ is zero:

$$4 + 2y = 0$$

$$y = -2$$

To find the corresponding value of x, substitute $y = -2$ into the given equation:

$$x^2 - (-2)^2 = 2x + 4(-2)$$

$$x^2 - 2x + 4 = 0$$

But this quadratic equation has no real solutions, which, in turn, implies that the given graph has no vertical tangents. The graph is shown in Figure 2.17.

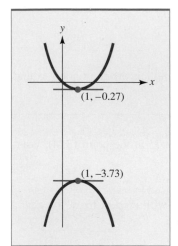

FIGURE 2.17 The graph of the equation $x^2 - y^2 = 2x + 4y$.

Application to Economics

Implicit differentiation is used in economics in both practical and theoretical work. In Section 4.3, you will see it used to derive certain theoretical relationships. A more practical application of implicit differentiation is given in Example 2.6.4, which is a preview of the discussion of level curves of a function of two variables given in Section 7.1.

EXAMPLE | 2.6.4

Suppose the output at a certain factory is $Q = 2x^3 + x^2y + y^3$ units, where x is the number of hours of skilled labor used and y is the number of hours of unskilled labor. The current labor force consists of 30 hours of skilled labor and 20 hours of unskilled labor. Use calculus to estimate the change in unskilled labor y that should be made to offset a 1-hour increase in skilled labor x so that output will be maintained at its current level.

Solution

The current level of output is the value of Q when $x = 30$ and $y = 20$. That is,

$$Q = 2(30)^3 + (30)^2(20) + (20)^3 = 80,000 \text{ units}$$

If output is to be maintained at this level, the relationship between skilled labor x and unskilled labor y is given by the equation

$$80,000 = 2x^3 + x^2y + y^3$$

which defines y implicitly as a function of x.

The goal is to estimate the change in y that corresponds to a 1-unit increase in x when x and y are related by this equation. As you saw in Section 2.5, the change in y caused by a 1-unit increase in x can be approximated by the derivative $\dfrac{dy}{dx}$. To find this derivative, use implicit differentiation. (Remember that the derivative of the constant 80,000 on the left-hand side is zero.)

$$0 = 6x^2 + x^2\frac{dy}{dx} + y\frac{d}{dx}(x^2) + 3y^2\frac{dy}{dx}$$

$$0 = 6x^2 + x^2\frac{dy}{dx} + 2xy + 3y^2\frac{dy}{dx}$$

$$-(x^2 + 3y^2)\frac{dy}{dx} = 6x^2 + 2xy$$

$$\frac{dy}{dx} = -\frac{6x^2 + 2xy}{x^2 + 3y^2}$$

Now evaluate this derivative when $x = 30$ and $y = 20$ to conclude that

$$\text{Change in } y \approx \frac{dy}{dx}\bigg|_{\substack{x=30 \\ y=20}} = \frac{6(30)^2 + 2(30)(20)}{(30)^2 + 3(20)^2} \approx -3.14 \text{ hours}$$

That is, to maintain the current level of output, unskilled labor should be decreased by approximately 3.14 hours to offset a 1-hour increase in skilled labor.

NOTE Did you notice that the solution to Example 2.6.4 contains an unnecessary step? The calculation of the current output to be 80,000 units could be omitted and the equation relating x and y written as

$$C = 2x^3 + x^2y + y^3$$

where C is a constant that stands for the (unspecified) current level of output. Since the derivative of any constant is zero, the differentiated equations will be the same. ∎

Related Rates

In certain practical problems, x and y are related by an equation and can be regarded as functions of a third variable t, which often represents time. Then implicit differentiation can be used to relate $\dfrac{dx}{dt}$ to $\dfrac{dy}{dt}$. This kind of problem is said to involve **related rates.** Here is a general procedure for analyzing related rates problems.

> ### A Procedure for Solving Related Rates Problems
> **1.** Draw a figure (if appropriate) and assign variables.
> **2.** Find a formula relating the variables.
> **3.** Use implicit differentiation to find how the rates are related.
> **4.** Substitute any given numerical information into the equation in step 3 to find the desired rate of change.

Here are four applied problems involving related rates.

EXAMPLE 2.6.5

The manager of a company determines that when q hundred units of a particular commodity are produced, the total cost of production is C thousand dollars, where $C^2 - 3q^3 = 4,275$. When 1,500 units are being produced, the level of production is increasing at the rate of 20 units per week. What is the total cost at this time and at what rate is it changing?

Solution

We want to find $\dfrac{dC}{dt}$ when $q = 15$ (1,500 units) and $\dfrac{dq}{dt} = 0.2$ (20 units per week with q measured in hundreds of units). Differentiating the equation $C^2 - 3q^3 = 4,275$ implicitly with respect to time, we get

$$2C\frac{dC}{dt} - 3\left[3q^2\frac{dq}{dt}\right] = 0$$

so that

$$2C\frac{dC}{dt} = 9q^2\frac{dq}{dt}$$

and

$$\frac{dC}{dt} = \frac{9q^2}{2C}\frac{dq}{dt}$$

When $q = 15$, the cost C satisfies

$$C^2 - 3(15)^3 = 4,275$$
$$C^2 = 4,275 + 3(15)^3 = 14,400$$
$$C = 120$$

and by substituting $q = 15$, $C = 120$, and $\dfrac{dq}{dt} = 0.2$ into the formula for $\dfrac{dC}{dt}$, we obtain

$$\frac{dC}{dt} = \left[\frac{9(15)^2}{2(120)}\right](0.2) = 1.6875$$

thousand dollars ($1,687.50) per week. To summarize, the cost of producing 1,500 units is $120,000 ($C = 120$) and at this level of production, total cost is increasing at the rate of $1,687.50 per week.

EXAMPLE | 2.6.6

A storm at sea has damaged an oil rig. Oil spills from the rupture at the constant rate of 60 ft³/min, forming a slick that is roughly circular in shape and 3 inches thick.

a. How fast is the radius of the slick increasing when the radius is 70 feet?

b. Suppose the rupture is repaired in such a way that the flow is shut off instantaneously. If the radius of the slick is increasing at the rate of 0.2 ft/min when the flow stops, what is the total volume of oil that spilled onto the sea?

Solution

We can think of the slick as a cylinder of oil of radius r feet and thickness $h = \dfrac{3}{12} = 0.25$ feet. Such a cylinder will have volume

$$V = \pi r^2 h = 0.25\pi r^2 \ \text{ft}^3$$

Differentiating implicitly in this equation with respect to time t, we get

$$\frac{dV}{dt} = 0.25\pi \left(2r \frac{dr}{dt} \right) = 0.5\pi r \frac{dr}{dt}$$

and since $\dfrac{dV}{dt} = 60$ at all times, we obtain the rate relationship

$$60 = 0.5\pi r \frac{dr}{dt}$$

a. We want to find $\dfrac{dr}{dt}$ when $r = 70$. Substituting into the rate relationship we have just obtained, we find that

$$60 = 0.5\pi (70) \frac{dr}{dt}$$

so that

$$\frac{dr}{dt} = \frac{60}{(0.5)\pi(70)} \approx 0.55$$

Thus, when the radius is 70 feet, it is increasing at about 0.55 ft/min.

b. We can compute the total volume of oil in the spill if we know the radius of the slick at the instant the flow stops. Since $\dfrac{dr}{dt} = 0.2$ at that instant, we have

$$60 = 0.5\pi r(0.2)$$

and the radius is

$$r = \frac{60}{0.5\pi(0.2)} \approx 191 \text{ feet}$$

Therefore, the total amount of oil spilled is

$$V = 0.25\pi(191)^2 \approx 28,652 \text{ ft}^3$$

(about 214,317 gallons).

EXAMPLE 2.6.7

A lake is polluted by waste from a plant located on its shore. Ecologists determine that when the level of pollutant is x parts per million (ppm), there will be F fish of a certain species in the lake, where

$$F = \frac{32,000}{3 + \sqrt{x}}$$

When there are 4,000 fish left in the lake, the pollution is increasing at the rate of 1.4 ppm/year. At what rate is the fish population changing at this time?

Solution

We want to find $\dfrac{dF}{dt}$ when $F = 4,000$ and $\dfrac{dx}{dt} = 1.4$. When there are 4,000 fish in the lake, the level of pollution x satisfies

$$4,000 = \frac{32,000}{3 + \sqrt{x}}$$
$$4,000(3 + \sqrt{x}) = 32,000$$
$$3 + \sqrt{x} = 8$$
$$\sqrt{x} = 5$$
$$x = 25$$

We find that

$$\frac{dF}{dx} = \frac{32,000(-1)}{(3 + \sqrt{x})^2}\left(\frac{1}{2}\frac{1}{\sqrt{x}}\right) = \frac{-16,000}{\sqrt{x}(3 + \sqrt{x})^2}$$

and according to the chain rule

$$\frac{dF}{dt} = \frac{dF}{dx}\frac{dx}{dt} = \left[\frac{-16,000}{\sqrt{x}(3 + \sqrt{x})^2}\right]\frac{dx}{dt}$$

Substituting $F = 4,000$, $x = 25$, and $\dfrac{dx}{dt} = 1.4$, we find that

$$\frac{dF}{dt} = \left[\frac{-16,000}{\sqrt{25}(3 + \sqrt{25})^2}\right](1.4) = -70$$

so the fish population is decreasing by 70 fish per year.

EXAMPLE 2.6.8

When the price of a certain commodity is p dollars per unit, the manufacturer is willing to supply x thousand units, where

$$x^2 - 2x\sqrt{p} - p^2 = 31$$

How fast is the supply changing when the price is $9 per unit and is increasing at the rate of 20 cents per week?

Solution

We know that when $p = 9$, $\dfrac{dp}{dt} = 0.20$. We are asked to find $\dfrac{dx}{dt}$ at this time. First, note that when $p = 9$, we have

$$x^2 - 2x\sqrt{9} - 9^2 = 31$$
$$x^2 - 6x - 112 = 0$$
$$(x + 8)(x - 14) = 0$$
$$x = 14 \quad (x = -8 \text{ has no practical value})$$

Next, we differentiate both sides of the supply equation implicitly with respect to time to obtain

$$2x\frac{dx}{dt} - 2\left[\left(\frac{dx}{dt}\right)\sqrt{p} + x\left(\frac{1}{2}\frac{1}{\sqrt{p}}\frac{dp}{dt}\right)\right] - 2p\frac{dp}{dt} = 0$$

Finally, by substituting $x = 14$, $p = 9$, and $\dfrac{dp}{dt} = 0.20$ into this rate equation and then solving for the required rate $\dfrac{dx}{dt}$, we get

$$2(14)\frac{dx}{dt} - 2\left[\sqrt{9}\frac{dx}{dt} + 14\left(\frac{1}{2}\frac{1}{\sqrt{9}}\right)(0.20)\right] - 2(9)(0.20) = 0$$

$$[28 - 2(3)]\frac{dx}{dt} = 2(14)\left(\frac{1}{2\sqrt{9}}\right)(0.20) + 2(9)(0.20)$$

$$\frac{dx}{dt} = \frac{14\left(\dfrac{1}{3}\right)(0.20) + 18(0.20)}{22}$$

$$\approx 0.206$$

Since the supply is given in terms of thousands of units, it follows that the supply is increasing at the rate of $0.206(1,000) = 206$ units per week.

PROBLEMS 2.6

In Problems 1 through 6, find $\dfrac{dy}{dx}$ in two ways:

   (a) by implicit differentiation
   (b) by differentiating an explicit formula for y.
In each case, show that the two answers are the same.

1. $2x + 3y = 7$ **2.** $xy = 4$

3. $xy + 2y = 3$ **4.** $x^2 + y^3 = 12$

5. $xy + 2y = x^2$ **6.** $x + \dfrac{1}{y} = 5$

In Problems 7 through 20, find $\dfrac{dy}{dx}$ by implicit differentiation.

7. $x^2 + y^2 = 25$ **8.** $x^2 + y = x^3 + y^2$

9. $x^3 + y^3 = xy$

10. $5x - x^2y^3 = 2y$

11. $y^2 + 2xy^2 - 3x + 1 = 0$

12. $\dfrac{1}{x} + \dfrac{1}{y} = 1$

13. $\sqrt{x} + \sqrt{y} = 1$

14. $\sqrt{2x} + y^2 = 4$

15. $xy - x = y + 2$

16. $y^2 + 3xy - 4x^2 = 9$

17. $(2x + y)^3 = x$

18. $(x - 2y)^2 = y$

19. $(x^2 + 3y^2)^5 = 2xy$

20. $(3xy^2 + 1)^4 = 2x - 3y$

In Problems 21 through 26, find the equation of the tangent line to the given curve at the specified point.

21. $x^2 = y^3$; $(8, 4)$

22. $\dfrac{1}{x} - \dfrac{1}{y} = 2$; $\left(\dfrac{1}{4}, \dfrac{1}{2}\right)$

23. $xy = 2$; $(2, 1)$

24. $x^2y^3 - 2xy = 6x + y + 1$; $(0, -1)$

25. $(1 - x + y)^3 = x + 7$; $(1, 2)$

26. $(x^2 + 2y)^3 = 2xy^2 + 64$; $(0, 2)$

In Problems 27 through 32, find all points (both coordinates) on the given curve where the tangent line is (a) horizontal and (b) vertical.

27. $x + y^2 = 9$

28. $xy = 16y^2$

29. $x^2 + xy + y = 3$

30. $\dfrac{y}{x} - \dfrac{x}{y} = 5$

31. $x^2 + xy + y^2 = 3$

32. $x^2 - xy + y^2 = 3$

In Problems 33 and 34, use implicit differentiation to find the second derivative $\dfrac{d^2y}{dx^2}$.

33. $x^2 + 3y^2 = 5$

34. $xy + y^2 = 1$

35. MANUFACTURING   The output at a certain plant is $Q = 0.08x^2 + 0.12xy + 0.03y^2$ units per day, where x is the number of hours of skilled labor used and y the number of hours of unskilled labor used. Currently, 80 hours of skilled labor and 200 hours of unskilled labor are used each day. Use calculus to estimate the change in unskilled labor that should be made to offset a 1-hour increase in skilled labor so that output will be maintained at its current level.

36. MANUFACTURING   The output of a certain plant is $Q = 0.06x^2 + 0.14xy + 0.05y^2$ units per day, where x is the number of hours of skilled labor used and y is the number of hours of unskilled labor used. Currently, 60 hours of skilled labor and 300 hours of unskilled labor are used each day. Use calculus to estimate the change in unskilled labor that should be made to offset a 1-hour increase in skilled labor so that output will be maintained at its current level.

37. SUPPLY RATE   When the price of a certain commodity is p dollars per unit, the manufacturer is willing to supply x hundred units, where
$$3p^2 - x^2 = 12$$
How fast is the supply changing when the price is \$4 per unit and is increasing at the rate of 87 cents per month?

38. DEMAND RATE   When the price of a certain commodity is p dollars per unit, customers demand x hundred units of the commodity, where
$$x^2 + 3px + p^2 = 79$$
How fast is the demand x changing with respect to time when the price is \$5 per unit and is decreasing at the rate of 30 cents per month?

39. GROWTH OF A TUMOR   A tumor is modeled as being roughly spherical, with radius R. If the radius of the tumor is currently $R = 0.54$ cm and is increasing at the rate of 0.13 cm per month, what is the corresponding rate of change of the volume $V = \dfrac{4}{3}\pi R^3$?

40. REFRIGERATION   An ice block used for refrigeration is modeled as a cube of side s. The block currently has volume 125,000 cm³ and is melting at the rate of 1,000 cm³ per hour.

a. What is the current length s of each side of the cube? At what rate is s currently changing with respect to time t?

b. What is the current rate of change of the surface area S of the block with respect to time? [*Note:* A cube of side s has volume $V = s^3$ and surface area $S = 6s^2$.]

41. **MEDICINE** A tiny spherical balloon is inserted into a clogged artery and is inflated at the rate of 0.002π mm³/min. How fast is the radius of the balloon growing when the radius is $R = 0.005$ mm? [*Note:* A sphere of radius R has volume $V = \frac{4}{3}\pi R^3$.]

42. **POLLUTION CONTROL** An environmental study for a certain community indicates that there will be $Q(p) = p^2 + 4p + 900$ units of a harmful pollutant in the air when the population is p thousand people. If the population is currently 50,000 and is increasing at the rate of 1,500 per year, at what rate is the level of pollution increasing?

43. **DEMAND RATE** When the price of a certain commodity is p dollars per unit, consumers demand x hundred units of the commodity, where
$$75x^2 + 17p^2 = 5,300$$
How fast is the demand x changing with respect to time when the price is \$7 and is decreasing at the rate of 75 cents per month? (That is, $\dfrac{dp}{dt} = -0.75$.)

44. **BOYLE'S LAW** Boyle's law states that when gas is compressed at constant temperature, the pressure P and volume V of a given sample satisfy the equation $PV = C$, where C is constant. Suppose that at a certain time the volume is 40 in.³, the pressure is 70 lb/in.², and the volume is increasing at the rate of 12 in.³/sec. How fast is the pressure changing at this instant? Is it increasing or decreasing?

45. **METABOLIC RATE** The *basal metabolic rate* is the rate of heat produced by an animal per unit time. Observations indicate that the basal metabolic rate of a warm-blooded animal of mass w kilograms (kg) is given by
$$M = 70w^{3/4} \text{ kilocalories per day}$$

a. Find the rate of change of the metabolic rate of an 80-kg cougar that is gaining mass at the rate of 0.8 kg per day.

b. Find the rate of change of the metabolic rate of a 50-kg ostrich that is losing mass at the rate of 0.5 kg per day.

46. **SPEED OF A LIZARD** Herpetologists have proposed using the formula $s = 1.1w^{0.2}$ to estimate the maximum sprinting speed s (meters per second) of a lizard of mass w (grams). At what rate is the maximum sprinting speed of an 11-gram lizard increasing if the lizard is growing at the rate of 0.02 grams per day?

47. **PRODUCTION** At a certain factory, output is given by $Q = 60\,K^{1/3}L^{2/3}$ units, where K is the capital investment (in thousands of dollars) and L is the size of the labor force, measured in worker-hours. If output is kept constant, at what rate is capital investment changing at a time when $K = 8$, $L = 1,000$, and L is increasing at the rate of 25 worker-hours per week?
[*Note:* Output functions such as the one in Problem 47 of the general form $Q = AK^\alpha L^{1-\alpha}$, where A and α are constants with $0 \le \alpha \le 1$, are called **Cobb-Douglas production functions.** Such functions appear in examples and exercises throughout this text, especially in Chapter 7.]

48. **WATER POLLUTION** A circular oil slick spreads in such a way that its radius is increasing at the rate of 20 ft/hr. How fast is the area of the slick changing when the radius is 200 feet?

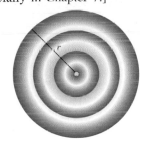

PROBLEM 48

49. A 6-foot-tall man walks at the rate of 4 ft/sec away from the base of a street light 12 feet above the ground. At what rate is the length of his shadow changing when he is 20 feet away from the base of the light?

12

6

20

PROBLEM 49

50. **CHEMISTRY** In an *adiabatic* chemical process, there is no net change (gain or loss) of heat. Suppose a container of oxygen is subjected to such a process. Then if the pressure on the oxygen is P and its volume is V, it can be shown that $PV^{1.4} = C$, where C is a constant. At a certain time, $V = 5$ m^3, $P = 0.6$ kg/m^2, and P is increasing at 0.23 kg/m^2 per sec. What is the rate of change of V? Is V increasing or decreasing?

51. **MANUFACTURING** At a certain factory, output Q is related to inputs x and y by the equation
$$Q = 2x^3 + 3x^2y^2 + (1 + y)^3$$
If the current levels of input are $x = 30$ and $y = 20$, use calculus to estimate the change in input y that should be made to offset a decrease of 0.8 unit in input x so that output will be maintained at its current level.

52. **LUMBER PRODUCTION** To estimate the amount of wood in the trunk of a tree, it is reasonable to assume that the trunk is a cutoff cone (see the figure). If the upper radius of the trunk is r, the lower radius is R, and the height is H, the volume of the wood is given by
$$V = \frac{\pi}{3}H(R^2 + rR + r^2)$$
Suppose r, R, and H are increasing at the respective rates of 4 in/yr, 5 in/yr, and 9 in/yr. At what rate is V increasing at the time when $r = 2$ feet, $R = 3$ feet, and $H = 15$ feet?

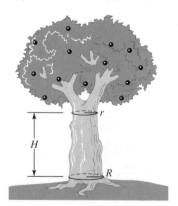

PROBLEM 52

53. **BLOOD FLOW** One of Poiseuille's laws (see Problem 57, Section 1.1) says that the speed of blood flowing under constant pressure in a blood vessel at a distance r from the center of the vessel is given by

$$v = \frac{K}{L}(R^2 - r^2)$$

where K is a positive constant, R is the radius of the vessel, and L is the length of the vessel.* Suppose the radius R and length L of the vessel change with time in such a way that the speed of blood flowing at the center is unaffected; that is, v does not change with time. Show that in this case, the relative rate of change of L with respect to time must be twice the relative rate of change of R.

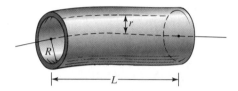

PROBLEM 53

54. **MANUFACTURING** At a certain factory, output Q is related to inputs u and v by the equation
$$Q = 3u^2 + \frac{2u + 3v}{(u + v)^2}$$
If the current levels of input are $u = 10$ and $v = 25$, use calculus to estimate the change in input v that should be made to offset a decrease of 0.7 unit in input u so that output will be maintained at its current level.

55. Show that the tangent line to the curve
$$\frac{x^2}{a^2} + \frac{y^2}{b^2} = 1$$
at the point (x_0, y_0) is
$$\frac{x_0 x}{a^2} + \frac{y_0 y}{b^2} = 1$$

56. Consider the equation $x^2 + y^2 = 6y - 10$.
   a. Show that there are no points (x, y) that satisfy this equation. [*Hint:* Complete the square.]
   b. Show that by applying implicit differentiation to the given equation, you obtain
$$\frac{dy}{dx} = \frac{x}{3 - y}$$
The point of this exercise is to show that one must be careful when applying implicit differentiation.

*E. Batschelet, *Introduction to Mathematics for Life Scientists,* 3rd ed., New York: Springer-Verlag, 1979, pp. 102–103.

Just because it is possible to find a derivative formally by implicit differentiation does not mean that the derivative has any meaning.

57. Prove the power rule $\frac{d}{dx}[x^n] = nx^{n-1}$ for the case where $n = r/s$ is a rational number. [*Hint:* Note that if $y = x^{r/s}$, then $y^s = x^r$ and use implicit differentiation.]

58. Use a graphing utility to graph the curve $5x^2 - 2xy + 5y^2 = 8$. Draw the tangent line to the curve at $(1, 1)$. How many horizontal tangent lines does the curve have? Find the equation of each horizontal tangent.

59. Use a graphing utility to graph the curve $11x^2 + 4xy + 14y^2 = 21$. Draw the tangent line to the curve at $(-1, 1)$. How many horizontal

tangent lines does the curve have? Find the equation of each horizontal tangent.

60. Answer the following questions about the curve $x^3 + y^3 = 3xy$ (called the **folium of Descartes**).
   a. Find the equation of each horizontal tangent to the curve.
   b. The curve intersects the line $y = x$ at exactly one point other than the origin. What is the equation of the tangent line at this point?
   c. Try to figure out a way to sketch the graph of the curve using your calculator (it won't be easy).

61. Use a graphing utility to graph the curve $x^2 + y^2 = \sqrt{x^2 + y^2} + x$. Find the equation of each horizontal tangent to the curve. (The curve is called a **cardioid**.)

Important Terms, Symbols, and Formulas

Secant line and tangent line   (98 and 100)

Average and instantaneous rates of change   (100)

Difference quotient
$$\frac{f(x + h) - f(x)}{h} \quad (101)$$

Derivative of $f(x)$: $f'(x) = \lim_{h \to 0} \dfrac{f(x + h) - f(x)}{h}$ (101)

Differentiation   (98)

Differentiable function   (101)

The derivative $f'(x_0)$ as slope of the tangent line to $y = f(x)$ at $(x_0, f(x_0))$ (102)

The derivative $f'(x_0)$ as the rate of change of $f(x)$ with respect to x when $x = x_0$ (102)

Derivative notation for $f(x)$: $f'(x)$ and $\dfrac{df}{dx}$ (105)

A differentiable function is continuous   (106)

The constant rule: $\dfrac{d}{dx}[c] = 0$ (113)

The power rule: $\dfrac{d}{dx}[x^n] = nx^{n-1}$ (113)

The constant multiple rule: $\dfrac{d}{dx}[cf(x)] = c\dfrac{df}{dx}$ (115)

The sum rule: $\dfrac{d}{dx}[f(x) + g(x)] = \dfrac{df}{dx} + \dfrac{dg}{dx}$ (115)

Relative rate of change of $Q(x)$: $\dfrac{Q'(x)}{Q(x)}$ (117)

Percentage rate of change of $Q(x)$: $\dfrac{100Q'(x)}{Q(x)}$ (117)

Rectilinear motion: position $s(t)$
velocity $v(t) = s'(t)$
acceleration $a(t) = v'(t)$ (118)

The motion of a projectile:

height $H(t) = -\dfrac{1}{2}gt^2 + V_0t + H_0$ (119)

The product rule:

$$\frac{d}{dx}[f(x)g(x)] = f(x)\frac{dg}{dx} + g(x)\frac{df}{dx} \quad (125)$$

The quotient rule: For $g(x) \neq 0$

$$\frac{d}{dx}\left[\frac{f(x)}{g(x)}\right] = \frac{g(x)\dfrac{df}{dx} - f(x)\dfrac{dg}{dx}}{g^2(x)} \quad (128)$$

The second derivative of $f(x)$: $f''(x) = [f'(x)]' = \dfrac{d^2f}{dx^2}$
gives the rate of change of $f'(x)$ (131)

Notation for the nth derivative of $f(x)$:

$$f^{(n)}(x) = \frac{d^nf}{dx^n} \quad (133)$$

The chain rule for $y = f(u(x))$: $\dfrac{dy}{dx} = \dfrac{dy}{du}\dfrac{du}{dx}$

or $[f(u(x))]' = f'(u)\,u'(x)$ (139)

The general power rule:

$\dfrac{d}{dx}[h(x)]^n = n[h(x)]^{n-1}\dfrac{dh}{dx}$ (143)

Marginal cost $C'(x_0)$ estimates $C(x_0 + 1) - C(x_0)$, the extra cost of producing the $(x_0 + 1)$st unit (152)

Marginal revenue $R'(x_0)$ estimates $R(x_0 + 1) - R(x_0)$, the extra revenue from producing the $(x_0 + 1)$st unit (153)

Marginal profit: $P'(x_0)$ estimates $P(x_0 + 1) - P(x_0)$, the extra profit from producing the $(x_0 + 1)$st unit (153)

Approximation by increments:

$f(x_0 + \Delta x) \approx f(x_0) + f'(x_0)\Delta x$ (155)

Propagated error (156)

Differential of $y = f(x)$ is $dy = f'(x)\,dx$ (158)

Implicit differentiation (164–165)

Related rates (168)

Checkup for Chapter 2

1. In each case, find the derivative $\dfrac{dy}{dx}$.

   a. $y = 3x^4 - 4\sqrt{x} + \dfrac{5}{x^2} - 7$

   b. $y = (3x^3 - x + 1)(4 - x^2)$

   c. $y = \dfrac{5x^2 - 3x + 2}{1 - 2x}$

   d. $y = (3 - 4x + 3x^2)^{3/2}$

2. Find the second derivative of the function $f(t) = t(2t + 1)^2$.

3. Find an equation for the tangent line to the curve $y = x^2 - 2x + 1$ at the point where $x = -1$.

4. Find the rate of change of the function $f(x) = \dfrac{x + 1}{1 - 5x}$ with respect to x when $x = 1$.

5. **PROPERTY TAX** Records indicate that x years after the year 2000, the average property tax on a four-bedroom home in a suburb of a major city was $T(x) = 3x^2 + 40x + 1{,}800$ dollars.

   a. At what rate was the property tax increasing with respect to time in 2003?

   b. At what percentage rate was the property tax increasing in 2003?

6. **MOTION ON A LINE** An object moves along a line in such a way that its position at time t is given by $s(t) = 2t^3 - 3t^2 + 2$ for $t \geq 0$.

   a. Find the velocity $v(t)$ and acceleration $a(t)$ of the object.

   b. When is the object stationary? When is it advancing? Retreating?

   c. What is the total distance traveled by the object for $0 \leq t \leq 2$?

7. **PRODUCTION COST** Suppose the cost of producing x units of a particular commodity is $C(x) = 0.04x^2 + 5x + 73$ hundred dollars.

   a. Use marginal cost to estimate the cost of producing the sixth unit.

   b. What is the actual cost of producing the sixth unit?

8. **INDUSTRIAL OUTPUT** At a certain factory, the daily output is $Q = 500L^{3/4}$ units, where L denotes the size of the labor force in worker-hours. Currently, 2,401 worker-hours of labor are used each day. Use calculus (increments) to estimate the effect on output of increasing the size of the labor force by 200 worker-hours from its current level.

9. **PEDIATRIC MEASUREMENT** Pediatricians use the formula $S = 0.2029w^{0.425}$ to estimate the surface area S (m^2) of a child 1 meter tall who weighs w kilograms (kg). A particular child weighs 30 kg and is gaining weight at the rate of 0.13 kg per week while remaining 1 meter tall. At what rate is this child's surface area changing?

10. **GROWTH OF A TUMOR** A cancerous tumor is roughly spherical in shape. Estimate the percentage error that can be allowed in the measurement of the radius r to ensure there will be no more than an 8% error in the calculation of volume using the formula $V = \dfrac{4}{3}\pi r^3$.

Review Problems

In Problems 1 and 2, use the definition of the derivative to find $f'(x)$.

1. $f(x) = x^2 - 3x + 1$

2. $f(x) = \dfrac{1}{x-2}$

In Problems 3 through 13, find the derivative of the given function.

3. $f(x) = 6x^4 - 7x^3 + 2x + \sqrt{2}$

4. $f(x) = x^3 - \dfrac{1}{3x^5} + 2\sqrt{x} - \dfrac{3}{x} + \dfrac{1-2x}{x^3}$

5. $y = \dfrac{2-x^2}{3x^2+1}$

6. $y = (x^3 + 2x - 7)(3 + x - x^2)$

7. $f(x) = (5x^4 - 3x^2 + 2x + 1)^{10}$

8. $f(x) = \sqrt{x^2+1}$

9. $y = \left(x + \dfrac{1}{x}\right)^2 - \dfrac{5}{\sqrt{3x}}$

10. $y = \left(\dfrac{x+1}{1-x}\right)^2$

11. $f(x) = (3x+1)\sqrt{6x+5}$

12. $f(x) = \dfrac{(3x+1)^3}{(1-3x)^4}$

13. $y = \sqrt{\dfrac{1-2x}{3x+2}}$

In Problems 14 through 17, find an equation for the tangent line to the graph of the given function at the specified point.

14. $f(x) = x^2 - 3x + 2$; $x = 1$

15. $f(x) = \dfrac{4}{x-3}$; $x = 1$

16. $f(x) = \dfrac{x}{x^2+1}$; $x = 0$

17. $f(x) = \sqrt{x^2+5}$; $x = -2$

18. In each of these cases, find the rate of change of $f(t)$ with respect to t at the given value of t.
    a. $f(t) = t^3 - 4t^2 + 5t\sqrt{t} - 5$ at $t = 4$
    b. $f(t) = \dfrac{2t^2 - 5}{1 - 3t}$ at $t = -1$

c. $f(t) = t^3(t^2 - 1)$ at $t = 0$
d. $f(t) = (t^2 - 3t + 6)^{1/2}$ at $t = 1$

19. In each of these cases, find the percentage rate of change of the function $f(t)$ with respect to t at the given value of t.
    a. $f(t) = t^2 - 3t + \sqrt{t}$ at $t = 4$
    b. $f(t) = t^2(3 - 2t)^3$ at $t = 1$
    c. $f(t) = \dfrac{1}{t+1}$ at $t = 0$

20. Use the chain rule to find $\dfrac{dy}{dx}$.
    a. $y = 5u^2 + u - 1$; $u = 3x + 1$
    b. $y = \dfrac{1}{u^2}$; $u = 2x + 3$

21. Use the chain rule to find $\dfrac{dy}{dx}$ for the given value of x.
    a. $y = u^3 - 4u^2 + 5u + 2$; $u = x^2 + 1$; for $x = 1$
    b. $y = \sqrt{u}$, $u = x^2 + 2x - 4$; for $x = 2$
    c. $y = \left(\dfrac{u-1}{u+1}\right)^{1/2}$, $u = \sqrt{x-1}$; for $x = \dfrac{34}{9}$

22. Find the second derivative of each of these functions:
    a. $f(x) = 6x^5 - 4x^3 + 5x^2 - 2x + \dfrac{1}{x}$
    b. $z = \dfrac{2}{1+x^2}$
    c. $y = (3x^2 + 2)^4$
    d. $f(x) = 2x(x+4)^3$
    e. $f(x) = \dfrac{x-1}{(x+1)^2}$

23. Find $\dfrac{dy}{dx}$ by implicit differentiation.
    a. $5x + 3y = 12$
    b. $x^2 y = 1$
    c. $(2x + 3y)^5 = x + 1$
    d. $(1 - 2xy^3)^5 = x + 4y$

24. Use implicit differentiation to find the slope of the line that is tangent to the given curve at the specified point.
    a. $xy^3 = 8$; $(1, 2)$
    b. $x^2 y - 2xy^3 + 6 = 2x + 2y$; $(0, 3)$

c. $x^2 + 2y^3 = \dfrac{3}{xy}$; $(1, 1)$

d. $y = \dfrac{x + y}{x - y}$; $(6, 2)$

25. Use implicit differentiation to find $\dfrac{d^2y}{dx^2}$ if $3x^2 - 2y^2 = 6$.

26. A projectile is launched vertically upward from ground level with an initial velocity of 160 ft/sec.

   a. When will the projectile hit the ground?

   b. What is the impact velocity?

   c. When will the projectile reach its maximum height? What is the maximum height?

27. **POPULATION GROWTH** Suppose that a 5-year projection of population trends suggests that t years from now, the population of a certain community will be P thousand, where
$$P(t) = -t^3 + 9t^2 + 48t + 200.$$

   a. At what rate will the population be growing 3 years from now?

   b. At what rate will the rate of population growth be changing with respect to time 3 years from now?

In Problems 28 and 29, s(t) denotes the position of an object moving along a line.

   (a) Find the velocity and acceleration of the object and describe its motion during the indicated time interval.

   (b) Compute the total distance traveled by the object during the indicated time interval.

28. $s(t) = 2t^3 - 21t^2 + 60t - 25$; $1 \le t \le 6$

29. $s(t) = \dfrac{2t + 1}{t^2 + 12}$; $0 \le t \le 4$

30. **RAPID TRANSIT** After x weeks, the number of people using a new rapid transit system was approximately $N(x) = 6x^3 + 500x + 8,000$.

   a. At what rate was the use of the system changing with respect to time after 8 weeks?

   b. By how much did the use of the system change during the eighth week?

31. **PRODUCTION** It is estimated that the weekly output at a certain plant is $Q(x) = 50x^2 + 9,000x$ units, where x is the number of workers employed at the plant. Currently there are 30 workers employed at the plant.

   a. Use calculus to estimate the change in the weekly output that will result from the addition of 1 worker to the force.

   b. Compute the actual change in output that will result from the addition of 1 worker.

32. **POPULATION** It is projected that t months from now, the population of a certain town will be $P(t) = 3t + 5t^{3/2} + 6,000$. At what percentage rate will the population be changing with respect to time 4 months from now?

33. **PRODUCTION** At a certain factory, the daily output is $Q(L) = 20,000L^{1/2}$ units, where L denotes the size of the labor force measured in worker-hours. Currently 900 worker-hours of labor are used each day. Use calculus to estimate the effect on output that will be produced if the labor force is cut to 885 worker-hours.

34. **GROSS DOMESTIC PRODUCT** The gross domestic product of a certain country was $N(t) = t^2 + 6t + 300$ billion dollars t years after 2000. Use calculus to predict the percentage change in the GDP during the second quarter of 2008.

35. **POLLUTION** The level of air pollution in a certain city is proportional to the square of the population. Use calculus to estimate the percentage by which the air pollution level will increase if the population increases by 5%.

36. **AIDS EPIDEMIC** In its early phase, specifically the period 1984–1990, the AIDS epidemic could be modeled* by the cubic function
$$C(t) = -170.36t^3 + 1,707.5t^2 + 1,998.4t + 4,404.8$$
for $0 \le t \le 6$, where C is the number of reported cases t years after the base year 1984.

   a. Compute and interpret the derivative $C'(t)$.

   b. At what rate was the epidemic spreading in the year 1984?

   c. At what percentage rate was the epidemic spreading in 1984? In 1990?

37. **POPULATION DENSITY** The formula $D = 36\,m^{-1.14}$ is sometimes used to determine the ideal population density D (individuals per square kilometer) for a large animal of mass m kilograms (kg).

   a. What is the ideal population density for humans, assuming that a typical human weighs about 70 kg?

Mortality and Morbidity Weekly Report, U.S. Centers for Disease Control, Vol. 40, No. 53, October 2, 1992.

b. The area of the United States is about 9.2 million square kilometers. What would the population of the United States have to be for the population density to be ideal?

c. Consider an island of area 3,000 km² Two hundred animals of mass $m = 30$ kg are brought to the island, and t years later, the population is given by
$$P(t) = 0.43t^2 + 13.37t + 200$$
How long does it take for the ideal population density to be reached? At what rate is the population changing when the ideal density is attained?

38. BACTERIAL GROWTH The population P of a bacterial colony t days after observation begins is modeled by the cubic function
$$P(t) = 1.035t^3 + 103.5t^2 + 6,900t + 230,000$$

a. Compute and interpret the derivative $P'(t)$.

b. At what rate is the population changing after 1 day? After 10 days?

c. What is the initial population of the colony? How long does it take for the population to double? At what rate is the population growing at the time it doubles?

39. PRODUCTION The output at a certain factory is $Q(L) = 600L^{2/3}$ units, where L is the size of the labor force. The manufacturer wishes to increase output by 1%. Use calculus to estimate the percentage increase in labor that will be required.

40. PRODUCTION The output Q at a certain factory is related to inputs x and y by the equation
$$Q = x^3 + 2xy^2 + 2y^3$$
If the current levels of input are $x = 10$ and $y = 20$, use calculus to estimate the change in input y that should be made to offset an increase of 0.5 in input x so that output will be maintained at its current level.

41. BLOOD FLOW Physiologists have observed that the flow of blood from an artery into a small capillary is given by the formula
$$F = kD^2\sqrt{A - C} \quad \text{(cm}^3\text{/sec)}$$
where D is the diameter of the capillary, A is the pressure in the artery, C is the pressure in the capillary, and k is a positive constant.

a. By how much is the flow of blood F changing with respect to pressure C in the capillary if A and D are kept constant? Does the flow increase or decrease with increasing C?

b. What is the percentage rate of change of flow F with respect to A if C and D are kept constant?

42. POLLUTION CONTROL It is estimated that t years from now, the population of a certain suburban community will be $p(t) = 10 - \dfrac{20}{(t + 1)^2}$ thousand. An environmental study indicates that the average daily level of carbon monoxide in the air will be $c(p) = 0.8\sqrt{p^2 + p + 139}$ units when the population is p thousand. At what percentage rate will the level of carbon monoxide be changing with respect to time 1 year from now?

43. You measure the radius of a circle to be 12 cm and use the formula $A = \pi r^2$ to calculate the area. If your measurement of the radius is accurate to within 3%, how accurate is your calculation of the area?

44. Estimate what will happen to the volume of a cube if the length of each side is decreased by 2%. Express your answer as a percentage.

45. PRODUCTION The output at a certain factory is $Q = 600K^{1/2}L^{1/3}$ units, where K denotes the capital investment and L is the size of the labor force. Estimate the percentage increase in output that will result from a 2% increase in the size of the labor force if capital investment is not changed.

46. BLOOD FLOW The speed of blood flowing along the central axis of a certain artery is $S(R) = 1.8 \times 10^5R^2$ centimeters per second, where R is the radius of the artery. A medical researcher measures the radius of the artery to be 1.2×10^{-2} centimeter and makes an error of 5×10^{-4} centimeter. Estimate the amount by which the calculated value of the speed of the blood will differ from the true speed if the incorrect value of the radius is used in the formula.

47. SIZE OF A TUMOR You measure the radius of a spherical tumor to be 1.2 cm and use the formula $V = \dfrac{4}{3}\pi r^3$ to calculate the volume. If your measurement of the radius is accurate to within 3%, how accurate is your calculation of the volume?

48. CARDIOVASCULAR SYSTEM One model of the cardiovascular system relates $V(t)$, the stroke volume of blood in the aorta at a time t during systole (the contraction phase), to the pressure $P(t)$ in the aorta during systole by the equation
$$V(t) = [C_1 + C_2P(t)]\left(\frac{3t^2}{T^2} - \frac{2t^3}{T^3}\right)$$

CHAPTER SUMMARY

where C_1 and C_2 are positive constants and T is the (fixed) time length of the systole phase.* Find a relationship between the rates $\dfrac{dV}{dt}$ and $\dfrac{dP}{dt}$.

49. **CONSUMER DEMAND**   When electric toasters are sold for p dollars apiece, local consumers will buy $D(p) = \dfrac{32{,}670}{2p + 1}$ toasters. It is estimated that t months from now, the unit price of the toasters will be $p(t) = 0.04t^{3/2} + 44$ dollars. Compute the rate of change of the monthly demand for toasters with respect to time 25 months from now. Will the demand be increasing or decreasing?

50. At noon, a truck is at the intersection of two roads and is moving north at 70 km/hr. An hour later, a car passes through the same intersection, traveling east at 105 km/hr. How fast is the distance between the car and truck changing at 2 P.M.?

51. **POPULATION GROWTH**   It is projected that t years from now, the population of a certain suburban community will be $P(t) = 20 - \dfrac{6}{t + 1}$ thousand. By approximately what percentage will the population grow during the next quarter year?

52. **WORKER EFFICIENCY**   An efficiency study of the morning shift at a certain factory indicates that an average worker arriving on the job at 8:00 A.M. will have produced $Q(t) = -t^3 + 9t^2 + 12t$ units t hours later.

   a. Compute the worker's rate of production $R(t) = Q'(t)$.

   b. At what rate is the worker's rate of production changing with respect to time at 9:00 A.M.?

   c. Use calculus to estimate the change in the worker's rate of production between 9:00 and 9:06 A.M.

   d. Compute the actual change in the worker's rate of production between 9:00 and 9:06 A.M.

53. **TRAFFIC SAFETY**   A car is traveling at 88 ft/sec when the driver applies the brakes to avoid hitting a child. After t seconds, the car is $s = 88t - 8t^2$ feet from the point where the brakes were applied. How long does it take for the car to come to a stop, and how far does it travel before stopping?

54. **CONSTRUCTION MATERIAL**   Sand is leaking from a bag in such a way that after t seconds, there are

$$S(t) = 50\left(1 - \frac{t^2}{15}\right)^3$$

pounds of sand left in the bag.

   a. How much sand was originally in the bag?

   b. At what rate is sand leaking from the bag after 1 second?

   c. How long does it take for all of the sand to leak from the bag? At what rate is the sand leaking from the bag at the time it empties?

55. **INFLATION**   It is projected that t months from now, the average price per unit for goods in a certain sector of the economy will be P dollars, where $P(t) = -t^3 + 7t^2 + 200t + 300$.

   a. At what rate will the price per unit be increasing with respect to time 5 months from now?

   b. At what rate will the rate of price increase be changing with respect to time 5 months from now?

   c. Use calculus to estimate the change in the rate of price increase during the first half of the sixth month.

   d. Compute the actual change in the rate of price increase during the first half of the sixth month.

56. **PRODUCTION COST**   At a certain factory, approximately $q(t) = t^2 + 50t$ units are manufactured during the first t hours of a production run, and the total cost of manufacturing q units is $C(q) = 0.1q^2 + 10q + 400$ dollars. Find the rate at which the manufacturing cost is changing with respect to time 2 hours after production commences.

57. **PRODUCTION COST**   It is estimated that the monthly cost of producing x units of a particular commodity is $C(x) = 0.06x + 3x^{1/2} + 20$ hundred dollars. Suppose production is decreasing at the rate of 11 units per month when the monthly production is 2,500 units. At what rate is the cost changing at this level of production?

58. Estimate the largest percentage error you can allow in the measurement of the radius of a sphere if you want the error in the calculation of its surface area using the formula $S = 4\pi r^2$ to be no greater than 8%.

59. A soccer ball made of leather 1/8 inch thick has an inner diameter of 8.5 inches. Estimate the volume of

its leather shell. [*Hint:* Think of the volume of the shell as a certain change ΔV in volume.]

60. A car traveling north at 60 mph and a truck traveling east at 45 mph leave an intersection at the same time. At what rate is the distance between them changing 2 hours later?

61. A child is flying a kite at a height of 80 feet above her hand. If the kite moves horizontally at a constant speed of 5 ft/sec, at what rate is string being paid out when the kite is 100 feet away from the child?

62. A person stands at the end of a pier 8 feet above the water and pulls in a rope attached to a buoy. If the rope is hauled in at the rate of 2 ft/min, how fast is the buoy moving in the water when it is 6 feet from the pier?

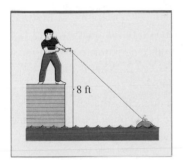

PROBLEM 62

63. A 10-foot-long ladder leans against the side of a wall. The top of the ladder is sliding down the wall at the rate of 3 ft/sec. How fast is the foot of the ladder moving away from the building when the top is 6 feet above the ground?

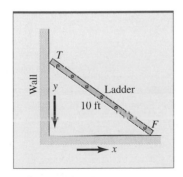

PROBLEM 63

64. A lantern falls from the top of a building in such a way that after t seconds, it is $h(t) = 150 - 16t^2$ feet above ground. A woman 5 feet tall originally stand-

ing directly under the lantern sees it start to fall and walks away at the constant rate of 5 ft/sec. How fast is the length of the woman's shadow changing when the lantern is 10 feet above the ground?

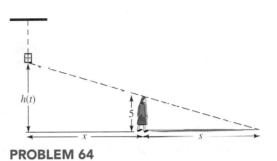

PROBLEM 64

65. A baseball diamond is a square, 90 feet on a side. A runner runs from second base to third at 20 ft/sec. How fast is the distance s between the runner and home base changing when he is 15 feet from third base?

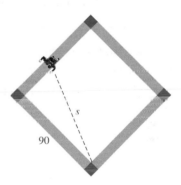

PROBLEM 65

66. **MANUFACTURING COST** Suppose the total manufacturing cost C at a certain factory is a function of the number q of units produced, which in turn is a function of the number t of hours during which the factory has been operating.

a. What quantity is represented by the derivative $\dfrac{dC}{dq}$? In what units is this quantity measured?

b. What quantity is represented by the derivative $\dfrac{dq}{dt}$? In what units is this quantity measured?

c. What quantity is represented by the product $\dfrac{dC}{dq}\dfrac{dq}{dt}$? In what units is this quantity measured?

67. An object projected from a point P moves along a straight line. It is known that the velocity of the object is directly proportional to the product of the time the object has been moving and the distance it has moved from P. It is also known that at the end of 5 seconds, the object is 20 feet from P and is moving at the rate of 4 ft/sec. Find the acceleration of the object at this time (when $t = 5$).

68. Find all the points (x, y) on the graph of the function $y = 4x^2$ with the property that the tangent to the graph at (x, y) passes through the point $(2, 0)$.

69. Suppose y is a linear function of x; that is, $y = mx + b$. What will happen to the percentage rate of change of y with respect to x as x increases without bound? Explain.

70. Find an equation for the tangent line to the curve

$$\frac{x^2}{a^2} - \frac{y^2}{b^2} = 1$$

at the point (x_0, y_0).

71. Let $f(x) = (3x + 5)(2x^3 - 5x + 4)$. Use a graphing utility to graph $f(x)$ and $f'(x)$ on the same set of coordinate axes. Use **TRACE** and **ZOOM** to find where $f'(x) = 0$.

72. Use a graphing utility to graph $f(x) = \dfrac{(2x + 3)}{(1 - 3x)}$ and $f'(x)$ on the same set of coordinate axes. Use **TRACE** and **ZOOM** to find where $f'(x) = 0$.

73. The curve $y^2(2 - x) = x^3$ is called a **cissoid.**

   a. Use a graphing utility to sketch the curve.

   b. Find an equation for the tangent line to the curve at all points where $x = 1$.

   c. What happens to the curve as x approaches 2 from the left?

   d. Does the curve have a tangent line at the origin? If so, what is its equation?

74. An object moves along a straight line in such a way that its position at time t is given by

   $$s(t) = t^{5/2}(0.73t^2 - 3.1t + 2.7) \qquad \text{for } 0 \le t \le 2$$

   a. Find the velocity $v(t)$ and the acceleration $a(t)$ and then use a graphing utility to graph $s(t)$, $v(t)$, and $a(t)$ on the same axes for $0 \le t \le 2$.

   b. Use your calculator to find a time when $v(t) = 0$ for $0 \le t \le 2$. What is the object's position at this time?

   c. When does the smallest value of $a(t)$ occur? Where is the object at this time and what is its velocity?

Finding Roots

1. The x intercepts of a function, also called **roots,** are an important feature of a function. Later we will see what characteristics of a function the roots of its derivative can tell us. For now, we explore the various ways a graphing calculator can find roots.

Check that your calculator is in function mode (Func) by accessing the **MODE** menu. Store $f(x) = x^2 - 2x - 10$ into Y1 of the equation editor (**Y=**). Graph using a standard **ZOOM** window. Find the x intercepts by tracing the graph or using the Zoom In option of **ZOOM.** Or press **CALC (2nd TRACE)** and select 2:zero. You will need to specify a left bound, right bound, and an estimate in the process. Verify algebraically that the negative root of this function is $x = 1 - \sqrt{11}$.

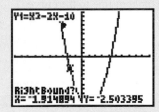 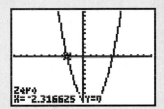

Another method for finding roots is to use the equation solver feature of your graphing calculator. Access the last item of your **MATH** key (0:Solver). Write the equation to be solved, in this case, Y1, into the equation solver, eqn: 0 = Y1, and then press **ENTER.** Suppose we are trying to locate the positive root for $f(x)$ shown in the previous graph. That root looks close to $x = 4$, so we use this value as a starting point. Next press the green **ALPHA** key and then the **ENTER** key **(SOLVE).** The resulting value $x = 4.317$ can be verified to be $x = 1 + \sqrt{11}$.

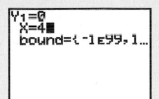 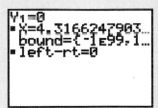

2. Another way to locate a root is Newton's method, an iterative root-finding algorithm that uses tangent line approximations to zero in on a root. See page 162.

Let us apply this method to $f(x) = x^2 - 2x - 10$. Place $f(x)$ into Y1 and write Y2 = nDeriv(Y1, X, X), which is the numerical derivative of Y1, obtained from item 8 of the **MATH** key. Newton's method requires a starting value. From the graph of $f(x)$ it appears that the positive root is smaller than $x = 5$. On a cleared homescreen **(QUIT = 2nd MODE)**, store the value 5 into X by writing 5 $\rightarrow$ X, and then write Newton's algorithm, X − Y1(X)/Y2(X) $\rightarrow$ X. Now press **ENTER** successively and observe the sequence of resulting values. Note how many iterations were needed so that two consecutive approximations agree to four (or more) decimal places. Try a starting value of $x = 4$. Do you construct a sequence of approximations to the same root? How about if you started at $x = -2$?

   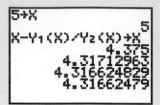   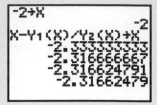

3. We can also use this equation solver feature to locate the roots of the derivative of a function. As an example, store $f(x) = x^2(3x - 1)$ into Y1 of the equation editor and write Y2 = nDeriv(Y1, X, X). Use a bold graphing style for Y2 and a small size **WINDOW,** $[-1, 1]0.2$ by $[-1, 1]0.2$. From the following graph, it is evident that one of the roots of the derivative is $x = 0$. The other root appears to be close to $x = 0.2$.

   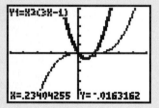

To find this other root, place Y2 into the equation solver position, eqn: 0= (use the up arrow to return to the equation solver screen). Use $x = 0.2$ as a starting point, and then press **ALPHA ENTER (SOLVE)** to obtain the second root of the derivative. What happens if you set the starting x value as -0.2?

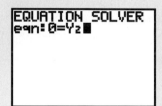   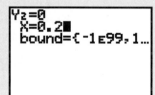   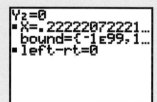

THINK ABOUT IT

PROBLEM

 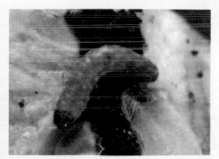

The **codling moth** is an insect that causes serious damage to apples. Adult codling moths emerge from their cocoons in the spring. They soon mate and the female lays as many as 130 tiny eggs on the leaves of apple trees. After the codling moth larva, also known as the common apple worm, hatches from its egg, it goes looking for an apple. The time between hatching and finding an apple is called the **searching period.** Once a codling moth finds an apple, it squirms into the apple and eats the fruit and seeds of the apple, thereby ruining it. After about 4 weeks, the codling moth backs out of the apple and crawls under the bark of the tree or into the soil where it forms a cocoon.

Observations regarding the behavior of codling moths indicate that the length of the searching period, $S(T)$, and the percentage of larvae that survive the searching period, $N(T)$, depend on the air temperature, T. Methods of data analysis (polynomial regression) applied to data recorded from observations suggest that if T is measured in degrees Celsius with $20 \leq T \leq 30$, then $S(T)$ and $N(T)$ may be modeled* by

$$S(T) = (-0.03T^2 + 1.6T - 13.65)^{-1} \text{ days}$$

and

$$N(T) = -0.85T^2 + 45.4T - 547$$

Use these formulas to answer the following questions.

Questions

1. What do these formulas for $S(T)$ and $N(T)$ predict for the length of searching period and percentage of larvae surviving the searching period when the air temperature is 25 degrees Celsius?

---

*P. L. Shaffer and H. J. Gold, "A Simulation Model of Population Dynamics of the Codling Moth *Cydia Pomonella*," *Ecological Modeling*, Vol. 30, 1985, pp. 247–274.

2.   Sketch the graph of $N(T)$ and determine the temperature at which the largest percentage of codling moth larvae survive. Then determine the temperature at which the smallest percentage of larvae survive. (Remember, $20 \le T \le 30$.)

3.   Find $\dfrac{dS}{dT}$, the rate of change of the searching period with respect to temperature T.

When does this rate equal zero? What (if anything) occurs when $\dfrac{dS}{dT} = 0$?

4.   Find $\dfrac{dN}{dS}$, the rate of change of the percentage of larvae surviving the searching period with respect to the length of the searching period using the chain rule,

$$\frac{dN}{dT} = \frac{dN}{dS}\frac{dS}{dT}$$

What (if any) information does this rate provide?

CHAPTER 3

Determining when an assembly line is working at peak efficiency is one application of the derivative.

ADDITIONAL APPLICATIONS OF THE DERIVATIVE

SECTION   3.1 | Increasing and Decreasing Functions; Relative Extrema

Intuitively, we regard a function $f(x)$ as *increasing* where the graph of f is rising, and *decreasing* where the graph is falling. For instance, the graph in Figure 3.1 shows the military spending by the former Soviet bloc countries during the crucial period 1990–1995 following the dissolution of the Soviet Union as a percentage of GDP (Gross Domestic Product). The shape of the graph suggests that the spending decreased dramatically from 1990 until 1992, before increasing slightly from 1992 to 1994, after which it decreased still more.

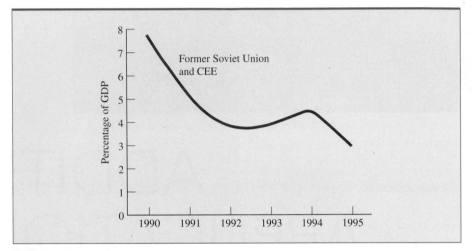

FIGURE 3.1   Military expenditure of former Soviet bloc countries as a percentage of GDP.

Here is a more formal definition of increasing and decreasing functions, which is illustrated graphically in Figure 3.2.

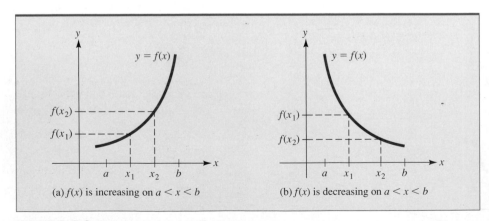

(a) $f(x)$ is increasing on $a < x < b$ (b) $f(x)$ is decreasing on $a < x < b$

FIGURE 3.2   Intervals of increase and decrease.

Increasing and Decreasing Functions ■ Let $f(x)$ be a function defined on the interval $a < x < b$, and let x_1 and x_2 be two numbers in the interval. Then

$f(x)$ is **increasing** on the interval if $f(x_2) > f(x_1)$ whenever $x_2 > x_1$.

$f(x)$ is **decreasing** on the interval if $f(x_2) < f(x_1)$ whenever $x_2 > x_1$.

As demonstrated in Figure 3.3a, if the graph of a function $f(x)$ has tangent lines with only positive slopes on the interval $a < x < b$, then the graph will be rising and $f(x)$ will be increasing on the interval. Since the slope of each such tangent line is given by the derivative $f'(x)$, it follows that $f(x)$ is increasing (graph rising) on intervals where $f'(x) > 0$. Similarly, $f(x)$ is decreasing (graph falling) on intervals where $f'(x) < 0$ (Figure 3.3b).

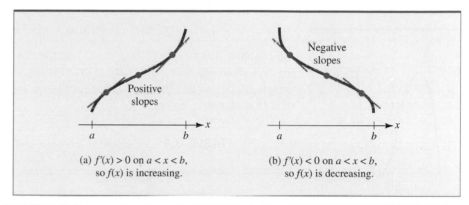

(a) $f'(x) > 0$ on $a < x < b$,
so $f(x)$ is increasing.

(b) $f'(x) < 0$ on $a < x < b$,
so $f(x)$ is decreasing.

FIGURE 3.3   Derivative criteria for increasing and decreasing functions.

Thanks to the intermediate value property (Section 1.6), we know that a continuous function cannot change sign without first becoming 0. This means that if we mark on a number line all numbers x where $f'(x)$ is discontinuous or $f'(x) = 0$, the line will be divided into intervals where $f'(x)$ does not change sign. Therefore, if we pick a test number c from each such interval and find that $f'(c) > 0$, we know that $f'(x) > 0$ for *all* numbers x in the interval and $f(x)$ must be increasing (graph rising) throughout the interval. Similarly, if $f'(c) < 0$, it follows that $f(x)$ is decreasing (graph falling) throughout the interval. These observations may be summarized as follows.

Procedure for Using the Derivative to Determine Intervals of Increase and Decrease for a Function f.

Step 1. Find all values of x for which $f'(x) = 0$ or $f'(x)$ is not continuous, and mark these numbers on a number line. This divides the line into a number of open intervals.

Step 2. Choose a test number c from each interval $a < x < b$ determined in step 1 and evaluate $f'(c)$. Then,

If $f'(c) > 0$, the function $f(x)$ is increasing (graph rising) on $a < x < b$.

If $f'(c) < 0$, the function $f(x)$ is decreasing (graph falling) on $a < x < b$.

This procedure is illustrated in Examples 3.1.1 and 3.1.2.

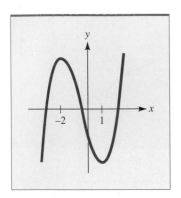

FIGURE 3.4 The graph of $f(x) = 2x^3 + 3x^2 - 12x - 7$.

EXAMPLE | 3.1.1

Find the intervals of increase and decrease for the function

$$f(x) = 2x^3 + 3x^2 - 12x - 7.$$

Solution

The derivative of $f(x)$ is

$$f'(x) = 6x^2 + 6x - 12 = 6(x + 2)(x - 1)$$

which is continuous everywhere, with $f'(x) = 0$ where $x = 1$ and $x = -2$. The numbers -2 and 1 divide the x axis into three open intervals; namely, $x < -2$, $-2 < x < 1$, and $x > 1$. Choose a test number c from each of these intervals; say, $c = -3$ from $x < -2$, $c = 0$ from $-2 < x < 1$, and $c = 2$ from $x > 1$. Then evaluate $f'(c)$ for each test number:

$$f'(-3) = 24 > 0 \qquad f'(0) = -12 < 0 \qquad f'(2) = 24 > 0$$

We conclude that $f'(x) > 0$ for $x < -2$ and for $x > 1$, so $f(x)$ is increasing (graph rising) on these intervals. Similarly, $f'(x) < 0$ on $-2 < x < 1$, so $f(x)$ is decreasing (graph falling) on this interval. These results are summarized in Table 3.1. The graph of $f(x)$ is shown in Figure 3.4.

TABLE 3.1 Intervals of Increase and Decrease for $f(x) = 2x^3 + 3x^2 - 12x - 7$

Interval	Test Number c	$f'(c)$	Conclusion	Direction of Graph
$x < -2$	-3	$f'(-3) > 0$	f is increasing	Rising
$-2 < x < 1$	0	$f'(0) < 0$	f is decreasing	Falling
$x > 1$	2	$f'(2) > 0$	f is increasing	Rising

NOTATION Henceforth, we shall indicate an interval where $f(x)$ is increasing by an "up arrow" ($\nearrow$) and an interval where $f(x)$ is decreasing by a "down arrow" ($\searrow$). Thus, the results in Example 3.1.1 can be represented by this arrow diagram:

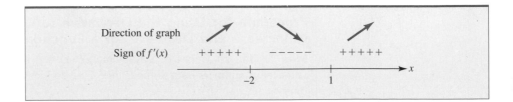

EXAMPLE | 3.1.2

Find the intervals of increase and decrease for the function

$$f(x) = \frac{x^2}{x - 2}$$

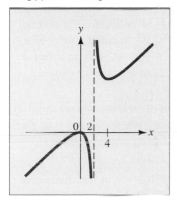

FIGURE 3.5   The graph of $f(x) = \dfrac{x^2}{x-2}$.

Solution

The function is defined for $x \neq 2$, and its derivative is

$$f'(x) = \frac{(x-2)(2x) - x^2(1)}{(x-2)^2} = \frac{x(x-4)}{(x-2)^2}$$

which is discontinuous at $x = 2$ and has $f'(x) = 0$ at $x = 0$ and $x = 4$. Thus, there are four intervals on which the sign of $f'(x)$ does not change; namely, $x < 0$, $0 < x < 2$, $2 < x < 4$, and $x > 4$. Choosing test numbers in these intervals (say, -2, 1, 3, and 5, respectively), we find that

$$f'(-2) = \frac{3}{4} > 0 \qquad f'(1) = -3 < 0 \qquad f'(3) = -3 < 0 \qquad f'(5) = \frac{5}{9} > 0$$

We conclude that $f(x)$ is increasing (graph rising) for $x < 0$ and for $x > 4$ and that it is decreasing (graph falling) for $0 < x < 2$ and for $2 < x < 4$. These results are summarized in the arrow diagram displayed next [the dashed vertical line indicates that $f(x)$ is not defined at $x = 2$].

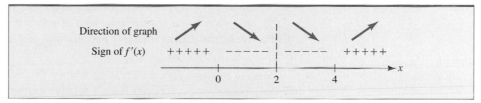

Intervals of increase and decrease for $f(x) = \dfrac{x^2}{x-2}$.

The graph of $f(x)$ is shown in Figure 3.5. Notice how the graph approaches the vertical line $x = 2$ as x tends toward 2. This behavior identifies $x = 2$ as a *vertical asymptote* of the graph of $f(x)$. We will discuss asymptotes in Section 3.3.

Relative Extrema

The simplicity of the graphs in Figures 3.4 and 3.5 may be misleading. A more general graph is shown in Figure 3.6. Note that "peaks" occur at C and E and "valleys" at B, D, and G, and while there are horizontal tangents at B, C, D, and G, no tangent

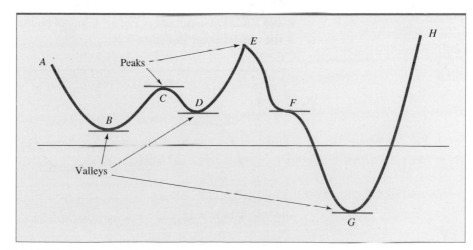

FIGURE 3.6   A graph with various kinds of "peaks" and "valleys."

can be drawn at the "sharp" point E. Moreover, there is a horizontal tangent at F that is neither a peak nor a valley. In this section and the next, you will see how the methods of calculus can be used to locate and identify the "peaks" and "valleys" of a graph. This, in turn, provides the basis for a curve sketching procedure and for methods of optimization.

Less informally, a "peak" on the graph of a function f is known as a **relative maximum** of f, and a "valley" is a **relative minimum.** Thus, a relative maximum is a point on the graph of f that is at least as high as any nearby point on the graph, while a relative minimum is at least as low as any nearby point. Collectively, relative maxima and minima are called **relative extrema.** In Figure 3.6, the relative maxima are located at C and E, and the relative minima are at B, D, and G. Note that a relative extremum does not have to be the highest or lowest point on the entire graph. For instance, in Figure 3.6, the lowest point on the graph is at the relative minimum G, but the highest point occurs at the right endpoint H. Here is a summary of this terminology.

Relative Extrema ■ The graph of the function $f(x)$ is said to have a *relative maximum* at $x = c$ if $f(c) \geq f(x)$ for all x in an interval $a < x < b$ containing c. Similarly, the graph has a *relative minimum* at $x = c$ if $f(c) \leq f(x)$ on such an interval. Collectively, the relative maxima and minima of f are called its *relative extrema.*

Since a function $f(x)$ is increasing when $f'(x) > 0$ and decreasing when $f'(x) < 0$, the only points where $f(x)$ can have a relative extremum are where $f'(x) = 0$ or $f'(x)$ does not exist. Such points are so important that we give them a special name.

Critical Numbers and Critical Points ■ A number c in the domain of $f(x)$ is called a **critical number** if either $f'(c) = 0$ or $f'(c)$ does not exist. The corresponding point $(c, f(c))$ on the graph of $f(x)$ is called a **critical point** for $f(x)$. *Relative extrema can only occur at critical points.*

It is important to note that while relative extrema occur at critical points, *not all critical points correspond to relative extrema.* For example, Figure 3.7 shows three different situations where $f'(c) = 0$, so a horizontal tangent line occurs at the critical point $(c, f(c))$. The critical point corresponds to a relative maximum in Figure 3.7a and to a relative minimum in Figure 3.7b, but neither kind of relative extremum occurs at the critical point in Figure 3.7c.

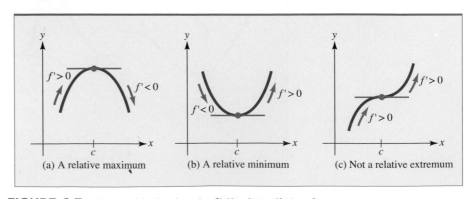

(a) A relative maximum   (b) A relative minimum   (c) Not a relative extremum

FIGURE 3.7 Three critical points $(c, f(c))$ where $f'(c) = 0$.

Three functions with critical points at which the derivative is undefined are shown in Figure 3.8. In Figure 3.8c, the tangent line is vertical at $(c, f(c))$, so the slope $f'(c)$ is undefined. In Figures 3.8a and 3.8b, no (unique) tangent line can be drawn at the "sharp" point $(c, f(c))$.

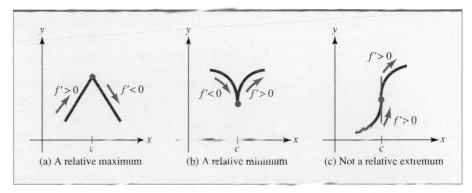

| (a) A relative maximum | (b) A relative minimum | (c) Not a relative extremum |

FIGURE 3.8   Three critical points $(c, f(c))$ where $f'(c)$ is undefined.

The First Derivative Test for Relative Extrema

Figures 3.7 and 3.8 also suggest a method for using the sign of the derivative to classify critical points as relative maxima, relative minima, or neither. Suppose c is a critical number of f and that $f'(x) > 0$ to the left of c, while $f'(x) < 0$ to the right. Geometrically, this means the graph of f goes up before the critical point $P(c, f(c))$ and then comes down, which implies that P is a relative maximum. Similarly, if $f'(x) < 0$ to the left of c and $f'(x) > 0$ to the right, the graph goes down before $P(c, f(c))$ and up afterward, so there must be a relative minimum at P. On the other hand, if the derivative has the same sign on both sides of c, then the graph either rises through P or falls through P, and no relative extremum occurs there. These observations can be summarized as follows.

The First Derivative Test for Relative Extrema   ▪   Let c be a critical number for $f(x)$ [that is, $f(c)$ is defined and either $f'(c) = 0$ or $f'(c)$ does not exist]. Then the critical point $P(c, f(c))$ is

A **relative maximum** if $f'(x) > 0$ to the left of c and $f'(x) < 0$ to the right of c

A **relative minimum** if $f'(x) < 0$ to the left of c and $f'(x) > 0$ to the right of c

Not a relative extremum if $f'(x)$ has the same sign on both sides of c

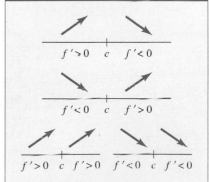

EXAMPLE | 3.1.3

Find all critical numbers of the function

$$f(x) = 2x^4 - 4x^2 + 3$$

and classify each critical point as a relative maximum, a relative minimum, or neither.

Solution

The polynomial $f(x)$ is defined for all x, and its derivative is

$$f'(x) = 8x^3 - 8x = 8x(x^2 - 1) = 8x(x - 1)(x + 1)$$

Since the derivative exists for all x, the only critical numbers are where $f'(x) = 0$; that is, $x = 0$, $x = 1$, and $x = -1$. These numbers divide the x axis into four intervals, on each of which the sign of the derivative does not change; namely, $x < -1$, $-1 < x < 0$, $0 < x < 1$, and $x > 1$. Choose a test number c in each of these intervals (say, -5, $-\frac{1}{2}$, $\frac{1}{4}$, and 2, respectively) and evaluate $f'(c)$ in each case:

$$f'(-5) = -960 < 0 \quad f'\left(-\frac{1}{2}\right) = 3 > 0 \quad f'\left(\frac{1}{4}\right) = -\frac{15}{8} < 0 \quad f'(2) = 48 > 0$$

Thus, the graph of f falls for $x < -1$ and for $0 < x < 1$, and rises for $-1 < x < 0$ and for $x > 1$, so there must be a relative maximum at $x = 0$ and relative minima at $x = -1$ and $x = 1$, as indicated in this arrow diagram.

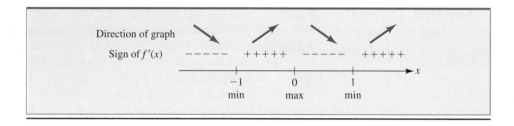

Applications   Once you determine the intervals of increase and decrease of a function f and find its relative extrema, you can obtain a rough sketch of the graph of the function. Here is a step-by-step description of the procedure for sketching the graph of a continuous function $f(x)$ using the derivative $f'(x)$. In Section 3.3, we will extend this procedure to cover the situation where $f(x)$ is discontinuous.

A Procedure for Sketching the Graph of a Continuous Function f(x) Using the Derivative f'(x)

Step 1. Determine the domain of $f(x)$. Set up a number line restricted to include only those numbers in the domain of $f(x)$.

Step 2. Find $f'(x)$ and mark each critical number on the restricted number line obtained in step 1. Then analyze the sign of the derivative to determine intervals of increase and decrease for $f(x)$ on the restricted number line.

Step 3. For each critical number c, find $f(c)$ and plot the critical point $P(c, f(c))$ on a coordinate plane, with a "cap" ⌢ at P if it is a relative maximum (⟋ ⟍), or a "cup" ⌣ if P is a relative minimum (⟍ ⟋). Plot intercepts and other key points that can be easily found.

Step 4. Sketch the graph of f as a smooth curve joining the critical points in such a way that it rises where $f'(x) > 0$, falls where $f'(x) < 0$, and has a horizontal tangent where $f'(x) = 0$.

EXPLORE!

Graph $f(x)$ from Example 3.1.4 in bold using the window $[-4.7, 4.7]1$ by $[-15, 45]5$. In the same window, graph $g(x) = x^4 + 8x^3 + 18x^2 + 2$, which is $f(x)$ with a change in the constant term from -8 to $+2$. What effect does this change have on the critical values?

EXAMPLE | 3.1.4

Sketch the graph of the function $f(x) = x^4 + 8x^3 + 18x^2 - 8$.

Solution

Since $f(x)$ is a polynomial, it is defined for all x. Its derivative is

$$f'(x) = 4x^3 + 24x^2 + 36x = 4x(x^2 + 6x + 9) = 4x(x + 3)^2$$

Since the derivative exists for all x, the only critical numbers are where $f'(x) = 0$; namely, at $x = 0$ and $x = -3$. These numbers divide the x axis into three intervals, on each of which the sign of the derivative $f'(x)$ does not change; namely, $x < -3$, $-3 < x < 0$, and $x > 0$. Choose a test number c in each interval (say, -5, -1, and 1, respectively), and determine the sign of $f'(c)$:

$$f'(-5) = -80 < 0 \qquad f'(-1) = -16 < 0 \qquad f'(1) = 64 > 0$$

Thus, the graph of f has horizontal tangents where x is -3 and 0, and the graph is falling (f decreasing) in the intervals $x < -3$ and $-3 < x < 0$ and is rising (f increasing) for $x > 0$, as indicated in this arrow diagram:

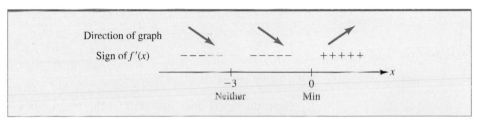

Interpreting the diagram, we see that the graph falls to a horizontal tangent at $x = -3$, then continues falling to the relative minimum at $x = 0$, after which it rises indefinitely. We find that $f(-3) = 19$ and $f(0) = -8$. To begin your sketch, plot a "cup" $\smile$ at the critical point $(0, -8)$ to indicate that a relative minimum occurs there (if it had been a relative maximum, you would have used a "cap" $\frown$), and a "twist" at $(-3, 19)$ to indicate a falling graph with a horizontal tangent at this point. This is shown in the preliminary graph in Figure 3.9a. Finally, complete the sketch by passing a smooth curve through the critical points in the directions indicated by the arrows, as shown in Figure 3.9b.

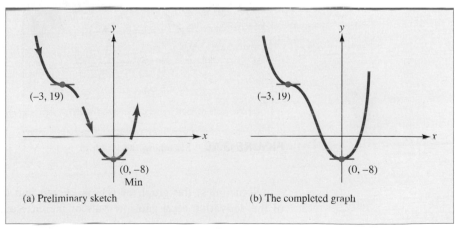

(a) Preliminary sketch   (b) The completed graph

FIGURE 3.9   The graph of $f(x) = x^4 + 8x^3 + 18x^2 - 8$.

EXAMPLE 3.1.5

Find the intervals of increase and decrease and the relative extrema of the function $g(t) = \sqrt{3 - 2t - t^2}$. Sketch the graph.

Just-In-Time Review

The product ab satisfies $ab \geq 0$ *only* when both $a \geq 0$ and $b \geq 0$ or when $a \leq 0$ and $b \leq 0$. If a and b have opposite signs, then $ab \leq 0$.

Solution

Since $\sqrt{u}$ is defined only for $u \geq 0$, the domain of g will be the set of all t such that $3 - 2t - t^2 \geq 0$. Factoring the expression $3 - 2t - t^2$, we get

$$3 - 2t - t^2 = (3 + t)(1 - t)$$

Note that $3 + t \geq 0$ when $t \geq -3$ and that $1 - t \geq 0$ when $t \leq 1$. We have $(3 + t)(1 - t) \geq 0$ when *both* terms are nonnegative; that is, when $t \geq -3$ *and* $t \leq 1$, or equivalently, when $-3 \leq t \leq 1$. We would also have $(3 + t)(1 - t) \geq 0$ if $3 + t \leq 0$ and $1 - t \leq 0$, but this is not possible (do you see why?). Thus, $g(t)$ is defined only for $-3 \leq t \leq 1$.

Next, using the chain rule, we compute the derivative of $g(t)$:

$$g'(t) = \frac{1}{2} \frac{1}{\sqrt{3 - 2t - t^2}} (-2 - 2t)$$

$$= \frac{-1 - t}{\sqrt{3 - 2t - t^2}}$$

Note that $g'(t)$ does not exist at the endpoints $t = -3$ and $t = 1$ of the domain of $g(t)$, and that $g'(t) = 0$ only when $t = -1$. Next, mark these three critical numbers on a number line restricted to the domain of g (namely, $-3 \leq t \leq 1$), and determine the sign of the derivative $g'(t)$ on the subintervals $-3 < t < -1$ and $-1 < t < 1$ to obtain the arrow diagram shown in Figure 3.10a. Finally, compute $g(-3) = g(1) = 0$ and $g(-1) = 2$, and observe that the arrow diagram suggests there is a relative maximum at $(-1, 2)$. The completed graph is shown in Figure 3.10b.

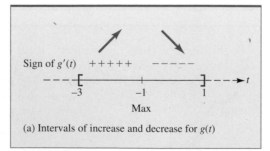

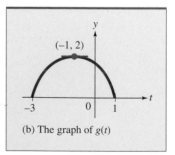

(a) Intervals of increase and decrease for $g(t)$ (b) The graph of $g(t)$

FIGURE 3.10 Sketching the graph of $g(t) = \sqrt{3 - 2t - t^2}$.

Sometimes, the graph of $f(x)$ is known and the relationship between the sign of the derivative $f'(x)$ and intervals of increase and decrease can be used to determine the general shape of the graph of $f'(x)$. The procedure is illustrated in Example 3.1.6.

EXAMPLE | 3.1.6

The graph of a function $f(x)$ is shown here. Sketch a possible graph for the derivative $f'(x)$.

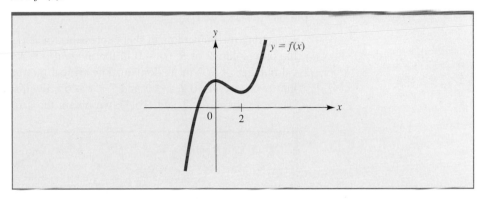

Solution

Since the graph of $f(x)$ is falling for $0 < x < 2$, we have $f'(x) < 0$ and the graph of $f'(x)$ is below the x axis on this interval. Similarly, for $x < 0$ and for $x > 2$, the graph of $f(x)$ is rising, so $f'(x) > 0$ and the graph of $f'(x)$ is above the x axis on both these intervals. The graph of $f(x)$ is "flat" (horizontal tangent line) at $x = 0$ and $x = 2$, so $f'(0) = f'(2) = 0$, and $x = 0$ and $x = 2$ are the x intercepts of the graph of $f'(x)$. Here is one possible graph that satisfies these conditions:

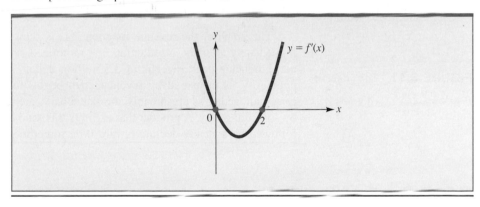

Curve sketching using the derivative $f'(x)$ will be refined by adding features using the second derivative $f''(x)$ in Section 3.2, and a general curve sketching procedure involving derivatives and limits will be presented in Section 3.3. The same reasoning used to analyze graphs can be applied to determining optimal values, such as minimum cost of a production process or maximal sustainable yield for a salmon fishery. Optimization is illustrated in Example 3.1.7 and is discussed in more detail in Sections 3.4 and 3.5.

EXAMPLE | 3.1.7

The revenue derived from the sale of a new kind of motorized skateboard t weeks after its introduction is given by

$$R(t) = \frac{63t - t^2}{t^2 + 63} \qquad 0 \le t \le 63$$

million dollars. When does maximum revenue occur? What is the maximum revenue?

EXPLORE!

Store $f(x) = x^3 - x^2 - 4x + 4$ into Y1 and $f'(x)$ (via the numerical derivative) into Y2 using the bold graphing style and a window size of $[-4.7, 4.7]1$ by $[-10, 10]2$. How do the relative extrema of $f(x)$ relate to the features of the graph of $f'(x)$? What are the largest and smallest values of $f(x)$ over the interval $[-2, 1]$?

Solution

Differentiating $R(t)$ by the quotient rule, we get

$$R'(t) = \frac{(t^2 + 63)(63 - 2t) - (63t - t^2)(2t)}{(t^2 + 63)^2} = \frac{-63(t - 7)(t + 9)}{(t^2 + 63)^2}$$

By setting the numerator in this expression for $R'(t)$ equal to 0, we find that $t = 7$ is the only solution of $R'(t) = 0$ in the interval $0 \le t \le 63$, and hence is the only critical number of $R(t)$ in its domain. The critical number divides the domain $0 \le t \le 63$ into two intervals, $0 \le t < 7$ and $7 < t \le 63$. Evaluating $R'(t)$ at test numbers in each interval (say, at $t = 1$ and $t = 9$), we obtain the arrow diagram shown here.

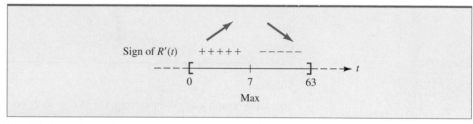

The arrow pattern indicates that the revenue increases to a maximum t at $t = 7$, after which it decreases. At the optimal time $t = 7$, the revenue produced is

$$R(7) = \frac{63(7) - (7)^2}{(7)^2 + 63} = 3.5 \text{ million dollars}$$

The graph of the revenue function $R(t)$ is shown in Figure 3.11. It suggests that immediately after its introduction, the motorized skateboard is a very popular product, producing peak revenue of 3.5 million dollars after only 7 weeks. However, its popularity, as measured by revenue, then begins to wane. After 63 weeks, revenue ceases altogether as presumably, the skateboards are taken off the shelves and replaced by something new. A product that exhibits this kind of revenue pattern, steep increase followed by a steady decline toward 0, is sometimes referred to as a "fad."

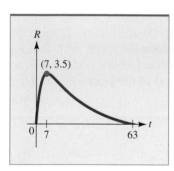

FIGURE 3.11   The graph of $R(t) = \dfrac{63t - t^2}{t^2 + 63}$ for $0 \le t \le 63$.

P R O B L E M S   3.1

In Problems 1 through 4, specify the intervals on which the derivative of the given function is positive and those on which it is negative.

1.
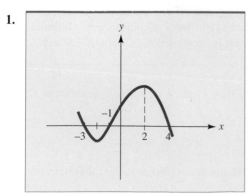

2.
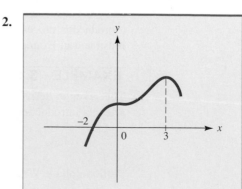

3.

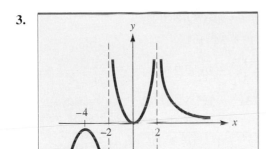

4.

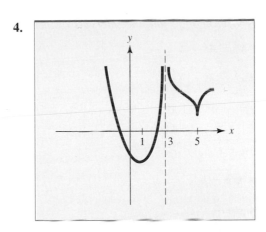

Each curve A, B, C, and D shown here is the graph of the derivative of a function whose graph is shown in one of Problems 5 through 8. Match each graph with that of its derivative.

5.

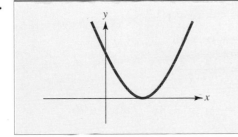

A

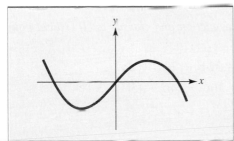

6.

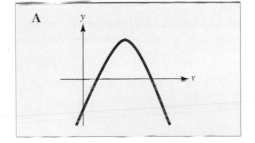

B

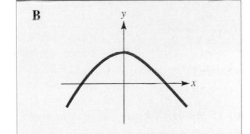

7.

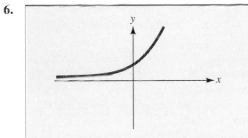

C

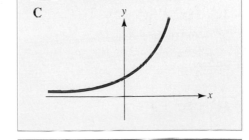

8.

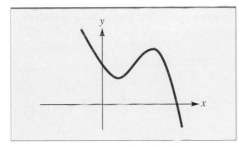

D

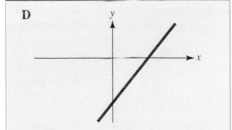

In Problems 9 through 22 find the intervals of increase and decrease for the given function.

9. $f(x) = x^2 - 4x + 5$

10. $f(t) = t^3 + 3t^2 + 1$

11. $f(x) = x^3 - 3x - 4$

12. $f(x) = \frac{1}{3}x^3 - 9x + 2$

13. $g(t) = t^5 - 5t^4 + 100$

14. $f(x) = 3x^5 - 5x^3$

15. $f(t) = \frac{1}{4 - t^2}$

16. $g(t) = \frac{1}{t^2 + 1} - \frac{1}{(t^2 + 1)^2}$

17. $h(u) = \sqrt{9 - u^2}$

18. $f(x) = \sqrt{6 - x - x^2}$

19. $F(x) = x + \frac{9}{x}$

20. $f(t) = \frac{t}{(t + 3)^2}$

21. $f(x) = \sqrt{x} + \frac{1}{\sqrt{x}}$

22. $G(x) = x^2 - \frac{1}{x^2}$

In Problems 23 through 34 determine the critical numbers of the given function and classify each critical point as a relative maximum, a relative minimum, or neither.

23. $f(x) = 3x^4 - 8x^3 + 6x^2 + 2$

24. $f(x) = 324x - 72x^2 + 4x^3$

25. $f(t) = 2t^3 + 6t^2 + 6t + 5$

26. $f(t) = 10t^6 + 24t^5 + 15t^4 + 3$

27. $g(x) = (x - 1)^5$

28. $F(x) = 3 - (x + 1)^3$

29. $f(t) = \frac{t}{t^2 + 3}$

30. $f(t) = t\sqrt{9 - t}$

31. $h(t) = \frac{t^2}{t^2 + t - 2}$

32. $g(x) = 4 - \frac{2}{x} + \frac{3}{x^2}$

33. $S(t) = (t^2 - 1)^4$

34. $F(x) = \frac{x^2}{x - 1}$

In Problems 35 through 44, use calculus to sketch the graph of the given function.

35. $f(x) = x^3 - 3x^2$

36. $f(x) = 3x^4 - 4x^3$

37. $f(x) = 3x^4 - 8x^3 + 6x^2 + 2$

38. $g(x) = 3 - (x + 1)^3$

39. $f(t) = 2t^3 + 6t^2 + 6t + 5$

40. $f(x) = x^3(x + 5)^2$

41. $g(t) = \frac{t}{t^2 + 3}$

42. $f(x) = \frac{x + 1}{x^2 + x + 1}$

43. $f(x) = 3x^5 - 5x^3 + 4$

44. $H(x) = \frac{1}{50}(3x^4 - 8x^3 - 90x^2 + 70)$

In Problems 45 through 48, the derivative of a function f(x) is given. In each case, find the critical numbers of f(x) and classify each as corresponding to a relative maximum, a relative minimum, or neither.

45. $f'(x) = x^2(4 - x^2)$

46. $f'(x) = \frac{x(2 - x)}{x^2 + x + 1}$

47. $f'(x) = \frac{(x + 1)^2(4 - 3x)^3}{(x^2 + 1)^2}$

48. $f'(x) = x^3(2x - 7)^2(x + 5)$

In Problems 49 through 52, the graph of a function f is given. In each case, sketch a possible graph for f'.

49.

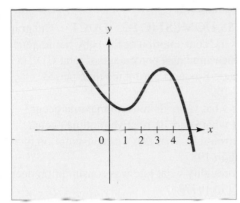

50.

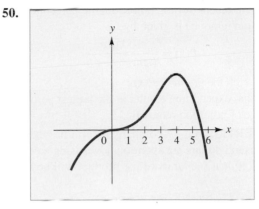

51.

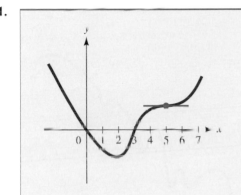

52.

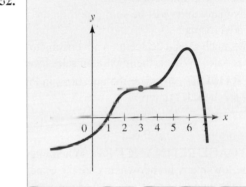

53. AVERAGE COST   The total cost of producing x units of a certain commodity is $C(x)$ thousand dollars, where

$$C(x) = x^3 - 20x^2 + 179x + 242$$

  a. Find $A'(x)$, where $A(x) = C(x)/x$ is the average cost function.

  b. For what values of x is $A(x)$ increasing? For what values is it decreasing?

  c. For what level of production x is average cost minimized? What is the minimum average cost?

54. MARGINAL ANALYSIS   The total cost of producing x units of a certain commodity is given by $C(x) = \sqrt{5x + 2} + 3$. Sketch the cost curve and find the marginal cost. Does marginal cost increase or decrease with increasing production?

55. MARGINAL ANALYSIS   Let $p = (10 - 3x)^2$ for $0 \le x \le 3$ be the price at which x hundred units of a certain commodity will be sold, and let $R(x) = xp(x)$ be the revenue obtained from the sale of the x units. Find the marginal revenue $R'(x)$ and sketch the revenue and marginal revenue curves on the same graph. For what level of production is revenue maximized?

56. PROFIT UNDER A MONOPOLY   To produce x units of a particular commodity, a monopolist has a total cost of

$$C(x) = 2x^2 + 3x + 5$$

and total revenue of $R(x) = xp(x)$, where $p(x) = 5 - 2x$ is the price at which the x units will be sold. Find the profit function $P(x) = R(x) - C(x)$ and sketch its graph. For what level of production is profit maximized?

57. MEDICINE   The concentration of a drug t hours after being injected into the arm of a patient is given by

$$C(t) = \frac{0.15t}{t^2 + 0.81}$$

Sketch the graph of the concentration function. When does the maximum concentration occur?

58. POLLUTION CONTROL Commissioners of a certain city determine that when x million dollars are spent on controlling pollution, the percentage of pollution removed is given by

$$P(x) = \frac{100\sqrt{x}}{0.04x^2 + 12}$$

   a. Sketch the graph of $P(x)$.

   b. What expenditure results in the largest percentage of pollution removal?

59. ADVERTISING A company determines that if x thousand dollars are spent on advertising a certain product, then $S(x)$ units of the product will be sold, where

$$S(x) = -2x^3 + 27x^2 + 132x + 207 \qquad 0 \le x \le 17$$

   a. Sketch the graph of $S(x)$.

   b. How many units will be sold if nothing is spent on advertising?

   c. How much should be spent on advertising to maximize sales? What is the maximum sales level?

60. ADVERTISING Answer the questions in Problem 59 for the sales function

$$S(x) = \frac{200x + 1{,}500}{0.02x^2 + 5}$$

61. MORTGAGE REFINANCING When interest rates are low, many homeowners take the opportunity to refinance their mortgages. As rates start to rise, there is often a flurry of activity as latecomers rush in to refinance while they still can do so profitably. Eventually, however, rates reach a level where refinancing begins to wane.

   Suppose in a certain community, there will be $M(r)$ thousand refinanced mortgages when the 30-year fixed mortgage rate is $r\%$, where

$$M(r) = \frac{1 + 0.05r}{1 + 0.004r^2} \qquad \text{for} \quad 1 \le r \le 8$$

   a. For what values of r is $M(r)$ increasing?

   b. For what interest rate r is the number of refinanced mortgages maximized? What is this maximum number?

62. POPULATION DISTRIBUTION A demographic study of a certain city indicates that $P(r)$ hundred people live r miles from the civic center, where

$$P(r) = \frac{5(3r + 1)}{r^2 + r + 2}$$

   a. What is the population at the city center?

   b. For what values of r is $P(r)$ increasing? For what values is it decreasing?

   c. At what distance from the civic center is the population largest? What is this largest population?

63. GROSS DOMESTIC PRODUCT The graph shows the consumption of the baby boom generation, measured as a percentage of total GDP (Gross Domestic Product) during the time period 1970–1997.

   a. At what years do relative maxima occur?

   b. At what years do relative minima occur?

   c. At roughly what rate was consumption increasing in 1987?

   d. At roughly what rate was consumption decreasing in 1972?

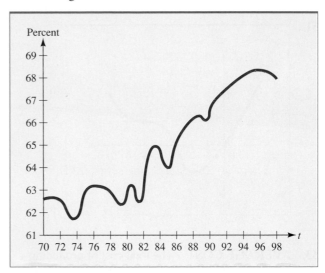

PROBLEM 63

Baby Boom Consumption as Percentage of GDP.
Source: Bureau of Economic Analysis.

64. DEPRECIATION The value V (in thousands of dollars) of an industrial machine is modeled by

$$V(N) = \left(\frac{3N + 430}{N + 1}\right)^{2/3}$$

where N is the number of hours the machine is used each day. Suppose further that usage varies with time in such a way that

$$N(t) = \sqrt{t^2 - 10t + 61}$$

where t is the number of months the machine has been in operation.

   a. Over what time interval is the value of the machine increasing? When is it decreasing?

   b. At what time t is the value of the machine the largest? What is this maximum value?

65. FISHERY MANAGEMENT The manager of a fishery determines that t weeks after 300 fish of a particular species are released into a pond, the average weight of an individual fish (in pounds) for the first 10 weeks will be

$$w(t) = 3 + t - 0.05t^2$$

He further determines that the proportion of the fish that are still alive after t weeks is given by

$$p(t) = \frac{31}{31 + t}$$

a. The expected yield $Y(t)$ of the fish after t weeks is the total weight of the fish that are still alive. Express $Y(t)$ in terms of $w(t)$ and $p(t)$ and sketch the graph of $Y(t)$ for $0 \le t \le 10$.

b. When is the expected yield $Y(t)$ the largest? What is the maximum yield?

66. FISHERY MANAGEMENT Suppose for the situation described in Problem 65, it costs the fishery $C(t) = 50 + 1.2t$ hundred dollars to maintain and monitor the pond for t weeks after the fish are released, and that each fish harvested after t weeks can be sold for $2.75 per pound.

a. If all fish that remain alive in the pond after t weeks are harvested, express the profit obtained by the fishery as a function of t.

 b. When should the fish be harvested in order to maximize profit? What is the maximum profit?

67. GROWTH OF A SPECIES The percentage of codling moth eggs* that hatch at a given temperature T (degrees Celsius) is given by

$$H(T) = -0.53T^2 + 25T - 209 \qquad \text{for } 15 \le T \le 30$$

Sketch the graph of the hatching function $H(T)$. At what temperature T ($15 \le T \le 30$) does the maximum percentage of eggs hatch? What is the maximum percentage? (See the Think About It article on page 185 for more information about the codling moth.)

68. Sketch a graph of a function that has all of the following properties:
a. $f'(0) = f'(1) = f'(2) = 0$
b. $f'(x) < 0$ when $x < 0$ and $x > 2$
c. $f'(x) > 0$ when $0 < x < 1$ and $1 < x < 2$

69. Sketch a graph of a function that has all of the following properties:
a. $f'(0) = f'(1) = f'(2) = 0$

b. $f'(x) < 0$ when $0 < x < 1$
c. $f'(x) > 0$ when $x < 0$, $1 < x < 2$, and $x > 2$

70. Sketch a graph of a function that has all of the following properties:
a. $f'(x) > 0$ when $x < -5$ and when $x > 1$
b. $f'(x) < 0$ when $-5 < x < 1$
c. $f(-5) = 4$ and $f(1) = -1$

71. Sketch a graph of a function that has all of the following properties:
a. $f'(x) < 0$ when $x < -1$
b. $f'(x) > 0$ when $-1 < x < 3$ and when $x > 3$
c. $f'(-1) = 0$ and $f'(3) = 0$

72. Find constants a, b, and c so that the graph of the function $f(x) = ax^2 + bx + c$ has a relative maximum at $(5, 12)$ and crosses the y axis at $(0, 3)$.

73. Find constants a, b, c, and d so the graph of the function $f(x) = ax^3 + bx^2 + cx + d$ will have a relative maximum at $(-2, 8)$ and a relative minimum at $(1, -19)$.

74. Sketch the graph of $f(x) = (x - 1)^{2/5}$. Explain why $f'(x)$ is not defined at $x = 1$.

75. Sketch the graph of $f(x) = 1 - x^{3/5}$.

76. Use calculus to prove that the relative extremum of the quadratic function

$$f(x) = ax^2 + bx + c$$

occurs when $x = -\dfrac{b}{2a}$. Where is $f(x)$ increasing and decreasing?

77. Use calculus to prove that the relative extremum of the quadratic function $y = (x - p)(x - q)$ occurs midway between its x intercepts.

 In Problems 78 through 81 use a graphing utility to sketch the graph of $f(x)$. Then find $f'(x)$ and graph it on the same axes (screen) as $f(x)$. Finally, use **TRACE, ZOOM,** *or other utility methods to find the values of x where $f'(x) = 0$.*

78. $f(x) = x^4 + 3x^3 - 9x^2 + 4$

79. $f(x) = (x^2 + x - 1)^3(x + 3)^2$

80. $f(x) = x^5 - 7.6x^3 + 2.1x^2 - 5$

81. $f(x) = (1 - x^{1/2})^{1/2}$

82. Use a graphing utility to sketch the graph of $f(x) = x^3 + 3x^2 - 5x + 11$. Then sketch the graph of $g(x) = (x + 1)^3 + 3(x + 1)^2 - 5(x + 1) + 11$ on the same axes. What function $h(x)$ would have a graph that is the same as that of $f(x)$, only shifted upward by 2 units and to the left by 3 units?

---

*P. L. Shaffer and H. J. Gold, "A Simulation Model of Population Dynamics of the Codling Moth *Cydia pomonelia*," *Ecological Modeling*, Vol. 30, 1985, pp. 247–274.

83. Let $f(x) = 4 + \sqrt{9 - 2x - x^2}$. Before actually graphing this function, what do you think the graph looks like? Now use a graphing utility to sketch the graph. Were you right?

84. Let $f(x) = x^3 - 6x^2 + 5x - 11$. Use a graphing utility to sketch the graph of $f(x)$. Then, on the same axes, sketch the graph of

$$g(x) = f(2x) = (2x)^3 - 6(2x)^2 + 5(2x) - 11$$

What (if any) relation exists between the two graphs?

SECTION 3.2 | Concavity and Points of Inflection

In Section 3.1, you saw how to use the sign of the derivative $f'(x)$ to determine where $f(x)$ is increasing and decreasing and where its graph has relative extrema. In this section, you will see that the second derivative $f''(x)$ also provides useful information about the graph of $f(x)$. By way of introduction, here is a brief description of a situation from industry that can be analyzed using the second derivative.

The number of units that a factory worker can produce in t hours after arriving at work is often given by a function $Q(t)$ like the one whose graph is shown in Figure 3.12. Notice that at first the graph is not very steep. The steepness increases, however, until the graph reaches a point of maximum steepness, after which the steepness begins to decrease. This reflects the fact that at first the worker's rate of production is low. The rate of production increases, however, as the worker settles into a routine and continues to increase until the worker is performing at maximum efficiency. Then fatigue sets in and the rate of production begins to decrease. The point P on the output curve where maximum efficiency occurs is known in economics as the **point of diminishing returns.**

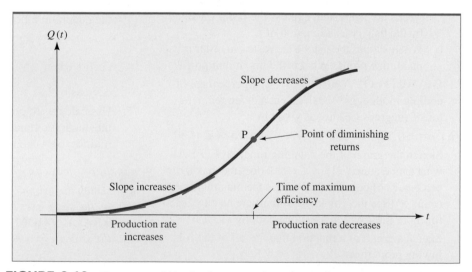

FIGURE 3.12   The output $Q(t)$ of a factory worker t hours after coming to work.

The behavior of the graph of this production function on either side of the point of diminishing returns can be described in terms of its tangent lines. To the left of this point, the slope of the tangent increases as t increases. To the right of this point, the slope of the tangent decreases as t increases. It is this increase and decrease of slopes that we shall examine in this section with the aid of the second derivative. (We shall return to the questions of worker efficiency and diminishing returns in Example 3.2.6, later in this section.)

Concavity   The increase and decrease of tangential slope will be described in terms of a graphical feature called **concavity.** Here is the definition we will use.

Concavity ▪ If the function $f(x)$ is differentiable on the interval $a < x < b$, then the graph of f is

concave upward on $a < x < b$ if f' is increasing on the interval

concave downward on $a < x < b$ if f' is decreasing on the interval

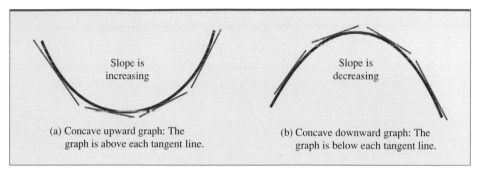

(a) Concave upward graph: The
graph is above each tangent line.

Slope is
increasing

(b) Concave downward graph: The
graph is below each tangent line.

Slope is
decreasing

FIGURE 3.13 Concavity.

Equivalently, a graph is concave upward on an interval if it lies above all its tangent lines on the interval (Figure 3.13a), and concave downward on an interval where it lies below all its tangent lines (Figure 3.13b). Informally, a concave upward portion of graph "holds water," while a concave downward portion "spills water," as illustrated in Figure 3.14.

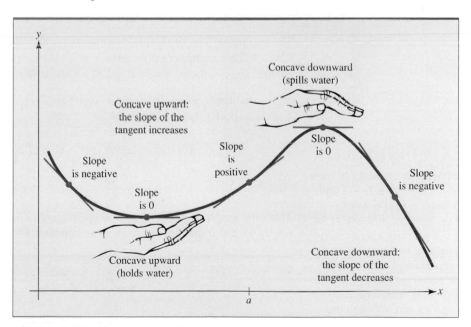

FIGURE 3.14 Concavity and the slope of the tangent.

Determining Intervals of Concavity Using the Sign of f''

There is a simple characterization of the concavity of the graph of a function $f(x)$ in terms of the second derivative $f''(x)$. In particular, in Section 3.1, we observed that a function $f(x)$ is increasing where its derivative is positive. Thus, the derivative function $f'(x)$ must be increasing where *its* derivative $f''(x)$ is positive. Suppose $f''(x) > 0$ on an interval $a < x < b$. Then $f'(x)$ is increasing, which in turn means that the graph of $f(x)$ is concave upward on this interval. Similarly, on an interval $a < x < b$, where $f''(x) < 0$, the derivative $f'(x)$ will be decreasing and the graph of $f(x)$ will be concave downward. Using these observations, we can modify the procedure for determining intervals of increase and decrease developed in Section 3.1 to obtain this procedure for determining intervals of concavity.

> ### Second Derivative Procedure for Determining Intervals of Concavity for a Function f
>
> **Step 1.** Find all values of x for which $f''(x) = 0$ or $f''(x)$ does not exist, and mark these numbers on a number line. This divides the line into a number of open intervals.
>
> **Step 2.** Choose a test number c from each interval $a < x < b$ determined in step 1 and evaluate $f''(c)$. Then,
>   If $f''(c) > 0$, the graph of $f(x)$ is concave upward on $a < x < b$.
>   If $f''(c) < 0$, the graph of $f(x)$ is concave downward on $a < x < b$.

EXAMPLE 3.2.1

Determine intervals of concavity for the function

$$f(x) = 2x^6 - 5x^4 + 7x - 3$$

Solution

We find that $f'(x) = 12x^5 - 20x^3 + 7$ and

$$f''(x) = 60x^4 - 60x^2 = 60x^2(x^2 - 1) = 60x^2(x - 1)(x + 1)$$

The second derivative $f''(x)$ is continuous for all x and $f''(x) = 0$ for $x = 0$, $x = 1$, and $x = -1$. These numbers divide the x axis into four intervals on which $f''(x)$ does not change sign; namely, $x < -1$, $-1 < x < 0$, $0 < x < 1$, and $x > 1$. Evaluating $f''(x)$ at test numbers in each of these intervals (say, at $x = -2$, $x = -\frac{1}{2}$, $x = \frac{1}{2}$, and $x = 5$, respectively), we find that

$$f''(-2) = 720 > 0 \qquad\qquad f''\left(\frac{-1}{2}\right) = \frac{-45}{4} < 0$$

$$f''\left(\frac{1}{2}\right) = \frac{-45}{4} < 0 \qquad\qquad f''(5) = 36{,}000 > 0$$

Thus, the graph of $f(x)$ is concave up for $x < -1$ and for $x > 1$ and concave down for $-1 < x < 0$ and for $0 < x < 1$, as indicated in this concavity diagram.

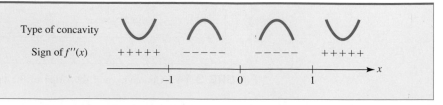

Intervals of concavity for $f(x) = 2x^6 - 5x^4 + 7x - 3$

The graph of $f(x)$ is shown in Figure 3.15.

EXPLORE!

Following Example 3.2.1, store $f(x) = 2x^6 - 5x^4 + 7x - 3$ into Y1 using a bold graphing style. Write Y2 = nDeriv(Y1, X, X) but deselect it by moving the cursor to the equal sign and pressing **ENTER**, which will define but not graph Y2. Write Y3 = nDeriv(Y2, X, X). Graph Y1 and Y3 using the window [−2.35, 2.35]1 by [−18, 8]2. How does the behavior of Y3, representing $f''(x)$, relate to the concavity of $f(x)$, specifically at X = −1, 0, and 1?

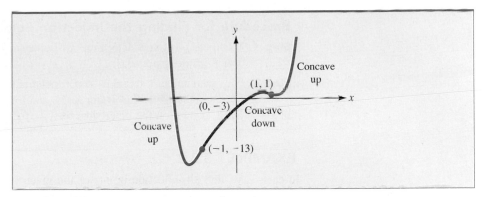

FIGURE 3.15   The graph of $f(x) = 2x^6 - 5x^4 + 7x - 3$.

NOTE   Do not confuse the concavity of a graph with its "direction" (rising or falling). A function f may be increasing or decreasing on an interval regardless of whether its graph is concave upward or concave downward on that interval. The four possibilities are illustrated in Figure 3.16.   ■

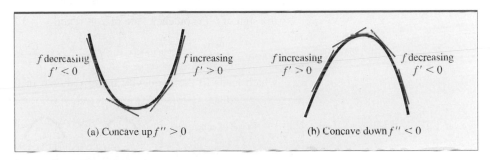

FIGURE 3.16   Possible combinations of increase, decrease, and concavity.

Inflection Points

A point $P(c, f(c))$ on the graph of a function f is called an *inflection point* of f if the concavity of the graph changes at P; that is, the graph of f is concave upward on one side of P and concave downward on the other side. Such transition points provide useful information about the graph of f. For instance, the concavity diagram for the function $f(x) = 2x^6 - 5x^4 + 7x - 3$ analyzed in Example 3.2.1 shows that the concavity of the graph of f changes from upward to downward at $x = -1$ and from downward to upward at $x = 1$, so the corresponding points $(-1, -13)$ and $(1, 1)$ on the graph are inflection points of f. Inflection points may also be of practical interest when interpreting a mathematical model based on f. For example, the point of diminishing returns on the production curve shown in Figure 3.12 is an inflection point.

At an inflection point $P(c, f(c))$, the graph of f can be neither concave upward ($f''(c) > 0$) nor concave downward ($f''(c) < 0$). Therefore, if $f''(c)$ exists, we must have $f''(c) = 0$. To summarize:

Inflection Points   ■   An inflection point (or point of inflection) is a point $(c, f(c))$ on the graph of a function f where the concavity changes. At such a point, either $f''(c) = 0$ or $f''(c)$ does not exist.

Procedure for Finding the Inflection Points for a Function *f*

Step 1. Compute $f''(x)$ and determine all points in the domain of f where either $f''(c) = 0$ or $f''(c)$ does not exist.

Step 2. For each number c found in step 1, determine the sign of $f''(x)$ to the left of $x = c$ and to the right $x = c$; that is, for $x < c$ and for $x > c$. If $f''(x) > 0$ on one side of $x = c$ and $f''(x) < 0$ on the other side, then $(c, f(c))$ is an inflection point for f.

EXAMPLE | 3.2.2

In each case, find all inflection points of the given function.

a. $f(x) = 3x^5 - 5x^4 - 1$ **b.** $g(x) = x^{1/3}$

Solution

a.   Note that $f(x)$ exists for all x and that

$$f'(x) = 15x^4 - 20x^3$$
$$f''(x) = 60x^3 - 60x^2 = 60x^2(x - 1)$$

Thus, $f''(x)$ is continuous for all x and $f''(x) = 0$ when $x = 0$ and $x = 1$. Testing the sign of $f''(x)$ on each side of $x = 0$ and $x = 1$ (say, at $x = -1, \frac{1}{2}$, and 2), we get

$$f''(-1) = -120 < 0 \quad f''\left(\frac{1}{2}\right) = \frac{-15}{2} < 0 \quad f''(2) = 240 > 0$$

which leads to the concavity pattern shown in this diagram:

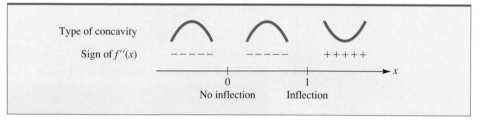

We see that the concavity does not change at $x = 0$ but changes from downward to upward at $x = 1$. Since $f(1) = -3$, it follows that $(1, -3)$ is an inflection point of f. The graph of f is shown in Figure 3.17a.

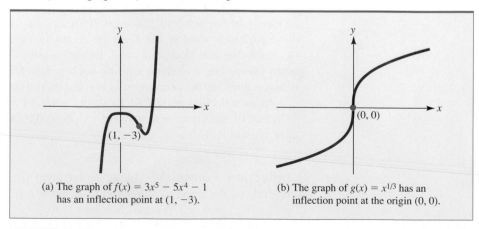

(a) The graph of $f(x) = 3x^5 - 5x^4 - 1$ has an inflection point at $(1, -3)$.

(b) The graph of $g(x) = x^{1/3}$ has an inflection point at the origin $(0, 0)$.

FIGURE 3.17   Two graphs with inflection points.

b.   The function $g(x)$ is continuous for all x and since

$$g'(x) = \frac{1}{3}x^{-2/3} \qquad \text{and} \qquad g''(x) = \frac{-2}{9}x^{-5/3}$$

it follows that $g''(x)$ is never 0 but does not exist when $x = 0$. Testing the sign of $g''(x)$ on each side of $x = 0$, we obtain the results displayed in this concavity diagram:

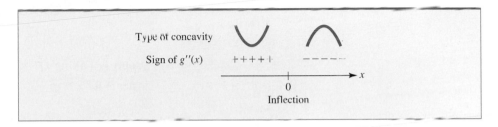

Since the concavity of the graph changes at $x = 0$ and $g(0) = 0$, there is an inflection point at the origin, $(0, 0)$. The graph of g is shown in Figure 3.17b.

NOTE   A function can have an inflection point only where it is continuous. In particular, if $f(c)$ is not defined, there cannot be an inflection point corresponding to $x = c$ even if $f''(x)$ changes sign at $x = c$. For example, if $f(x) = 1/x$, then $f''(x) = 2/x^3$, so $f''(x) < 0$ if $x < 0$ and $f''(x) > 0$ if $x > 0$. The concavity changes from downward to upward at $x = 0$ (see Figure 3.18a) but there is no inflection point at $x = 0$ since $f(0)$ is not defined.

Moreover, just knowing that $f(c)$ is defined and that $f''(c) = 0$ does not guarantee that $(c, f(c))$ is an inflection point. For instance, if $f(x) = x^4$, then $f(0) = 0$ and $f''(x) = 12x^2$ so $f''(0) = 0$. However, $f''(x) > 0$ for any number $x \neq 0$, so the graph of f is always concave upward, and there is no inflection point at $(0, 0)$ (see Figure 3.18b).

Do you think that if $f(c)$ is defined and $f''(c) = 0$, then you can at least conclude that either an inflection point or a relative extremum occurs at $x = c$? For the answer, see Problem 67.   ∎

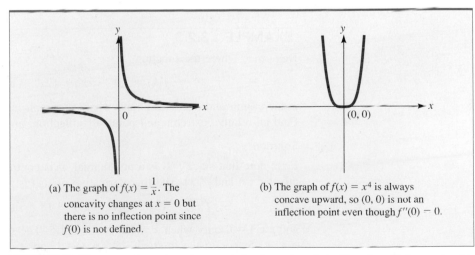

(a) The graph of $f(x) = \dfrac{1}{x}$. The concavity changes at $x = 0$ but there is no inflection point since $f(0)$ is not defined.

(b) The graph of $f(x) = x^4$ is always concave upward, so $(0, 0)$ is not an inflection point even though $f''(0) = 0$.

FIGURE 3.18   A graph need not have an inflection point where $f'' = 0$ or f'' does not exist.

Curve Sketching with the Second Derivative

Geometrically, inflection points occur at "twists" on a graph. Here is a summary of the graphical possibilities.

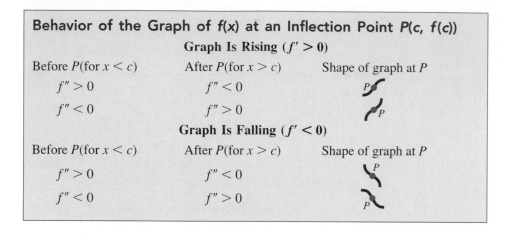

Behavior of the Graph of $f(x)$ at an Inflection Point $P(c, f(c))$		
Graph Is Rising ($f' > 0$)		
Before P (for $x < c$)	After P (for $x > c$)	Shape of graph at P
$f'' > 0$	$f'' < 0$	
$f'' < 0$	$f'' > 0$	
Graph Is Falling ($f' < 0$)		
Before P (for $x < c$)	After P (for $x > c$)	Shape of graph at P
$f'' > 0$	$f'' < 0$	
$f'' < 0$	$f'' > 0$	

Figure 3.19 shows several ways that inflection points can occur on a graph.

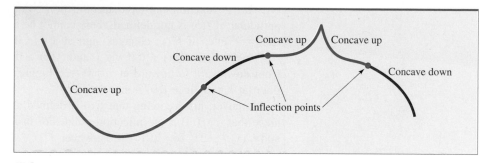

FIGURE 3.19   A graph showing concavity and inflection points.

By adding the criteria for concavity and inflection points to the first derivative methods discussed in Section 3.1, you can sketch a variety of graphs with considerable detail. Here is an example.

EXAMPLE | 3.2.3

Determine where the function

$$f(x) = 3x^4 - 2x^3 - 12x^2 + 18x + 15$$

is increasing and decreasing, and where its graph is concave up and concave down. Find all relative extrema and points of inflection, and sketch the graph.

Solution

First, note that since $f(x)$ is a polynomial, it is continuous for all x, as are the derivatives $f'(x)$ and $f''(x)$. The first derivative of $f(x)$ is

$$f'(x) = 12x^3 - 6x^2 - 24x + 18 = 6(x - 1)^2(2x + 3)$$

and $f'(x) = 0$ only when $x = 1$ and $x = -1.5$. The sign of $f'(x)$ does not change for $x < -1.5$, nor in the interval $-1.5 < x < 1$, nor for $x > 1$. Evaluating $f'(x)$ at test numbers in each interval (say, at -2, 0, and 3), you obtain the arrow diagram shown. Note that there is a relative minimum at $x = -1.5$ but no extremum at $x = 1$.

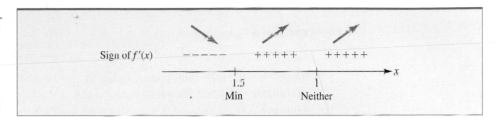

The second derivative is

$$f''(x) = 36x^2 - 12x - 24 = 12(x - 1)(3x + 2)$$

and $f''(x) = 0$ only when $x = 1$ and $x = \dfrac{-2}{3}$. The sign of $f''(x)$ does not change on each of the intervals $x < \dfrac{-2}{3}$, $\dfrac{-2}{3} < x < 1$, and $x > 1$. Evaluating $f''(x)$ at test numbers in each interval, you obtain the concavity diagram shown.

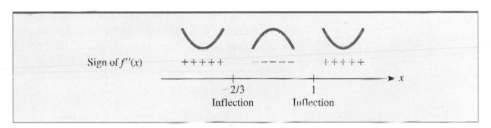

The patterns in these two diagrams suggest that there is a relative minimum at $x = -1.5$ and inflection points at $x = \dfrac{-2}{3}$ and $x = 1$ (since the concavity changes at both points).

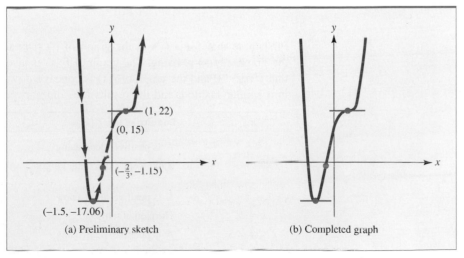

FIGURE 3.20   The graph of $f(x) = 3x^4 - 2x^3 - 12x^2 + 18x + 15$.

To sketch the graph, you find that

$$f(-1.5) = -17.06 \qquad f\left(\frac{-2}{3}\right) = -1.15 \qquad f(1) = 22$$

and plot a "cup" ⌣ at $(-1.5, -17.06)$ to mark the relative minimum located there. Likewise, plot twists ∫ at $\left(\frac{-2}{3}, -1.15\right)$ and ⌐ at $(1, 22)$ to mark the shape of the graph near the inflection points. Using the arrow and concavity diagrams, you get the preliminary diagram shown in Figure 3.20a. Finally, complete the sketch as shown in Figure 3.20b by passing a smooth curve through the minimum point, the inflection points, and the y intercept $(0, 15)$.

Sometimes you are given the graph of a derivative function $f'(x)$ and asked to analyze the graph of $f(x)$ itself. For instance, it would be quite reasonable for a manufacturer who knows the marginal cost $C'(x)$ associated with producing x units of a particular commodity to want to know as much as possible about the total cost $C(x)$. The following example illustrates a procedure for carrying out this kind of analysis.

EXAMPLE | 3.2.4

The graph of the derivative $f'(x)$ of a function $f(x)$ is shown in the figure. Find intervals of increase and decrease and concavity for $f(x)$ and locate all relative extrema and inflection points. Then sketch a curve that has all these features.

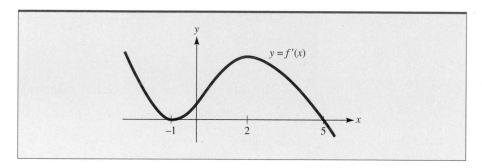

Solution

First, note that for $x < -1$, the graph of $f'(x)$ is above the x axis, so $f'(x) > 0$ and the graph of $f(x)$ is rising. The graph of $f'(x)$ is also falling for $x < -1$, which means that $f''(x) < 0$ and the graph of $f(x)$ is concave down. Other intervals can be analyzed in a similar fashion, and the results are summarized in the table.

x	Feature of $y = f'(x)$	Feature of $y = f(x)$
$x < -1$	f' is positive; decreasing	f is increasing; concave down
$x = -1$	x intercept; horizontal tangent	Horizontal tangent; possible inflection point ($f'' = 0$)
$-1 < x < 2$	f' is positive; increasing	f is increasing; concave up
$x = 2$	Horizontal tangent	Possible inflection point
$2 < x < 5$	f' is positive; decreasing	f is increasing; concave down
$x = 5$	x intercept	Horizontal tangent
$x > 5$	f' is negative; decreasing	f is decreasing; concave down

Since the concavity changes at $x = -1$ (down to up), an inflection point occurs there, along with a horizontal tangent. At $x = 2$, there is also an inflection point (concavity changes from up to down), but no horizontal tangent. The graph of $f(x)$ is rising to the left of $x = 5$ and falling to the right, so there must be a relative maximum at $x = 5$.

One possible graph that has all the features required for $y = f(x)$ is shown in Figure 3.21. Note, however, that since you are not given the values of $f(-1)$, $f(2)$, and $f(5)$, many other graphs will also satisfy the requirements.

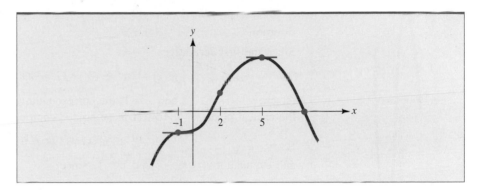

FIGURE 3.21   A possible graph for $y = f(x)$.

The Second Derivative Test

The second derivative can also be used to classify a critical point of a function as a relative maximum or minimum. Here is a statement of the procedure, which is known as the **second derivative test.**

> **The Second Derivative Test**   ■   Suppose $f''(x)$ exists on an open interval containing $x = c$ and that $f'(c) = 0$.
>
> If $f''(c) > 0$, then f has a relative minimum at $x = c$.
>
> If $f''(c) < 0$, then f has a relative maximum at $x = c$.
>
> However, if $f''(c) = 0$ or if $f''(c)$ does not exist, the test is inconclusive and f may have a relative maximum, a relative minimum, or no relative extremum at all at $x = c$.

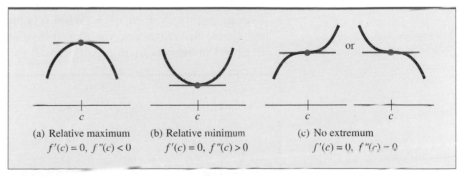

(a) Relative maximum
$f'(c) = 0$, $f''(c) < 0$

(b) Relative minimum
$f'(c) = 0$, $f''(c) > 0$

(c) No extremum
$f'(c) = 0$, $f''(c) = 0$

FIGURE 3.22   The second derivative test.

To see why the second derivative test works, look at Figure 3.22, which shows four possibilities that can occur when $f'(c) = 0$. Figure 3.22a suggests that at a relative maximum, the graph of f must be concave downward, so $f''(c) < 0$. Likewise, at a relative minimum (Figure 3.22b), the graph of f must be concave upward, so $f''(c) > 0$. On the other hand, Figure 3.22c suggests that if a point where $f'(c) = 0$ is not a relative extremum, then it must be an inflection point and $f''(c) = 0$ (if $f''(c)$ is defined). It follows that if $f'(c) = 0$ and $f''(c) < 0$, then $P(c, f(c))$ must be a relative maximum, while if $f'(c) = 0$ and $f''(c) > 0$, the corresponding critical point must be a relative minimum.

The use of the second derivative test is illustrated in Example 3.2.5.

EXAMPLE 3.2.5

Find the critical points of $f(x) = 2x^3 + 3x^2 - 12x - 7$ and use the second derivative test to classify each critical point as a relative maximum or minimum.

Solution

Since the first derivative

$$f'(x) = 6x^2 + 6x - 12 = 6(x + 2)(x - 1)$$

is zero when $x = -2$ and $x = 1$, the corresponding points $(-2, 13)$ and $(1, -14)$ are the critical points of f. To test these points, compute the second derivative

$$f''(x) = 12x + 6$$

and evaluate it at $x = -2$ and $x = 1$. Since

$$f''(-2) = -18 < 0$$

it follows that the critical point $(-2, 13)$ is a relative maximum, and since

$$f''(1) = 18 > 0$$

it follows that the critical point $(1, -14)$ is a relative minimum. For reference, the graph of f is sketched in Figure 3.23.

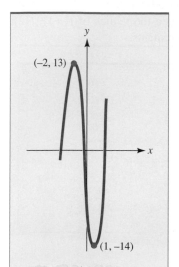

FIGURE 3.23 The graph of $f(x) = 2x^3 + 3x^2 - 12x - 7$.

NOTE Although it was easy to use the second derivative test to classify critical points in Example 3.2.5, the test does have some limitations. For some functions, the work involved in computing the second derivative is time-consuming, which may diminish the appeal of the test. Moreover, the test applies only to critical points at which the derivative is zero and not to those where the derivative is undefined. Finally, if both $f'(c)$ and $f''(c)$ are zero, the second derivative test tells you nothing about the nature of the critical point. This is illustrated in Figure 3.24, which shows the graphs of three functions whose first and second derivatives are both zero when $x = 0$. When it is inconvenient or impossible to apply the second derivative test, you may still be able to use the first derivative test described in Section 3.1 to classify critical points. ■

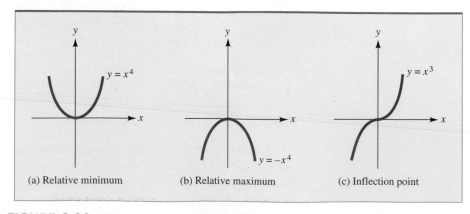

(a) Relative minimum

(b) Relative maximum

(c) Inflection point

FIGURE 3.24 Three functions whose first and second derivatives are zero at $x = 0$.

In Example 3.2.6, we return to the questions of worker efficiency and diminishing returns examined in the illustration at the beginning of this section. Our goal is to maximize a worker's *rate* of production; that is, the derivative of the worker's output. Hence, we will set to zero the *second* derivative of output and find an inflection point of the output function, which we interpret as the point of diminishing returns for the output.

EXAMPLE 3.2.6

An efficiency study of the morning shift at a factory indicates that an average worker who starts at 8:00 A.M. will have produced $Q(t) = -t^3 + 9t^2 + 12t$ units t hours later. At what time during the morning is the worker performing most efficiently?

Solution

The worker's rate of production is the derivative of the output $Q(t)$; that is,

$$R(t) = Q'(t) = -3t^2 + 18t + 12$$

Assuming that the morning shift runs from 8:00 A.M. until noon, the goal is to find the largest rate $R(t)$ for $0 \le t \le 4$. The derivative of the rate function is

$$R'(t) = Q''(t) = -6t + 18$$

which is zero when $t = 3$, positive for $0 < t < 3$, and negative for $3 < t < 4$, as indicated in the arrow diagram shown.

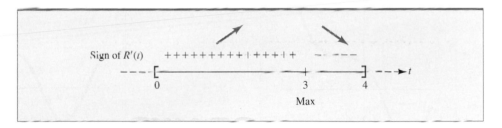

Thus, the rate of production $R(t)$ increases for $0 < t < 3$, decreases for $3 < t < 4$, and has its maximum value when $t = 3$ (11:00 A.M.) This means that the output function $Q(t)$ has an inflection point at $t = 3$, since $Q''(t) = R'(t)$ changes sign at that time. The graph of the production function $Q(t)$ is shown in Figure 3.25a, while that of the production rate function $R(t)$ is shown in Figure 3.25b.

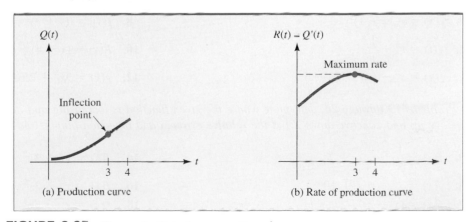

(a) Production curve   (b) Rate of production curve

FIGURE 3.25 The production of an average worker.

PROBLEMS 3.2

In Problems 1 through 4, determine where the second derivative of the function is positive and where it is negative.

1.

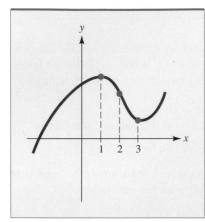

2.

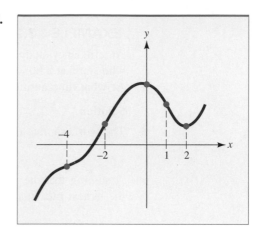

3.

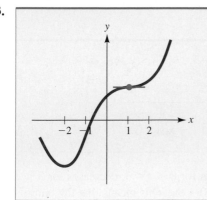

4.

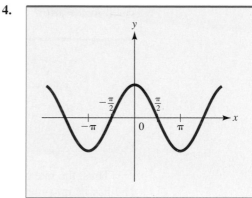

In Problems 5 through 12, determine where the graph of the given function is concave upward and concave downward. Find the coordinates of all inflection points.

5. $f(x) = x^3 + 3x^2 + x + 1$

6. $f(x) = x^4 - 4x^3 + 10x - 9$

7. $f(x) = x(2x + 1)^2$

8. $f(s) = s(s + 3)^2$

9. $g(t) = t^2 - \dfrac{1}{t}$

10. $F(x) = (x - 4)^{7/3}$

11. $f(x) = x^4 - 6x^3 + 7x - 5$

12. $g(x) = 3x^5 - 25x^4 + 11x - 17$

In Problems 13 through 26, determine where the given function is increasing and decreasing, and where its graph is concave up and concave down. Find the relative extrema and inflection points, and sketch the graph of the function.

13. $f(x) = \dfrac{1}{3}x^3 - 9x + 2$

14. $f(x) = x^3 + 3x^2 + 1$

15. $f(x) = x^4 - 4x^3 + 10$

16. $f(x) = x^3 - 3x^2 + 3x + 1$

17. $f(x) = (x - 2)^3$

18. $f(x) = x^5 - 5x$

19. $f(x) = (x^2 - 5)^3$

20. $f(x) = (x - 2)^4$

21. $f(s) = 2s(s + 4)^3$

22. $f(x) = (x^2 - 3)^2$

23. $g(x) = \sqrt{x^2 + 1}$

24. $f(x) = \dfrac{x^2}{x^2 + 3}$

25. $f(x) = \dfrac{1}{x^2 + x + 1}$

26. $f(x) = x^4 + 6x^3 - 24x^2 + 24$

In Problems 27 through 38 use the second derivative test to find the relative maxima and minima of the given function.

27. $f(x) = x^3 + 3x^2 + 1$

28. $f(x) = x^4 - 2x^2 + 3$

29. $f(x) = (x^2 - 9)^2$

30. $f(x) = x + \dfrac{1}{x}$

31. $f(x) = 2x + 1 + \dfrac{18}{x}$

32. $f(x) = \dfrac{x^2}{x - 2}$

33. $f(x) = x^2(x - 5)^2$

34. $f(x) = \left(\dfrac{x}{x + 1}\right)^2$

35. $h(t) = \dfrac{2}{1 + t^2}$

36. $f(s) = \dfrac{s + 1}{(s - 1)^2}$

37. $f(x) = \dfrac{(x - 2)^3}{x^2}$

38. $h(t) = \dfrac{(t + 3)^3}{(t - 1)^2}$

In Problems 39 through 42, the second derivative $f''(x)$ of a function is given. In each case, use this information to determine where the graph of $f(x)$ is concave upward and concave downward and find all values of x for which an inflection point occurs. [You are not required to find $f(x)$ or the y coordinates of the inflection points.]

39. $f''(x) = x^2(x - 3)(x - 1)$

40. $f''(x) = x^3(x^2 + 2x - 3)$

41. $f''(x) = (x - 1)^{1/3}$

42. $f''(x) = \dfrac{x^2 + x - 2}{x^2 + 4}$

In Problems 43 through 46, the first derivative $f'(x)$ of a certain function $f(x)$ is given. In each case,
  (a) Find intervals on which f is increasing and decreasing.
  (b) Find intervals on which the graph of f is concave up and concave down.
  (c) Find the x coordinates of the relative extrema and inflection points of f.
  (d) Sketch a possible graph for $f(x)$.

43. $f'(x) = x^2 - 4x$

44. $f'(x) = x^2 - 2x - 8$

45. $f'(x) = 5 - x^2$

46. $f'(x) = x(1 - x)$

47. Sketch the graph of a function that has all of the following properties:
  a. $f'(x) > 0$ when $x < -1$ and when $x > 3$
  b. $f'(x) < 0$ when $-1 < x < 3$
  c. $f''(x) < 0$ when $x < 2$
  d. $f''(x) > 0$ when $x > 2$

48. Sketch the graph of a function f that has all of the following properties:
  a. The graph has discontinuities at $x = -1$ and at $x = 3$
  b. $f'(x) > 0$ for $x < 1, x \neq -1$
  c. $f'(x) < 0$ for $x > 1, x \neq 3$
  d. $f''(x) > 0$ for $x < -1$ and $x > 3$ and $f''(x) < 0$ for $-1 < x < 3$
  e. $f(0) = 0 = f(2), f(1) = 3$

In Problems 49 through 52 the graph of a derivative function $y = f'(x)$ is given. Describe the function $f(x)$ and sketch a possible graph of $y = f(x)$.

49.
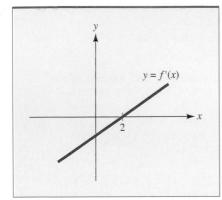

50.
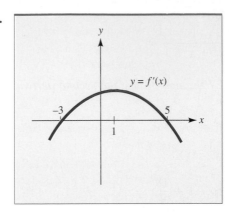

51.
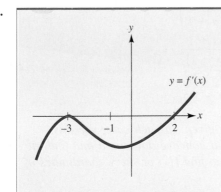

52.
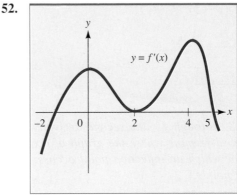

53. MARGINAL ANALYSIS   The cost of producing *x* units of a commodity per week is

$$C(x) = 0.3x^3 - 5x^2 + 28x + 200$$

a. Find the marginal cost $C'(x)$, and sketch its graph along with the graph of $C(x)$ on the same coordinate plane.

b. Find all values of *x* where $C''(x) = 0$. How are these levels of production related to the graph of the marginal cost?

54. SALES   A company estimates that when *x* thousand dollars are spent on the marketing of a certain product, $Q(x)$ units of the product will be sold, where

$$Q(x) = -4x^3 + 252x^2 - 3,200x + 17,000$$

for $10 \le x \le 40$. Sketch the graph of $Q(x)$. Where does the graph have an inflection point? What is the significance of the marketing expenditure that corresponds to this point?

55. WORKER EFFICIENCY   An efficiency study of the morning shift (from 8:00 A.M. to 12:00 noon) at a factory indicates that an average worker who arrives on the job at 8:00 A.M. will have produced *Q* units *t* hours later, where $Q(t) = -t^3 + \dfrac{9}{2}t^2 + 15t$.

a. At what time during the morning is the worker performing most efficiently?

b. At what time during the morning is the worker performing least efficiently?

56. WORKER EFFICIENCY   An efficiency study of the morning shift (from 8:00 A.M. to 12:00 noon) at a certain factory indicates that an average worker who arrives on the job at 8:00 A.M. will have assembled $Q(t) = -t^3 + 6t^2 + 15t$ transistor radios *t* hours later.

a. At what time during the morning is the worker performing most efficiently?

b. At what time during the morning is the worker performing least efficiently?

57. **POPULATION GROWTH** A 5-year projection of population trends suggests that t years from now, the population of a certain community will be $P(t) = -t^3 + 9t^2 + 48t + 50$ thousand.

    a. At what time during the 5-year period will the population be growing most rapidly?

    b. At what time during the 5-year period will the population be growing least rapidly?

    c. At what time is the rate of population growth changing most rapidly?

58. **ADVERTISING** The manager of the Footloose sandal company determines that t months after initiating an advertising campaign, $S(t)$ hundred pairs of sandals will be sold, where

$$S(t) = \frac{3}{t + 2} - \frac{12}{(t + 2)^2} + 5$$

    a. Find $S'(t)$ and $S''(t)$.

    b. At what time will sales be maximized? What is the maximum level of sales?

    c. The manager plans to terminate the advertising campaign when the sales rate is minimized. When does this occur? What are the sales level and sales rate at this time?

59. **SPREAD OF A DISEASE** An epidemiologist determines that a particular epidemic spreads in such a way that t weeks after the outbreak, N hundred new cases will be reported, where

$$N(t) = \frac{5t}{12 + t^2}$$

    a. Find $N'(t)$ and $N''(t)$.

    b. At what time is the epidemic at its worst? What is the maximum number of reported new cases?

    c. Health officials declare the epidemic to be under control when the rate of reported new cases is minimized. When does this occur? What number of new cases will be reported at that time?

60. **GOVERNMENT SPENDING** During a recession, Congress decides to stimulate the economy by providing funds to hire unemployed workers for government projects. Suppose that t months after the stimulus program begins, there are $N(t)$ thousand people unemployed, where

$$N(t) = -t^3 + 45t^2 + 408t + 3078$$

    a. What is the maximum number of unemployed workers? When does the maximum level of unemployment occur?

b. In order to avoid overstimulating the economy (and inducing inflation), a decision is made to terminate the stimulus program as soon as the rate of unemployment begins to decline. When does this occur? At this time, how many people are unemployed?

61. **HOUSING STARTS** Suppose that in a certain community, there will be $M(r)$ thousand new houses built when the 30-year fixed mortgage rate is $r\%$, where

$$M(r) = \frac{1 + 0.02r}{1 + 0.009r^2}$$

    a. Find $M'(r)$ and $M''(r)$.

    b. Sketch the graph of the construction function $M(r)$.

    c. At what rate of interest r is the rate of construction of new houses minimized?

62. Water is poured at a constant rate into the vase shown in the accompanying figure. Let $h(t)$ be the height of the water in the vase at time t (assume the vase is empty when $t = 0$). Sketch a rough graph of the function $h(t)$. In particular, what happens when the water level reaches the neck of the vase?

PROBLEM 62

63. **POPULATION GROWTH** Studies show that when environmental factors impose an upper bound on the possible size of a population $P(t)$, the population often tends to grow in such a way that the percentage rate of change of $P(t)$ satisfies

$$\frac{100 P'(t)}{P(t)} = A - BP(t)$$

where A and B are positive constants. Where does the graph of $P(t)$ have an inflection point? What is the significance of this point? (Your answer will be in terms of A and B.)

64. **THE SPREAD OF AN EPIDEMIC** Let $Q(t)$ denote the number of people in a city of population N_0 who have been infected with a certain disease t days after the beginning of an epidemic. Studies

indicate that the rate $R(Q)$ at which an epidemic spreads is jointly proportional to the number of people who have contracted the disease and the number who have not, so $R(Q) = kQ(N_0 - Q)$. Sketch the graph of the rate function, and interpret your graph. In particular, what is the significance of the highest point on the graph of $R(Q)$?

65. **TISSUE GROWTH** Suppose a particular tissue culture has area $A(t)$ at time t and a potential maximum area M. Based on properties of cell division, it is reasonable to assume that the area A grows at a rate jointly proportional to $\sqrt{A(t)}$ and $M - A(t)$; that is,

$$\frac{dA}{dt} = k\sqrt{A(t)}\,[M - A(t)]$$

where k is a positive constant.
   a. Let $R(t) = A'(t)$ be the rate of tissue growth. Show that $R'(t) = 0$ when $A(t) = M/3$.
   b. Is the rate of tissue growth greatest or least when $A(t) = M/3$? [*Hint:* Use the first derivative test or the second derivative test.]
   c. Based on the given information and what you discovered in part (a), what can you say about the graph of $A(t)$?

66. Use calculus to show that the graph of the quadratic function $y = ax^2 + bx + c$ is concave upward if a is positive and concave downward if a is negative.

67. Let $f(x) = x^4 + x$. Show that even though $f''(0) = 0$, the graph of f has neither a relative extremum nor an inflection point where $x = 0$. Sketch the graph of $f(x)$.

68. If $f(x)$ and $g(x)$ are continuous functions that both have an inflection point at $x = c$, is it true that the sum $h(x) = f(x) + g(x)$ must also have an inflection point at $x = c$? Either explain why it must always

be true or find functions $f(x)$ and $g(x)$ for which it is false.

69. Given the function $f(x) = 2x^3 + 3x^2 - 12x - 7$, complete these steps:
   a. Graph using $[-10, 10]1$ by $[-10, 10]1$ and $[-10, 10]1$ by $[-20, 20]2$.
   b. Fill in this table:

x	-4	-2	-1	0	1	2
$f(x)$						
$f'(x)$						
$f''(x)$						

   c. Approximate the x intercept and y intercept to two decimal places.
   d. Find the relative maximum and relative minimum points.
   e. Find the intervals over which $f(x)$ is increasing.
   f. Find the intervals over which $f(x)$ is decreasing.
   g. Find any inflection points.
   h. Find the intervals over which the graph of $f(x)$ is concave upward.
   i. Find the intervals over which the graph of $f(x)$ is concave downward.
   j. Show that the concavity changes from upward to downward, or vice versa, when x moves from a little less than the point of inflection to a little greater than the point of inflection.
   k. Find the largest and smallest values for this function for $-4 \le x \le 2$.

70. Repeat Problem 69 for the function
$$f(x) = 3.7x^4 - 5.03x^3 + 2x^2 - 0.7$$

SECTION 3.3 | Curve Sketching

Just-In-Time Review

It is important to remember that ∞ is *not* a number. It is used only to represent a process of unbounded growth or the result of such growth.

So far in this chapter, you have seen how to use the derivative $f'(x)$ to determine where the graph of $f(x)$ is rising and falling and to use the second derivative $f''(x)$ to determine the concavity of the graph. While these tools are adequate for locating the high and low points of a graph and for sculpting its twists and turns, there are other graphical features that are best described using limits.

Recall from Section 1.5 that a limit of the form $\lim\limits_{x \to +\infty} f(x)$ or $\lim\limits_{x \to -\infty} f(x)$, in which the independent variable x either increases or decreases without bound, is called a *limit at infinity*. On the other hand, if the functional values $f(x)$ themselves grow without bound as x approaches a number c, we say that $f(x)$ has an *infinite limit* at $x = c$ and write $\lim\limits_{x \to c} f(x) = +\infty$ if $f(x)$ increases indefinitely as x approaches c or

$\lim\limits_{x \to c} f(x) = -\infty$ if $f(x)$ decreases indefinitely. Collectively, limits at infinity and infinite limits are referred to as **limits involving infinity.** Our first goal in this section is to see how limits involving infinity may be interpreted as graphical features. This information will then be combined with the derivative methods of Sections 3.1 and 3.2 to form a general procedure for sketching graphs.

Vertical Asymptotes

Limits involving infinity can be used to describe graphical features called *asymptotes.* In particular, the graph of a function $f(x)$ is said to have a **vertical asymptote** at $x = c$ if $f(x)$ increases or decreases without bound as x tends toward c, from either the right or the left.

For instance, consider the rational function

$$f(x) = \frac{x + 1}{x - 2}$$

As x approaches 2 from the left ($x < 2$), the functional values decrease without bound, but they increase without bound if the approach is from the right ($x > 2$). This behavior is illustrated in the table and demonstrated graphically in Figure 3.26.

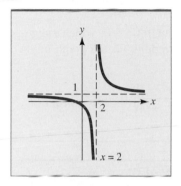

FIGURE 3.26 The graph of $f(x) = \dfrac{x + 1}{x - 2}$.

x	1.95	1.97	1.99	1.999	2	2.001	2.005	2.01
$f(x) = \dfrac{x + 1}{x - 2}$	−59	−99	−299	−2,999	Undefined	3,001	601	301

The behavior in this example can be summarized as follows using the one-sided limit notation introduced in Section 1.6:

$$\lim_{x \to 2^-} \frac{x + 1}{x - 2} = -\infty \qquad \text{and} \qquad \lim_{x \to 2^+} \frac{x + 1}{x - 2} = +\infty$$

In a similar fashion, we use the limit notation to define the concept of vertical asymptote.

Vertical Asymptotes ■ The line $x = c$ is a *vertical asymptote* of the graph of $f(x)$ if either

$$\lim_{x \to c^-} f(x) = +\infty \quad (\text{or } -\infty)$$

or

$$\lim_{x \to c^+} f(x) = +\infty \quad (\text{or } -\infty)$$

In general, a rational function $R(x) = \dfrac{p(x)}{q(x)}$ has a vertical asymptote $x = c$ whenever $q(c) = 0$ but $p(c) \neq 0$. Here is an example of a function with a vertical asymptote.

EXAMPLE 3.3.1

Determine all vertical asymptotes of the graph of

$$f(x) = \frac{x^2 - 9}{x^2 + 3x}$$

Solution

Let $p(x) = x^2 - 9$ and $q(x) = x^2 + 3x$ be the numerator and denominator of $f(x)$, respectively. Then $q(x) = 0$ when $x = -3$ and when $x = 0$. However, for $x = -3$, we also have $p(-3) = 0$ and

$$\lim_{x \to -3} \frac{x^2 - 9}{x^2 + 3x} = \lim_{x \to -3} \frac{x - 3}{x} = 2$$

This means that the graph of $f(x)$ has a "hole" at the point $(-3, 2)$ and $x = -3$ is *not* a vertical asymptote of the graph.

On the other hand, for $x = 0$ we have $q(0) = 0$ but $p(0) \neq 0$, which suggests that the y axis (the vertical line $x = 0$) is a vertical asymptote for the graph of $f(x)$. This asymptotic behavior is verified by noting that

$$\lim_{x \to 0^-} \frac{x^2 - 9}{x^2 + 3x} = +\infty \qquad \text{and} \qquad \lim_{x \to 0^+} \frac{x^2 - 9}{x^2 + 3x} = -\infty$$

The graph of $f(x)$ is shown in Figure 3.27.

EXPLORE!

Referring to Example 3.3.1, store $f(x) = (x^2 - 9)/(x^2 + 3x)$ into Y1 and graph using the expanded decimal window $[-4.7, 4.7]1$ by $[-6.2, 6.2]1$. Use the **TRACE** feature to confirm that $f(x)$ is not defined at X = -3 and X = 0. How does your calculator indicate these undefined points?

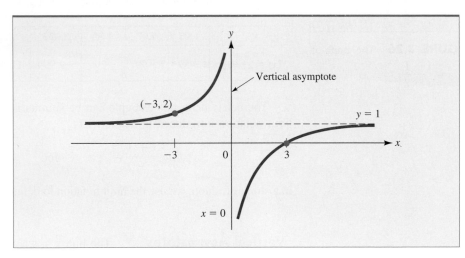

FIGURE 3.27 The graph of $f(x) = \dfrac{x^2 - 9}{x^2 + 3x}$.

Horizontal Asymptotes

In Figure 3.27, note that the graph approaches the horizontal line $y = 1$ as x increases or decreases without bound; that is,

$$\lim_{x \to -\infty} \frac{x^2 - 9}{x^2 + 3x} = 1 \qquad \text{and} \qquad \lim_{x \to +\infty} \frac{x^2 - 9}{x^2 + 3x} = 1$$

In general, when a function $f(x)$ tends toward a finite value b as x either increases or decreases without bound (or both as in our example), then the horizontal line $y = b$ is called a **horizontal asymptote** of the graph of $f(x)$. Here is a definition.

Horizontal Asymptotes ■ The horizontal line $y = b$ is called a *horizontal asymptote* of the graph of $y = f(x)$ if

$$\lim_{x \to -\infty} f(x) = b \qquad \text{or} \qquad \lim_{x \to +\infty} f(x) = b$$

EXAMPLE | 3.3.2

Determine all horizontal asymptotes of the graph of

$$f(x) = \frac{x^2}{x^2 + x + 1}$$

Solution

Dividing each term in the rational function $f(x)$ by x^2 (the highest power of x in the denominator), we find that

$$\lim_{x \to +\infty} f(x) = \lim_{x \to +\infty} \frac{x^2}{x^2 + x + 1} = \lim_{x \to +\infty} \frac{x^2/x^2}{x^2/x^2 + x/x^2 + 1/x^2}$$

$$= \lim_{x \to +\infty} \frac{1}{1 + 1/x + 1/x^2} = \frac{1}{1 + 0 + 0} = 1 \qquad \begin{array}{l} \textit{reciprocal power} \\ \textit{rule} \end{array}$$

and similarly,

$$\lim_{x \to -\infty} f(x) = \lim_{x \to -\infty} \frac{x^2}{x^2 + x + 1} = 1$$

Thus, the graph of $f(x)$ has $y = 1$ as a horizontal asymptote. The graph is shown in Figure 3.28.

Just-In-Time Review

Recall the reciprocal power rules for limits (pg. 66 in Section 1.5):

$$\lim_{x \to +\infty} \frac{A}{x^k} = 0$$

and

$$\lim_{x \to -\infty} \frac{A}{x^k} = 0$$

for constants A and k, with $k > 0$ and x^k defined for all x.

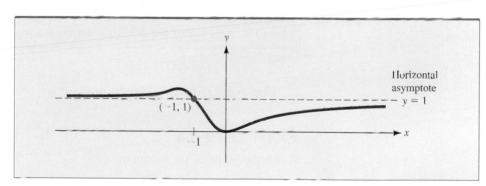

FIGURE 3.28 The graph of $f(x) = \dfrac{x^2}{x^2 + x + 1}$.

NOTE The graph of a function $f(x)$ can never cross a vertical asymptote $x = c$ because at least one of the one-sided limits $\lim_{x \to c^-} f(x)$ and $\lim_{x \to c^+} f(x)$ must be infinite. However, it is quite possible for a graph to cross its horizontal asymptotes. For instance, in Example 3.3.2, the graph of $y = \dfrac{x^2}{x^2 + x + 1}$ crosses the horizontal asymptote $y = 1$ at the point where

$$\frac{x^2}{x^2 + x + 1} = 1$$
$$x^2 = x^2 + x + 1$$
$$x = -1$$

that is, at the point $(-1, 1)$. ∎

A General Graphing Procedure

We now have the tools we need to describe a general procedure for sketching a variety of graphs.

A General Procedure for Sketching the Graph of $f(x)$

Step 1. Find the domain of $f(x)$ (that is, where $f(x)$ is defined).

Step 2. Find and plot all intercepts. The y intercept (where $x = 0$) is usually easy to find, but the x intercepts (where $f(x) = 0$) may require a calculator.

Step 3. Determine all vertical and horizontal asymptotes of the graph. Draw the asymptotes in a coordinate plane.

Step 4. Find $f'(x)$ and use it to determine the critical numbers of $f(x)$ and intervals of increase and decrease.

Step 5. Determine all relative extrema (both coordinates). Plot each relative maximum with a "cap" ($\frown$) and each relative minimum with a "cup" ($\smile$).

Step 6. Find $f''(x)$ and use it to determine intervals of concavity and points of inflection. Plot each inflection point with a "twist" to suggest the shape of the graph near the point.

Step 7. You now have a preliminary graph, with asymptotes in place, intercepts plotted, arrows indicating the direction of the graph, and "caps," "cups," and "twists" suggesting the shape at key points. Plot additional points if needed, and complete the sketch by joining the plotted points in the directions indicated. Be sure to remember that the graph cannot cross a vertical asymptote.

Here is a step-by-step analysis of the graph of a rational function.

EXAMPLE │ 3.3.3

Sketch the graph of the function

$$f(x) = \frac{x}{(x + 1)^2}$$

Solution

Steps 1 and 2. The function is defined for all x except $x = -1$, and the only intercept is the origin $(0, 0)$.

Step 3. The line $x = -1$ is a vertical asymptote of the graph of $f(x)$ since $f(x)$ decreases indefinitely as x approaches -1 from either side; that is,

$$\lim_{x \to -1^-} \frac{x}{(x + 1)^2} = \lim_{x \to -1^+} \frac{x}{(x + 1)^2} = -\infty$$

Moreover, since

$$\lim_{x \to -\infty} \frac{x}{(x + 1)^2} = \lim_{x \to +\infty} \frac{x}{(x + 1)^2} = 0$$

the line $y = 0$ (the x axis) is a horizontal asymptote. Draw dashed lines $x = -1$ and $y = 0$ on a coordinate plane.

Step 4. Applying the quotient rule, compute the derivative of $f(x)$:

$$f'(x) = \frac{(x + 1)^2(1) - x[2(x + 1)(1)]}{(x + 1)^4} = \frac{1 - x}{(x + 1)^3}$$

Since $f'(1) = 0$, it follows that $x = 1$ is a critical number. Note that even though $f'(-1)$ does not exist, $x = -1$ is not a critical number since it is not in the domain of $f(x)$. Place $x = 1$ and $x = -1$ on a number line with a dashed vertical line at $x = -1$ to indicate the vertical asymptote there. Then evaluate $f'(x)$ at appropriate test numbers (say, at -2, 0, and 3) to obtain the arrow diagram shown.

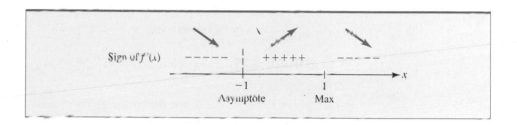

Step 5. The arrow pattern in the diagram obtained in step 4 indicates there is a relative maximum at $x = 1$. Since $f(1) = \frac{1}{4}$, we plot a "cap" at $\left(1, \frac{1}{4}\right)$.

Step 6. Apply the quotient rule again to get

$$f''(x) = \frac{2(x - 2)}{(x + 1)^4}$$

Since $f''(x) = 0$ at $x = 2$ and $f''(x)$ does not exist at $x = -1$, plot -1 and 2 on a number line and check the sign of $f''(x)$ on the intervals $x < -1$, $-1 < x < 2$, and $x > 2$ to obtain the concavity diagram shown.

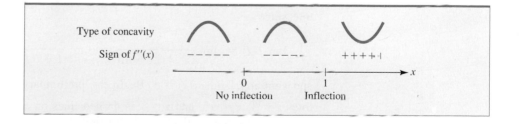

Note that the concavity changes at $x = 2$. Since $f(2) = \frac{2}{9}$, plot a "twist" $\sim$ at $\left(2, \frac{2}{9}\right)$ to indicate the inflection point there.

Step 7. The preliminary graph is shown in Figure 3.29a. Note that the vertical asymptote (dashed line) breaks the graph into two parts. Join the features in each separate part by a smooth curve to obtain the completed graph shown in Figure 3.29b.

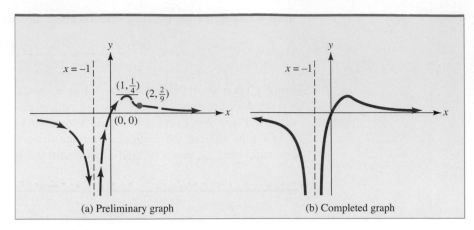

(a) Preliminary graph   (b) Completed graph

FIGURE 3.29 The graph of $f(x) = \dfrac{x}{(x + 1)^2}$.

In Example 3.3.4, we sketch the graph of a more complicated rational function using a condensed form of the step-by-step solution featured in Example 3.3.3.

EXPLORE!

Store the function $f(x)$ in Example 3.3.4 into the equation editor. How does the graph of $f(x)$ behave as you trace out points with large x values?

EXAMPLE | 3.3.4

Sketch the graph of

$$f(x) = \frac{3x^2}{x^2 + 2x - 15}$$

Solution

Since $x^2 + 2x - 15 = (x + 5)(x - 3)$, the function $f(x)$ is defined for all x except $x = -5$ and $x = 3$. The only intercept is the origin $(0, 0)$.

We see that $x = 3$ and $x = -5$ are vertical asymptotes because if we write $f(x) = p(x)/q(x)$, where $p(x) = 3x^2$ and $q(x) = x^2 + 2x - 15$, then $q(3) = 0$ and $q(-5) = 0$, while $p(3) \neq 0$ and $p(-5) \neq 0$. Moreover, $y = 3$ is a horizontal asymptote since

$$\lim_{x \to +\infty} f(x) = \lim_{x \to +\infty} \frac{3x^2}{x^2 + 2x - 15} = \lim_{x \to +\infty} \frac{3}{1 + 2/x - 15/x^2} = \frac{3}{1 + 0 - 0} = 3$$

and similarly, $\lim_{x \to -\infty} f(x) = 3$. Begin the preliminary sketch by drawing the asymptotes $x = 3$, $x = -5$, and $y = 3$ as dashed lines on a coordinate plane.

Next, use the quotient rule to obtain

$$f'(x) = \frac{(x^2 + 2x - 15)(6x) - (2x + 2)(3x^2)}{(x^2 + 2x - 15)^2} = \frac{6x(x - 15)}{(x^2 + 2x - 15)^2}$$

We see that $f'(x) = 0$ when $x = 0$ and $x = 15$ and that $f'(x)$ does not exist when $x = -5$ and $x = 3$. Put $x = -5, 0, 3,$ and 15 on a number line and obtain the arrow diagram shown in Figure 3.30a by determining the sign of $f'(x)$ at appropriate test numbers (say, at $-7, -1, 2, 5,$ and 20). Interpreting the arrow diagram, we see that there is a relative maximum at $x = 0$ and a relative minimum at $x = 15$. Since $f(0) = 0$ and $f(15) \approx 2.81$, we plot a "cap" at $(0, 0)$ on our preliminary graph and a "cup" at $(15, 2.81)$.

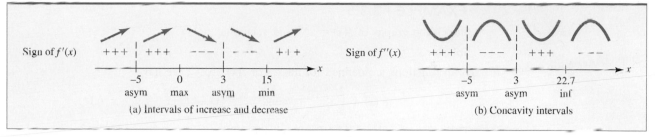

FIGURE 3.30 Arrow and concavity diagrams for $f(x) = \dfrac{3x^2}{x^2 + 2x - 15}$.

Apply the quotient rule again to get

$$f''(x) = \frac{-6(2x^3 - 45x^2 - 225)}{(x^2 + 2x - 15)^3}$$

We see that $f''(x)$ does not exist when $x = -5$ and $x = 3$, and that $f''(x) = 0$ when

$$2x^3 - 45x^2 - 225 = 0$$
$$x \approx 22.7 \qquad \textit{found by calculator}$$

Put $x = -5$, 3, and 22.7 on a number line and obtain the concavity diagram shown in Figure 3.30b by determining the sign of $f''(x)$ at appropriate test numbers (say, at -6, 0, 4, and 25). The concavity changes at all three subdivision numbers, but only $x = 22.7$ corresponds to an inflection point since $x = -5$ and $x = 3$ are not in the domain of $f(x)$. We find that $f(22.7) \approx 2.83$ and plot a "twist" ( ) at (22.7, 2.83).

The preliminary graph is shown in Figure 3.31a. Note that the two vertical asymptotes break the graph into three parts. Join the features in each separate part by a smooth curve to obtain the completed graph shown in Figure 3.31b.

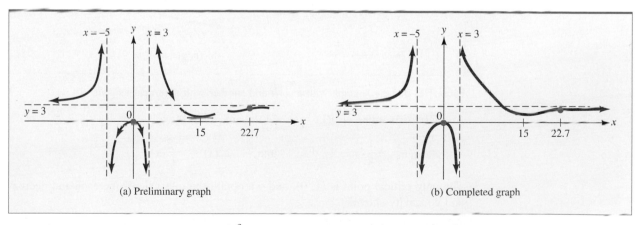

FIGURE 3.31 The graph of $f(x) = \dfrac{3x^2}{x^2 + 2x - 15}$.

When $f'(c)$ does not exist at a number $x = c$ in the domain of $f(x)$, there are several possibilities for the graph of $f(x)$ at the point $(c, f(c))$. Two such cases are examined in Example 3.3.5.

EXAMPLE | 3.3.5

Sketch the graphs of $f(x) = x^{2/3}$ and $g(x) = (x - 1)^{1/3}$.

Solution

Both functions are defined for all x. For $f(x) = x^{2/3}$, we have

$$f'(x) = \frac{2}{3}x^{-1/3} \quad \text{and} \quad f''(x) = \frac{-2}{9}x^{-4/3} \quad x \neq 0$$

The only critical point is $(0, 0)$ and the intervals of increase and decrease and concavity are as shown:

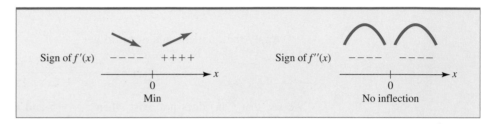

Interpreting these diagrams, we conclude that the graph of $f(x)$ is concave downward for all $x \neq 0$ but is falling for $x < 0$ while rising for $x > 0$. Thus, the graph of $f(x)$ has a relative minimum at the origin $(0, 0)$ and its shape there is $\vee$, called a "cusp." The graph of $f(x)$ is shown in Figure 3.32a.

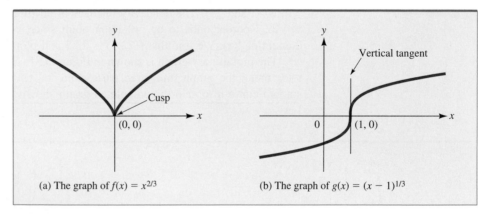

(a) The graph of $f(x) = x^{2/3}$ (b) The graph of $g(x) = (x - 1)^{1/3}$

FIGURE 3.32 A graph with a cusp and another with a vertical tangent.

The derivatives of $g(x) = (x - 1)^{1/3}$ are given by

$$g'(x) = \frac{1}{3}(x - 1)^{-2/3} \quad \text{and} \quad g''(x) = \frac{-2}{9}(x - 1)^{-5/3} \quad x \neq 1$$

The only critical point is $(1, 0)$, and we obtain the intervals of increase and decrease and concavity shown:

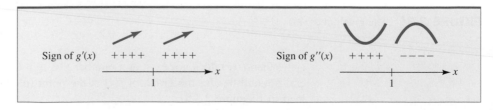

Thus, the graph of $g(x)$ is rising for all $x \neq 1$ but is concave upward for $x < 1$ and concave downward for $x > 1$. This means that $(1, 0)$ is an inflection point, but in addition, note that

$$\lim_{x \to 1} g'(x) = \lim_{x \to 1^+} g'(x) = +\infty$$

Geometrically, this means that as x approaches 0 from either side, the tangent line at $(x, g(x))$ becomes steeper and steeper (with positive slope). This can be interpreted as saying that the graph of $g(x)$ has a tangent line at $(1, 0)$ with "infinite" slope, that is, a **vertical tangent** line. The graph of $g(x)$ is shown in Figure 3.32b.

Sometimes, it is useful to represent observations about a quantity in graphical form. This procedure is illustrated in Example 3.3.6.

EXAMPLE | 3.3.6

The population of a community is 230,000 in 1990 and increases at an increasing rate for 5 years, reaching the 300,000 level in 1995. It then continues to rise, but at a decreasing rate until it peaks out at 350,000 in 2002. After that, the population decreases, at a decreasing rate for 3 years to 320,000 and then at an increasing rate, approaching 280,000 in the long run. Represent this information in graphical form.

Solution

Let $P(t)$ denote the population of the community t years after the base year 1990, where P is measured in units of 10,000 people. Since the population increases at an increasing rate for 5 years from 230,000 to 300,000, the graph of $P(t)$ rises from $(0, 23)$ to $(5, 30)$ and is concave upward for $0 < t < 5$. The population then continues to increase until 2002, but at a decreasing rate, until it reaches a maximum value of 350,000. That is, the graph continues to rise from $(5, 30)$ to a high point at $(12, 35)$ but is now concave downward. Since the graph changes concavity at $x = 5$ (from up to down), it has an inflection point at $(5, 30)$.

For the next 3 years, the population decreases at a decreasing rate, so the graph of $P(t)$ is falling and concave downward for $12 < t < 15$. Since the population continues to decrease from the 320,000 level reached in 2005 but at an increasing rate, the graph falls from the point $(15, 32)$ for $t > 15$ but is concave upward. The change in concavity at $x = 15$ (from down to up) means that $(15, 32)$ is another inflection point.

The statement that "the population decreases at an increasing rate" for $t > 15$ means that the population is changing at a negative rate, which is becoming less negative with increasing time. In other words, the decline in population "slows down" after 2005. This, coupled with the statement that the population "approaches 280,000 in the long run," suggests that the population curve $y = P(t)$ "flattens out" and approaches $y = 28$ asymptotically as $t \to +\infty$.

These observations are summarized in Table 3.2 and are represented in graphical form in Figure 3.33.

TABLE 3.2 Behavior of a Population $P(t)$

Time Period	The function $P(t)$ is . . .	and the graph of $P(t)$ is . . .
$t = 0$	$P(0) = 23$	at the point $(0, 23)$
$0 < t < 5$	increasing at an increasing rate	rising and concave upward
$t = 5$	$P(5) = 30$	at the inflection point $(5, 30)$
$5 < t < 12$	increasing at a decreasing rate	rising and concave downward
$t = 12$	$P(12) = 35$	at the high point $(12, 35)$
$12 < t < 15$	decreasing at a decreasing rate	falling and concave downward
$t = 15$	$P(15) = 32$	at the inflection point $(15, 32)$
$t > 15$	decreasing at an increasing rate and gradually tending toward 28	falling and concave upward asymptotically approaching $y = 28$

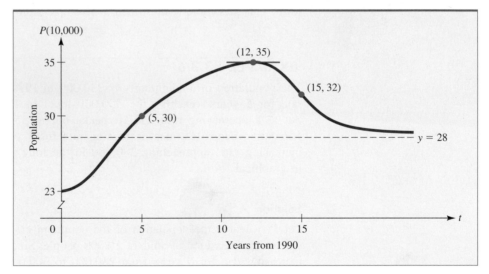

FIGURE 3.33 Graph of a population function.

PROBLEMS 3.3

In Problems 1 through 8, name the vertical and horizontal asymptotes of the given graph.

1.

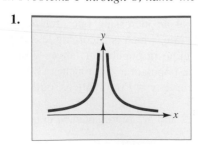

2.

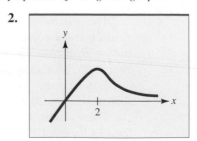

3.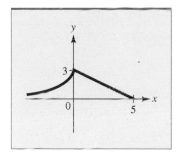

4.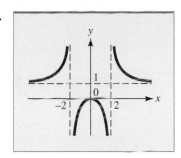

5.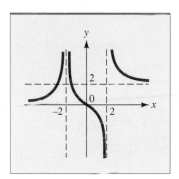

6.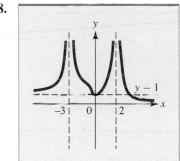

7.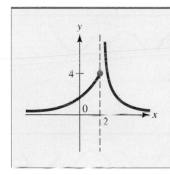

8.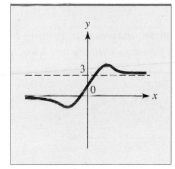

In Problems 9 through 16, find all vertical and horizontal asymptotes of the graph of the given function.

9. $f(x) = \dfrac{3x - 1}{x + 2}$

10. $f(x) = \dfrac{x}{2 - x}$

11. $f(x) = \dfrac{x^2 + 2}{x^2 + 1}$

12. $f(t) = \dfrac{t + 2}{t^2}$

13. $f(t) = \dfrac{t^2 + 3t - 5}{t^2 - 5t + 6}$

14. $g(x) = \dfrac{5x^2}{x^2 - 3x + 4}$

15. $h(x) = \dfrac{1}{x} - \dfrac{1}{x - 1}$

16. $g(t) = \dfrac{t}{\sqrt{t^2 - 4}}$

In Problems 17 through 32, sketch the graph of the given function.

17. $f(x) = x^3 + 3x^2 - 2$

18. $f(x) = x^5 - 5x^4 + 93$

19. $f(x) = x^4 + 4x^3 + 4x^2$

20. $f(x) = 3x^4 - 4x^2 + 3$

21. $f(x) = (2x - 1)^2(x^2 - 9)$

22. $f(x) = x^3 - 3x^4$

23. $f(x) = \dfrac{1}{2x + 3}$

24. $f(x) = \dfrac{x + 3}{x - 5}$

25. $f(x) = x - \dfrac{1}{x}$

26. $f(x) = \dfrac{x^2}{x + 2}$

27. $f(x) = \dfrac{1}{x^2 - 9}$

28. $f(x) = \dfrac{1}{\sqrt{1 - x^2}}$

29. $f(x) = \dfrac{x^2 - 9}{x^2 + 1}$

30. $f(x) = \dfrac{1}{\sqrt{x}} - \dfrac{1}{x}$

31. $f(x) = x^{3/2}$

32. $f(x) = x^{4/3}$

In Problems 33 through 38, diagrams indicating intervals of increase or decrease and concavity are given. Sketch a possible graph for a function with these characteristics.

33.

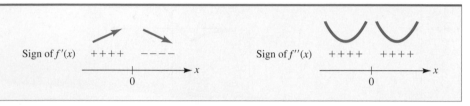

34.

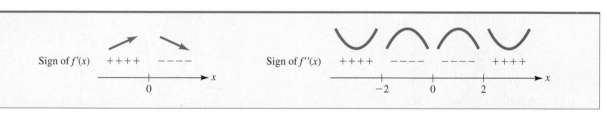

35.

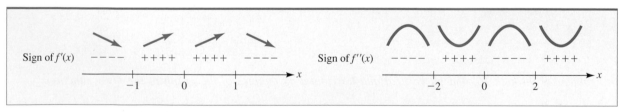

36.

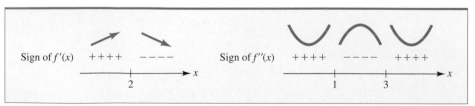

37.

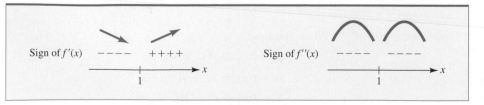

38.

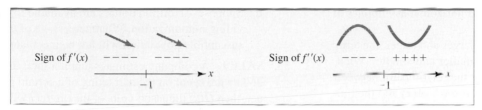

In Problems 39 through 42, the derivative $f'(x)$ of a differentiable function $f(x)$ is given. In each case,
(a) Find intervals of increase and decrease for $f(x)$.
(b) Determine values of x for which relative maxima and minima occur on the graph of $f(x)$.
(c) Find $f''(x)$ and determine intervals of concavity for the graph of $f(x)$.
(d) For what values of x do inflection points occur on the graph of $f(x)$?

39. $f'(x) = x^3(x - 2)^2$

40. $f'(x) = \dfrac{x + 2}{(x - 1)^2}$

41. $f'(x) = \dfrac{x + 3}{(x - 2)^2}$

42. $f'(x) = x^2(x + 1)^3$

43. Find constants A and B so that the graph of the function

$$f(x) = \frac{Ax - 3}{5 + Bx}$$

will have $x = 2$ as a vertical asymptote and $y = 4$ as a horizontal asymptote. Once you find A and B, sketch the graph of $f(x)$.

44. CONCENTRATION OF DRUG   A patient is given an injection of a particular drug at noon, and samples of blood are taken at regular intervals to determine the concentration of drug in the patient's system. It is found that the concentration increases at an increasing rate with respect to time until 1 P.M., and for the next 3 hours, continues to increase but at a decreasing rate until the peak concentration is reached at 4 P.M. The concentration then decreases at a decreasing rate until 5 P.M., after which it decreases at an increasing rate toward zero. Sketch a possible graph for the concentration of drug $C(t)$ as a function of time.

45. BACTERIAL POPULATION   The population of a bacterial colony increases at an increasing rate for 1 hour, after which it continues to increase but at a rate that gradually decreases toward zero. Sketch a possible graph for the population $P(t)$ as a function of time t.

46. EPIDEMIOLOGY   Epidemiologists studying a contagious disease observe that the number of newly infected people increases at an increasing rate during the first 3 years of the epidemic. At that time, a new drug is introduced, and the number of infected people declines at a decreasing rate. Two years after its introduction, the drug begins to lose effectiveness. The number of new cases continues to decline for 1 more year but at an increasing rate before rising again at an increasing rate. Draw a possible graph for the number of new cases $N(t)$ as a function of time.

47. ADOPTION OF TECHNOLOGY   Draw a possible graph for the percentage of households adopting a new type of consumer electronic technology if the percentage grows at an increasing rate for the first 2 years, after which the rate of increase declines, with the market penetration of the technology eventually approaching 90%.

48. EXPERIMENTAL PSYCHOLOGY   To study the rate at which animals learn, a psychology student performed an experiment in which a rat was sent repeatedly through a laboratory maze. Suppose the time required for the rat to traverse the maze on the nth trial was approximately $f(n) = 3 + \dfrac{12}{n}$ minutes.

   a. Graph the function $f(n)$.
   b. What portion of the graph is relevant to the practical situation under consideration?
   c. What happens to the graph as n increases without bound? Interpret your answer in practical terms.

49. AVERAGE COST   The total cost of producing x units of a particular commodity is C thousand dollars, where $C(x) = 3x^2 + x + 48$, and the average cost is

$$A(x) = \frac{C(x)}{x} = 3x + 1 + \frac{48}{x}$$

a. Find all vertical and horizontal asymptotes of the graph of $A(x)$.

b. Note that as x gets larger and larger, the term $48/x$ in $A(x)$ gets smaller and smaller. What does this say about the relationship between the average cost curve $y = A(x)$ and the line $y = 3x + 1$?

c. Sketch the graph of $A(x)$, incorporating the result of part (b) in your sketch. [*Note:* The line $y = 3x + 1$ is called an *oblique* (or *slant*) *asymptote* of the graph.]

50. **INVENTORY COST**   A manufacturer estimates that if each shipment of raw materials contains x units, the total cost in dollars of obtaining and storing the year's supply of raw materials will be

$$C(x) = 2x + \frac{80{,}000}{x}.$$

a. Find all vertical and horizontal asymptotes of the graph of $C(x)$.

b. Note that as x gets larger and larger, the term $\dfrac{80{,}000}{x}$ in $C(x)$ gets smaller and smaller. What does this say about the relationship between the cost curve $y = C(x)$ and the line $y = 2x$?

c. Sketch the graph of $C(x)$, incorporating the result of part (b) in your sketch. [*Note:* The line $y = 2x$ is called an *oblique* (or *slant*) *asymptote* of the graph.]

51. **DISTRIBUTION COST**   The number of worker-hours W required to distribute new telephone books to $x\%$ of the households in a certain community is modeled by the function

$$W(x) = \frac{200x}{100 - x}$$

a. Sketch the graph of $W(x)$.

b. Suppose only 1,500 worker-hours are available for distributing telephone books. What percentage of households do not receive new books?

52. **IMMUNIZATION**   During a nationwide program to immunize the population against a new strain of influenza, public health officials determined that the cost of inoculating $x\%$ of the susceptible population would be approximately

$$C(x) = \frac{1.7x}{100 - x}$$

million dollars.

a. Sketch the graph of the cost function $C(x)$.

b. Suppose 40 million dollars are available for providing immunization. What percentage of the susceptible population will not be inoculated?

53. **SALES**   A company estimates that if x thousand dollars are spent on the marketing of a certain product, then $Q(x)$ thousand units of the product will be sold, where

$$Q(x) = \frac{7x}{27 + x^2}$$

a. Sketch the graph of the sales function $Q(x)$.

b. For what marketing expenditure x are sales maximized? What is the maximum sales level?

c. For what value of x is the sales rate minimized?

54. **PRODUCTION**   A business manager determines that t months after production begins on a new product, the number of units produced will be P million per month, where

$$P(t) = \frac{t}{(t + 1)^2}$$

a. Find $P'(t)$ and $P''(t)$.

b. Sketch the graph of $P(t)$.

c. What happens to production in the long run (as $t \to \infty$)?

55. **POLITICAL POLLING**   A poll commissioned by a politician estimates that t days after she comes out in favor of a controversial bill, the percentage of her constituency (those who support her at the time she declares her position on the bill) that still supports her is given by

$$S(t) = \frac{100(t^2 - 3t + 25)}{t^2 + 7t + 25}$$

The vote is to be taken 10 days after she announces her position.

a. Sketch the graph of $S(t)$ for $0 \le t \le 10$.

b. When is her support at its lowest level? What is her minimum support level?

c. The derivative $S'(t)$ may be thought of as an approval rate. Is her approval rate positive or negative when the vote is taken? Is the approval rate increasing or decreasing at this time? Interpret your results.

56. **ADVERTISING**   A manufacturer of motorcycles estimates that if x thousand dollars are spent on advertising, then for $x > 0$,

$$M(x) = 2{,}300 + \frac{125}{x} - \frac{500}{x^2}$$

cycles will be sold.

a. Sketch the graph of the sales function $M(x)$.

b. What level of advertising expenditure results in maximum sales? What is the maximum sales level?

57. **COST MANAGEMENT** A company uses a truck to deliver its products. To estimate costs, the manager models gas consumption by the function

$$G(x) = \frac{1}{2,000}\left(\frac{800}{x} + 5x\right)$$

gal/mile, assuming that the truck is driven at a constant speed of x miles per hour for $x \geq 5$. The driver is paid $18 per hour to drive the truck 400 miles, and gasoline costs $2.95 per gallon. Highway regulations require $30 \leq x \leq 65$.

a. Find an expression for the total cost $C(x)$ of the trip. Sketch the graph of $C(x)$ for the legal speed interval $30 \leq x \leq 65$.

b. What legal speed will minimize the total cost of the trip? What is the minimal total cost?

58. Let $f(x) = x^{1/3}(x - 4)$.

a. Find $f'(x)$ and determine intervals of increase and decrease for $f(x)$. Locate all relative extrema on the graph of $f(x)$.

b. Find $f''(x)$ and determine intervals of concavity for $f(x)$. Find all inflection points on the graph of $f(x)$.

c. Find all intercepts for the graph of $f(x)$. Does the graph have any asymptotes?

d. Sketch the graph of $f(x)$.

59. Repeat Problem 58 for the function

$$f(x) = x^{7/3}(2x - 5)$$

60. Repeat Problem 58 for the function

$$f(x) = \frac{x + 9.4}{25 - 1.1x - x^2}$$

61. Let $f(x) = \frac{x - 1}{x^2 - 1}$ and let $g(x) = \frac{x - 1.01}{x^2 - 1}$.

 a. Use a graphing utility to sketch the graph of $f(x)$. What happens at $x = 1$?

b. Sketch the graph of $g(x)$. Now what happens at $x = 1$?

SECTION 3.4 Optimization

You have already seen several situations where the methods of calculus were used to determine the largest or smallest value of a function of interest (for example, maximum profit or minimum cost). In most such optimization problems, the goal is to find the absolute maximum or absolute minimum of a particular function on some relevant interval. The absolute maximum of a function on an interval is the largest value of the function on that interval and the absolute minimum is the smallest value. Here is a definition of absolute extrema.

> **Absolute Maxima and Minima of a Function** ■ Let f be a function defined on an interval I that contains the number c. Then
>
> $f(c)$ is the *absolute maximum* of f on I if $f(c) \geq f(x)$ for all x in I
>
> $f(c)$ is the *absolute minimum* of f on I if $f(c) \leq f(x)$ for all x in I
>
> Collectively, absolute maxima and minima are called *absolute extrema*.

Absolute extrema often coincide with relative extrema but not always. For example, in Figure 3.34 the absolute maximum and relative maximum on the interval $a \leq x \leq b$ are the same, but the absolute minimum occurs at the left endpoint, $x = a$.

In this section, you will learn how to find absolute extrema of functions on intervals. We begin by considering intervals $a \leq x \leq b$ that are "closed" in the sense they include both endpoints, a and b. It can be shown that a function continuous on such an interval has both an absolute maximum and an absolute minimum on the interval.

EXPLORE!

Use a graphing calculator to graph $f(x) = \dfrac{x^3}{\sqrt{x+2}}$ with the modified decimal window [0, 4.7]1 by [0, 60]5. Trace or use other utility methods to find the absolute maximum and absolute minimum of $f(x)$ over the interval [1, 3].

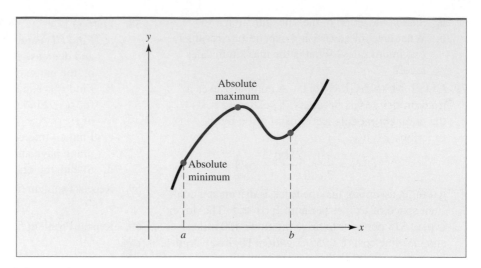

FIGURE 3.34 Absolute extrema.

Moreover, each absolute extremum can occur either at an endpoint of the interval (at a or b) or at a critical number c between a and b (that is, $a < c < b$) (Figure 3.35).

The Extreme Value Property ■ A function $f(x)$ that is continuous on the closed interval $a \le x \le b$ attains its absolute extrema on the interval either at an endpoint of the interval (a or b) or at a critical number c such that $a < c < b$.

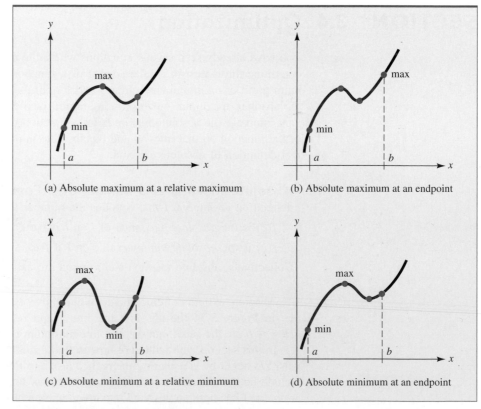

FIGURE 3.35 Absolute extrema of a continuous function on $a \le x \le b$.

Thanks to the extreme value property, you can find the absolute extrema of a continuous function on a closed interval $a \leq x \leq b$ by using this straightforward procedure.

How to Find the Absolute Extrema of a Continuous Function f on a Closed Interval $a \leq x \leq b$

Step 1. Find all critical numbers of f in the open interval $a < x < b$.

Step 2. Compute $f(x)$ at the critical numbers found in step 1 and at the endpoints $x = a$ and $x = b$.

Step 3. Interpretation: The largest and smallest values found in step 2 are, respectively, the absolute maximum and absolute minimum values of $f(x)$ on the closed interval $a \leq x \leq b$.

The procedure is illustrated in Examples 3.4.1 through 3.4.3.

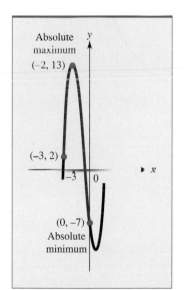

FIGURE 3.36 The absolute extrema on $-3 \leq x \leq 0$ for $y = 2x^3 + 3x^2 - 12x - 7$

EXAMPLE 3.4.1

Find the absolute maximum and absolute minimum of the function

$$f(x) = 2x^3 + 3x^2 - 12x - 7$$

on the interval $-3 \leq x \leq 0$.

Solution

From the derivative

$$f'(x) = 6x^2 + 6x - 12 = 6(x + 2)(x - 1)$$

we see that the critical numbers are $x = -2$ and $x = 1$. Of these, only $x = -2$ lies in the interval $-3 \leq x \leq 0$. Compute $f(x)$ at $x = -2$ and at the endpoints $x = -3$ and $x = 0$.

$$f(-2) = 13 \qquad f(-3) = 2 \qquad f(0) = -7$$

Compare these values to conclude that the absolute maximum of f on the interval $-3 \leq x \leq 0$ is $f(-2) = 13$ and the absolute minimum is $f(0) = -7$.

Notice that we did not have to classify the critical points or draw the graph to locate the absolute extrema. The sketch in Figure 3.36 is presented only for the sake of illustration.

EXAMPLE 3.4.2

For several weeks, the highway department has been recording the speed of freeway traffic flowing past a certain downtown exit. The data suggest that between 1:00 and 6:00 P.M. on a normal weekday, the speed of the traffic at the exit is approximately $S(t) = t^3 - 10.5t^2 + 30t + 20$ miles per hour, where t is the number of hours past noon. At what time between 1:00 and 6:00 P.M. is the traffic moving the fastest, and at what time is it moving the slowest?

Solution

The goal is to find the absolute maximum and absolute minimum of the function $S(t)$ on the interval $1 \le t \le 6$. From the derivative

$$S'(t) = 3t^2 - 21t + 30 = 3(t^2 - 7t + 10) = 3(t - 2)(t - 5)$$

we get the critical numbers $t = 2$ and $t = 5$, both of which lie in the interval $1 \le t \le 6$.

Compute $S(t)$ for these values of t and at the endpoints $t = 1$ and $t = 6$ to get

$$S(1) = 40.5 \qquad S(2) = 46 \qquad S(5) = 32.5 \qquad S(6) = 38$$

Since the largest of these values is $S(2) = 46$ and the smallest is $S(5) = 32.5$, we can conclude that the traffic is moving fastest at 2:00 P.M., when its speed is 46 miles per hour, and slowest at 5:00 P.M., when its speed is 32.5 miles per hour. For reference, the graph of S is sketched in Figure 3.37.

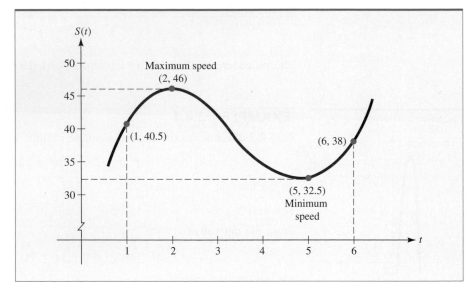

FIGURE 3.37   Traffic speed $S(t) = t^3 - 10.5t^2 + 30t + 20$.

EXAMPLE | 3.4.3

When you cough, the radius of your trachea (windpipe) decreases, affecting the speed of the air in the trachea. If r_0 is the normal radius of the trachea, the relationship between the speed S of the air and the radius r of the trachea during a cough is given by a function of the form $S(r) = ar^2(r_0 - r)$, where a is a positive constant.* Find the radius r for which the speed of the air is greatest.

Solution

The radius r of the contracted trachea cannot be greater than the normal radius r_0 or less than zero. Hence, the goal is to find the absolute maximum of $S(r)$ on the interval $0 \le r \le r_0$.

---

*Philip M. Tuchinsky, "The Human Cough," *UMAP Modules 1976: Tools for Teaching*, Lexington, MA: Consortium for Mathematics and Its Application, Inc., 1977.

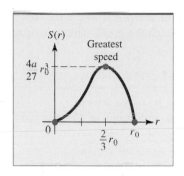

FIGURE 3.38 The speed of air during a cough

$$S(r) = ar^2(r_0 - r)$$

First differentiate $S(r)$ with respect to r using the product rule and factor the derivative as follows (note that a and r_0 are constants):

$$S'(r) = -ar^2 + (r_0 - r)(2ar) = ar[-r + 2(r_0 - r)] = ar(2r_0 - 3r)$$

Then set the factored derivative equal to zero and solve to get the critical numbers:

$$ar(2r_0 - 3r) = 0$$

$$r = 0 \quad \text{or} \quad r - \frac{2}{3}r_0$$

Both of these values of r lie in the interval $0 \le r \le r_0$, and one is actually an endpoint of the interval. Compute $S(r)$ for these two values of r and for the other endpoint $r = r_0$ to get

$$S(0) = 0 \qquad S\left(\frac{2}{3}r_0\right) = \frac{4a}{27}r_0^3 \qquad S(r_0) = 0$$

Compare these values and conclude that the speed of the air is greatest when the radius of the contracted trachea is $\frac{2}{3}r_0$; that is, when it is two-thirds the radius of the uncontracted trachea.

A graph of the function $S(r)$ is given in Figure 3.38. Note that the r intercepts of the graph are obvious from the factored function $S(r) = ar^2(r_0 - r)$. Notice also that the graph has a horizontal tangent when $r = 0$, reflecting the fact that $S'(0) = 0$.

More General Optimization

When the interval on which you wish to maximize or minimize a continuous function is not of the form $a \le x \le b$, the procedure illustrated in Examples 3.4.1 through 3.4.3 no longer applies. This is because there is no longer any guarantee that the function actually has an absolute maximum or minimum on the interval in question. On the other hand, if an absolute extremum does exist and the function is continuous on the interval, the absolute extremum will still occur at a relative extremum or endpoint contained in the interval. Several possibilities for functions on unbounded intervals are illustrated in Figure 3.39.

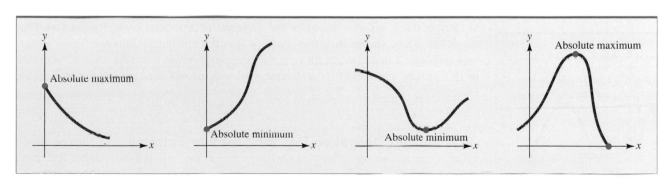

FIGURE 3.39 Extrema for functions defined on unbounded intervals.

To find the absolute extrema of a continuous function on an interval that is not of the form $a \le x \le b$, you still evaluate the function at all the critical points and endpoints that are contained in the interval. However, before you can draw any final conclusions, you must find out if the function actually has relative extrema on the interval. One way to do this is to use the first derivative to determine where the function is increasing and where it is decreasing and then to sketch the graph. The technique is illustrated in Example 3.4.4.

EXAMPLE | 3.4.4

If they exist, find the absolute maximum and absolute minimum of the function $f(x) = x^2 + \dfrac{16}{x}$ on the interval $x > 0$.

Solution

The function is continuous on the interval $x > 0$ since its only discontinuity occurs at $x = 0$. The derivative is

$$f'(x) = 2x - \frac{16}{x^2} = \frac{2x^3 - 16}{x^2} = \frac{2(x^3 - 8)}{x^2}$$

which is zero when

$$x^3 - 8 = 0 \qquad x^3 = 8 \qquad \text{or} \qquad x = 2$$

Since $f'(x) < 0$ for $0 < x < 2$ and $f'(x) > 0$ for $x > 2$, the graph of f is decreasing for $0 < x < 2$ and increasing for $x > 2$, as shown in Figure 3.40. It follows that

$$f(2) = 2^2 + \frac{16}{2} = 12$$

is the absolute minimum of f on the interval $x > 0$ and that there is no absolute maximum.

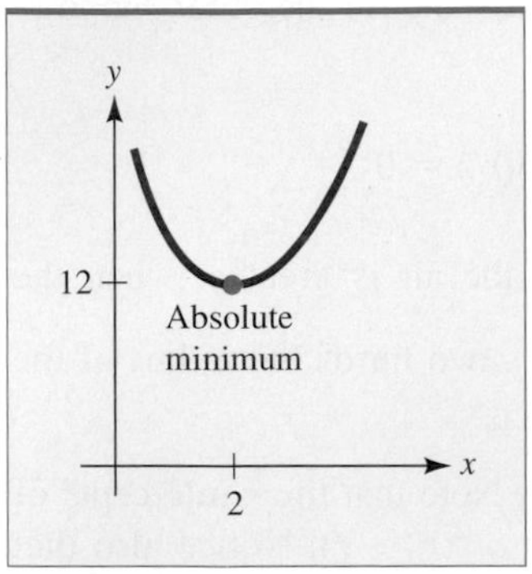

FIGURE 3.40 The function $f(x) = x^2 + \dfrac{16}{x}$ on the interval $x > 0$.

The procedure illustrated in Example 3.4.4 can be used whenever we wish to find the largest or smallest value of a function f that is continuous on an interval I on which it has *exactly one* critical number c. In particular, if this condition is satisfied and $f(x)$ has a *relative* maximum (minimum) at $x = c$, it also has an *absolute* maximum (minimum) there. To see why, suppose the graph has a relative minimum at $x = c$. Then the graph is always falling before c and always rising after c, since to change direction would require the presence of a second critical point (see Figure 3.41). Thus, the relative minimum is also the absolute minimum. These observations suggest that any test for relative extrema becomes a test for absolute extrema in this special case. Here is a statement of the second derivative test for absolute extrema.

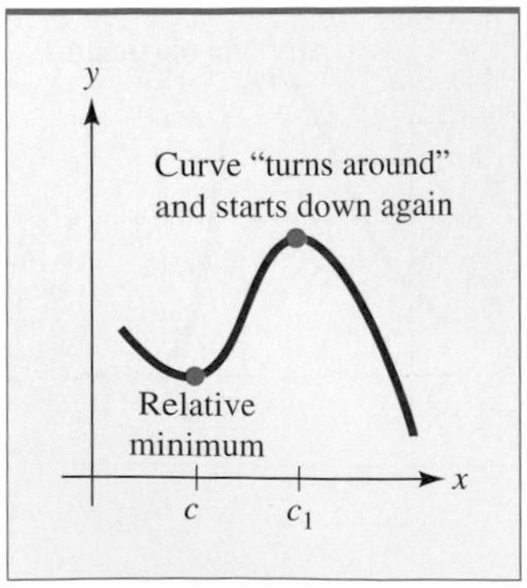

FIGURE 3.41 The relative minimum is not the absolute minimum because of the effect of another critical point.

> **The Second Derivative Test for Absolute Extrema** ■ Suppose that $f(x)$ is continuous on an interval I where $x = c$ is the only critical number and that $f'(c) = 0$. Then,
>
> if $f''(c) > 0$, the absolute minimum of $f(x)$ on I is $f(c)$
>
> if $f''(c) < 0$, the absolute maximum of $f(x)$ on I is $f(c)$

Example 3.4.5 illustrates how the second derivative test for absolute extrema can be used in practice.

EXAMPLE | 3.4.5

A manufacturer estimates that when q thousand units of a particular commodity are produced each month, the total cost will be $C(q) = 0.4q^2 + 3q + 40$ thousand dollars, and all q units can be sold at a price of $p(q) = 22.2 - 1.2q$ dollars per unit.

a. Determine the level of production that results in maximum profit. What is the maximum profit?

b. At what level of production is the average cost per unit $A(q) = \dfrac{C(q)}{q}$ minimized?

c. At what level of production is the average cost equal to the marginal cost $C'(q)$?

Solution

a. The revenue is

$$R(q) = qp(q) = q(22.2 - 1.2q) = -1.2q^2 + 22.2q$$

thousand dollars, so the profit is

$$\begin{aligned} P(q) = R(q) - C(q) &= -1.2q^2 + 22.2q - (0.4q^2 + 3q + 40) \\ &= -1.6q^2 + 19.2q - 40 \end{aligned}$$

thousand dollars. We have

$$\begin{aligned} P'(q) &= -1.6(2q) + 19.2 = -3.2q + 19.2 \\ &= 0 \end{aligned}$$

when

$$-3.2q + 19.2 = 0$$
$$q = \frac{19.2}{3.2} = 6$$

Since $P''(q) = -3.2$, it follows that $P''(6) < 0$, and the second derivative test tells us that maximum profit occurs when $q = 6$ (thousand) units are produced. The maximum profit is

$$P(6) = -1.6(6)^2 + 19.2(6) - 40$$
$$= 17.6$$

thousand dollars ($17,600). The graph of the profit function is shown in Figure 3.42a.

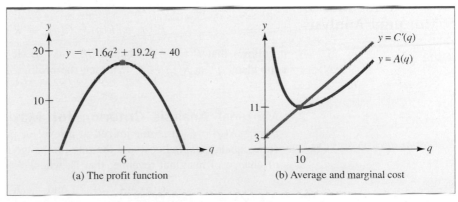

(a) The profit function   (b) Average and marginal cost

FIGURE 3.42   Graphs of profit, average cost, and marginal cost for Example 3.4.5.

b. The average cost is

$$A(q) = \frac{C(q)}{q} = \frac{0.4q^2 + 3q + 40}{q} \qquad \frac{\textit{thousand dollars}}{\textit{thousand units}}$$

$$= 0.4q + 3 + \frac{40}{q} \qquad \frac{\textit{dollars}}{\textit{unit}}$$

for $q > 0$ (the level of production cannot be negative or zero). We find

$$A'(q) = 0.4 - \frac{40}{q^2} = \frac{0.4q^2 - 40}{q^2}$$

which is 0 for $q > 0$ only when $q = 10$. Since

$$A''(q) = \frac{80}{q^3} > 0 \quad \text{when } q > 0$$

it follows from the second derivative test for absolute extrema that average cost $A(q)$ is minimized when $q = 10$ (thousand) units. The minimal average cost is

$$A(10) = 0.4(10) + 3 + \frac{40}{10} = 11 \qquad \textit{dollars/unit}$$

c. The marginal cost is $C'(q) = 0.8q + 3$, and it equals average cost when

$$0.8q + 3 = 0.4q + 3 + \frac{40}{q}$$

$$0.4q = \frac{40}{q}$$

$$0.4q^2 = 40$$

$$q = 10 \text{ (thousand) units}$$

which equals the optimal level of production in part (b). The graphs of the marginal cost $C'(q)$ and average cost $A(q) = \dfrac{C(q)}{q}$ are shown in Figure 3.42b.

Two General Principles of Marginal Analysis

If the revenue derived from the sale of q units is $R(q)$ and the cost of producing those units is $C(q)$, then the profit is $P(q) = R(q) - C(q)$. Since

$$P'(q) = [R(q) - C(q)]' = R'(q) - C'(q)$$

it follows that $P'(q) = 0$ when $R'(q) = C'(q)$. If it is also true that $P''(q) < 0$, or equivalently, that $R''(q) < C''(q)$, then the profit will be maximized.

Marginal Analysis Criterion for Maximum Profit ■ The profit $P(q) = R(q) - C(q)$ is maximized at a level of production q where marginal revenue equals marginal cost and the rate of change of marginal cost exceeds the rate of change of marginal revenue; that is, where

$$R'(q) = C'(q) \qquad \text{and} \qquad R''(q) < C''(q)$$

For instance, in Example 3.4.5, the revenue is $R(q) = -1.2q^2 + 22.2q$, and the cost is $C(q) = 0.4q^2 + 3q + 40$, so the marginal revenue is $R'(q) = -2.4q + 22.2$ and the marginal cost is $C'(q) = 0.8q + 3$. Thus, marginal revenue equals marginal cost when

$$R'(q) = C'(q)$$
$$-2.4q + 22.2 = 0.8q + 3$$
$$3.2q = 19.2$$
$$q = 6$$

which is the level of production for maximum profit found in part (a) of Example 3.4.5. Note that $R'' < C''$ is also satisfied since $R'' = -2.4$ and $C'' = 0.8$.

In part (c) of Example 3.4.5, you found that marginal cost equals average cost at the level of production where average cost is minimized. This, too, is no accident. To see why, let $C(q)$ be the cost of producing q units of a commodity. Then the average cost per unit is $A(q) = \dfrac{C(q)}{q}$, and by applying the quotient rule, you get

$$A'(q) = \frac{qC'(q) - C(q)}{q^2}$$

Thus, $A'(q) = 0$ when the numerator on the right is zero. That is, when

$$qC'(q) = C(q)$$

or equivalently, when

$$C'(q) = \underbrace{\frac{C(q)}{q}}_{\substack{\text{average} \\ \text{cost}}} = A(q)$$

$$\underbrace{}_{\substack{\text{marginal} \\ \text{cost}}}$$

To show that average cost is actually minimized where average cost equals marginal cost, it is necessary to make a few reasonable assumptions about total cost (see Problem 52).

Marginal Analysis Criterion for Minimal Average Cost ■ Average cost is minimized at the level of production where average cost equals marginal cost; that is, when $A(q) = C'(q)$.

Here is an informal explanation of the relationship between average and marginal cost that is often given in economics texts. The marginal cost (MC) is approximately the same as the cost of producing one additional unit. If the additional unit costs less to produce than the average cost (AC) of the existing units (if MC < AC), then this less-expensive unit will cause the average cost per unit to decrease. On the other hand, if the additional unit costs more than the average cost of the existing units (if MC > AC), then this more-expensive unit will cause the average cost per unit to increase. However, if the cost of the additional unit is equal to the average cost of the existing units (if MC = AC), then the average cost will neither increase nor decrease, which means (AC)′ = 0.

The relation between average cost and marginal cost can be generalized to apply to any pair of average and marginal quantities. The only possible modification involves the nature of the critical point that occurs when the average quantity equals the marginal quantity. For example, average revenue usually has a relative *maximum* (instead of a minimum) when average revenue equals marginal revenue.

Price Elasticity of Demand

In Section 1.1, we introduced the economic concept of demand as a means for expressing the relationship between the unit price p of a commodity and the number of units q that will be demanded (that is, produced and purchased) by consumers at that price. Previously, we have expressed unit price p as a function of production level q, but for the present discussion, it is more convenient to turn things around and write the demand relationship as $q = D(p)$.

In general, an increase in the unit price of a commodity will result in decreased demand, but the sensitivity or responsiveness of demand to a change in price varies from one product to another. For instance, the demand for products such as soap, flashlight batteries, or salt will not be much affected by a small percentage change in unit price, while a comparable percentage change in the price of airline tickets or home loans can affect demand dramatically.

Sensitivity of demand is commonly measured by the ratio of the percentage rate of change in quantity demanded to the percentage rate of change in price. This is approximately the same as the change in demand produced by a 1% change in unit price. Recall from Section 2.2 that the percentage rate of change of a quantity $Q(x)$ is given by $\dfrac{100Q'(x)}{Q(x)}$. In particular, if the demand function $q = D(p)$ is differentiable, then

$$\begin{bmatrix} \textbf{Percentage rate of} \\ \textbf{change of demand } \boldsymbol{q} \end{bmatrix} = \frac{100\dfrac{dq}{dp}}{q}$$

and

$$\begin{bmatrix} \textbf{Percentage rate of} \\ \textbf{change of price } \boldsymbol{p} \end{bmatrix} = \frac{100\dfrac{dp}{dp}}{p} = \frac{100}{p}$$

Thus, sensitivity to change in price is measured by the ratio

$$\frac{\textbf{Percentage rate of change in } \boldsymbol{q}}{\textbf{Percentage rate of change in } \boldsymbol{p}} = \frac{\dfrac{100\dfrac{dq}{dp}}{q}}{\dfrac{100}{p}} = \frac{p}{q}\frac{dq}{dp}$$

which, in economics, is called the *price elasticity of demand*. To summarize:

Price Elasticity of Demand ■ If $q = D(p)$ units of a commodity are demanded by the market at a unit price p, where D is a differentiable function, then the **price elasticity of demand** for the commodity is given by

$$E(p) = \frac{p}{q}\frac{dq}{dp}$$

and has the interpretation

$$E(p) \approx \left[\begin{array}{l}\text{percentage rate of change in demand } q \\ \text{produced by a 1\% rate of change in price } p\end{array}\right]$$

NOTE Since demand q decreases as the unit price p increases, we have $\dfrac{dq}{dp} < 0$. Therefore, since $q > 0$ and $p > 0$, it follows that the price elasticity of demand will be negative; that is,

$$E(p) = \frac{p}{q}\frac{dq}{dp} < 0$$

For instance, to say that a particular commodity has a price elasticity of demand of -0.5 at a certain unit price p means that a 10% rise in price for the commodity will result in a decline of approximately 5% in the number of units demanded (sold). These ideas are illustrated further in Example 3.4.6. ■

EXAMPLE | 3.4.6

Suppose the demand q and price p for a certain commodity are related by the linear equation $q - 240 - 2p$ (for $0 \le p \le 120$).

a. Express the elasticity of demand as a function of p.

b. Calculate the elasticity of demand when the price is $p = 100$. Interpret your answer.

c. Calculate the elasticity of demand when the price is $p = 50$. Interpret your answer.

d. At what price is the elasticity of demand equal to -1? What is the economic significance of this price?

Solution

a. The elasticity of demand is

$$E(p) = \frac{p}{q}\frac{dq}{dp} = \frac{p}{q}(-2) = \frac{-2p}{240 - 2p} = \frac{-p}{120 - p}$$

b. When $p = 100$, the elasticity of demand is

$$E(100) = \frac{-100}{120 - 100} - 5$$

That is, when the price is $p = 100$, a 1% increase in price will produce a decrease in demand of approximately 5%.

c. When $p = 50$, the elasticity of demand is

$$E(50) = \frac{-50}{120 - 50} \approx -0.71$$

That is, when the price is $p = 50$, a 1% increase in price will produce a decrease in demand of approximately 0.71%.

d. The elasticity of demand will be equal to -1 when

$$-1 = \frac{-p}{120 - p} \qquad 120 - p = p \qquad 2p = 120 \qquad \text{or} \qquad p = 60$$

At this price, a 1% increase in price will result in a decrease in demand of approximately the same percentage.

There are three levels of elasticity, depending on whether $|E(p)|$ is greater than, less than, or equal to 1. Here is a description and economic interpretation of each level.

Levels of Elasticity

$|E(p)| > 1$ **Elastic demand.** The percentage decrease in demand is greater than the percentage increase in price that caused it. Thus, demand is relatively sensitive to changes in price.

$|E(p)| < 1$ **Inelastic demand.** The percentage decrease in demand is less than the percentage increase in price that caused it. When this occurs, demand is relatively insensitive to changes in price.

$|E(p)| = 1$ **Demand is of unit elasticity (or unitary).** The percentage changes in price and demand are (approximately) equal.

For instance, in Example 4.6b, we found that $E(100) = -5$. Thus, the demand is elastic with respect to price when $p = 100$ since $|E(100)| = 5 > 1$. In part (c) of the same example, we found that $|E(50)| = |-0.71| < 1$, so the demand is inelastic when $p = 50$. Finally, in part (d), we found $|E(60)| = |-1| = 1$, which means that the demand is of unit elasticity when $p = 60$.

The level of elasticity of demand for a commodity gives useful information about the total revenue R obtained from the sale of q units of the commodity at p dollars per unit. Assuming that the demand q is a differentiable function of unit price p, the revenue is $R(p) = pq(p)$ and by differentiating implicitly with respect to p, we find that

$$\frac{dR}{dp} = p\frac{dq}{dp} + q \qquad\qquad \textit{by the product rule}$$

To get the elasticity $E(p) = \dfrac{p}{q}\dfrac{dq}{dp}$ into the picture, simply multiply the expression on the right-hand side by $\dfrac{q}{q}$ as follows:

$$\frac{dR}{dp} = \frac{q}{q}\left(p\frac{dq}{dp} + q\right) = q\left(\frac{p}{q}\frac{dq}{dp} + 1\right) = q[E(p) + 1]$$

Suppose demand is elastic so that $|E(p)| > 1$. It follows that $E(p) < -1$ [since $E(p) < 0$] so that $[E(p) + 1] < 0$ and

$$\frac{dR}{dp} = q(p)\,[E(p) + 1] < 0$$

which means that the result of a small increase in price will be to decrease revenue. Similarly, when demand is inelastic, we have $-1 < E(p) < 0$ so that $[E(p) + 1] > 0$ and $\dfrac{dR}{dp} > 0$. In this case, a small increase in price results in increased revenue. If the demand is of unit elasticity, then $E(p) = -1$ so $\dfrac{dR}{dp} = 0$ and a small increase in price leaves revenue approximately unchanged (neither increased nor decreased). These observations are summarized as follows.

Levels of Elasticity and the Effect on Revenue

If demand is **elastic** ($|E(p)| > 1$), revenue $R = pq(p)$ decreases as price p increases.

If demand is **inelastic** ($|E(p)| < 1$), revenue R increases as price p increases.

If demand is of **unit elasticity** ($|E(p)| = 1$), revenue is unaffected by a small change in price.

The relationship between revenue and price is shown in Figure 3.43. Note that the revenue curve is rising where demand is inelastic, falling where demand is elastic, and has a horizontal tangent line where the demand is of unit elasticity.

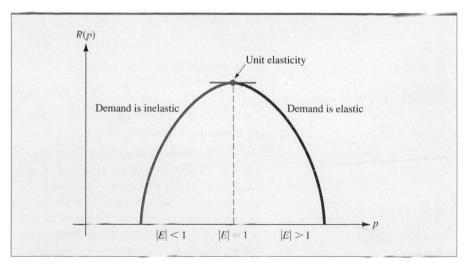

FIGURE 3.43   Revenue as a function of price.

The relationship between elasticity of demand and total revenue is illustrated in Example 3.4.7.

EXAMPLE | 3.4.7

The manager of a bookstore determines that when a certain new novel is priced at p dollars per copy, the daily demand will be $q = 300 - p^2$ copies, where $0 \le p \le \sqrt{300}$.

a. Determine where the demand is elastic, inelastic, and of unit elasticity with respect to price.

b. Interpret the results of part (a) in terms of the behavior of total revenue as a function of price.

Solution

a. The elasticity of demand is

$$E(p) = \frac{p}{q}\frac{dq}{dp} = \frac{p}{300 - p^2}(-2p) = \frac{-2p^2}{300 - p^2}$$

and since $0 \le p \le \sqrt{300}$,

$$|E(p)| = \frac{2p^2}{300 - p^2}$$

The demand is of unit elasticity when $|E| = 1$; that is, when

$$\frac{2p^2}{300 - p^2} = 1$$
$$2p^2 = 300 - p^2$$
$$3p^2 = 300$$
$$p = \pm 10$$

of which only $p = 10$ is in the relevant interval $0 \le p \le \sqrt{300}$. If $0 \le p < 10$, then

$$|E| = \frac{2p^2}{300 - p^2} < \frac{2(10)^2}{300 - (10)^2} = 1$$

so the demand is inelastic. Likewise, if $10 < p < \sqrt{300}$, then

$$|E| = \frac{2p^2}{300 - p^2} > \frac{2(10)^2}{300 - (10)^2} = 1$$

and the demand is elastic.

b. The total revenue, $R = pq$, increases when demand is inelastic; that is, when $0 \le p < 10$. For this range of prices, a specified percentage increase in price results in a smaller percentage decrease in demand, so the bookstore will take in more money for each increase in price up to $10 per copy.

However, for the price range $10 < p \le \sqrt{300}$, the demand is elastic, so the revenue is decreasing. If the book is priced in this range, a specified percentage increase in price results in a larger percentage decrease in demand. Thus, if the bookstore increases the price beyond $10 per copy, it will lose revenue.

This means that the optimal price is $10 per copy, which corresponds to unit elasticity. The graphs of the demand and revenue functions are shown in Figure 3.44.

EXPLORE!

Suppose the demand/price equation in Example 3.4.7 is $q = 300 - ap^2$. Describe how the price is affected when the demand is of unit elasticity for $a = 0, 1, 3,$ or 5 by examining the graph of $x(300 - ax^2)$ for these values of a. Use a viewing rectangle of [0, 20]1 by [0, 3,000]500.

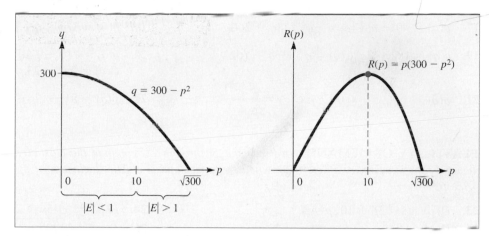

FIGURE 3.44 Demand and revenue curves for Example 3.4.7.

PROBLEMS 3.4

In Problems 1 through 16, find the absolute maximum and absolute minimum (if any) of the given function on the specified interval.

1. $f(x) = x^2 + 4x + 5; -3 \le x \le 1$

2. $f(x) = x^3 + 3x^2 + 1; -3 \le x < 2$

3. $f(x) = \frac{1}{3}x^3 - 9x + 2; 0 \le x < 2$

4. $f(x) = x^5 - 5x^4 + 1; 0 \le x \le 5$

5. $f(t) = 3t^5 - 5t^3; -2 \le t \le 0$

6. $f(x) = 10x^6 + 24x^5 + 15x^4 + 3; -1 \le x \le 1$

7. $f(x) = (x^2 - 4)^5; -3 \le x \le 2$

8. $f(t) = \frac{t^2}{t - 1}; -2 \le t \le -\frac{1}{2}$

9. $g(x) = x + \frac{1}{x}; \frac{1}{2} \le x \le 3$

10. $g(x) = \frac{1}{x^2 - 9}; 0 \le x \le 2$

11. $f(u) = u + \frac{1}{u}; u > 0$

12. $f(u) = 2u + \frac{32}{u}; u > 0$

13. $f(x) = \frac{1}{x}; x > 0$

14. $f(x) = \frac{1}{x^2}; x > 0$

15. $f(x) = \frac{1}{x + 1}; x \ge 0$

16. $f(x) = -\frac{1}{(x + 1)^2}; x \ge 0$

MAXIMUM PROFIT AND MINIMUM AVERAGE COST In Problems 17 through 22, you are given the price $p(q)$ at which q units of a particular commodity can be sold and the total cost $C(q)$ of producing the q units. In each case:

(a) Find the revenue function $R(q)$, the profit function $P(q)$, the marginal revenue $R'(q)$, and marginal cost $C'(q)$. Sketch the graphs of $P(q)$, $R'(q)$, and $C'(q)$ on the same coordinate axes and determine the level of production q where $P(q)$ is maximized.

(b) Find the average cost $A(q) = C(q)/q$ and sketch the graphs of $A(q)$, and the marginal cost $C'(q)$ on the same axes. Determine the level of production q at which $A(q)$ is minimized.

17. $p(q) = 49 - q$; $C(q) = \dfrac{1}{8}q^2 + 4q + 200$

18. $p(q) = 37 - 2q$; $C(q) = 3q^2 + 5q + 75$

19. $p(q) = 180 - 2q$; $C(q) = q^3 + 5q + 162$

20. $p(q) = 710 - 1.1q^2$;
$C(q) = 2q^3 - 23q^2 + 90.7q + 151$

21. $p(q) = 1.0625 - 0.0025q$; $C(q) = \dfrac{q^2 + 1}{q + 3}$

22. $p(q) = 81 - 3q$; $C(q) = \dfrac{q + 1}{q + 3}$

ELASTICITY OF DEMAND *In Problems 23 through 28, compute the elasticity of demand for the given demand function $D(p)$ and determine whether the demand is elastic, inelastic, or of unit elasticity at the indicated price p.*

23. $D(p) = -1.3p + 10$; $p = 4$

24. $D(p) = -1.5p + 25$; $p = 12$

25. $D(p) = 200 - p^2$; $p = 10$

26. $D(p) = \sqrt{400 - 0.01p^2}$; $p = 120$

27. $D(p) = \dfrac{3,000}{p} - 100$; $p = 10$

28. $D(p) = \dfrac{2,000}{p^2}$; $p = 5$

29. AVERAGE REVENUE Suppose the total revenue in dollars from the sale of q units of a certain commodity is $R(q) = -2q^2 + 68q - 128$.
  a. At which level of sales is the average revenue per unit equal to the marginal revenue?
  b. Verify that the average revenue is increasing if the level of sales is less than the level in part (a) and decreasing if the level of sales is greater than the level in part (a).
  c. On the same set of axes, graph the relevant portions of the average and marginal revenue functions.

30. At what point does the tangent to the curve $y = 2x^3 - 3x^2 + 6x$ have the smallest slope? What is the slope of the tangent at this point?

31. For what value of x in the interval $-1 \le x \le 4$ is the graph of the function

$$f(x) = 2x^2 - \frac{1}{3}x^3$$

steepest? What is the slope of the tangent at this point?

In Problems 32 through 38 solve the practical optimization problem and use one of the techniques from this section to verify that you have actually found the desired absolute extremum.

32. BROADCASTING An all-news radio station has made a survey of the listening habits of local residents between the hours of 5:00 P.M. and midnight. The survey indicates that the percentage of

the local adult population that is tuned in to the station x hours after 5:00 P.M. is

$$f(x) = \frac{1}{8}(-2x^3 + 27x^2 - 108x + 240)$$

  a. At what time between 5:00 P.M. and midnight are the most people listening to the station? What percentage of the population is listening at this time?
  b. At what time between 5:00 P.M. and midnight are the fewest people listening? What percentage of the population is listening at this time?

33. GROUP MEMBERSHIP Suppose that x years after its founding in 1992, a certain national consumers' association had a membership of

$$f(x) = 100(2x^3 - 45x^2 + 264x)$$

  a. At what time between 1992 and 2006 was the membership of the association largest? What was the membership at that time?
  b. At what time between 1992 and 2006 was the membership of the association smallest? What was the membership at that time?

34. SPEED OF FLIGHT In a model* developed by C. J. Pennycuick, the power P required by a bird to maintain flight is given by the formula

$$P = \frac{w^2}{2\rho Sv} + \frac{1}{2}\rho Av^3$$

---
*C. J. Pennycuick, "The Mechanics of Bird Migration," *Ibis* III, 1969, pp. 525–556.

where v is the relative speed of the bird, w is the weight, ρ is the density of air, and S and A are positive constants associated with the bird's size and shape. What relative speed v will minimize the power required by the bird?

35. **LEARNING** In a learning model, two responses (A and B) are possible for each of the series of observations. If there is a probability p of getting response A in any individual observation, the probability of getting response A exactly n times in a series of m observations is $F(p) = p^n(1 - p)^{m-n}$. The **maximum likelihood estimate** is the value of p that maximizes $F(p)$ for $0 \le p \le 1$. For what value of p does this occur?

36. **GROWTH OF A SPECIES** More on codling moths*. The percentage of codling moths that survive the pupa stage at a given temperature T (degrees Celsius) is modeled by the formula

$$P(T) = -1.42T^2 + 68T - 746$$
$$\text{for } 20 \le T \le 30$$

Find the temperatures at which the greatest and smallest percentage of moths survive.

37. **BLOOD CIRCULATION** Poiseuille's law asserts that the speed of blood that is r centimeters from the central axis of an artery of radius R is $S(r) = c(R^2 - r^2)$, where c is a positive constant. Where is the speed of the blood greatest?

38. **POLITICS** A poll indicates that x months after a particular candidate for public office declares her candidacy, she will have the support of $S(x)$ percent of the voters, where

$$S(x) = \frac{1}{29}(-x^3 + 6x^2 + 63x + 1,080)$$
$$\text{for } 0 \le x \le 12$$

If the election is held in November, when should the politician announce her candidacy? Should she expect to win if she needs at least 50% of the vote?

39. **ELASTICITY OF DEMAND** When a particular commodity is priced at p dollars per unit, consumers demand q units, where p and q are related by the equation $q^2 + 3pq = 22$.
    a. Find the elasticity of demand for this commodity.
    b. For a unit price of $3, is the demand elastic, inelastic, or of unit elasticity?

40. **ELASTICITY OF DEMAND** When an electronics store prices a certain brand of stereo at p hundred dollars per set, it is found that q sets will be sold each month, where $q^2 + 2p^2 = 41$.
    a. Find the elasticity of demand for the stereos.
    b. For a unit price of $p = 4$ ($400), is the demand elastic, inelastic, or of unit elasticity?

41. **DEMAND FOR ART** An art gallery offers 50 prints by a famous artist. If each print in the limited edition is priced at p dollars, it is expected that $q = 500 - 2p$ prints will be sold.
    a. What limitations are there on the possible range of the price p?
    b. Find the elasticity of demand. Determine the values of p for which the demand is elastic, inelastic, and of unit elasticity.
    c. Interpret the results of part (b) in terms of the behavior of the total revenue as a function of unit price p.
    d. If you were the owner of the gallery, what price would you charge for each print? Explain the reasoning behind your decision.

42. **DEMAND FOR AIRLINE TICKETS** An airline determines that when a round-trip ticket between Los Angeles and San Francisco costs p dollars ($0 \le p \le 160$), the daily demand for tickets is $q = 256 - 0.01p^2$.
    a. Find the elasticity of demand. Determine the values of p for which the demand is elastic, inelastic, and of unit elasticity.
    b. Interpret the results of part (a) in terms of the behavior of the total revenue as a function of unit price p.
    c. What price would you advise the airline to charge for each ticket? Explain your reasoning.

43. **ORNITHOLOGY** According to the results[†] of Tucker and Schmidt-Koenig, the energy expended by a certain species of parakeet is given by

$$E(v) = \frac{1}{v}[0.074(v - 35)^2 + 22]$$

where v is the bird's velocity (in km/hr).

    a. What velocity minimizes energy expenditure?
    b. Read an article on how mathematical methods can be used to study animal behavior, and write

*P. L. Shaffer and H. J. Gold, "A Simulation Model of Population Dynamics of the Codling Moth, *Cydia pomonelia*," *Ecological Modeling*, Vol. 30. 1985, pp 247–274.

†V. A. Tucker and K. Schmidt-Koenig, "Flight Speeds of Birds in Relation to Energetics and Wind Directions," *The Auk*, Vol. 88 1971, pp. 97–107.

a paragraph on whether you think such methods are valid. You may wish to begin with the reference cited in this problem.

44. **NATIONAL CONSUMPTION**   Assume that total national consumption is given by a function $C(x)$, where x is the total national income. The derivative $C'(x)$ is called the **marginal propensity to consume.** Then $S = x - C$ represents total national savings, and $S'(x)$ is called **marginal propensity to save.** Suppose the consumption function is $C(x) = 8 - 0.8x - 0.8\sqrt{x}$. Find the marginal propensity to consume, and determine the value of x that results in the smallest total savings.

45. **SENSITIVITY TO DRUGS**   Body reaction to drugs is often modeled* by an equation of the form

$$R(D) = D^2\left(\frac{C}{2} - \frac{D}{3}\right)$$

where D is the dosage and C (a constant) is the maximum dosage that can be given. The rate of change of $R(D)$ with respect to D is called the **sensitivity.**
   a. Find the value of D for which the sensitivity is the greatest. What is the greatest sensitivity? (Express your answer in terms of C.)
   b. What is the reaction (in terms of C) when the dosage resulting in greatest sensitivity is used?

46. **AERODYNAMICS**   In designing airplanes, an important feature is the so-called drag factor; that is, the retarding force exerted on the plane by the air. One model measures drag by a function of the form

$$F(v) = Av^2 + \frac{B}{v^2}$$

where A and B are positive constants. It is found experimentally that drag is minimized when $v = 160$ mph. Use this information to find the ratio $\dfrac{B}{A}$.

47. **ELECTRICITY**   When a resistor of R ohms is connected across a battery with electromotive force E volts and internal resistance r ohms, a current of I amperes will flow, generating P watts of power, where

$$I = \frac{E}{r + R} \qquad \text{and} \qquad P = I^2R$$

Assuming r is constant, what choice of R results in maximum power?

48. **SURVIVAL OF AQUATIC LIFE**   It is known that a quantity of water that occupies 1 liter at 0°C will occupy

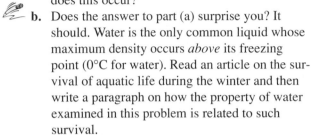

$$V(T) = \left(\frac{-6.8}{10^8}\right)T^3 + \left(\frac{8.5}{10^6}\right)T^2 - \left(\frac{6.4}{10^5}\right)T + 1$$

liters when the temperature is $T°C$, for $0 \le T \le 30$.

   a. Use a graphing utility to graph $V(T)$ for $0 \le T \le 10$. The density of water is maximized when $V(T)$ is minimized. At what temperature does this occur?
   b. Does the answer to part (a) surprise you? It should. Water is the only common liquid whose maximum density occurs *above* its freezing point (0°C for water). Read an article on the survival of aquatic life during the winter and then write a paragraph on how the property of water examined in this problem is related to such survival.

49. **BLOOD PRODUCTION**   A useful model for the production $p(x)$ of blood cells involves a function of the form

$$p(x) = \frac{Ax}{B + x^m}$$

where x is the number of cells present, and A, B, and m are positive constants.[†]
   a. Find the rate of blood production $R(x) = p'(x)$ and determine where $R(x) = 0$.
   b. Find the rate at which $R(x)$ is changing with respect to x and determine where $R'(x) = 0$.
   c. If $m > 1$, does the nonzero critical number you found in part (b) correspond to a relative maximum or a relative minimum? Explain.

50. **AMPLITUDE OF OSCILLATION**   In physics, it can be shown that a particle forced to oscillate in a resisting medium has amplitude $A(r)$ given by

$$A(r) = \frac{1}{(1 - r^2)^2 + kr^2}$$

where r is the ratio of the forcing frequency to the natural frequency of oscillation and k is a positive constant that measures the damping effect of the resisting medium. Show that $A(r)$ has exactly one positive critical number. Does it correspond to a relative maximum or a relative minimum? Can anything be said about the *absolute* extrema of $A(r)$?

---

*R. M. Thrall et al., *Some Mathematical Models in Biology,* U. of Michigan, 1967.

[†]M. C. Mackey and L. Glass, "Oscillations and Chaos in Physiological Control Systems," *Science,* Vol. 197, pp 287–289.

51. RESPIRATION Physiologists define the flow F of air in the trachea by the formula $F = SA$, where S is the speed of the air and A is the area of a cross section of the trachea.

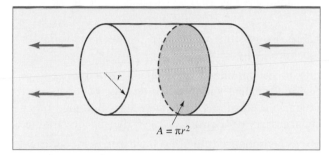

$A = \pi r^2$

PROBLEM 51

  a. Assume the trachea has a circular cross section with radius r. Use the formula for the speed of air in the trachea during a cough given in Example 3.4.3 to express air flow F in terms of r.

  b. Find the radius r for which the flow is greatest.

52. MARGINAL ANALYSIS Suppose $q > 0$ units of a commodity are produced at a total cost of $C(q)$ dollars and an average cost of $A(q) = \dfrac{C(q)}{q}$. In this section, we showed that $q = q_c$ satisfies $A'(q_c) = 0$ if and only if $C'(q_c) = A(q_c)$; that is, when marginal cost equals average cost. The purpose of this problem is to show that $A(q)$ is *minimized* when $q = q_c$.

  a. Generally speaking, the cost of producing a commodity increases at an increasing rate as more and more goods are produced. Using this economic principle, what can be said about the sign of $C''(q)$ as q increases?

  b. Show that $A''(q_c) > 0$ if and only if $C''(q_c) > 0$. Then use part (a) to argue that average cost $A(q)$ is minimized when $q = q_c$.

53. ELASTICITY AND REVENUE Suppose the demand for a certain commodity is given by $q = b - ap$, where a and b are positive constants, and $0 \le p \le b/a$.

  a. Express elasticity of demand as a function of p.

  b. Show that the demand is of unit elasticity at the midpoint $p = b/2a$ of the interval $0 \le p \le b/a$.

  c. For what values of p is the demand elastic? Inelastic?

54. INCOME ELASTICITY OF DEMAND **Income elasticity of demand** is defined to be the percentage change in quantity purchased divided by the percentage change in real income.

  a. Write a formula for income elasticity of demand E in terms of real income I and quantity purchased Q.

  b. In the United States, which would you expect to be greater, the income elasticity of demand for cars or for food? Explain your reasoning.

  c. What do you think is meant by a *negative* income elasticity of demand? Which of the following goods would you expect to have $E < 0$: used clothing, personal computers, bus tickets, refrigerators, used cars? Explain your reasoning.

  d. Read an article on income elasticity of demand and write a paragraph on why the income elasticity of demand for food is much larger in a developing country than in a country such as the United States or Japan.*

55. ELASTICITY Suppose that the demand equation for a certain commodity is $q = \dfrac{a}{p^m}$, where a and m are positive constants. Show that the elasticity of demand is equal to $-m$ for all values of p. Interpret this result.

*You may find the following text a helpful starting point for your research: Campbell R. McConnell and Stanley L. Brue, *Microeconomics,* 12th ed., New York: McGraw-Hill, 1993.

SECTION 3.5 Additional Applied Optimization

In Section 3.4, you saw a number of applications in which a formula was given and it was required to determine either a maximum or a minimum value. In practice, things are often not that simple, and it is necessary to first gather information about a quantity of interest, then formulate and analyze an appropriate mathematical model.

In this section, you will learn how to combine the techniques of model-building from Section 1.4 with the optimization techniques of Section 3.4. Here is a procedure for dealing with such problems.

Guidelines for Solving Optimization Problems

Step 1. Begin by deciding precisely what you want to maximize or minimize. Once this has been done, assign names to all variables of interest. It may help to pick letters that suggest the nature or role of the variable, such as "R" for "revenue."

Step 2. After assigning variables, express the relationships between the variables in terms of equations or inequalities. A figure may help.

Step 3. Express the quantity to be optimized (maximized or minimized) in terms of just one variable (the independent variable). To do this, you may need to use one or more of the available equations from step 2 to eliminate other variables. Also determine any necessary restrictions on the independent variable.

Step 4. If $f(x)$ is the quantity to be optimized, find $f'(x)$ and determine all critical numbers of f. Then find the required maximum or minimum value using the methods of Section 3.4 (the extreme value theorem or the second derivative test for absolute extrema). Remember, you may have to check the value of $f(x)$ at endpoints of an interval.

Step 5. Interpret your results in terms of appropriate physical, geometric, or economic quantities.

This procedure is illustrated in Examples 3.5.1 through 3.5.3.

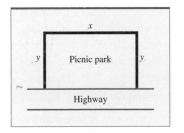

FIGURE 3.45 Rectangular picnic park.

EXAMPLE | 3.5.1

The highway department is planning to build a picnic park for motorists along a major highway. The park is to be rectangular with an area of 5,000 square yards and is to be fenced off on the three sides not adjacent to the highway. What is the least amount of fencing required for this job? How long and wide should the park be for the fencing to be minimized?

Solution

Step 1. Draw the picnic area, as in Figure 3.45. Let x (yards) be the length of the park (the side parallel to the highway) and let y (yards) be the width.

Step 2. Since the park is to have area 5,000 square yards, we must have $xy = 5,000$.

Step 3. The length of the fencing is $F = x + 2y$, where x and y are both positive, since otherwise there would be no park. Since

$$xy = 5,000 \qquad \text{or} \qquad y = \frac{5,000}{x}$$

we can eliminate y from the formula for F to obtain a formula in terms of x alone:

$$F(x) = x + 2y = x + 2\left(\frac{5,000}{x}\right) = x + \frac{10,000}{x} \qquad \text{for } x > 0$$

Step 4. The derivative of $F(x)$ is

$$F'(x) = 1 - \frac{10,000}{x^2}$$

and we get the critical numbers for $F(x)$ by setting $F'(x) = 0$ and solving for x:

$$F'(x) = 1 - \frac{10{,}000}{x^2} = 0$$

$$\Rightarrow \quad \frac{x^2 - 10{,}000}{x^2} = 0 \qquad \text{put the fraction over a common denominator}$$

$$x^2 = 10{,}000 \qquad \text{equate the numerator to 0}$$

$$x = 100 \qquad \text{reject } -100 \text{ since } x > 0$$

Since $x = 100$ is the only critical number in the interval $x > 0$, we can apply the second derivative test for absolute extrema. The second derivative of $F(x)$ is

$$F''(x) = \frac{20{,}000}{x^3}$$

so $F''(100) > 0$ and an absolute minimum of $F(x)$ occurs where $x = 100$. For reference, the graph of the fencing function $F(x)$ is shown in Figure 3.46.

Step 5. We have shown the minimal amount of fencing is

$$F(100) = 100 + \frac{10{,}000}{100} = 200 \text{ yards}$$

which is achieved when the park is $x = 100$ yards long and

$$y = \frac{5{,}000}{100} = 50 \text{ yards}$$

wide.

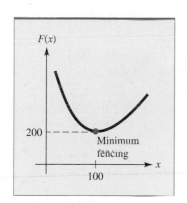

FIGURE 3.46 The graph of $F(x) = x + \dfrac{10{,}000}{x}$ for $x > 0$.

EXAMPLE | 3.5.2

A cylindrical can is to be constructed to hold a fixed volume of liquid. The cost of the material used for the top and bottom of the can is 3 cents per square inch, and the cost of the material used for the curved side is 2 cents per square inch. Use calculus to derive a simple relationship between the radius and height of the can that is the least costly to construct.

Solution

Let r denote the radius, h the height, C the cost (in cents), and V the (fixed) volume. The goal is to minimize total cost, which comes from three sources:

$$\text{Cost} = \text{cost of top} + \text{cost of bottom} + \text{cost of side}$$

where, for each component of the cost,

$$\text{Cost} = (\text{cost per square inch})(\text{area})$$

Hence, $\text{Cost of top} = \text{cost of bottom} = 3(\pi r^2)$

and $\text{Cost of side} = 2(2\pi r h) = 4\pi r h$

so the total cost is

$$C = \underbrace{3\pi r^2}_{top} + \underbrace{3\pi r^2}_{bottom} + \underbrace{4\pi r h}_{sides} = 6\pi r^2 + 4\pi r h$$

Before you can apply calculus, you must write the cost in terms of just one variable. To do this, use the fact that the can is to have a fixed volume V_0 and solve the equation $V_0 = \pi r^2 h$ for h to get

$$h = \frac{V_0}{\pi r^2}$$

Then substitute this expression for h into the formula for C to express the cost in terms of r alone:

$$C(r) = 6\pi r^2 + 4\pi r\left(\frac{V_0}{\pi r^2}\right) = 6\pi r^2 + \frac{4V_0}{r}$$

The radius r can be any positive number, so the goal is to find the absolute minimum of $C(r)$ for $r > 0$. Differentiating $C(r)$, you get

$$C'(r) = 12\pi r - \frac{4V_0}{r^2} \qquad \textit{remember } V_0 \textit{ is a constant}$$

Since $C'(r)$ exists for all $r > 0$, any critical number $r = R$ must satisfy $C'(R) = 0$; that is,

Just-In-Time Review

A cylinder of radius r and height h has lateral (curved) area $A = 2\pi rh$ and volume $V = \pi r^2 h$.

$$C'(R) = 12\pi R - \frac{4V_0}{R^2} = 0$$

$$12\pi R = \frac{4V_0}{R^2}$$

$$R^3 = \frac{4V_0}{12\pi}$$

$$R = \sqrt[3]{\frac{V_0}{3\pi}}$$

If H is the height of the can that corresponds to the radius R, then $V_0 = \pi r^2 H$, and since R must satisfy

$$12\pi R = \frac{4V_0}{R^2}$$

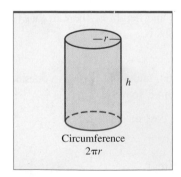

Circumference
$2\pi r$

you find that

$$12\,\pi R = \frac{4(\pi R^2 H)}{R^2} = 4\pi H$$

or

$$H = \frac{12\pi R}{4\pi} = 3R$$

Finally, note that the second derivative of $C(r)$ satisfies

$$C''(r) = 12\pi + \frac{8V_0}{r^3} > 0 \qquad \text{for all } r > 0$$

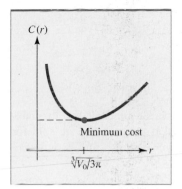

FIGURE 3.47 The cost function $C(r) = 6\pi r^2 + \dfrac{4V}{r}$ for $r > 0$.

Therefore, since $r = R$ is the only critical number for $C(r)$ and since $C''(R) > 0$, the second derivative test for absolute extrema assures you that when the total cost of constructing the can is minimized, the height of the can is three times its radius. The graph of the cost function $C(r)$ is shown in Figure 3.47.

EXAMPLE | 3.5.3

A manufacturer can produce premium-grade blank videotape cassettes at a cost of $2 apiece. The cassettes have been selling for $5 apiece, and at this price, consumers have been buying 4,000 cassettes a month. The manufacturer is planning to raise the price of the cassettes and estimates that for each $1 increase in the price, 400 fewer cassettes will be sold each month. At what price should the manufacturer sell the cassettes to maximize profit?

Solution

Let x denote the new price at which the cassettes will be sold and $P(x)$ the corresponding profit. The goal is to maximize the profit. As in Example 1.4.5 of Section 1.4, begin by stating the formula for profit in words.

$$\text{Profit} = (\text{number of cassettes sold})(\text{profit per cassette})$$

Since 4,000 cassettes are sold each month when the price is $5 and 400 fewer will be sold each month for each $1 increase in the price, it follows that

$$\text{Number of cassettes sold} = 4,000 - 400(\text{number of } \$1 \text{ increases})$$

The number of $1 increases in the price is the difference $x - 5$ between the new and old selling prices. Hence,

$$
\begin{aligned}
\text{Number of cassettes sold} &= 4,000 - 400(x - 5) \\
&= 400[10 - (x - 5)] \\
&= 400(15 - x)
\end{aligned}
$$

The profit per cassette is simply the difference between the selling price x and the cost $2. That is,

$$\text{Profit per cassette} = x - 2$$

Putting it all together,

$$P(x) = 400(15 - x)(x - 2)$$

The goal is to find the absolute maximum of the profit function $P(x)$. To determine the relevant interval for this problem, note that since the new price x is to be at least as high as the old price $5, we must have $x \geq 5$. On the other hand, the number of cassettes sold is $400(15 - x)$, which will be negative if $x > 15$. If you assume that the manufacturer will not price the cassettes so high that no one buys them, you can restrict the optimization problem to the closed interval $5 \leq x \leq 15$.

To find the critical points, compute the derivative using the product and constant multiple rules to get

$$
\begin{aligned}
P'(x) &= 400[(15 - x)(1) + (x - 2)(-1)] \\
&= 400(15 - x - x + 2) = 400(17 - 2x)
\end{aligned}
$$

which is zero when

$$17 - 2x = 0 \qquad \text{or} \qquad x = 8.5$$

Comparing the values of the profit function

$$P(5) = 12{,}000 \qquad P(8.5) = 16{,}900 \qquad \text{and} \qquad P(15) = 0$$

at this critical value and at the endpoints of the interval, you can conclude that the maximum possible profit is \$16,900, which will be generated if the cassettes are sold for \$8.50 apiece. For reference, the graph of the profit function is shown in Figure 3.48.

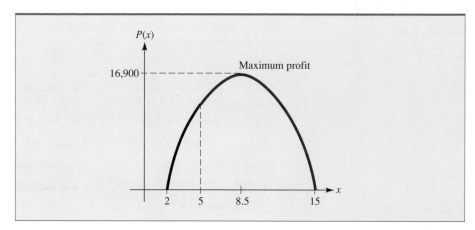

FIGURE 3.48   The profit function $P(x) = 400(15 - x)(x - 2)$.

NOTE   **An Alternative Solution to the Profit Maximization Problem in Example 3.5.3**
Since the number of cassettes sold in Example 3.5.3 is described in terms of the number N of \$1 increases in price, you may wish to use N as the independent variable in your solution rather than the new price itself. With this choice, you find that

$$\text{Number of cassettes} = 4{,}000 - 400N$$
$$\text{Profit per cassette} = (N + 5) - 2 = N + 3$$

Thus, the total profit is

$$P(N) = (4{,}000 - 400N)(N + 3) = 400(10 - N)(N + 3)$$

and the relevant interval is $0 \le N \le 10$ (Do you see why?). The absolute maximum in this case will occur when $N = 3.5$ (provide the details); that is, when the old price is increased from 5 to $5 + 3.5 = 8.5$ dollars (\$8.50). As you would expect, this is the same as the result obtained using price as the independent variable in Example 3.5.3.   ∎

In modern cities, where residential areas are often developed in the vicinity of industrial plants, it is important to carefully monitor and control the emission of pollutants. In Example 3.5.4, we examine a modeling problem in which calculus is used to determine the location in a community where pollution is minimized.

EXPLORE!

Refer to Example 3.5.4. Store $P(x) = 75/x + 300/(15 - x)$ into Y1 and graph using the modified decimal window [0, 14]1 by [0, 350]10. Now use the TRACE to move the cursor from X = 1 to 14 and confirm the location of minimal pollution. To view the behavior of the derivative $P'(x)$, write Y2 = nDeriv(Y1, X, X) and graph using the window [0, 14]1 by [−75, 300]10. What do you observe?

EXAMPLE | 3.5.4

Two industrial plants, A and B, are located 15 miles apart, and emit 75 ppm (parts per million) and 300 ppm of particulate matter, respectively. Each plant is surrounded by a restricted area of radius 1 mile in which no housing is allowed, and the concentration of pollutant arriving at any other point Q from each plant decreases with the reciprocal of the distance between that plant and Q. Where should a house be located on a road joining the two plants to minimize the total pollution arriving from both plants?

Solution

Suppose a house H is located x miles from plant A and hence, $15 - x$ miles from plant B, where x satisfies $1 \leq x \leq 14$ since there is a 1-mile restricted area around each plant (see Figure 3.49). Since the concentration of particulate matter arriving at H from each plant decreases with the reciprocal of the distance from the plant to H, the concentration of pollutant from plant A is $75/x$ and from plant B is $300/(15 - x)$. Thus, the total concentration of particulate matter arriving at H is given by the function

$$P(x) = \underbrace{\frac{75}{x}}_{\substack{pollution \\ from\ A}} + \underbrace{\frac{300}{15 - x}}_{\substack{pollution \\ from\ B}}$$

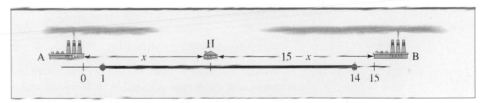

FIGURE 3.49　Pollution at a house located between two industrial plants.

To minimize the total pollution $P(x)$, we first find the derivative $P'(x)$ and solve $P'(x) = 0$. Applying the quotient rule and chain rule, we obtain

$$P'(x) = \frac{-75}{x^2} + \frac{-300(-1)}{(15 - x)^2} = \frac{-75}{x^2} + \frac{300}{(15 - x)^2}$$

Solving $P'(x) = 0$, we find that

$$\frac{-75}{x^2} + \frac{300}{(15 - x)^2} = 0$$

$$\frac{75}{x^2} = \frac{300}{(15 - x)^2}$$

$75(15 - x)^2 = 300x^2$	*cross multiply*
$x^2 - 30x + 225 = 4x^2$	*divide by 75 and expand*
$3x^2 + 30x - 225 = 0$	*combine terms*
$3(x - 5)(x + 15) = 0$	*factor*
$x = 5, x = -15$	

Just-In-Time **Review**

Cross-multiplying means that if

$$\frac{A}{B} = \frac{C}{D}$$

then

$$AD = CB$$

The only critical number in the allowable interval $1 \leq x \leq 14$ is $x = 5$. Evaluating $P(x)$ at $x = 5$ and at the two endpoints of the interval, $x = 1$ and $x = 14$, we get

$$P(1) \approx 96.43 \text{ ppm}$$
$$P(5) = 45 \text{ ppm}$$
$$P(14) \approx 305.36 \text{ ppm}$$

Thus, total pollution is minimized when the house is located 5 miles from plant A.

EXAMPLE | 3.5.5

A cable is to be run from a power plant on one side of a river 900 meters wide to a factory on the other side, 3,000 meters downstream. The cost of running the cable under the water is $5 per meter, while the cost over land is $4 per meter. What is the most economical route over which to run the cable?

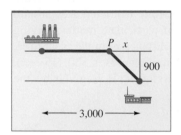

FIGURE 3.50 Relative positions of factory, river, and power plant.

Solution

To help you visualize the situation, begin by drawing a diagram as shown in Figure 3.50. (Notice that in drawing the diagram in Figure 3.50, we have already assumed that the cable should be run in a *straight line* from the power plant to some point P on the opposite bank. Do you see why this assumption is justified?)

The goal is to minimize the cost of installing the cable. Let C denote this cost and represent C as follows:

$$C = 5(\text{number of meters of cable under water})$$
$$+ \ 4(\text{number of meters of cable over land})$$

Since you wish to describe the optimal route over which to run the cable, it will be convenient to choose a variable in terms of which you can easily locate the point P. Two reasonable choices for the variable x are illustrated in Figure 3.51.

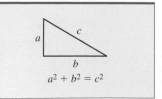

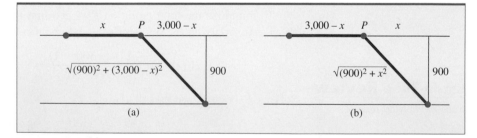

FIGURE 3.51 Two choices for the variable x.

Before plunging into the calculations, take a minute to decide which choice of variables is more advantageous. In Figure 3.51a, the distance across the water from the power plant to the point P is (by the Pythagorean theorem) $\sqrt{(900)^2 + (3,000 - x)^2}$, and the corresponding total cost function is

$$C(x) = 5\sqrt{(900)^2 + (3,000 - x)^2} + 4x$$

In Figure 3.51b, the distance across the water is $\sqrt{(900)^2 + x^2}$, and the total cost function is

$$C(x) = 5\sqrt{(900)^2 + x^2} + 4(3{,}000 - x)$$

The second function is the more attractive since the term $3{,}000 - x$ is merely multiplied by 4, while in the first function it is squared and appears under the radical. Hence, you should choose x as in Figure 3.51b and work with the total cost function

$$C(x) = 5\sqrt{(900)^2 + x^2} + 4(3{,}000 - x)$$

Since the distances x and $3{,}000 - x$ cannot be negative, the relevant interval is $0 \le x \le 3{,}000$, and your goal is to find the absolute minimum of the function $C(x)$ on this closed interval. To find the critical values, compute the derivative

$$C'(x) = \frac{5}{2}[(900)^2 + x^2]^{-1/2}(2x) - 4 = \frac{5x}{\sqrt{(900)^2 + x^2}} - 4$$

and set it equal to zero to get

$$\frac{5x}{\sqrt{(900)^2 + x^2}} - 4 = 0 \quad \text{or} \quad \sqrt{(900)^2 + x^2} = \frac{5}{4}x$$

Square both sides of the equation and solve for x to get

$$(900)^2 + x^2 = \frac{25}{16}x^2$$

$$x^2 - \frac{25}{16}x^2 - (900)^2 \qquad\qquad \textit{subtract } \frac{25x^2}{16} \textit{ and } (900)^2 \textit{ from each side}$$

$$-\frac{9}{16}x^2 = -(900)^2 \qquad\qquad \textit{combine terms on the left}$$

$$x^2 - \frac{16}{9}(900)^2 \qquad\qquad \textit{cross multiply by } -\frac{16}{9}$$

$$x = \pm\frac{4}{3}(900) = \pm 1{,}200 \qquad\qquad \textit{take square roots on each side}$$

Since only the positive value $x = 1{,}200$ is in the interval $0 \le x \le 3{,}000$, compute $C(x)$ at this critical value and at the endpoints $x = 0$ and $x = 3{,}000$. Since

$$C(0) = 5\sqrt{(900)^2 + 0} + 4(3{,}000 - 0) = 16{,}500$$
$$C(1{,}200) = 5\sqrt{(900)^2 + (1{,}200)^2} + 4(3{,}000 - 1{,}200) = 14{,}700$$
$$C(3{,}000) = 5\sqrt{(900)^2 + (3{,}000)^2} + 4(3{,}000 - 3{,}000) = 15{,}660$$

it follows that the minimal installation cost is \$14,700, which will occur if the cable reaches the opposite bank 1,200 meters downstream from the power plant.

In Example 3.5.6, the function to be maximized has practical meaning only when its independent variable is a whole number. However, the optimization procedure leads to a fractional value of this variable, and additional analysis is needed to obtain a meaningful solution.

EXAMPLE | 3.5.6

A bus company will charter a bus that holds 50 people to groups of 35 or more. If a group contains exactly 35 people, each person pays $60. In large groups, everybody's fare is reduced by $1 for each person in excess of 35. Determine the size of the group for which the bus company's revenue will be greatest.

Solution

Let R denote the bus company's revenue. Then,

$$R = \text{(number of people in the group)(fare per person)}$$

You could let x denote the total number of people in the group, but it is slightly more convenient to let x denote the number of people in excess of 35. Then,

$$\text{Number of people in the group} = 35 + x$$

and

$$\text{Fare per person} = 60 - x$$

so the revenue function is

$$R(x) = (35 + x)(60 - x)$$

Since x represents the number of people in excess of 35 but less than 50, you want to maximize $R(x)$ for a positive integer x in the interval $0 \leq x \leq 15$ (see Figure 3.52a). However, to use the methods of calculus, consider the *continuous* function $R(x) = (35 + x)(60 - x)$ defined on the entire interval $0 \leq x \leq 15$ (see Figure 3.52b).

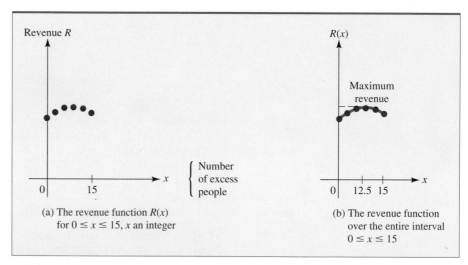

(a) The revenue function $R(x)$ for $0 \leq x \leq 15$, x an integer

(b) The revenue function over the entire interval $0 \leq x \leq 15$

FIGURE 3.52 The revenue function $R(x) = (35 + x)(60 - x)$.

The derivative is

$$R'(x) = (35 + x)(-1) + (60 - x)(1) = 25 - 2x$$

which is zero when $x = 12.5$. Since

$$R(0) = 2,100 \qquad R(12.5) = 2,256.25 \qquad R(15) = 2,250$$

it follows that the absolute maximum of $R(x)$ on the interval $0 \leq x \leq 15$ occurs when $x = 12.5$.

But x represents a certain number of people and must be a whole number. Hence, $x = 12.5$ cannot be the solution to this practical optimization problem. To find the optimal *integer* value of x, observe that R is increasing for $0 < x < 12.5$ and decreasing for $x > 12.5$, as shown in this diagram (see also Figure 3.52b).

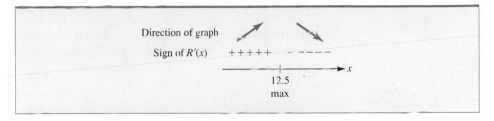

It follows that the optimal integer value of x is either $x = 12$ or $x = 13$. Since

$$R(12) = 2{,}256 \qquad \text{and} \qquad R(13) = 2{,}256$$

you can conclude that the bus company's revenue will be greatest when the group contains either 12 or 13 people in excess of 35; that is, for groups of 47 or 48. The revenue in either case will be $2,256.

Inventory Control

Inventory control is an important consideration in business. In particular, for each shipment of raw materials, a manufacturer must pay an ordering fee to cover handling and transportation. When the raw materials arrive, they must be stored until needed, and storage costs result. If each shipment of raw materials is large, few shipments will be needed, so ordering costs will be low, while storage costs will be high. On the other hand, if each shipment is small, ordering costs will be high because many shipments will be needed, but storage costs will be low. Example 3.5.7 shows how the methods of calculus can be used to determine the shipment size that minimizes total cost.

EXAMPLE | 3.5.7

A bicycle manufacturer buys 6,000 tires a year from a distributor. The ordering fee is $20 per shipment, the storage cost is 96 cents per tire per year, and each tire costs $5.75. Suppose that the tires are used at a constant rate throughout the year and that each shipment arrives just as the preceding shipment is being used up. How many tires should the manufacturer order each time to minimize cost?

Solution

The goal is to minimize the total cost, which can be written as

$$\text{Total cost} = \text{storage cost} + \text{ordering cost} + \text{purchase cost}$$

Let x denote the number of tires in each shipment and $C(x)$ the corresponding total cost in dollars. Then,

$$\text{Ordering cost} = (\text{ordering cost per shipment})(\text{number of shipments})$$

Since 6,000 tires are ordered during the year and each shipment contains x tires, the number of shipments is $\dfrac{6{,}000}{x}$ and so

$$\text{Ordering cost} = 20\left(\frac{6{,}000}{x}\right) = \frac{120{,}000}{x}$$

Moreover,

$$\text{Purchase cost} = (\text{total number of tires ordered})(\text{cost per tire})$$
$$= 6{,}000(5.75) = 34{,}500$$

The storage cost is slightly more complicated. When a shipment arrives, all x tires are placed in storage and then withdrawn for use at a constant rate. The inventory decreases linearly until there are no tires left, at which time the next shipment arrives. The situation is illustrated in Figure 3.53a. This is sometimes called a **just-in-time inventory pattern.**

EXPLORE!

Refer to Example 3.5.7 and the cost function $C(t)$, but vary the ordering fee $q = \$20$. Graph the cost functions for $q = 10$, 15, 20, and 25 using a viewing window of [0, 6,000]500 by [30,000, 40,000]5,000 followed by a window of [0, 6,000] 500 by [34,500, 35,500]5,000. Describe the difference in the graphs for these values of q. Find the minimum cost in each case. Describe how the minimum changes with the changing values of q.

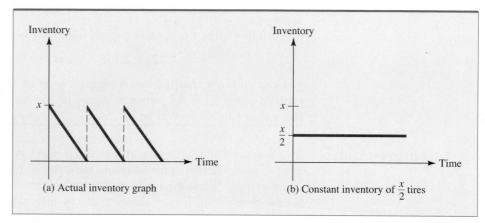

FIGURE 3.53   Inventory graphs.

The average number of tires in storage during the year is $x/2$, and the total yearly storage cost is the same as if $x/2$ tires were kept in storage for the entire year (Figure 3.53b). (This assertion, although reasonable, is not really obvious, and you have every right to be unconvinced. In Chapter 5, you will learn how to prove this fact mathematically using integral calculus.) It follows that

$$\text{Storage cost} = (\text{average number of tires stored})(\text{storage cost per tire})$$

$$= \frac{x}{2}(0.96) = 0.48x$$

Putting it all together, the total cost is

$$C(x) = \underbrace{0.48x}_{\substack{storage \\ cost}} + \underbrace{\frac{120{,}000}{x}}_{\substack{ordering \\ cost}} + \underbrace{34{,}500}_{\substack{cost\ of \\ purchase}}$$

and the goal is to find the absolute minimum of $C(x)$ on the interval

$$0 < x \le 6{,}000$$

The derivative of $C(x)$ is

$$C'(x) = 0.48 - \frac{120{,}000}{x^2}$$

which is zero when

$$x^2 = \frac{120,000}{0.48} = 250,000 \qquad \text{or} \qquad x = \pm 500$$

Since $x = 500$ is the only critical number in the relevant interval $0 < x \leq 6,000$, you can apply the second derivative test for absolute extrema. You find that the second derivative of the cost function is

$$C''(x) = \frac{240,000}{x^3}$$

which is positive when $x > 0$. Hence, the absolute minimum of the total cost $C(x)$ on the interval $0 < x \leq 6,000$ occurs when $x = 500$; that is, when the manufacturer orders the tires in lots of 500. For reference, the graph of the total cost function is sketched in Figure 3.54.

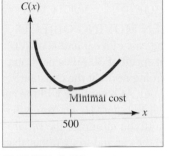

FIGURE 3.54 Total cost $C(x) = 0.48x + \dfrac{120,000}{x} + 34,500$.

NOTE Since the derivative of the (constant) purchase price $34,500 in Example 3.5.7 was zero, this component of the total cost had no bearing on the optimization problem. In general, economists distinguish between **fixed costs** (such as the total purchase price) and **variable costs** (such as the storage and ordering costs). To minimize total cost, it is sufficient to minimize the sum of all the variable components of cost. ∎

PROBLEMS 3.5

1. What number exceeds its square by the largest amount? [*Hint:* Find the number x that maximizes $f(x) = x - x^2$.]

2. What number is exceeded by its square root by the largest amount?

3. Find two positive numbers whose sum is 50 and whose product is as large as possible.

4. Find two positive numbers x and y whose sum is 30 and are such that xy^2 is as large as possible.

5. **RETAIL SALES** A store has been selling a popular computer game at the price of $40 per unit, and at this price, players have been buying 50 units per month. The owner of the store wishes to raise the price of the game and estimates that for each $1 increase in price, 3 fewer units will be sold each month. If each unit costs the store $25, at what price should the game be sold to maximize profit?

6. **RETAIL SALES** A bookstore can obtain a certain gift book from the publisher at a cost of $3 per book. The bookstore has been offering the book at a price of $15 per copy and, at this price, has been selling 200 copies a month. The bookstore is planning to lower its price to stimulate sales and estimates that for each $1 reduction in the price, 20 more books will be sold each month. At what price should the bookstore sell the book to generate the greatest possible profit?

7. **AGRICULTURAL YIELD** A Florida citrus grower estimates that if 60 orange trees are planted, the average yield per tree will be 400 oranges. The average yield will decrease by 4 oranges per tree for each additional tree planted on the same acreage. How many trees should the grower plant to maximize the total yield?

8. **HARVESTING** Farmers can get $2 per bushel for their potatoes on July first, and after that, the

price drops by 2 cents per bushel per day. On July first, a farmer has 80 bushels of potatoes in the field and estimates that the crop is increasing at the rate of 1 bushel per day. When should the farmer harvest the potatoes to maximize revenue?

9. **FENCING** A city recreation department plans to build a rectangular playground having an area of 3,600 square meters and surround it by a fence. How can this be done using the least amount of fencing?

10. **FENCING** There are 320 yards of fencing available to enclose a rectangular field. How should this fencing be used so that the enclosed area is as large as possible?

11. Prove that of all rectangles with a given perimeter, the square has the largest area.

12. Prove that of all rectangles with a given area, the square has the smallest perimeter.

13. A rectangle is inscribed in a right triangle, as shown in the accompanying figure. If the triangle has sides of length 5, 12, and 13, what are the dimensions of the inscribed rectangle of greatest area?

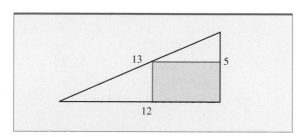

PROBLEM 13

14. A triangle is positioned with its hypotenuse on a diameter of a circle, as shown in the accompanying figure. If the circle has radius 4, what are the dimensions of the triangle of greatest area?

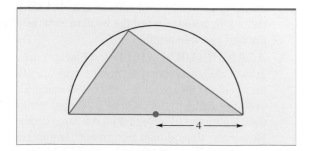

PROBLEM 14

15. **CONSTRUCTION COST** A carpenter has been asked to build an open box with a square base. The

sides of the box will cost \$3 per square meter, and the base will cost \$4 per square meter. What are the dimensions of the box of greatest volume that can be constructed for \$48?

16. **CONSTRUCTION COST** A closed box with a square base is to have a volume of 250 cubic meters. The material for the top and bottom of the box costs \$2 per square meter, and the material for the sides costs \$1 per square meter. Can the box be constructed for less than \$300?

17. **DISTANCE BETWEEN MOVING OBJECTS** A truck is 300 miles due east of a car and is traveling west at the constant speed of 30 miles per hour. Meanwhile, the car is going north at the constant speed of 60 miles per hour. At what time will the car and truck be closest to each other? [*Hint:* You will simplify the calculation if you minimize the *square* of the distance between the car and truck rather than the distance itself. Can you explain why this simplification is justified?]

18. **SPY STORY** It is noon, and the spy is back from space (see Problem 70 in Section 2.2) and driving a jeep through the sandy desert in the tiny principality of Alta Loma. He is 32 kilometers from the nearest point on a straight paved road. Down the road 16 kilometers is an abandoned power plant where a group of rival spies are holding captive his superior, code name "N." If the spy doesn't arrive with a ransom by 12:50 P.M., the bad guys have threatened to do N in. The jeep can travel at 48 km/hr in the sand and at 80 km/hr on the paved road. Can the spy make it in time, or is this the end of N? [*Hint:* The goal is to minimize time, which is distance divided by speed.]

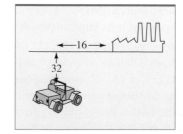

PROBLEM 18

19. **PACKAGING** Use the fact that 12 fluid ounces is approximately 6.89π cubic inches to find the dimensions of the 12-ounce soda can that can be constructed using the least amount of metal. Compare these dimensions with those of one of the soda

cans in your refrigerator. What do you think accounts for the difference?

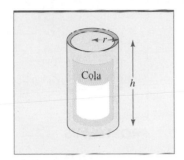

PROBLEM 19

20. **INSTALLATION COST**   A cable is to be run from a power plant on one side of a river 1,200 meters wide to a factory on the other side, 1,500 meters downstream. The cost of running the cable under the water is $25 per meter, while the cost over land is $20 per meter. What is the most economical route over which to run the cable?

21. **INSTALLATION COST**   Find the most economical route in Problem 20 if the power plant is 2,000 meters downstream from the factory.

22. **PACKAGING**   A cylindrical can is to hold 4π cubic inches of frozen orange juice. The cost per square inch of constructing the metal top and bottom is twice the cost per square inch of constructing the cardboard side. What are the dimensions of the least expensive can?

23. **PACKAGING**   What is the maximum possible volume of a cylindrical can with no top that can be made from 27π square inches of metal?

24. **PACKAGING**   A cylindrical can (with top) is to be constructed using a fixed amount of metal. Use calculus to derive a simple relationship between the radius and height of the can having the greatest volume.

25. **CONSTRUCTION COST**   A cylindrical container with no top is to be constructed to hold a fixed volume of liquid. The cost of the material used for the bottom is 3 cents per square inch, and that for the curved side is 2 cents per square inch. Use calculus to derive a simple relationship between the radius and height of the least expensive container.

26. **POSTER DESIGN**   A printer receives an order to produce a rectangular poster containing 25 square centimeters of print surrounded by margins of 2 centimeters on each side and 4 centimeters on the top and bottom. What are the dimensions of the smallest piece of paper that can be used to make the poster? [*Hint:* An unwise choice of variables will make the calculations unnecessarily complicated.]

27. **PRODUCTION COST**   A plastics firm has received an order from the city recreation department to manufacture 8,000 special Styrofoam kickboards for its summer swimming program. The firm owns 10 machines, each of which can produce 30 kickboards an hour. The cost of setting up the machines to produce the kickboards is $20 per machine. Once the machines have been set up, the operation is fully automated and can be overseen by a single production supervisor earning $15 per hour.

   a. How many of the machines should be used to minimize the cost of production?

   b. How much will the supervisor earn during the production run if the optimal number of machines is used?

   c. How much will it cost to set up the optimal number of machines?

28. **INVENTORY**   An electronics firm uses 600 cases of transistors each year. Each case costs $1,000. The cost of storing one case for a year is 90 cents, and the ordering fee is $30 per shipment. How many cases should the firm order each time to keep total cost at a minimum? (Assume that the transistors are used at a constant rate throughout the year and that each shipment arrives just as the preceding shipment is being used up.)

29. **INVENTORY**   A store expects to sell 800 bottles of perfume this year. The perfume costs $20 per bottle, the ordering fee is $10 per shipment, and the cost of storing the perfume is 40 cents per bottle per year. The perfume is consumed at a constant rate throughout the year, and each shipment arrives just as the preceding shipment is being used up.

   a. How many bottles should the store order in each shipment to minimize total cost?

   b. How often should the store order the perfume?

30. **INVENTORY**   A manufacturer of medical monitoring devices uses 36,000 cases of transistors per year. The ordering cost is $54 per shipment, and the annual cost of storage is $1.20 per case.

The transistors are used at a constant rate throughout the year, and each shipment arrives just as the preceding shipment is being used up. How many cases should be ordered in each shipment in order to minimize total cost?

31. **ACCOUNT MANAGEMENT** Tom requires $10,000 of spending money each year, which he withdraws from his savings account by making N equal withdrawals. Each withdrawal incurs a transaction fee of $8, and money in his account earns interest at the annual rate of 4%.

   a. The total cost C of managing the account is the transaction cost plus the loss of interest due to withdrawn funds. Express C as a function of N. [*Hint:* You may need the fact that $1 + 2 + \ldots + N = N(N + 1)/2$.]

   b. How many withdrawals should Tom make each year in order to minimize the total transaction cost $C(N)$?

32. **PRODUCTION COST** Each machine at a certain factory can produce 50 units per hour. The setup cost is $80 per machine, and the operating cost is $5 per hour. How many machines should be used to produce 8,000 units at the least possible cost? (Remember that the answer should be a whole number.)

33. **RETAIL SALES** A retailer has bought several cases of a certain imported wine. As the wine ages, its value initially increases, but eventually the wine will pass its prime and its value will decrease. Suppose that x years from now, the value of a case will be changing at the rate of $53 - 10x$ dollars per year. Suppose, in addition, that storage rates will remain fixed at $3 per case per year. When should the retailer sell the wine to obtain the greatest possible profit?

34. **CONSTRUCTION** An open box is to be made from a square piece of cardboard, 18 inches by 18 inches, by removing a small square from each corner and folding up the flaps to form the sides. What are the dimensions of the box of greatest volume that can be constructed in this way?

35. **POSTAL REGULATIONS** According to postal regulations, the girth plus length of parcels sent by fourth-class mail may not exceed 108 inches. What is the largest possible volume of a rectangular parcel with two square sides that can be sent by fourth-class mail?

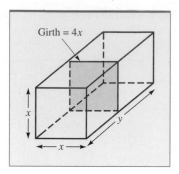

PROBLEM 35

36. **POSTAL REGULATIONS** Refer to Problem 35. What is the largest volume of a cylindrical parcel that can be sent by fourth-class mail?

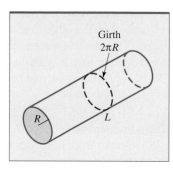

PROBLEM 36

37. **MINIMAL COST** A manufacturer finds that in producing x units per day (for $0 < x < 100$), three different kinds of cost are involved:

   a. A fixed cost of $1,200 per day in wages

   b. A production cost of $1.20 per day for each unit produced

   c. An ordering cost of $100/x^2$ dollars per day. Express the total cost as a function of x and determine the level of production that results in minimal total cost.

38. **TRANSPORTATION COST** For speeds between 40 and 65 miles per hour, a truck gets $480/x$ miles per gallon when driven at a constant speed of x miles per hour. Diesel gasoline costs $2.23 per gallon, and the driver is paid $15.10 per hour. What is the most economical constant speed between 40 and 65 miles per hour at which to drive the truck?

39. **AVIAN BEHAVIOR** Homing pigeons will rarely fly over large bodies of water unless forced to do so, presumably because it requires more energy to maintain altitude in flight in the heavy air over

cool water.* Suppose a pigeon is released from a boat B floating on a lake 5 miles from a point A on the shore and 13 miles from the pigeon's loft L, as shown in the accompanying figure. Assuming the pigeon requires twice as much energy to fly over water as over land, what path should it follow to minimize the total energy expended in flying from the boat to its loft? Assume the shoreline is straight and describe your path as a line from B to a point P on the shore followed by a line from P to L.

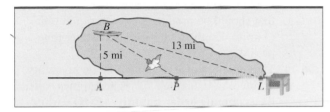

PROBLEM 39

40. **CONSTRUCTION** The strength of a rectangular beam is proportional to the product of its width and the square of its depth. Find the dimensions of the strongest beam that can be cut from a wooden log of diameter 15 inches.

41. **CONSTRUCTION** The stiffness of a rectangular beam is proportional to the product of its width w and the cube of its depth h. Find the dimensions of the stiffest beam that can be cut from a wooden log of diameter 15 inches. (Note the accompanying figure.)

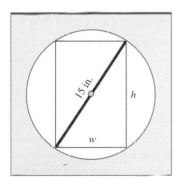

PROBLEM 41

42. **TRANSPORTATION COST** A truck is hired to transport goods from a factory to a warehouse. The driver's wages are figured by the hour and so are in-

versely proportional to the speed at which the truck is driven. The amount of gasoline used is directly proportional to the speed at which the truck is driven, and the price of gasoline remains constant during the trip. Show that the total cost is smallest at the speed for which the driver's wages are equal to the cost of the gasoline used.

43. **URBAN PLANNING** Two industrial plants, A and B, are located 18 miles apart, and each day, respectively emit 80 ppm (parts per million) and 720 ppm of particulate matter. Plant A is surrounded by a restricted area of radius 1 mile, while the restricted area around plant B has a radius of 2 miles. The concentration of particulate matter arriving at any other point Q from each plant decreases with the reciprocal of the distance between that plant and Q. Where should a house be located on a road joining the two plants to minimize the total concentration of particulate matter arriving from both plants?

44. **URBAN PLANNING** In Problem 43, suppose the concentration of particulate matter arriving at a point Q from each plant decreases with the reciprocal of the *square* of the distance between that plant and Q. With this alteration, now where should a house be located to minimize the total concentration of particulate matter arriving from both plants?

45. **OPTIMAL SETUP COST** Suppose that at a certain factory, setup cost is directly proportional to the number N of machines used and operating cost is inversely proportional to N. Show that when the total cost is minimal, the setup cost is equal to the operating cost.

46. **RECYCLING** To raise money, a service club has been collecting used bottles that it plans to deliver to a local glass company for recycling. Since the project began 80 days ago, the club has collected 24,000 pounds of glass for which the glass company currently offers 1 cent per pound. However, because bottles are accumulating faster than they can be recycled, the company plans to reduce by 1 cent each day the price it will pay for 100 pounds of used glass. Assume that the club can continue to collect bottles at the same rate and that transportation costs make more than one trip to the glass company unfeasible. What is the most advantageous time for the club to conclude its project and deliver the bottles?

*E. Batschelet, *Introduction to Mathematics for Life Sciences,* 3rd ed., New York: Springer-Verlag, 1979, pp. 276–277.

47. INSTALLATION COST For the summer, the company installing the cable in Example 3.5.5 has hired Frank Kornercutter as a consultant. Frank, recalling a problem from first-year calculus, asserts that no matter how far downstream the factory is located (beyond 1,200 meters), it would be most economical to have the cable reach the opposite bank 1,200 meters downstream from the power plant. The supervisor, amused by Frank's naivete, replies, "Any fool can see that if the factory is farther away, the cable should reach the opposite bank farther downstream. It's just common sense!" Of course, Frank is no common fool, but is he right? Why?

48. CONSTRUCTION As part of a construction project, it is necessary to carry a pipe around a corner as shown in the accompanying figure. What is the length of the longest pipe that will fit horizontally?

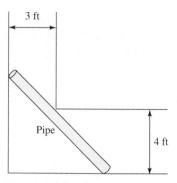

3 ft

4 ft

Pipe

PROBLEM 48

49. PRODUCTION COST A manufacturing firm receives an order for q units of a certain commodity. Each of the firm's machines can produce n units per hour. The setup cost is s dollars per machine, and the operating cost is p dollars per hour.

  a. Derive a formula for the number of machines that should be used to keep total cost as low as possible.

  b. Prove that when the total cost is minimal, the cost of setting up the machines is equal to the cost of operating the machines.

50. INVENTORY The inventory model analyzed in Example 3.5.7 is not the only such model possible. Suppose a company must supply N units per time period at a uniform rate. Assume that the storage cost per unit is D_1 dollars per time period and that the setup (ordering) cost is D_2 dollars. If production is at a uniform rate of m units per time period

(with no items in inventory at the end of each period), it can be shown that the total storage cost is

$$C_1 = \frac{D_1 x}{2}\left(1 - \frac{N}{m}\right)$$

where x is the number of items produced in each run.

  a. Show that the total average cost per period is

$$C = \frac{D_1 x}{2}\left(1 - \frac{N}{m}\right) + \frac{D_2 N}{x}$$

  b. Find an expression for the number of items that should be produced in each run in order to minimize the total average cost per time period.

 c. The optimum quantity found in the inventory problem in Example 3.5.7 is sometimes called the **economic order quantity (EOQ),** while the optimum found in part (b) of this problem is called the **economic production quantity (EPQ).** Modern inventory management goes far beyond the simple conditions in the EOQ and EPQ models, but elements of these models are still very important. For instance, the just-in-time inventory pattern described in Example 3.5.7 fits well with the production management philosophy of the Japanese. Read an article on Japanese production methods and write a paragraph on why the Japanese regard using space for the storage of materials as undesirable.*

51. EFFECT OF TAXATION ON A MONOPOLY A **monopolist** is a manufacturer who can manipulate the price of a commodity and usually does so with an eye toward maximizing profit. When the government taxes output, the tax effectively becomes an additional cost item, and the monopolist is forced to decide how much of the tax to absorb and how much to pass on to the consumer.

Suppose a particular monopolist estimates that when x units are produced, the total cost will be $C(x) = \frac{7}{8}x^2 + 5x + 100$ dollars and the market price of the commodity will be $p(x) = 15 - \frac{3}{8}x$

---

*You may wish to begin your search with the text by Philip E. Hicks, *Industrial Engineering and Management: A New Perspective,* New York: McGraw-Hill, 1994, pp. 144–170.

dollars per unit Further assume that the government imposes a tax of t dollars on each unit produced.

a. Show that profit is maximized when

$$x = \frac{2}{5}(10 - t).$$

b. Suppose the government assumes that the monopolist will always act so as to maximize total profit. What value of t should be chosen to guarantee maximum total tax revenue?

c. If the government chooses the optimum rate of taxation found in part (b), how much of this tax will be absorbed by the monopolist and how much will be passed on to the consumer?

 d. Read an article on taxation and write a paragraph on how it affects consumer spending.*

52. AVERAGE PRODUCTIVITY The output Q at a certain factory is a function of the number L of worker-hours of labor that are used. Use calculus to prove that when the average output per worker-hour is greatest, the average output is equal to the marginal output per worker-hour. [*Hint:* The marginal output per worker-hour is the derivative of output Q with respect to labor L.] You may assume without proof that the critical point of the average output function is actually the desired absolute maximum.

---

*You may wish to begin your research by consulting the following references: Robert Eisner, *The Misunderstood Economics: What Counts and How to Count It*, Boston, MA: Harvard Business School Press, 1994, pp. 196–199; and Robert H. Frank, *Microeconomics and Behavior*, 2nd ed., New York: McGraw-Hill, 1994, pp. 656–657.

Important Terms, Symbols, and Formulas

f is increasing $f'(x) > 0$ (189)

f is decreasing $f'(x) < 0$ (189)

Critical point: $(c, f(c))$, where $f'(c) = 0$ or $f'(c)$ does not exist   (192)

Relative maxima and minima   (192)

First derivative test for relative extrema: (193)
   If $f'(c) = 0$ or $f'(c)$ does not exist, then

Point of diminishing returns   (204)

Concavity: (205)
   Upward if $f'(x)$ is increasing; that is, $f''(x) > 0$
   Downward if $f'(x)$ is decreasing; that is, $f''(x) < 0$

Inflection point: A point on a graph where the concavity changes   (207)

Second derivative test for relative extrema: (213)
   If $f'(c) = 0$, then

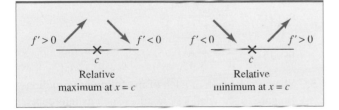

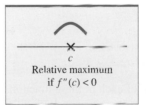

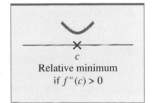

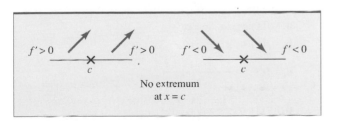

Vertical asymptote   (221)

Horizontal asymptote   (222)

Absolute maxima and minima   (235)

Extreme value property: (236)
   Absolute extrema of a continuous function on a closed interval $a \le x \le b$ occur at critical numbers in $a < x < b$ or at endpoints of the interval (a or b).

CHAPTER SUMMARY

Second derivative test for absolute extrema: (240)
   If $f(x)$ has only one critical number $x = c$ on an interval
   I, then $f(c)$ is an absolute maximum on I if $f''(c) < 0$ and
   an absolute minimum if $f''(c) > 0$.

Profit $P(q) = R(q) - C(q)$ is maximized when marginal
   revenue equals marginal cost: $R'(q) = C'(q)$.   (242)

Average cost $A(q) = \dfrac{C(q)}{q}$ is minimized when average

cost equals marginal cost: $A(q) = C'(q)$.   (243)

Elasticity of demand $q = D(p)$: $E(p) = \dfrac{p}{q}\dfrac{dq}{dp}$ (245)

Levels of elasticity of demand: (246)
$$\text{inelastic if } |E(p)| < 1$$
$$\text{elastic if } |E(p)| > 1$$
$$\text{unit elasticity if } |E(p)| = 1$$

Inventory models   (263)

Checkup for Chapter 3

1. Of the two curves shown here, one is the graph of a function $f(x)$ and the other is the graph of its derivative $f'(x)$.
    Determine which graph is the derivative and give reasons for your decision.

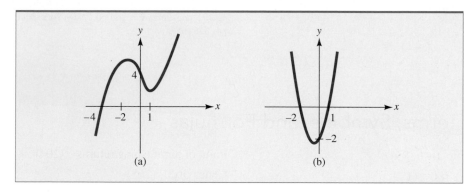

(a)   (b)

2. Find intervals of increase and decrease for each of
    these functions and determine whether each critical
    number corresponds to a relative maximum, a rela-
    tive minimum, or neither.

    a. $f(x) = -x^4 + 4x^3 + 5$

    b. $f(t) = 2t^3 - 9t^2 + 12t + 5$

    c. $g(t) = \dfrac{t}{t^2 + 9}$

    d. $g(x) = \dfrac{4 - x}{x^2 + 9}$

3. Determine where the graph of each of these functions
    is concave upward and concave downward. Find the
    x (or t) coordinate of each point of inflection.

    a. $f(x) = 3x^5 - 10x^4 + 2x - 5$

    b. $f(x) = 3x^5 + 20x^4 - 50x^3$

    c. $f(t) = \dfrac{t^2}{t - 1}$

    d. $g(t) = \dfrac{3t^2 + 5}{t^2 + 3}$

4. Determine all vertical and horizontal asymptotes for
    the graph of each of these functions.

    a. $f(x) = \dfrac{2x - 1}{x + 3}$

    b. $f(x) = \dfrac{x}{x^2 - 1}$

    c. $f(x) = \dfrac{x^2 + x - 1}{2x^2 + x - 3}$

    d. $f(x) = \dfrac{1}{x} - \dfrac{1}{\sqrt{x}}$

5. Sketch the graph of each of the following functions.
    Be sure to show all key features such as intercepts,
    asymptotes, high and low points, points of inflec-
    tion, cusps, and vertical tangents.

    a. $f(x) = 3x^4 - 4x^3$

    b. $f(x) = x^4 - 3x^3 + 3x^2 + 1$

    c. $f(x) = \dfrac{x^2 + 2x + 1}{x^2}$

    d. $f(x) = \dfrac{1 - 2x}{(x - 1)^2}$

6. Sketch the graph of a function $f(x)$ with all of these properties:
   a. $f'(x) > 0$ for $x < 0$ and for $0 < x < 2$
   b. $f'(x) < 0$ for $x > 2$
   c. $f'(0) = f'(2) = 0$
   d. $f''(x) < 0$ for $x < 0$ and for $x > 1$
   e. $f''(x) > 0$ for $0 < x < 1$
   f. $f(-1) = f(4) = 0; f(0) = 1, f(1) = 2, f(2) = 3$

7. **WORKER EFFICIENCY**   A postal clerk comes to work at 6 A.M. and t hours later has sorted approximately $f(t) = -t^3 + 7t^2 + 200t$ letters. At what time during the period from 6 A.M. to 10 A.M. is the clerk performing at peak efficiency?

8. **MAXIMIZING PROFIT**   A manufacturer can produce CD players at a cost of $20 apiece. It is estimated that if the CD players are sold for x dollars apiece, consumers will buy $120 - x$ of them each month. What unit price should the manufacturer charge to maximize profit?

9. **CONCENTRATION OF DRUG**   The concentration of a drug in a patient's bloodstream t hours after it is injected is given by

$$C(t) = \frac{0.05t}{t^2 + 27}$$

   milligrams per cubic centimeter.

   a. Sketch the graph of the concentration function.
   b. At what time is the concentration decreasing most rapidly?
   c. What happens to the concentration in the long run (as $t \to +\infty$)?

10. **BACTERIAL POPULATION**   A bacterial colony is estimated to have a population of

$$P(t) = \frac{15t^2 + 10}{t^3 + 6} \quad \text{million}$$

   t hours after the introduction of a toxin.

   a. What is the population at the time the toxin is introduced?
   b. When does the largest population occur? What is the largest population?
   c. Sketch the graph of the population curve. What happens to the population in the long run (as $t \to +\infty$)?

Review Problems

In Problems 1 through 10, determine intervals of increase and decrease and intervals of concavity for the given function. Then sketch the graph of the function. Be sure to show all key features such as intercepts, asymptotes, high and low points, points of inflection, cusps, and vertical tangents.

1. $f(x) = -2x^3 + 3x^2 + 12x - 5$

2. $f(x) = x^2 - 6x + 1$

3. $f(x) = 3x^3 - 4x^2 - 12x + 17$

4. $f(x) = x^3 - 3x^2 + 2$

5. $f(t) = 3t^5 - 20t^3$

6. $f(x) = \dfrac{x^2 + 3}{x - 1}$

7. $g(t) = \dfrac{t^2}{t + 1}$

8. $G(x) = (2x - 1)^2(x - 3)^3$

9. $F(x) = 2x + \dfrac{8}{x} + 2$

10. $f(x) = \dfrac{1}{x^3} + \dfrac{2}{x^2} + \dfrac{1}{x}$

In Problems 11 and 12, one of the two curves shown is the graph of a certain function f(x) and the other is the graph of its derivative f'(x). Determine which curve is the graph of the derivative and give reasons for your decision.

11.

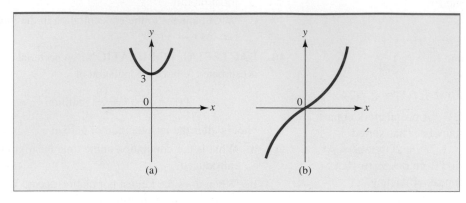

12.

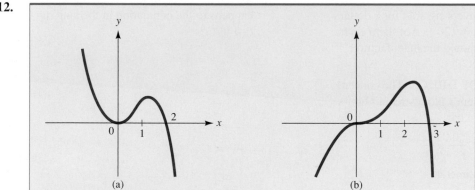

In Problems 13 through 16, the derivative f'(x) of a function is given. Use this information to classify each critical number of f(x) as a relative maximum, a relative minimum, or neither.

13. $f'(x) = x^3(2x - 3)^2(x + 1)^5(x - 7)$

14. $f'(x) = \dfrac{x^2 + 2x - 3}{x^2(x^2 + 1)}$

15. $f'(x) = \dfrac{x(x - 2)^2}{x^4 + 1}$

16. $f'(x) = \sqrt[3]{x}(3 - x)(x + 1)^2$

In Problems 17 through 20, sketch the graph of a function f that has all of the given properties.

17.
  a. $f'(x) > 0$ when $x < 0$ and when $x > 5$
  b. $f'(x) < 0$ when $0 < x < 5$
  c. $f''(x) > 0$ when $-6 < x < -3$ and when $x > 2$
  d. $f''(x) < 0$ when $x < -6$ and when $-3 < x < 2$

18.
  a. $f'(x) > 0$ when $x < -2$ and when $-2 < x < 3$
  b. $f'(x) < 0$ when $x > 3$
  c. $f'(-2) = 0$ and $f'(3) = 0$

19.
  a. $f'(x) > 0$ when $1 < x < 2$
  b. $f'(x) < 0$ when $x < 1$ and when $x > 2$
  c. $f''(x) > 0$ for $x < 2$ and for $x > 2$
  d. $f'(1) = 0$ and $f'(2)$ is undefined.

20.
  a. $f'(x) > 0$ when $x < 1$
  b. $f'(x) < 0$ when $x > 1$
  c. $f''(x) > 0$ when $x < 1$ and when $x > 1$
  d. $f'(1)$ is undefined.

In Problems 21 through 24, find all critical numbers for the given function f(x) and use the second derivative test to determine which (if any) critical points are relative maxima or relative minima.

21. $f(x) = -2x^3 + 3x^2 + 12x - 5$

22. $f(x) = x(2x - 3)^2$

23. $f(x) = \dfrac{x^2}{x + 1}$

24. $f(x) = \dfrac{1}{x} - \dfrac{1}{(x + 3)}$

In Problems 25 through 28, find the absolute maximum and the absolute minimum values (if any) of the given function on the specified interval.

25. $f(x) = -2x^3 + 3x^2 + 12x - 5; \; -3 \le x \le 3$

26. $f(t) = -3t^4 + 8t^3 - 10; \; 0 \le t \le 3$

27. $g(s) = \dfrac{s^2}{s + 1}; \; -\dfrac{1}{2} \le s \le 1$

28. $f(x) = 2x + \dfrac{8}{x} + 2; \; x > 0$

29. The first derivative of a certain function is $f'(x) = x(x - 1)^2$.

   a. On what intervals is f increasing? Decreasing?

   b. On what intervals is the graph of f concave up? Concave down?

   c. Find the x coordinates of the relative extrema and inflection points of f.

   d. Sketch a possible graph of $f(x)$.

30. OPTIMAL DESIGN A farmer wishes to enclose a rectangular pasture with 320 feet of fence. What dimensions give the maximum area if

   a. the fence is on all four sides of the pasture?

   b. the fence is on three sides of the pasture and the fourth side is bounded by a wall?

31. PROFIT A manufacturer can produce radios at a cost of $5 apiece and estimates that if they are sold for x dollars apiece, consumers will buy $20 - x$ radios a day. At what price should the manufacturer sell the radios to maximize profit?

32. TRAFFIC CONTROL It is estimated that between the hours of noon and 7:00 P.M., the speed of highway traffic flowing past a certain downtown exit is approximately $S(t) = t^3 - 9t^2 + 15t + 45$ miles per hour, where t is the number of hours past noon. At what time between noon and 7:00 P.M. is the traffic moving the fastest, and at what time between noon and 7:00 P.M. is it moving the slowest?

33. SPREAD OF A RUMOR The rate at which a rumor spreads through a community of P people is jointly proportional to the number of people N who have heard the rumor and the number who have not. Show that the rumor is spreading most rapidly when half the people have heard it.

34. CONSTRUCTION COST A box with a rectangular base is to be constructed of material costing $2/in.^{2} for the sides and bottom and $3/in.^{2} for the top. If the box is to have volume 1,215 in.^{3} and the length of its base is to be twice its width, what dimensions of the box will minimize its cost of construction? What is the minimal cost?

35. CONSTRUCTION COST A cylindrical container with no top is to be constructed for a fixed amount of money. The cost of the material used for the bottom is 3 cents per square inch, and the cost of the material used for the curved side is 2 cents per square inch. Use calculus to derive a simple relationship between the radius and height of the container having the greatest volume.

36. PROFIT A manufacturer has been selling lamps at $6 apiece, and at this price, consumers have been buying 3,000 lamps per month. The manufacturer wishes to raise the price and estimates that for each $1 increase in the price, 1,000 fewer lamps will be sold each month. The manufacturer can produce the lamps at a cost of $4 per lamp. At what price should the manufacturer sell the lamps to generate the greatest possible profit?

37. PROFIT A baseball card store can obtain Mel Schlabotnik rookie cards at a cost of $5 per card. The store has been offering the cards at $10 apiece and, at this price, has been selling 25 cards per month. The store is planning to lower the price to stimulate sales and estimates that for each 25-cent reduction in the price, 5 more cards will be sold each month. At what price should the cards be sold in order to maximize total monthly profit?

38. MARGINAL ANALYSIS A manufacturer estimates that if x units of a particular commodity are produced, the total cost will be $C(x)$ dollars, where

$$C(x) = x^3 - 24x^2 + 350x + 338$$

a. At what level of production will the marginal cost $C'(x)$ be minimized?

b. At what level of production will the average cost $A(x) = \dfrac{C(x)}{x}$ be minimized?

39. COST ANALYSIS It is estimated that the cost of constructing an office building that is n floors high is $C(n) = 2n^2 + 500n + 600$ thousand dollars. How many floors should the building have in order to minimize the average cost per floor? (Remember that your answer should be a whole number.)

40. CONSTRUCTION You wish to use 300 meters of fencing to surround two identical adjacent rectangular plots, as shown in the accompanying figure. How should you do this to make the combined area of the plots as large as possible?

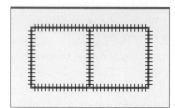

PROBLEM 40

41. MINIMIZING TRAVEL TIME You are standing on the bank of a river that is 1 mile wide and want to get to a town on the opposite bank, 1 mile upstream. You plan to row on a straight line to some point P on the opposite bank and then walk the remaining distance along the bank. To what point P should you row to reach the town in the shortest possible time if you can row at 4 miles per hour and walk at 5 miles per hour?

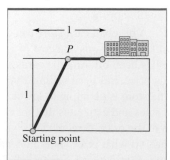

PROBLEM 41

42. REAL ESTATE DEVELOPMENT A real estate developer estimates that if 60 luxury houses are

built in a certain area, the average profit will be $47,500 per house. The average profit will decrease by $500 per house for each additional house built in the area. How many houses should the developer build to maximize the total profit? (Remember, the answer must be an integer.)

43. INVENTORY A manufacturing firm has received an order to make 400,000 souvenir medals commemorating the 200th anniversary of the Louisiana Purchase. The firm owns 20 machines, each of which can produce 200 medals per hour. The cost of setting up the machines to produce the medals is $80 per machine, and the total operating cost is $5.76 per hour. How many machines should be used to minimize the cost of producing the 400,000 medals? (Remember, the answer must be an integer.)

44. ELASTICITY OF DEMAND Suppose that the demand equation for a certain commodity is $q = 60 - 0.1p$ (for $0 \leq p \leq 600$).

a. Express the elasticity of demand as a function of p.

b. Calculate the elasticity of demand when the price is $p = 200$. Interpret your answer.

c. At what price is the elasticity of demand equal to -1?

45. ELASTICITY OF DEMAND Suppose that the demand equation for a certain commodity is $q = 200 - 2p^2$ (for $0 \leq p \leq 10$).

a. Express the elasticity of demand as a function of p.

b. Calculate the elasticity of demand when the price is $p = 6$. Interpret your answer.

c. At what price is the elasticity of demand equal to -1?

46. ELASTICITY AND REVENUE Suppose that $q = 500 - 2p$ units of a certain commodity are demanded when p dollars per unit are charged, for $0 \leq p \leq 250$.

a. Determine where the demand is elastic, inelastic, and of unit elasticity with respect to price.

b. Use the results of part (a) to determine the intervals of increase and decrease of the revenue function and the price at which revenue is maximized.

c. Find the total revenue function explicitly and use its first derivative to determine its intervals of increase and decrease and the price at which revenue is maximized.

d. Graph the demand and revenue functions.

47. DEMAND FOR CRUISE TICKETS A cruise line estimates that when each deluxe view stateroom on a particular cruise is priced at p thousand dollars, then q tickets for staterooms will be demanded by travelers, where $q = 300 - 0.7p^2$.

a. Find the elasticity of demand for the stateroom tickets.

b. When the price is $8,000 ($p = 8$) per stateroom, should the cruise line raise or lower the price in order to increase total revenue?

48. Find constants A, B, and C so that the function $f(x) = Ax^3 + Bx^2 + C$ will have a relative extremum at $(2, 11)$ and an inflection point at $(1, 5)$. Sketch the graph of f.

49. **ARCHITECTURE**   A window is in the form of an equilateral triangle surmounted on a rectangle. The rectangle is of clear glass and transmits twice as much light as does the triangle, which is made of stained glass. If the entire window has a perimeter of 20 feet, find the dimensions of the window that will admit the most light.

50. **WORKER EFFICIENCY**   An efficiency study of the morning shift at a certain factory indicates that an average worker who is on the job at 8:00 A.M. will have assembled $f(x) = -x^3 + 6x^2 + 15x$ transistor radios x hours later. The study indicates further that after a 15-minute coffee break the worker can assemble $g(x) = -\frac{1}{3}x^3 + x^2 + 23x$ radios in x hours. Determine the time between 8:00 A.M. and noon at which a 15-minute coffee break should be scheduled so that the worker will assemble the maximum number of radios by lunchtime at 12:15 P.M. [*Hint:* If the coffee break begins x hours after 8:00 A.M., $4 - x$ hours will remain after the break.]

51. **PRODUCTION CONTROL**   A toy manufacturer produces an inexpensive doll (Floppsy) and an expensive doll (Moppsy) in units of x hundreds and y hundreds, respectively. Suppose that it is possible to produce the dolls in such a way that $y = \dfrac{82 - 10x}{10 - x}$ for $0 \leq x \leq 8$ and that the company receives *twice* as much for selling a Moppsy doll as for selling a Floppsy doll. Find the level of production (both x and y) for which the total revenue derived from selling these dolls is maximized. You can assume that the company sells every doll it produces.

52. **CONSTRUCTION COST**   Oil from an offshore rig located 3 miles from the shore is to be pumped to a location on the edge of the shore that is 8 miles east of the rig. The cost of constructing a pipe in the ocean

from the rig to the shore is 1.5 times more expensive than the cost of construction on land. How should the pipe be laid to minimize cost?

53. **INVENTORY**   Through its franchised stations, an oil company gives out 16,000 road maps per year. The cost of setting up a press to print the maps is $100 for each production run. In addition, production costs are 6 cents per map and storage costs are 20 cents per map per year. The maps are distributed at a uniform rate throughout the year and are printed in equal batches timed so that each arrives just as the preceding batch has been used up. How many maps should the oil company print in each batch to minimize cost?

54. **MARINE BIOLOGY**   When a fish swims upstream at a speed v against a constant current v_w, the energy it expends in traveling to a point upstream is given by a function of the form

$$E(v) = \frac{Cv^k}{v - v_w}$$

where $C > 0$ and $k > 2$ is a number that depends on the species of fish involved.*

a. Show that $E(v)$ has exactly one critical number. Does it correspond to a relative maximum or a relative minimum?

b. Note that the critical number in part (a) depends on k. Let $F(k)$ be the critical number. Sketch the graph of $F(k)$. What can be said about $F(k)$ if k is very large?

55. **INVENTORY**   A manufacturing firm receives raw materials in equal shipments arriving at regular intervals throughout the year. The cost of storing the raw materials is directly proportional to the size of each shipment, while the total yearly ordering cost is inversely proportional to the shipment size. Show that the total cost is lowest when the total storage cost and total ordering cost are equal.

56. **PHYSICAL CHEMISTRY**   In physical chemistry, it is shown that the pressure P of a gas is related to the volume V and temperature T by *van der Waals' equation*:

$$\left(P + \frac{a}{V^2}\right)(V - b) = nRT$$

where a, b, n, and R are constants. The *critical temperature* T_c of the gas is the highest temperature at which the gaseous and liquid phases can exist as separate phases.

---

*E. Batschelet, *Introduction to Mathematics for Life Scientists*, 2nd ed., New York: Springer-Verlag, 1976, p. 280.

a. When $T = T_c$, the pressure P is given as a function $P(V)$ of volume alone. Sketch the graph of $P(V)$.

b. The critical volume V_c is the volume for which $P'(V_c) = 0$ and $P''(V_c) = 0$. Show that $V_c = 3b$.

c. Find the critical pressure $P_c = P(V_c)$ and then T_c in terms of a, b, n, and R.

57. CRYSTALLOGRAPHY   A fundamental problem in crystallography is the determination of the **packing fraction** of a crystal lattice, which is the fraction of space occupied by the atoms in the lattice, assuming that the atoms are hard spheres. When the lattice contains exactly two different kinds of atoms, it can be shown that the packing fraction is given by the formula*

$$f(x) = \frac{K(1 + c^2 x^3)}{(1 + x)^3}$$

where $x = \dfrac{r}{R}$ is the ratio of the radii, r and R, of the two kinds of atoms in the lattice, and c and K are positive constants.

a. The function $f(x)$ has exactly one critical number. Find it and use the second derivative test to classify it as a relative maximum or a relative minimum.

b. The numbers c and K and the domain of $f(x)$ depend on the cell structure in the lattice. For ordinary rock salt: $c = 1$, $K = \dfrac{2\pi}{3}$, and the domain is the interval $(\sqrt{2} - 1) \le x \le 1$. Find the largest and smallest values of $f(x)$.

c. Repeat part (b) for β-cristobalite, for which $c = \sqrt{2}$, $K = \dfrac{\sqrt{3}\pi}{16}$, and the domain is $0 \le x \le 1$.

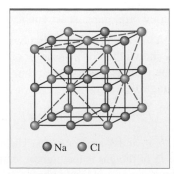

● Na   ● Cl

PROBLEM 57

d. What can be said about the packing fraction $f(x)$ if r is much larger than R? Answer this question by computing $\lim\limits_{x \to \infty} f(x)$

 e. Read the article on which this problem is based, and write a paragraph on how packing factors are computed and used in crystallography.

58. VOTING PATTERNS   After a presidential election, the proportion $h(p)$ of seats in the House of Representatives won by the party of the winning presidential candidate may be modeled by the "cube rule":

$$h(p) = \frac{p^3}{p^3 + (1 - p)^3} \qquad \text{for } 0 \le p \le 1$$

where p is the proportion of the popular vote received by the winning presidential candidate.

a. Find $h'(p)$ and $h''(p)$.

b. Sketch the graph of $h(p)$.

c. In 1964, the Democrat Lyndon Johnson received 61% of the popular vote. What percentage of seats in the House does the cube rule predict should have gone to Democrats? (They actually won 72%.)

 d. For the most part, the cube rule has been extremely accurate for presidential elections since 1900. Use the Internet to research the actual proportions that occurred during these elections, and write a paragraph on your findings.

59. ELIMINATION OF HAZARDOUS WASTE   Certain hazardous waste products have the property that as the concentration of substrate (the substance undergoing change by enzymatic action) increases, there is a toxic inhibition effect. A mathematical model for this behavior is the **Haldane equation**[†]

$$R(S) = \frac{cS}{a + S + bS^2}$$

where R is the specific growth rate of the substance (the rate at which cells divide); S is the substrate concentration; and a, b, and c are positive constants.

a. Sketch the graph of $R(S)$. Does the graph appear to have a highest point? A lowest point? A point of inflection? What happens to the growth rate as S grows larger and larger? Interpret your observations.

 b. Read an article on hazardous waste management, and write a paragraph on how mathemati-

---

*John C. Lewis and Peter P. Gillis, "Packing Factors in Diatomic Crystals," *American Journal of Physics*, Vol. 61, No. 5, 1993, pp. 434–438.

[†]Michael D. La Grega, Philip L. Buckingham, and Jeffrey C. Evans, *Hazardous Waste Management*, New York: McGraw-Hill, 1994, p. 578.

cal models are used to develop methods for eliminating waste. A good place to start is the reference cited here.

60. **FIRE FIGHTING** If air resistance is neglected, it can be shown that the stream of water emitted by a fire hose will have height

$$y = -16(1 + m^2)\left(\frac{x}{v}\right)^2 + mx$$

feet above a point located x feet from the nozzle, where m is the slope of the nozzle and v is the velocity of the stream of water as it leaves the nozzle. Assume v is constant.

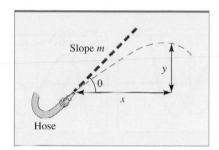

PROBLEM 60

a. Suppose m is also constant. What is the maximum height reached by the stream of water? How far away from the nozzle does the stream reach (that is, what is x when $y = 0$)?

b. If m is allowed to vary, find the slope that allows a firefighter to spray water on a fire from the greatest distance.

c. Suppose the firefighter is $x - x_0$ feet from the base of a building. If m is allowed to vary, what is the highest point on the building that the firefighter can reach with the water from her hose?

61. **URBAN CRIME** A study* of urban crime in Detroit published in 1977 indicated that the average number of monthly property crimes N_1 and personal crimes (such as rape or murder) N_2 were related to the retail price of heroin p (dollars per gram) by the equations

$$N_1 = 3{,}351p^{0.287} \qquad \text{and} \qquad N_2 = 207.8p^{0.349}$$

For simplicity, we assume that there is no overlap between the two kinds of crime.

---

*The formulas given in this problem appear on page 170 of the UMAP module cited in part (e). They were originally presented by L. P. Silverman and N. L. Spruill in the article, "Urban Crime and the Price of Heroin," *Journal of Urban Economics*, Vol. 4, No. 1, 1977, pp. 80–103.

a. Find the elasticity of N_1 with respect to p. By what approximate percentage do property crimes N_1 increase if the price of heroin rises by 2%? By 5%?

b. Answer the questions in part (a) for personal crimes N_2.

c. Let $N = N_1 + N_2$ denote total crime and suppose the price of heroin is \$75 per gram. Find the elasticity of N with respect to p and use it to estimate the percentage of increase in total crime to be expected if the price of heroin increases by 5 percent.

d. For what price p of heroin will a 17 percent increase in the price result in an approximate 5 percent increase in total crime?

e. Read the article by Yves Nievergelt with the intriguing title, "Price Elasticity of Demand: Gambling, Heroin, Marijuana, Whiskey, Prostitution, and Fish," *UMAP Modules 1987: Tools for Teaching,* Arlington, MA: Consortium for Mathematics and Its Applications, Inc., 1988, pp. 153–181. Then write a paragraph on what you think about this type of socioeconomic analysis.

62. **BOTANY** An experimental garden plot contains N annual plants, each of which produces S seeds that are dropped within the same plot. A botanical model measures the number of offspring plants $A(N)$ that survive until the next year by the function

$$A(N) = \frac{NS}{1 + (cN)^p}$$

where c and p are positive constants.

a. For what value of N is $A(N)$ maximized? What is the maximum value? Express your answer in terms of S, c, and p.

b. For what value of N is the offspring survival rate $A'(N)$ minimized?

c. The function $F(N) = A(N)/N$ is called the *net reproductive rate*. It measures the number of surviving offspring per plant. Find $F'(N)$ and use it to show that the greater the number of plants, the lower the number of surviving offspring per plant. This is called the principle of *density-dependent mortality*.

Complete solutions for all EXPLORE! boxes throughout the text can be accessed at the book specific website, www.mhhe.com/hoffmann.

Solution for Explore! on Page 190

Place the function $f(x) = 2x^3 + 3x^2 - 12x - 7$ into Y1, using a bold graphing style, and store $f'(x)$ into Y2, representing the numerical derivative. Using a modified decimal window $[-4.7, 4.7]1$ by $[-20, 20]2$, we obtain the following graphs. Using the tracing feature of your graphing calculator, you can locate where the derivative $f'(x)$ intersects the x axis (the y value displayed in scientific notation $2E^-6 = 0.000002$ can be considered negligible).

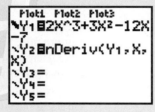 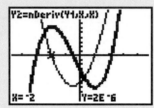 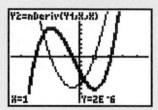

You will observe that the graph of $f(x)$ in bold is decreasing between $x = -2$ and $x = 1$, the same interval over which $f'(x)$ is negative. Note that the graph of $f'(x)$ is below the x axis in this interval. Where $f(x)$ is increasing, $f'(x)$ is positive, specifically, for $x < -2$ and $x > 1$. It appears that where the derivative is zero, $x = -2$ and 1, the graph of $f(x)$ has turns or bends in the curve, indicating possible high or low points.

Solution for Explore! on Page 191

Store $f(x) = x^2/(x - 2)$ into Y1 and $g(x) = x^2/(x - 4)$ into Y2 with a bold graphing style. Use the graphing window $[-9.4, 9.4]1$ by $[-20, 30]5$ to obtain the graphs shown here. How do the two graphs differ? The relative maximum (on the left branch of the graph before the asymptote) occurs at the origin for both graphs. But the relative minimum (on the right branch of the graph) appears to shift from $x = 4$ to $x = 8$. Note that in both graphs the relative maximum is actually smaller than any point in the right branch of the respective graph. The intervals over which $g(x)$ decreases are $(0, 4)$ and $(4, 8)$.

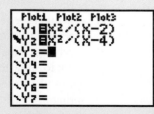 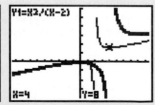 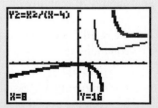

Solution for Explore! on Page 211

Place $f(x) = 3x^4 - 2x^3 - 12x^2 + 18x + 15$ into Y1 of your equation editor using a bold graphing style, and let Y2 and Y3 be the first and second derivatives of Y1, respectively, as shown here. Now deselect Y3 and only consider the graphs of Y1 and Y2. Use the modified decimal window given here. The critical values of $f(x)$ are located at the zeros of $f'(x)$, namely, $x = -1.5$ and 1, the latter a double root.

 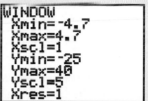 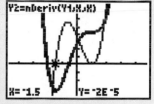

Next select (turn on) Y3 and deselect (turn off) Y2, as shown in the following window. The graph of Y3, the second derivative of $f(x)$, shows that there are two inflection points, at $x = -2/3$ and $x = 1$. The root finding feature of your graphing calculator can verify that these are the roots of $f''(x)$, coinciding with the locations at which $f(x)$ changes concavity.

 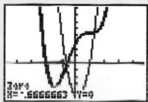 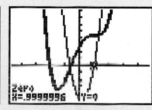

Solution for Explore! on Page 226

Store $f(x) = 3x^2/(x^2 + 2x - 15)$ into Y1 of your equation editor. As you trace the graph for large x values, $f(x)$ hovers below but appears to approach the value 3, visually demonstrating that $y = 3$ is a horizontal asymptote. Do you think that there is a relative minimum on the right branch of the graph of $f(x)$? Also how far out on the x axis do you need to trace to obtain a y value above 2.9? The table on the far right should help.

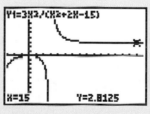 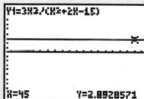 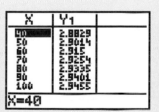

EXPLORE! UPDATE

THINK ABOUT IT

MODELING AN EPIDEMIC: DEATHS DUE TO AIDS IN THE UNITED STATES

Mathematical models are used extensively to study infectious disease. In particular, during the past 20 years, many different mathematical models for the number of deaths from AIDS per year have been developed. In this essay, we will describe some of the simplest such models and discuss their accuracy. Before doing so, however, we will provide some background.

From 1980 until 1995, the number of people dying from AIDS in the United States increased rapidly from close to zero to more than 51,000. In the mid-1990s it appeared that the number of these deaths would continue to increase each year. However, with the advent of new treatments and drugs, the number of annual deaths from AIDS in the United States has steadily decreased, with less than 16,000 such deaths reported in 2001. That said, the future trend for the total number of annual AIDS deaths in the United States is not clear. The virus that causes AIDS has developed resistance to some of the drugs that have been effective in treating it, and in addition, there are indications that many susceptible people are not taking adequate precautions to prevent the spread of the disease. Will these factors result in an increase in AIDS-related deaths, or will the many new drugs currently being developed result in a continuation of the current downward trend?

Year	United States AIDS Deaths
1989	27,408
1990	31,120
1991	36,175
1992	40,587
1993	45,850
1994	50,842
1995	51,670
1996	38,296
1997	22,245
1998	18,823
1999	18,249
2000	16,672
2001	15,603

The total number of deaths from AIDS in the United States each year from 1989 to 2001 is shown in the table. (SOURCE: Centers for Disease Control & Prevention, National Center for HIV, STD, and TB Prevention.)

The goal in modeling the number of AIDS deaths in the United States per year is to find a relatively simple function $f(t)$ that provides a close approximation to the number of AIDS deaths in the United States, in the year t after the year 1989. One of the simplest approaches to construct such functions is to use *polynomial regression,* a technique that produces best-fit polynomials of specified degree that are good approximations to observed data. When using polynomial regression, we first need to specify the degree of the polynomial we want to use to approximate the number of AIDS deaths per year in the United States. The higher the degree, the better the approximation, but the more complicated the function. Suppose we decide to use a polynomial of degree 3, so that $f(t)$ has the general form $f(t) = at^3 + bt^2 + ct + d$. The coefficients $a, b, c,$ and d of the best-fitting cubic curve we seek are then determined by requiring the sum of squares of the vertical distances between data points in the table and corresponding points on the cubic polynomial $y = f(t)$ to be as small as possible. Calculus methods for carrying out this optimization procedure are developed in Section 7.4. In practice, however, regression polynomials are almost always found using a computer or graphing calculator. There is a detailed description of how to use your calculator for finding regression polynomials in the Calculator Introduction at the front of the text.

It is important to realize that there are no underlying biological reasons why a polynomial should provide a good approximation to the number of AIDS deaths in the United States during a year. Nevertheless, we begin with such models because they are easy to construct. More sophisticated models, including some that are based on biological reasoning, can be constructed using exponential functions studied in Chapter 4.

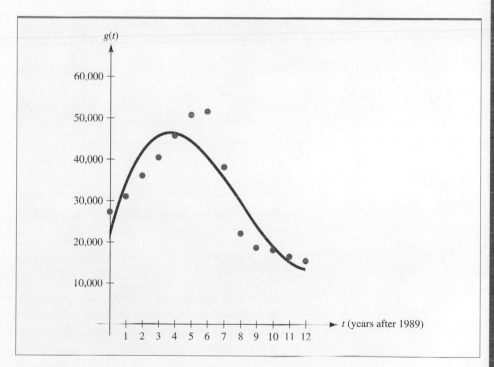

FIGURE 1   The best fitting cubic curve.

We will take a look at both the degree three (cubic) and the degree four (quartic) polynomials produced with polynomial regression using our 13 points (t_j, d_j), where d_j is the number of AIDS deaths in the United States that occurred in the year

j after 1989. (Our data corresponds to $t_j = 0, 1, \ldots, 12$.) Computations show that the best-fit cubic approximation to our data is

$$g(t) = 107.0023t^3 - 2,565.0889t^2 + 14,682.6031t + 21,892.5055$$

This cubic polynomial is graphed together with the data points in Figure 1. Similar computations show that the best-fit quartic approximation to our data is

$$h(t) = 29.6957t^4 - 605.6941t^3 + 2,801.3456t^2 + 1,599.5330t + 26,932.2873$$

which is graphed together with the given data points in Figure 2. These best-fitting graphs can then be used to analyze the given data and to make predictions. A few examples of this kind of analysis are provided in the accompanying questions.

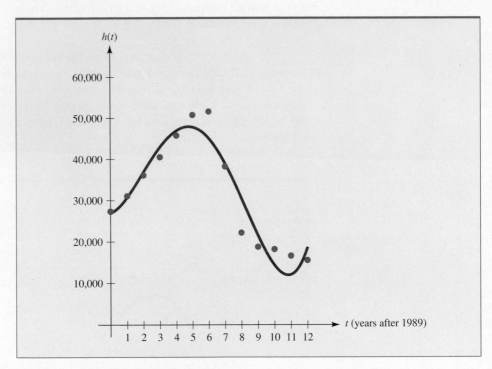

FIGURE 2   The best-fitting quartic curve.

Questions

1. For which time period does the best-fit cubic polynomial $g(t)$ do the best job of approximating the number of AIDS deaths in the United States? For what time period does it do the poorest job? Answer the same questions for the best-fit quartic polynomial $h(t)$.

2. Find all critical numbers of the best-fit cubic polynomial $g(t)$ for the given time interval $0 \le t \le 12$. What are the largest and smallest number of deaths due to AIDS in the United States predicted by this model for the period 1989–2001? Answer the same questions for the best-fit quartic polynomial $h(t)$. Comment on these predictions.

3. Estimate the number of AIDS deaths in the United States in 2002 projected by the best-fit cubic approximation $g(t)$ and the best-fit quartic approximation $h(t)$. If you can find the actual data, determine which produces the better prediction. What do these two models predict about the number of AIDS deaths in the United States in the years beyond 2002?

4. Why would a linear equation, such as that produced by linear regression, not do a good job of approximating the number of AIDS deaths in the United States from 1989 until 2001?

 5. Use your graphing calculator to generate a best-fit quadratic (degree 2) polynomial $Q(t)$ for the given data. Using your $Q(t)$, answer questions 2 and 3. Comment on how well your polynomial fits the given data.

6. Could we ever find a polynomial that exactly agrees with the number of U.S. AIDS deaths for every year from 1989 until 2001? If there is such a polynomial, how large must we take the degree to be sure that we have found it?

CHAPTER 4

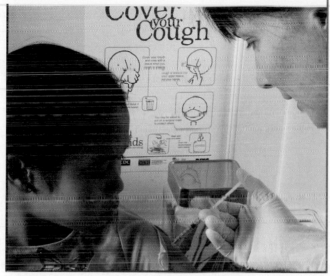

Exponential functions can be used for describing the effect of medication.

EXPONENTIAL AND LOGARITHMIC FUNCTIONS

SECTION 4.1 | Exponential Functions

A population $Q(t)$ is said to grow **exponentially** if whenever it is measured at equally spaced time intervals, the population at the end of any particular interval is a fixed constant (greater than 1) times the population at the end of the previous interval. For instance, according to the United Nations, in the year 2000, the population of the world was 6.1 billion people and was growing at an annual rate of about 1.4%. If this pattern were to continue, then every year, the population would be 1.014 times the population of the previous year. Thus, if $P(t)$ is the world population (in billions) t years after the base year 2000, the population would grow as follows:

2000	$P(0) = 6.1$
2001	$P(1) = 6.1(1.014) = 6.185$
2002	$P(2) = 6.185(1.014) = [6.1(1.014)](1.014) = 6.1(1.014)^2 = 6.272$
2003	$P(3) = 6.272(1.014) = [6.1(1.014)^2](1.014) = 6.1(1.014)^3 = 6.360$
$\vdots$	$\vdots$
$2000 + t$	$P(t) = 6.1(1.014)^t$

The graph of $P(t)$ is shown in Figure 4.1a. Notice that according to this model, the world population grows gradually at first but doubles after about 50 years (to 12.22 billion in 2050).

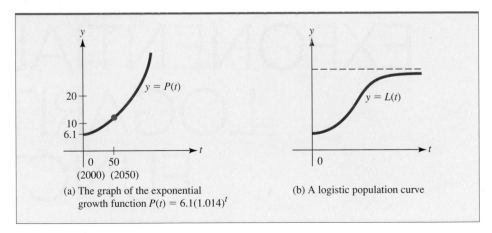

(a) The graph of the exponential growth function $P(t) = 6.1(1.014)^t$

(b) A logistic population curve

FIGURE 4.1 Two models for population growth.

Exponential population models are sometimes referred to as *Malthusian,* after Thomas Malthus (1766–1834), a British economist who predicted mass starvation would result if a population grows exponentially while the food supply grows at a constant rate (linearly). Fortunately, world population does not continue to grow exponentially as predicted by Malthus's model, and models that take into account various restrictions on the growth rate actually provide more accurate predictions. The population curve that results from one such model, the so-called *logistic* model, is shown in Figure 4.1b. Note how the logistic growth curve rises steeply at first, like an exponential curve, but then eventually turns over and flattens out as environmental factors act to brake the growth rate. We will examine logistic curves in Section 4.4 (Example 4.4.5) and again in Chapter 6 as part of a more detailed study of population models.

A function of the general form $f(x) = b^x$, where b is a positive number, is called an **exponential function.** Such functions can be used to describe exponential and logistic growth and a variety of other important quantities. For instance, exponential functions are used in demography to forecast population size, in finance to calculate the value of investments, in archaeology to date ancient artifacts, in psychology to study learning patterns, and in industry to estimate the reliability of products.

In this section, we will explore the properties of exponential functions and introduce a few basic models in which such functions play a prominent role. Additional applications such as the logistic model are examined in subsequent sections.

Review of Exponential Notation and Properties

Working with exponential functions requires the use of exponential notation and the algebraic laws of exponents. Worked examples and practice problems involving this notation can be found in the Algebra Review in Appendix A. Here is a brief summary of the major facts you will need.

Definition of b^n for Rational Values of n (and $b > 0$) ▪ Integer powers: If n is a positive integer,

$$b^n = \underbrace{b \cdot b \cdots b}_{n \; factors}$$

Fractional powers: If n and m are positive integers,

$$b^{n/m} = (\sqrt[m]{b})^n = \sqrt[m]{b^n}$$

where $\sqrt[m]{b}$ denotes the positive mth root.

Negative powers: $b^{-n} = \dfrac{1}{b^n}$

Zero power: $b^0 = 1$

For example,

$$3^4 = 3 \cdot 3 \cdot 3 \cdot 3 = 81 \qquad 3^{-4} = \frac{1}{3^4} = \frac{1}{81}$$

$$4^{1/2} = \sqrt{4} = 2 \qquad 4^{3/2} = (\sqrt{4})^3 = 2^3 = 8$$

$$4^{-3/2} = \frac{1}{4^{3/2}} = \frac{1}{8} \qquad 27^{-2/3} = \frac{1}{(\sqrt[3]{27})^2} = \frac{1}{3^2} = \frac{1}{9}$$

EXPLORE!

Graph $y = (-1)^x$ using a modified decimal window [−4.7, 4.7]1 by [−1, 7]1. Why does the graph appear with dotted points? Next set b to decimal values, $b = 0.5, 0.25,$ and 0.1, and graph $y = b^x$ for each case. Explain the behavior of these graphs.

We know what is meant by b^r for any rational number r, but if we try to graph $y = b^x$ there will be a "hole" in the graph for each value of x that is not rational, such as $x = \sqrt{2}$. However, using methods beyond the scope of this book, it can be shown that because there are "so many" rational numbers, there can be only one unbroken curve that passes through all the points (r, b^r) for r rational. In other words, there exists a unique continuous function $f(x)$ that is defined for all real numbers x and is equal to b^r when $x = r$ is rational. It is this function that we define as $f(x) = b^x$.

Exponential Functions ▪ If b is a positive number other than 1 ($b > 0$, $b \neq 1$), there is a unique function called the *exponential function* with base b that is defined by

$$f(x) = b^x \qquad \text{for every real number } x$$

To get a feeling for the appearance of the graph of an exponential function, consider Example 4.1.1.

EXAMPLE │ 4.1.1

Sketch the graphs of $y = 2^x$ and $y = \left(\frac{1}{2}\right)^x$.

Solution

Begin by constructing a table of values for $y = 2^x$ and $y = \left(\frac{1}{2}\right)^x$:

x	−15	−10	−1	0	1	3	5	10	15
$y = 2^x$	0.00003	0.001	0.5	1	2	8	32	1,024	32,768
$y = \left(\frac{1}{2}\right)^x$	32,768	1,024	2	1	0.5	0.125	0.313	0.001	0.00003

The pattern of values in this table suggests that the functions $y = 2^x$ and $y = \left(\frac{1}{2}\right)^x$ have the following features:

The function $y = 2^x$

always increasing

$$\lim_{x \to -\infty} 2^x = 0$$

$$\lim_{x \to +\infty} 2^x = +\infty$$

The function $y = \left(\frac{1}{2}\right)^x$

always decreasing

$$\lim_{x \to -\infty} \left(\frac{1}{2}\right)^x = +\infty$$

$$\lim_{x \to +\infty} \left(\frac{1}{2}\right)^x = 0$$

Using this information, we sketch the graphs shown in Figure 4.2. Notice that each graph has (0, 1) as its y intercept, has the x axis as a horizontal asymptote, and appears to be concave upward for all x. The graphs also appear to be reflections of one another in the y axis. You are asked to verify this observation in Problem 62.

EXPLORE!

Graph $y = b^x$, for $b = 1, 2, 3$, and 4, using the modified decimal window [−4.7, 4.7]1 by [−1, 7]1. Explain what you observe. Conjecture and then check where the graph of $y = 4^x$ will lie relative to that of $y = 2^x$ and $y = 6^x$. Where does $y = e^x$ lie, assuming e is a value between 2 and 3?

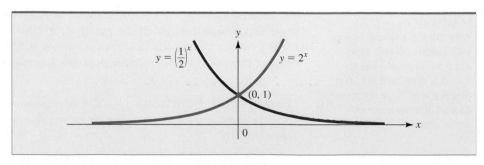

FIGURE 4.2 The graphs of $y = 2^x$ and $y = \left(\frac{1}{2}\right)^x$.

Figure 4.3 shows graphs of various members of the family of exponential functions $y = b^x$. Notice that the graphs of $y = 2^x$ and $y = \left(\dfrac{1}{2}\right)^x$ shown in Figure 4.2 are typical of those of $y = b^x$ for $b > 1$ and $0 < b < 1$, respectively.

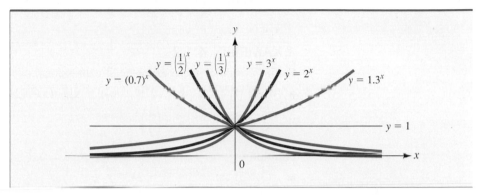

FIGURE 4.3   Graphs of the exponential form $y = b^x$.

NOTE   Students often confuse the *power* function $p(x) = x^b$ with the *exponential* function $f(x) = b^x$. Remember that in x^b, the variable x is the base and the exponent b is constant, while in b^x, the base b is constant and the variable x is the exponent. The graphs of $y = x^2$ and $y = 2^x$ are shown in Figure 4.4. Notice that after the crossover point $(4, 16)$, the exponential curve $y = 2^x$ rises much more steeply than the power curve $y = x^2$. For instance, when $x = 10$, the y value on the power curve is $y = 10^2 = 100$, while the corresponding y value on the exponential curve is $y = 2^{10} = 1{,}024$.   ■

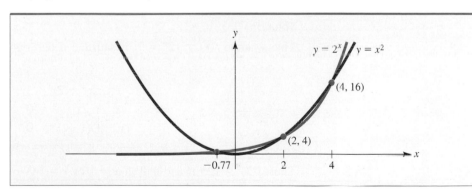

FIGURE 4.4   Comparing the power curve $y = x^2$ with the exponential curve $y = 2^x$.

In general, exponential functions obey these rules.

Basic Properties of Exponential Functions   ■   For bases a, b ($a > 0$, $b > 0$) and any real numbers x, y, we have

The **equality rule:** $b^x = b^y$ if and only if $x = y$

The **product rule:** $b^x b^y = b^{x+y}$

The **quotient rule:** $\dfrac{b^x}{b^y} = b^{x-y}$

(continued)

The **power rule:** $(b^x)^y = b^{xy}$

The **multiplication rule:** $(ab)^x = a^x b^x$

The **division rule:** $\left(\dfrac{a}{b}\right)^x = \dfrac{a^x}{b^x}$

EXAMPLE | 4.1.2

Evaluate each of these exponential expressions:

a. $(3)^2(3)^3$ **b.** $(2^3)^2$ **c.** $(5^{1/3})(2^{1/3})$

d. $\dfrac{2^3}{2^5}$ **e.** $\left(\dfrac{4}{7}\right)^3$

Solution

a. $(3)^2(3)^3 = 3^{2+3} = 3^5 = 243$

b. $(2^3)^2 = 2^{(3)(2)} = 2^6 = 64$

c. $(5^{1/3})(2^{1/3}) = [(5)(2)]^{1/3} = 10^{1/3} = \sqrt[3]{10}$

d. $\dfrac{2^3}{2^5} = 2^{3-5} = 2^{-2} = \dfrac{1}{4}$

e. $\left(\dfrac{4}{7}\right)^3 = \dfrac{4^3}{7^3} = \dfrac{64}{343}$

EXAMPLE | 4.1.3

If $f(x) = 5^{x^2 + 2x}$, find all values of x such that $f(x) = 125$.

Solution

The equation $f(x) = 125 = 5^3$ is satisfied if and only if

$$5^{x^2 + 2x} = 5^3$$
$$x^2 + 2x = 3 \qquad \textit{since } b^x = b^y \textit{ only when } x = y$$
$$x^2 + 2x - 3 = 0$$
$$(x - 1)(x + 3) = 0 \qquad \textit{factor}$$
$$x = 1, \; x = -3$$

Thus, $f(x) = 125$ if and only if $x = 1$ or $x = -3$.

The Natural Exponential Base e

In algebra, it is common practice to use the base $b = 10$ or, in some cases, $b = 2$, but in calculus, it turns out to be more convenient to use a number denoted by e and defined by the limit

$$e = \lim_{n \to +\infty} \left(1 + \frac{1}{n}\right)^n$$

"Hold on!" you say, "That limit has to be 1, since $1 + \dfrac{1}{n}$ certainly tends to 1 as n increases without bound, and $1^n = 1$ for any n." Not so. The limit process does not work this way, as you can see from this table:

n	10	100	1,000	10,000	100,000
$\left(1 + \frac{1}{n}\right)^n$	2.59374	2.70481	2.71692	2.71815	2.71827

EXPLORE!

 Store $\left(1 + \frac{1}{x}\right)^x$ into Y1 of the function editor and examine its graph. Trace the graph to the right for large values of x. What number is y approaching as x gets larger and larger? Try using the table feature of the graphing calculator, setting both the initial value and the incremental change first to 10, then successively to 1,000 and 100,000. Estimate the limit to five decimal places. Now do the same as x approaches $-\infty$ and observe this limit.

The number e is one of the most important numbers in all mathematics, and its value has been computed with great precision. To five decimal places, its value is

$$e = \lim_{n \to +\infty} \left(1 + \frac{1}{n}\right)^n = 2.71828\ldots$$

To compute e^N for a particular number N, you will use a table of exponential values or, more likely, the "e^X" key on your calculator. For example, to find $e^{1.217}$ press the e^X key, and then enter the number 1.217 to obtain $e^{1.217} \approx 3.37704$.

Continuous Compounding of Interest

The number e is called the "natural exponential base," but it may seem anything but "natural" to you. As an illustration of how this number appears in practical situations, we use it to describe the accounting practice known as continuous compounding of interest.

First, let us review the basic ideas behind compound interest. Suppose a sum of money is invested and the interest is compounded only once. If P is the initial investment (the *principal*) and r is the interest rate (expressed as a decimal), the balance B after the interest is added will be

$$B = P + Pr = P(1 + r) \quad \text{dollars}$$

That is, to compute the balance at the end of an interest period, you multiply the balance at the beginning of the period by the expression $1 + r$.

At most banks, interest is compounded more than once a year. The interest that is added to the account during one period will itself earn interest during the subsequent periods. If the annual interest rate is r and interest is compounded k times per year, then the year is divided into k equal compounding periods and the interest rate in each period is $\frac{r}{k}$. Hence, the balance at the end of the first period is

$$P_1 = P + \underbrace{P\left(\frac{r}{k}\right)}_{\text{principal}} = \underbrace{P\left(1 + \frac{r}{k}\right)}_{\text{interest}}$$

principal   interest

At the end of the second period, the balance is

$$P_2 = P_1 + P_1\left(\frac{r}{k}\right) = P_1\left(1 + \frac{r}{k}\right)$$

$$= \left[P\left(1 + \frac{r}{k}\right)\right]\left(1 + \frac{r}{k}\right) = P\left(1 + \frac{r}{k}\right)^2$$

and, in general, the balance at the end of the mth period is

$$P_m = P\left(1 + \frac{r}{k}\right)^m$$

Since there are k periods in a year, the balance after 1 year is

$$P\left(1 + \frac{r}{k}\right)^k$$

At the end of t years, interest has been compounded kt times and the balance is given by the function

$$B(t) = P\left(1 + \frac{r}{k}\right)^{kt}$$

To summarize:

EXPLORE!

Suppose you have $1,000 to invest. Which is the better investment, 5% compounded monthly for 10 years or 6% compounded quarterly for 10 years? Write the expression 1,000(1 + R/K)^(K∗T) on the Home Screen and evaluate after storing appropriate values for *R*, *K*, and *T*.

Compound Interest   ■   If P dollars are invested at an annual interest rate r (expressed as a decimal) and interest is compounded k times per year, the balance (future value) $B(t)$ after t years will be

$$B(t) = P\left(1 + \frac{r}{k}\right)^{kt} \quad \text{dollars}$$

   As the frequency with which interest is compounded increases, the corresponding balance $B(t)$ also increases. Hence, a bank that compounds interest frequently may attract more customers than one that offers the same interest rate but compounds interest less often. But what happens to the balance at the end of t years as the compounding frequency increases without bound? More specifically, what will the balance be at the end of t years if interest is compounded not quarterly, not monthly, not daily, but continuously? In mathematical terms, this question is equivalent to asking what happens to the expression $P\left(1 + \frac{r}{k}\right)^{kt}$ as k increases without bound. The answer turns out to involve the number e. Here is the argument.

   To simplify the calculation, let $n = \frac{k}{r}$. Then, $k = nr$ and so

$$P\left(1 + \frac{r}{k}\right)^{kt} = P\left(1 + \frac{1}{n}\right)^{nrt} = P\left[\left(1 + \frac{1}{n}\right)^n\right]^{rt}.$$

Since n increases without bound as k does, and since $\left(1 + \frac{1}{n}\right)^n$ approaches e as n increases without bound, it follows that the balance after t years is

$$B(t) = \lim_{k \to +\infty} P\left(1 + \frac{r}{k}\right)^{kt} = P\left[\lim_{n \to +\infty}\left(1 + \frac{1}{n}\right)^n\right]^{rt} = Pe^{rt}$$

To summarize:

Continuously Compounded Interest   ■   If P dollars are invested at an annual interest rate r (expressed as a decimal) and interest is compounded continuously, the balance (*future value*) $B(t)$ after t years will be

$$B(t) = Pe^{rt} \quad \text{dollars}$$

EXPLORE!

Write the expressions

$$P*(1 + R/K)^{\wedge}(K*T)$$

and

$$P*e^{\wedge}(R*T)$$

on the Home Screen and compare these two expressions for the *P*, *R*, *T*, and *K* values in Example 4.1.4. Repeat using the same values for *P*, *R*, and *K*, but with *T* = 15 years.

EXAMPLE | 4.1.4

Suppose $1,000 is invested at an annual interest rate of 6%. Compute the balance after 10 years if the interest is compounded

a. Quarterly   **b.** Monthly   **c.** Daily   **d.** Continuously

Solution

a. To compute the balance after 10 years if the interest is compounded quarterly, use the formula $B(t) = P\left(1 + \dfrac{r}{k}\right)^{kt}$, with $t = 10$, $P = 1,000$, $r = 0.06$, and $k = 4$:

$$B(10) = 1,000\left(1 + \frac{0.06}{4}\right)^{40} \approx \$1,814.02$$

b. This time, take $t = 10$, $P = 1,000$, $r = 0.06$, and $k = 12$ to get

$$B(10) = 1,000\left(1 + \frac{0.06}{12}\right)^{120} \approx \$1,819.40$$

c. Take $t = 10$, $P = 1,000$, $r = 0.06$, and $k = 365$ to obtain

$$B(10) = 1,000\left(1 + \frac{0.06}{365}\right)^{3,650} \approx \$1,822.03$$

d. For continuously compounded interest use the formula $B(t) = Pe^{rt}$, with $t = 10$, $P = 1,000$, and $r = 0.06$:

$$B(10) = 1,000e^{0.6} \approx \$1,822.12$$

This value, $1,822.12, is an upper bound for the possible balance. No matter how often interest is compounded, $1,000 invested at an annual interest rate of 6% cannot grow to more than $1,822.12 in 10 years.

Present Value

The quantity $B(t)$ that we have called the "balance at the end of t years when P dollars are invested" is sometimes called the *future value* of P dollars in t years. Turning things around, the *present value* of B dollars payable in t years is the amount P that should be invested today (as principal) to obtain a balance $B(t)$ at the end of the t-year period.

To derive a formula for present value, we need only solve an appropriate future value formula for P. In particular, if the investment is compounded k times per year at an annual rate of r, then

$$B = P\left(1 + \frac{r}{k}\right)^{kt}$$

Just-In-Time **Review**

If $C \neq 0$, the equation

$$A = PC$$

can be solved for P by multiplying by

$$\frac{1}{C} = C^{-1}$$

so

$$P = AC^{-1}$$

and the present value of B dollars in t years is obtained by multiplying both sides of the equation by $\left(1 + \dfrac{r}{k}\right)^{-kt}$:

$$P = B\left(1 + \frac{r}{k}\right)^{kt}$$

Likewise, if the compounding is continuous, then

$$B = Pe^{rt}$$

and the present value is given by

$$P = Be^{-rt}$$

To summarize:

Present Value ■ The **present value** of B dollars in t years invested at the annual rate r compounded k times per year is given by

$$P = B\left(1 + \frac{r}{k}\right)^{-kt}$$

If interest is compounded continuously at the same rate, the present value in t years is given by

$$P = Be^{-rt}$$

EXPLORE!

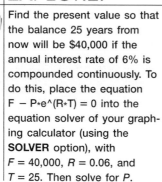

Find the present value so that the balance 25 years from now will be $40,000 if the annual interest rate of 6% is compounded continuously. To do this, place the equation $F - P*e^{\wedge}(R*T) = 0$ into the equation solver of your graphing calculator (using the **SOLVER** option), with $F = 40,000$, $R = 0.06$, and $T = 25$. Then solve for P.

EXAMPLE 4.1.5

Sue is about to enter college. When she graduates 4 years from now, she wants to take a trip to Europe that she estimates will cost $5,000. How much should she invest now at 7% to have enough for the trip if interest is compounded:

a. Quarterly **b.** Continuously

Solution

The required future value is $F = \$5,000$ in $t = 4$ years with $r = 0.07$.

a. If the compounding is quarterly, then $k = 4$ and the present value is

$$P = 5,000\left(1 + \frac{0.07}{4}\right)^{-4(4)} = \$3,788.08$$

b. For continuous compounding, the present value is

$$P = 5,000e^{-0.07(4)} = \$3,778.92$$

Thus, Sue would have to invest about $9 more if interest is compounded quarterly than if the compounding is continuous.

Exponential Growth and Decay

A quantity $Q(t)$ that obeys a law of the form $Q(t) = Q_0e^{kt}$ with $k > 0$ is said to experience **exponential growth.** Similarly, if $Q(t) = Q_0e^{-kt}$ for $k > 0$, the quantity Q experiences **exponential decay.** Many important quantities in business and economics, and the physical, social, and biological sciences can be modeled in terms of exponential growth or decay. For example, if interest is compounded continuously, the future value $B(t) = Pe^{rt}$ grows exponentially, and in the absence of restrictions, population increases exponentially. Radioactive substances decay exponentially, as do the present value of an investment, the amount of drug in a person's bloodstream, and sales of certain commodities once advertising is discontinued.

Typical exponential growth and decay graphs are shown in Figure 4.5. It is customary to display such graphs only for nonnegative time ($t \geq 0$). Note that in both cases, when $t = 0$

$$Q(0) = Q_0e^0 = Q_0$$

so the graph "begins" at Q_0.

Exponential Growth and Decay ■ A quantity $Q(t)$ *grows exponentially* if $Q(t) = Q_0 e^{kt}$ for $k > 0$ and *decays exponentially* if $Q(t) = Q_0 e^{-kt}$ for $k > 0$.

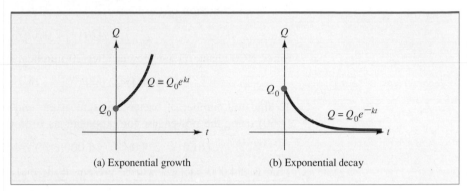

FIGURE 4.5 Exponential change.

EXAMPLE | 4.1.6

The concentration of drug in a patient's bloodstream t hours after an injection is given by

$$C(t) = 3.97e^{-0.9t}$$

milligrams per milliliter (mg/ml). What is the average rate of change of drug concentration during the second hour?

Solution

During the second hour (that is, from $t = 1$ to $t = 2$), the concentration changes by $C(2) - C(1)$, so the average rate of change during this time period is given by

$$A = \frac{C(2) - C(1)}{2 - 1}$$
$$= \frac{3.97e^{-0.9(2)} - 3.97e^{-0.9(1)}}{1} = 3.97(e^{-1.8} - e^{-0.9})$$
$$= 3.97(0.1653 - 0.4066)$$
$$- -0.9580$$

Thus, the concentration of drug decreases at the average rate of roughly 0.96 mg/ml per hour during the second hour after the injection.

EXPLORE!

Refer to Example 4.1.7. In the equation solver of your graphing calculator place the equation 0 = Q − 2,000e^(K∗T). With Q = 6,000 and T = 20, find the value of K. Then change T to 60 and find the corresponding value for Q.

EXAMPLE | 4.1.7

Biologists have determined that when sufficient space and nutrients are available, the number of bacteria in a culture grows exponentially. Suppose that 2,000 bacteria are initially present in a certain culture and that 6,000 are present 20 minutes later. How many bacteria will be present at the end of 1 hour?

Solution

Let $Q(t)$ denote the number of bacteria present after t minutes. Since the number of bacteria grows exponentially and since 2,000 bacteria were initially present, you know that Q is a function of the form

$$Q(t) = 2,000e^{kt}$$

Since 6,000 bacteria are present after 20 minutes, it follows that

$$6,000 = 2,000e^{20k} \qquad \text{or} \qquad e^{20k} = 3$$

To find the number of bacteria present at the end of 1 hour (60 minutes), compute $Q(60)$ using the power law for exponents as follows:

$$Q(60) = 2,000e^{60k} = 2,000(e^{20k})^3 = 2,000(3)^3 = 54,000$$

That is, 54,000 bacteria will be present at the end of 1 hour.

NOTE In Example 4.1.7, you used the given condition $e^{20k} = 3$ to solve the exponential equation $2,000e^{60k} = 54,000$. In general, however, you will solve such an equation for k using **logarithms,** which will be developed in Section 4.2. ∎

PROBLEMS 4.1

 In Problems 1 and 2, use your calculator to find the indicated power of e. (Round off your answers to three decimal places.)

1. $e^2, e^{-2}, e^{0.05}, e^{-0.05}, e^0, e, \sqrt{e}$, and $\dfrac{1}{\sqrt{e}}$

2. $e^3, e^{-1}, e^{0.01}, e^{-0.1}, e^2, e^{-1/2}, e^{1/3}$, and $\dfrac{1}{\sqrt[3]{e}}$

3. Sketch the curves $y = 3^x$ and $y = 4^x$ on the same set of axes.

4. Sketch the curves $y = \left(\dfrac{1}{3}\right)^x$ and $y = \left(\dfrac{1}{4}\right)^x$ on the same set of axes.

In Problems 5 through 12, evaluate the given expressions.

5. a. $27^{2/3}$

    b. $\left(\dfrac{1}{9}\right)^{3/2}$

6. a. $(-128)^{3/7}$

    b. $\left(\dfrac{27}{64}\right)^{2/3}\left(\dfrac{64}{25}\right)^{3/2}$

7. a. $8^{2/3} + 16^{3/4}$

    b. $\left(\dfrac{27 + 36}{121}\right)^{3/2}$

8. a. $(2^3 - 3^2)^{11/7}$

    b. $(27^{2/3} + 8^{4/3})^{-3/2}$

9. a. $(3^3)(3^{-2})$

    b. $(4^{2/3})(2^{2/3})$

10. a. $\dfrac{5^2}{5^3}$

    b. $\left(\dfrac{\pi^2}{\sqrt{\pi}}\right)^{4/3}$

11. a. $(3^2)^{5/2}$

    b. $(e^2e^{3/2})^{4/3}$

12. a. $\dfrac{(3^{1.2})(3^{2.7})}{3^{4.1}}$

    b. $\left(\dfrac{16}{81}\right)^{1/4}\left(\dfrac{125}{8}\right)^{-2/3}$

In Problems 13 through 18, use the properties of exponents to simplify the given expressions.

13. **a.** $(27x^6)^{2/3}$
 b. $(8x^2y^3)^{1/3}$

14. **a.** $(x^{1/3})^{3/2}$
 b. $(x^{2/3})^{-3/4}$

15. **a.** $\dfrac{(x+y)^0}{(x^2y^3)^{1/6}}$
 b. $(x^{1.1}y^2)(x^2+y^3)^0$

16. **a.** $(-2t^{-3})(3t^{2/3})$
 b. $(t^{-2/3})(t^{3/4})$

17. **a.** $(t^{5/6})^{-6/5}$
 b. $(t^{-3/2})^{-2/3}$

18. **a.** $(x^2y^{-3}z)^3$
 b. $\left(\dfrac{x^3y^{-2}}{z^4}\right)^{1/6}$

In Problems 19 through 24, find all real numbers x that satisfy the given equation.

19. $4^{2x-1} = 16$

20. $3^x 2^{2x} = 144$

21. $2^{3-x} = 4^x$

22. $4^x(\frac{1}{2})^{3x} = 8$

23. $(2.14)^{x-1} = (2.14)^{1-x}$

24. $(3.2)^{2x-3} = (3.2)^{2-x}$

In Problems 25 and 26, find the values of the constants C and b so that the curve $y = Cb^x$ contains the indicated points.

25. $(2, 12)$ and $(3, 24)$

26. $(2, 3)$ and $(3, 9)$

27. **COMPOUND INTEREST**   Suppose $1,000 is invested at an annual interest rate of 7%. Compute the balance after 10 years if the interest is compounded:
 a. Annually
 b. Quarterly
 c. Monthly
 d. Continuously

28. **COMPOUND INTEREST**   Suppose $5,000 is invested at an annual interest rate of 10%. Compute the balance after 10 years if the interest is compounded:
 a. Annually
 b. Semiannually
 c. Daily (using 365 days per year)
 d. Continuously

29. **PRESENT VALUE**   How much money should be invested today at 7% compounded quarterly so that it will be worth $5,000 in 5 years?

30. **PRESENT VALUE**   How much money should be invested today at an annual interest rate of 7% compounded continuously so that 20 years from now it will be worth $20,000?

31. **PRESENT VALUE**   How much money should be invested now at 7% to obtain $9,000 in 5 years if interest is compounded:

 a. Quarterly
 b. Continuously

32. **PRESENT VALUE**   What is the present value of $10,000 over a 5-year period of time if interest is compounded continuously at an annual rate of 7%? What is the present value of $20,000 under the same conditions?

33. **POPULATION GROWTH**   It is projected that t years from now, the population of a certain country will be $P(t) = 50e^{0.02t}$ million.
 a. What is the current population?
 b. What will the population be 30 years from now?

34. **POPULATION GROWTH**   It is estimated that the population of a certain country grows exponentially. If the population was 60 million in 1997 and 90 million in 2002, what will the population be in 2012?

35. Find $f(2)$ if $f(x) = e^{kx}$ and $f(1) = 20$.

36. Find $f(9)$ if $f(x) = e^{kx}$ and $f(3) = 2$.

37. Find $f(8)$ if $f(x) = A2^{kx}$, $f(0) = 20$, and $f(2) = 40$.

38. **COMPOUND INTEREST**   A sum of money is invested at a certain fixed interest rate, and the interest is compounded continuously. After 10 years, the money has doubled. How will the balance at the end of 20 years compare with the initial investment?

39. **COMPOUND INTEREST** A sum of money is invested at a certain fixed interest rate, and the interest is compounded quarterly. After 15 years, the money has doubled. How will the balance at the end of 30 years compare with the initial investment?

40. **GROWTH OF GDP** The Gross Domestic Product (GDP) of a certain country was $500 billion at the beginning of the year 2000 and increases at the rate of 2.7% per year.
    a. Express the GDP of this country as a function of the number of years t after 2000.
    b. What does this formula predict the GDP of the country will be at the beginning of the year 2010?

41. **POPULATION GROWTH** The size of a bacterial population $P(t)$ grows at the rate of 3.1% per day. If the initial population is 10,000, what is the population after 10 days?

42. **REAL ESTATE INVESTMENT** In 1626, Peter Minuit traded trinkets worth $24 to a tribe of Native Americans for land on Manhattan Island. Assume that in 1990 the same land was worth $25.2 billion. If the sellers in this transaction had invested their $24 at 7% annual interest compounded continuously during the entire 364-year period, who would have gotten the better end of the deal? By how much?

43. **GROWTH OF BACTERIA** The following data were compiled by a researcher during the first 10 minutes of an experiment designed to study the growth of bacteria:

Number of minutes	0	10
Number of bacteria	5,000	8,000

Assuming that the number of bacteria grows exponentially, how many bacteria will be present after 30 minutes?

44. **RETAIL SALES** The total number of hamburgers sold by a national fast-food chain is growing exponentially. If 4 billion had been sold by 1995 and 12 billion had been sold by 2000, how many will have been sold by 2005?

45. **SALES** Once the initial publicity surrounding the release of a new book is over, sales of the hardcover edition tend to decrease exponentially. At the time publicity was discontinued, a certain book was experiencing sales of 25,000 copies per month. One month later, sales of the book had dropped to 10,000 copies per month. What will the sales be after 1 more month?

46. **COMPOUND INTEREST** A bank pays 5% interest compounded quarterly, and a savings institution pays 4.9% interest compounded continuously. Over a 1-year period, which account pays more interest? What about a 5-year period?

47. **POPULATION DENSITY** The population density x miles from the center of a certain city is $D(x) = 12e^{-0.07x}$ thousand people per square mile.
    a. What is the population density at the center of the city?
    b. What is the population density 10 miles from the center of the city?

48. **RADIOACTIVE DECAY** The amount of a sample of a radioactive substance remaining after t years is given by a function of the form $Q(t) = Q_0e^{-0.0001t}$. At the end of 5,000 years, 200 grams of the substance remain. How many grams were present initially?

49. **RADIOACTIVE DECAY** A radioactive substance decays exponentially. If 500 grams of the substance were present initially and 400 grams are present 50 years later, how many grams will be present after 200 years?

50. **LEARNING** According to the Ebbinghaus model, the fraction $F(t)$ of subject matter you will remember from this course t months after the final exam can be estimated by the formula
$$F(t) = B + (1 - B)e^{-kt}$$
where B is the fraction of the material you will never forget and k is a constant that depends on the quality of your memory. Suppose you are tested and it is found that $B = 0.3$ and $k = 0.2$. What fraction of the material will you remember one month after the class ends? What fraction will you remember after one year?

51. **PRODUCT RELIABILITY** A statistical study indicates that the fraction of the electric toasters manufactured by a certain company that are still in working condition after t years of use is approximately $f(t) = e^{-0.2t}$.
    a. What fraction of the toasters can be expected to work for at least three years?
    b. What fraction of the toasters can be expected to fail before one year of use?
    c. What fraction of the toasters can be expected to fail during the third year of use?

52. **BUSINESS TRAINING** A company organizes a training program in which it is determined that after t weeks, the average trainee produces

$$P(t) = 50(1 - e^{-0.15t}) \text{ units}$$

while a typical new worker without special training produces

$$W(t) = \sqrt{150t} \text{ units}$$

   a. How many units does the average trainee produce during the third week of the training period?
   b. Explain how the function $F(t) = P(t) - W(t)$ can be used to evaluate the effectiveness of the training program. Is the program effective if it lasts just five weeks? What if it lasts at least seven weeks? Explain your reasoning.

53. **AQUATIC PLANT LIFE** Plant life exists only in the top 10 meters of a lake or sea, primarily because the intensity of sunlight decreases exponentially with depth. Specifically, the **Bouguer-Lambert law** says that a beam of light that strikes the surface of a body of water with intensity I_0 will have intensity I at a depth of x meters, where $I = I_0 e^{-kx}$ with $k > 0$. The constant k, called the **absorption coefficient,** depends on the wavelength of the light and the density of the water. Suppose a beam of sunlight is only 10% as intense at a depth of 3 meters as at the surface. How intense is the beam at a depth of 1 meter? (Express your answer in terms of I_0.)

54. **EFFECTIVE INTEREST RATE** When a bank offers interest at an annual rate r (expressed as a decimal) and compounds interest more than once a year, the total interest earned during the year is greater than $100r\%$ of the balance at the beginning of that year. The actual percentage by which the balance grows during a year is called the **effective interest rate,** while the advertised rate r is called the **nominal interest rate.** In other words, the effective interest rate is the simple rate that is equivalent to the nominal compound interest rate.

   a. If interest is compounded k times a year, show that the effective rate is $(1 + i)^k - 1$, where $i = \dfrac{r}{k}$.

   b. If interest is compounded continuously, show that the effective rate is $e^r - 1$.

In Problems 55 and 56, you will need the formulas for effective rate of interest derived in Problem 54.

55. **EFFECTIVE INTEREST RATE** If a bank offers interest at a nominal rate of 6%, how much greater is the effective rate if interest is compounded continuously than if the compounding is quarterly?

56. **EFFECTIVE INTEREST RATE** Which investment has the greater effective rate: 8.2% compounded quarterly, or 8.1% compounded continuously?

57. **LINGUISTICS** **Glottochronology** is the methodology used by linguists to determine how many years have passed since two modern languages "branched" from a common ancestor. Experiments suggest that if N words are in common use at a base time $t = 0$, then the number $N(t)$ of them still in use with essentially the same meaning t thousand years later is given by the so-called **fundamental glottochronology equation***

$$N(t) = N_0 e^{-0.217t}$$

   a. Out of a set of 500 basic words used in classical Latin in 200 B.C., how many would you expect to be still in use in modern Italian in the year 2010?

   b. The research of C. W. Feng and M. Swadesh indicated that out of a set of 210 words commonly used in classical Chinese in 950 A.D., 167 were still in use in modern Mandarin in 1950. Is this the same number that the fundamental glottochronology equation would predict? How do you account for the difference?

 c. Read an article on glottochronology and write a paragraph on its methodology. You may wish to begin your research by reading the article cited in this problem.

Source: Anthony LoBello and Maurice D. Weir, "Glottochronology: An Application of Calculus to Linguistics," *UMAP Modules 1982: Tools for Teaching*, Lexington, MA: Consortium for Mathematics and Its Applications, Inc., 1983.

Amortization of Debt ■ If a loan of A dollars is amortized over n years at an annual interest rate r (expressed as a decimal) compounded monthly, the monthly payments are given by

$$M = \frac{Ai}{1 - (1 + i)^{-12n}}$$

where $i = \dfrac{r}{12}$ is the monthly interest rate. Use this formula in Problems 58 through 61.

58. FINANCE PAYMENTS Determine the monthly car payment for a new car costing $15,675, if there is a down payment of $4,000 and the car is financed over a 5-year period at an annual rate of 6% compounded monthly.

59. MORTGAGE PAYMENTS A home loan is made for $150,000 at 9% annual interest, compounded monthly, for 30 years. What is the monthly mortgage payment on this loan?

60. MORTGAGE PAYMENTS Suppose a family figures it can handle monthly mortgage payments of no more than $1,200. What is the largest amount of money they can borrow, assuming the lender is willing to amortize over 30 years at 8% annual interest compounded monthly?

61. TRUTH IN LENDING You are selling your car for $6,000. A potential buyer says, "I will pay you $1,000 now for the car and pay off the rest at 12% interest with monthly payments for 3 years. Let's see . . . 12% of the $5,000 is $600 and $5,600 divided by 36 months is $155.56, but I'll pay you $160 per month for the trouble of carrying the loan. Is it a deal?"

a. If this deal sounds fair to you, I have a perfectly lovely bridge I think you should consider as your next purchase. If not, explain why the deal is fishy and compute a fair monthly payment (assuming you still plan to amortize the debt of $5,000 over 3 years at 12%).

b. Read an article on truth in lending and think up some examples of plausible yet shady deals, such as the proposed used-car transaction in this problem.

62. Two graphs $y = f(x)$ and $y = g(x)$ are reflections of one another in the y axis if whenever (a, b) is a point on one of the graphs, then $(-a, b)$ is a point on the other, as indicated in the accompanying figure. Use this criterion to show that the graphs of $y = b^x$ and $y = \left(\dfrac{1}{b}\right)^x$ for $b > 0$, $b \ne 1$ are reflections of one another in the y axis.

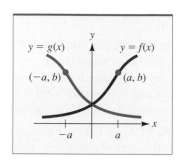

PROBLEM 62

63. Complete the following table for $f(x) = \dfrac{1}{2}\left(\dfrac{1}{4}\right)^x$.

x	-2.2	-1.5	0	1.5	2.3
$f(x)$					

64. Program a computer or use a calculator to evaluate $\left(1 + \dfrac{1}{n}\right)^n$ for $n = 1{,}000, 2{,}000, \ldots, 50{,}000$.

65. Program a computer or use a calculator to evaluate $\left(1 + \dfrac{1}{n}\right)^n$ for $n = -1{,}000, -2{,}000, \ldots, -50{,}000$.

On the basis of these calculations, what can you conjecture about the behavior of $\left(1 + \dfrac{1}{n}\right)^n$ as n *decreases* without bound?

66. Program a computer or use a calculator to estimate $\displaystyle \lim_{n \to +\infty}\left(1 + \frac{3}{n}\right)^{2n}$.

67. Program a computer or use a calculator to estimate $\displaystyle \lim_{n \to +\infty}\left(2 - \frac{5}{2n}\right)^{n/3}$.

SECTION 4.2 | Logarithmic Functions

Suppose you invest $1,000 at 8% compounded continuously and wish to know how much time must pass for your investment to double in value to $2,000. According to the formula derived in Section 4.1, the value of your account after t years will be $1,000e^{0.08t}$, so to find the doubling time for your account, you must solve for t in the equation

$$1,000e^{0.08t} = 2,000$$

or, by dividing both sides by 1,000,

$$e^{0.08t} = 2$$

We will answer the question about doubling time in Example 4.2.10. Solving an exponential equation such as this involves using *logarithms*, which reverse the process of exponentiation. Logarithms play an important role in a variety of applications, such as measuring the capacity of a transmission channel and in the famous Richter scale for measuring earthquake intensity. In this section, we examine the basic properties of logarithmic functions and a few applications. We begin with a definition.

Logarithmic Functions ■ If x is a positive number, then the **logarithm** of x to the base b ($b > 0$, $b \neq 1$), denoted $\log_b x$, is the number y such that $b^y = x$; that is,

$$y = \log_b x \quad \text{if and only if} \quad b^y = x \quad \text{for } x > 0$$

EXAMPLE | 4.2.1

Evaluate

a. $\log_{10} 1,000$ **b.** $\log_2 32$ **c.** $\log_5\left(\dfrac{1}{125}\right)$

Solution

a. $\log_{10} 1,000 = 3$ since $10^3 = 1,000$.

b. $\log_2 32 = 5$ since $2^5 = 32$.

c. $\log_5 \dfrac{1}{125} = -3$ since $5^{-3} = \dfrac{1}{125}$.

EXAMPLE | 4.2.2

Solve each of the following equations for x:

a. $\log_4 x = \dfrac{1}{2}$ **b.** $\log_{64} 16 = x$ **c.** $\log_x 27 = 3$

Solution

a. By definition, $\log_4 x = \dfrac{1}{2}$ is equivalent to $x = 4^{1/2} = 2$.

b. $\log_{64} 16 = x$ means

$$16 = 64^x$$
$$2^4 = (2^6)^x = 2^{6x}$$
$$4 = 6x \qquad\qquad b^m = b^n \text{ implies } m = n$$
$$x = \frac{2}{3}$$

c. $\log_x 27 = 3$ means

$$x^3 = 27$$
$$x = (27)^{1/3} = 3$$

Properties of Logarithms

Logarithms were introduced in the 17th century as a computational device, primarily because they can be used to convert expressions involving products and quotients to much simpler expressions involving sums and differences. Here are the rules for logarithms that allow such simplification.

> **Properties of Logarithms** ■ Let b $(b > 0, b \neq 1)$ be any logarithmic base. Then,
>
> $$\log_b 1 = 0 \qquad \text{and} \qquad \log_b b = 1$$
>
> and if u and v are any positive numbers, we have
>
> The **equality rule** $\qquad \log_b u = \log_b v$ if and only if $u = v$
>
> The **product rule** $\qquad \log_b (uv) = \log_b u + \log_b v$
>
> The **power rule** $\qquad \log_b u^r = r \log_b u$ for any real number r
>
> The **quotient rule** $\qquad \log_b \left(\dfrac{u}{v}\right) = \log_b u - \log_b v$
>
> The **inversion rule** $\qquad \log_b b^u = u$

All these logarithmic rules follow from corresponding exponential rules. For example,

$$\log_b 1 = 0 \qquad \text{since} \qquad b^0 = 1$$
$$\log_b b = 1 \qquad \text{since} \qquad b^1 = b$$

To prove the equality rule, let

$$m = \log_b u \quad \text{and} \quad n = \log_b v$$

so that by definition,

$$b^m = u \quad \text{and} \quad b^n = v$$

Therefore, if

$$\log_b u = \log_b v$$

then $m = n$ so

$$b^m = b^n \qquad \textit{equality rule for exponentials}$$

or, equivalently,

$$u = v$$

as stated in the equality rule for logarithms. Similarly, to prove the product rule for logarithms, note that

$$
\begin{aligned}
\log_b u + \log_b v &= m + n \\
&= \log_b (b^{m+n}) \quad \textit{definition of logarithm} \\
&= \log_b (b^m b^n) \quad \textit{product rule for exponentials} \\
&= \log_b (uv) \qquad \textit{since } b^m = u \textit{ and } b^n = v
\end{aligned}
$$

Proofs of the power rule and the quotient rule are left as exercises (see Problem 74). Table 4.1 displays the correspondence between basic properties of exponential and logarithmic functions.

TABLE 4.1　Comparison of Properties of Exponential and Logarithmic Functions

Exponential Property	Logarithmic Property
$b^x b^y = b^{x+y}$	$\log_b (xy) = \log_b x + \log_b y$
$\dfrac{b^x}{b^y} = b^{x-y}$	$\log_b \left(\dfrac{x}{y} \right) = \log_b x - \log_b y$
$b^{xy} = (b^x)^y$	$\log_b x^p = p \log_b x$

EXAMPLE │ 4.2.3

Use logarithm rules to rewrite each of the following expressions in terms of $\log_5 2$ and $\log_5 3$.

a. $\log_5 \left(\dfrac{5}{3} \right)$　　　　　**b.** $\log_5 8$　　　　　**c.** $\log_5 36$

Solution

a.
$$
\begin{aligned}
\log_5 \left(\frac{5}{3} \right) &= \log_5 5 - \log_5 3 \qquad \textit{quotient rule} \\
&= 1 - \log_5 3 \qquad\quad \textit{since } \log_5 5 = 1
\end{aligned}
$$

b. $\log_5 8 = \log_5 2^3 = 3 \log_5 2 \qquad \textit{power rule}$

c.
$$
\begin{aligned}
\log_5 36 &= \log_5 (2^2 3^2) \\
&= \log_5 2^2 + \log_5 3^2 \qquad \textit{product rule} \\
&= 2 \log_5 2 + 2 \log_5 3 \quad \textit{power rule}
\end{aligned}
$$

EXAMPLE | 4.2.4

Use logarithm rules to expand each of the following expressions.

a. $\log_3 (x^3 y^{-4})$ **b.** $\log_2 \left(\dfrac{y^5}{x^2} \right)$ **c.** $\log_7 (x^3 \sqrt{1 - y^2})$

Solution

a. $\log_3 (x^3 y^{-4}) = \log_3 x^3 + \log_3 y^{-4}$ *product rule*

$\qquad\qquad = 3 \log_3 x + (-4) \log_3 y$ *power rule*

$\qquad\qquad = 3 \log_3 x - 4 \log_3 y$

b. $\log_2 \left(\dfrac{y^5}{x^2} \right) = \log_2 y^5 - \log_2 x^2$ *quotient rule*

$\qquad\qquad = 5 \log_2 y - 2 \log_2 x$ *power rule*

c. $\log_7 (x^3 \sqrt{1 - y^2}) = \log_7 [x^3 (1 - y^2)^{1/2}]$

$\qquad\qquad = \log_7 x^3 + \log_7 (1 - y^2)^{1/2}$ *product rule*

$\qquad\qquad = 3 \log_7 x + \dfrac{1}{2} \log_7 (1 - y^2)$ *power rule*

$\qquad\qquad = 3 \log_7 x + \dfrac{1}{2} \log_7 [(1 - y)(1 + y)]$ *factor* $1 - y^2$

$\qquad\qquad = 3 \log_7 x + \dfrac{1}{2} [\log_7 (1 - y) + \log_7 (1 + y)]$ *product rule*

$\qquad\qquad = 3 \log_7 x + \dfrac{1}{2} \log_7 (1 - y) + \dfrac{1}{2} \log_7 (1 + y)$

Graphs of Logarithmic Functions

There is an easy way to obtain the graph of the logarithmic function $y = \log_b x$ from the graph of the exponential function $y = b^x$. The idea is that since $y = \log_b x$ is equivalent to $x = b^y$, the graph of $y = \log_b x$ is the same as the graph of $y = b^x$ with the roles of x and y reversed. That is, if (u, v) is a point on the curve $y = \log_b x$, then $v = \log_b u$, or equivalently, $u = b^v$, which means that (v, u) is on the graph of $y = b^x$. But, as illustrated in Figure 4.6a, the points (u, v) and (v, u) are mirror images of one another in the line $y = x$ (see Problem 75). Thus, the graph of $y = \log_b x$ can be obtained by simply *reflecting* the graph of $y = b^x$ in the line $y = x$, as shown in Figure 4.6b for the case where $b > 1$. To summarize:

> ### Relationship Between the Graphs of $y = \log_b x$ and $y = b^x$ ▪
> The graphs of $y = \log_b x$ and $y = b^x$ are mirror images of one another in the line $y = x$. Therefore, the graph of $y = \log_b x$ can be obtained by reflecting the graph of $y = b^x$ in the line $y = x$.

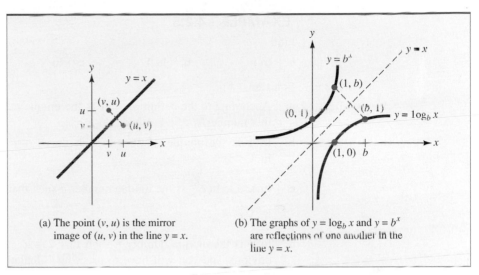

(a) The point (v, u) is the mirror image of (u, v) in the line $y = x$.

(b) The graphs of $y = \log_b x$ and $y = b^x$ are reflections of one another in the line $y = x$.

FIGURE 4.6   The graph of $y = \log_b x$ for $b > 1$ is obtained by reflecting the graph of $y = b^x$ in the line $y = x$.

Figure 4.6b reveals these facts about the logarithmic function $y = \log_b x$ for $b > 1$ and its graph:

1. The function $y = \log_b x$ is always increasing and its graph is always concave downward for $x > 0$.

2. The graph of $y = \log_b x$ has $(1, 0)$ as its only intercept and the y axis as a vertical asymptote.

3. The function $y = \log_b x$ increases without bound as x increases indefinitely and decreases without bound as x approaches 0 from the right; that is,

$$\lim_{x \to +\infty} \log_b x = +\infty \quad \text{and} \quad \lim_{x \to 0^+} \log_b x = -\infty$$

What (if anything) do you think is different about the function $y = \log_b x$ and its graph for the case where $0 < b < 1$? You are asked to explore this question in Problem 76.

The Natural Logarithm

In calculus, the most frequently used logarithmic base is e. In this case, the logarithm $\log_e x$ is called the **natural logarithm** of x and is denoted by $\ln x$ (read as "el en x"); that is, for $x > 0$

$$y = \ln x \quad \text{if and only if} \quad e^y = x$$

To evaluate $\ln a$ for a particular number $a > 0$, use the **LN** key on your calculator. For example, to find $\ln(2.714)$, you would press the **LN** key, and then enter the number 2.714 to get

$$\ln(2.714) = 0.9984 \quad \text{(to four decimal places)}$$

Here is an example illustrating the computation of natural logarithms.

EXAMPLE | 4.2.5

Find

a. $\ln e$ **b.** $\ln 1$ **c.** $\ln \sqrt{e}$ **d.** $\ln 2$

Solution

a. According to the definition, $\ln e$ is the unique number c such that $e = e^c$. Clearly this number is $c = 1$. Hence, $\ln e = 1$.

b. $\ln 1$ is the unique number c such that $1 = e^c$. Since $e^0 = 1$, it follows that $\ln 1 = 0$.

c. $\ln \sqrt{e} = \ln e^{1/2}$ is the unique number c such that $e^{1/2} = e^c$; that is, $c = \dfrac{1}{2}$. Hence, $\ln \sqrt{e} = \dfrac{1}{2}$.

d. $\ln 2$ is the unique number c such that $2 = e^c$. The value of this number is not obvious, and you will have to use your calculator to find that $\ln 2 \approx 0.69315$.

Since the natural logarithm $y = \ln x$ is the logarithmic function used most frequently in calculus, it is helpful to have this list of its properties.

> **Properties of the Natural Logarithm $y = \ln x$** ■ For positive numbers u and v:
>
> The **equality rule** $\ln u = \ln v$ if and only if $u = v$
>
> The **product rule** $\ln(uv) = \ln u + \ln v$
>
> The **power rule** $\ln u^r = r \ln u$ where r is a real number
>
> The **quotient rule** $\ln\left(\dfrac{u}{v}\right) = \ln u - \ln v$
>
> **Special values** $\ln 1 = 0$ and $\ln e = 1$

EXPLORE!

Store $y = e^x$ into Y1, using a bold graphing style, and $y = x$ into Y2. Set a decimal window. Since $y = \ln x$ is equivalent to $e^y = x$, we can graph $y = \ln x$ as the inverse relation of $e^x = y$ using option **8: DrawInv** of the **DRAW (2nd PRGM)** key and writing **DrawInv Y1** on the Home Screen.

The graph of $y = \ln x$ is obtained by reflecting the graph of $y = e^x$ in the line $y = x$, as shown in Figure 4.7.

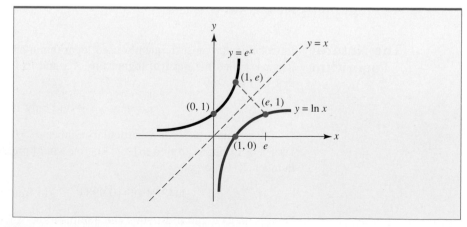

FIGURE 4.7   The graph of $y = \ln x$.

EXAMPLE | 4.2.6

a. Find $\ln \sqrt{ab}$ if $\ln a = 3$ and $\ln b = 7$.

b. Show that $\ln \dfrac{1}{x} = -\ln x$.

c. Find x if $2^x = e^3$.

Solution

a. $\ln \sqrt{ab} = \ln (ab)^{1/2} = \dfrac{1}{2} \ln ab = \dfrac{1}{2}(\ln a + \ln b) = \dfrac{1}{2}(3 + 7) = 5$

b. $\ln \dfrac{1}{x} = \ln 1 - \ln x = 0 - \ln x = -\ln x$

c. Take the natural logarithm of each side of the equation $2^x = e^3$ and solve for x to get

$$x \ln 2 = 3 \quad \text{or} \quad x = \frac{3}{\ln 2} \approx 4.33$$

The Relationship Between e^x and $\ln x$

Example 4.2.7 establishes two important identities that show how logarithmic and exponential functions have a certain "neutralizing" effect on each other.

EXAMPLE | 4.2.7

Simplify the following expressions:

a. $e^{\ln x}$ (for $x > 0$) **b.** $\ln e^x$

Solution

a. According to the definition, $\ln x$ is the unique number b for which $x = e^b$. Hence, $e^{\ln x} = e^b = x$.

b. Similarly, $\ln e^x$ is the unique number b for which $e^x = e^b$. Clearly, this number b is x itself. Hence, $\ln e^x = x$.

The two identities derived in Example 4.2.7 show that the composite functions $\ln e^x$ and $e^{\ln x}$ leave the variable x unchanged. In general, two functions f and g for which $f(g(x)) = x$ and $g(f(x)) = x$ are said to be **inverses** of one another. Thus, the exponential function $y = e^x$ and the logarithmic function $y = \ln x$ are inverses.

The Inverse Relationship Between e^x and $\ln x$ ■

$$e^{\ln x} = x \quad \text{for } x > 0 \quad \text{and} \quad \ln e^x = x \quad \text{for all } x$$

Example 4.2.8 illustrates how you can use the inverse relationship between e^x and $\ln x$ to solve equations.

EXPLORE!

Solve the equation

$$3 - e^x = \ln(x^2 + 1)$$

by placing the left side of the equation into Y1 and the right side into Y2. Use Standard Window (**ZOOM** 6) to find the x values of the intersection points.

EXAMPLE | 4.2.8

Solve each of the following equations for x:

a. $3 = e^{20x}$ **b.** $2 \ln x = 1$

Solution

a. Take the natural logarithm of each side of the equation to get

$$\ln 3 = \ln e^{20x} \quad \text{or} \quad \ln 3 = 20x$$

Solve for x, using a calculator, to find $\ln 3$:

$$x = \frac{\ln 3}{20} \approx \frac{1.0986}{20} \approx 0.0549$$

b. First isolate $\ln x$ on the left side of the equation by dividing both sides by 2:

$$\ln x = \frac{1}{2}$$

Then apply the exponential function to both sides of the equation to get

$$e^{\ln x} = e^{1/2} \quad \text{or} \quad x = e^{1/2} = \sqrt{e} \approx 1.6487$$

EXPLORE!

Put the function $y = 10^x$ into Y1 and graph using a bold graphing style. Then put $y = x$ into Y2, $y = \ln x$ into Y3, and $y = \log x$ into Y4. What can you conclude from this series of graphs?

You have already seen how to use the **LN** key to compute natural logarithms, and most calculators have a **LOG** key for computing logarithms to base 10, but what about logarithms to bases other than e or 10? To be specific, suppose you wish to calculate the logarithmic number $c = \log_b a$. You have

$$c = \log_b a$$
$$b^c = a \qquad \qquad \textit{definition of the logarithm}$$
$$\ln b^c = \ln a$$
$$c \ln b = \ln a \qquad \qquad \textit{power rule}$$
$$c = \frac{\ln a}{\ln b}$$

Thus, the logarithm $\log_b a$ can be computed by finding the ratio of two natural logarithms, $\ln a$ and $\ln b$. To summarize:

EXPLORE!

Store $f(x) = B^x$ into Y1 and then $g(x) = \log_B x$ into Y2 as $\ln(x)/\ln(B)$. Experiment with different values of $1 < B < 2$, using the **STO▶** key, to determine for which B values these two functions intersect, touch at a point, or are separated, as shown in Figure 4.6b.

Conversion Formula for Logarithms ▪ If a and b are positive numbers with $b \neq 1$, then

$$\log_b a = \frac{\ln a}{\ln b}$$

EXAMPLE | 4.2.9

Find $\log_5 3.172$.

Solution

Using the conversion formula, you find

$$\log_5 3.172 = \frac{\ln 3.172}{\ln 5} \approx \frac{1.1544}{1.6094} \approx 0.7172$$

Doubling Time In the introductory paragraph at the beginning of this section, you were asked how long it would take for a particular investment to double in value. This question is answered in Example 4.2.10.

EXAMPLE | 4.2.10

If $1,000 is invested at 8% annual interest, compounded continuously, how long will it take for the investment to double? Would the doubling time change if the principal were something other than $1,000?

Solution

With a principal of $1,000, the balance after t years is $B(t) = 1,000 e^{0.08t}$, so the investment doubles when $B(t) = \$2,000$; that is, when

$$2,000 = 1,000 e^{0.08t}$$

Dividing by 1,000 and taking the natural logarithm on each side of the equation, we get

$$2 = e^{0.08t}$$
$$\ln 2 = 0.08t$$
$$t = \frac{\ln 2}{0.08} \approx 8.66 \text{ years}$$

If the principal had been P_0 dollars instead of $1,000, the doubling time would satisfy

$$2P_0 = P_0 e^{0.08t}$$
$$2 = e^{0.08t}$$

which is exactly the same equation we had with $P_0 = \$1,000$, so once again, the doubling time is 8.66 years.

EXPLORE!

Use the equation solver of your graphing calculator with the equation F − P∗e^(R∗T) = 0 to determine how long it will take for $2,500 to double at 8.5% compounded continuously.

The situation illustrated in Example 4.2.10 applies to any quantity $Q(t) = Q_0 e^{kt}$ that grows exponentially with time t. In particular, since at time $t = 0$, we have $Q(0) = Q_0 e^0 = Q_0$, the quantity doubles when

$$2Q_0 = Q_0 e^{kt}$$
$$2 = e^{kt}$$
$$\ln 2 = kt$$
$$t = \frac{\ln 2}{k}$$

To summarize:

Doubling Time ▪ A quantity $Q(t) = Q_0 e^{kt}$ ($k > 0$) undergoing exponential growth doubles when $t = d$, where

$$d = \frac{\ln 2}{k}$$

Half-Life   It has been experimentally determined that most radioactive substances decay exponentially, so that the amount of a sample of initial size Q_0 that is present after t years is given by a function of the form $Q(t) = Q_0 e^{-kt}$. The positive constant k measures the rate of decay, but this rate is usually given by specifying the amount of time $t = h$ required for half a given sample to decay. This time is called the **half-life** of the radioactive substance. Example 4.2.11 shows how half-life is related to k.

EXAMPLE | 4.2.11

Show that a radioactive substance that decays according to the formula $Q(t) = Q_0 e^{-kt}$ has half-life $h = \dfrac{\ln 2}{k}$.

Solution

The goal is to find the value of t for which $Q(h) = \dfrac{1}{2} Q_0$; that is,

$$\frac{1}{2} Q_0 = Q_0 e^{-kh}$$

Divide by Q_0 and take the natural logarithm of each side to get

$$\ln \frac{1}{2} = -kh$$

Thus, the half-life is

$$h = \frac{\ln \dfrac{1}{2}}{-k}$$

$$= \frac{-\ln 2}{-k} = \frac{\ln 2}{k} \qquad since \; \ln \frac{1}{2} = \ln 2^{-1} = -\ln 2$$

as required.

Carbon Dating   In 1960, W. F. Libby won a Nobel prize for his discovery of **carbon dating,** a technique for determining the age of certain fossils and artifacts. Here is an outline of the technique.*

The carbon dioxide in the air contains the radioactive isotope ^{14}C ("carbon 14") as well as the stable isotope ^{12}C ("carbon 12"). Living plants absorb carbon dioxide from the air, which means that the ratio of ^{14}C to ^{12}C in a living plant (or in an animal that eats plants) is the same as that in the air itself. When a plant or an animal dies, the absorption of carbon dioxide ceases. The ^{12}C already in the plant or animal remains the same as at the time of death while the ^{14}C decays, and the ratio of ^{14}C to ^{12}C decreases exponentially. It is reasonable to assume that the ratio R_0 of ^{14}C to ^{12}C in the atmosphere is the same today as it was in the past, so that the ratio of ^{14}C to ^{12}C in a sample (e.g., a fossil or an artifact) is given by a function of the form

---

*For instance, see Raymond J. Cannon, "Exponential Growth and Decay," *UMAP Modules 1977: Tools for Teaching,* Lexington, MA: Consortium for Mathematics and Its Applications, Inc., 1978. More advanced dating procedures are discussed in Paul J. Campbell, "How Old Is the Earth?" *UMAP Modules 1992: Tools for Teaching,* Lexington, MA: Consortium for Mathematics and Its Applications, Inc., 1993.

$R(t) = R_0 e^{-kt}$. The half-life of ^{14}C is 5,730 years. By comparing $R(t)$ to R_0, archaeologists can estimate the age of the sample. Example 4.2.12 illustrates the dating procedure.

EXAMPLE | 4.2.12

An archaeologist has found a fossil in which the ratio of ^{14}C to ^{12}C is $\frac{1}{5}$ the ratio found in the atmosphere. Approximately how old is the fossil?

Solution

The age of the fossil is the value of t for which $R(t) = \frac{1}{5} R_0$; that is, for which

$$\frac{1}{5} R_0 = R_0 e^{-kt}$$

Dividing by R_0 and taking logarithms, you find that

$$\frac{1}{5} = e^{-kt}$$

$$\ln \frac{1}{5} = -kt$$

and

$$t = \frac{-\ln \frac{1}{5}}{k} = \frac{\ln 5}{k}$$

In Example 4.2.11, you found that the half-life h satisfies $h = \frac{\ln 2}{k}$, and since ^{14}C has half-life $h = 5,730$ years, you have

$$k = \frac{\ln 2}{h} = \frac{\ln 2}{5,730} \approx 0.000121$$

Therefore, the age of the fossil is

$$t = \frac{\ln 5}{k} = \frac{\ln 5}{0.000121} \approx 13,300$$

That is, the fossil is approximately 13,300 years old.

Exponential Curve Fitting

In our final example, you will see how to use logarithms to find an exponential function that satisfies certain specified conditions.

EXAMPLE | 4.2.13

The population density x miles from the center of a city is given by a function of the form $Q(x) = Ae^{-kx}$. Find this function if it is known that the population density at the center of the city is 15,000 people per square mile and the density 10 miles from the center is 9,000 people per square mile.

EXPLORE!

Refer to Example 4.2.13. Place $0 = Q - A*e^{\wedge}(-K*X)$ into the equation solver of your graphing calculator. Find the distance from the center of the city if $Q = 13,500$ people per square mile. Recall from the example that the density at city center is 15,000 and the density 10 miles from city center is 9,000 people per square mile.

Solution

For simplicity, express the density in units of 1,000 people per square mile. The fact that $Q(0) = 15$ tells you that $A = 15$. The fact that $Q(10) = 9$ means that

$$9 = 15e^{-10k} \qquad \text{or} \qquad \frac{3}{5} = e^{-10k}$$

Taking the logarithm of each side of this equation, you get

$$\ln \frac{3}{5} = -10k \qquad \text{or} \qquad k = -\frac{\ln 3/5}{10} \approx 0.051$$

Hence the exponential function for the population density is $Q(x) = 15e^{-0.051x}$.

PROBLEMS 4.2

In Problems 1 and 2, use your calculator to find the indicated natural logarithms.

1. Find $\ln 1$, $\ln 2$, $\ln e$, $\ln 5$, $\ln \frac{1}{5}$, and $\ln e^2$. What happens if you try to find $\ln 0$ or $\ln(-2)$? Why?

2. Find $\ln 7$, $\ln \frac{1}{3}$, $\ln e^{-3}$, $\ln \frac{1}{e^{2.1}}$, and $\ln \sqrt[5]{e}$. What happens if you try to find $\ln(-7)$ or $\ln(-e)$?

In Problems 3 through 8, evaluate the given expression using properties of the logarithm.

3. $\ln e^3$

4. $\ln \sqrt{e}$

5. $e^{\ln 5}$

6. $e^{2 \ln 3}$

7. $e^{3 \ln 2 - 2 \ln 5}$

8. $\ln \dfrac{e^3 \sqrt{e}}{e^{1/3}}$

In Problems 9 through 12, use logarithm rules to rewrite each expression in terms of $\log_3 2$ and $\log_3 5$.

9. $\log_3 270$

10. $\log_3 (2.5)$

11. $\log_3 100$

12. $\log_3 \left(\dfrac{64}{125}\right)$

In Problems 13 through 20, use logarithm rules to simplify each expression.

13. $\log_2 (x^4 y^3)$

14. $\log_3 (x^5 y^{-2})$

15. $\ln \sqrt[3]{x^2 - x}$

16. $\ln(x^2 \sqrt{4 - x^2})$

17. $\ln \left[\dfrac{x^2(3 - x)^{2/3}}{\sqrt{x^2 + x + 1}}\right]$

18. $\ln \left[\dfrac{1}{x} + \dfrac{1}{x^2}\right]$

19. $\ln(x^3 e^{-x^2})$

20. $\ln \left[\dfrac{\sqrt[4]{x}}{x^3 \sqrt{1 - x^2}}\right]$

In Problems 21 through 36, solve the given equation for x.

21. $4^x = 53$

22. $\log_2 x = 4$

23. $\log_3 (2x - 1) = 2$

24. $3^{2x-1} = 17$

25. $2 = e^{0.06x}$

26. $\frac{1}{2} Q_0 = Q_0 e^{-1.2x}$

27. $3 = 2 + 5e^{-4x}$

28. $-2 \ln x = b$

29. $-\ln x = \frac{t}{50} + C$

30. $5 = 3 \ln x - \frac{1}{2} \ln x$

31. $\ln x = \frac{1}{3}(\ln 16 + 2 \ln 2)$

32. $\ln x = 2(\ln 3 - \ln 5)$

33. $3^x = e^2$

34. $a^k = e^{kx}$

35. $\dfrac{25 e^{0.1x}}{e^{0.1x} + 3} = 10$

36. $\dfrac{5}{1 + 2e^{-x}} = 3$

37. If $\log_2 x = 5$, what is $\ln x$?

38. If $\log_{10} x = -3$, what is $\ln x$?

39. If $\log_5 (2x) = 7$, what is $\ln x$?

40. If $\log_3 (x - 5) = 2$, what is $\ln x$?

41. Find $\ln \dfrac{1}{\sqrt{ab^3}}$ if $\ln a = 2$ and $\ln b = 3$.

42. Find $\dfrac{1}{a} \ln \left(\dfrac{\sqrt{b}}{c} \right)^a$ if $\ln b = 6$ and $\ln c = -2$.

43. **COMPOUND INTEREST** How quickly will money double if it is invested at an annual interest rate of 6% compounded continuously?

44. **COMPOUND INTEREST** How quickly will money double if it is invested at an annual interest rate of 7% compounded continuously?

45. **COMPOUND INTEREST** Money deposited in a certain bank doubles every 13 years. The bank compounds interest continuously. What annual interest rate does the bank offer?

46. **TRIPLING TIME** How long will it take for a quantity of money A_0 to triple in value if it is invested at an annual interest rate r compounded continuously?

47. **TRIPLING TIME** If an account that earns interest compounded continuously takes 12 years to double in value, how long will it take to triple in value?

48. **CONCENTRATION OF DRUG** A drug is injected into a patient's bloodstream and t seconds later, the concentration of the drug is C grams per cubic centimeter (g/cm^3), where
$$C(t) = 0.1(1 + 3e^{-0.03t})$$
   a. What is the drug concentration after 10 seconds?
   b. How long does it take for the drug concentration to reach 0.12 g/cm^3?

49. **CONCENTRATION OF DRUG** The concentration of a drug in a patient's kidneys at time t (seconds) is C grams per cubic centimeter (g/cm^3), where
$$C(t) = 0.4(2 - 0.13e^{-0.02t})$$
   a. What is the drug concentration after 20 seconds? After 60 seconds?
   b. How long does it take for the drug concentration to reach 0.75 g/cm^3?

50. **RADIOACTIVE DECAY** The amount of a certain radioactive substance remaining after t years is given by a function of the form $Q(t) = Q_0 e^{-0.003t}$. Find the half-life of the substance.

51. **RADIOACTIVE DECAY** Radium decays exponentially. Its half-life is 1,690 years. How long will it take for a 50-gram sample of radium to be reduced to 5 grams?

52. **ADVERTISING** The editor at a major publishing house estimates that if x thousand complimentary copies are distributed to instructors, the first-year sales of a new text will be approximately $f(x) = 20 - 12e^{-0.03x}$ thousand copies. According to this estimate, approximately how many complimentary copies should the editor send out to generate first-year sales of 12,000 copies?

53. **GROWTH OF BACTERIA** A medical student studying the growth of bacteria in a certain culture has compiled these data:

Number of minutes	0	20
Number of bacteria	6,000	9,000

Use these data to find an exponential function of the form $Q(t) = Q_0 e^{kt}$ expressing the number of bacteria in the culture as a function of time. How many bacteria are present after 1 hour?

54. **GROSS DOMESTIC PRODUCT** An economist has compiled these data on the Gross Domestic Product (GDP) of a certain country:

Year	1990	2000
GDP (in billions)	100	180

Use these data to predict the GDP in the year 2010 if the GDP is growing:
a. Linearly   b. Exponentially

55. **WORKER EFFICIENCY** An efficiency expert hired by a manufacturing firm has compiled these data relating workers' output to their experience:

Experience t (months)	0	6
Output Q (units per hour)	300	410

Suppose output Q is related to experience t by a function of the form $Q(t) = 500 - Ae^{-kt}$. Find the function of this form that fits the data. What output is expected from a worker with 1 year's experience?

56. **ARCHAEOLOGY** An archaeologist has found a fossil in which the ratio of ^{14}C to ^{12}C is $\frac{1}{3}$ the ratio found in the atmosphere. Approximately how old is the fossil?

57. **ARCHAEOLOGY** Tests of an artifact discovered at the Debert site in Nova Scotia show that 28% of the original ^{14}C is still present. Approximately how old is the artifact?

58. **ARCHAEOLOGY** The Dead Sea Scrolls were written on parchment in about 100 B.C. What percentage of the original ^{14}C in the parchment remained when the scrolls were discovered in 1947?

59. **ART FORGERY** A forged painting allegedly painted by Rembrandt in 1640 is found to have 99.7% of its original ^{14}C. When was it actually painted? What percentage of the original ^{14}C should remain if it were legitimate?

60. **ARCHAEOLOGY** In 1389, Pierre d'Arcis, the bishop of Troyes, wrote a memo to the pope, accusing a colleague of passing off "a certain cloth, cunningly painted" as the burial shroud of Jesus Christ. Despite this early testimony of forgery, the image on the cloth is so compelling that many people regard it as a sacred relic. Known as the Shroud of Turin, the cloth was subjected to carbon dating in 1988. If authentic, the cloth would have been approximately 1,960 years old at that time.
a. If the Shroud were actually 1,960 years old, what percentage of the ^{14}C would have remained?
b. Scientists determined that 92.3% of the Shroud's original ^{14}C remained. Based on this information alone, what was the likely age of the Shroud in 1988?

61. **COOLING** Instant coffee is made by adding boiling water (212°F) to coffee mix. If the air temperature is 70°F, Newton's law of cooling says that after t minutes, the temperature of the coffee will be given by a function of the form $f(t) = 70 - Ae^{-kt}$. After cooling for 2 minutes, the coffee is still 15°F too hot to drink, but 2 minutes later it is just right. What is this "ideal" temperature for drinking?

62. **DEMOGRAPHICS** The world's population grows at the rate of approximately 2% per year. If it is assumed that the population growth is exponential, then the population t years from now will be given by a function of the form $P(t) = P_0 e^{0.02t}$, where P_0 is the current population. (This formula is derived in Chapter 6.) Assuming that this model of population growth is correct, how long will it take for the world's population to double?

63. **SPY STORY** Having ransomed his superior in Problem 18 of Section 3.5, the spy returns home, only to learn that his best friend, Sigmund ("Siggy") Leiter, has been murdered. The police say that Siggy's body was discovered at 1 P.M. on Thursday, stuffed in a freezer where the temperature was 10°F. He is also told that the temperature of the corpse at the time of discovery was 40°F, and he remembers that t hours after death a body has temperature

$$T = T_a + (98.6 - T_a)(0.97)^t$$

where T_a is the air temperature adjacent to the body. The spy knows the dark deed was done by either Ernst Stavro Blohardt or André Scélérat. If Blohardt was in jail until noon on Wednesday and Scélérat was seen in Las Vegas from noon Wednesday until Friday, who "iced" Siggy, and when?

64. **SOUND LEVELS** A **decibel,** named for Alexander Graham Bell, is the smallest increase of the loudness of sound that is detectable by the human

ear. In physics, it is shown that when two sounds of intensity I_1 and I_2 (watts/cm^3) occur, the difference in loudness is D decibels, where

$$D = 10 \log_{10}\left(\frac{I_1}{I_2}\right)$$

When sound is rated in relation to the threshold of human hearing ($I_0 = 10^{-12}$), the level of normal conversation is about 60 decibels, while a rock concert may be 50 times as loud (110 decibels).
a. How much more intense is the rock concert than normal conversation?
b. The threshold of pain is reached at a sound level roughly 10 times as loud as a rock concert. What is the decibel level of the threshold of pain?

65. **SEISMOLOGY** The magnitude formula for the Richter scale is

$$R = \frac{2}{3}\log_{10}\left(\frac{E}{E_0}\right)$$

where E is the energy released by the earthquake (in joules), and $E_0 = 10^{4.4}$ joules is the energy released by a small reference earthquake used as a standard of measurement.
a. The 1906 San Francisco earthquake released approximately 5.96×10^{16} joules of energy. What was its magnitude on the Richter scale?
b. How much energy was released by the Indian earthquake of 1993, which measured 6.4 on the Richter scale?

66. **LEARNING** In an experiment designed to test short-term memory,* L. R. Peterson and M. J. Peterson found that the probability $p(t)$ of a subject recalling a pattern of numbers and letters t seconds after being given the pattern is
$$p(t) = 0.89[0.01 + 0.99(0.85)^t]$$
a. What is the probability that the subject can recall the pattern immediately ($t = 0$)?
b. How much time passes before $p(t)$ drops to 0.5?
c. Sketch the graph of $p(t)$.

67. **SEISMOLOGY** On the Richter scale, the magnitude R of an earthquake of intensity I is given by
$$R = \frac{\ln I}{\ln 10}.$$

---
*L. R. Peterson and M. J. Peterson, "Short-Term Retention of Individual Verbal Items," *Journal of Experimental Psychology,* Vol. 58, 1959, pp. 193–198.

a. Find the intensity of the 1906 San Francisco earthquake, which measured $R = 8.3$ on the Richter scale.
b. How much more intense was the San Francisco earthquake of 1906 than the devastating 1995 earthquake in Kobe, Japan, which measured $R = 7.1$?

68. **ENTOMOLOGY** It is determined that the volume of the yolk of a house fly egg shrinks according to the formula $V(t) = 5e^{-1.3t}$ mm^3 (cubic millimeters), where t is the number of days from the time the egg is produced. The egg hatches after 4 days.
a. What is the volume of the yolk when the egg hatches?
b. Sketch the graph of the volume of the yolk over the time period $0 \le t \le 4$.
c. Find the half-life of the volume of the yolk; that is, the time it takes for the volume of the yolk to shrink to half its original size.

69. **RADIOLOGY** Radioactive iodine ^{133}I has a half-life of 20.9 hours. If injected into the bloodstream, the iodine accumulates in the thyroid gland.
a. After 24 hours, a medical technician scans a patient's thyroid gland to determine whether thyroid function is normal. If the thyroid has absorbed all of the iodine, what percentage of the original amount should be detected?
b. A patient returns to the medical clinic 25 hours after having received an injection of ^{133}I. The medical technician scans the patient's thyroid gland and detects the presence of 41.3% of the original iodine. How much of the original ^{133}I remains in the rest of the patient's body?

70. **AIR PRESSURE** The air pressure $f(s)$ at a height of s meters above sea level is given by
$$f(s) = e^{-0.000125s} \text{ atmospheres}$$
a. The atmospheric pressure outside an airplane is 0.25 atmosphere. How high is the plane?
b. A mountain climber decides she will wear an oxygen mask once she has reached an altitude of 7,000 meters. What is the atmospheric pressure at this altitude?

71. **POPULATION GROWTH** Based on the estimate that there are 10 billion acres of arable land on the earth and that each acre can produce enough food to feed 4 people, some demographers believe that the earth can support a population of no more than 40 billion people. The population of the earth

was approximately 3 billion in 1960 and 4 billion in 1975. If the population of the earth were growing exponentially, when would it reach the theoretical limit of 40 billion?

72. **POPULATION GROWTH**   According to a logistic model based on the assumption that the earth can support no more than 40 billion people, the world's population (in billions) t years after 1960 is given by a function of the form $P(t) = \dfrac{40}{1 + Ce^{-kt}}$, where C and k are positive constants. Find the function of this form that is consistent with the fact that the world's population was approximately 3 billion in 1960 and 4 billion in 1975. What does your model predict for the population in the year 2000? Check the accuracy of the model by consulting an almanac.

73. **ALLOMETRY**   Suppose that for the first 6 years of a moose's life, its shoulder height $H(t)$ and tip-to-tip antler length $A(t)$ increase with time t (years) according to the formulas $H(t) = 125e^{0.08t}$ and $A(t) = 50e^{0.16t}$, where H and A are both measured in centimeters (cm).
   a. On the same graph, sketch $y = H(t)$ and $y = A(t)$ for the applicable period $0 \le t \le 6$.
   b. Express antler length A as a function of height H. [*Hint:* First take logarithms on both sides of the equation $H = 125e^{0.08t}$ to express time t in terms of H and then substitute into the formula for $A(t)$.]

74. In each case, use one of the laws of exponents to prove the indicated law of logarithms.
   a. The quotient rule: $\ln \dfrac{u}{v} = \ln u - \ln v$
   b. The power rule: $\ln u^r = r \ln u$

75. Show that the reflection of the point (a, b) in the line $y = x$ is (b, a). [*Hint:* Show that the line joining (a, b) and (b, a) is perpendicular to $y = x$ and that the distance from (a, b) to $y = x$ is the same as the distance from $y = x$ to (b, a).]

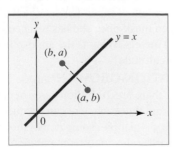

PROBLEM 75

76. Sketch the graph of $y = \log_b x$ for $0 < b < 1$ by reflecting the graph of $y = b^x$ in the line $y = x$. Then answer these questions:
   a. Is the graph of $y = \log_b x$ rising or falling for $x > 0$?
   b. Is the graph concave upward or concave downward for $x > 0$?
   c. What are the intercepts of the graph? Does the graph have any horizontal or vertical asymptotes?
   d. What can be said about
$$\lim_{x \to +\infty} \log_b x \quad \text{and} \quad \lim_{x \to 0^+} \log_b x?$$

77. Show that if y is a power function of x, so that $y = Cx^k$ where C and k are constants, then $\ln y$ is a linear function of $\ln x$. (*Hint:* Take the logarithm on both sides of the equation $y = Cx^k$.)

78. Use the graphing utility of your calculator to graph $y = 10^x$, $y = x$, and $y = \log_{10} x$ on the same coordinate axes (use $[-5, 5]1$ by $[-5, 5]1$). How are these graphs related?

In Problems 79 through 82 solve for x.

79. $x = \ln(3.42 \times 10^{-8.1})$

80. $3{,}500e^{0.31x} = \dfrac{e^{-3.5x}}{1 + 257e^{-1.1x}}$

81. $e^{0.113x} + 4.72 = 7.031 - x$

82. $\ln(x + 3) - \ln x = 5 \ln(x^2 - 4)$

83. Let a and b be any positive numbers other than 1.
   a. Show that $(\log_a b)(\log_b a) = 1$.
   b. Show that $\log_a x = \dfrac{\log_b x}{\log_b a}$ for any number $x > 0$.

SECTION 4.3 | Differentiation of Logarithmic and Exponential Functions

Exponential functions and logarithmic functions both have simple derivatives. In this section, we will obtain derivative formulas for these functions and use them to analyze a few practical problems. We begin with the derivative formula for $\ln x$.

The Derivative of $\ln x$

$$\frac{d}{dx}(\ln x) = \frac{1}{x} \quad \text{for } x > 0$$

The derivation of this formula is given at the end of this section, after we examine a variety of examples involving its use.

EXPLORE!

Graph $y = \ln x$ using a modified decimal window, $[-0.7, 8.7]1$ by $[-3.1, 3.1]1$. Choose a value of x and construct the tangent line to the curve at this x. Observe how close the slope of the tangent line is to $1/x$. Repeat this for several additional values of x.

EXAMPLE | 4.3.1

Differentiate the function $f(x) = x \ln x$.

Solution

Combine the product rule with the formula for the derivative of $\ln x$ to get

$$f'(x) = x\left(\frac{1}{x}\right) + \ln x = 1 + \ln x$$

Using the rules for logarithms can simplify the differentiation of complicated expressions. In Example 4.3.2, we use the power rule for logarithms before differentiating.

EXAMPLE | 4.3.2

Differentiate $f(x) = \dfrac{\ln \sqrt[3]{x^2}}{x^4}$.

Solution

First, since $\sqrt[3]{x^2} = x^{2/3}$, the power rule for logarithms allows us to write

$$f(x) = \frac{\ln \sqrt[3]{x^2}}{x^4} = \frac{\ln x^{2/3}}{x^4} = \frac{\frac{2}{3}\ln x}{x^4}$$

Then, by the quotient rule, we find

$$f'(x) = \frac{2}{3}\left[\frac{x^4(\ln x)' - (x^4)'\ln x}{(x^4)^2}\right]$$

$$= \frac{2}{3}\left[\frac{x^4\left(\dfrac{1}{x}\right) - 4x^3 \ln x}{x^8}\right]$$

$$= \frac{2}{3}\left[\frac{1 - 4\ln x}{x^5}\right] \qquad \textit{cancel common } x^3 \textit{ terms}$$

EXAMPLE | 4.3.3

Differentiate $g(t) = (t + \ln t)^{3/2}$.

Solution

The function has the form $g(t) = u^{3/2}$, where $u = t + \ln t$, and by applying the general power rule, we find

$$g'(t) = \frac{d}{dt}(u^{3/2}) = \frac{3}{2} u^{1/2} \frac{du}{dt}$$

$$= \frac{3}{2} (t + \ln t)^{1/2} \frac{d}{dt}(t + \ln t)$$

$$= \frac{3}{2} (t + \ln t)^{1/2} \left(1 + \frac{1}{t}\right)$$

If $f(x) = \ln u(x)$, where $u(x)$ is a differentiable function of x, then the chain rule yields the following formula for $f'(x)$.

The Chain Rule for Logarithmic Functions   ■   If $u(x)$ is a differentiable function of x, then

$$\frac{d}{dx}[\ln u(x)] = \frac{1}{u(x)} \frac{du}{dx}$$

EXAMPLE | 4.3.4

Differentiate the function $f(x) = \ln(2x^3 + 1)$.

Solution

Here, we have $f(x) = \ln u$, where $u(x) = 2x^3 + 1$. Thus,

$$f'(x) = \frac{1}{u} \frac{du}{dx} = \frac{1}{2x^3 + 1} \frac{d}{dx}(2x^3 + 1)$$

$$= \frac{2(3x^2)}{2x^3 + 1} = \frac{6x^2}{2x^3 + 1}$$

EXAMPLE | 4.3.5

Find an equation for the tangent line to the graph of $f(x) = x - \ln \sqrt{x}$ at the point where $x = 1$.

Solution

When $x = 1$, we have

$$y = f(1) = 1 - \ln(\sqrt{1}) = 1 - 0 = 1$$

so the point of tangency is (1, 1). To find the slope of the tangent line at this point, we first write

$$f(x) = x - \ln \sqrt{x} = x - \frac{1}{2} \ln x$$

and compute the derivative

$$f'(x) = 1 - \frac{1}{2}\left(\frac{1}{x}\right) = 1 - \frac{1}{2x}$$

Thus, the tangent line passes through the point (1, 1) with slope

$$f'(1) = 1 - \frac{1}{2(1)} = \frac{1}{2}$$

so it has the equation

$$\frac{y - 1}{x - 1} = \frac{1}{2} \qquad \qquad point\text{-}slope\ formula$$

or equivalently,

$$y = \frac{1}{2}x + \frac{1}{2}$$

EXAMPLE 4.3.6

A manufacturer determines that x units of a particular luxury item will be sold when the price is $p(x) = 112 - x \ln x^3$ hundred dollars per unit.

a. Find the revenue and marginal revenue functions.

b. Use marginal analysis to estimate the revenue obtained from producing the fifth unit. What is the actual revenue obtained from producing the fifth unit?

Solution

a. The revenue is

$$R(x) = xp(x) = x(112 - x \ln x^3) = 112x - x^2(3 \ln x)$$

hundred dollars, and the marginal revenue is

$$R'(x) = 112 - 3\left[x^2\left(\frac{1}{x}\right) + (2x) \ln x \right] = 112 - 3x - 6x \ln x$$

b. The revenue obtained from producing the fifth unit is estimated by the marginal revenue evaluated at $x = 4$; that is, by

$$R'(4) = 112 - 3(4) - 6(4) \ln(4) \approx 66.73$$

hundred dollars ($6,673) for the additional unit. The actual revenue obtained from producing the fifth unit is

$$R(5) - R(4) = [112(5) - 3(5)^2 \ln 5] - [112(4) - 3(4)^2 \ln 4]$$
$$= 439.29 - 381.46 = 57.83$$

hundred dollars ($5,783).

Exponential Functions

To find the formula for the derivative of e^x, differentiate both sides of the equation

$$\ln e^x = x$$

with respect to x, using the chain rule for logarithms to differentiate $\ln e^x$. You get

$$\frac{\frac{d}{dx}(e^x)}{e^x} = 1 \qquad \text{or} \qquad \frac{d}{dx}(e^x) = e^x$$

EXPLORE!

Graph $y = e^x$ using a modified decimal window, $[-0.7, 8.7]1$ by $[-0.1, 6.1]1$. Trace the curve to any value of x and determine the value of the derivative at this point. Observe how close the derivative value is to the y coordinate of the graph. Repeat this for several values of x.

The Derivative of the Exponential Function

$$\frac{d}{dx}(e^x) = e^x \qquad \text{for every real number } x$$

The fact that e^x is its own derivative means that at each point $P(c, e^c)$ on the curve $y = e^x$, the slope is equal to e^c, the y coordinate of P (see Figure 4.8). This is one of the most important reasons for using e as the base for exponential functions in calculus.

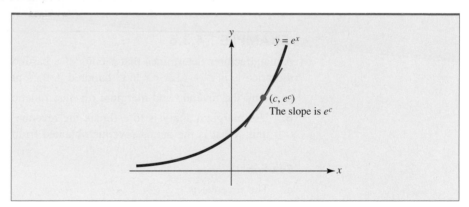

FIGURE 4.8   At each point $P(c, e^c)$ on the graph of $y = e^x$, the slope equals e^c.

By using the chain rule in conjunction with the differentiation formula

$$\frac{d}{dx}(e^x) = e^x$$

we obtain this formula for differentiating general exponential functions.

The Chain Rule for Exponential Functions   ▪   If $u(x)$ is a differentiable function of x, then

$$\frac{d}{dx}(e^{u(x)}) = e^{u(x)}\frac{du}{dx}$$

EXAMPLE 4.3.7

Differentiate the function $f(x) = e^{x^2+1}$.

Solution

Using the chain rule with $u = x^2 + 1$, we find

$$f'(x) = e^{x^2+1}\left[\frac{d}{dx}(x^2 + 1)\right] = 2xe^{x^2+1}$$

EXAMPLE 4.3.8

Differentiate the function

$$f(x) = \frac{e^{-3x}}{x^2 + 1}$$

Solution

Using the chain rule together with the quotient rule, you get

$$f'(x) = \frac{(x^2 + 1)(-3e^{-3x}) - (2x)e^{-3x}}{(x^2 + 1)^2}$$

$$= e^{-3x}\left[\frac{-3(x^2 + 1) - 2x}{(x^2 + 1)^2}\right] = e^{-3x}\left[\frac{-3x^2 - 2x - 3}{(x^2 + 1)^2}\right]$$

EXAMPLE 4.3.9

Find the largest and the smallest values of the function $f(x) = xe^{2x}$ on the interval $-1 \le x \le 1$.

Solution

By the product rule

$$f'(x) = x\frac{d}{dx}(e^{2x}) + e^{2x}\frac{d}{dx}(x) = x(2e^{2x}) + e^{2x}(1) = (2x + 1)e^{2x}$$

so $f'(x) = 0$ when

$$(2x + 1)e^{2x} = 0$$
$$2x + 1 = 0 \qquad \text{since } e^{2x} > 0 \text{ for all } x$$
$$x = -\frac{1}{2}$$

Evaluating $f(x)$ at the critical number $x = -\frac{1}{2}$ and at the endpoints of the interval, $x = -1$ and $x = 1$, we find that

$$f(-1) = (-1)e^{-2} \approx -0.135$$
$$f\left(-\frac{1}{2}\right) = \left(-\frac{1}{2}\right)e^{-1} \approx -0.184 \qquad \textit{minimum}$$
$$f(1) = (1)e^2 \approx 7.389 \qquad \textit{maximum}$$

Thus, $f(x)$ has its largest value 7.389 at $x = 1$ and its smallest value -0.184 at $x = -\dfrac{1}{2}$.

Exponential Growth and Decay

Recall (from Section 4.1) that a quantity $Q(x) = Q_0 e^{kx}$ is said to **grow exponentially** if $k > 0$ and **decay exponentially** if $k < 0$. In both cases, notice that

$$Q'(x) = Q_0(e^{kx})(k) = k(Q_0 e^{kx}) = kQ(x)$$

and that the percentage rate of change of Q is given by

$$\frac{100Q'(x)}{Q(x)} = \frac{100(kQ)}{Q} = 100k$$

To summarize:

> **Exponential Growth and Decay** ■ If $Q(x) = Q_0 e^{kx}$, then $Q'(x) = kQ(x)$ and $Q(x)$ changes at the percentage rate $100k$.

NOTE Observe that this is consistent with what you already know about compound interest; namely, that if interest is compounded continuously, the balance after t years is $B(t) = Pe^{rt}$, where r is the interest rate expressed as a decimal. ■

EXAMPLE | 4.3.10

A demographer studying a certain community uses an exponential model $P(t) = P_0 e^{kt}$ for the population, where t is the number of years after 1990. If the population was 100,000 in 1990 and 145,000 in 2003, at what annual percentage rate is the population increasing?

Solution

For simplicity, measure the population $P(t)$ in thousands of individuals. Since the population is 100,000 when $t = 0$ (1990) and 145,000 when $t = 13$ (2003), we have

$$P_0 = P(0) = 100$$

and

$$P(13) = 100e^{k(13)} = 145$$

$$e^{k(13)} = \frac{145}{100} = 1.45$$

$$\ln(e^{k(13)}) = \ln(1.45) \qquad \text{\textit{take logarithms on both sides}}$$
$$k(13) = 0.3716 \qquad\qquad \text{\textit{use} } \ln(e^x) = x$$

$$k = \frac{0.3716}{13} = 0.0286$$

Thus, the population is increasing at the annual rate of $100k = 2.86\%$.

In Example 4.3.11, we use marginal analysis to examine exponential demand for a certain commodity and to determine the price at which the revenue obtained from satisfying the demand is maximized.

EXAMPLE | 4.3.11

A manufacturer determines that $D(p) = 5,000e^{-0.02p}$ units of a particular commodity will be demanded (sold) when the price is p dollars per unit.

a. Find the elasticity of demand for this commodity. For what values of p is the demand elastic, inelastic, and of unit elasticity?

b. If the price is increased by 3% from $40, what is the expected effect on demand?

c. Find the revenue $R(p)$ obtained by selling $q = D(p)$ units at p dollars per unit. For what value of p is the revenue maximized?

Solution

a. According to the formula derived in Section 3.4, the elasticity of demand is given by

$$F(p) = \frac{p}{q}\frac{dq}{dp}$$

$$= \left(\frac{p}{5,000e^{-0.02p}}\right)[5,000e^{-0.02p}(-0.02)]$$

$$= \frac{p[5,000(-0.02)e^{-0.02p}]}{5,000e^{-0.02p}} = -0.02p$$

You find that

$$|E(p)| = |-0.02p| = 0.02p$$

so

demand is **of unit elasticity** when $|E(p)| = 0.02p = 1$; that is, when $p = 50$

demand is **elastic** when $|E(p)| = 0.02p > 1$; or $p > 50$

demand is **inelastic** when $|E(p)| = 0.02p < 1$; or $p < 50$

The graph of the demand function, showing levels of elasticity, is displayed in Figure 4.9a.

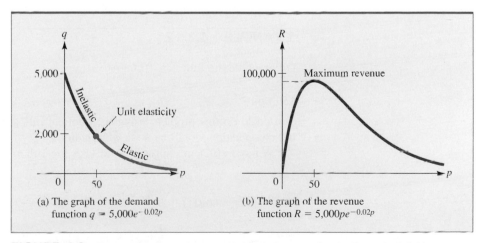

(a) The graph of the demand function $q = 5,000e^{-0.02p}$

(b) The graph of the revenue function $R = 5,000pe^{-0.02p}$

FIGURE 4.9 Demand and revenue curves for the commodity in Example 4.3.11.

b. When $p = 40$, the demand is

$$q(40) = 5{,}000e^{-0.02(40)} \approx 2{,}247 \text{ units}$$

and the elasticity of demand is

$$E(p) = -0.02(40) = -0.8$$

Thus, an increase of 1% in price from $p = \$40$ will result in a decrease in the quantity demanded by approximately 0.8%. Consequently, an increase of 3% in price, from $40 to $41.20, results in a decrease in demand of approximately $2{,}247[3(0.008)] = 54$ units, from 2,247 to 2,193 units.

c. The revenue function is

$$R(p) = pq = 5{,}000pe^{-0.02p}$$

for $p \geq 0$ (only nonnegative prices have economic meaning), with derivative

$$R'(p) = 5{,}000(-0.02pe^{-0.02p} + e^{-0.02p})$$
$$= 5{,}000(1 - 0.02p)e^{-0.02p}$$

Since $e^{-0.02p}$ is always positive, $R'(p) = 0$ if and only if

$$1 - 0.02p = 0 \qquad \text{or} \qquad p = \frac{1}{0.02} = 50$$

To verify that $p = 50$ actually gives the absolute maximum, note that

$$R''(p) = 5{,}000(0.0004p - 0.04)e^{-0.02p}$$

so

$$R''(50) = 5{,}000[0.0004(50) - 0.04]e^{-0.02(50)} \approx -37 < 0$$

Thus, the second derivative test tells you that the absolute maximum of $R(p)$ does indeed occur when $p = 50$ (see Figure 4.9b).

Logarithmic Differentiation

Differentiating a function that involves products, quotients, or powers can often be simplified by first taking the logarithm of the function. This technique, called **logarithmic differentiation,** is illustrated in Example 4.3.12.

EXAMPLE 4.3.12

Differentiate the function $f(x) = \dfrac{\sqrt[3]{x + 1}}{(1 - 3x)^4}$.

Solution

You could do this problem using the quotient rule and the chain rule, but the resulting computation would be somewhat tedious. (Try it!)

A more efficient approach is to take logarithms of both sides of the expression for f:

$$\ln f(x) = \ln\left[\frac{\sqrt[3]{x + 1}}{(1 - 3x)^4}\right] = \ln \sqrt[3]{(x + 1)} - \ln (1 - 3x)^4$$

$$= \frac{1}{3} \ln (x + 1) - 4 \ln (1 - 3x)$$

(Notice that by introducing the logarithm, you eliminate the quotient, the cube root, and the fourth power.)

Now use the chain rule for logarithms to differentiate both sides of this equation to get

$$\frac{f'(x)}{f(x)} = \frac{1}{3}\frac{1}{(x+1)} - 4\left(\frac{-3}{1-3x}\right) = \frac{1}{3}\frac{1}{x+1} + \frac{12}{1-3x}$$

so that

$$f'(x) = f(x)\left[\frac{1}{3}\frac{1}{x+1} + \frac{12}{1-3x}\right]$$

$$= \left[\frac{\sqrt[3]{x+1}}{(1-3x)^4}\right]\left[\frac{1}{3}\frac{1}{x+1} + \frac{12}{1-3x}\right]$$

Occasionally, you may need to differentiate an exponential function $y = b^x$ or a logarithmic function $y = \log_b x$ with base $b \neq e$. Example 4.3.13 shows how logarithmic differentiation can be used to handle such cases. In Problem 82, you are asked to find general formulas for such derivatives.

EXAMPLE | 4.3.13

Differentiate each of these functions:

a. $y = 2^x$

b. $y = \log_3 x$

Solution

a. To differentiate $y = 2^x$, we use logarithmic differentiation as follows:

$$y = 2^x$$

$\ln y = x(\ln 2)$ *take natural logarithms on both sides*

$\dfrac{1}{y}\dfrac{dy}{dx} = \ln 2$ *differentiate both sides*

$\dfrac{dy}{dx} = (\ln 2)\,y$ *multiply both sides by y*

$\dfrac{dy}{dx} = (\ln 2)\,2^x$ *substitute $y = 2^x$*

b. In this case, we find

$$y = \log_3 x$$

$3^y = x$ *definition of logarithm*

$y \ln 3 = \ln x$ *take natural logarithms on both sides*

$(\ln 3)\dfrac{dy}{dx} = \dfrac{1}{x}$ *differentiate both sides*

$\dfrac{dy}{dx} = \dfrac{1}{x(\ln 3)}$ *divide by $\ln 3$*

If $Q(x)$ is a differentiable function of x, note that

$$\frac{d}{dx}(\ln Q) = \frac{Q'(x)}{Q(x)}$$

where the ratio on the right is the relative rate of change of $Q(x)$. That is, *the relative rate of change of a quantity $Q(x)$ can be computed by finding the derivative of* $\ln Q$. This special kind of logarithmic differentiation can be used to simplify the computation of various growth rates, as illustrated in Example 4.3.14.

EXAMPLE | 4.3.14

A country exports three goods, wheat W, steel S, and oil O. Suppose at a particular time $t = t_0$, the revenue (in billions of dollars) derived from each of these goods is

$$W(t_0) = 4 \qquad S(t_0) = 7 \qquad O(t_0) = 10$$

and that S is growing at 8%, O is growing at 15%, while W is declining at 3%. At what relative rate is total export revenue growing at this time?

Solution

Let $R = W + S + O$. At time $t = t_0$, we know that

$$R(t_0) = W(t_0) + S(t_0) + O(t_0) = 4 + 7 + 10 = 21$$

and

$$\frac{W'(t_0)}{W(t_0)} = -0.03 \qquad \frac{S'(t_0)}{S(t_0)} = 0.08 \qquad \frac{O'(t_0)}{O(t_0)} = 0.15$$

so that

$$W'(t_0) = -0.03W(t_0) \qquad S'(t_0) = 0.08S(t_0) \qquad O'(t_0) = 0.15\,O(t_0)$$

Thus, at $t = t_0$, the relative rate of growth of R is

$$\frac{R'(t_0)}{R(t_0)} = \frac{d(\ln R)}{dt} = \frac{d}{dt}\left[\ln(W + S + O)\right]\Big|_{t=t_0}$$

$$= \frac{[W'(t_0) + S'(t_0) + O'(t_0)]}{[W(t_0) + S(t_0) + O(t_0)]}$$

$$= \frac{-0.03W(t_0) + 0.08S(t_0) + 0.15\,O(t_0)}{W(t_0) + S(t_0) + O(t_0)}$$

$$= \frac{-0.03W(t_0) + 0.08S(t_0) + 0.15\,O(t_0)}{R(t_0)}$$

$$= \frac{-0.03W(t_0)}{R(t_0)} + \frac{0.08S(t_0)}{R(t_0)} + \frac{0.15\,O(t_0)}{R(t_0)}$$

$$= \frac{-0.03(4)}{21} + \frac{0.08(7)}{21} + \frac{0.15(10)}{21}$$

$$\approx 0.0924$$

That is, at time $t = t_0$, the total revenue obtained from the three exported goods is increasing at the rate of 9.24%.

The Proof That

$$\frac{d}{dx} (\ln x) = \frac{1}{x}$$

The proof that $\frac{d}{dx} (\ln x) = \frac{1}{x}$ is based on the fact that $\left(1 + \frac{1}{n}\right)^n$ approaches e as n increases (or decreases) without bound. To derive the formula for the derivative of $f(x) = \ln x$, form the difference quotient and rewrite it using the properties of logarithms as follows:

$$\frac{f(x + h) - f(x)}{h} = \frac{\ln (x + h) - \ln x}{h} \qquad \text{for fixed } x > 0$$

$$= \frac{1}{h} \ln\left(\frac{x + h}{x}\right)$$

$$= \ln\left(\frac{x + h}{x}\right)^{1/h}$$

$$= \ln\left(1 + \frac{h}{x}\right)^{1/h}$$

To find the derivative of $\ln x$, let h approach zero in the simplified difference quotient. This will be easier to do if you first let $n = \frac{x}{h}$. Then,

$$\frac{h}{x} = \frac{1}{n} \qquad \text{and} \qquad \frac{1}{h} = \frac{n}{x}$$

and so

$$\frac{f(x + h) - f(x)}{h} = \ln\left(1 + \frac{1}{n}\right)^{n/x} = \ln\left[\left(1 + \frac{1}{n}\right)^n\right]^{1/x}$$

As h approaches zero, $n = \frac{x}{h}$ increases or decreases without bound, depending on the sign of h. Since

$$\lim_{n \to \infty} \left(1 + \frac{1}{n}\right)^n = e$$

it follows that

$$\frac{d}{dx} (\ln x) = \lim_{h \to 0} \frac{f(x + h) - f(x)}{h}$$

$$= \lim_{n \to \infty} \ln\left[\left(1 + \frac{1}{n}\right)^n\right]^{1/x}$$

$$= \ln\left[\lim_{n \to \infty} \left(1 + \frac{1}{n}\right)^n\right]^{1/x}$$

$$= \ln e^{1/x} = \frac{1}{x}$$

as claimed.

PROBLEMS 4.3

In Problems 1 through 34 differentiate the given function.

1. $f(x) = e^{5x}$

2. $f(x) = 3e^{4x+1}$

3. $f(x) = xe^x$

4. $f(x) = \dfrac{e^x}{x}$

5. $f(x) = 30 + 10e^{-0.05x}$

6. $f(x) = e^{x^2+2x-1}$

7. $f(x) = (x^2 + 3x + 5)e^{6x}$

8. $f(x) = xe^{-x^2}$

9. $f(x) = (1 - 3e^x)^2$

10. $f(x) = \sqrt{1 + e^x}$

11. $f(x) = e^{\sqrt{3x}}$

12. $f(x) = e^{1/x}$

13. $f(x) = \ln x^3$

14. $f(x) = \ln 2x$

15. $f(x) = x^2 \ln x$

16. $f(x) = x \ln \sqrt{x}$

17. $f(x) = \sqrt[3]{e^{2x}}$

18. $f(x) = \dfrac{\ln x}{x}$

19. $f(x) = \ln\left(\dfrac{x + 1}{x - 1}\right)$

20. $f(x) = e^x \ln x$

21. $f(x) = e^{-2x} + x^3$

22. $f(t) = t^2 \ln \sqrt[3]{t}$

23. $g(s) = (e^s + s + 1)(2e^{-s} + s)$

24. $F(x) = \ln(2x^3 - 5x + 1)$

25. $h(t) = \dfrac{e^t + t}{\ln t}$

26. $g(u) = \ln(u^2 - 1)^3$

27. $f(x) = \dfrac{e^x + e^{-x}}{2}$

28. $h(x) = \dfrac{e^{-x}}{x^2}$

29. $f(t) = \sqrt{\ln t + t}$

30. $f(x) = \dfrac{e^x + e^{-x}}{e^x - e^{-x}}$

31. $f(x) = \ln(e^{-x} + x)$

32. $f(s) = e^{s+\ln s}$

33. $g(u) = \ln(u + \sqrt{u^2 + 1})$

34. $L(x) = \ln\left[\dfrac{x^2 + 2x - 3}{x^2 + 2x + 1}\right]$

In Problems 35 through 38, find the largest and smallest values of the given function over the prescribed closed, bounded interval.

35. $f(x) = \dfrac{\ln(x + 1)}{x + 1}$ for $0 \le x \le 2$

36. $F(x) = e^{x^2-2x}$ for $0 \le x \le 2$

37. $g(t) = t^{3/2}e^{-2t}$ for $0 \le t \le 1$

38. $h(s) = 2s \ln s - s^2$ for $0.5 \le s \le 2$

In Problems 39 through 44, find an equation for the tangent line to $y = f(x)$ at the specified point.

39. $f(x) = xe^{-x}$; where $x = 0$

40. $f(x) = (x + 1)e^{-2x}$; where $x = 0$

41. $f(x) = \dfrac{e^{2x}}{x^2}$; where $x = 1$

42. $f(x) = \dfrac{\ln x}{x}$; where $x = 1$

43. $f(x) = x^2 \ln \sqrt{x}$; where $x = 1$

44. $f(x) = x - \ln x$; where $x = e$

In Problems 45 through 48, find the second derivative of the given function.

45. $f(x) = e^{2x} + 2e^{-x}$

46. $f(x) = \ln(2x) + x^2$

47. $f(t) = t^2 \ln t$

48. $g(t) = t^2 e^{-t}$

In Problems 49 through 56, use logarithmic differentiation to find the derivative $f'(x)$.

49. $f(x) = (2x + 3)^2(x - 5x^2)^{1/2}$

50. $f(x) = x^2 e^{-x}(3x + 5)^3$

51. $f(x) = \dfrac{(x + 2)^5}{\sqrt[6]{3x - 5}}$

52. $f(x) = \sqrt[4]{\dfrac{2x + 1}{1 - 3x}}$

53. $f(x) = (x + 1)^3(6 - x)^2\sqrt[3]{2x + 1}$

54. $f(x) = \dfrac{e^{-3x}\sqrt{2x - 5}}{(6 - 5x)^4}$

55. $f(x) = 5^{x^2}$

56. $f(x) = \log_2(\sqrt{x})$

MARGINAL ANALYSIS   *In Problems 57 through 60, the demand function $q = D(p)$ for a particular commodity is given in terms of a price p per unit at which all q units can be sold. In each case:*

> (a) Find the elasticity of demand and determine the values of p for which the demand is elastic, inelastic, and of unit elasticity.
>
> (b) If the price is increased by 2% from \$15, what is the approximate effect on demand?
>
> (c) Find the revenue $R(p)$ obtained by selling q units at the unit price p. For what value of p is revenue maximized?

57. $D(p) = 3,000e^{-0.04p}$

58. $D(p) = 10,000e^{-0.025p}$

59. $D(p) = 5,000(p + 11)e^{-0.1p}$

60. $D(p) = \dfrac{10,000e^{-p/10}}{p + 1}$

MARGINAL ANALYSIS   *In Problems 61 through 64, the cost $C(x)$ of producing x units of a particular commodity is given. In each case:*

> (a) Find the marginal cost $C'(x)$.
>
> (b) Determine the level of production x for which the average cost $A(x) = C(x)/x$ is minimized.

61. $C(x) = e^{0.2x}$

62. $C(x) = 100e^{0.01x}$

63. $C(x) = 12\sqrt{x}\, e^{x/10}$

64. $C(x) = x^2 + 10xe^{-x}$

65. **POPULATION GROWTH**   It is projected that t years from now, the population of a certain country will become $P(t) = 50e^{0.02t}$ million.

   a. At what rate will the population be changing with respect to time 10 years from now?

   b. At what percentage rate will the population be changing with respect to time t years from now? Does this percentage rate depend on t or is it constant?

66. **COMPOUND INTEREST**   Money is deposited in a bank offering interest at an annual rate of 6% compounded continuously. Find the percentage rate of change of the balance with respect to time.

67. **DEPRECIATION**   A certain industrial machine depreciates so that its value after t years becomes $Q(t) = 20,000e^{-0.4t}$ dollars.

   a. At what rate is the value of the machine changing with respect to time after 5 years?

   b. At what percentage rate is the value of the machine changing with respect to time after t years? Does this percentage rate depend on t or is it constant?

68. **COOLING**   A cool drink is removed from a refrigerator on a hot summer day and placed in a room whose temperature is 30° Celsius. According to Newton's law of cooling, the temperature of the drink t minutes later is given by a function of the form $f(t) = 30 - Ae^{-kt}$. Show that the rate of change of the temperature of the drink with respect to time is proportional to the difference between the temperature of the room and that of the drink.

69. **MARGINAL ANALYSIS**   The mathematics editor at a major publishing house estimates that if x thousand complimentary copies are distributed to professors, the first-year sales of a certain new text will be $f(x) = 20 - 15e^{-0.2x}$ thousand copies. Currently, the editor is planning to distribute 10,000 complimentary copies.

   a. Use marginal analysis to estimate the increase in first-year sales that will result if 1,000 additional complimentary copies are distributed.

   b. Calculate the actual increase in first-year sales that will result from the distribution of the additional 1,000 complimentary copies. Is the estimate in part (a) a good one?

70. ECOLOGY In a model developed by John Helms,* the water evaporation $E(T)$ for a ponderosa pine is given by

$$E(T) = 4.6e^{17.3T/(T+237)}$$

where T (degrees Celsius) is the surrounding air temperature.

   a. What is the rate of evaporation when $T = 30°C$?
   b. What is the percentage rate of evaporation? At what temperature does the percentage rate of evaporation first drop below 0.5?

71. LEARNING According to the Ebbinghaus model (recall Problem 50, Section 4.1), the fraction $F(t)$ of subject matter you will remember from this course t months after the final exam can be estimated by the formula $F(t) = B + (1 - B)e^{-kt}$, where B is the fraction of the material you will never forget and k is a constant that depends on the quality of your memory.

   a. Find $F'(t)$ and explain what this derivative represents.
   b. Show that $F'(t)$ is proportional to $F - B$ and interpret this result. [*Hint:* What does $F - B$ represent in terms of what you remember?]
   c. Sketch the graph of $F(t)$ for the case where $B = 0.3$ and $k = 0.2$.

72. ALCOHOL ABUSE CONTROL Suppose the percentage of alcohol in the blood t hours after consumption is given by

$$C(t) = 0.12te^{-t/2}$$

   a. At what rate is the blood alcohol level changing at time t?
   b. How much time passes before the blood alcohol level begins to decrease?
   c. Suppose the legal limit for blood alcohol is 0.04%. How much time must pass before the blood alcohol reaches this level? At what rate is the blood alcohol level decreasing when it reaches the legal limit?

73. CONSUMER EXPENDITURE The demand for a certain commodity is $D(p) = 3,000\,e^{-0.01p}$ units per month when the market price is p dollars per unit.

   a. At what rate is the consumer expenditure $E(p) = pD(p)$ changing with respect to price p?
   b. At what price does consumer expenditure stop increasing and begin to decrease?
   c. At what price does the *rate* of consumer expenditure begin to increase? Interpret this result.

---

*John A. Helms, "Environmental Control of Net Photosynthesis in Naturally Growing Pinus Ponderosa Nets," *Ecology,* Winter, 1972, p. 92.

74. LEARNING In an experiment to test memory learning, a subject is confronted by a series of tasks, and it is found that t minutes after the experiment begins, the number of tasks successfully completed is

$$R(t) = \frac{15(1 - e^{-0.01t})}{1 + 1.5e^{-0.01t}}$$

   a. For what values of t is the learning function $R(t)$ increasing? For what values is it decreasing?
   b. When is the rate of change of the learning function $R(t)$ increasing? When is it decreasing? Interpret your results.

75. ENDANGERED SPECIES An international agency determines that the number of individuals of an endangered species that remain in the wild t years after a protection policy is instituted may be modeled by

$$N(t) = \frac{600}{1 + 3e^{-0.02t}}$$

   a. At what rate is the population changing at time t? When is the population increasing? When is it decreasing?
   b. When is the rate of change of the population increasing? When is it decreasing? Interpret your results.
   c. What happens to the population in the "long run" (as $t \rightarrow +\infty$)?

76. ENDANGERED SPECIES The agency in Problem 75 studies a second endangered species but fails to receive funding to develop a policy of protection. The population of the species is modeled by

$$N(t) = \frac{30 + 500e^{-0.3t}}{1 + 5e^{-0.3t}}$$

   a. At what rate is the population changing at time t? When is the population increasing? When is it decreasing?
   b. When is the rate of change of the population increasing? When is it decreasing? Interpret your results.
   c. What happens to the population in the "long run" (as $t \rightarrow +\infty$)?

77. PLANT GROWTH Two plants grow in such a way that t days after planting, they are $P_1(t)$ and $P_2(t)$ centimeters tall, respectively, where

$$P_1(t) = \frac{21}{1 + 25e^{-0.3t}} \quad \text{and} \quad P_2(t) = \frac{20}{1 + 17e^{-0.6t}}$$

   a. At what rate is the first plant growing at time $t = 10$ days? Is the rate of growth of the second plant increasing or decreasing at this time?

b. At what time do the two plants have the same height? What is this height? Which plant is growing more rapidly when they have the same height?

78. **PER CAPITA GROWTH** The national income $I(t)$ of a particular country is increasing by 2.3% per year, while the population $P(t)$ of the country is decreasing at the annual rate of 1.75%. The per capita income C is defined to be $C(t) = \dfrac{I(t)}{P(t)}$.

   a. Find the derivative of $\ln C(t)$.

   b. Use the result of part (a) to determine the percentage rate of growth of per capita income.

79. **REVENUE GROWTH** A country exports electronic components E and textiles T. Suppose at a particular time $t = t_0$, the revenue (in billions of dollars) derived from each of these goods is

   $$E(t_0) = 11 \quad \text{and} \quad T(t_0) = 8$$

   and that E is growing at 9%, while T is declining at 2%. At what relative rate is total export revenue $R = E + T$ changing at this time?

80. A quantity grows so that $Q(t) = Q_0 \dfrac{e^{kt}}{t}$. Find the percentage rate of change of Q with respect to t.

81. **POPULATION GROWTH** It is projected that x years from now the population of a certain town will be approximately $P(x) = 5,000\sqrt{x^2 + 4x + 19}$. At what percentage rate will the population be changing with respect to time 3 years from now?

82. **DERIVATIVES OF b^x AND $\log_b x$ FOR BASE $b \neq e$** Let b be a positive number other than 1 $(b > 0, b \neq 1)$.

   a. Show that

   $$\frac{d}{dx}(b^x) = (\ln b)b^x$$

   b. Show that

   $$\frac{d}{dx}(\log_b x) = \frac{1}{(\ln b)}\frac{1}{x}$$

In Problems 83 through 86, use the formulas in Problem 78 to differentiate the given function.

83. $f(x) = \dfrac{2^x}{x}$

84. $f(x) = x^2 3^{x^2}$

85. $f(x) = x \log_{10} x$

86. $f(x) = \dfrac{\log_2 x}{\sqrt{x}}$

87. Use a numerical differentiation utility to find $f'(c)$, where $c = 0.65$ and

   $$f(x) = \ln\left[\frac{\sqrt[3]{x+1}}{(1+3x)^4}\right]$$

   Then use a graphing utility to sketch the graph of $f(x)$ and to draw the tangent line at the point where $x = c$.

88. Repeat Problem 87 with the function

   $$f(x) = (3.7x^2 - 2x + 1)e^{-3x+2}$$

   and $c = -2.17$.

SECTION 4.4 Additional Exponential Models

Earlier in this chapter, we examined applied topics such as continuous compounding, exponential growth and decay, and carbon dating. In this section, we introduce several additional models from a variety of areas such as business and economics, biology, psychology, demography, and sociology. We begin with two examples illustrating the particular issues that arise in sketching exponential and logarithmic graphs.

Curve Sketching

As when graphing polynomials or rational functions, the key to graphing a function $f(x)$ involving e^x or $\ln x$ is to use the derivative $f'(x)$ to find intervals of increase and decrease and then the second derivative $f''(x)$ to determine concavity.

EXAMPLE 4.4.1

Sketch the graph of $f(x) = x^2 - 8 \ln x$.

Solution

The function $f(x)$ is defined only for $x > 0$. Its derivative is

$$f'(x) = 2x - \frac{8}{x} = \frac{2x^2 - 8}{x}$$

and $f'(x) = 0$ if and only if $2x^2 = 8$ or $x = 2$ (since $x > 0$). Testing the sign of $f'(x)$ for $0 < x < 2$ and for $x > 2$, you obtain the intervals of increase and decrease shown in the figure.

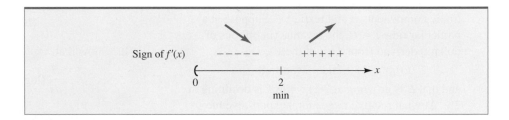

Notice that the arrow pattern indicates there is a relative minimum at $x = 2$, and since $f(2) = 2^2 - 8 \ln 2 \approx -1.5$, the minimum point is $(2, -1.5)$.

The second derivative

$$f''(x) = 2 + \frac{8}{x^2}$$

satisfies $f''(x) > 0$ for all $x > 0$, so the graph of $f(x)$ is always concave up and there are no inflection points.

Checking for asymptotes, you find

$$\lim_{x \to 0^+} (x^2 - 8 \ln x) = +\infty \qquad \text{and} \qquad \lim_{x \to +\infty} (x^2 - 8 \ln x) = +\infty$$

so the y axis ($x = 0$) is a vertical asymptote, but there is no horizontal asymptote. The x intercepts are found by using your calculator to solve the equation

$$x^2 - 8 \ln x = 0$$
$$x \approx 1.2 \qquad \text{and} \qquad x \approx 2.9$$

To summarize, the graph falls (down from the vertical asymptote) to the minimum at $(2, -1.5)$, after which it rises indefinitely, while maintaining a concave up shape. It passes through the x intercept $(1.2, 0)$ on the way down, and $(2.9, 0)$ as it rises. The graph is shown in Figure 4.10.

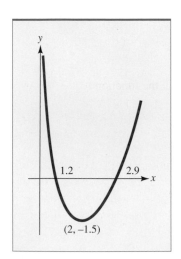

FIGURE 4.10 The graph of $f(x) = x^2 - 8 \ln x$.

EXAMPLE 4.4.2

Determine where the function

$$f(x) = \frac{1}{\sqrt{2\pi}} e^{-x^2/2}$$

is increasing and decreasing, and where its graph is concave up and concave down. Find the relative extrema and inflection points and draw the graph.

Solution

The first derivative is

$$f'(x) = \frac{-x}{\sqrt{2\pi}} e^{-x^2/2}$$

Since $e^{-x^2/2}$ is always positive, $f'(x)$ is zero if and only if $x = 0$. Since $f(0) = \frac{1}{\sqrt{2\pi}} \approx 0.4$, the only critical point is $(0, 0.4)$. By the product rule, the second derivative is

$$f''(x) = \frac{x^2}{\sqrt{2\pi}} e^{-x^2/2} - \frac{1}{\sqrt{2\pi}} e^{-x^2/2} = \frac{1}{\sqrt{2\pi}}(x^2 - 1)e^{-x^2/2}$$

which is zero if $x = \pm 1$. Since

$$f(1) = \frac{e^{1/2}}{\sqrt{2\pi}} \approx 0.24 \qquad \text{and} \qquad f(-1) = \frac{e^{-1/2}}{\sqrt{2\pi}} \approx 0.24$$

the potential inflection points are $(1, 0.24)$ and $(-1, 0.24)$.

Plot the critical points and then check the signs of the first and second derivatives on each of the intervals defined by the x coordinates of these points:

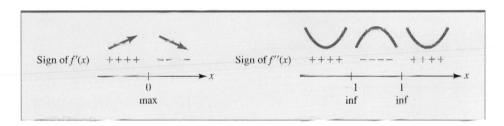

The arrow pattern indicates there is a relative maximum at $(0, 0.4)$, and since the concavity changes at $x = -1$ (from up to down) and at $x = 1$ (from down to up), both $(-1, 0.24)$ and $(1, 0.24)$ are inflection points.

Complete the graph, as shown in Figure 4.11, by connecting the key points with a curve of appropriate shape on each interval. Notice that the graph has no x intercepts since $e^{-x^2/2}$ is always positive and that the graph approaches the x axis as a horizontal asymptote since $e^{-x^2/2}$ approaches zero as $|x|$ increases without bound.

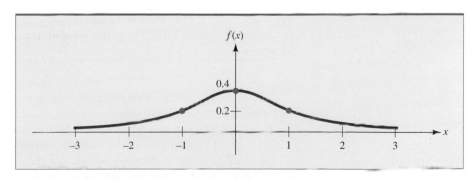

FIGURE 4.11 The standard normal density function: $f(x) = \frac{1}{\sqrt{2\pi}} e^{-x^2/2}$.

NOTE The function $f(x) = \dfrac{1}{\sqrt{2\pi}}e^{-x^2/2}$ whose graph was sketched in Example 4.4.2 is known as the **standard normal probability density function** and plays a vital role in probability and statistics. The famous "bell-shape" of the graph is used by physicists and social scientists to describe the distributions of IQ scores, measurements on large populations of living organisms, the velocity of a molecule in a gas, and numerous other important phenomena. ∎

Optimal Holding Time

Suppose you own an asset whose value increases with time. The longer you hold the asset, the more it will be worth, but there may come a time when you could do better by selling the asset and reinvesting the proceeds. Economists determine the optimal time for selling by maximizing the present value of the asset in relation to the prevailing rate of interest, compounded continuously. The application of this criterion is illustrated in Example 4.4.3.

EXPLORE!

Store the function

$P(x) = 20,000*e^\wedge(\sqrt{x} - 0.07x)$

from Example 4.4.3 into Y1 of the equation editor of your graphing calculator. Find an appropriate window in order to view the graph and determine its maximum value.

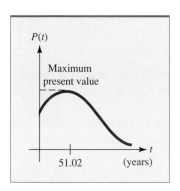

FIGURE 4.12 Present value $P(t) = 20,000e^{\sqrt{t} - 0.07t}$.

EXAMPLE │ 4.4.3

Suppose you own a parcel of land whose market price t years from now will be $V(t) = 20,000e^{\sqrt{t}}$ dollars. If the prevailing interest rate remains constant at 7% compounded continuously, when will the present value of the market price of the land be greatest?

Solution

In t years, the market price of the land will be $V(t) = 20,000e^{\sqrt{t}}$. The present value of this price is

$$P(t) = V(t)e^{-0.07t} = 20,000e^{\sqrt{t}}e^{-0.07t} = 20,000e^{\sqrt{t} - 0.07t}$$

The goal is to maximize $P(t)$ for $t \geq 0$. The derivative of P is

$$P'(t) = 20,000e^{\sqrt{t} - 0.07t}\left(\frac{1}{2\sqrt{t}} - 0.07\right)$$

Thus, $P'(t)$ is undefined when $t = 0$ and $P'(t) = 0$ when

$$\frac{1}{2\sqrt{t}} - 0.07 = 0 \qquad \text{or} \qquad t = \left[\frac{1}{2(0.07)}\right]^2 \approx 51.02$$

Since $P'(t)$ is positive if $0 < t < 51.02$ and negative if $t > 51.02$, it follows that the graph of P is increasing for $0 < t < 51.02$ and decreasing for $t > 51.02$, as shown in Figure 4.12. Thus, the present value is maximized in approximately 51 years.

NOTE The optimization criterion used in Example 4.4.3 is not the only way to determine the optimal holding time. For instance, you may decide to sell the asset when the percentage rate of growth of the asset's value just equals the prevailing rate of interest (7% in the example). Which criterion seems more reasonable to you? Actually, it does not matter, for the two criteria yield exactly the same result! A proof of this equivalence is outlined in Problem 52. ∎

Learning Curves

The graph of a function of the form $Q(t) = B - Ae^{-kt}$, where A, B, and k are positive constants, is sometimes called a **learning curve**. The name arose when psychologists discovered that for $t \geq 0$, functions of this form often realistically model the relationship between the efficiency with which an individual performs a task and the amount of training time or experience the "learner" has had.

To sketch the graph of $Q(t) = B - Ae^{-kt}$ for $t \geq 0$, note that

$$Q'(t) = -Ae^{-kt}(-k) = Ake^{-kt}$$

and

$$Q''(t) = Ake^{-kt}(-k) = -Ak^2e^{-kt}$$

Since A and k are positive, it follows that $Q'(t) > 0$ and $Q''(t) < 0$ for all t, so the graph of $Q(t)$ is always rising and is always concave down. Furthermore, the vertical (Q axis) intercept is $Q(0) = B - A$, and $Q = B$ is a horizontal asymptote since

$$\lim_{t \to +\infty} Q(t) = \lim_{t \to +\infty} (B - Ae^{-kt}) = B - 0 = B$$

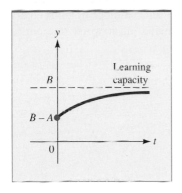

FIGURE 4.13 A learning curve $y = B - Ae^{-kt}$.

A graph with these features is sketched in Figure 4.13. The behavior of the learning curve as $t \to +\infty$ reflects the fact that "in the long run," an individual approaches his or her learning "capacity," and additional training time will result in only marginal improvement in performance efficiency.

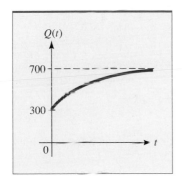

FIGURE 4.14 Worker efficiency
$$Q(t) = 700 - 400e^{-0.5t}$$

EXAMPLE 4.4.4

The rate at which a postal clerk can sort mail is a function of the clerk's experience. Suppose the postmaster of a large city estimates that after t months on the job, the average clerk can sort $Q(t) = 700 - 400e^{-0.5t}$ letters per hour.

a. How many letters can a new employee sort per hour?

b. How many letters can a clerk with 6 months' experience sort per hour?

c. Approximately how many letters will the average clerk ultimately be able to sort per hour?

Solution

a. The number of letters a new employee can sort per hour is

$$Q(0) = 700 - 400e^0 = 300$$

b. After 6 months, the average clerk can sort

$$Q(6) = 700 - 400e^{-0.5(6)} = 700 - 400e^{-3} \approx 680 \quad \text{letters per hour}$$

c. As t increases without bound, $Q(t)$ approaches 700. Hence, the average clerk will ultimately be able to sort approximately 700 letters per hour. The graph of the function $Q(t)$ is sketched in Figure 4.14.

Logistic Curves

The graph of a function of the form $Q(t) = \dfrac{B}{1 + Ae^{-Bkt}}$, where A, B, and k are positive constants, is called a **logistic curve**. A typical logistic curve is shown in Figure 4.15. Notice that it rises steeply like an exponential curve at first, and then turns over and flattens out, approaching a horizontal asymptote in much the same way as a learning curve. The asymptotic line represents a "saturation level" for the quantity represented by the logistic curve and is called the **carrying capacity** of the quantity. For instance, in population models, the carrying capacity represents the maximum number of individuals the environment can support, while in a logistic model for the spread of an epidemic, the carrying capacity is the total number of individuals susceptible to the disease, say, those who are unvaccinated or, at worst, the entire community.

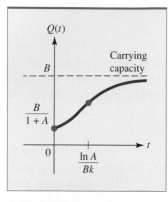

FIGURE 4.15 A logistic curve $Q(t) = \dfrac{B}{1 + Ae^{-Bkt}}$.

To sketch the graph of $Q(t) = \dfrac{B}{1 + Ae^{-Bkt}}$ for $t \geq 0$, note that

$$Q'(t) = \frac{AB^2 ke^{-Bkt}}{(1 + Ae^{-Bkt})^2}$$

and

$$Q''(t) = \frac{AB^3 k^2 e^{-Bkt}(-1 + Ae^{-Bkt})}{(1 + Ae^{-Bkt})^3}$$

Verify these formulas and also verify the fact that $Q'(t) > 0$ for all t, which means the graph of $Q(t)$ is always rising. The equation $Q''(t) = 0$ has exactly one solution; namely, when

$$-1 + Ae^{-Bkt} = 0$$

$$e^{-Bkt} = \frac{1}{A} \qquad \textit{add 1 to both sides and divide by A}$$

$$-Bkt = \ln\left(\frac{1}{A}\right) = -\ln A \qquad \textit{take logarithms on both sides}$$

$$t = \frac{\ln A}{Bk} \qquad \textit{divide both sides by } -Bk$$

As shown in this diagram, there is an inflection point at $t = \dfrac{\ln A}{Bk}$ since the concavity changes there (from up to down).

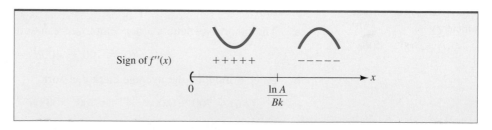

The vertical intercept of the logistic curve is

$$Q(0) = \frac{B}{1 + Ae^0} = \frac{B}{1 + A}$$

Since $Q(t)$ is defined for all $t \geq 0$, the logistic curve has no vertical asymptotes, but $y = B$ is a horizontal asymptote since

$$\lim_{t \to +\infty} Q(t) = \lim_{t \to +\infty} \frac{B}{1 + Ae^{-Bkt}} = \frac{B}{1 + A(0)} = B$$

To summarize, as shown in Figure 4.15, the logistic curve begins at $Q(0) = \dfrac{B}{1 + A}$, rises sharply (concave up) until it reaches the inflection point at $t = \dfrac{\ln A}{Bk}$, and then flattens out as it continues to rise (concave down) toward the horizontal asymptote $y = B$. Thus, B is the carrying capacity of the quantity $Q(t)$ represented by the logistic curve, and the inflection point at $t = \dfrac{\ln A}{Bk}$ can be interpreted as a point of *diminishing growth*.

Logistic curves often provide accurate models of population growth when environmental factors such as restricted living space, inadequate food supply, or urban pollution impose an upper bound on the possible size of the population. Logistic curves are also often used to describe the dissemination of privileged information or rumors in a community, where the restriction is the number of individuals susceptible to receiving such information. Here is an example in which a logistic curve is used to describe the spread of an infectious disease.

EXPLORE!

Graph the function in Example 4.4.5 using the window [0, 10]1 by [0, 25]5. Trace out this function for large values of x. What do you observe? Determine a graphic way to find out when 90% of the population has caught the disease.

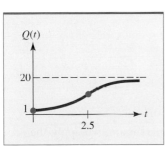

FIGURE 4.16 The spread of an epidemic $Q(t) = \dfrac{20}{1 + 19e^{-1.2t}}$.

EXAMPLE   4.4.5

Public health records indicate that t weeks after the outbreak of a certain form of influenza, approximately $Q(t) = \dfrac{20}{1 + 19e^{-1.2t}}$ thousand people had caught the disease.

a. How many people had the disease when it broke out? How many had it 2 weeks later?

b. At what time does the rate of infection begin to decline?

c. If the trend continues, approximately how many people will eventually contract the disease?

Solution

a. Since $Q(0) = \dfrac{20}{1 + 19} = 1$, it follows that 1,000 people initially had the disease. When $t = 2$,

$$Q(2) = \frac{20}{1 + 19e^{-1.2(2)}} \approx 7.343$$

so about 7,343 had contracted the disease by the second week.

b. The rate of infection begins to decline at the inflection point on the graph of $Q(t)$. By comparing the given formula with the logistic formula $Q(t) = \dfrac{B}{1 + Ae^{-Bkt}}$, you find that $B = 20$, $A = 19$, and $Bk = 1.2$. Thus, the inflection point occurs when

$$t = \frac{\ln A}{Bk} = \frac{\ln 19}{1.2} \approx 2.454$$

so the epidemic begins to fade about 2.5 weeks after it starts.

c. Since $Q(t)$ approaches 20 as t increases without bound, it follows that approximately 20,000 people will eventually contract the disease. For reference, the graph is sketched in Figure 4.16.

Optimal Age for Reproduction

An organism such as Pacific salmon or bamboo that breeds only once during its lifetime is said to be *semelparous*. Biologists measure the per capita rate of reproduction of such an organism by the function*

$$R(x) = \frac{\ln [p(x)f(x)]}{x}$$

where $p(x)$ is the likelihood of an individual organism surviving to age x and $f(x)$ is the number of female births to an individual reproducing at age x. The larger the value of $R(x)$, the more offspring will be produced. Hence, the age at which $R(x)$ is maximized is regarded as the optimal age for reproduction.

EXAMPLE 4.4.6

Suppose that for a particular semelparous organism, the likelihood of an individual surviving to age x (years) is given by $p(x) = e^{-0.15x}$ and that the number of female births at age x is $f(x) = 3x^{0.85}$. What is the optimal age for reproduction?

Solution

The per capita rate of increase function for this model is

$$R(x) = \frac{\ln [e^{-0.15x}(3x^{0.85})]}{x}$$

$$= x^{-1}(\ln e^{-0.15x} + \ln 3 + \ln x^{0.85}) \qquad \text{\textit{product rule for logarithms}}$$

$$= x^{-1}(-0.15x + \ln 3 + 0.85 \ln x) \qquad \text{\textit{power rule for logarithms}}$$

$$= -0.15 + (\ln 3 + 0.85 \ln x)x^{-1}$$

Differentiating $R(x)$ by the product rule, we find that

$$R'(x) = 0 + (\ln 3 + 0.85 \ln x)(-x^{-2}) + \left[0.85\left(\frac{1}{x}\right)\right]x^{-1}$$

$$= \frac{-\ln 3 - 0.85 \ln x + 0.85}{x^2}$$

Therefore, $R'(x) = 0$ when

$$-\ln 3 - 0.85 \ln x + 0.85 = 0$$

$$\ln x = \frac{0.85 - \ln 3}{0.85} \approx -0.2925$$

$$x = e^{-0.2925} \approx 0.7464$$

To show that this critical number corresponds to a maximum, we apply the second derivative test. We find that

$$R''(x) = \frac{1.7 \ln x + 2 \ln 3 - 2.55}{x^3}$$

(supply the details), and since

$$R''(0.7464) \approx -2.0440 < 0$$

it follows that the largest value of $R(x)$ occurs when $x \approx 0.7464$. Thus, the optimal age for an individual organism to reproduce is when it is 0.7464 years old (approximately 9 months).

---

*Adapted from Claudia Neuhauser, *Calculus for Biology and Medicine*, Upper Saddle River, New Jersey: Prentice-Hall, 2000, p. 199 (Problem 22).

PROBLEMS 4.4

Each of the curves shown in Problems 1 through 4 is the graph of one of the six functions listed here. In each case, match the given curve to the proper function.

$$f_1(x) = 2 - e^{-2x} \qquad f_2(x) = x \ln x^5$$

$$f_3(x) = \frac{2}{1 - e^{-x}} \qquad f_4(x) = \frac{2}{1 + e^{-x}}$$

$$f_5(x) = \frac{\ln x^5}{x} \qquad f_6(x) = (x - 1)e^{-2x}$$

1.
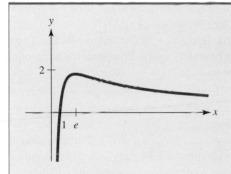

2.
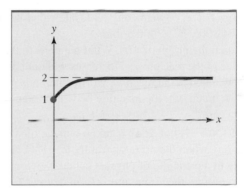

3.
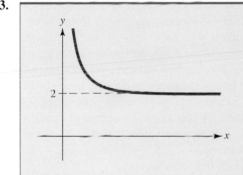

4.
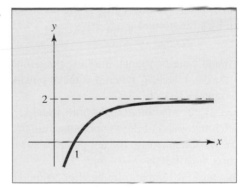

In Problems 5 through 20, determine where the given function is increasing and decreasing and where its graph is concave upward and concave downward. Sketch the graph of the function. Show as many key features as possible (high and low points, points of inflection, vertical and horizontal asymptotes, intercepts, cusps, vertical tangents).

5. $f(t) = 2 + e^t$

6. $g(x) = 3 + e^{-x}$

7. $g(x) = 2 - 3e^x$

8. $f(t) = 3 - 2e^t$

9. $f(x) = \dfrac{2}{1 + 3e^{-2x}}$

10. $h(t) = \dfrac{2}{1 + 3e^{2t}}$

11. $f(x) = xe^x$

12. $f(x) = xe^{-x}$

13. $f(x) = xe^{2-x}$

14. $f(x) = e^{-x^2}$

15. $f(x) = x^2 e^{-x}$

16. $f(x) = e^x + e^{-x}$

17. $f(x) = \dfrac{6}{1 + e^{-x}}$

18. $f(x) = x - \ln x$ (for $x > 0$)

19. $f(x) = (\ln x)^2$ (for $x > 0$)

20. $f(x) = \dfrac{\ln x}{x}$ (for $x > 0$)

21. PRODUCT RELIABILITY A manufacturer of toys has found that the fraction of its plastic battery-operated toy oil tankers that sink in fewer than t days is approximately $f(t) = 1 - e^{-0.03t}$.

a. Sketch this reliability function. What happens to the graph as t increases without bound?

b. What fraction of the tankers can be expected to float for at least 10 days?

c. What fraction of the tankers can be expected to sink between the 15th and 20th days?

22. DEPRECIATION When a certain industrial machine has become t years old, its resale value will be $V(t) = 4,800e^{-t/5} + 400$ dollars.

a. Sketch the graph of $V(t)$. What happens to the value of the machine as t increases without bound?

b. How much is the machine worth when it is new?

c. How much will the machine be worth after 10 years?

23. COOLING A hot drink is taken outside on a cold winter day when the air temperature is $-5°C$. According to a principle of physics called Newton's law of cooling, the temperature T (in degrees Celsius) of the drink t minutes after being taken outside is given by a function of the form

$$T(t) = -5 + Ae^{-kt}$$

where A and k are constants. Suppose the temperature of the drink is $80°C$ when it is taken outside and 20 minutes later, is $25°C$.

a. Use this information to determine A and k.

b. Sketch the graph of the temperature function $T(t)$. What happens to the temperature as t increases indefinitely $(t \to +\infty)$?

c. What will the temperature be after 30 minutes?

d. When will the temperature reach $0°C$?

24. POPULATION GROWTH It is estimated that t years from now, the population of a certain country will be $P(t) = \dfrac{20}{2 + 3e^{-0.06t}}$ million.

a. Sketch the graph of $P(t)$.

b. What is the current population?

c. What will be the population 50 years from now?

d. What will happen to the population in the long run?

25. THE SPREAD OF AN EPIDEMIC Public health records indicate that t weeks after the outbreak of a certain form of influenza, approximately $f(t) = \dfrac{2}{1 + 3e^{-0.8t}}$ thousand people had caught the disease.

a. Sketch the graph of $f(t)$.

b. How many people had the disease initially?

c. How many had caught the disease by the end of 3 weeks?

d. If the trend continues, approximately how many people in all will contract the disease?

26. RECALL FROM MEMORY Psychologists believe that when a person is asked to recall a set of facts, the number of facts recalled after t minutes is given by a function of the form $Q(t) = A(1 - e^{-kt})$, where k is a positive constant and A is the total number of relevant facts in the person's memory.

a. Sketch the graph $Q(t)$.

b. What happens to the graph as t increases without bound? Explain this behavior in practical terms.

27. EFFICIENCY The daily output of a worker who has been on the job for t weeks is given by a function of the form $Q(t) = 40 - Ae^{-kt}$. Initially the worker could produce 20 units a day, and after 1 week the worker can produce 30 units a day. How many units will the worker produce per day after 3 weeks?

28. ADVERTISING When professors select texts for their courses, they usually choose from among the books already on their shelves. For this reason, most publishers send complimentary copies of new texts to professors teaching related courses. The mathematics editor at a major publishing house estimates that if x thousand complimentary copies are distributed, the first-year sales of a certain new mathematics text will be approximately $f(x) = 20 - 15e^{-0.2x}$ thousand copies.

a. Sketch this sales function.

b. How many copies can the editor expect to sell in the first year if no complimentary copies are sent out?

c. How many copies can the editor expect to sell in the first year if 10,000 complimentary copies are sent out?

d. If the editor's estimate is correct, what is the most optimistic projection for the first-year sales of the text?

29. MARGINAL ANALYSIS The economics editor at a major publishing house estimates that if x thousand complimentary copies are distributed to professors, the first-year sales of a certain new text will be $f(x) = 15 - 20e^{-0.3x}$ thousand copies. Currently, the editor is planning to distribute 9,000 complimentary copies.

a. Use marginal analysis to estimate the increase in first-year sales that will result if 1,000 additional complimentary copies are distributed.

b. Calculate the actual increase in first-year sales that will result from the distribution of the additional 1,000 complimentary copies. Is the estimate in part (a) a good one?

30. MARGINAL ANALYSIS A business estimates that when x thousand people are employed, its profit will be $P(x)$ million dollars, where

$$P(x) = 10 + \ln\left(\frac{x}{25}\right) - 12x^2$$

for $x > 0$. What level of employment maximizes profit? What is the maximum profit?

31. CHILDHOOD LEARNING A psychologist measures a child's capability to learn and remember by the function

$$L(t) = \frac{\ln(t + 1)}{t + 1}$$

where t is the child's age in years, for $0 \le t \le 5$. Answer these questions about this model.

a. At what age does a child have the greatest learning capability?

b. At what age is a child's learning capability increasing most rapidly?

32. AEROBIC RATE The aerobic rating of a person x years old is modeled by the function

$$A(x) = \frac{110(\ln x - 2)}{x} \quad \text{for } x \ge 10$$

a. At what age is a person's aerobic rating largest?

b. At what age is a person's aerobic rating decreasing most rapidly?

33. STOCK SPECULATION In a classic paper on the theory of conflict,* L. F. Richardson claimed that the proportion p of a population advocating war or other aggressive action at a time t satisfies

$$p(t) = \frac{Ce^{kt}}{1 + Ce^{kt}}$$

where k and C are positive constants. Speculative day-trading in the stock market can be regarded as "aggressive action." Suppose that initially, 1/200 of total daily market volume is attributed to

_____
*Richardson's original work appeared in *Generalized Foreign Politics,* Monograph Supplement 23 of the *British Journal of Psychology* (1939). His work was also featured in the text *Mathematical Models of Arms Control and Disarmament* by T. L. Saaty, New York: John Wiley & Sons, 1968.

day-trading and that 4 weeks later, the proportion is 1/100. When will the proportion be increasing most rapidly? What will the proportion be at that time?

34. WORLD POPULATION According to a certain logistic model, the world's population (in billions) t years after 1960 will be approximately

$$P(t) = \frac{40}{1 + 12e^{-0.08t}}$$

a. If this model is correct, at what rate was the world's population increasing with respect to time in the year 2000? At what *percentage* rate was the population increasing at this time?

b. When will the population be growing most rapidly?

c. Sketch the graph of $P(t)$. What feature occurs on the graph at the time found in part (b)? What happens to $P(t)$ "in the long run"?

 d. Do you think such a population model is reasonable? Why or why not?

35. MARGINAL ANALYSIS A manufacturer can produce VCRs at a cost of $125 apiece and estimates that if they are sold for x dollars apiece, consumers will buy approximately $1,000e^{-0.02x}$ each week.

a. Express the profit P as a function of x. Sketch the graph of $P(x)$.

b. At what price should the manufacturer sell the VCRs to maximize profit?

36. THE SPREAD OF AN EPIDEMIC An epidemic spreads through a community so that t weeks after its outbreak, the number of people who have been infected is given by a function of the form

$$f(t) = \frac{B}{1 + Ce^{-kt}}, \text{ where } B \text{ is the number of resi-}$$

dents in the community who are susceptible to the disease. If 1/5 of the susceptible residents were infected initially and 1/2 had been infected by the end of the fourth week, what fraction of the susceptible residents will have been infected by the end of the eighth week?

37. OPTIMAL HOLDING TIME Suppose you own a parcel of land whose value t years from now will be $V(t) = 8,000e^{\sqrt{t}}$ dollars. If the prevailing interest rate remains constant at 6% per year compounded continuously, when should you sell the land to maximize its present value?

38. OPTIMAL HOLDING TIME Suppose your family owns a rare book whose value t years from now will be $V(t) = 200e^{\sqrt{2t}}$ dollars. If the prevailing interest rate remains constant at 6% per year compounded continuously, when will it be most advantageous for your family to sell the book and invest the proceeds?

39. **OPTIMAL AGE FOR REPRODUCTION** Suppose that for a particular semelparous organism, the likelihood of an individual surviving to age x years is $p(x) = e^{-0.2x}$ and that the number of female births to an individual at age x is $f(x) = 5x^{0.9}$. What is the ideal age for reproduction for an individual organism of this species? (See Example 4.4.6.)

40. **OPTIMAL AGE FOR REPRODUCTION** Suppose an environmental change affects the organism in Problem 39 in such a way that an individual is only half as likely as before to survive to age x years. If the number of female births $f(x)$ remains the same, how is the ideal age for reproduction affected by this change?

41. **RESPONSE TO STIMULATION** According to **Hoorweg's law,** when a nerve is stimulated by discharges from an electrical condenser of capacitance C, the electrical energy required to elicit a minimal response (a muscle contraction) is given by

$$E(C) = C\left(aR + \frac{b}{C}\right)^2$$

where a, b, and R are positive constants.

  a. For what value of C is the energy $E(C)$ minimized? How do you know you have found the minimum value? (Your answer will be in terms of a, b, and R.)

  b. Another model for E has the form

$$E(C) = mCe^{k/C}$$

  where m and k are constants. What must m and k be for the two models to have the same minimum value for the same value of C?

42. **FISHERY MANAGEMENT** The manager of a fishery determines that t weeks after 3,000 fish of a particular species are hatched, the average weight of an individual fish will be $w(t) = 0.8te^{-0.05t}$ pounds, for $0 \le t \le 20$. Moreover, the proportion of the fish that will still be alive after t weeks is estimated to be

$$p(t) = \frac{10}{10 + t}$$

  a. The expected yield $E(t)$ from harvesting after t weeks is the total weight of the fish that are still alive. Express $E(t)$ in terms of $w(t)$ and $p(t)$.

  b. For what value of t is the expected yield $E(t)$ the largest? What is the maximum expected yield?

  c. Sketch the yield curve $y = E(t)$ for $0 \le t \le 20$.

43. **FISHERY MANAGEMENT** The manager of a fishery determines that t days after 1,000 fish of a particular species are released into a pond, the average weight of an individual fish will be $w(t)$ pounds and the proportion of the fish still alive after t days will be $p(t)$, where

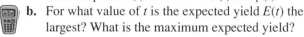

$$w(t) = \frac{10}{1 + 15e^{-0.05t}} \quad \text{and} \quad p(t) = e^{-0.01t}$$

  a. The expected yield $E(t)$ from harvesting after t days is the total weight of the fish that are still alive. Express $E(t)$ in terms of $w(t)$ and $p(t)$.

  b. For what value of t is the expected yield $E(t)$ the largest? What is the maximum expected yield?

  c. Sketch the yield curve $y = E(t)$.

44. **OPTIMAL HOLDING TIME** Suppose you own a stamp collection that is currently worth $1,200 and whose value increases linearly at the rate of $200 per year. If the prevailing interest rate remains constant at 8% per year compounded continuously, when will it be most advantageous for you to sell the collection and invest the proceeds?

45. **THE SPREAD OF A RUMOR** A traffic accident was witnessed by 10% of the residents of a small town, and 25% of the residents had heard about the accident 2 hours later. Suppose the number $N(t)$ of residents who had heard about the accident t hours after it occurred is given by a function of the form

$$N(t) = \frac{B}{1 + Ce^{-kt}}$$

where B is the population of the town and C and k are constants.

  a. Use this information to find C and k.

  b. How long does it take for half the residents of the town to know about the accident?

  c. When is the news about the accident spreading most rapidly? [*Hint:* See Example 4.4.5(b).]

46. **EFFECT OF A TOXIN** A medical researcher determines that t hours from the time a toxin is introduced to a bacterial colony, the population will be

$$P(t) = 10,000(7 + 15e^{-0.05t} + te^{-0.05t})$$

  a. What is the population at the time the toxin is introduced?

  b. When does the maximum bacterial population occur? What is the maximum population?

  c. What eventually happens to the bacterial population as $t \to +\infty$?

47. CORPORATE ORGANIZATION   A Gompertz

curve is the graph of a function of the general form

$$N(t) = CA^{B^t}$$

where A, B, and C are constants. Such curves are used by psychologists and others to describe such things as learning and growth within an organization.[*]

a. Suppose the personnel director of a large corporation conducts a study that indicates that after t years, the corporation will have

$$N(t) = 500(0.03)^{(0.4)^t}$$

employees. How many employees are there originally (at time $t = 0$)? How many are there after 5 years? When will there be 300 employees? How many employees will there be "in the long run"?

b. Sketch the graph of $N(t)$. Then, on the same graph sketch the graph of the Gompertz function

$$F(t) = 500(0.03)^{-(0.4)^t}$$

How would you describe the relationship between the two graphs?

48. LEARNING THEORY   In a learning model proposed by C. L. Hull, the habit strength H of an individual is related to the number r of reinforcements by the equation

$$H(r) = M(1 - e^{-kr})$$

a. Sketch the graph of $H(r)$. What happens to $H(r)$ as $r \to +\infty$?

b. Show that if the number of reinforcements is doubled from r to $2r$, the habit strength is multiplied by $1 + e^{-kr}$.

49. CONCENTRATION OF DRUG   A function of the form $C(t) = Ate^{-kt}$, where A and k are positive constants, is called a **surge function** and is sometimes used to model the concentration of drug in a patient's bloodstream t hours after the drug is administered. Assume $t \geq 0$.

a. Find $C'(t)$ and determine the time interval where the drug concentration is increasing and where it is decreasing. For what value of t is the concentration maximized? What is the maximum concentration? (Your answers will be in terms of A and k.)

b. Find $C''(t)$ and determine time intervals of concavity for the graph of $C(t)$. Find all points of inflection and explain what is happening to the rate of change of drug concentration at the times that correspond to the inflection points.

c. Sketch the graph of $C(t) = te^{-kt}$ for $k = 0.2$, $k = 0.5$, $k = 1.0$, and $k = 2.0$. Describe how the shape of the graph changes as k increases.

50. CONCENTRATION OF DRUG   The concentration of a certain drug in a patient's bloodstream t hours after being administered orally is assumed to be given by the surge function $C(t) = Ate^{-kt}$, where C is measured in micrograms of drug per milliliter of blood. Monitoring devices indicate that a maximum concentration of 5 occurs 20 minutes after the drug is administered.

a. Use this information to find A and k.

b. What is the drug concentration in the patient's blood after 1 hour?

c. At what time after the maximum concentration occurs will the concentration be half the maximum?

d. If you double the time in part (c), will the resulting concentration of drug be 1/4 the maximum? Explain.

51. BUREAUCRATIC GROWTH   Parkinson's law[†] states that in any administrative department not actively at war, the staff will grow by about 6% per year, regardless of need.

a. Parkinson applied his law to the size of the British Colonial Office. He noted that the Colonial Office had 1,139 staff members in 1947. How many staff members did the law predict for the year 1954? (The actual number was 1,661.)

b. Based on Parkinson's law, how long should it take for a staff to double in size?

 c. Read about Parkinson's law and write a paragraph about whether or not you think it is valid in today's world. You may wish to begin your research with the book cited in this problem.

52. OPTIMAL HOLDING TIME   Let $V(t)$ be the value of an asset t years from now and assume that the prevailing annual interest rate remains fixed at r (expressed as a decimal) compounded continuously.

---

[*]A discussion of models of Gompertz curves and other models of differential growth can be found in an article by Roger V. Jean titled, "Differential Growth, Huxley's Allometric Formula, and Sigmoid Growth," *UMAP Modules 1983: Tools for Teaching,* Lexington, MA: Consortium for Mathematics and Its Applications, Inc., 1984.

[†]C. N. Parkinson, *Parkinson's Law,* Boston, MA: Houghton-Mifflin, 1957.

---

a. Show that the present value of the asset $P(t) = V(t)e^{-rt}$ has a critical number where $V'(t) = V(t)r$. (Using economic arguments, it can be shown that the critical number corresponds to a maximum.)

b. Explain why the present value of $V(t)$ is maximized at a value of t where the percentage rate of change (expressed in decimal form) equals r.

53. **CANCER RESEARCH** In Problem 64, Exercise set 2.3, you were given a function to model the production of blood cells. Such models are useful in the study of leukemia and other so-called *dynamical diseases* in which certain physiological systems begin to behave erratically. An alternative model* for blood cell production developed by A. Lasota involves the exponential production function

$$p(x) = Ax^s e^{-sx/r}$$

where A, s, and r are positive constants and x is the number of granulocytes (a type of white blood cell) present.

a. Find the blood cell level x that maximizes the production function $p(x)$. How do you know the optimum level is a maximum?

b. If $s > 1$, show that the graph of $p(x)$ has two inflection points. Sketch the graph. Give a physical interpretation of the inflection points.

c. Sketch the graph of $p(x)$ in the case where $0 \leq s \leq 1$. What is different in this case?

 d. Read an article on mathematical methods in the study of dynamical diseases and write a paragraph on these methods. A good place to start is the article referenced in this problem.

54. **MORTALITY RATES** An actuary measures the probability that a person in a certain population will die at age x by the formula

$$P(x) = \lambda^2 x e^{-\lambda x}$$

where λ is a constant such that $0 < \lambda < e$.

a. Find the maximum value of $P(x)$ in terms of λ.

b. Sketch the graph of $P(x)$.

c. Read an article about actuarial formulas of this kind. Write a paragraph on what is represented by the number λ.

55. **THE SPREAD OF AN EPIDEMIC** An epidemic spreads throughout a community so that

t weeks after its outbreak, the number of residents who have been infected is given by a function of the form $f(t) = \dfrac{A}{1 + Ce^{-kt}}$, where A is the total number of susceptible residents. Show that the epidemic is spreading most rapidly when half of the susceptible residents have been infected.

56. Use the graphing utility of your calculator to sketch the graph of $f(x) = x(e^{-x} + e^{-2x})$. Use **ZOOM** and **TRACE** to find the largest value of $f(x)$. What happens to $f(x)$ as $x \to +\infty$?

57. **MARKET RESEARCH** A company is trying to use television advertising to expose as many people as possible to a new product in a large metropolitan area with 2 million possible viewers. A model for the number of people N (in millions) who are aware of the product after t days is found to be

$$N = 2(1 - e^{-0.037t})$$

Use a graphing utility to graph this function. What happens as $t \to +\infty$? [Suggestion: Set the range on your viewing screen to [0, 200]10 by [0, 3]1.]

58. **OPTIMAL HOLDING TIME** Suppose you win a parcel of land whose market value t years from now is estimated to be $V(t) = 20,000te^{\sqrt{0.4t}}$ dollars. If the prevailing interest rate remains constant at 7% compounded continuously, when will it be most advantageous to sell the land? (Use a graphing utility and **ZOOM** and **TRACE** to make the required determination.)

59. **DRUG CONCENTRATION** In a classic paper,† E. Heinz modeled the concentration $y(t)$ of a drug injected into the body intramuscularly by the function

$$y(t) = \frac{c}{b - a}(e^{-at} - e^{-bt}) \quad t \geq 0$$

where t is the number of hours after the injection and a, b, and c are positive constants, with $b > a$.

a. When does the maximum concentration occur? What happens to the concentration "in the long run"?

b. Sketch the graph of $y(t)$.

 c. Write a paragraph on the reliability of the Heinz model. In particular, is it more reliable when t is small or large? You may wish to begin your research with the article cited in this problem.

---

*W. B. Gearhart and M. Martelli, "A Blood Cell Population Model, Dynamical Diseases, and Chaos," *UMAP Modules 1990: Tools for Teaching,* Arlington, MA: Consortium for Mathematics and Its Applications, Inc., 1991.

†E. Heinz, "Probleme bei der Diffusion kleiner Substanzmengen innerhalb des menschlichen Körpers," *Biochem. Z.,* Vol. 319, 1949, pp. 482–492.

60. STRUCTURAL DESIGN When a chain, a telephone line, or a TV cable is strung between supports, the curve it forms is called a **catenary**. A typical catenary curve is
$$y = 0.125(e^{4x} + e^{-4x})$$
  a. Sketch this catenary curve.

  b. Catenary curves are important in architecture. Read an article on the Gateway Arch to the West in St. Louis, Missouri, and write a paragraph on the use of the catenary shape in its design.*

61. ACCOUNTING The **double declining balance** formula in accounting is
$$V(t) = V_0\left(1 - \frac{2}{L}\right)^t$$

---

*A good place to start is the article by William V. Thayer, "The St. Louis Arch Problem," *UMAP Modules 1983: Tools for Teaching*, Lexington, MA: Consortium for Mathematics and Its Applications, Inc., 1984.

where $V(t)$ is the value after t years of an article that originally cost V_0 dollars and L is a constant, called the "useful life" of the article.

  a. A refrigerator costs $875 and has a useful life of 8 years. What is its value after 5 years? What is its annual rate of depreciation?
  b. In general, what is the percentage rate of change of $V(t)$?

62. SPREAD OF DISEASE In the Think About It essay at the end of Chapter 3, we examined several models associated with the AIDS epidemic. Using a data analysis program, we obtain the function
$$C(t) = 456 + 1{,}234te^{-0.137t}$$
as a model for the number of cases of AIDS reported t years after the base year of 1990,
  a. According to this model, in what year will the largest number of cases be reported? What will the maximum number of reported cases be?
  b. When will the number of reported cases be the same as the number reported in 1990?

Important Terms, Symbols, and Formulas

Exponential function: (288)
$$f(x) = b^x \quad b > 0, b \neq 1$$
The graph of $y = b^x$ for $b > 1$: (290, 291)

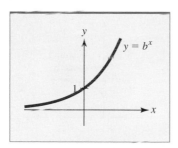

Properties of exponential functions: (291, 292)
$$b^x = b^y \quad \text{if and only if } x = y$$
$$b^x b^y = b^{x+y}$$
$$\frac{b^x}{b^y} = b^{x-y}$$
$$(b^x)^y = b^{xy}$$
$$b^0 = 1$$
The natural exponential base: (292)
$$e = \lim_{n \to +\infty}\left(1 + \frac{1}{n}\right)^n = 2.71828\ldots$$

Compounding interest k times per year at an annual rate r for t years:
  Future value of P dollars is $B = P\left(1 + \dfrac{r}{k}\right)^{kt}$ (294)

  Present value of B dollars is $P = B\left(1 + \dfrac{r}{k}\right)^{-kt}$ (296)

Continuous compounding at an annual rate r for t years:
  Future value of P dollars is $B = Pe^{rt}$ (294)
  Present value of B dollars is $P = Be^{-rt}$ (296)
Logarithmic function: (303)
$$y = \log_b x \quad \text{if and only if } b^y = x.$$
Properties of logarithms: (304)
$$\log_b u = \log_b v \quad \text{if and only if } u = v$$
$$\log_b uv = \log_b u + \log_b v$$
$$\log_b\left(\frac{u}{v}\right) = \log_b u - \log_b v$$
$$\log_b u^r = r \log_b u$$
$$\log_b 1 = 0 \text{ and } \log_b b = 1$$
$$\log_b b^u = u$$

CHAPTER SUMMARY

Comparison of Properties of Exponential and Logarithmic Functions (305)

Exponential Property	Logarithmic Property
$b^x b^y = b^{x+y}$	$\log_b(xy) = \log_b x + \log_b y$
$\dfrac{b^x}{b^y} = b^{x-y}$	$\log_b\left(\dfrac{x}{y}\right) = \log_b x - \log_b y$
$b^{xp} = (b^x)^p$	$\log_b x^p = p \log_b x$

The graph of $y = \log_b x$ for $b > 1$: (307)

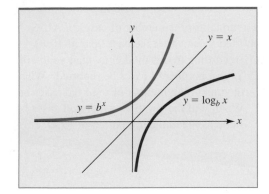

The natural logarithm: (307)
$$y = \ln x \quad \text{if and only if} \quad e^y = x$$

Inversion formulas: (309)
$$e^{\ln x} = x \quad \text{for } x > 0$$
and
$$\ln e^x = x \quad \text{for all } x$$

Conversion formula for logarithms (310)
$$\log_b a = \frac{\ln a}{\ln b}$$

Exponential growth: (297)

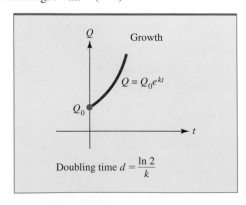

Exponential decay: (297)

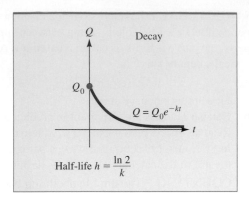

Carbon dating (312)

Derivatives of exponential functions: (322)
$$\frac{d}{dx}(e^x) = e^x \quad \text{and} \quad \frac{d}{dx}[e^{u(x)}] = e^{u(x)} \frac{du}{dx}$$

Derivatives of logarithmic functions: (319, 320)
$$\frac{d}{dx}(\ln x) = \frac{1}{x} \quad \text{and} \quad \frac{d}{dx}[\ln u(x)] = \frac{1}{u(x)} \frac{du}{dx}$$

Logarithmic differentiation: (326)

Relative rate of change of $Q(x)$: (328)
$$\frac{Q'(x)}{Q(x)} = \frac{d}{dx}(\ln Q)$$

Optimal holding time (336)

Learning curve $y = B - Ae^{-kt}$ (337)

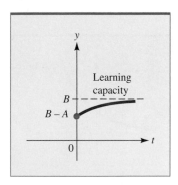

Logistic curve $y = \dfrac{B}{1 + Ae^{-Bkt}}$ (338)

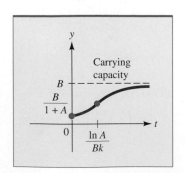

Checkup for Chapter 4

1. Evaluate each of these expressions:

   a. $\dfrac{(3^{-2})(9^2)}{(27)^{2/3}}$

   b. $\sqrt[3]{(25)^{1.5}\left(\dfrac{8}{27}\right)}$

   c. $\log_2 4 + \log_4 16^{-1}$

   d. $\left(\dfrac{8}{27}\right)^{-2/3}\left(\dfrac{16}{81}\right)^{3/2}$

2. Simplify each of these expressions:

   a. $(9x^4 y^2)^{3/2}$

   b. $(3x^2 y^{4/3})^{-1/2}$

   c. $\left(\dfrac{y}{x}\right)^{3/2}\left(\dfrac{x^{2/3}}{y^{1/6}}\right)^2$

   d. $\left(\dfrac{x^{0.2} y^{-1.2}}{x^{1.5} y^{0.4}}\right)^5$

3. Find all real numbers x that satisfy each of these equations.

   a. $4^{2x-x^2} = \dfrac{1}{64}$

   b. $e^{1/x} = 4$

   c. $\log_4 x^2 = 2$

   d. $\dfrac{25}{1 + 2e^{-0.5t}} = 3$

4. In each case, find the derivative $\dfrac{dy}{dx}$. (In some cases, it may help to use logarithmic differentiation.)

   a. $y = \dfrac{e^x}{x^2 - 3x}$

   b. $y = \ln(x^3 + 2x^2 - 3x)$

   c. $y = x^3 \ln x$

   d. $y = \dfrac{e^{-2x}(2x - 1)^3}{1 - x^2}$

5. In each of these cases, determine where the given function is increasing and decreasing and where its graph is concave upward and concave downward. Sketch the graph, showing as many key features as possible (high and low points, points of inflection, asymptotes, intercepts, cusps, vertical tangents).

   a. $y = x^2 e^{-x}$

   b. $y = \dfrac{\ln \sqrt{x}}{x^2}$

   c. $y = \ln(\sqrt{x} - x)^2$

   d. $y = \dfrac{4}{1 + e^{-x}}$

6. If you invest \$2,000 at 5% compounded continuously, how much will your account be worth in 3 years? How long does it take before your account is worth \$3,000?

7. **PRICE ANALYSIS** A product is introduced and t months later, its unit price is $p(t)$ hundred dollars, where

   $$p = \dfrac{\ln(t + 1)}{t + 1} + 5$$

   a. For what values of t is the price increasing? When is it decreasing?

   b. When is the price decreasing most rapidly?

   c. What happens to the price in the long run (as $t \to +\infty$)?

8. **MAXIMIZING REVENUE** It is determined that q units of a commodity can be sold when the price is p hundred dollars per unit, where

   $$q(p) = 1,000(p + 2)e^{-p}$$

   a. Verify that the demand function $q(p)$ decreases as p increases for $p \geq 0$.

   b. For what price p is revenue $R = pq$ maximized? What is the maximum revenue?

9. **CARBON DATING** An archaeological artifact is found to have 45% of its original ^{14}C. How old is the artifact? (Use 5,730 years as the half-life of ^{14}C.)

10. **BACTERIAL GROWTH** A toxin is introduced into a bacterial colony, and t hours later, the population is given by

    $$N(t) = 10,000(8 + t)e^{-0.1t}$$

    a. What was the population when the toxin was introduced?

    b. When is the population maximized? What is the maximum population?

    c. What happens to the population in the long run (as $t \to +\infty$)?

Review Problems

In Problems 1 through 4, sketch the graph of the given exponential or logarithmic function without using calculus.

1. $f(x) = 5^x$

2. $f(x) = -2e^{-x}$

3. $f(x) = \ln x^2$

4. $f(x) = \log_3 x$

5. **a.** Find $f(4)$ if $f(x) = Ae^{-kx}$ and $f(0) = 10$, $f(1) = 25$.

  b. Find $f(3)$ if $f(x) = Ae^{kx}$ and $f(1) = 3$, $f(2) = 10$.

  c. Find $f(9)$ if $f(x) = 30 + Ae^{-kx}$ and $f(0) = 50$, $f(3) = 40$.

  d. Find $f(10)$ if $f(t) = \dfrac{6}{1 + Ae^{-kt}}$ and $f(0) = 3$, $f(5) = 2$.

6. Evaluate the following expressions without using tables or a calculator.

  a. $\ln e^5$

  b. $e^{\ln 2}$

  c. $e^{3\ln 4 - \ln 2}$

  d. $\ln 9e^2 + \ln 3e^{-2}$

In Problems 7 through 13, find all real numbers x that satisfy the given equation.

7. $8 = 2e^{0.04x}$

8. $5 = 1 + 4e^{-6x}$

9. $4 \ln x = 8$

10. $5^x = e^3$

11. $\log_9 (4x - 1) = 2$

12. $\ln (x - 2) + 3 = \ln (x + 1)$

13. $e^{2x} + e^x - 2 = 0$ [*Hint:* Let $u = e^x$.]

In Problems 14 through 29, find the derivative $\dfrac{dy}{dx}$. In some of these problems, you may need to use implicit differentiation or logarithmic differentiation.

14. $y = 2e^{3x+5}$

15. $y = x^2 e^{-x}$

16. $y = \ln \sqrt{x^2 + 4x + 1}$

17. $y = x \ln x^2$

18. $y = \dfrac{x}{\ln 2x}$

19. $y = \log_3 (x^2)$

20. $y = \dfrac{e^{3x}}{e^{3x} + 2}$

21. $y = \dfrac{e^{-x} + e^x}{1 + e^{-2x}}$

22. $y = (1 + e^{-x})^{4/5}$

23. $y = \ln (e^{-2x} + e^{-x})$

24. $y = \ln\left(\dfrac{e^{3x}}{1 + x}\right)$

25. $y = \dfrac{e^{-x}}{x + \ln x}$

26. $xe^{-y} + ye^{-x} = 3$

27. $ye^{x - x^2} = x + y$

28. $y = \dfrac{e^{-2x}(2 - x^3)^{3/2}}{\sqrt{1 + x^2}}$

29. $y = \dfrac{(x^2 + e^{2x})^3 e^{-2x}}{(1 + x - x^2)^{2/3}}$

In Problems 30 through 37, determine where the given function is increasing and decreasing, and where its graph is concave upward and concave downward. Sketch the graph, showing as many key features as possible (high and low points, points of inflection, asymptotes, intercepts, cusps, vertical tangents).

30. $f(x) = xe^{-2x}$

31. $f(x) = e^x - e^{-x}$

32. $f(x) = \dfrac{4}{1 + e^{-x}}$

33. $f(t) = t + e^{-t}$

34. $g(t) = \dfrac{\ln(t+1)}{t+1}$

35. $F(u) = u^2 + 2\ln(u+2)$

36. $f(u) = e^{2u} + e^{-u}$

37. $G(x) = \ln(e^{-2x} + e^{-x})$

In Problems 38 through 41, find the largest and smallest values of the given function over the prescribed closed, bounded interval.

38. $f(x) = \dfrac{e^{-x/2}}{x^2}$ for $-5 \le x \le -1$

39. $f(x) = \ln(4x - x^2)$ for $1 \le x \le 3$

40. $g(t) = \dfrac{\ln(\sqrt{t})}{t^2}$ for $1 \le t \le 2$

41. $h(t) = (e^{-t} + e^t)^5$ for $-1 \le t \le 1$

In Problems 42 through 45, find an equation for the tangent line to the given curve at the specified point.

42. $y = (x^2 - x)e^{-x}$ where $x = 0$

43. $y = x\ln x^2$ where $x = 1$

44. $y = (x + \ln x)^3$ where $x = 1$

45. $y = x^3 e^{2-x}$ where $x = 2$

46. EXPONENTIAL GROWTH The value of a certain industrial machine is decreasing exponentially. If the machine was originally worth \$50,000 and was worth \$20,000 when it was 5 years old, how much will it be worth when it is 10 years old?

47. SALES FROM ADVERTISING It is estimated that if x thousand dollars are spent on advertising, approximately $Q(x) = 50 - 40e^{-0.1x}$ thousand units of a certain commodity will be sold.

   a. Sketch the sales curve for $x \ge 0$.

   b. How many units will be sold if no money is spent on advertising?

   c. How many units will be sold if \$8,000 is spent on advertising?

   d. How much should be spent on advertising to generate sales of 35,000 units?

   e. According to this model, what is the most optimistic sales projection?

48. WORKER PRODUCTION An employer determines that the daily output of a worker on the job for t weeks is $Q(t) = 120 - Ae^{-kt}$ units. Initially, the worker can produce 30 units per day, and after 8 weeks, can produce 80 units per day. How many units can the worker produce per day after 4 weeks?

49. POPULATION GROWTH It is estimated that t years from now the population of a certain country will be P million people, where

$$P(t) = \dfrac{30}{1 + 2e^{-0.05t}}$$

   a. Sketch the graph of $P(t)$.

   b. What is the current population?

   c. What will be the population in 20 years?

   d. What happens to the population in "the long run"?

50. COMPOUND INTEREST How quickly will \$2,000 grow to \$5,000 when invested at an annual interest rate of 8% if interest is compounded:

   a. Quarterly

   b. Continuously

51. EFFECTIVE RATE OF INTEREST Which investment has the greater effective interest rate: 8.25% per year compounded quarterly or 8.20% per year compounded continuously? (Recall Problem 54, Section 4.1.)

52. COMPOUND INTEREST How much should you invest now at an annual interest rate of 6.25% so that your balance 10 years from now will be \$2,000 if interest is compounded:

   a. Monthly

   b. Continuously

53. PRESENT VALUE Find the present value of \$8,000 payable 10 years from now if the annual interest rate is 6.25% and interest is compounded:

   a. Semiannually

   b. Continuously

CHAPTER SUMMARY

54. COMPOUND INTEREST A bank compounds interest continuously. What (nominal) interest rate does it offer if $1,000 grows to $2,054.44 in 12 years?

55. COMPOUND INTEREST A certain bank offers an interest rate of 6% per year compounded annually. A competing bank compounds its interest continuously. What (nominal) interest rate should the competing bank offer so that the effective interest rates of the two banks will be equal?

56. BACTERIAL GROWTH The number of bacteria in a certain culture grows exponentially. If 5,000 bacteria were initially present and 8,000 were present 10 minutes later, how long will it take for the number of bacteria to double?

57. AIR POLLUTION An environmental study of a certain suburban community suggests that t years from now, the average level of carbon monoxide in the air will be $Q(t) = 4e^{0.03t}$ parts per million.

   a. At what rate will the carbon monoxide level be changing with respect to time 2 years from now?

   b. At what percentage rate will the carbon monoxide level be changing with respect to time t years from now? Does this percentage rate of change depend on t or is it constant?

58. PROFIT A manufacturer can produce cameras at a cost of $40 apiece and estimates that if they are sold for p dollars apiece, consumers will buy approximately $D(p) = 800e^{-0.01p}$ cameras per week. At what price should the manufacturer sell the cameras to maximize profit?

59. OPTIMAL HOLDING TIME Suppose you own an asset whose value t years from now will be $V(t) = 2,000e^{\sqrt{2t}}$ dollars. If the prevailing interest rate remains constant at 5% per year compounded continuously, when will it be most advantageous to sell the collection and invest the proceeds?

60. RULE OF 70 Investors are often interested in knowing how long it takes for a particular investment to double. A simple means for making this determination is the "rule of 70," which says: The doubling time of an investment with an annual interest rate r (expressed as a decimal) compounded continuously is given by $d = \dfrac{70}{r}$.

   a. For interest rate r, use the formula $B = Pe^{rt}$ to find the doubling time for $r = 4, 6, 9, 10,$ and 12. In each case, compare the value with the value obtained from the rule of 70.

 b. Some people prefer a "rule of 72" and others use a "rule of 69." Test these alternative rules as in part (a) and write a paragraph on which rule you would prefer to use.

61. RADIOACTIVE DECAY A radioactive substance decays exponentially with half-life λ. Suppose the amount of the substance initially present (when $t = 0$) is Q_0.

   a. Show that the amount of the substance that remains after t years will be $Q(t) = Q_0 e^{-(\ln 2/\lambda)t}$.

   b. Find a number k so that the amount in part (a) can be expressed as $Q(t) = Q_0(0.5)^{kt}$.

62. ANIMAL DEMOGRAPHY A naturalist at an animal sanctuary has determined that the function

$$f(x) = \frac{4e^{-(\ln x)^2}}{\sqrt{\pi}\,x}$$

provides a good measure of the number of animals in the sanctuary that are x years old. Sketch the graph of $f(x)$ for $x > 0$ and find the most "likely" age among the animals (that is, the age for which $f(x)$ is largest).

63. CARBON DATING "Ötzi the Iceman" is the name given a neolithic corpse found frozen in an Alpine glacier in 1991. He was originally thought to be from the Bronze Age because of the hatchet he was carrying. However, the hatchet proved to be made of copper rather than bronze. Read an article on the Bronze Age and determine the least age of the Ice Man assuming that he dates before the Bronze Age. What is the *largest* percentage of ^{14}C that can remain in a sample taken from his body?

64. FICK'S LAW Fick's law* says that $f(t) = C(1 - e^{-kt})$, where $f(t)$ is the concentration of solute inside a cell at time t, C is the (constant) concentration of solute surrounding the cell, and k is a positive constant. Suppose that for a particular cell, the concentration on the inside of the cell after 2 hours is 0.8% of the concentration outside the cell.

   a. What is k?

   b. What is the percentage rate of change of $f(t)$ at time t?

   c. Write a paragraph on the role played by Fick's law in ecology.

---

*Fick's law plays an important role in ecology. For instance, see M.D. LaGrega, P. L. Buckingham, and J. C. Evans, *Hazardous Waste Management,* New York: McGraw-Hill, 1994, pp. 95, 464, and especially p. 813, where the authors discuss contaminant transport through landfill.

65. COOLING A child falls into a lake where the water temperature is $-3°C$. Her body temperature after t minutes in the water is $T(t) = 35e^{-0.32t}$. She will lose consciousness when her body temperature reaches $27°C$. How long do rescuers have to save her? How fast is her body temperature dropping at the time it reaches $27°C$?

66. FORENSIC SCIENCE The temperature T of the body of a murder victim found in a room where the air temperature is $20°C$ is given by

$$T(t) = 20 + 17e^{-0.07t} \quad °C$$

where t is the number of hours after the victim's death.

a. Graph the body temperature $T(t)$ for $t \geq 0$. What is the horizontal asymptote of this graph and what does it represent?

b. What is the temperature of the victim's body after 10 hours? How long does it take for the body's temperature to reach $25°C$?

c. Abel Baker is a clerk in the firm of Dewey, Cheatum, and Howe. He comes to work early one morning and finds the corpse of his boss, Will Cheatum, draped across his desk. He calls the police, and at 8 A.M., they determine that the temperature of the corpse is $33°C$. Since the last item entered on the victim's notepad was, "Fire that idiot, Baker," Abel is considered the prime suspect. Actually, Abel is bright enough to have been reading this text in his spare time. He glances at the thermostat to confirm that the room temperature is $20°C$. For what time will he need an alibi in order to establish his innocence?

67. CONCENTRATION OF DRUG Suppose that t hours after an antibiotic is administered orally, its concentration in the patient's bloodstream is given by a surge function of the form $C(t) = Ate^{-kt}$, where A and k are positive constants and C is measured in micrograms per milliliter of blood. Blood samples are taken periodically, and it is determined that the maximum concentration of drug occurs 2 hours after it is administered and is 10 micrograms per milliliter.

a. Use this information to determine A and k.

b. A new dose will be administered when the concentration falls to 1 microgram per milliliter. When does this occur?

68. CHEMICAL REACTION RATE The effect of temperature on the reaction rate of a chemical reaction is given by the **Arrhenius equation**

$$k = Ae^{-E_0/RT}$$

where k is the rate constant, T (in degrees kelvin) is the temperature, and R is the gas constant. The quantities A and E_0 are fixed once the reaction is specified. Let k_1 and k_2 be the reaction rate constants associated with temperatures T_1 and T_2.

Find an expression for $\ln\left(\dfrac{k_1}{k_2}\right)$ in terms of E_0, R, T_1, and T_2.

69. OZONE DEPLETION It is known that fluorocarbons have the effect of depleting ozone in the upper atmosphere. Suppose it is found that the amount Q of ozone in the atmosphere is depleted by 0.15% per year, so that after t years, the amount of original ozone Q_0 that remains is

$$Q = Q_0 e^{-0.0015t}$$

a. At what percentage rate is the ozone level decreasing at time t?

b. How many years will it take for 10% of the ozone to be depleted? At what percentage rate is the ozone level decreasing at this time?

70. THE SPREAD OF AN EPIDEMIC Public health records indicate that t weeks after the outbreak of a certain form of influenza, approximately

$$Q(t) = \frac{80}{4 + 76e^{-1.2t}}$$

thousand people had caught the disease. At what rate was the disease spreading at the end of the second week? At what time is the disease spreading most rapidly?

71. ACIDITY (pH) OF A SOLUTION In chemistry, the acidity of a solution is measured by its pH value, which is defined by $pH = -\log_{10}[H_3O^+]$, where $[H_3O^+]$ is the hydronium ion concentration (moles/liter) of the solution. On average, milk has a pH value that is three times the pH value of a lime, which in turn has half the pH value of an orange. If the average pH of an orange is 3.2, what is the average hydronium ion concentration of a lime?

72. CARBON DATING A Cro-Magnon cave painting at Lascaux, France, is approximately 15,000 years old. Approximately what ratio of ^{14}C to ^{12}C would you expect to find in a fossil found in the same cave as the painting?

73. MORTALITY RATES It is sometimes useful for actuaries to be able to project mortality rates within a given population. A formula sometimes used for computing the mortality rate $D(t)$ for women in the age group 25–29 is

$$D(t) = (D_0 - 0.00046)e^{-0.162t} + 0.00046$$

where t is the number of years after a fixed base year and D_0 is the mortality rate when $t = 0$.

a. Suppose the initial mortality rate of a particular group is 0.008 (8 deaths per 1,000 women). What is the mortality rate of this group 10 years later? What is the rate 25 years later?

b. Sketch the graph of the mortality function $D(t)$ for the group in part (a) for $0 \leq t \leq 25$.

74. GROSS DOMESTIC PRODUCT The Gross Domestic Product (GDP) of a certain country was 100 billion dollars in 1990 and 165 billion dollars in 2000. Assuming that the GDP is growing exponentially, what will it be in the year 2010?

75. ARCHAEOLOGY "Lucy," the famous prehuman whose skeleton was discovered in Africa, has been found to be approximately 3.8 million years old.

a. Approximately what percentage of original ^{14}C would you expect to find if you tried to apply carbon dating to Lucy? Why would this be a problem if you were actually trying to "date" Lucy?

b. In practice, carbon dating works well only for relatively "recent" samples—those that are no more than approximately 50,000 years old. For older samples, such as Lucy, variations on carbon dating have been developed, such as potassium-argon and rubidium-strontium dating. Read an article on alternative dating methods and write a paragraph on how they are used.*

76. RADIOLOGY The radioactive isotope gallium-67 (^{67}Ga), used in the diagnosis of malignant tumors, has a half-life of 46.5 hours. If we start with 100 milligrams of the isotope, how many milligrams will be left after 24 hours? When will there be only 25 milligrams left? Answer these questions by first using a graphing utility to graph an appropriate exponential function and then using the **TRACE** and **ZOOM** features.

77. A population model developed by the U.S. Census Bureau uses the formula

$$P(t) = \frac{202.31}{1 + e^{3.938 - 0.314t}}$$

to estimate the population of the United States (in millions) for every tenth year from the base year

1790. Thus, for instance, $t = 0$ corresponds to 1790, $t = 1$ to 1800, $t = 10$ to 1890, and so on. The figures exclude Alaska and Hawaii.

a. Use this formula to compute the population of the United States for the years 1790, 1800, 1830, 1860, 1880, 1900, 1920, 1940, 1960, 1980, 1990, and 2000.

b. Sketch the graph of $P(t)$. When does this model predict that the population of the United States will be increasing most rapidly?

c. Use an almanac or some other source to find the actual population figures for the years listed in part (a). Does the given population model seem to be accurate? Write a paragraph describing some possible reasons for any major differences between the predicted population figures and the actual census figures.

78. Use a graphing utility to graph $y = 2^{-x}$, $y = 3^{-x}$, $y = 5^{-x}$, and $y = (0.5)^{-x}$ on the same set of axes. How does a change in base affect the graph of the exponential function? (*Suggestion:* Use the graphing window $[-3, 3]1$ by $[-3, 3]1$.)

79. Use a graphing utility to draw the graphs of $y = \sqrt{3^x}$, $y = \sqrt{3^{-x}}$, and $y = 3^{-x}$ on the same set of axes. How do these graphs differ? (*Suggestion:* Use the graphing window $[-3, 3]1$ by $[-3, 3]1$.)

80. Use a graphing utility to draw the graphs of $y = 3^x$ and $y = 4 - \ln \sqrt{x}$ on the same axes. Then use **TRACE** and **ZOOM** to find all points of intersection of the two graphs.

81. Solve this equation with three decimal place accuracy:

$$\log_5 (x + 5) - \log_2 x = 2 \log_{10} (x^2 + 2x)$$

82. Use a graphing utility to draw the graphs of

$$y = \ln (1 + x^2) \quad \text{and} \quad y = \frac{1}{x}$$

on the same axes. Do these graphs intersect?

83. Make a table for the quantities $(\sqrt{n})^{\sqrt{n+1}}$ and $(\sqrt{n + 1})^{\sqrt{n}}$, with $n = 8, 9, 12, 20, 25, 31, 37, 38, 43, 50, 100,$ and $1,000$. Which of the two quantities seems to be larger? Do you think this inequality holds for all $n \geq 8$?

---

*A good place to start your research is the article by Paul J. Campbell, "How Old Is the Earth?", *UMAP Modules 1992: Tools for Teaching*, Arlington, MA: Consortium for Mathematics and Its Applications, 1993.

Complete solutions for all EXPLORE! boxes throughout the text can be accessed at the book specific website, www.mhhe.com/hoffmann.

**Solution for Explore!
on Page 290**

One method to display all the desired graphs is to list the desired bases in the function form. First, write Y1 = {1, 2, 3, 4}^X into the equation editor of your graphing calculator. Observe that for $b > 1$, the functions increase exponentially, with steeper growth for larger bases. Also all the curves pass through the point $(0, 1)$. Why? Now try Y1 = {2, 4, 6}^X. Note that $y = 4^x$ lies between $y = 2^x$ and $y = 6^x$. Likewise the graph of $y = e^x$ would lie between $y = 2^x$ and $y = 3^x$.

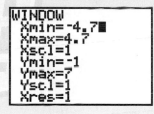 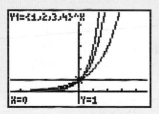 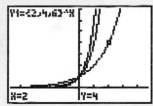

**Solution for Explore!
on Page 310**

Store $f(x) = B^x$ into Y1 and $g(x) = \log_B x$ into Y2 as $\ln(x)/\ln(B)$. Experimenting with different values of B, we find this: For $B < e^{1/e} \approx 1.444668$, the two curves intersect in two places (where, in terms of B?), for $B = e^{1/e}$ they touch only at one place, and for $B > e^{1/e}$, there is no intersection. (See Classroom Capsules, "An Overlooked Calculus Question," *The College Mathematics Journal*, Vol. 33, No. 5, November 2002.)

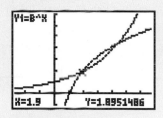 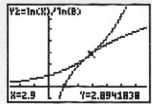 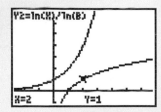

**Solution for Explore!
on Page 326**

Following Example 4.3.11, store the function $f(x) = Axe^{-Bx}$ into Y1 of the equation editor. We attempt to find the maximum of $f(x)$ in terms of A and B. We can set $A = 1$ and vary the value of B (say, 1, 0.5, and 0.01). Then we can fix B to be 0.1 and let A vary (say, 1, 10, 100).

For instance, when $A = 1$ and $B = 1$, the maximal y value occurs at $x = 1$ (see the figure on the left). When $A = 1$ and $B = 0.1$, it occurs at $x = 10$ (middle figure).

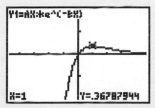 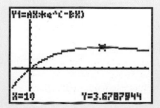 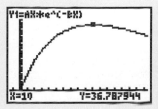

When $A = 10$ and $B = 0.1$, this maximum remains at $x = 10$ (figure on the right). In this case, the y coordinate of the maximum increases by the A factor. In general, it can be shown that the x value of the maximal point is just $1/B$. The A factor does not change the location of the x value of the maximum, but it does affect the y value as a multiplier. To confirm this analytically, set the derivative of $y = Axe^{-Bx}$ equal to zero and solve for the location of the maximal point.

Solution for Explore! on Page 335

Following Example 4.4.2, store $f(x)$ into Y1 and $f'(x)$ into Y2 (but deselected), and $f''(x)$ into Y3 in bold, using the window $[-4.7, 4.7]1$ by $[-0.5, 0.5]0.1$. Using the **TRACE** or the root finding feature of the calculator, you can determine that the two x intercepts of $f''(x)$ are located at $x = -1$ and $x = 1$. Since the second derivative $f''(x)$ represents the concavity function of $f(x)$, we know that at these values $f(x)$ changes concavity. At the inflection point $(-1, 0.242)$, $f(x)$ changes concavity from positive (concave upward) to negative (concave downward). At $x = 1$, $y = 0.242$, concavity changes from downward to upward.

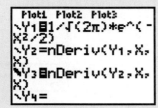 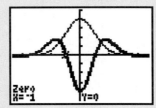 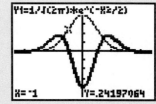

Solution for Explore! on Page 339

Following Example 4.4.5, store $Q(t) = 20/(1 + 19e^{-1.2t})$ into Y1 and graph using the window $[0, 10]1$ by $[0, 25]5$. We can trace the function for large values of the independent variable, time. As x approaches 10 (weeks), the function attains a value close to 20,000 people infected ($Y > 19.996$ thousand). Since 90% of the population is 18,000, by setting Y2 = 18 and using the intersection feature of the calculator, you can determine that 90% of the population becomes infected after $x = 4.28$ weeks (about 30 days).

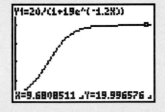 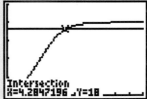

THINK ABOUT IT

MODELING WITH THE ARRHENIUS LAW

It's a warm summer evening, but how warm? Can you determine outside temperature without a thermometer? People have long used the rate at which crickets chirp to estimate the temperature. The kind of cricket matters since worldwide, there are roughly 1,000 different species of cricket, all with different chirping behavior. The snowy tree cricket (*Oecanthus fultoni*) is a good species for estimating temperature. It is common in the United States and much of Canada, occurring in every state but Florida, Montana, Hawaii, and Alaska. Moreover, its chirps are slow enough to count, individuals chirp synchronously, and it chirps at a rate that does not depend much on factors other than temperature. You can listen to the chirping of this cricket at different temperatures via our text specific website at www.mhhe.com/hoffmann.

Several different rules of thumb are used to estimate the temperature by counting the chirps of this cricket. One of the more accurate of these rules says that the number of chirps in 13 seconds plus 40 is a good estimate of T_F, the temperature in degrees Fahrenheit. To express this rule as a formula, note that if c is the number of chirps in a minute, then $\dfrac{13c}{60}$ is the number of chirps in a 13-second interval, so the temperature T_F can be estimated using the equation

$$T_F = \left(\frac{13}{60}\right)c + 40$$

Turning things around, we can express the number of chirps per minute as a function of temperature by solving this equation for c in terms of T_F to obtain

$$c = \left(\frac{60}{13}\right)(T_F - 40)$$

Part of this essay was adapted from a chapter from the General Chemistry website of Dartmouth College, available at http://www.dartmouth.edu/~genchem/0102/spring/6winn/cricket.html.

Figure 4.17 shows this line along with a number of data points (T_F, c), where in each case, c is the number of chirps per minute observed to occur when the temperature is T_F.

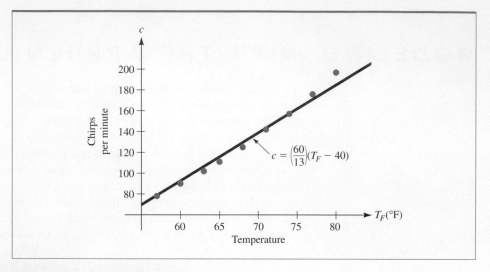

FIGURE 4.17   The number of cricket chirps per minute c is approximately linear as a function of temperature T_F.

Although Figure 4.17 indicates that the linear equation $c = \left(\dfrac{60}{13}\right)(T_F - 40)$ fits the observed data fairly well, it appears that the number of chirps moves away from the line at higher temperatures. Can we find a curve that fits the data even better than the line? One possible approach for achieving a better fit is to develop a model based on chemical reaction rates. This is reasonable since chemical reaction rates increase with temperature, as do processes that are based on chemical reactions, including many biological phenomena such as cricket chirping. We will use a result called the **Arrhenius law,** which says that the rate k of a chemical reaction is given by the function

$$k = Ae^{-E_a/RT}$$

where A is a constant called the *frequency factor*, E_a is the *activation energy*, R is the *universal gas constant*, and T is the *temperature* in degrees kelvin, which is related to degrees Fahrenheit by the formula

$$T = \frac{1}{9}(5T_F + 2{,}297)$$

If we regard the number of cricket chirps per minute c as a chemical reaction rate, the particular function that best fits the data shown in Figure 4.17 is

$$c = 25.98 \cdot 10^{10} e^{-6{,}290/T}$$

which can be found using the regression (least-squares) techniques discussed later in this text, in Section 7.4. When this curve is drawn on a graph with the data in Figure 4.17, we obtain the graph shown in Figure 4.18, which suggests that the model based on the Arrhenius law fits the observed data better than the linear model, especially for higher temperatures.

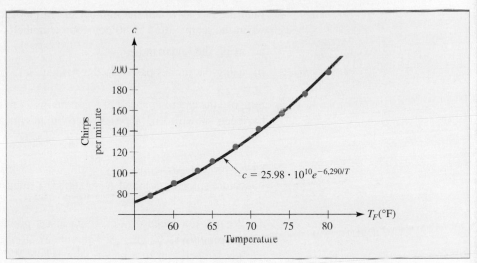

FIGURE 4.18 The cricket chirp data is better approximated by an exponential curve.

Although cricket chirping is more complicated than a single chemical reaction, we have seen that the exponential model based on the Arrhenius law provides a more accurate estimate of the rate of cricket chirping as a function of temperature than a linear model. It may be possible to find models that are even more accurate, but they would probably have to be much more complicated.

Besides cricket chirping, biological phenomena that have been successfully modeled using the Arrhenius law include the rate at which ants creep, the rates fireflies flash, the heartbeat rate of terrapins, alpha brain wave frequencies, and the rate people forget information. A discussion of such applications may be found in the article *Unconventional Applications of the Arrhenius Law,* by Keith J. Laidler, *Journal of Chemical Education,* Vol. 49, No. 5, May 1972, pp. 343–344.

Questions

1. Find the temperature corresponding to a snowy tree cricket chirping at a rate of 100 chirps per minute

   a. using the linear model.

   b. using the model based on the Arrhenius law.

2. Find the rate of chirping of a snowy tree cricket when the temperature is 75 degrees Fahrenheit that is predicted

   a. by the linear model.

   b. by the model based on the Arrhenius law.

3. Show that the logarithm of the reaction rate k in the Arrhenius law is a linear function of the reciprocal of the temperature T. What is the slope of this line?

4. Find the rate at which the number of chirps of a snowy tree cricket is changing when the temperature is 80 degrees Fahrenheit

   a. using the linear model.

   b. using the model based on the Arrhenius law.

5. Suppose the creeping speed of a certain insect in centimeters per second (cm/sec) has been shown to be approximated by an equation

   $$s = Ae^{-E_0/RT}$$

   based on the Arrhenius law with $A = 685$ cm/sec, $E_0 = 10.1$ (measured in kilocalories per mole), and $R = 0.008314$ (measured in kilojoules per mole per degree kelvin).

   a. What is the speed of a creeping insect predicted by this model when the temperature is 60 degrees Fahrenheit? (Remember to convert the Fahrenheit temperature to a kelvin temperature.)

   b. What is the temperature in degrees Fahrenheit predicted by this model when insects of this kind creep at 10 cm/sec?

6. Find the rate at which the creeping speed of the insects in Question 5 is changing when the temperature is 70 degrees Fahrenheit.

CHAPTER 5

Computing area under a curve, like the area of the region spanned by the scaffolding under the roller coaster track, is an application of integration.

INTEGRATION

SECTION   5.1 | Antidifferentiation: The Indefinite Integral

How can a known rate of inflation be used to determine future prices? What is the velocity of an object moving along a straight line with known acceleration? How can knowing the rate at which a population is changing be used to predict future population levels? In all these situations, the derivative (rate of change) of a quantity is known and the quantity itself is required. Here is the terminology we will use in connection with obtaining a function from its derivative.

Antidifferentiation   ■   A function $F(x)$ is said to be an *antiderivative* of $f(x)$ if

$$F'(x) = f(x)$$

for every x in the domain of $f(x)$. The process of finding antiderivatives is called *antidifferentiation* or *indefinite integration*.

| **NOTE**   Sometimes we write the equation

$$F'(x) = f(x)$$

as

$$\frac{dF}{dx} = f(x) \quad ■$$

Later in this section, you will learn techniques you can use to find antiderivatives. Once you have found what you believe to be an antiderivative of a function, you can always check your answer by differentiating. You should get the original function back. Here is an example.

EXAMPLE | 5.1.1

Verify that $F(x) = \dfrac{1}{3}x^3 + 5x + 2$ is an antiderivative of $f(x) = x^2 + 5$.

Solution

$F(x)$ is an antiderivative of $f(x)$ if and only if $F'(x) = f(x)$. Differentiate F and you will find that

$$F'(x) = \frac{1}{3}(3x^2) + 5$$
$$= x^2 + 5 = f(x)$$

as required.

The General Antiderivative of a Function

A function has more than one antiderivative. For example, one antiderivative of the function $f(x) = 3x^2$ is $F(x) = x^3$, since

$$F'(x) = 3x^2 = f(x)$$

but so are $x^3 + 12$ and $x^3 - 5$ and $x^3 + \pi$, since

$$\frac{d}{dx}(x^3 + 12) = 3x^2 \qquad \frac{d}{dx}(x^3 - 5) = 3x^2 \qquad \frac{d}{dx}(x^3 + \pi) = 3x^2$$

In general, if F is one antiderivative of f, then so is any function of the form $G(x) = F(x) + C$, for constant C since

$$
\begin{aligned}
G'(x) &= [F(x) + C]' \\
&= F'(x) + C' && \textit{sum rule for derivatives} \\
&= F'(x) + 0 && \textit{derivative of a constant is 0} \\
&= f(x) && \textit{since } F \textit{ is an antiderivative of } f
\end{aligned}
$$

Conversely, it can be shown that if F and G are both antiderivatives of f, then $G(x) = F(x) + C$, for some constant C (Problem 62). To summarize:

Fundamental Property of Antiderivatives   ■   If $F(x)$ is an antiderivative of the continuous function $f(x)$, then any other antiderivative of $f(x)$ has the form $G(x) = F(x) + C$ for some constant C.

There is a simple geometric interpretation for the fundamental property of antiderivatives. If F and G are both antiderivatives of f, then

$$G'(x) = F'(x) = f(x)$$

This means that the slope $F'(x)$ of the tangent line to $y = F(x)$ at the point $(x, F(x))$ is the same as the slope $G'(x)$ of the tangent line to $y = G(x)$ at $(x, G(x))$. Since the slopes are equal, it follows that the tangent lines at $(x, F(x))$ and $(x, G(x))$ are parallel, as shown in Figure 5.1a. Since this is true for all x, the entire curve $y = G(x)$ must be parallel to the curve $y = F(x)$, so that

$$y = G(x) = F(x) + C$$

In general, the collection of graphs of all antiderivatives of a given function f is a family of parallel curves that are vertical translations of one another. This is illustrated in Figure 5.1b for the family of antiderivatives of $f(x) = 3x^2$.

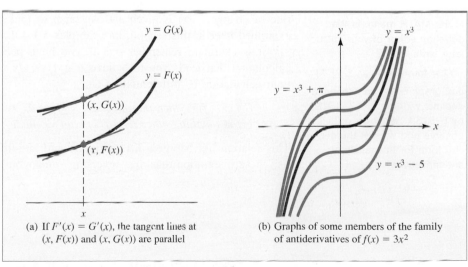

(a) If $F'(x) = G'(x)$, the tangent lines at $(x, F(x))$ and $(x, G(x))$ are parallel

(b) Graphs of some members of the family of antiderivatives of $f(x) = 3x^2$

FIGURE 5.1   Graphs of antiderivatives of a function f form a family of parallel curves.

The Indefinite Integral

You have just seen that if $F(x)$ is one antiderivative of the continuous function $f(x)$, then all such antiderivatives may be characterized by $F(x) + C$ for constant C. The family of all antiderivatives of $f(x)$ is written

$$\int f(x)\,dx = F(x) + C$$

and is called the **indefinite integral** of $f(x)$. The integral is "indefinite" because it involves a constant C that can take on any value. In Section 5.3, we introduce a **definite integral** that has a specific numerical value and is used to represent a variety of quantities, such as area, average value, present value of an income flow, and cardiac output, to name a few. The connection between definite and indefinite integrals is made in Section 5.3 through a result so important that it is referred to as the **fundamental theorem of calculus.**

In the context of the indefinite integral $\int f(x)\,dx = F(x) + C$, the **integral symbol** is $\int$, the function $f(x)$ is called the **integrand,** C is the **constant of integration,** and dx is a differential that indicates x is the **variable of integration.** These features are displayed in this diagram for the indefinite integral of $f(x) = 3x^2$:

integrand ——⌐   ⌐— *constant of integration*

$$\int 3x^2\,dx = x^3 + C$$

integral symbol ——⌐   ⌐———*variable of integration*

For any differentiable function F, we have

$$\int F'(x)\,dx = F(x) + C$$

since by definition, $F(x)$ is an antiderivative of $F'(x)$. Equivalently,

$$\int \frac{dF}{dx}\,dx = F(x) + C$$

This property of indefinite integrals is especially useful in applied problems where a rate of change $F'(x)$ is given and we wish to find $F(x)$. Several such problems are examined later in this section, in Examples 5.1.4 through 5.1.8.

It is useful to remember that if you have performed an indefinite integration calculation that leads you to believe that $\int f(x)\,dx = G(x) + C$, then you can check your calculation by differentiating $G(x)$:

If $G'(x) = f(x)$, then the integration $\int f(x)\,dx = G(x) + C$ is correct, but if $G'(x)$ is anything other than $f(x)$, you've made a mistake.

This relationship between differentiation and antidifferentiation enables us to establish these integration rules by "reversing" analogous differentiation rules.

Just-In-Time **Review**

Recall that differentials were introduced in Section 2.5.

EXPLORE!

Most graphing calculators allow the construction of an antiderivative through its numerical integral, **fnInt**(expression, variable, lower limit, upper limit), found in the **MATH** menu. In the equation editor of your calculator write

$Y1 = fnInt(2X, X, \{0, 1, 2\}, X)$

and graph using an expanded decimal window, $[-4.7, 4.7]1$ by $[-5, 5]1$. What do you observe and what is the general form for this family of antiderivatives?

Rules for Integrating Common Functions

The **constant rule:** $\displaystyle\int k\,dx = kx + C$ for constant k

The **power rule:** $\displaystyle\int x^n\,dx = \frac{x^{n+1}}{n+1} + C$ for all $n \neq -1$

The **logarithmic rule:** $\displaystyle\int \frac{1}{x}\,dx = \ln|x| + C$ for all $x \neq 0$

The **exponential rule:** $\displaystyle\int e^{kx}\,dx = \frac{1}{k}e^{kx} + C$ for constant $k \neq 0$

To verify the power rule, it is enough to show that the derivative of $\dfrac{x^{n+1}}{n+1}$ is x^n:

$$\frac{d}{dx}\left(\frac{x^{n+1}}{n+1}\right) = \frac{1}{n+1}[(n+1)x^n] = x^n$$

For the logarithmic rule, if $x > 0$, then $|x| = x$ and

$$\frac{d}{dx}(\ln|x|) = \frac{d}{dx}(\ln x) = \frac{1}{x}$$

If $x < 0$, then $-x > 0$ and $\ln|x| = \ln(-x)$, and it follows from the chain rule that

$$\frac{d}{dx}(\ln|x|) = \frac{d}{dx}[\ln(-x)] = \frac{1}{(-x)}(-1) = \frac{1}{x}$$

Thus, for all $x \neq 0$,

$$\frac{d}{dx}(\ln|x|) = \frac{1}{x}$$

so

$$\int \frac{1}{x}\,dx = \ln|x| + C$$

You are asked to verify the constant rule and exponential rule in Problem 64.

NOTE   Notice that the logarithm rule "fills the gap" in the power rule; namely, the case where $n = -1$. You may wish to blend the two rules into this combined form:

$$\int x^n\,dx = \begin{cases} \dfrac{x^{n+1}}{n+1} + C & \text{if } n \neq -1 \\[2mm] \ln|x| + C & \text{if } n = -1 \end{cases} \qquad \blacksquare$$

EXPLORE!

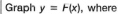

Graph $y = F(x)$, where

    $F(x) = \ln|x| = \ln(\text{abs}(x))$

in bold and $f(x) = 1/x$ in the regular graphing style using a decimal graphing window. At any point $x \neq 0$, show that the derivative of $F(x)$ is equal to the value of $f(x)$ at that particular point, confirming that $F(x)$ is the antiderivative of $f(x)$.

EXAMPLE | 5.1.2

Find these integrals:

a. $\displaystyle\int 3\,dx$ **b.** $\displaystyle\int x^{17}\,dx$ **c.** $\displaystyle\int \frac{1}{\sqrt{x}}\,dx$ **d.** $\displaystyle\int e^{-3x}\,dx$

Solution

a. Use the constant rule with $k = 3$: $\int 3\,dx = 3x + C$

b. Use the power rule with $n = 17$: $\int x^{17}\,dx = \frac{1}{18}x^{18} + C$

c. Use the power rule with $n = -\frac{1}{2}$: Since $n + 1 = \frac{1}{2}$,

$$\int \frac{dx}{\sqrt{x}} = \int x^{-1/2}\,dx = \frac{1}{1/2}x^{1/2} + C = 2\sqrt{x} + C$$

d. Use the exponential rule with $k = -3$:

$$\int e^{-3x}\,dx = \frac{1}{-3}e^{-3x} + C$$

Example 5.1.2 illustrates how certain basic functions can be integrated, but what about combinations of functions, such as the polynomial $x^5 + 2x^3 + 7$ or an expression like $5e^{-x} + \sqrt{x}$? Here are algebraic rules that will enable you to handle such expressions in a natural fashion.

Algebraic Rules for Indefinite Integration

The **constant multiple rule:** $\int kf(x)\,dx = k\int f(x)\,dx$ for constant k

The **sum rule:** $\int [f(x) + g(x)]\,dx = \int f(x)\,dx + \int g(x)\,dx$

The **difference rule:** $\int [f(x) - g(x)]\,dx = \int f(x)\,dx - \int g(x)\,dx$

To prove the constant multiple rule, note that if $\dfrac{dF}{dx} = f(x)$, then

$$\frac{d}{dx}[kF(x)] = k\frac{dF}{dx} = kf(x)$$

which means that

$$\int kf(x)\,dx = k\int f(x)\,dx$$

The sum and difference rules can be established in a similar fashion.

EXAMPLE | 5.1.3

Find the following integrals:

a. $\int (2x^5 + 8x^3 - 3x^2 + 5)\,dx$

b. $\displaystyle\int\left(\frac{x^3 + 2x - 7}{x}\right) dx$

c. $\displaystyle\int (3e^{-5t} + \sqrt{t}) \, dt$

Solution

a. By using the power rule in conjunction with the sum and difference rules and the multiple rule, you get

$$\int (2x^5 + 8x^3 - 3x^2 + 5) \, dx = 2\int x^5 \, dx + 8\int x^3 \, dx - 3\int x^2 \, dx + \int 5 \, dx$$

$$= 2\left(\frac{x^6}{6}\right) + 8\left(\frac{x^4}{4}\right) - 3\left(\frac{x^3}{3}\right) + 5x + C$$

$$= \frac{1}{3}x^6 + 2x^4 - x^3 + 5x + C$$

b. There is no "quotient rule" for integration, but at least in this case, you can still divide the denominator into the numerator and then integrate using the method in part (a):

$$\int \left(\frac{x^3 + 2x - 7}{x}\right) dx = \int \left(x^2 + 2 - \frac{7}{x}\right) dx$$

$$= \frac{1}{3}x^3 + 2x - 7 \ln |x| + C$$

c. $\displaystyle\int (3e^{-5t} + \sqrt{t}) \, dt = \int (3e^{-5t} + t^{1/2}) \, dt$

$$= 3\left(\frac{1}{-5}e^{-5t}\right) + \frac{1}{3/2}t^{3/2} + C = \frac{-3}{5}e^{-5t} + \frac{2}{3}t^{3/2} + C$$

EXPLORE!

Refer to Example 5.1.4. Store the function $f(x) = 3x^2 + 1$ into Y1. Graph using a bold graphing style and the window [0, 2.35]0.5 by [−2, 12]1. Place into Y2 the family of antiderivatives

$$F(x) = x^3 + x + L1$$

where L1 is the list of integer values −5 to 5. Which of these antiderivatives passes through the point (2, 6)? Repeat this exercise for $f(x) = 3x^2 − 2$.

EXAMPLE 5.1.4

Find the function $f(x)$ whose tangent has slope $3x^2 + 1$ for each value of x and whose graph passes through the point (2, 6).

Solution

The slope of the tangent at each point $(x, f(x))$ is the derivative $f'(x)$. Thus,

$$f'(x) = 3x^2 + 1$$

and so $f(x)$ is the antiderivative

$$f(x) = \int f'(x) \, dx = \int (3x^2 + 1) \, dx = x^3 + x + C$$

To find C, use the fact that the graph of f passes through (2, 6). That is, substitute $x = 2$ and $f(2) = 6$ into the equation for $f(x)$ and solve for C to get

$$6 = (2)^3 + 2 + C \qquad \text{or} \qquad C = -4$$

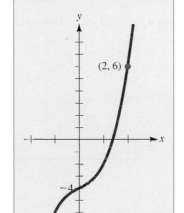

The graph of $y = x^3 + x - 4$.

Thus, the desired function is $f(x) = x^3 + x - 4$. The graph of this function is shown in the accompanying figure.

Applied Initial Value Problems

A **differential equation** is an equation that involves differentials or derivatives. Such equations are of great importance in modeling and occur in a variety of applications. An **initial value problem** is a problem that involves solving a differential equation subject to a specified initial condition. For instance, in Example 5.1.4, we were required to find $y = f(x)$ so that

$$\frac{dy}{dx} = 3x^2 + 1 \quad \text{subject to the condition } y = 6 \text{ when } x = 2$$

We solved this initial value problem by finding the antiderivative

$$y = \int (3x^2 + 1)\, dx = x^3 + x + C$$

and then using the initial condition to evaluate C. The same approach is used in Examples 5.1.5 through 5.1.8 to solve a selection of applied initial value problems from business, economics, biology, and physics. Similar initial value problems appear in examples and exercises throughout this chapter.

EXAMPLE | 5.1.5

A manufacturer has found that marginal cost is $3q^2 - 60q + 400$ dollars per unit when q units have been produced. The total cost of producing the first 2 units is $900. What is the total cost of producing the first 5 units?

Solution

Recall that the marginal cost is the derivative of the total cost function $C(q)$. Thus,

$$\frac{dC}{dq} = 3q^2 - 60q + 400$$

and so $C(q)$ must be the antiderivative

$$C(q) = \int \frac{dC}{dq}\, dq = \int (3q^2 - 60q + 400)\, dq = q^3 - 30q^2 + 400q + K$$

for some constant K. (The letter K was used for the constant to avoid confusion with the cost function C.)

The value of K is determined by the fact that $C(2) = 900$. In particular,

$$900 = (2)^3 - 30(2)^2 + 400(2) + K \quad \text{or} \quad K = 212$$

Hence,
$$C(q) = q^3 - 30q^2 + 400q + 212$$

and the cost of producing the first 5 units is

$$C(5) = (5)^3 - 30(5)^2 + 400(5) + 212 = \$1,587$$

EXAMPLE | 5.1.6

The population $P(t)$ of a bacterial colony t hours after observation begins is found to be changing at the rate

$$\frac{dP}{dt} = 200e^{0.1t} + 150e^{-0.03t}$$

If the population was 200,000 bacteria when observations began, what will the population be 12 hours later?

Solution

The population $P(t)$ is found by antidifferentiating $\dfrac{dP}{dt}$ as follows:

$$P(t) = \int \frac{dP}{dt}\, dt = \int (200\,e^{0.1t} + 150\,e^{-0.03t})\, dt$$

$$= \frac{200\,e^{0.1t}}{0.1} + \frac{150\,e^{-0.03t}}{-0.03} + C \qquad \text{exponential and sum rules}$$

$$= 2,000\,e^{0.1t} - 5,000\,e^{-0.03t} + C$$

Since the population is 200,000 when $t = 0$, we have

$$P(0) = 200,000 = 2,000\,e^0 - 5,000\,e^0 + C$$

$$= -3,000 + C$$

so $C = 203,000$ and

$$P(t) = 2,000e^{0.1t} - 5,000e^{-0.03t} + 203,000$$

Thus, after 12 hours, the population is

$$P(12) = 2,000e^{0.1(12)} - 5,000e^{-0.03(12)} + 203,000$$

$$\approx 206,152$$

EXAMPLE 5.1.7

A retailer receives a shipment of 10,000 kilograms of rice that will be used up over a 5-month period at the constant rate of 2,000 kilograms per month. If storage costs are 1 cent per kilogram per month, how much will the retailer pay in storage costs over the next 5 months?

Solution

Let $S(t)$ denote the total storage cost (in dollars) over t months. Since the rice is used up at a constant rate of 2,000 kilograms per month, the number of kilograms of rice in storage after t months is 10,000 - 2,000t. Therefore, since storage costs are 1 cent per kilogram per month, the rate of change of the storage cost with respect to time is

$$\frac{dS}{dt} = \left(\begin{array}{c}\text{monthly cost}\\ \text{per kilogram}\end{array}\right)\left(\begin{array}{c}\text{number of}\\ \text{kilograms}\end{array}\right) = 0.01(10,000 - 2,000t)$$

It follows that $S(t)$ is an antiderivative of

$$0.01(10,000 - 2,000t) = 100 - 20t$$

That is, $S(t) = \displaystyle\int \frac{dS}{dt}\, dt = \int (100 - 20t)\, dt$

$$= 100t - 10t^2 + C$$

for some constant C. To determine C, use the fact that at the time the shipment arrives (when $t = 0$) there is no cost, so that

$$0 = 100(0) - 10(0)^2 + C \qquad \text{or} \qquad C = 0$$

Hence, $$S(t) = 100t - 10t^2$$

and the total storage cost over the next 5 months will be

$$S(5) = 100(5) - 10(5)^2 = \$250$$

Motion along a Line

Recall from Section 2.2 that if an object moving along a straight line is at the position $s(t)$ at time t, then its velocity is given by $v = \dfrac{ds}{dt}$ and its acceleration by $a = \dfrac{dv}{dt}$. Turning things around, if the acceleration of the object is given, then its velocity and position can be found by integration. Here is an example.

EXPLORE!

Refer to Example 5.1.8. Graph the position function $s(t)$ in the equation editor of your calculator as Y1 = $-11x^2 + 66x$, using the window [0, 9.4]1 by [0, 200]10. Locate the stopping time and the corresponding position on the graph. Work the problem again for a car traveling 60 mph (88 ft/sec). In this case, what is happening at the 3-sec mark?

EXAMPLE │ 5.1.8

A car is traveling along a straight, level road at 45 miles per hour (66 feet per second) when the driver is forced to apply the brakes to avoid an accident. If the brakes supply a constant deceleration of 22 ft/sec^2 (feet per second, per second), how far does the car travel before coming to a complete stop?

Solution

Let $s(t)$ denote the distance traveled by the car in t seconds after the brakes are applied. Since the car decelerates at 22 ft/sec^2, it follows that $a(t) = -22$; that is,

$$\frac{dv}{dt} = a(t) = -22$$

Integrating, you find that the velocity at time t is given by

$$v(t) = \int \frac{dv}{dt}\, dt = \int -22\, dt = -22t + C_1$$

To evaluate C_1, note that $v = 66$ when $t = 0$ so that

$$66 = v(0) = -22(0) + C_1$$

and $C_1 = 66$. Thus, the velocity at time t is $v(t) = -22t + 66$.

Next, to find the distance $s(t)$, begin with the fact that

$$\frac{ds}{dt} = v(t) = -22t + 66$$

and use integration to show that

$$s(t) = \int \frac{ds}{dt}\, dt = \int (-22t + 66)\, dt = -11t^2 + 66t + C_2$$

Since $s(0) = 0$ (do you see why?), it follows that $C_2 = 0$ and

$$s(t) = -11t^2 + 66t$$

Finally, to find the stopping distance, note that the car stops when $v(t) = 0$, and this occurs when

$$v(t) = -22t + 66 = 0$$

Solving this equation, you find that the car stops after 3 seconds of deceleration, and in that time it has traveled

$$s(3) = -11(3)^2 + 66(3) = 99 \text{ feet}$$

PROBLEMS | 5.1

In Problems 1 through 30, find the indicated integral. Check your answers by differentiation.

1. $\int -3\, dx$

2. $\int dx$

3. $\int x^5\, dx$

4. $\int \sqrt{t}\, dt$

5. $\int \frac{1}{x^2}\, dx$

6. $\int 3e^x\, dx$

7. $\int \frac{2}{\sqrt{t}}\, dt$

8. $\int x^{-0.3}\, dx$

9. $\int u^{-2/5}\, du$

10. $\int \left(\frac{1}{x^2} - \frac{1}{x^3} \right) dx$

11. $\int (3t^2 - \sqrt{5t} + 2)\, dt$

12. $\int (x^{1/3} - 3x^{-2/3} + 6)\, dx$

13. $\int (3\sqrt{y} - 2y^{-3})\, dy$

14. $\int \left(\frac{1}{2y} - \frac{2}{y^2} + \frac{3}{\sqrt{y}} \right) dy$

15. $\int \left(\frac{e^x}{2} + x\sqrt{x} \right) dx$

16. $\int \left(\sqrt{x^3} - \frac{1}{2\sqrt{x}} + \sqrt{2} \right) dx$

17. $\int u^{1.1} \left(\frac{1}{3u} - 1 \right) du$

18. $\int \left(2e^u + \frac{6}{u} + \ln 2 \right) du$

19. $\int \left(\frac{x^2 + 2x + 1}{x^2} \right) dx$

20. $\int \frac{x^2 + 3x - 2}{\sqrt{x}}\, dx$

21. $\int (x^3 - 2x^2) \left(\frac{1}{x} - 5 \right) dx$

22. $\int y^3 \left(2y + \frac{1}{y} \right) dy$

23. $\int \sqrt{t}(t^2 - 1)\, dt$

24. $\int x(2x + 1)^2\, dx$

25. $\int (e^t + 1)^2\, dt$

26. $\int e^{-0.02t}(e^{-0.13t} + 4)\, dt$

27. $\int \left(\frac{1}{3y} - \frac{5}{\sqrt{y}} + e^{-y/2} \right) dy$

28. $\int \frac{1}{x}(x + 1)^2\, dx$

29. $\int t^{-1/2}(t^2 - t + 2)\, dt$

30. $\int \ln(e^{-x^2})\, dx$

In Problems 31 through 34, solve the given initial value problem for $y = f(x)$.

31. $\dfrac{dy}{dx} = 3x - 2$ where $y = 2$ when $x = -1$

32. $\dfrac{dy}{dx} = e^{-x}$ where $y = 3$ when $x = 0$

33. $\dfrac{dy}{dx} = \dfrac{2}{x} - \dfrac{1}{x^2}$ where $y = -1$ when $x = 1$

34. $\dfrac{dy}{dx} = \dfrac{x + 1}{\sqrt{x}}$ where $y = 5$ when $x = 4$

In Problems 35 through 40, the slope f'(x) at each point (x, y) on a curve y = f(x) is given along with a particular point (a, b) on the curve. Use this information to find f(x).

35. $f'(x) = 4x + 1$; $(1, 2)$

36. $f'(x) = 3x^2 + 6x - 2$; $(0, 6)$

37. $f'(x) = x^3 - \dfrac{2}{x^2} + 2$; $(1, 3)$

38. $f'(x) = x^{-1/2} + x$; $(1, 2)$

39. $f'(x) = e^{-x} + x^2$; $(0, 4)$

40. $f'(x) = \dfrac{3}{x} - 4$; $(1, 0)$

41. POPULATION GROWTH It is estimated that t months from now the population of a certain town will be changing at the rate of $4 + 5t^{2/3}$ people per month. If the current population is 10,000, what will be the population 8 months from now?

42. MARGINAL PROFIT A manufacturer estimates marginal revenue to be $R'(q) = 100q^{-1/2}$ dollars per unit when the level of production is q units. The corresponding marginal cost has been found to be $0.4q$ dollars per unit. Suppose the manufacturer's profit is $520 when the level of production is 16 units. What is the manufacturer's profit when the level of production is 25 units?

43. NET CHANGE IN A BIOMASS A biomass is growing at the rate of $M'(t) = 0.5e^{0.2t}$ g/hr. By how much does the mass change during the second hour?

44. TREE GROWTH An environmentalist finds that a certain type of tree grows in such a way that its height $h(t)$ after t years is changing at the rate of

$$h'(t) = 0.2t^{2/3} + \sqrt{t} \quad \text{ft/yr}$$

If the tree was 2 feet tall when it was planted, how tall will it be in 27 years?

45. MARGINAL COST A manufacturer estimates that the marginal cost of producing q units of a certain commodity is $C'(q) = 3q^2 - 24q + 48$ dollars per unit. If the cost of producing 10 units is $5,000, what is the cost of producing 30 units?

46. MARGINAL REVENUE The marginal revenue derived from producing q units of a certain commodity is $R'(q) = 4q - 1.2q^2$ dollars per unit. If the revenue derived from producing 20 units is $30,000, how much revenue should be expected from producing 40 units?

47. LEARNING Bob is taking a learning test in which the time he takes to memorize items from a given list is recorded. Let $M(t)$ be the number of items he can memorize in t minutes. His learning rate is found to be

$$M'(t) = 0.4t - 0.005t^2$$

a. How many items can Bob memorize during the first 10 minutes?

b. How many additional items can he memorize during the next 10 minutes (from time $t = 10$ to $t = 20$)?

48. SALES The monthly sales at an import store are currently $10,000 but are expected to be declining at the rate of

$$S'(t) = -10t^{2/5} \quad \text{dollars per month}$$

t months from now. The store is profitable as long as the sales level is above $8,000 per month.

a. Find a formula for the expected sales in t months.

b. What sales figure should be expected 2 years from now?

 c. For how many months will the store remain profitable?

49. ADVERTISING After initiating an advertising campaign in an urban area, a satellite dish provider estimates that the number of new subscribers will grow at a rate given by

$$N'(t) = 154t^{2/3} + 37 \quad \text{subscribers per month}$$

where t is the number of months after the advertising begins. How many new subscribers should be expected 8 months from now?

50. ENDANGERED SPECIES A conservationist finds that the population $P(t)$ of a certain endangered species is growing at a rate given by $P'(t) = 0.51e^{-0.03t}$, where t is the number of years after records began to be kept.

a. If the population is $P_0 = 500$ now (at time $t = 0$), what will it be in 10 years?

b. Read an article on endangered species and write a paragraph on the use of mathematical models in studying populations of such species.*

51. DEFROSTING A roast is removed from the freezer of a refrigerator and left on the counter to defrost. The temperature of the roast was $-4°C$

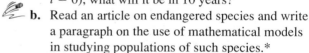

---

*You may wish to begin your research with the journal *Ecology*.

when it was removed from the freezer and t hours later, was increasing at the rate of
$$T'(t) = 7e^{-0.35t} \quad °C/hr$$

a. Find a formula for the temperature of the roast after t hours.
b. What is the temperature after 2 hours?
c. Assume the roast is defrosted when its temperature reaches 10°C. How long does it take for the roast to defrost?

52. **MARGINAL REVENUE**   Suppose it has been determined that the marginal revenue associated with the production of x units of a particular commodity is $R'(x) = 240 - 4x$ dollars per unit. What is the revenue function $R(x)$? You may assume $R(0) = 0$. What price will be paid for each unit when the level of production is $x = 5$ units?

53. **MARGINAL PROFIT**   The marginal profit of a certain commodity is $P'(q) = 100 - 2q$ when q units are produced. When 10 units are produced, the profit is $700.
a. Find the profit function $P(q)$.
b. What production level q results in maximum profit? What is the maximum profit?

54. **PRODUCTION**   At a certain factory, when K thousand dollars are invested in the plant, the production Q is changing at a rate given by
$$Q'(K) = 200K^{-2/3}$$
units per thousand dollars invested. When $8,000 are invested, the level of production is 5,500 units.
a. Find a formula for the level of production Q to be expected when K thousand dollars are invested.
b. How many units will be produced when $27,000 are invested?
c. What capital investment K is required to produce 7,000 units?

55. **MARGINAL PROPENSITY TO CONSUME**
Suppose the consumption function for a particular country is $c(x)$, where x is national disposable income. Then the **marginal propensity to consume** is $c'(x)$. Suppose x and c are both measured in billions of dollars and
$$c'(x) = 0.9 + 0.3\sqrt{x}$$
If consumption is 10 billion dollars when $x = 0$, find $c(x)$.

56. **SPY STORY**   Our spy, intent on avenging the death of Siggy Leiter (Problem 63 in Section 4.2), is driving a sports car toward the lair of the fiend who

killed his friend. To remain as inconspicuous as possible, he is traveling at the legal speed of 60 mph (88 feet per second) when suddenly, he sees a camel in the road, 199 feet in front of him. It takes him 00.7 seconds to react to the crisis. Then he hits the brakes, and the car decelerates at the constant rate of 28 ft/sec² (28 feet per second, per second). Does he stop before hitting the camel?

57. **LEARNING**   Let $f(x)$ represent the total number of items a subject has memorized x minutes after being presented with a long list of items to learn. Psychologists refer to the graph of $y = f(x)$ as a **learning curve** and to $f'(x)$ as the **learning rate.** The time of **peak efficiency** is the time when the learning rate is maximized. Suppose the learning rate is
$$f'(x) = 0.1(10 + 12x - 0.6x^2) \quad \text{for } 0 \le x < 25$$
a. When does peak efficiency occur? What is the learning rate at peak efficiency?
b. What is $f(x)$?
c. What is the largest number of items memorized by the subject?

58. **CANCER THERAPY**   A new medical procedure is applied to a cancerous tumor with volume 30 cm³, and t days later the volume is found to be changing at the rate
$$V'(t) = 0.15 - 0.09e^{0.006t} \quad cm^3/day$$
a. Find a formula for the volume of the tumor after t days.
b. What is the volume after 60 days? After 120 days?
c. For the procedure to be successful, it should take no longer than 90 days for the tumor to begin to shrink. Based on this criterion, does the procedure succeed?

59. **MARGINAL ANALYSIS**   A manufacturer estimates marginal revenue to be $200q^{-1/2}$ dollars per unit when the level of production is q units. The corresponding marginal cost has been found to be $0.4q$ dollars per unit. If the manufacturer's profit is $2,000 when the level of production is 25 units, what is the profit when the level of production is 36 units?

60. **CORRECTION FACILITY MANAGEMENT**
Statistics compiled by the local department of corrections indicate that x years from now the number of inmates in county prisons will be increasing at the rate of $280e^{0.2x}$ per year. Currently, 2,000 inmates are housed in county prisons. How many inmates should the county expect 10 years from now?

61. FLOW OF BLOOD One of Poiseuille's laws for the flow of blood in an artery says that if $v(r)$ is the velocity of flow r cm from the central axis of the artery, then the velocity decreases at a rate proportional to r. That is,

$$v'(r) = -ar$$

where a is a positive constant.* Find an expression for $v(r)$. Assume $v(R) = 0$, where R is the radius of the artery.

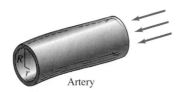

PROBLEM 61

62. If $H'(x) = 0$ for all real numbers x, what must be true about the graph of $H(x)$? Explain how your observation can be used to show that if $G'(x) = F'(x)$ for all x, then $G(x) = F(x) + C$ for constant C.

63. DISTANCE AND VELOCITY An object is moving so that its velocity after t minutes is $v(t) = 3 + 2t + 6t^2$ meters per minute. How far does the object travel during the second minute?

64. a. Prove the constant rule: $\int k \, dx = kx + C$.

    b. Prove the exponential rule: $\int e^{kx} \, dx = \dfrac{1}{k} e^{kx} + C$.

---

*E. Batschelet, *Introduction to Mathematics for Life Scientists*, 2nd ed., New York: Springer-Verlag, 1979, pp. 101–103.

65. What is $\int b^x \, dx$ for base b ($b > 0$, $b \neq 1$)? [*Hint:* Recall that $b^x = e^{x \ln b}$.]

66. It is estimated that x months from now, the population of a certain town will be changing at the rate of $P'(x) = 2 + 1.5\sqrt{x}$ people per month. The current population is 5,000.

    a. Find a function $P(x)$ that satisfies these conditions. Use the graphing utility of your calculator to graph this function.

    b. Use **TRACE** and **ZOOM** to determine the level of population 9 months from now. When will the population be 7,590?

    c. Suppose the current population were 2,000 (instead of 5,000). Sketch the graph of $P(x)$ with this assumption. Then sketch the graph of $P(x)$ assuming current populations of 4,000 and 6,000. What is the difference between the graphs?

67. A car traveling at 67 ft/sec decelerates at the constant rate of 23 ft/sec² when the brakes are applied.

    a. Find the velocity $v(t)$ of the car t seconds after the brakes are applied. Then find its distance $s(t)$ from the point where the brakes are applied.

    b. Use the graphing utility of your calculator to sketch the graphs of $v(t)$ and $s(t)$ on the same screen (use [0, 5]1 by [0, 200]10).

    c. Use **TRACE** and **ZOOM** to determine when the car comes to a complete stop and how far it travels in that time. How fast is the car traveling when it has traveled 45 feet?

SECTION 5.2 Integration by Substitution

The majority of functions that occur in practical situations can be differentiated by applying rules and formulas such as those you learned in Chapter 2. Integration, however, is at least as much an art as a science, and many integrals that appear deceptively simple may actually require a special technique or clever insight.

For example, we easily find that

$$\int x^7 \, dx = \frac{1}{8} x^8 + C$$

by applying the power rule, but suppose we wish to compute

$$\int (3x + 5)^7 \, dx$$

We could proceed by expanding the integrand $(3x + 5)^7$ and then integrating term by term, but the algebra involved in this approach is daunting. Instead, we make the change of variable

$$u = 3x + 5 \qquad \text{so that} \qquad du = 3\,dx \quad \text{or} \quad dx = \frac{1}{3}\,du$$

Just-In-Time Review

Recall that the differential of $y = f(x)$ is $dy = f'(x)\,dx$.

Then, by substituting these quantities into the given integral, we get

$$\int (3x + 5)^7\,dx = \int u^7\!\left(\frac{1}{3}\,du\right)$$

$$= \frac{1}{3}\!\left(\frac{1}{8}u^8\right) + C = \frac{1}{24}u^8 + C \quad \text{power rule}$$

$$= \frac{1}{24}(3x + 5)^8 + C \qquad\qquad \text{since } u = 3x + 5$$

We can check this computation by differentiating using the chain rule (Section 2.4):

$$\frac{d}{dx}\!\left[\frac{1}{24}(3x + 5)^8\right] = \frac{1}{24}[8(3x + 5)^7(3)] = (3x + 5)^7$$

which verifies that $\dfrac{1}{24}(3x + 5)^8$ is indeed an antiderivative of $(3x + 5)^7$.

The change of variable procedure we have just demonstrated is called **integration by substitution,** and it amounts to reversing the chain rule for differentiation. To see how, consider an integral that can be written as

$$\int f(x)\,dx = \int g(u(x))u'(x)\,dx$$

Suppose G is an antiderivative of g, so that $G' = g$. Then, according to the chain rule

$$\frac{d}{dx}[G(u(x))] = G'(u(x))\ u'(x)$$

$$= g(u(x))\ u'(x) \quad \text{since } G' = g$$

Therefore, by integrating both sides of this equation with respect to x, we find that

$$\int f(x)\,dx = \int g(u(x))u'(x)\,dx$$

$$= \int \left(\frac{d}{dx}[G(u(x))]\right)dx$$

$$= G(u(x)) + C \qquad\qquad \text{since } \int G' = G$$

In other words, once we have an antiderivative for $g(u)$, we also have one for $f(x)$.

A useful device for remembering the substitution procedure is to think of $u = u(x)$ as a change of variable whose differential $du = u'(x)\,dx$ can be manipulated algebraically. Then

$$\int f(x)\,dx = \int g(u(x))\,u'(x)\,dx$$

$$= \int g(u)\,du \qquad\qquad \textit{substitute } du \textit{ for } u'(x)\,dx$$

$$= G(u) + C \qquad\qquad \textit{where } G \textit{ is an antiderivative of } g$$

$$= G(u(x)) + C \qquad\quad \textit{substitute } u(x) \textit{ for } u$$

Here is a step-by-step procedure for integrating by substitution.

Using Substitution to Integrate $\int f(x)\,dx$

Step 1. Choose a substitution $u = u(x)$ that "simplifies" the integrand $f(x)$.

Step 2. Express the entire integral in terms of u and $du = u'(x)\,dx$. This means that *all* terms involving x and dx must be transformed to terms involving u and du.

Step 3. When step 2 is complete, the given integral should have the form

$$\int f(x)\,dx = \int g(u)\,du$$

If possible, evaluate this transformed integral by finding an antiderivative $G(u)$ for $g(u)$.

Step 4. Replace u by $u(x)$ in $G(u)$ to obtain an antiderivative $G(u(x))$ for $f(x)$, so that

$$\int f(x)\,dx = G(u(x)) + C$$

An old saying goes, "The first step in making rabbit stew is to catch a rabbit." Likewise, the first step in integrating by substitution is to find a suitable change of variable $u = u(x)$ that simplifies the integrand of the given integral $\int f(x)\,dx$ without adding undesired complexity when dx is replaced by $du = u'(x)\,dx$. Here are a few guidelines for choosing $u(x)$:

1. If possible, try to choose u so that $u'(x)$ is part of the integrand $f(x)$.

2. Consider choosing u as the part of the integrand that makes $f(x)$ difficult to integrate directly, such as the quantity inside a radical, the denominator of a fraction, or the exponent of an exponential function.

3. Don't "oversubstitute." For instance, in our introductory example $\int (3x + 5)^7\,dx$, a common mistake is to use $u = (3x + 5)^7$. This certainly simplifies the integrand, but then $du = 7(3x + 5)^6(3)\,dx$, and you are left with a transformed integral that is more complicated than the original.

4. Persevere. If you try a substitution that does not result in a transformed integral you can evaluate, try a different substitution.

Examples 5.2.1 through 5.2.6 illustrate how substitutions are chosen and used in various kinds of integrals.

EXAMPLE | 5.2.1

Find $\int \sqrt{2x + 7}\, dx$.

Solution

We choose $u = 2x + 7$ and obtain

$$du = 2\, dx \qquad \text{so that} \qquad dx = \frac{1}{2}\, du$$

Then the integral becomes

$$\int \sqrt{2x + 7}\, dx = \int \sqrt{u}\left(\frac{1}{2}\, du\right)$$

$$= \frac{1}{2}\int u^{1/2}\, du \qquad\qquad\qquad \text{since } \sqrt{u} = u^{1/2}$$

$$= \frac{1}{2}\frac{u^{3/2}}{3/2} + C = \frac{1}{3}u^{3/2} + C \quad \text{power rule}$$

$$= \frac{1}{3}(2x + 7)^{3/2} + C \qquad\qquad \text{substitute } 2x + 7 \text{ for } u$$

EXAMPLE | 5.2.2

Find $\int 8x(4x^2 - 3)^5\, dx$.

Solution

First, note that the integrand $8x(4x^2 - 3)^5$ is a product in which one of the factors, $8x$, is the derivative of an expression, $4x^2 - 3$, that appears in the other factor. This suggests that you make the substitution

$$u = 4x^2 - 3 \qquad \text{with} \qquad du = 4(2x\, dx) = 8x\, dx$$

to obtain

$$\int 8x(4x^2 - 3)^5\, dx = \int (4x^2 - 3)^5(8x\, dx)$$

$$= \int u^5\, dx$$

$$= \frac{1}{6}u^6 + C \qquad\qquad \text{power rule}$$

$$= \frac{1}{6}(4x^2 - 3)^6 + C \quad \text{substitute } 4x^2 - 3 \text{ for } u$$

EXAMPLE | 5.2.3

Find $\int x^3 e^{x^4+2}\, dx$.

Solution

If the integrand of an integral contains an exponential function, it is often useful to substitute for the exponent. In this case, we choose

$$u = x^4 + 2 \qquad \text{so that} \qquad du = 4x^3\, dx$$

and

$$\int x^3 e^{x^4 + 2}\, dx = \int e^{x^4 + 2}(x^3\, dx)$$

$$= \int e^u\left(\frac{1}{4}\, du\right) \qquad \text{since } du = 4x^3\, dx$$

$$= \frac{1}{4}e^u + C \qquad \text{exponential rule}$$

$$= \frac{1}{4}e^{x^4 + 2} + C \qquad \text{substitute } x^4 + 2 \text{ for } u$$

EXAMPLE | 5.2.4

Find $\displaystyle\int \frac{x}{x - 1}\, dx$.

Solution

Following our guidelines, we substitute for the denominator of the integrand, so that $u = x - 1$ and $du = dx$. Since $u = x - 1$, we also have $x = u + 1$. Thus,

$$\int \frac{x}{x - 1}\, dx = \int \frac{u + 1}{u}\, du$$

$$= \int \left[1 + \frac{1}{u}\right] du \qquad \text{divide}$$

$$= u + \ln |u| + C \qquad \text{constant and logarithmic rules}$$

$$= x - 1 + \ln |x - 1| + C \qquad \text{substitute } x - 1 \text{ for } u$$

EXAMPLE | 5.2.5

Find $\displaystyle\int \frac{3x + 6}{\sqrt{2x^2 + 8x + 3}}\, dx$.

Solution

This time, our guidelines suggest substituting for the quantity inside the radical in the denominator; that is,

$$u = 2x^2 + 8x + 3 \qquad du = (4x + 8)\, dx$$

At first glance, it may seem that this substitution fails, since $du = (4x + 8)\, dx$ appears quite different from the term $(3x + 6)\, dx$ in the integral. However, note that

$$(3x + 6)\, dx = 3(x + 2)\, dx = \frac{3}{4}(4)[(x + 2)\, dx]$$

$$= \frac{3}{4}[(4x + 8)\, dx] = \frac{3}{4}\, du$$

Substituting, we find that

$$\int \frac{3x + 6}{\sqrt{2x^2 + 8x + 3}} \, dx = \int \frac{1}{\sqrt{2x^2 + 8x + 3}} [(3x + 6) \, dx]$$

$$= \int \frac{1}{\sqrt{u}} \left(\frac{3}{4} \, du \right) = \frac{3}{4} \int u^{-1/2} \, du$$

$$= \frac{3}{4} \left(\frac{u^{1/2}}{1/2} \right) + C = \frac{3}{2} \sqrt{u} + C$$

$$= \frac{3}{2} \sqrt{2x^2 + 8x + 3} + C \qquad \begin{array}{l} \textit{substitute} \\ u = 2x^2 + 8x + 3 \end{array}$$

EXAMPLE | 5.2.6

Find $\displaystyle\int \frac{(\ln x)^2}{x} \, dx$.

Solution

Because

$$\frac{d}{dx}(\ln x) = \frac{1}{x}$$

the integrand

$$\frac{(\ln x)^2}{x} = (\ln x)^2 \left(\frac{1}{x} \right)$$

is a product in which one factor $\dfrac{1}{x}$ is the derivative of an expression $\ln x$ that appears in the other factor. This suggests substituting $u = \ln x$ with $du = \dfrac{1}{x} \, dx$ so that

$$\int \frac{(\ln x)^2}{x} \, dx = \int (\ln x)^2 \left(\frac{1}{x} \, dx \right)$$

$$= \int u^2 \, du = \frac{1}{3} u^3 + C$$

$$= \frac{1}{3} (\ln x)^3 + C \qquad \textit{substitute } \ln x \textit{ for } u$$

Sometimes an integral "looks" like it should be evaluated using a substitution but closer examination reveals a more direct approach. Consider Example 5.2.7.

EXAMPLE | 5.2.7

Find $\displaystyle\int e^{5x+2} \, dx$.

Solution

You can certainly handle this integral using the substitution

$$u = 5x + 2 \qquad du = 5\, dx$$

but it is not really necessary since $e^{5x+2} = e^{5x}e^2$, and e^2 is a constant. Thus,

$$\int e^{5x+2}\, dx = \int e^{5x}e^2\, dx$$

$$= e^2 \int e^{5x}\, dx \qquad \textit{factor constant } e^2 \textit{ outside integral}$$

$$= e^2 \left[\frac{e^{5x}}{5} \right] + C \quad \textit{exponential rule}$$

$$= \frac{1}{5}e^{5x+2} + C \quad \textit{since } e^2 e^{5x} = e^{5x+2}$$

In Example 5.2.7, we used algebra to put the integrand into a form where substitution was not necessary. In Examples 5.2.8 and 5.2.9, we use algebra as a first step, before making a substitution.

EXAMPLE | 5.2.8

Find $\displaystyle\int \frac{x^2 + 3x + 5}{x + 1}\, dx.$

Solution

There is no easy way to approach this integral as it stands (remember, there is no "quotient rule" for integration). However, suppose we simply divide the denominator into the numerator:

$$
\begin{array}{r}
x + 2 \\
x + 1 \overline{\big)\; x^2 + 3x + 5} \\
-x(x + 1) \\
\hline
2x + 5 \\
-2(x + 1) \\
\hline
3
\end{array}
$$

that is,

$$\frac{x^2 + 3x + 5}{x + 1} = x + 2 + \frac{3}{x + 1}$$

We can integrate $x + 2$ directly using the power rule. For the term $\dfrac{3}{x + 1}$, we use the substitution $u = x + 1$; $du = dx$:

$$\int \frac{x^2 + 3x + 5}{x + 1}\, dx = \int \left[x + 2 + \frac{3}{x + 1} \right] dx$$

$$= \int x\, dx + \int 2\, dx + \int \frac{3}{u}\, du \qquad \begin{array}{l} u = x + 1 \\ du = dx \end{array}$$

$$= \frac{1}{2}x^2 + 2x + 3 \ln |u| + C$$

$$= \frac{1}{2}x^2 + 2x + 3 \ln |x + 1| + C \qquad \textit{substitute } x + 1 \textit{ for } u$$

EXAMPLE | 5.2.9

Find $\displaystyle\int \frac{1}{1 + e^{-x}} \, dx$.

Solution

You may try to substitute $w = 1 + e^{-x}$. However, this is a dead end because $dw = -e^{-x} \, dx$ but there is no e^{-x} term in the numerator of the integrand. Instead, note that

$$\frac{1}{1 + e^{-x}} = \frac{1}{1 + \dfrac{1}{e^x}} = \frac{1}{\dfrac{e^x + 1}{e^x}}$$

$$= \frac{e^x}{e^x + 1}$$

Now, if you substitute $u = e^x + 1$ with $du = e^x \, dx$ into the given integral, you get

$$\int \frac{1}{1 + e^{-x}} \, dx = \int \frac{e^x}{e^x + 1} \, dx = \int \frac{1}{e^x + 1}(e^x \, dx)$$

$$= \int \frac{1}{u} \, du$$

$$= \ln |u| + C$$

$$= \ln |e^x + 1| + C \qquad \textit{substitute } e^x + 1 \textit{ for } u$$

When Substitution Fails

The method of substitution does not always succeed. In Example 5.2.10, we consider an integral very similar to the one in Example 5.2.3 but just enough different so no substitution will work.

Just-In-Time Review

Notice that if $u = x^4 + 2$, then $x^4 = u - 2$ so

$$x = (u - 2)^{1/4} = \sqrt[4]{u - 2}$$

EXAMPLE | 5.2.10

Evaluate $\displaystyle\int x^4 e^{x^4 + 2} \, dx$.

Solution

The natural substitution is $u = x^4 + 2$, as in Example 5.2.3. As before, you find $du = 4x^3 \, dx$, so $x^3 \, dx = \dfrac{1}{4} \, du$, but this integrand involves x^4, not x^3. The "extra" factor of x satisfies $x = \sqrt[4]{u - 2}$, so when the substitution is made, you have

$$\int x^4 e^{x^4 + 2} \, dx = \int x e^{x^4 + 2}(x^3 \, dx) = \int \sqrt[4]{u - 2} \, e^u \, du$$

which is hardly an improvement on the original integral! Try a few other possible substitutions (say, $u = x^2$ or $u = x^3$) to convince yourself that nothing works.

An Application Involving Substitution

EXAMPLE 5.2.11

The price p (dollars) of each unit of a particular commodity is estimated to be changing at the rate

$$\frac{dp}{dx} = \frac{-135x}{\sqrt{9 + x^2}}$$

where x (hundred) units is the consumer demand (the number of units purchased at that price). Suppose 400 units ($x = 4$) are demanded when the price is $30 per unit.

a. Find the demand function $p(x)$.

b. At what price will 300 units be demanded? At what price will no units be demanded?

c. How many units are demanded at a price of $20 per unit?

Solution

a. The price per unit demanded $p(x)$ is found by integrating $p'(x)$ with respect to x. To perform this integration, use the substitution

$$u = 9 + x^2, \qquad du = 2x\, dx, \qquad x\, dx = \frac{1}{2} du$$

to get

$$p(x) = \int \frac{-135x}{\sqrt{9 + x^2}}\, dx = \int \frac{-135}{u^{1/2}}\left(\frac{1}{2}\right) du$$

$$= \frac{-135}{2} \int u^{-1/2}\, du$$

$$= \frac{-135}{2}\left(\frac{u^{1/2}}{1/2}\right) + C$$

$$= -135\sqrt{9 + x^2} + C \qquad \text{substitute } 9 + x^2 \text{ for } u$$

Since $p = 30$ when $x = 4$, you find that

$$30 = -135\sqrt{9 + 4^2} + C$$
$$C = 30 + 135\sqrt{25} = 705$$

so

$$p(x) = -135\sqrt{9 + x^2} + 705$$

b. When 300 units are demanded, $x = 3$ and the corresponding price is

$$p(3) = -135\sqrt{9 + 3^2} + 705 = \$132.24 \text{ per unit}$$

No units are demanded when $x = 0$ and the corresponding price is

$$p(0) = -135\sqrt{9 + 0} + 705 = \$300 \text{ per unit}$$

c. To determine the number of units demanded at a unit price of $20 per unit, you need to solve the equation

$$-135\sqrt{9 + x^2} + 705 = 20$$
$$135\sqrt{9 + x^2} = 685$$
$$\sqrt{9 + x^2} = \frac{685}{135}$$

$$9 + x^2 \approx 25.75 \qquad \textit{square both sides}$$
$$x^2 \approx 16.75$$
$$x \approx 4.09$$

That is, roughly 409 units will be demanded when the price is \$20 per unit.

PROBLEMS 5.2

In Problems 1 and 2, fill in the table by specifying the substitution you would choose to find each of the four given integrals.

1.

Integral	Substitution *u*
a. $\int (3x + 4)^{5/2}\, dx$	
b. $\int \dfrac{4}{3 - x}\, dx$	
c. $\int t e^{2 - t^2}\, dt$	
d. $\int t(2 + t^2)^3\, dt$	

2.

Integral	Substitution *u*
a. $\int \dfrac{3}{(2x - 5)^4}\, dx$	
b. $\int x^2 e^{-x^3}\, dx$	
c. $\int \dfrac{e^t}{e^t + 1}\, dt$	
d. $\int \dfrac{t + 3}{\sqrt[3]{t^2 + 6t + 5}}\, dt$	

In Problems 3 through 36, find the indicated integral and check your answer by differentiation.

3. $\displaystyle\int (2x + 6)^5\, dx$

4. $\displaystyle\int e^{5x}\, dx$

5. $\displaystyle\int \sqrt{4x - 1}\, dx$

6. $\displaystyle\int \frac{1}{3x + 5}\, dx$

7. $\displaystyle\int e^{1 - x}\, dx$

8. $\displaystyle\int [(x - 1)^5 + 3(x - 1)^2 + 5]\, dx$

9. $\displaystyle\int x e^{x^2}\, dx$

10. $\displaystyle\int 2x e^{x^2 - 1}\, dx$

11. $\displaystyle\int t(t^2 + 1)^5\, dt$

12. $\displaystyle\int 3t\sqrt{t^2 + 8}\, dt$

13. $\displaystyle\int x^2(x^3 + 1)^{3/4}\, dx$

14. $\displaystyle\int x^5 e^{1 - x^6}\, dx$

15. $\displaystyle\int \frac{2y^4}{y^5 + 1}\, dy$

16. $\displaystyle\int \frac{y^2}{(y^3 + 5)^2}\, dy$

17. $\displaystyle\int (x + 1)(x^2 + 2x + 5)^{12}\, dx$

18. $\displaystyle\int (3x^2 - 1)e^{x^3 - x}\, dx$

19. $\displaystyle\int \frac{3x^4 + 12x^3 + 6}{x^5 + 5x^4 + 10x + 12}\, dx$

20. $\displaystyle\int \frac{10x^3 - 5x}{\sqrt{x^4 - x^2 + 6}}\, dx$

21. $\displaystyle\int \frac{3u - 3}{(u^2 - 2u + 6)^2}\, du$

22. $\displaystyle\int \frac{6u - 3}{4u^2 - 4u + 1}\, du$

23. $\displaystyle\int \frac{\ln 5x}{x}\, dx$

24. $\displaystyle\int \frac{1}{x \ln x}\, dx$

25. $\displaystyle\int \frac{1}{x(\ln x)^2}\, dx$

26. $\displaystyle\int \frac{\ln x^2}{x}\, dx$

27. $\displaystyle\int \frac{2x \ln (x^2 + 1)}{x^2 + 1}\, dx$

28. $\displaystyle\int \frac{e^{\sqrt{x}}}{\sqrt{x}}\, dx$

29. $\displaystyle\int \frac{e^x + e^{-x}}{e^x - e^{-x}}\, dx$

30. $\displaystyle\int e^{-x}(1 + e^{2x})\, dx$

31. $\displaystyle\int \frac{x}{2x + 1}\, dx$

32. $\displaystyle\int \frac{t - 1}{t + 1}\, dt$

33. $\displaystyle\int x\sqrt{2x + 1}\, dx$

34. $\displaystyle\int \frac{x}{\sqrt[3]{4 - 3x}}\, dx$

35. $\displaystyle\int \frac{1}{\sqrt{x}(\sqrt{x} + 1)}\, dx$

36. $\displaystyle\int \frac{1}{x^2}\left(\frac{1}{x} - 1\right)^{2/3} dx$

[*Hint:* Let $u = \sqrt{x} + 1$.]

$\left[\textit{Hint:} \text{ Let } u = \dfrac{1}{x} - 1.\right]$

In Problems 37 through 40, solve the given initial value problem for $y = f(x)$.

37. $\dfrac{dy}{dx} = \dfrac{1}{x + 1}$　where $y = 1$ when $x = 0$

38. $\dfrac{dy}{dx} = e^{2-x}$　where $y = 0$ when $x = 2$

39. $\dfrac{dy}{dx} = \dfrac{x + 2}{x^2 + 4x + 5}$　where $y = 3$ when $x = -1$

40. $\dfrac{dy}{dx} = \dfrac{\ln\sqrt{x}}{x}$　where $y = 2$ when $x = 1$

In Problems 41 through 44, the slope $f'(x)$ at each point (x, y) on a curve $y = f(x)$ is given, along with a particular point (a, b) on the curve. Use this information to find $f(x)$.

41. $f'(x) = (1 - 2x)^{3/2}$; $(0, 0)$

42. $f'(x) = x\sqrt{x^2 + 5}$; $(2, 10)$

43. $f'(x) = xe^{4 - x^2}$; $(-2, 1)$

44. $f'(x) = \dfrac{2x}{1 + 3x^2}$; $(0, 5)$

In Problems 45 through 48, the velocity $v(t) = x'(t)$ at time t of an object moving along the x axis is given, along with the initial position $x(0)$ of the object. In each case, find:

(a) *The position $x(t)$ at time t.*

(b) *The position of the object at time $t = 4$.*

(c) *The time when the object is at $x = 3$.*

45. $x'(t) = -2(3t + 1)^{1/2}$; $x(0) = 4$

46. $x'(t) = \dfrac{-1}{1 + 0.5t}$; $x(0) = 5$

47. $x'(t) = \dfrac{1}{\sqrt{2t + 1}}$; $x(0) = 0$

48. $x'(t) = \dfrac{-2t}{(1 + t^2)^{3/2}}$; $x(0) = 4$

49. MARGINAL COST At a certain factory, the marginal cost is $3(q - 4)^2$ dollars per unit when the level of production is q units.
   a. Express the total production cost in terms of the overhead (the cost of producing 0 units) and the number of units produced.
   b. What is the cost of producing 14 units if the overhead is $436?

50. DEPRECIATION The resale value of a certain industrial machine decreases at a rate that depends on its age. When the machine is t years old, the rate at which its value is changing is $-960e^{-t/5}$ dollars per year.
   a. Express the value of the machine in terms of its age and initial value.
   b. If the machine was originally worth $5,200, how much will it be worth when it is 10 years old?

51. TREE GROWTH A tree has been transplanted and after x years is growing at the rate of
$$1 + \frac{1}{(x + 1)^2}$$ meters per year. After 2 years it has reached a height of 5 meters. How tall was it when it was transplanted?

52. RETAIL PRICES In a certain section of the country, it is estimated that t weeks from now, the price of chicken will be increasing at the rate of $p'(t) = 3\sqrt{t + 1}$ cents per kilogram per week. If chicken currently costs $3 per kilogram, what will it cost 8 weeks from now?

53. REVENUE The marginal revenue from the sale of x units of a particular commodity is estimated to be
$$R'(x) = 50 + 3.5xe^{-0.01x^2}$$ dollars per unit
where $R(x)$ is revenue in dollars.
   a. Find $R(x)$, assuming that $R(0) = 0$.
   b. What revenue should be expected from the sale of 1,000 units?

54. WATER POLLUTION An oil spill in the ocean is roughly circular in shape, with radius $R(t)$ feet t minutes after the spill begins. The radius is increasing at the rate
$$R'(t) = \frac{21}{0.07t + 5} \text{ ft/min}$$
   a. Find an expression for the radius $R(t)$, assuming that $R = 0$ when $t = 0$.
   b. What is the area $A = \pi R^2$ of the spill after 1 hour?

55. DRUG CONCENTRATION The concentration $C(t)$ in milligrams per cubic centimeter (mg/cm^3) of a drug in a patient's bloodstream is 0.5 mg/cm^3 immediately after an injection and t minutes later is decreasing at the rate
$$C'(t) = \frac{-0.01e^{0.01t}}{(e^{0.01t} + 1)^2} \text{ mg/cm}^3 \text{ per minute}$$
A new injection is given when the concentration drops below 0.05 mg/cm^3.
   a. Find an expression for $C(t)$.
   b. What is the concentration after 1 hour? After 3 hours?
    c. Use the graphing utility of your calculator with **TRACE** and **ZOOM** to determine how much time passes before the next injection is given.

56. LAND VALUE It is estimated that x years from now, the value $V(x)$ of an acre of farmland will be increasing at the rate of
$$V'(x) = \frac{0.4x^3}{\sqrt{0.2x^4 + 8,000}}$$
dollars per year. The land is currently worth $500 per acre.
   a. Find $V(x)$.
   b. How much will the land be worth in 10 years?
    c. Use the graphing utility of your calculator with **TRACE** and **ZOOM** to determine how long it will take for the land to be worth $1,000 per acre.

57. AIR POLLUTION In a certain suburb of Los Angeles, the level of ozone $L(t)$ at 7:00 A.M. is 0.25 parts per million (ppm). A 12-hour weather forecast predicts that the ozone level t hours later will be changing at the rate of
$$L'(t) = \frac{0.24 - 0.03t}{\sqrt{36 + 16t - t^2}}$$
parts per million per hour (ppm/hr).
   a. Express the ozone level $L(t)$ as a function of t. When does the peak ozone level occur? What is the peak level?
    b. Use the graphing utility of your calculator to sketch the graph of $L(t)$ and use **TRACE** and **ZOOM** to answer the questions in part (a). Then determine at what other time the ozone level will be the same as it is at 11:00 A.M.

58. SUPPLY The owner of a fast-food chain determines that if x thousand units of a new meal item are supplied, then the marginal price at that level of supply is given by

$$p'(x) = \frac{x}{(x+3)^2} \quad \text{dollars per meal}$$

where $p(x)$ is the price (in dollars) per unit at which all x meal units will be sold. Currently, 5,000 units are being supplied at a price of $2.20 per unit.

a. Find the supply (price) function $p(x)$.

b. If 10,000 meal units are supplied to restaurants in the chain, what unit price should be charged so that all the units will be sold?

59. **DEMAND**   The manager of a shoe store determines that the price p (dollars) for each pair of a popular brand of sports sneakers is changing at the rate of

$$p'(x) = \frac{-300x}{(x^2 + 9)^{3/2}}$$

when x (hundred) pairs are demanded by consumers. When the price is $75 per pair, 400 pairs ($x = 4$) are demanded by consumers.

a. Find the demand (price) function $p(x)$.

b. At what price will 500 pairs of sneakers be demanded? At what price will no sneakers be demanded?

c. How many pairs will be demanded at a price of $90 per pair?

60. **MARGINAL PROFIT**   A company determines that the marginal revenue from the production of x units is $R'(x) = 7 - 3x - 4x^2$ hundred dollars per unit, and the corresponding marginal cost is $C'(x) = 5 + 2x$ hundred dollars per unit. By how much does the profit change when the level of production is raised from 5 to 9 units?

61. **MARGINAL PROFIT**   Repeat Problem 60 for marginal revenue $R'(x) = \dfrac{11 - x}{\sqrt{14 - x}}$ and for the marginal cost $C'(x) = 2 + x + x^2$.

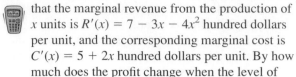

62. Find $\displaystyle\int x^3(4 - x^2)^{-1/2}\, dx$. [*Hint:* Substitute $u = 4 - x^2$ and use the fact that $x^2 = 4 - u$.]

63. Find $\displaystyle\int x^{1/3}(x^{2/3} + 1)^{3/2}\, dx$. [*Hint:* Substitute $u = x^{2/3} + 1$ and use $x^{2/3} = u - 1$.]

64. Find $\displaystyle\int e^{-x}(1 + e^x)^2\, dx$. [*Hint:* Is it better to set $u = 1 + e^x$ or $u = e^x$? Or is it better to not even use the method of substitution?]

65. Find $\displaystyle\int \frac{e^{2x}}{1 + e^x}\, dx$. [*Hint:* Let $u = 1 + e^x$.]

SECTION 5.3 | The Definite Integral and the Fundamental Theorem of Calculus

Suppose a real estate agent wants to evaluate an unimproved parcel of land that is 100 feet wide and is bounded by streets on three sides and by a stream on the fourth side. The agent determines that if a coordinate system is set up as shown in Figure 5.2, the stream can be described by the curve $y = x^3 + 1$, where x and y are measured in hundreds of feet. If the area of the parcel is A square feet and the agent estimates its land is worth $12 per square foot, then the total value of the parcel is $12A$ dollars. If the parcel were rectangular in shape or triangular or even trapezoidal, its area A could be found by substituting into a well-known formula, but the upper boundary of the parcel is curved, so how can the agent find the area and hence the total value of the parcel?

Our goal in this section is to show how area under a curve, such as the area A in our real estate example, can be expressed as a limit of a sum of terms called a **definite integral.** We will then introduce a result called the **fundamental theorem of calculus** that allows us to compute *definite* integrals and thus find area and other quantities by using the *indefinite* integration (antidifferentiation) methods of Sections 5.1 and 5.2. In Example 5.3.3, we will illustrate this procedure by expressing the area A in our real estate example as a definite integral and evaluating it using the fundamental theorem of calculus.

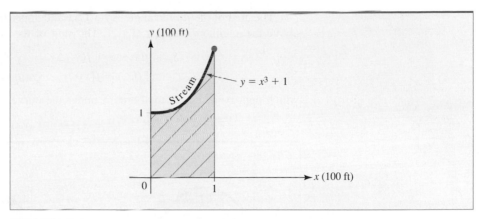

FIGURE 5.2   Determining land value by finding the area under a curve.

Area as the Limit of a Sum

Consider the area of the region under the curve $y = f(x)$ over an interval $a \leq x < b$, where $f(x) \geq 0$ and f is continuous, as illustrated in Figure 5.3. To find this area, we will follow a useful general policy:

> *When faced with something you don't know how to handle, try to relate it to something you do know how to handle.*

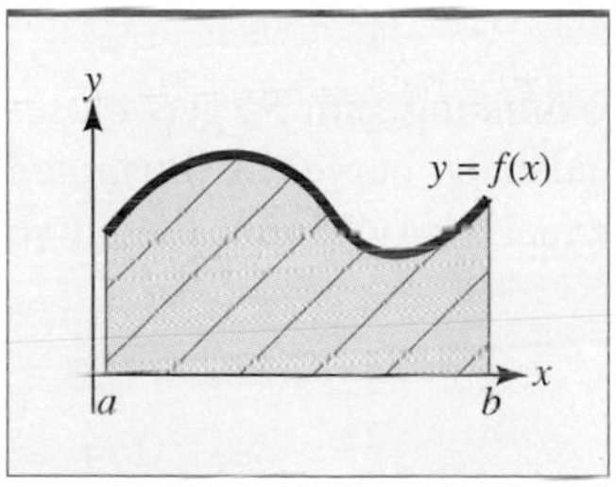

In this particular case, we may not know the area under the given curve, but we do know how to find the area of a rectangle. Thus, we proceed by subdividing the region into a number of rectangular regions and then approximate the area A under the curve $y = f(x)$ by adding the areas of the approximating rectangles.

To be more specific, begin the approximation by dividing the interval $a \leq x \leq b$ into n equal subintervals, each of length $\Delta x = (b - a)/n$, and let x_j denote the left endpoint of the jth subinterval, for $j = 1, 2, \ldots, n$. Then draw n rectangles such that the jth rectangle has the jth subinterval as its base and $f(x_j)$ as its height. The approximation scheme is illustrated in Figure 5.4.

FIGURE 5.3   The region under the curve $y = f(x)$ over the interval $a \leq x \leq b$.

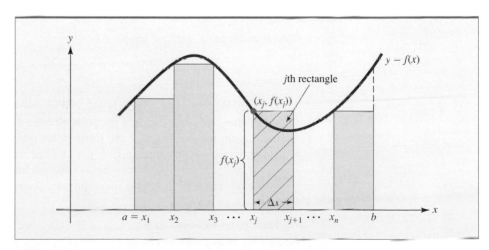

FIGURE 5.4   An approximation of area under a curve by rectangles.

The area of the jth rectangle is $f(x_j)\,\Delta x$ and approximates the area under the curve above the subinterval $x_j \le x \le x_{j+1}$. The sum of the areas of all n rectangles is

$$S_n = f(x_1)\Delta x + f(x_2)\Delta x + \cdots + f(x_n)\Delta x$$
$$= [f(x_1) + f(x_2) + \cdots + f(x_n)]\Delta x$$

which approximates the total area A under the curve.

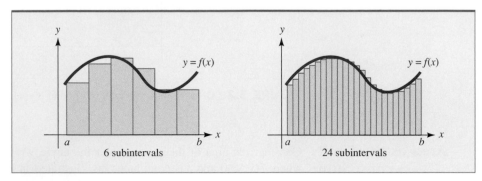

FIGURE 5.5   The approximation improves as the number of subintervals increases.

As the number of subintervals n increases, the approximating sum S_n gets closer and closer to what we intuitively think of as the area under the curve, as illustrated in Figure 5.5. Therefore, it is reasonable to define the actual area A under the curve as the limit of the sums. To summarize:

Area Under a Curve   ■   Let $f(x)$ be continuous and satisfy $f(x) \ge 0$ on the interval $a \le x \le b$. Then the region under the curve $y = f(x)$ over the interval $a \le x \le b$ has area

$$A = \lim_{n \to +\infty} [f(x_1) + f(x_2) + \cdots + f(x_n)]\Delta x$$

where x_j is the left endpoint of the jth subinterval if the interval $a \le x \le b$ is divided into n equal parts, each of length $\Delta x = \dfrac{b - a}{n}$.

NOTE   At this point, you may ask, "Why use the left endpoint of the subintervals rather than, say, the right endpoint or even the midpoint?" The answer is that there is no reason we can't use those other points to compute the height of our approximating rectangles. In fact, the interval $a \le x \le b$ can be subdivided arbitrarily and arbitrary points chosen in each subinterval, and the result will still be the same. However, proving this equivalence is difficult, well beyond the scope of this text.   ■

Here is an example in which area is computed as the limit of a sum and then checked using a geometric formula.

EXAMPLE 5.3.1

Let R be the region under the graph of $f(x) = 2x + 1$ over the interval $1 < x \le 3$, as shown in Figure 5.6a. Compute the area of R as the limit of a sum.

Solution

The region R is shown in Figure 5.6 with six approximating rectangles, each of width $\Delta x = \dfrac{3 - 1}{6} = \dfrac{1}{3}$. The left endpoints in the partition of $1 < x \le 3$ are $x_1 = 1$, $x_2 = 1 + \dfrac{1}{3} = \dfrac{4}{3}$, and similarly, $x_3 = \dfrac{5}{3}$, $x_4 = 2$, $x_5 = \dfrac{7}{3}$, and $x_6 = \dfrac{8}{3}$. The corresponding values of $f(x) = 2x + 1$ are given in the following table:

x_j	1	$\frac{4}{3}$	$\frac{5}{3}$	2	$\frac{7}{3}$	$\frac{8}{3}$
$f(x_j) = 2x_j + 1$	3	$\frac{11}{3}$	$\frac{13}{3}$	5	$\frac{17}{3}$	$\frac{19}{3}$

Thus, the area A of the region R is approximated by the sum

$$S = \left(3 + \frac{11}{3} + \frac{13}{3} + 5 + \frac{17}{3} + \frac{19}{3}\right)\left(\frac{1}{3}\right) = \frac{28}{3} \approx 9.333$$

Just-In-Time Review

A trapezoid is a four-sided polygon with at least two parallel sides. Its area is

$$A = \frac{1}{2}(s_1 + s_2)h$$

where s_1 and s_2 are the lengths of the two parallel sides and h is the distance between them.

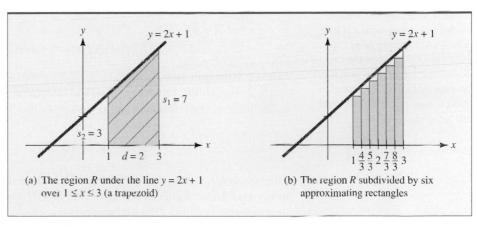

(a) The region R under the line $y = 2x + 1$ over $1 \le x \le 3$ (a trapezoid)

(b) The region R subdivided by six approximating rectangles

FIGURE 5.6 Approximating the area under a line with rectangles.

If you continue to subdivide the region R using more and more rectangles, the corresponding approximating sums S_n approach the actual area A of the region. The sum we have already computed for $n = 6$ is listed in the following table, along with those for $n = 10, 20, 50, 100$, and 500. (If you have access to a computer or a programmable calculator, see if you can write a program for generating any such sum for given n.)

Number of partitions n	6	10	20	50	100	500
Approximating sum S_n	9.333	9.600	9.800	9.920	9.960	9.992

The numbers on the bottom line of this table seem to be approaching 10 as n gets larger and larger. Thus, it is reasonable to conjecture that the region R has area

$$A = \lim_{n \to +\infty} S_n = 10$$

Notice in Figure 5.6a that the region R is a trapezoid of width $d = 3 - 1 = 2$ with parallel sides of lengths

$$s_1 = 2(3) + 1 = 7 \qquad \text{and} \qquad s_2 = 2(1) + 1 = 3$$

Such a trapezoid has area

$$A = \frac{1}{2}(s_1 + s_2)d = \frac{1}{2}(7 + 3)(2) = 10$$

the same result we just obtained using the limit of a sum procedure.

The Definite Integral

Area is just one of many quantities that can be expressed as the limit of a sum. To handle all such cases, including those for which $f(x) \geq 0$ is *not* required and left endpoints are not used, we require the terminology and notation introduced in the following definition.

EXPLORE!

Place into Y1 the function $f(x) = -x^2 + 4x - 3$ and view its graph using the window [0, 4.7]1 by [−0.5, 1.5]0.5. Visually estimate the area under the curve from $x = 2$ to $x = 3$, using triangles or rectangles. Now use the numerical integration feature of your graphing calculator (**CALC** key, option 7). How far off were you and why?

The Definite Integral ▪ Let $f(x)$ be a function that is continuous on the interval $a \leq x \leq b$. Subdivide the interval $a \leq x \leq b$ into n equal parts, each of width $\Delta x = \dfrac{b - a}{n}$, and choose a number x_k from the kth subinterval for $k = 1, 2, \ldots, n$. Form the sum

$$[f(x_1) + f(x_2) + \cdots + f(x_n)]\Delta x$$

called a **Riemann sum.**

Then the **definite integral** of f on the interval $a \leq x \leq b$, denoted by $\int_a^b f(x)\,dx$, is the limit of the Riemann sum as $n \to +\infty$; that is,

$$\int_a^b f(x)\,dx = \lim_{n \to +\infty} [f(x_1) + f(x_2) + \cdots + f(x_n)]\Delta x$$

The function $f(x)$ is called the **integrand,** and the numbers a and b are called the **lower and upper limits of integration,** respectively. The process of finding a definite integral is called **definite integration.**

Surprisingly, the fact that $f(x)$ is continuous on $a \leq x \leq b$ turns out to be enough to guarantee that the limit used to define the definite integral $\int_a^b f(x)\,dx$ exists and is the same regardless of how the subinterval representatives x_k are chosen.

The symbol $\int_a^b f(x)\,dx$ used for the definite integral is essentially the same as the symbol $\int f(x)\,dx$ for the indefinite integral, even though the definite integral is a specific number while the indefinite integral is a family of functions, the antiderivatives of f. You will soon see that these two apparently very different concepts are intimately related. Here is a compact form for the definition using the integral notation.

Area as a Definite Integral ■ If $f(x)$ is continuous and $f(x) \geq 0$ on the interval $a \leq x \leq b$, then the region R under the curve $y = f(x)$ over the interval $a \leq x \leq b$ has area A given by the definite integral $A = \int_a^b f(x)\, dx$.

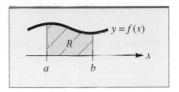

The Fundamental Theorem of Calculus

If computing the limit of a sum were the only way of evaluating a definite integral, the integration process probably would be little more than a mathematical novelty. Fortunately, there is an easier way of performing this computation, thanks to this remarkable result connecting the definite integral to antidifferentiation.

The Fundamental Theorem of Calculus ■ If the function $f(x)$ is continuous on the interval $a \leq x \leq b$, then

$$\int_a^b f(x)\, dx = F(b) - F(a)$$

where $F(x)$ is any antiderivative of $f(x)$ on $a \leq x \leq b$.

A special case of the fundamental theorem of calculus is verified at the end of this section. When applying the fundamental theorem, we use the notation

$$F(x)\Big|_a^b = F(b) - F(a)$$

Thus,

$$\int_a^b f(x)\, dx = F(x)\Big|_a^b = F(b) - F(a)$$

NOTE You may wonder how the fundamental theorem of calculus can promise that if $F(x)$ is *any* antiderivative of $f(x)$, then

$$\int_a^b f(x)\, dx = F(b) - F(a)$$

To see why this is true, suppose $G(x)$ is another such antiderivative. Then $G(x) = F(x) + C$ for some constant C, so $F(x) = G(x) - C$ and

$$\int_a^b f(x)\, dx = F(b) - F(a)$$

$$= [G(b) - C] - [G(a) - C]$$
$$= G(b) - G(a)$$

since the C's cancel. Thus, the valuation is the same regardless of which antiderivative is used. ■

In Example 5.3.2, we demonstrate the computational value of the fundamental theorem of calculus by using it to compute the same area we estimated as the limit of a sum in Example 5.3.1.

EXAMPLE 5.3.2

Use the fundamental theorem of calculus to find the area of the region under the line $y = 2x + 1$ over the interval $1 \le x \le 3$.

Solution

Since $f(x) = 2x + 1$ satisfies $f(x) \ge 0$ on the interval $1 \le x \le 3$, the area is given by the definite integral $A = \displaystyle\int_{1}^{3} (2x + 1)\, dx$. Since an antiderivative of $f(x) = 2x + 1$ is $F(x) = x^2 + x$, the fundamental theorem of calculus tells us that

$$A = \int_{1}^{3} (2x + 1)\, dx = x^2 + x \Big|_{1}^{3}$$

$$= [(3)^2 + (3)] - [(1)^2 + (1)] = 10$$

as estimated in Example 5.3.1.

EXPLORE!

Refer to Example 5.3.3. Use the numerical integration feature of your calculator to confirm numerically that

$$\int_{0}^{1} (x^3 + 1)\, dx = 1.25$$

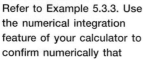

EXAMPLE 5.3.3

Find the area of the parcel of land described in the introduction to this section; that is, the area under the curve $y = x^3 + 1$ over the interval $0 \le x \le 1$, where x and y are in hundreds of feet. If the land in the parcel is appraised at \$12 per square foot, what is the total value of the parcel?

Solution

The area of the parcel is given by the definite integral

$$A = \int_{0}^{1} (x^3 + 1)\, dx$$

Since an antiderivative of $f(x) = x^3 + 1$ is $F(x) = \dfrac{1}{4}x^4 + x$, the fundamental theorem of calculus tells us that

$$A = \int_{0}^{1} (x^3 + 1)\, dx = \frac{1}{4}x^4 + x \Big|_{0}^{1}$$

$$= \left[\frac{1}{4}(1)^4 + 1\right] - \left[\frac{1}{4}(0)^4 + 0\right] = \frac{5}{4}$$

Because x and y are measured in hundreds of feet, the total area is

$$\frac{5}{4} \times 100 \times 100 = 12{,}500 \text{ ft}^2$$

and since the land in the parcel is worth \$12 per square foot, the total value of the parcel is

$$V = (\$12/\text{ft}^3)(12{,}500 \text{ ft}^2) = \$150{,}000$$

EXAMPLE 5.3.4

Evaluate the definite integral $\displaystyle\int_{0}^{1} (e^{-x} + \sqrt{x})\, dx$.

Just-In-Time **Review**

When the fundamental theorem of calculus

$$\int_a^b f(x)\, dx = F(b) - F(a)$$

is used to evaluate a definite integral, remember to compute both $F(b)$ and $F(a)$, even when $a = 0$.

Solution

An antiderivative of $f(x) = e^{-x} + \sqrt{x}$ is $F(x) = -e^{-x} + \dfrac{2}{3}x^{3/2}$, so the definite integral is

$$\int_0^1 (e^{-x} + \sqrt{x})\, dx = \left(-e^{-x} + \frac{2}{3}x^{3/2} \right)\Bigg|_0^1$$

$$= \left[-e^{-1} + \frac{2}{3}(1)^{3/2} \right] - \left[-e^0 + \frac{2}{3}(0) \right]$$

$$= \frac{5}{3} - \frac{1}{e} \approx 1.299$$

Our definition of the definite integral was motivated by computing area, which is a nonnegative quantity. However, since the definition does not require $f(x) \geq 0$, it is quite possible for a definite integral to be negative, as illustrated in Example 5.3.5.

EXAMPLE | 5.3.5

Evaluate $\displaystyle\int_1^4 \left(\frac{1}{x} - x^2 \right) dx$.

Solution

An antiderivative of $f(x) = \dfrac{1}{x} - x^2$ is $F(x) = \ln |x| - \dfrac{1}{3}x^3$, so we have

$$\int_1^4 \left(\frac{1}{x} - x^2 \right) dx = \left(\ln |x| - \frac{1}{3}x^3 \right)\Bigg|_1^4$$

$$= \left[\ln 4 - \frac{1}{3}(4)^3 \right] - \left[\ln 1 - \frac{1}{3}(1)^3 \right]$$

$$= \ln 4 - 21 \approx -19.6137$$

Integration Rules   This list of rules can be used to simplify the computation of definite integrals.

Rules for Definite Integrals

Let f and g be any functions continuous on $a \leq x \leq b$. Then,

1. Constant multiple rule: $\displaystyle\int_a^b k\, f(x)\, dx = k \int_a^b f(x)\, dx \qquad$ for constant k

2. Sum rule: $\displaystyle\int_a^b [f(x) + g(x)]\, dx = \int_a^b f(x)\, dx + \int_a^b g(x)\, dx$

3. Difference rule: $\displaystyle\int_a^b [f(x) - g(x)]\, dx = \int_a^b f(x)\, dx - \int_a^b g(x)\, dx$

4. $\displaystyle\int_a^a f(x)\, dx = 0$

5. $\displaystyle\int_b^a f(x)\, dx = -\int_a^b f(x)\, dx$

6. Subdivision rule: $\displaystyle\int_a^b f(x)\, dx = \int_a^c f(x)\, dx + \int_c^b f(x)\, dx$

Rules 4 and 5 are really special cases of the definition of the definite integral. The first three rules can be proved by using the fundamental theorem of calculus along with an analogous rule for indefinite integrals. For instance, to verify the constant multiple rule, suppose $F(x)$ is an antiderivative of $f(x)$. Then, according to the constant multiple rule for indefinite integrals, $kF(x)$ is an antiderivative of $kf(x)$ and the fundamental theorem of calculus tells us that

$$\int_a^b kf(x)\, dx = kF(x)\Big|_a^b$$
$$= kF(b) - kF(a) = k[F(b) - F(a)]$$
$$= k\int_a^b f(x)\, dx$$

You are asked to verify the sum rule using similar reasoning in Problem 64.

In the case where $f(x) \geq 0$ on the interval $a \leq x \leq b$, the subdivision rule is a geometric reflection of the fact that the area under the curve $y = f(x)$ over the interval $a \leq x \leq b$ is the sum of the areas under $y = f(x)$ over the subintervals $a \leq x \leq c$ and $c \leq x \leq b$, as illustrated in Figure 5.7. However, it is important to remember that the subdivision rule is true even if $f(x)$ does *not* satisfy $f(x) \geq 0$ on $a \leq x \leq b$.

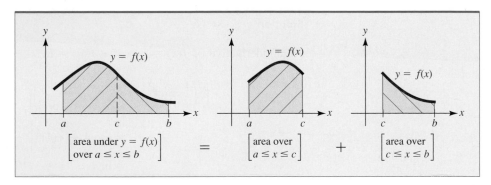

FIGURE 5.7   The subdivision rule for definite integrals (case where $f(x) \geq 0$).

EXAMPLE   5.3.6

Let $f(x)$ and $g(x)$ be functions that are continuous on the interval $-2 \leq x \leq 5$ and that satisfy

$$\int_{-2}^5 f(x)\, dx = 3 \qquad \int_{-2}^5 g(x)\, dx = -4 \qquad \int_3^5 f(x)\, dx = 7$$

Use this information to evaluate each of these definite integrals:

a. $\displaystyle\int_{-2}^5 [2f(x) - 3g(x)]\, dx$ **b.** $\displaystyle\int_{-2}^3 f(x)\, dx$

Solution

a. By combining the difference rule and constant multiple rule and substituting the given information, we find that

$$\int_{-2}^5 [2f(x) - 3g(x)]\, dx = \int_{-2}^5 2f(x)\, dx - \int_{-2}^5 3g(x)\, dx \quad \textit{difference rule}$$

$$= 2 \int_2^5 f(x) \, dx - 3 \int_{-2}^5 g(x) \, dx \qquad \textit{constant multiple rule}$$

$$= 2(3) - 3(-4) = 18 \qquad \textit{substitute given information}$$

b. According to the subdivision rule

$$\int_{-2}^5 f(x) \, dx = \int_2^3 f(x) \, dx + \int_3^5 f(x) \, dx$$

Solving this equation for the required integral $\int_{-2}^3 f(x) \, dx$ and substituting the given information, we obtain

$$\int_{-2}^3 f(x) \, dx = \int_{-2}^5 f(x) \, dx - \int_3^5 f(x) \, dx$$

$$= 3 - 7 = -4$$

Substituting in a Definite Integral

When using a substitution $u = g(x)$ to evaluate a definite integral $\int_a^b f(x) \, dx$, you can proceed in either of these two ways:

1. Use the substitution to find an antiderivative $F(x)$ for $f(x)$ and then evaluate the definite integral using the fundamental theorem of calculus.

2. Use the substitution to express the integrand and dx in terms of u and du and to replace the original limits of integration, a and b, with transformed limits $c = g(a)$ and $d = g(b)$. The original integral can then be evaluated by applying the fundamental theorem of calculus to the transformed definite integral.

These procedures are illustrated in Examples 5.3.7 and 5.3.8.

Just-In-Time Review

Only one member of the family of antiderivatives of $f(x)$ is needed for evaluating $\int_a^b f(x) \, dx$ by the fundamental theorem of calculus. Therefore, the "+C" may be left out of intermediate integrations.

EXAMPLE | 5.3.7

Evaluate $\int_0^1 8x(x^2 + 1)^3 \, dx$.

Solution

The integrand is a product in which one factor $8x$ is a constant multiple of the derivative of an expression $x^2 + 1$ that appears in the other factor. This suggests that you let $u = x^2 + 1$. Then $du = 2x \, dx$, and so

$$\int 8x(x^2 + 1)^3 \, dx = \int 4u^3 \, du = u^4$$

The limits of integration, 0 and 1, refer to the variable x and not to u. You can, therefore, proceed in one of two ways. Either you can rewrite the antiderivative in terms of x, or you can find the values of u that correspond to $x = 0$ and $x = 1$.

If you choose the first alternative, you find that

$$\int 8x(x^2 + 1)^3 \, dx = u^4 = (x^2 + 1)^4$$

and so

$$\int_0^1 8x(x^2 + 1)^3 \, dx = (x^2 + 1)^4 \Big|_0^1 = 16 - 1 - 15$$

If you choose the second alternative, use the fact that $u = x^2 + 1$ to conclude that $u = 1$ when $x = 0$ and $u = 2$ when $x = 1$. Hence,

$$\int_0^1 8x(x^2 + 1)^3 \, dx = \int_1^2 4u^3 \, du = u^4 \Big|_1^2 = 16 - 1 = 15$$

EXPLORE!

Refer to Example 5.3.8. Use a graphing utility with the window [0, 3]1 by [−4, 1]1 to graph $f(x) = \ln x/x$. Explain in terms of area why the integral of $f(x)$ over $\dfrac{1}{4} \leq x \leq 2$ is negative.

EXAMPLE 5.3.8

Evaluate $\displaystyle\int_{1/4}^2 \left(\dfrac{\ln x}{x}\right) dx$.

Solution

Let $u = \ln x$, so $du = \dfrac{1}{x} \, dx$. Then

$$\int \frac{\ln x}{x} \, dx = \int \ln x \left(\frac{1}{x} \, dx\right) = \int u \, du$$

$$= \frac{1}{2} u^2 = \frac{1}{2}(\ln x)^2$$

Thus,

$$\int_{1/4}^2 \frac{\ln x}{x} \, dx = \left[\frac{1}{2}(\ln x)^2\right]\Bigg|_{1/4}^2 = \frac{1}{2}(\ln 2)^2 - \frac{1}{2}\left(\ln \frac{1}{4}\right)^2$$

$$= \frac{-3}{2}(\ln 2)^2 \approx -0.721$$

Alternatively, use the substitution $u = \ln x$ to transform the limits of integration:

$$\text{when } x = \frac{1}{4}, \text{ then } u = \ln \frac{1}{4}$$

$$\text{when } x = 2, \text{ then } u = \ln 2$$

Substituting, we find

$$\int_{1/4}^2 \frac{\ln x}{x} \, dx = \int_{\ln 1/4}^{\ln 2} u \, du = \frac{1}{2} u^2 \Bigg|_{\ln 1/4}^{\ln 2}$$

$$= \frac{1}{2}(\ln 2)^2 - \frac{1}{2}\left(\ln \frac{1}{4}\right)^2 \approx -0.721$$

Net Change   In certain applications, we are given the rate of change $Q'(x)$ of a quantity $Q(x)$ and required to compute the **net change** $Q(b) - Q(a)$ in $Q(x)$ as x varies from $x = a$ to $x = b$. We did this in Section 5.1 by solving initial value problems (recall Examples 5.1.5 through 5.1.8). However, since $Q(x)$ is an antiderivative of $Q'(x)$, the fundamental theorem of calculus allows us to compute net change by the following definite integration formula.

> **Net Change** ■ If $Q'(x)$ is continuous on the interval $a \leq x \leq b$, then the **net change** in $Q(x)$ as x varies from $x = a$ to $x = b$ is given by
>
> $$Q(b) - Q(a) = \int_a^b Q'(x) \, dx$$

Here are two examples involving net change.

EXAMPLE | 5.3.9

At a certain factory, the marginal cost is $3(q - 4)^2$ dollars per unit when the level of production is q units. By how much will the total manufacturing cost increase if the level of production is raised from 6 units to 10 units?

Solution

Let $C(q)$ denote the total cost of producing q units. Then the marginal cost is the derivative $\dfrac{dC}{dq} = 3(q - 4)^2$, and the increase in cost if production is raised from 6 units to 10 units is given by the definite integral

$$C(10) - C(6) = \int_6^{10} \frac{dC}{dq}\, dq$$

$$= \int_6^{10} 3(q - 4)^2\, dq = (q - 4)^3 \Big|_6^{10}$$

$$= (10 - 4)^3 - (6 - 4)^3$$

$$= \$208$$

EXAMPLE | 5.3.10

A protein with mass m (grams) disintegrates into amino acids at a rate given by

$$\frac{dm}{dt} = \frac{-30}{(t + 3)^2} \quad \text{g/hr}$$

What is the net change in mass of the protein during the first 2 hours?

Solution

The net change is given by the definite integral

$$m(2) - m(0) = \int_0^2 \frac{dm}{dt}\, dt = \int_0^2 \frac{-30}{(t + 3)^2}\, dt$$

If we substitute $u = t + 3$, $du = dt$, and change the limits of integration accordingly ($t = 0$ becomes $u = 3$ and $t = 2$ becomes $u = 5$), we find

$$m(2) - m(0) = \int_0^2 \frac{-30}{(t + 3)^2}\, dt = \int_3^5 -30u^{-2}\, du$$

$$= -30\left(\frac{u^{-1}}{-1}\right)\Big|_3^5 = 30\left[\frac{1}{5} - \frac{1}{3}\right]$$

$$= -4$$

Thus, the mass of the protein has a net decrease of 4 g over the first 2 hours.

Area Justification of the Fundamental Theorem of Calculus

We close this section with a justification of the fundamental theorem of calculus for the case where $f(x) \geq 0$. In this case, the definite integral $\int_a^b f(x)\, dx$ represents the area under the curve $y = f(x)$ over the interval $[a, b]$. For fixed x between a and b, let $A(x)$ denote the area under $y = f(x)$ over the interval $[a, x]$. Then the difference quotient of $A(x)$ is

$$\frac{A(x + h) - A(x)}{h}$$

and the expression $A(x + h) - A(x)$ in the numerator is just the area under the curve $y = f(x)$ between x and $x + h$. If h is small, this area is approximately the same as the area of the rectangle with height $f(x)$ and width h as indicated in Figure 5.8. That is,

$$A(x + h) - A(x) \approx f(x)h$$

or, equivalently,

$$\frac{A(x + h) - A(x)}{h} \approx f(x)$$

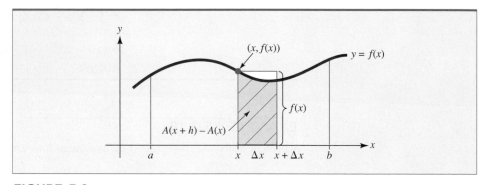

FIGURE 5.8   The area $A(x + h) - A(x)$.

As h approaches 0, the error in the approximation approaches 0, and it follows that

$$\lim_{h \to 0} \frac{A(x + h) - A(x)}{h} = f(x)$$

But by the definition of the derivative,

$$\lim_{h \to 0} \frac{A(x + h) - A(x)}{h} = A'(x)$$

so that

$$A'(x) = f(x)$$

In other words, $A(x)$ is an antiderivative of $f(x)$.

Suppose $F(x)$ is any other antiderivative of $f(x)$. Then, according to the fundamental property of antiderivatives (Section 5.1), we have

$$A(x) = F(x) + C$$

for some constant C and all x in the interval $a \leq x \leq b$. Since $A(x)$ represents the area under $y = f(x)$ between a and x, it is certainly true that $A(a)$, the area between a and a, is 0, so that

$$A(a) = 0 = F(a) + C$$

and $C = -F(a)$. The area under $y = f(x)$ between $x = a$ and $x = b$ is $A(b)$, which satisfies

$$A(b) = F(b) + C = F(b) - F(a)$$

Finally, since the area under $y = f(x)$ above $a \le x \le b$ is also given by the definite integral $\int_a^b f(x)\,dx$, it follows that

$$\int_a^b f(x)\,dx = A(b) = F(b) - F(a)$$

as claimed in the fundamental theorem of calculus.

PROBLEMS 5.3

In Problems 1 through 30, evaluate the given definite integral using the fundamental theorem of calculus.

1. $\displaystyle\int_{-1}^{2} 5\,dx$

2. $\displaystyle\int_{-2}^{1} \pi\,dx$

3. $\displaystyle\int_{0}^{5} (3x + 2)\,dx$

4. $\displaystyle\int_{1}^{4} (5 - 2t)\,dt$

5. $\displaystyle\int_{-1}^{1} 3t^4\,dt$

6. $\displaystyle\int_{1}^{4} 2\sqrt{u}\,du$

7. $\displaystyle\int_{-1}^{1} (2u^{1/3} - u^{2/3})\,du$

8. $\displaystyle\int_{4}^{9} x^{-3/2}\,dx$

9. $\displaystyle\int_{0}^{1} e^{-x}(4 - e^x)\,dx$

10. $\displaystyle\int_{-1}^{1} \left(\frac{1}{e^x} - \frac{1}{e^{-x}}\right) dx$

11. $\displaystyle\int_{0}^{1} (x^4 + 3x^3 + 1)\,dx$

12. $\displaystyle\int_{-1}^{0} (-3x^5 - 3x^2 + 2x + 5)\,dx$

13. $\displaystyle\int_{2}^{5} (2 + 2t + 3t^2)\,dt$

14. $\displaystyle\int_{1}^{9} \left(\sqrt{t} - \frac{4}{\sqrt{t}}\right) dt$

15. $\displaystyle\int_{1}^{3} \left(1 + \frac{1}{x} + \frac{1}{x^2}\right) dx$

16. $\displaystyle\int_{0}^{\ln 2} (e^t - e^{-t})\,dt$

17. $\displaystyle\int_{-3}^{-1} \frac{t + 1}{t^3}\,dt$

18. $\displaystyle\int_{1}^{6} x^2(x - 1)\,dx$

19. $\displaystyle\int_{1}^{2} (2x - 4)^4\,dx$

20. $\displaystyle\int_{-3}^{0} (2x + 6)^4\,dx$

21. $\displaystyle\int_{0}^{4} \frac{1}{\sqrt{6t + 1}}\,dt$

22. $\displaystyle\int_{1}^{2} \frac{x^2}{(x^3 + 1)^2}\,dx$

23. $\displaystyle\int_{0}^{1} (x^3 + x)\sqrt{x^4 + 2x^2 + 1}\,dx$

24. $\displaystyle\int_{0}^{1} \frac{6t}{t^2 + 1}\,dt$

25. $\displaystyle\int_2^{e+1} \frac{x}{x-1}\,dx$

26. $\displaystyle\int_1^2 (t+1)(t-2)^6\,dt$

27. $\displaystyle\int_1^{e^2} \frac{(\ln x)^2}{x}\,dx$

28. $\displaystyle\int_e^{e^2} \frac{1}{x\ln x}\,dx$

29. $\displaystyle\int_{1/3}^{1/2} \frac{e^{1/x}}{x^2}\,dx$

30. $\displaystyle\int_1^4 \frac{(\sqrt{x}-1)^{3/2}}{\sqrt{x}}\,dx$

In Problems 31 through 36, f(x) and g(x) are functions that are continuous on the interval $-3 \le x \le 2$ and satisfy

$$\int_{-3}^2 f(x)\,dx = 5 \qquad \int_{-3}^2 g(x)\,dx = -2 \qquad \int_{-3}^1 f(x)\,dx = 0 \qquad \int_{-3}^1 g(x)\,dx = 4$$

In each case, use this information along with rules for definite integrals to evaluate the indicated integral.

31. $\displaystyle\int_{-3}^2 [-2f(x)+5g(x)]\,dx$

32. $\displaystyle\int_1^2 g(x)\,dx$

33. $\displaystyle\int_4^4 g(x)\,dx$

34. $\displaystyle\int_2^{-3} f(x)\,dx$

35. $\displaystyle\int_1^2 [3f(x)+2g(x)]\,dx$

36. $\displaystyle\int_{-3}^1 [2f(x)+3g(x)]\,dx$

In Problems 37 through 42, find the area of the region R that lies under the given curve $y = f(x)$ over the indicated interval $a \le x \le b$.

37. Under $y = x^4$, over $-1 \le x \le 2$

38. Under $y = \sqrt{x}\,(x+1)$, over $0 \le x \le 4$

39. Under $y = e^{2x}$, over $0 \le x \le \ln 3$

40. Under $y = \dfrac{3}{x}$, over $1 \le x \le e^2$

41. Under $y = (3x+4)^{1/2}$, over $0 \le x \le 4$

42. Under $y = xe^{-x^2}$, over $0 \le x \le 3$

43. LAND VALUES   It is estimated that t years from now the value of a certain parcel of land will be increasing at the rate of $V'(t)$ dollars per year. Find an expression for the amount by which the value of the land will increase during the next 5 years.

44. ADMISSION TO EVENTS   The promoters of a county fair estimate that t hours after the gates open at 9:00 A.M., visitors will be entering the fair at the rate of $N'(t)$ people per hour. Find an expression for the number of people who will enter the fair between 11:00 A.M. and 1:00 P.M.

45. STORAGE COST   A retailer receives a shipment of 12,000 pounds of soybeans that will be used at a constant rate of 300 pounds per week. If the cost of storing the soybeans is 0.2 cent per pound per week, how much will the retailer have to pay in storage costs over the next 40 weeks?

46. WATER POLLUTION   It is estimated that t years from now the population of a certain lakeside community will be changing at the rate of $0.6t^2 + 0.2t + 0.5$ thousand people per year. Environmentalists have found that the level of pollution in the lake increases at the rate of approximately 5 units per 1,000 people. By how much will the pollution in the lake increase during the next 2 years?

47. AIR POLLUTION   An environmental study of a certain community suggests that t years from now the level $L(t)$ of carbon monoxide in the air will be changing at the rate of $L'(t) = 0.1t + 0.1$ parts per million (ppm) per year. By how much will the pollution level change during the next 3 years?

48. MARGINAL COST The marginal cost of producing a certain commodity is $C'(q) = 6q + 1$ dollars per unit when q units are being produced.
  a. What is the total cost of producing the first 10 units?
  b. What is the cost of producing the *next* 10 units?

49. NET GROWTH OF POPULATION A study indicates that t months from now the population of a certain town will be growing at the rate of $P'(t) = 5 + 3t^{2/3}$ people per month. By how much will the population of the town increase over the next 8 months?

50. OIL PRODUCTION A certain oil well that yields 400 barrels of crude oil a month will run dry in 2 years. The price of crude oil is currently $25 per barrel and is expected to rise at a constant rate of 3 cents per barrel per month. If the oil is sold as soon as it is extracted from the ground, what will be the total future revenue from the well?

51. FARMING It is estimated that t days from now a farmer's crop will be increasing at the rate of $0.3t^2 + 0.6t + 1$ bushels per day. By how much will the value of the crop increase during the next 5 days if the market price remains fixed at $3 per bushel?

52. SALES REVENUE It is estimated that the demand for a manufacturer's product is increasing exponentially at the rate of 2% per year. If the current demand is 5,000 units per year and if the price remains fixed at $400 per unit, how much revenue will the manufacturer receive from the sale of the product over the next 2 years?

53. PRODUCTION Bejax Corporation has set up a production line to manufacture a new type of cellular telephone. The rate of production of the telephones is
$$\frac{dP}{dt} = 1,500\left(2 - \frac{t}{2t + 5}\right) \quad \text{units/month}$$
How many telephones are produced during the third month?

54. ADVERTISING An advertising agency begins a campaign to promote a new product and determines that t days later, the number of people $N(t)$ who have heard about the product is changing at a rate given by
$$N'(t) = 5t^2 - \frac{0.04t}{t^2 + 3} \quad \text{people per day}$$

How many people learn about the product during the first week? During the second week?

55. CONCENTRATION OF DRUG The concentration of a drug in a patient's bloodstream t hours after an injection is decreasing at the rate
$$C'(t) = \frac{-0.33t}{\sqrt{0.02t^2 + 10}} \quad \text{mg/cm}^3 \text{ per hour}$$
By how much does the concentration change over the first 4 hours after the injection?

56. ENDANGERED SPECIES A study conducted by an environmental group in the year 2000 determined that t years later, the population of a certain endangered bird species will be decreasing at the rate of $P'(t) = -0.75t\sqrt{10 - 0.2t}$ individuals per year. By how much is the population expected to change during the decade 2000–2010?

57. DEPRECIATION The resale value of a certain industrial machine decreases over a 10-year period at a rate that changes with time. When the machine is x years old, the rate at which its value is changing is $220(x - 10)$ dollars per year. By how much does the machine depreciate during the second year?

58. WATER CONSUMPTION The city manager of Paloma Linda estimates that water is being consumed by his community at the rate of $C'(t) = 10 + 0.3e^{0.03t}$ billion gallons per year, where $C(t)$ is the water consumption t years after the year 2000. How much water will be consumed by the community during the decade 2000–2010?

59. CHANGE IN BIOMASS A protein with mass m (grams) disintegrates into amino acids at a rate given by
$$\frac{dm}{dt} = \frac{-2}{t + 1} \quad \text{g/hr}$$
How much more protein is there after 2 hours than after 5 hours?

60. CHANGE IN BIOMASS Answer the question in Problem 59 if the rate of disintegration is given by
$$\frac{dm}{dt} = -(0.1t + e^{0.1t})$$

61. RATE OF LEARNING In a learning experiment, subjects are given a series of facts to memorize, and it is determined that t minutes after the experiment begins, the average subject is learning at the rate

$$L'(t) = \frac{4}{\sqrt{t+1}} \text{ facts per minute}$$

where $L(t)$ is the total number of facts memorized by time t. About how many facts does the typical subject learn during the second 5 minutes (between $t = 5$ and $t = 10$)?

62. **DISTANCE AND VELOCITY**   A driver, traveling at a constant speed of 45 mph, decides to speed up in such a way that her velocity t hours later is $v(t) = 45 + 12t$ mph. How far does she travel in the first 2 hours?

63. **PROJECTILE MOTION**   A ball is thrown upward from the top of a building, and t seconds later has velocity $v(t) = -32t + 80$ ft/sec. What is the difference in the ball's position after 3 seconds?

64. Verify the sum rule for definite integrals; that is, if $f(x)$ and $g(x)$ are continuous on the interval $a \le x \le b$, then

$$\int_a^b [f(x) + g(x)]\, dx = \int_a^b f(x)\, dx + \int_a^b g(x)\, dx$$

65. You have seen that the definite integral can be used to compute the area under a curve, but the "area as an integral" formula works both ways.

a. Compute $\int_0^1 \sqrt{1 - x^2}\, dx$. [*Hint:* Note that the integral is part of the area under the circle $x^2 + y^2 = 1$.]

b. Compute $\int_1^2 \sqrt{2x - x^2}\, dx$. [*Hint:* Describe the graph of $y = \sqrt{2x - x^2}$ and look for a geometric solution as in part (a).]

66. Given the function of $f(x) = 2\sqrt{x} + \dfrac{1}{x+1}$, approximate the value of the integral $\int_0^2 f(x)\, dx$ by completing these steps:

a. Find the numbers x_1, x_2, x_3, x_4, and x_5 that subdivide the interval $0 \le x \le 2$ into four equal subintervals. Use these numbers to form four rectangles that approximate the area under the curve $y = f(x)$ over $0 \le x \le 2$.

b. Estimate the value of the given integral by computing the sum of the areas of the four approximating rectangles in part (a).

c. Repeat steps (a) and (b) with eight subintervals instead of four.

SECTION 5.4 Applying Definite Integration: Area Between Curves and Average Value

We have seen that area can be expressed as a special kind of limit of a sum called a definite integral and then computed by applying the fundamental theorem of calculus. This procedure, called **definite integration,** was introduced through area because area is easy to visualize, but there are many applications other than area in which the integration process plays an important role.

In this section, we extend the ideas introduced in Section 5.3 to find the area between two curves and the average value of a function. As part of our study of area between curves, we will examine an important socioeconomic device called a Lorentz curve, which is used to measure relative wealth within a society.

Applying the Definite Integral   Intuitively, definite integration can be thought of as a process that "accumulates" an infinite number of small pieces of a quantity to obtain the total quantity. Here is a step-by-step description of how to use this process in applications.

A Procedure for Using Definite Integration in Applications

To use definite integration to "accumulate" a quantity Q over an interval $a \leq x \leq b$, proceed as follows:

Step 1. Divide the interval $a \leq x \leq b$ into n equal subintervals, each of length $\Delta x = \dfrac{b - a}{n}$. Choose a number x_j from the jth subinterval, for $j = 1, 2, \ldots, n$.

Step 2. Approximate small parts of the quantity Q by products of the form $f(x_j) \Delta x$, where $f(x)$ is an appropriate function that is continuous on $a \leq x \leq b$.

Step 3. Add the individual approximating products to estimate the total quantity Q by the Riemann sum

$$[f(x_1) + f(x_2) + \cdots + f(x_n)] \Delta x$$

Step 4. Make the approximation in step 3 exact by taking the limit of the Riemann sum as $n \to +\infty$ to express Q as a definite integral; that is,

$$Q = \lim_{n \to +\infty} [f(x_1) + f(x_2) + \cdots + f(x_n)] \Delta x = \int_a^b f(x)\, dx$$

Then use the fundamental theorem of calculus to compute $\int_a^b f(x)\, dx$ and thus to obtain the required quantity Q.

Just-In-Time **Review**

There is nothing special about using "j" for the index in the summation notation. The most commonly used indices are i, j, and k.

NOTATION: *The Summation Notation* We can use *summation notation* to represent the Riemann sums that occur when quantities are modeled using definite integration. Specifically, to describe the sum

$$a_1 + a_2 + \cdots + a_n$$

it suffices to specify the general term a_j in the sum and to indicate that n terms of this form are to be added, starting with the term where $j = 1$ and ending with $j = n$. For this purpose, it is customary to use the uppercase Greek letter sigma (Σ) and to write the sum as $\displaystyle\sum_{j=1}^{n} a_j$. Thus,

$$\sum_{j=1}^{n} a_j = a_1 + a_2 + \cdots + a_n$$

In particular, the Riemann sum

$$[f(x_1) + f(x_2) + \cdots + f(x_n)] \Delta x$$

can be written in the compact form

$$\sum_{j=1}^{n} f(x_j) \Delta x$$

Thus, the limit statement

$$\lim_{n \to +\infty} [f(x_1) + f(x_2) + \cdots + f(x_n)] \Delta x = \int_a^b f(x)\, dx$$

used to define the definite integral can be expressed as

$$\lim_{n \to +\infty} \sum_{j=1}^{n} f(x_j) \Delta x = \int_{a}^{b} f(x)\,dx$$

A review of the summation notation including examples may be found in Appendix A. ∎

Area Between Two Curves

In certain practical applications, you may find it useful to represent a quantity of interest in terms of area between two curves. First, suppose that f and g are continuous, nonnegative [that is, $f(x) \geq 0$ and $g(x) \geq 0$], and satisfy $f(x) \geq g(x)$ on the interval $a \leq x \leq b$, as shown in Figure 5.9a.

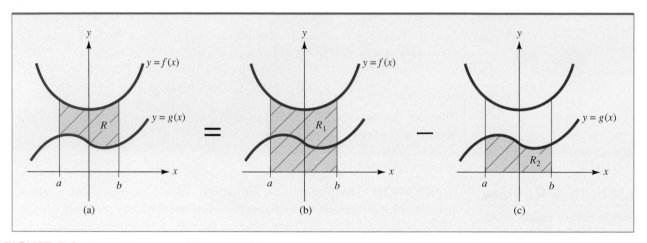

FIGURE 5.9 Area of R = area of R_1 − area of R_2.

Then, to find the area of the region R between the curves $y = f(x)$ and $y = g(x)$ over the interval $a \leq x \leq b$, we simply subtract the area under the lower curve $y = g(x)$ (Figure 5.9c) from the area under the upper curve $y = f(x)$ (Figure 5.9b), so that

$$\text{Area of } R = [\text{area under } y = f(x)] - [\text{area under } y = g(x)]$$

$$= \int_{a}^{b} f(x)\,dx - \int_{a}^{b} g(x)\,dx = \int_{a}^{b} [f(x) - g(x)]\,dx$$

This formula still applies whenever $f(x) \geq g(x)$ on the interval $a \leq x \leq b$, even when the curves $y = f(x)$ and $y = g(x)$ are not always both above the x axis. We will show that this is true by using the procedure for applying definite integration described on page 403.

Step 1. Subdivide the interval $a \leq x \leq b$ into n equal subintervals, each of width $\Delta x = \dfrac{b - a}{n}$. For $j = 1, 2, \ldots, n$, let x_j be the left endpoint of the jth subinterval.

Step 2. Construct approximating rectangles of width Δx and height $f(x_j) - g(x_j)$. This is possible since $f(x) \geq g(x)$ on $a \leq x \leq b$, which guarantees that the height is nonnegative; that is $f(x_j) - g(x_j) \geq 0$. For $j = 1, 2, \ldots, n$, the area $[f(x_j) - g(x_j)]\Delta x$ of the jth rectangle you have just constructed is approximately the same as the area between the two curves over the jth subinterval, as shown in Figure 5.10a.

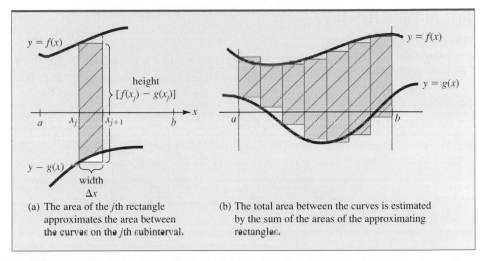

(a) The area of the jth rectangle approximates the area between the curves on the jth subinterval.

(b) The total area between the curves is estimated by the sum of the areas of the approximating rectangles.

FIGURE 5.10   Computing area between curves by definite integration.

Step 3. Add the individual approximating areas $[f(x_j) - g(x_j)]\Delta x$ to estimate the total area A between the two curves over the interval $a \leq x \leq b$ by the Riemann sum

$$A \approx [f(x_1) - g(x_1)]\Delta x + [f(x_2) - g(x_2)]\Delta x + \cdots + [f(x_n) - g(x_n)]\Delta x$$

$$= \sum_{j=1}^{n} [f(x_j) - g(x_j)]\Delta x$$

(See Figure 5.10b.)

Step 4. Make the approximation exact by taking the limit of the Riemann sum in step 3 as $n \to +\infty$ to express the total area A between the curves as a definite integral; that is,

$$A = \lim_{n \to +\infty} \sum_{j=1}^{n} [f(x_j) - g(x_j)]\Delta x = \int_a^b [f(x) - g(x)]\, dx$$

To summarize:

> **The Area Between Two Curves**   ▪   If $f(x)$ and $g(x)$ are continuous with $f(x) \geq g(x)$ on the interval $a \leq x \leq b$, then the area A between the curves $y = f(x)$ and $y = g(x)$ over the interval is given by
>
> $$A = \int_a^b [f(x) - g(x)]\, dx$$

EXAMPLE 5.4.1

Find the area of the region R enclosed by the curves $y = x^3$ and $y = x^2$

Solution

To find the points where the curves intersect, solve the equations simultaneously as follows:

$$x^3 = x^2$$

$$x^3 - x^2 = 0 \quad \textit{subtract } x^2 \textit{ from both sides}$$

Just-In-Time **Review**

Note that $x^2 \geq x^3$ for $0 \leq x \leq 1$.
For example,

$$\left(\frac{1}{3}\right)^2 > \left(\frac{1}{3}\right)^3$$

$$x^2(x - 1) = 0 \qquad \textit{factor out } x^2$$
$$x = 0,\ 1 \qquad uv = 0 \textit{ if and only if } u = 0 \textit{ or } v = 0$$

The corresponding points $(0, 0)$ and $(1, 1)$ are the only points of intersection.

The region R enclosed by the two curves is bounded above by $y = x^2$ and below by $y = x^3$, over the interval $0 \leq x \leq 1$ (see Figure 5.11). The area of this region is given by the integral

$$A = \int_0^1 (x^2 - x^3)\, dx = \frac{1}{3}x^3 - \frac{1}{4}x^4 \Big|_0^1$$

$$= \left[\frac{1}{3}(1)^3 - \frac{1}{4}(1)^4\right] - \left[\frac{1}{3}(0)^3 - \frac{1}{4}(0)^4\right] = \frac{1}{12}$$

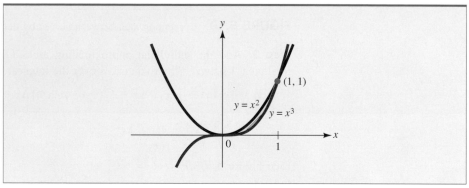

FIGURE 5.11 The region enclosed by the curves $y = x^2$ and $y = x^3$.

In certain applications, you may need to find the area A between the two curves $y = f(x)$ and $y = g(x)$ over an interval $a \leq x \leq b$, where $f(x) \geq g(x)$ for $a \leq x \leq c$ but $g(x) \geq f(x)$ for $c \leq x \leq b$. In this case, we have

$$A = \underbrace{\int_a^c [f(x) - g(x)]\, dx}_{f(x) \,\geq\, g(x) \text{ on } a \,\leq\, x \,\leq\, c} \quad + \quad \underbrace{\int_c^b [g(x) - f(x)]\, dx}_{g(x) \,\geq\, f(x) \text{ on } c \,\leq\, x \,\leq\, b}$$

Consider Example 5.4.2.

EXPLORE!

Refer to Example 5.4.2. Set Y1 = 4X and Y2 = X³ + 3X² in the equation editor of your graphing calculator. Graph using the window [−6, 2]1 by [−25, 10]5. Determine the points of intersection of the two curves. Another view of the area between the two curves is to set Y3 = Y2 − Y1, deselect (turn off) Y1 and Y2, and graph using [−4.5, 1.5]0.5 by [−5, 15]5. Numerical integration can be applied to this difference function.

EXAMPLE | 5.4.2

Find the area of the region enclosed by the line $y = 4x$ and the curve $y = x^3 + 3x^2$.

Solution

To find where the line and curve intersect, solve the equations simultaneously as follows:

$$x^3 + 3x^2 = 4x$$
$$x^3 + 3x^2 - 4x = 0 \qquad \textit{subtract } 4x \textit{ from both sides}$$
$$x(x^2 + 3x - 4) = 0 \qquad \textit{factor out } x$$

$$x(x - 1)(x + 4) = 0 \qquad \text{factor } x^2 + 3x - 4$$

$$x = 0, 1, -4 \qquad uv = 0 \text{ if and only if } u = 0 \text{ or } v = 0$$

The corresponding points of intersection are $(0, 0)$, $(1, 4)$, and $(-4, -16)$. The curve and the line are sketched in Figure 5.12.

Over the interval $-4 \leq x \leq 0$, the curve is above the line, so $x^3 + 3x^2 \geq 4x$, and the region enclosed by the curve and line has area

$$A_1 = \int_{-4}^{0} |(x^3 + 3x^2) - 4x| \, dx = \frac{1}{4}x^4 + x^3 - 2x^2 \bigg|_{-4}^{0}$$

$$= \left[\frac{1}{4}(0)^4 + (0)^3 - 2(0)^2 \right] - \left[\frac{1}{4}(-4)^4 + (-4)^3 - 2(-4)^2 \right] = 32$$

Over the interval $0 \leq x \leq 1$, the line is above the curve and the enclosed region has area

$$A_2 = \int_{0}^{1} [4x - (x^3 + 3x^2)] \, dx = 2x^2 - \frac{1}{4}x^4 - x^3 \bigg|_{0}^{1}$$

$$= \left[2(1)^2 - \frac{1}{4}(1)^4 - (1)^3 \right] - \left[2(0)^2 - \frac{1}{4}(0)^4 - (0)^3 \right] = \frac{3}{4}$$

Therefore, the total area enclosed by the line and the curve is given by the sum

$$A = A_1 + A_2 = 32 + \frac{3}{4} = 32.75$$

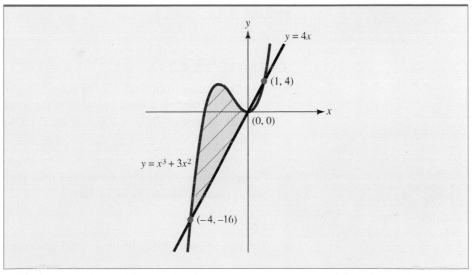

FIGURE 5.12   The region enclosed by the line $y = 4x$ and the curve $y = x^3 + 3x^2$.

Net Excess Profit   The area between curves can sometimes be used as a way of measuring the amount of a quantity that has been accumulated during a particular procedure. For instance, suppose that t years from now, two investment plans will be generating profit $P_1(t)$

and $P_2(t)$, respectively, and that their respective rates of profitability, $P_1'(t)$ and $P_2'(t)$, are expected to satisfy $P_2'(t) \geq P_1'(t)$ for the next N years; that is, over the time interval $0 \leq t \leq N$. Then $E(t) = P_2(t) - P_1(t)$ represents the **excess profit** of plan 2 over plan 1 at time t, and the **net excess profit** $NE = E(N) - E(0)$ over the time interval $0 \leq t \leq N$ is given by the definite integral

$$NE = E(N) - E(0) = \int_0^N E'(t)\, dt$$

$$= \int_0^N [P_2'(t) - P_1'(t)]\, dt \qquad \begin{array}{l} \textit{since } E'(t) = [P_2(t) - P_1(t)]' \\ \qquad\quad = P_2'(t) - P_1'(t) \end{array}$$

This integral can be interpreted geometrically as the area between the rate of profitability curves $y = P_1'(t)$ and $y = P_2'(t)$ as shown in Figure 5.13. Example 5.4.3 illustrates the computation of net excess profit.

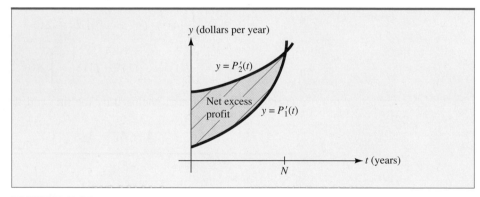

FIGURE 5.13   Net excess profit as the area between rate of profitability curves.

EXAMPLE | 5.4.3

Suppose that t years from now, one investment will be generating profit at the rate of $P_1'(t) = 50 + t^2$ hundred dollars per year, while a second investment will be generating profit at the rate of $P_2'(t) = 200 + 5t$ hundred dollars per year.

a. For how many years does the rate of profitability of the second investment exceed that of the first?

b. Compute the net excess profit for the time period determined in part (a). Interpret the net excess profit as an area.

Solution

a. The rate of profitability of the second investment exceeds that of the first until

$$
\begin{aligned}
P_1'(t) &= P_2'(t) \\
50 + t^2 &= 200 + 5t \\
t^2 - 5t - 150 &= 0 \qquad && \textit{subtract } 200 + 5t \textit{ from both sides} \\
(t - 15)(t + 10) &= 0 \qquad && \textit{factor} \\
t &= 15, -10 \qquad && \textit{since } uv = 0 \textit{ if and only if } u = 0 \textit{ or } v = 0 \\
t &= 15 \text{ years} \qquad && \textit{reject the negative time } t = -10
\end{aligned}
$$

b.   The excess profit of plan 2 over plan 1 is $E(t) = P_2(t) - P_1(t)$, and the net excess profit NE over the time period $0 \le t \le 15$ determined in part (a) is given by the definite integral

$$NE = E(15) - E(0) = \int_0^{15} E'(t)\,dt \qquad \text{\textit{fundamental theorem of calculus}}$$

$$= \int_0^{15} [P_2'(t) - P_1'(t)]\,dt \qquad \text{\textit{since } } E(t) = P_2(t) - P_1(t)$$

$$= \int_0^{15} [(200 + 5t) - (50 + t^2)]\,dt$$

$$= \int_0^{15} [150 + 5t - t^2]\,dt \qquad \text{\textit{combine terms}}$$

$$= \left[150t + 5\left(\frac{1}{2}t^2\right) - \left(\frac{1}{3}t^3\right) \right]\Big|_0^{15}$$

$$- \left[150(15) + \frac{5}{2}(15)^2 - \frac{1}{3}(15)^3 \right] - \left[150(0) + \frac{5}{2}(0)^2 - \frac{1}{3}(0)^3 \right]$$

$$= 1{,}687.50 \text{ hundred dollars}$$

Thus, the net excess profit is $168,750.

The graphs of the rate of profitability functions $P_1'(t)$ and $P_2'(t)$ are shown in Figure 5.14. The net excess profit

$$NE = \int_0^{15} [P_2'(t) - P_1'(t)]\,dt$$

can be interpreted as the area of the (shaded) region between the rate of profitability curves over the interval $0 \le t \le 15$.

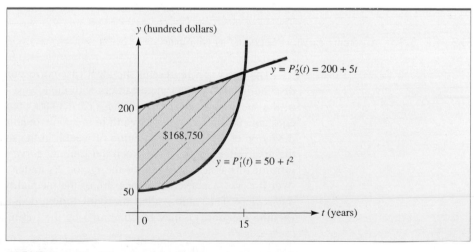

FIGURE 5.14   Net excess profit for one investment plan over another.

Lorentz Curves   Area also plays an important role in the study of **Lorentz curves,** a device used by both economists and sociologists to measure the percentage of a society's wealth that is possessed by a given percentage of its people. To be more specific, the Lorentz curve for a particular society's economy is the graph of the function $L(x)$, which denotes the fraction of total annual national income earned by the lowest-paid $100x\%$ of the wage-earners in the society, for $0 \leq x \leq 1$. For instance, if the lowest-paid 30% of all wage-earners receive 23% of the society's total income, then $L(0.3) = 0.23$.

Note that $L(x)$ is an increasing function on the interval $0 \leq x \leq 1$ and has these properties:

1. $0 \leq L(x) \leq 1$ because $L(x)$ is a percentage
2. $L(0) = 0$ because no wages are earned when no wage-earners are employed
3. $L(1) = 1$ because 100% of wages are earned by 100% of the wage-earners
4. $L(x) \leq x$ because the lowest-paid $100x\%$ of wage-earners cannot receive more than $100x\%$ of total income

A typical Lorentz curve is shown in Figure 5.15a.

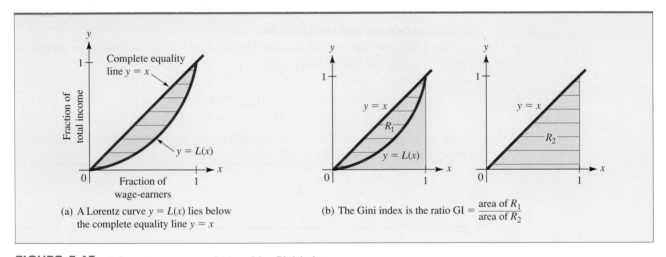

(a) A Lorentz curve $y = L(x)$ lies below the complete equality line $y = x$

(b) The Gini index is the ratio $\text{GI} = \dfrac{\text{area of } R_1}{\text{area of } R_2}$

FIGURE 5.15   A Lorentz curve $y = L(x)$ and its Gini index.

The line $y = x$ represents the ideal case corresponding to complete equality in the distribution of income (wage-earners with the lowest $100x\%$ of income receive $100x\%$ of the society's wealth). The closer a particular Lorentz curve is to this line, the more equitable the distribution of wealth in the corresponding society. We represent the total deviation of the actual distribution of wealth in the society from complete equality by the area of the region R_1 between the Lorentz curve $y = L(x)$ and the line $y = x$. The ratio of this area to the area of the region R_2 under the complete equality line $y = x$ over $0 \leq x \leq 1$ is used as a measure of the inequality in the distribution of wealth in the society. This ratio, called the **Gini index,** denoted GI (also called the **index of income inequality**), may be computed by the formula

$$\text{GI} = \frac{\text{area of } R_1}{\text{area of } R_2} = \frac{\text{area between } y = L(x) \text{ and } y = x}{\text{area under } y = x \text{ over } 0 \leq x \leq 1}$$

$$= \frac{\displaystyle\int_0^1 [x - L(x)]\, dx}{\displaystyle\int_0^1 x\, dx} = \frac{\displaystyle\int_0^1 [x - L(x)]\, dx}{1/2}$$

$$= 2 \int_0^1 [x - L(x)]\, dx$$

(see Figure 5.15b). To summarize:

Gini Index   ■   If $y = L(x)$ is the equation of a Lorentz curve, then the inequality in the corresponding distribution of wealth is measured by the *Gini index,* which is given by the formula

$$\text{Gini index} = 2 \int_0^1 [x - L(x)]\, dx$$

The Gini index always lies between 0 and 1. An index of 0 corresponds to total equity in the distribution of income, while an index of 1 corresponds to total inequity (all income belongs to 0% of the population). The smaller the index, the more equitable the distribution of income, and the larger the index, the more the wealth is concentrated in only a few hands. Example 5.4.4 illustrates how Lorentz curves and the Gini index can be used to compare the relative equity of income distribution for two professions.

EXAMPLE | 5.4.4

A governmental agency determines that the Lorentz curves for the distribution of income for dentists and contractors in a certain state are given by the functions

$$L_1(x) = x^{1.7} \quad \text{and} \quad L_2(x) = 0.8x^2 + 0.2x$$

respectively. For which profession is the distribution of income more fairly distributed?

Solution

The respective Gini indices are

$$G_1 = 2 \int_0^1 (x - x^{1.7})\, dx = 2 \left(\frac{x^2}{2} - \frac{x^{2.7}}{2.7} \right) \Big|_0^1 = 0.2593$$

and

$$G_2 = 2 \int_0^1 [x - (0.8x^2 + 0.2x)]\, dx$$

$$= 2 \left[-0.8 \left(\frac{x^3}{3} \right) + 0.8 \left(\frac{x^2}{2} \right) \right] \Big|_0^1 = 0.2667$$

Since the Gini index for dentists is smaller, it follows that in this state, the incomes of dentists are more evenly distributed than those of contractors.

Using the Gini index, we can see how the distribution of income in the United States compares to that in other countries. Table 5.1 lists the Gini index for selected industrial and developing nations. Note that with an index of 0.46, the distribution of income in the United States is about the same as that of Thailand, is less equitable than the United Kingdom, Germany, or Denmark, but much more equitable than Brazil or Panama. (Is there anything you know about the socio-political nature of these countries that would explain the difference in income equity?)

TABLE 5.1 Gini Indices for Selected Countries

Country	Gini Index
United States	0.460
Brazil	0.601
Canada	0.315
Denmark	0.247
Germany	0.281
Japan	0.350
South Africa	0.584
Panama	0.568
Thailand	0.462
United Kingdom	0.326

SOURCE: David C. Colander, *Economics,* 4th ed., Boston: McGraw-Hill, 2001, p. 435.

Average Value of a Function

As a second illustration of how definite integration can be used in applications, we will compute the **average value of a function,** which is of interest in a variety of situations. First, let us take a moment to clarify our thinking about what we mean by "average value." A teacher who wants to compute the average score on an examination simply adds all the individual scores and then divides by the number of students taking the exam, but how should one go about finding, say, the average pollution level in a city during the daytime hours? The difficulty is that since time is continuous, there are "too many" pollution levels to add up in the usual way, so how should we proceed?

Consider the general case in which we wish to find the average value of the function $f(x)$ over an interval $a \leq x \leq b$ on which f is continuous. We begin by subdividing the interval $a \leq x \leq b$ into n equal parts, each of length $\Delta x = (b - a)/n$. If x_j is a number taken from the jth subinterval for $j = 1, 2, \ldots, n$, then the average of the corresponding functional values $f(x_1), f(x_2), \ldots, f(x_n)$ is

$$
\begin{aligned}
V_n &= \frac{f(x_1) + f(x_2) + \cdots + f(x_n)}{n} \\
&= \frac{b - a}{b - a} \left[\frac{f(x_1) + f(x_2) + \cdots + f(x_n)}{n} \right] \qquad \text{\textit{multiply and divide by} } (b - a) \\
&= \frac{1}{b - a} [f(x_1) + f(x_2) + \cdots + f(x_n)]\left(\frac{b - a}{n}\right) \text{\textit{factor out the expression} } \frac{b - a}{n} \\
&= \frac{1}{b - a} [f(x_1) + f(x_2) + \cdots + f(x_n)]\Delta x \qquad \text{\textit{since} } \Delta x = \frac{b - a}{n} \\
&= \frac{1}{b - a} \sum_{j=1}^{n} f(x_j)\Delta x
\end{aligned}
$$

which we recognize as a Riemann sum.

If we refine the partition of the interval $a \leq x \leq b$ by taking more and more subdivision points, then V_n becomes more and more like what we may intuitively think of as the average value V of $f(x)$ over the entire interval $a \leq x \leq b$. Thus, it is reasonable to *define* the average value V as the limit of the Riemann sum V_n as

EXPLORE!

Suppose you wish to calculate the average value of $f(x) = x^3 - 6x^2 + 10x - 1$ over the interval [1, 4]. Store $f(x)$ in Y1 and obtain its graph using the window [0, 4.7]1 by [−2, 8]1. Shade the region under the curve over the interval [1, 4] and compute its area A. Set Y2 equal to the constant function $\dfrac{A}{b - a} = \dfrac{A}{3}$.

This is the average value. Plot Y2 and Y1 on the same screen. At what number(s) between 1 and 4 does $f(x)$ equal the average value?

$n \to +\infty$; that is, as the definite integral

$$V = \lim_{n \to +\infty} V_n = \lim_{n \to +\infty} \frac{1}{b-a} \sum_{j=1}^{n} f(x_j)\Delta x$$

$$- \frac{1}{b-a} \int_a^b f(x)\,dx$$

To summarize:

The Average Value of a Function ■ Let $f(x)$ be a function that is continuous on the interval $a \le x \le b$. Then the *average value* V of $f(x)$ over $a \le x \le b$ is given by the definite integral

$$V = \frac{1}{b-a} \int_a^b f(x)\,dx$$

EXAMPLE | 5.4.5

A manufacturer determines that t months after introducing a new product, the company's sales will be $S(t)$ thousand dollars, where

$$S(t) = \frac{750t}{\sqrt{4t^2 + 25}}$$

What are the average monthly sales of the company over the first 6 months after the introduction of the new product?

Solution

The average monthly sales V over the time period $0 \le t \le 6$ is given by the integral

$$V = \frac{1}{6-0} \int_0^6 \frac{750t}{\sqrt{4t^2 + 25}}\,dt$$

To evaluate this integral, make the substitution

$$u = 4t^2 + 25 \qquad \text{limits of integration:}$$
$$du = 4(2t\,dt) \qquad \text{if } t = 0, \text{ then } u = 4(0)^2 + 25 = 25$$
$$t\,dt = \frac{1}{8}\,du \qquad \text{if } t = 6, \text{ then } u = 4(6)^2 + 25 = 169$$

You obtain

$$V = \frac{1}{6} \int_0^6 \frac{750}{\sqrt{4t^2 + 25}}\,(t\,dt)$$

$$= \frac{1}{6} \int_{25}^{169} \frac{750}{\sqrt{u}}\left(\frac{1}{8}\,du\right) = \frac{750}{6(8)} \int_{25}^{169} u^{-1/2}\,du$$

$$= \frac{750}{6(8)} \left(\frac{u^{1/2}}{1/2}\right)\bigg|_{25}^{169} = \frac{750(2)}{6(8)} [(169)^{1/2} - (25)^{1/2}]$$

$$- 250$$

Thus, for the 6-month period immediately after the introduction of the new product, the company's sales average $250,000 per month.

EXAMPLE | 5.4.6

A researcher models the temperature T (in °C) during the time period from 6 A.M. to 6 P.M. in a certain northern city by the function

$$T(t) = 3 - \frac{1}{3}(t - 4)^2 \qquad \text{for } 0 \le t \le 12$$

where t is the number of hours after 6 A.M.

a. What is the average temperature in the city during the workday, from 8 A.M. to 5 P.M.?

b. At what time (or times) during the workday is the temperature in the city the same as the average temperature found in part (a)?

Solution

a. Since 8 A.M. and 5 P.M. are, respectively, $t = 2$ and $t = 11$ hours after 6 A.M., we want to compute the average of the temperature $T(t)$ for $2 \le t \le 11$, which is given by the definite integral

$$\begin{aligned} T_{\text{ave}} &= \frac{1}{11 - 2} \int_2^{11} \left[3 - \frac{1}{3}(t - 4)^2 \right] dt \\ &= \frac{1}{9} \left[3t - \frac{1}{3}\frac{1}{3}(t - 4)^3 \right]\Big|_2^{11} \\ &= \frac{1}{9} \left[3(11) - \frac{1}{9}(11 - 4)^3 \right] - \frac{1}{9} \left[3(2) - \frac{1}{9}(2 - 4)^3 \right] \\ &= -\frac{4}{3} \approx -1.33 \end{aligned}$$

Thus, the average temperature during the workday is approximately $-1.33°C$ (or $29.6°F$).

b. We want to find a time $t = t_a$ with $2 \le t_a \le 11$ such that $T(t_a) = -\frac{4}{3}$. Solving this equation, we find that

$$3 - \frac{1}{3}(t_a - 4)^2 = -\frac{4}{3}$$

$$-\frac{1}{3}(t_a - 4)^2 = -\frac{4}{3} - 3 = -\frac{13}{3} \qquad \text{subtract 3 from both sides}$$

$$(t_a - 4)^2 = (-3)\left(-\frac{13}{3}\right) = 13 \qquad \text{multiply both sides by } -3$$

$$t_a - 4 = \pm\sqrt{13} \qquad \text{take square roots on both sides}$$

$$t_a = 4 \pm \sqrt{13}$$

$$\approx 0.39 \quad \text{or} \quad 7.61$$

Since $t = 0.39$ is outside the time interval $2 \le t_a \le 11$ (8 A.M. to 5 P.M.), it follows that the temperature in the city is the same as the average temperature only when $t = 7.61$; that is, at approximately 1:37 P.M.

Just-In-Time Review

Fahrenheit temperature F is related to Celsius temperature C by the formula

$$F = \frac{9}{5}C + 32$$

Just-In-Time Review

Since there are 60 minutes in an hour, 0.61 hour is the same as $0.61(60) \approx 37$ minutes. Thus, 7.61 hours after 6 A.M. is 37 minutes past 1 P.M. or 1.37 P.M.

Two Interpretations of Average Value

The average value of a function has several useful interpretations. First, note that if $f(x)$ is continuous on the interval $a \leq x \leq b$ and $F(x)$ is any antiderivative of $f(x)$ over the same interval, then the average value V of $f(x)$ over the interval satisfies

$$V = \frac{1}{b-a} \int_a^b f(x)\, dx$$

$$= \frac{1}{b-a} [F(b) - F(a)] \qquad \textit{fundamental theorem of calculus}$$

$$= \frac{F(b) - F(a)}{b - a}$$

We recognize this difference quotient as the average rate of change of $F(x)$ over $a \leq x \leq b$ (see Section 2.1). Thus, we have this interpretation:

Rate Interpretation of Average Value ▪ The average value of a function $f(x)$ over an interval $a \leq x \leq b$ where $f(x)$ is continuous is the same as the average rate of change of any antiderivative $F(x)$ of $f(x)$ over the same interval.

For instance, since the total cost $C(x)$ of producing x units of a commodity is an antiderivative of marginal cost $C'(x)$, it follows that the *average rate of change of cost over a range of production $a \leq x \leq b$ equals the average value of the marginal cost over the same range.*

The average value of a function $f(x)$ on an interval $a \leq x \leq b$ where $f(x) \geq 0$ can also be interpreted geometrically by rewriting the integral formula for average value

$$V = \frac{1}{b-a} \int_a^b f(x)\, dx$$

in the form

$$(b-a)V = \int_a^b f(x)\, dx$$

In the case where $f(x) \geq 0$ on the interval $a \leq x \leq b$, the integral on the right can be interpreted as the area under the curve $y = f(x)$ over $a \leq x \leq b$, and the product on the left as the area of a rectangle of height V and width $b - a$ equal to the length of the interval. In other words:

Geometric Interpretation of Average Value ▪ The average value V of $f(x)$ over an interval $a \leq x \leq b$ where $f(x)$ is continuous and satisfies $f(x) \geq 0$ is equal to the height of a rectangle whose base is the interval and whose area is the same as the area under the curve $y = f(x)$ over $a \leq x \leq b$.

This geometric interpretation is illustrated in Figure 5.16.

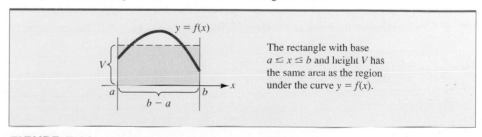

FIGURE 5.16 Geometric interpretation of average value V.

PROBLEMS 5.4

In Problems 1 through 4, find the area of the shaded region.

1.

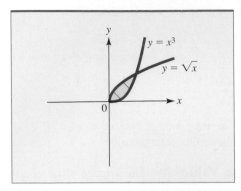

2.

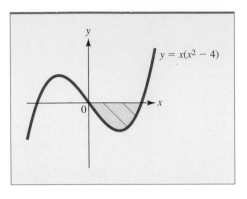

3.

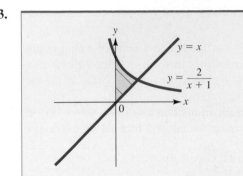

4.

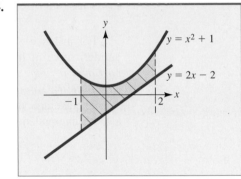

In Problems 5 through 17, sketch the given region R and then find its area.

5. R is the region bounded by the lines $y = x$, $y = -x$, and $x = 1$.

6. R is the region bounded by the curves $y = x^2$, $y = -x^2$, and the line $x = 1$.

7. R is the region bounded by the x axis and the curve $y = -x^2 + 4x - 3$.

8. R is the region bounded by the curves $y = e^x$, $y = e^{-x}$, and the line $x = \ln 2$.

9. R is the region bounded by the curve $y = x^2 - 2x$ and the x axis. [*Hint:* Note that the region is below the x axis.]

10. R is the region bounded by the curve $y = \dfrac{1}{x^2}$ and the lines $y = x$ and $y = \dfrac{x}{8}$.

11. R is the region bounded by the curves $y = x^2 - 2x$ and $y = -x^2 + 4$.

12. R is the region between the curve $y = x^3$ and the line $y = 9x$, for $x \geq 0$.

13. R is the region between the curves $y = x^3 - 3x^2$ and $y = x^2 + 5x$.

14. R is the triangle bounded by the line $y = 4 - 3x$ and the coordinate axes.

15. R is the triangle with vertices $(-4, 0)$, $(2, 0)$, and $(2, 6)$.

16. R is the rectangle with vertices $(1, 0)$, $(-2, 0)$, $(-2, 5)$, and $(1, 5)$.

17. R is the trapezoid bounded by the lines $y = x + 6$ and $x = 2$ and the coordinate axes.

In Problems 18 through 23, find the average value of the given function $f(x)$ over the specified interval $a \leq x \leq b$.

18. $f(x) = x^2 - 3x + 5$ over $-1 \leq x \leq 2$

19. $f(x) = 1 - x^2$ over $-3 \leq x \leq 3$

20. $f(x) = e^{2x} + e^{-x}$ over $0 \leq x \leq \ln 2$

21. $f(x) = e^{-x}(4 - e^{2x})$ over $-1 \leq x \leq 1$

22. $f(x) = \dfrac{x + 1}{x^2 + 2x + 6}$ over $-1 \leq x \leq 1$

23. $f(x) = \dfrac{e^x - e^{-x}}{e^x + e^{-x}}$ over $0 \leq x \leq \ln 3$

In Problems 24 through 27, find the average value V of the given function over the specified interval. In each case, sketch the graph of the function along with the rectangle whose base is the given interval and whose height is the average value V.

24. $f(x) = x$ over $0 \leq x \leq 4$

25. $f(x) = 2x - x^2$ over $0 \leq x \leq 2$

26. $g(t) = e^{-2t}$ over $-1 \leq t \leq 2$

27. $h(u) = \dfrac{1}{u}$ over $2 \leq u \leq 4$

LORENTZ CURVES   *In Problems 28 through 33, find the Gini index for the given Lorentz curve.*

28. $L(x) = x^2$

29. $L(x) = x^3$

30. $L(x) = 0.7x^2 + 0.3x$

31. $L(x) = 0.55x^2 + 0.45x$

32. $L(x) = \dfrac{e^x - 1}{e - 1}$

33. $L(x) = \dfrac{2}{3}x^{3.7} + \dfrac{1}{3}x$

34. TEMPERATURE   Records indicate that t hours past midnight, the temperature at the local airport was $f(t) = -0.3t^2 + 4t + 10$ degrees Celsius. What was the average temperature at the airport between 9:00 A.M. and noon?

35. FOOD PRICES   Records indicate that t months after the beginning of the year, the price of ground beef in local supermarkets was

$$P(t) = 0.09t^2 - 0.2t + 1.6$$

dollars per pound. What was the average price of ground beef during the first 3 months of the year?

36. EFFICIENCY   After t months on the job, a postal clerk can sort $Q(t) = 700 - 400e^{-0.5t}$ letters per hour. What is the average rate at which the clerk sorts mail during the first 3 months on the job?

37. BACTERIAL GROWTH   The number of bacteria present in a certain culture after t minutes of an experiment was $Q(t) = 2,000e^{0.05t}$. What was the average number of bacteria present during the first 5 minutes of the experiment?

38. AVERAGE VALUE OF INVESTMENT   Tom invests $5,000 in an account that pays 6% compounded continuously. What is the average value of his investment over the next 10 years?

39. INVENTORY   An inventory of 60,000 kilograms of a certain commodity is used at a constant rate and is exhausted after 1 year. What is the average inventory for the year?

40. INVESTMENT   Suppose that t years from now, one investment plan will be generating profit at the rate of $P_1'(t) = 100 + t^2$ hundred dollars per year, while a second investment will be generating profit at the rate of $P_2'(t) = 220 + 2t$ hundred dollars per year.

   a. For how many years does the rate of profitability of the second investment exceed that of the first?

   b. Compute the net excess profit assuming that you invest in the second plan for the time period determined in part (a).

   c. Sketch the rate of profitability curves $y = P_1'(t)$ and $y = P_2'(t)$ and shade the region whose area represents the net excess profit computed in part (b).

41. INVESTMENT   Answer the questions in Problem 40 for two investments with respective rates of profitability $P_1'(t) = 130 + t^2$ hundred dollars per year and $P_2'(t) = 306 + 5t$ hundred dollars per year.

42. **INVESTMENT** Answer the questions in Problem 40 for two investments with respective rates for profitability $P_1'(t) = 60e^{0.12t}$ thousand dollars per year and $P_2'(t) = 160e^{0.08t}$ thousand dollars per year.

43. **INVESTMENT** Answer the questions in Problem 40 for two investments with respective rates of profitability $P_1'(t) = 90e^{0.1t}$ thousand dollars per year and $P_2'(t) = 140e^{0.07t}$ thousand dollars per year.

44. **EFFICIENCY** After t hours on the job, one factory worker is producing $Q_1'(t) = 60 - 2(t - 1)^2$ units per hour, while a second worker is producing $Q_2'(t) = 50 - 5t$ units per hour.
    a. If both arrive on the job at 8:00 A.M., how many more units will the first worker have produced by noon than the second worker?
    b. Interpret the answer in part (a) as the area between two curves.

45. **AVERAGE POPULATION** The population of a certain community t years after the year 1990 is given by

$$P(t) = \frac{e^{0.2t}}{4 + e^{0.2t}} \quad \text{million people}$$

What was the average population of the community during the decade from 1990 to 2000?

46. **AVERAGE COST** The cost of producing x units of a new product is $C(x) = 3x\sqrt{x} + 10$ hundred dollars. What is the average cost of producing the first 81 units?

47. **AVERAGE DRUG CONCENTRATION** A patient is injected with a drug, and t hours later, the concentration of the drug remaining in the patient's bloodstream is given by

$$C(t) = \frac{3t}{(t^2 + 36)^{3/2}} \quad \text{mg/cm}^3$$

What is the average concentration of drug during the first 8 hours after the injection?

48. **AVERAGE PRODUCTION** A company determines that if L worker-hours of labor are employed, then Q units of a particular commodity will be produced, where

$$Q(L) = 500L^{2/3}$$

    a. What is the average production as labor varies from 1,000 to 2,000 worker-hours?
    b. What labor level between 1,000 and 2,000 worker-hours results in the average production found in part (a)?

49. **TEMPERATURE** A researcher models the temperature T (in °C) during the time period from 6 A.M. to 6 P.M. in a certain northern city by the function

$$T(t) = 3 - \frac{1}{3}(t - 5)^2 \quad \text{for } 0 \le t \le 12$$

where t is the number of hours after 6 A.M.
    a. What is the average temperature in the city during the workday, from 8 A.M. to 5 P.M.?
    b. At what time (or times) during the workday is the temperature in the city the same as the average temperature found in part (a)?

50. **ADVERTISING** An advertising firm is hired to promote a new television show for 3 weeks before its debut and 2 weeks afterward. After t weeks of the advertising campaign, it is found that $P(t)$ percent of the viewing public is aware of the show, where

$$P(t) = \frac{59t}{0.7t^2 + 16} + 6$$

    a. What is the average percentage of the viewing public that is aware of the show during the 5 weeks of the advertising campaign?
    b. At what time during the first 5 weeks of the campaign is the percentage of viewers the same as the average percentage found in part (a)?

51. **TRAFFIC MANAGEMENT** For several weeks, the highway department has been recording the speed of freeway traffic flowing past a certain downtown exit. The data suggest that between the hours of 1:00 and 6:00 P.M. on a normal weekday, the speed of traffic at the exit is approximately $S(t) = t^3 - 10.5t^2 + 30t + 20$ miles per hour, where t is the number of hours past noon.
    a. Compute the average speed of the traffic between the hours of 1:00 and 6:00 P.M.
     b. At what time between 1:00 and 6:00 P.M. is the traffic speed at the exit the same as the average speed found in part (a)?

52. **AVERAGE AEROBIC RATING** The aerobic rating of a person x years old is given by

$$A(x) = \frac{110(\ln x - 2)}{x} \quad \text{for } x \ge 10$$

What is a person's average aerobic rating from age 15 to age 25? From age 60 to age 70?

53. **THERMAL EFFECT OF FOOD** Normally, the metabolism of an organism functions at an essentially constant rate, called the *basal metabolic rate* of the organism. However, the metabolic rate may increase or decrease depending on the activity of the organism. In particular, after ingesting nutrients, the organism often experiences a surge in its metabolic rate, which then gradually returns to the basal level.

Michelle has just finished her Thanksgiving dinner, and her metabolic rate has surged from its basal level M_0 as she "works off" the meal over the next 12 hours. Suppose that t hours after the meal, her metabolic rate is given by

$$M(t) = M_0 + 50te^{-0.1t^2} \qquad 0 \le t \le 12$$

kilojoules per hour (kJ/hr).
   a. What is Michelle's average metabolic rate over the 12-hour period?
   b. Sketch the graph of $M(t)$. What is the peak metabolic rate and when does it occur? [*Note:* Both the graph and the peak rate will involve M_0.]

54. **DISTRIBUTION OF INCOME** In a certain state, it is found that the distribution of income for lawyers is given by the Lorentz curve $y = L_1(x)$, where

$$L_1(x) = \frac{4}{5}x^2 + \frac{1}{5}x$$

while that of surgeons is given by $y = L_2(x)$, where

$$L_2(x) = \frac{5}{8}x^4 + \frac{3}{8}x$$

Compute the Gini index for each Lorentz curve. Which profession has the more equitable income distribution?

55. **DISTRIBUTION OF INCOME** Suppose a study indicates that the distribution of income for professional baseball players is given by the Lorentz curve $y = L_1(x)$, where

$$L_1(x) = \frac{2}{3}x^3 + \frac{1}{3}x$$

while those of professional football players and basketball players are $y = L_2(x)$ and $y = L_3(x)$, respectively, where

$$L_2(x) = \frac{5}{6}x^2 + \frac{1}{6}x$$

and

$$L_3(x) = \frac{3}{5}x^4 + \frac{2}{5}x$$

Find the Gini index for each professional sport and determine which has the most equitable income distribution. Which has the least equitable distribution?

56. **COMPARATIVE GROWTH** The population of a third world country grows exponentially at the unsustainable rate of

$$P_1'(t) = 10e^{0.02t} \text{ thousand people per year}$$

where t is the number of years after 2000. A study indicates that if certain socioeconomic changes are instituted in this country, then the population will instead grow at the restricted rate

$$P_2'(t) = 10 + 0.02t + 0.002t^2$$

thousand people per year. How much smaller will the population of this country be in the year 2010 if the changes are made than if they are not?

57. **COMPARATIVE GROWTH** A second study of the country in Problem 56 indicates that natural restrictive forces are at work that make the actual rate of growth

$$P_3'(t) = \frac{20e^{0.02t}}{1 + e^{0.02t}}$$

instead of the exponential rate $P_1'(t) = 10e^{0.02t}$. If this rate is correct, how much smaller will the population be in the year 2010 than if the exponential rate were correct?

58. **REAL ESTATE** The territory occupied by a certain community is bounded on one side by a river and on all other sides by mountains, forming the shaded region shown in the accompanying figure. If a coordinate system is set up as indicated, the mountainous boundary is given roughly by the curve $y = 4 - x^2$, where x and y are measured in miles. What is the total area occupied by the community?

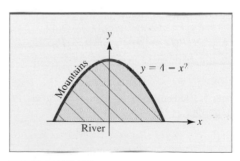

PROBLEM 58

59. **REAL ESTATE EVALUATION** A square cabin with a plot of land adjacent to a lake is shown in the accompanying figure. If a coordinate system is set up as indicated, with distances given in yards, the lakefront boundary of the property is part of the curve $y = 10e^{0.04x}$. Assuming that the cabin costs $2,000 per square yard and the land in the plot outside the cabin (the shaded region in the figure) costs $800 per square yard, what is the total cost of this vacation property?

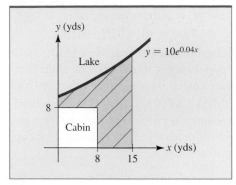

PROBLEM 59

60. **VOLUME OF BLOOD DURING SYSTOLE** A model* of the cardiovascular system relates the stroke volume $V(t)$ of blood in the aorta at time t during systole (the contraction phase) to the pressure $P(t)$ in the aorta at the same time by the equation

$$V(t) = [C_1 + C_2P(t)]\left(\frac{3t^2}{T^2} - \frac{2t^3}{T^3}\right)$$

where C_1 and C_2 are positive constants and T is the period of the systolic phase (a fixed time). Assume that aortic pressure $P(t)$ rises at a constant rate from P_0 when $t = 0$ to P_1 when $t = T$.

a. Show that

$$P(t) = \left(\frac{P_1 - P_0}{T}\right)t + P_0$$

---

*J. G. Defares, J. J. Osborn, and H. H. Hura, *Acta Physiol. Pharm. Neerl.*, Vol. 12, 1963, pp. 189–265.

b. Find the average volume of blood in the aorta during the systolic phase ($0 \le t \le T$). [*Note:* Your answer will be in terms of $C_1, C_2, P_0, P_1,$ and T.]

61. **REACTION TO MEDICATION** In certain biological models, the human body's reaction to a particular kind of medication is measured by a function of the form

$$F(M) = \frac{1}{3}(kM^2 - M^3) \qquad 0 \le M \le k$$

where k is a positive constant and M is the amount of medication absorbed in the blood. The sensitivity of the body to the medication is measured by the derivative $S = F'(M)$.

a. Show that the body is most sensitive to the medication when $M = \dfrac{k}{3}$.

b. What is the average reaction to the medication for $0 \le M \le \dfrac{k}{3}$?

62. Use the graphing utility of your calculator to draw the graphs of the curves $y = x^2e^{-x}$ and $y = \dfrac{1}{x}$ on the same screen. Use **ZOOM** and **TRACE** or some other feature of your calculator to find where the curves intersect. Then compute the area of the region bounded by the curves using the numeric integration feature.

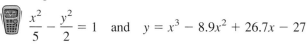

63. Repeat Problem 62 for the curves

$$\frac{x^2}{5} - \frac{y^2}{2} = 1 \quad \text{and} \quad y = x^3 - 8.9x^2 + 26.7x - 27$$

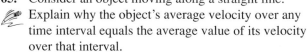

64. Show that the average value V of a continuous function $f(x)$ over the interval $a \le x \le b$ may be computed as the slope of the line joining the points $(a, F(a))$ and $(b, F(b))$ on the curve $y = F(x)$, where $F(x)$ is any antiderivative of $f(x)$ over $a \le x \le b$.

65. Consider an object moving along a straight line. Explain why the object's average velocity over any time interval equals the average value of its velocity over that interval.

SECTION 5.5 | Additional Applications to Business and Economics

In this section, we examine several important applications of definite integration to business and economics, such as future and present value of an income flow, consumers' willingness to spend, and consumers' and producers' surplus. We begin by showing how integration can be used to measure the value of an asset.

Useful Life of a Machine

Suppose that t years after being put into use, a machine has generated total revenue $R(t)$ and that the total cost of operating and servicing the machine up to this time is $C(t)$. Then the total profit generated by the machine up to time t is $P(t) = R(t) - C(t)$. Profit declines when costs accumulate at a higher rate than revenue; that is, when $C'(t) > R'(t)$. Thus, a manager may consider disposing of the machine at the time T when $C'(T) = R'(T)$, and for this reason, the time period $0 \leq t \leq T$ is called the **useful life** of the machine. The net profit over the useful life of the machine provides the manager with a measure of its value.

EXPLORE!

Refer to Example 5.5.1. Suppose a new cost rate function $C'_{now}(t) = 2,000 + 6t^2$ is in place. Compare its useful life and net profit with those of the original cost rate function. Use the window [0, 20]5 by [−2,000, 8,000]1,000.

EXAMPLE | 5.5.1

Suppose that when it is t years old, a particular industrial machine is generating revenue at the rate $R'(t) = 5,000 - 20t^2$ dollars per year and that operating and servicing costs related to the machine are accumulating at the rate $C'(t) = 2,000 + 10t^2$ dollars per year.

a. What is the useful life of this machine?

b. Compute the net profit generated by the machine over its period of useful life.

Solution

a. To find the machine's useful life T, solve

$$C'(t) = R'(t)$$
$$2,000 + 10t^2 = 5,000 - 20t^2$$
$$30t^2 = 3,000$$
$$t^2 = 100$$
$$t = 10$$

Thus, the machine has a useful life of $T = 10$ years.

b. Since profit $P(t)$ is given by $P(t) = R(t) - C(t)$, we have $P'(t) = R'(t) - C'(t)$, and the net profit generated by the machine over its useful life $0 \leq t \leq 10$ is

$$NP = P(10) - P(0) = \int_0^{10} P'(t)\, dt$$

$$= \int_0^{10} [R'(t) - C'(t)]\, dt$$

$$= \int_0^{10} [(5,000 - 20t^2) - (2,000 + 10t^2)]\, dt$$

$$= \int_0^{10} [3,000 - 30t^2]\, dt$$

$$= 3,000t - 10t^3 \Big|_0^{10} = \$20,000$$

The rate of revenue and rate of cost curves are sketched in Figure 5.17. The net earnings of the machine over its useful life are represented by the area of the (shaded) region between the two rate curves.

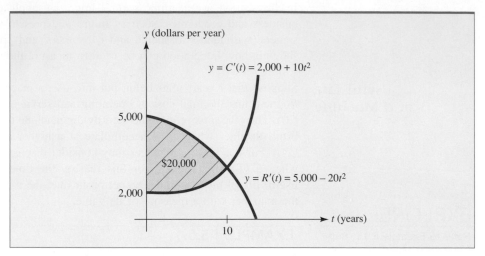

FIGURE 5.17 Net profit over the useful life of a machine.

Future Value and Present Value of an Income Flow

In our next application, we consider a stream of income transferred continuously into an account in which it earns interest over a specified time period (the **term**). Then the **future value of the income stream** is the total amount (money transferred into the account plus interest) that is accumulated in this way during the specified term.

The calculation of the amount of an income stream is illustrated in Example 5.5.2. The strategy is to approximate the continuous income stream by a sequence of discrete deposits called an **annuity**. The amount of the approximating annuity is a certain sum whose limit (a definite integral) is the future value of the income stream.

EXAMPLE | 5.5.2

Money is transferred continuously into an account at the constant rate of $1,200 per year. The account earns interest at the annual rate of 8% compounded continuously. How much will be in the account at the end of 2 years?

Solution

Recall from Section 4.1 that P dollars invested at 8% compounded continuously will be worth $Pe^{0.08t}$ dollars t years later.

To approximate the future value of the income stream, divide the 2-year time interval $0 \leq t \leq 2$ into n equal subintervals of length Δt years and let t_j denote the beginning of the jth subinterval. Then, during the jth subinterval (of length Δt years),

$$\text{Money deposited} = (\text{dollars per year})(\text{number of years}) = 1{,}200\,\Delta t$$

If all of this money were deposited at the beginning of the subinterval (at time t_j), it would remain in the account for $2 - t_j$ years and therefore would grow to $(1{,}200\,\Delta t)e^{0.08(2 - t_j)}$ dollars. Thus,

$$\begin{array}{c}\text{Future value of money deposited} \\ \text{during } j\text{th subinterval}\end{array} \approx 1{,}200e^{0.08(2 - t_j)}\Delta t$$

The situation is illustrated in Figure 5.18.

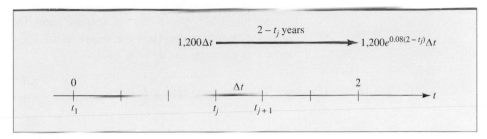

FIGURE 5.18   The (approximate) future value of the money deposited during the jth subinterval.

The future value of the entire income stream is the sum of the future values of the money deposited during each of the n subintervals. Hence,

$$\text{Future value of income stream} \approx \sum_{j=1}^{n} 1,200 e^{0.08(2-t_j)} \Delta t$$

(Note that this is only an approximation because it is based on the assumption that all $1,200 \, \Delta t_n$ dollars are deposited at time t_j rather than continuously throughout the jth subinterval.)

As n increases without bound, the length of each subinterval approaches zero and the approximation approaches the true future value of the income stream. Hence,

$$\begin{aligned}
\text{Future value of income stream} &= \lim_{n \to +\infty} \sum_{j=1}^{n} 1,200 e^{0.08(2-t_j)} \Delta t \\
&= \int_0^2 1,200 e^{0.08(2-t)} \, dt = 1,200 e^{0.16} \int_0^2 e^{-0.08t} \, dt \\
&= -\frac{1,200}{0.08} e^{0.16} (e^{-0.08t}) \Big|_0^2 = -15,000 e^{0.16}(e^{-0.16} - 1) \\
&= -15,000 + 15,000 e^{0.16} \approx \$2,602.66
\end{aligned}$$

By generalizing the reasoning illustrated in Example 5.5.2, we are led to this integration formula for the future value of an income stream with rate of flow given by $f(t)$ for a term of T years:

$$\begin{aligned}
FV &= \int_0^T f(t) \, e^{r(T-t)} \, dt \\
&= \int_0^T f(t) \, e^{rT} e^{-rt} \, dt \\
&= e^{rT} \int_0^T f(t) \, e^{-rt} \, dt \quad \textit{factor constant } e^{rT} \textit{ outside integral}
\end{aligned}$$

The first and last forms of the formula for future value are both listed next for future reference.

Future Value of an Income Stream ■ Suppose money is being transferred continuously into an account over a time period $0 \le t \le T$ at a rate given by the function $f(t)$ and that the account earns interest at an annual rate r compounded continuously. Then the future value FV of the income stream over the term T is given by the definite integral

$$FV = \int_0^T f(t)\, e^{r(T-t)}\, dt = e^{rT} \int_0^T f(t)\, e^{-rt}\, dt$$

In Example 5.5.2, we had $f(t) = 1{,}200$, $r = 0.08$, and $T = 2$, so that

$$FV = e^{0.08(2)} \int_0^2 1{,}200 e^{-0.08t}\, dt$$

The **present value** of an income stream generated at a continuous rate $f(t)$ over a specified term of T years is the amount of money A that must be deposited now at the prevailing interest rate to generate the same income as the income stream over the same T-year period. Since A dollars invested at an annual interest rate r compounded continuously will be worth Ae^{rT} dollars in T years, we must have

$$Ae^{rT} = e^{rT} \int_0^T f(t)\, e^{-rt}\, dt$$

$$A = \int_0^T f(t)\, e^{-rt}\, dt \qquad \textit{divide both sides by } e^{rT}$$

To summarize:

Present Value of an Income Stream ■ The **present value** PV of an income stream that is deposited continuously at the rate $f(t)$ into an account that earns interest at an annual rate r compounded continuously for a term of T years is given by

$$PV = \int_0^T f(t)\, e^{-rt}\, dt$$

Example 5.5.3 illustrates how present value can be used in making certain financial decisions.

EXAMPLE 5.5.3

Jane is trying to decide between two investments. The first costs \$1,000 and is expected to generate a continuous income stream at the rate of $f_1(t) = 3{,}000 e^{0.03t}$ dollars per year. The second investment costs \$4,000 and is estimated to generate income at the constant rate of $f_2(t) = 4{,}000$ dollars per year. If the prevailing annual interest rate remains fixed at 5% compounded continuously over the next 5 years, which investment is better over this time period?

Solution

The net value of each investment over the 5-year time period is the present value of the investment less its initial cost. For each investment, we have $r = 0.05$ and $T = 5$.

For the first investment:

$$PV - cost = \int_0^5 (3,000e^{0.03t})e^{-0.05t}\, dt - 1,000$$

$$= 3,000 \int_0^5 e^{0.03t - 0.05t}\, dt - 1,000$$

$$= 3,000 \int_0^5 e^{-0.02t}\, dt - 1,000$$

$$= 3,000 \left(\frac{e^{-0.02t}}{-0.02} \right) \Big|_0^5 - 1,000$$

$$= -150,000[e^{-0.02(5)} - e^0] - 1,000$$

$$= 13,274.39$$

For the second investment:

$$PV - cost = \int_0^5 (4,000)e^{-0.05t}\, dt - 4,000$$

$$= 4,000 \left(\frac{e^{-0.05t}}{-0.05} \right) \Big|_0^5 - 4,000$$

$$= -80,000[e^{-0.05(5)} - e^0] - 4,000$$

$$= 13,695.94$$

Thus, the net value of the first investment is $13,274.39 and of the second is $13,695.04. The second investment is slightly better.

The Consumers' Demand Curve and Willingness to Spend

In studying consumer behavior, economists often assume that the price a consumer or group of consumers is willing to pay to buy an additional unit of a commodity is a function of the number of units of the commodity that the consumer or group has already bought.

For example, a family might be willing to spend up to $200 to own one television set. For the convenience of having two sets (to eliminate the conflict between *Monday Night Football* and *American Idol,* for instance), the family might be willing to spend an additional $100 to buy a second set. Since there would be very little use for more than two sets, they might be willing to spend no more than $30 for a third set.

A function $p = D(q)$ giving the price per unit that consumers are willing to pay to get the qth unit of a commodity is known in economics as the **consumers' demand function** for the commodity. As illustrated in Figure 5.19, the consumers' demand function is usually a decreasing function of q. That is, the price that consumers are willing to pay to get one additional unit usually decreases as the number of units already bought increases.

The consumers' demand function $p = D(q)$ can also be thought of as the rate of change with respect to q of the total amount $A(q)$ that consumers are willing to spend for q units; that is, dA/dq. Integrating, you find that the total amount that consumers are willing to pay for q_0 units of the commodity is given by

$$A(q_0) - A(0) = \int_0^{q_0} \frac{dA}{dq}\, dq = \int_0^{q_0} D(q)\, dq$$

In this context, economists call $A(q)$ the **total willingness to spend** and $D(q) = A'(q)$ the **marginal willingness to spend.** In geometric terms, the total willingness to spend for q_0 units is the area under the demand curve $p = D(q)$ between $q = 0$ and $q = q_0$ (see Figure 5.19).

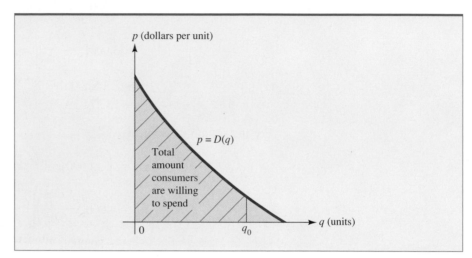

FIGURE 5.19   The amount consumers are willing to spend is the area under the demand curve.

EXPLORE!

In Example 5.5.4, change $D(q)$ to $D_{\text{new}}(q) = 4(23 - q^2)$. Will the amount of money consumers are willing to spend to obtain 3 units of the commodity increase or decrease? Graph $D_{\text{new}}(q)$ in bold to compare with the graph of $D(q)$, using the viewing window $[0, 5]1$ by $[0, 150]10$.

EXAMPLE   5.5.4

Suppose that the consumers' demand function for a certain commodity is $D(q) = 4(25 - q^2)$ dollars per unit.

a. Find the total amount of money consumers are willing to spend to get 3 units of the commodity.

b. Sketch the demand curve and interpret the answer to part (a) as an area.

Solution

a. Since the demand function $D(q) = 4(25 - q^2)$, measured in dollars per unit, is the rate of change with respect to q of consumers' willingness to spend, the total amount that consumers are willing to spend to get 3 units of the commodity is given by the definite integral

$$A(3) = \int_0^3 D(q)\, dq = 4 \int_0^3 (25 - q^2)\, dq$$

$$= 4\left(25q - \frac{1}{3}q^3\right)\Big|_0^3 = \$264$$

b. The consumers' demand curve is sketched in Figure 5.20. In geometric terms, the total amount, $264, that consumers are willing to spend to get 3 units of the commodity is the area under the demand curve from $q = 0$ to $q = 3$.

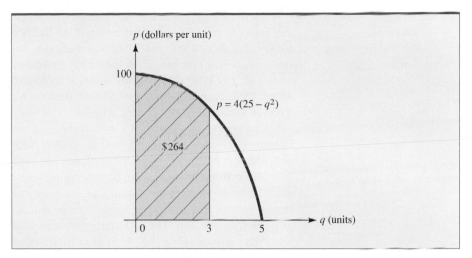

FIGURE 5.20 Consumers' willingness to spend for 3 units when demand is given by $D(q) = 4(25 - q^2)$.

Consumers' and Producers' Surplus

In a competitive economy, the total amount that consumers actually spend on a commodity is usually less than the total amount they would have been willing to spend. The difference between the two amounts can be thought of as a savings realized by consumers and is known in economics as the **consumers' surplus.** That is,

$$\begin{bmatrix} \text{Consumers'} \\ \text{surplus} \end{bmatrix} = \begin{pmatrix} \text{total amount consumers} \\ \text{would be willing to spend} \end{pmatrix} - \begin{pmatrix} \text{actual consumer} \\ \text{expenditure} \end{pmatrix}$$

Market conditions determine the price per unit at which a commodity is sold. Once the price, say p_0, is known, the demand equation $p = D(q)$ determines the number of units q_0 that consumers will buy. The actual consumer expenditure for q_0 units of the commodity at the price of p_0 dollars per unit is $p_0 q_0$ dollars. The consumers' surplus is calculated by subtracting this amount from the total amount consumers would have been willing to spend to get q_0 units of the commodity.

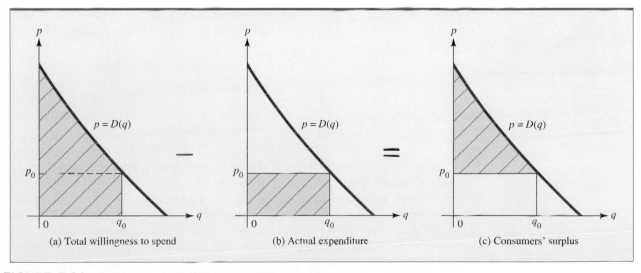

FIGURE 5.21 Geometric interpretation of consumers' surplus.

To get a better feel for the concept of consumers' surplus, consider once again the example of the family that was willing to spend $200 for its first television set, $100 for a second set, and $30 for a third set. Suppose the market price for television sets is $100 per set. Then the family would buy only two sets and would spend a total of $2 \times \$100 = \$200$. This is less than the $\$200 + \$100 = \$300$ that the family would have been willing to spend to get the two sets. The savings of $\$300 - \$200 = \$100$ is the family's consumers' surplus.

Consumers' surplus has a simple geometric interpretation, which is illustrated in Figure 5.21. The symbols p_0 and q_0 denote the market price and corresponding demand, respectively. Figure 5.21a shows the region under the demand curve from $q = 0$ to $q = q_0$. Its area, as we have seen, represents the total amount that consumers are willing to spend to get q_0 units of the commodity. The rectangle in Figure 5.21b has an area of $p_0 q_0$ and hence represents the actual consumer expenditure for q_0 units at p_0 dollars per unit. The difference between these two areas (Figure 5.21c) represents the consumers' surplus. That is, consumers' surplus CS is the area of the region between the demand curve $p = D(q)$ and the horizontal line $p = p_0$ and hence is equal to the definite integral

$$\text{CS} = \int_0^{q_0} [D(q) - p_0] \, dq = \int_0^{q_0} D(q) \, dq - \int_0^{q_0} p_0 \, dq$$

$$= \int_0^{q_0} D(q) \, dq - p_0 q \Big|_0^{q_0}$$

$$= \int_0^{q_0} D(q) \, dq - p_0 q_0$$

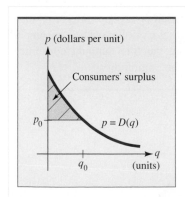

Consumers' Surplus ■ If q_0 units of a commodity are sold at a price of p_0 per unit and if $p = D(q)$ is the consumers' demand function for the commodity, then

$$\begin{bmatrix} \text{Consumers'} \\ \text{surplus} \end{bmatrix} = \begin{bmatrix} \text{total amount consumers} \\ \text{are willing to spend} \\ \text{for } q_0 \text{ units} \end{bmatrix} - \begin{bmatrix} \text{actual consumer} \\ \text{expenditure} \\ \text{for } q_0 \text{ units} \end{bmatrix}$$

$$\text{CS} = \int_0^{q_0} D(q) \, dq - p_0 q_0$$

Producers' surplus is the other side of the coin from consumers' surplus. In particular, a **supply function** $p = S(q)$ gives the price per unit that producers are willing to accept in order to supply q units to the marketplace. However, any producer who is willing to accept less than $p_0 = S(q_0)$ dollars for q_0 units gains from the fact that the price is p_0. Then producers' surplus is the difference between what producers would be willing to accept for supplying q_0 units and the price they actually receive. Reasoning as we did with consumers' surplus, we obtain the following formula for producers' surplus.

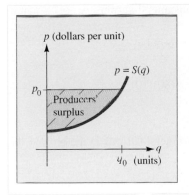

Producers' Surplus ■ If q_0 units of a commodity are sold at a price of p_0 dollars per unit and $p = S(q)$ is the producers' supply function for the commodity, then

$$\begin{bmatrix} \text{Producers'} \\ \text{surplus} \end{bmatrix} - \begin{bmatrix} \text{actual consumer} \\ \text{expenditure} \\ \text{for } q_0 \text{ units} \end{bmatrix} - \begin{bmatrix} \text{total amount producers} \\ \text{receive when } q_0 \\ \text{units are supplied} \end{bmatrix}$$

$$PS - p_0 q_0 \qquad \int_0^{q_0} S(q)\, dq$$

EXAMPLE 5.5.5

A tire manufacturer estimates that q (thousand) radial tires will be purchased (demanded) by wholesalers when the price is

$$p = D(q) = -0.1q^2 + 90$$

dollars per tire, and the same number of tires will be supplied when the price is

$$p = S(q) = 0.2q^2 + q + 50$$

dollars per tire.

a. Find the equilibrium price (where supply equals demand) and the quantity supplied and demanded at that price.

b. Determine the consumers' and producers' surplus at the equilibrium price.

Solution

a. The supply and demand curves are shown in Figure 5.22. Supply equals demand when

$$-0.1q^2 + 90 = 0.2q^2 + q + 50$$
$$0.3q^2 + q - 40 = 0$$
$$q = 10 \quad (\text{reject } q \approx -13.33)$$

and $p = -0.1(10)^2 + 90 - 80$ dollars per tire. Thus, equilibrium occurs at a price of \$80 per tire, and then 10,000 tires are supplied and demanded.

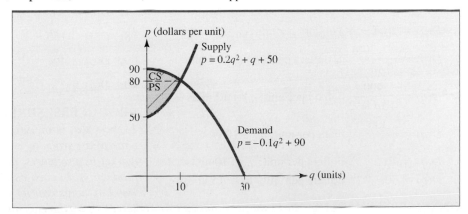

FIGURE 5.22 Consumers' surplus and producers' surplus for the demand and supply functions in Example 5.5.5.

b. Using $p_0 = 80$ and $q_0 = 10$, we find that the consumers' surplus is

$$CS = \int_0^{10} (-0.1q^2 + 90)\, dq - (80)(10)$$

$$= \left[-0.1\left(\frac{q^3}{3}\right) + 90q \right]\Big|_0^{10} - (80)(10)$$

$$\approx 866.67 - 800 = 66.67$$

or \$66,670 (since $q_0 = 10$ is really 10,000). The consumers' surplus is the area of the shaded region labeled CS in Figure 5.22.

The producers' surplus is

$$PS = (80)(10) - \int_0^{10} (0.2q^2 + q + 50)\, dq$$

$$= (80)(10) - \left[0.2\left(\frac{q^3}{3}\right) + \left(\frac{q^2}{2}\right) + 50q \right]\Big|_0^{10}$$

$$\approx 800 - 616.67 = 183.33$$

or \$183,330. The producers' surplus is the area of the shaded region labeled PS in Figure 5.22.

PROBLEMS 5.5

CONSUMERS' WILLINGNESS TO SPEND *For the consumers' demand functions $D(q)$ in Problems 1 through 6:*

(a) Find the total amount of money consumers are willing to spend to get q_0 units of the commodity.

(b) Sketch the demand curve and interpret the consumers' willingness to spend in part (a) as an area.

1. $D(q) = 2(64 - q^2)$ dollars per unit; $q_0 = 6$ units

2. $D(q) = \dfrac{300}{(0.1q + 1)^2}$ dollars per unit; $q_0 = 5$ units

3. $D(q) = \dfrac{400}{0.5q + 2}$ dollars per unit; $q_0 = 12$ units

4. $D(q) = \dfrac{300}{4q + 3}$ dollars per unit; $q_0 = 10$ units

5. $D(q) = 40e^{-0.05q}$ dollars per unit; $q_0 = 10$ units

6. $D(q) = 50e^{-0.04q}$ dollars per unit; $q_0 = 15$ units

CONSUMERS' SURPLUS *In Problems 7 through 10, $p = D(q)$ is the price (dollars per unit) at which q units of a particular commodity will be demanded by*

the market (that is, all q units will be sold at this price), and q_0 is a specified level of production. In each case, find the price $p_0 = D(q_0)$ at which q_0 units will be demanded and compute the corresponding consumers' surplus CS. Sketch the demand curve $y = D(q)$ and shade the region whose area represents the consumers' surplus.

7. $D(q) = 2(64 - q^2)$; $q_0 = 3$ units

8. $D(q) = 150 - 2q - 3q^2$; $q_0 = 6$ units

9. $D(q) = 40e^{-0.05q}$; $q_0 = 5$ units

10. $D(q) = 75e^{-0.04q}$; $q_0 = 3$ units

PRODUCERS' SURPLUS *In Problems 11 through 14, $p = S(q)$ is the price (dollars per unit) at which q units of a particular commodity will be supplied to the market by producers, and q_0 is a specified level of production. In each case, find the price $p_0 = S(q_0)$ at which q_0 units will be supplied and compute the corresponding producers' surplus PS. Sketch the supply curve $y = S(q)$ and shade the region whose area represents the producers' surplus.*

11. $S(q) = 0.3q^2 + 30$; $q_0 = 4$ units

12. $S(q) = 0.5q + 15$; $q_0 = 5$ units

13. $S(q) = 10 + 15e^{0.03q}$; $q_0 = 3$ units

14. $S(q) = 17 + 11e^{0.01q}$; $q_0 = 7$ units

CONSUMERS' AND PRODUCERS' SURPLUS AT EQUILIBRIUM
In Problems 15 through 19, the demand and supply functions, $D(q)$ and $S(q)$, for a particular commodity are given. Specifically, q thousand units of the commodity will be demanded (sold) at a price of $p = D(q)$ dollars per unit, while q thousand units will be supplied by producers when the price is $p = S(q)$ dollars per unit. In each case:
  (a) Find the equilibrium price p_e (where supply equals demand).
  (b) Find the consumers' surplus and the producers' surplus at equilibrium.

15. $D(q) = 131 - \frac{1}{3}q^2$; $S(q) = 50 + \frac{2}{3}q^2$

16. $D(q) = 65 - q^2$; $S(q) = \frac{1}{3}q^2 + 2q + 5$

17. $D(q) = -0.3q^2 + 70$; $S(q) = 0.1q^2 + q + 20$

18. $D(q) = \sqrt{245 - 2q}$; $S(q) = 5 + q$

19. $D(q) = \frac{16}{q + 2} - 3$; $S(q) = \frac{1}{3}(q + 1)$

20. **PROFIT OVER THE USEFUL LIFE OF A MACHINE** Suppose that when it is t years old, a particular industrial machine generates revenue at the rate $R'(t) = 6,025 - 8t^2$ dollars per year and that operating and servicing costs accumulate at the rate $C'(t) = 4,681 + 13t^2$ dollars per year.
  a. How many years pass before the profitability of the machine begins to decline?
  b. Compute the net profit generated by the machine over its useful lifetime.
  c. Sketch the revenue rate curve $y = R'(t)$ and the cost rate curve $y = C'(t)$ and shade the region whose area represents the net profit computed in part (b).

21. **PROFIT OVER THE USEFUL LIFE OF A MACHINE** Answer the questions in Problem 20 for a machine that generates revenue at the rate $R'(t) = 7,250 - 18t^2$ dollars per year and for which costs accumulate at the rate $C'(t) = 3,620 + 12t^2$ dollars per year.

22. **FUND-RAISING** It is estimated that t weeks from now, contributions in response to a fund-raising campaign will be coming in at the rate of $R'(t) = 5,000e^{-0.2t}$ dollars per week, while campaign expenses are expected to accumulate at the constant rate of $676 per week.
  a. For how many weeks does the rate of revenue exceed the rate of cost?
  b. What net earnings will be generated by the campaign during the period of time determined in part (a)?
  c. Interpret the net earnings in part (b) as an area between two curves.

23. **FUND-RAISING** Answer the questions in Problem 22 for a charity campaign in which contributions are made at the rate of $R'(t) = 6,537e^{-0.3t}$ dollars per week and expenses accumulate at the constant rate of $593 per week.

24. **THE AMOUNT OF AN INCOME STREAM** Money is transferred continuously into an account at the constant rate of $2,400 per year. The account earns interest at the annual rate of 6% compounded continuously. How much will be in the account at the end of 5 years?

25. **THE AMOUNT OF AN INCOME STREAM** Money is transferred continuously into an account at the constant rate of $1,000 per year. The account earns interest at the annual rate of 10% compounded continuously. How much will be in the account at the end of 10 years?

26. **CONSTRUCTION DECISION** Magda wants to expand and renovate her import store, and is presented with two plans for making the improvements. The first plan will cost her $40,000 and the second will cost only $25,000. However, she expects the improvements resulting from the first plan to provide income at the continuous rate of $10,000 per year, while the income flow from the second plan provides $8,000 per year. Which plan will result in more net income over the next 3 years if the prevailing rate of interest is 5% per year compounded continuously?

27. **RETIREMENT ANNUITY** At age 25, Tom starts making annual deposits of $2,500 into an IRA account that pays interest at an annual rate of 5% compounded continuously. Assuming that his payments are made as a continuous income flow, how much money will be in his account if he retires at age 60? At age 65?

28. RETIREMENT ANNUITY When she is 30, Sue starts making annual deposits of $2,000 into a bond fund that pays 8% annual interest compounded continuously. Assuming that her deposits are made as a continuous income flow, how much money will be in her account if she retires at age 55?

29. THE PRESENT VALUE OF AN INVESTMENT An investment will generate income continuously at the constant rate of $1,200 per year for 5 years. If the prevailing annual interest rate remains fixed at 5% compounded continuously, what is the present value of the investment?

30. THE PRESENT VALUE OF A FRANCHISE The management of a national chain of fast-food outlets is selling a 10-year franchise in Cleveland, Ohio. Past experience in similar localities suggests that t years from now the franchise will be generating profit at the rate of $f(t) = 10,000$ dollars per year. If the prevailing annual interest rate remains fixed at 4% compounded continuously, what is the present value of the franchise?

31. INVESTMENT ANALYSIS Adam is trying to choose between two investment opportunities. The first will cost $50,000 and is expected to produce income at the continuous rate of $15,000 per year. The second will cost $30,000 and is expected to produce income at the rate of $9,000 per year. If the prevailing rate of interest stays constant at 6% per year compounded continuously, which investment is better over the next 5 years?

32. INVESTMENT ANALYSIS Kevin spends $4,000 for an investment that generates a continuous income stream at the rate of $f_1(t) = 3,000$ dollars per year. His friend, Molly, makes a separate investment that also generates income continuously, but at a rate of $f_2(t) = 2,000e^{0.04t}$ dollars per year. The couple discovers that their investments have exactly the same net value over a 4-year period. If the prevailing annual interest rate stays fixed at 5% compounded continuously, how much did Molly pay for her investment?

33. CONSUMERS' SURPLUS A manufacturer of machinery parts determines that q units of a particular piece will be sold when the price is $p = 110 - q$ dollars per unit. The total cost of producing those q units is $C(q)$ dollars, where

$$C(q) = q^3 - 25q^2 + 2q + 3,000$$

a. How much profit is derived from the sale of the q units at p dollars per unit? [*Hint:* First find the revenue $R = pq$; then profit = revenue − cost.]

b. For what value of q is profit maximized?

c. Find the consumers' surplus for the level of production q_0 that corresponds to maximum profit.

34. CONSUMERS' SURPLUS Repeat Problem 33 for $p = 124 - 2q$ and $C(q) = 2q^3 - 59q^2 + 4q + 7,600$.

35. DEPLETION OF ENERGY RESOURCES Oil is being pumped from an oil field t years after its opening at the rate of $P'(t) = 1.3e^{0.04t}$ billion barrels per year. The field has a reserve of 20 billion barrels, and the price of oil holds steady at $56 per barrel.

a. Find $P(t)$, the amount of oil pumped from the field at time t. How much oil is pumped from the field during the first 3 years of its operation? The next 3 years?

b. For how many years T does the field operate before it runs dry?

c. If the prevailing annual interest rate stays fixed at 5% compounded continuously, what is the present value of the continuous income stream $V = 56P'(t)$ over the period of operation of the field $0 \leq t \leq T$?

d. If the owner of the oil field decides to sell on the first day of operation, do you think the present value determined in part (c) would be a fair asking price? Explain your reasoning.

36. DEPLETION OF ENERGY RESOURCES Answer the questions in Problem 35 for another oil field with a pumping rate of $P'(t) = 1.5e^{0.03t}$ and with a reserve of 16 billion barrels. You may assume that the price of oil is still $56 per barrel and that the prevailing annual interest rate is 5%.

37. DEPLETION OF ENERGY RESOURCES Answer the questions in Problem 35 for an oil field with a pumping rate of $P'(t) = 1.2e^{0.02t}$ and with a reserve of 12 billion barrels. Assume that the prevailing interest rate is 5% as before, but that the price of oil after t years is given by $A(t) = 56e^{0.015t}$.

38. LOTTERY PAYOUT A state lottery pays the winner of a $2 million lottery in 10 annual installments of $200,000 apiece. What is the present value of this income stream if the prevailing rate of interest is 5% per year, compounded continuously?

39. LOTTERY PAYOUT The winner of a state lottery is offered a choice of either receiving $10 million now as a lump sum or of receiving A dollars a year for the next 6 years as a continuous income stream. If the prevailing annual interest rate is 5% compounded continuously and the two payouts are worth the same, what is A?

40. SPORTS CONTRACTS A star baseball free agent is the object of a bidding war between two rival teams. The first team offers a 3 million dollar signing bonus and a 5-year contract guaranteeing him 8 million dollars this year and an increase of 3% per year for the remainder of the contract. The second team offers $9 million per year for 5 years with no incentives. If the prevailing annual interest rate stays fixed at 4% compounded continuously, which offer is worth more? [*Hint:* Assume that with both offers, the salary is paid as a continuous income stream.]

41. PRESENT VALUE OF AN INVESTMENT An investment produces a continuous income stream at the rate of $A(t)$ thousand dollars per year at time t, where

$$A(t) = 10e^{1-0.05t}$$

The prevailing rate of interest is 5% per year compounded continuously.

a. What is the future value of the investment over a term of 5 years ($0 \le t \le 5$)?

b. What is the present value of the investment over the time period $1 \le t \le 3$?

42. PROFIT FROM AN INVENTION A marketing survey indicates that t months after a new type of computerized air purifier is introduced to the market, sales will be generating profit at the rate of $P'(t)$ thousand dollars per month, where

$$P'(t) = \frac{500[1.4 - \ln(0.5t + 1)]}{t + 2}$$

a. When is the rate of profitability positive and when is it negative? When is the rate increasing and when is it decreasing?

b. At what time $t = t_m$ is monthly profit maximized? Find the net change in profit over the time period $0 \le t \le t_m$.

c. It costs the manufacturer $100,000 to develop the purifier product, so $P(0) = -100$. Use this information along with integration to find $P(t)$.

d. Sketch the graph of $P(t)$ for $t \ge 0$. A "fad" is a product that gains rapid success in the market, then just as quickly, fades from popularity. Based on the graph $P(t)$, would you call the purifiers a "fad"? Explain.

43. TOTAL REVENUE Consider the following problem: "A certain oil well that yields 300 barrels of crude oil a month will run dry in 3 years. It is estimated that t months from now the price of crude oil will be $P(t) = 18 + 0.3\sqrt{t}$ dollars per barrel. If the oil is sold as soon as it is extracted from the ground, what will be the total future revenue from the well?"

a. Solve the problem using definite integration. [*Hint:* Divide the 3-year (36-month) time interval $0 \le t \le 36$ into n equal subintervals of length Δt and let t_j denote the beginning of the jth subinterval. Find an expression that estimates the revenue $R(t_j)$ obtained during the jth subinterval. Then express the total revenue as the limit of a sum.]

 b. Read an article on the petroleum industry and write a paragraph on mathematical methods of modeling oil production.*

44. INVENTORY STORAGE COSTS A manufacturer receives N units of a certain raw material that are initially placed in storage and then withdrawn and used at a constant rate until the supply is exhausted 1 year later. Suppose storage costs remain fixed at p dollars per unit per year. Use definite integration to find an expression for the total storage cost the manufacturer will pay during the year. [*Hint:* Let $Q(t)$ denote the number of units in storage after t years and find an expression for $Q(t)$. Then subdivide the interval $0 \le t \le 1$ into n equal subintervals and express the total storage cost as the limit of a sum.]

---

*A good place to start is the article by J. A. Weyland and D. W. Ballew, "A Relevant Calculus Problem: Estimation of U.S. Oil Reserves," *The Mathematics Teacher*, Vol. 69, 1976, pp. 125–126.

SECTION 5.6 | Additional Applications to the Life and Social Sciences

We have already seen how definite integration can be used to compute quantities of interest in the social and life sciences, such as net change, average value, and the Gini index of a Lorentz curve. In this section, we examine several additional such applications, including survival and renewal within a group, blood flow through an artery, and cardiac output. We shall also discuss how volume can be computed using integration and used for purposes such as measuring the size of a lake or a tumor.

Survival and Renewal

In Example 5.6.1, a **survival function** gives the fraction of individuals in a group or population that can be expected to remain in the group for any specified period of time. A **renewal function** giving the rate at which new members arrive is also known, and the goal is to predict the size of the group at some future time. Problems of this type arise in many fields, including sociology, ecology, demography, and even finance, where the "population" is the number of dollars in an investment account and "survival and renewal" refer to features of an investment strategy.

EXAMPLE | 5.6.1

A new county mental health clinic has just opened. Statistics from similar facilities suggest that the fraction of patients who will still be receiving treatment at the clinic t months after their initial visit is given by the function $f(t) = e^{-t/20}$. The clinic initially accepts 300 people for treatment and plans to accept new patients at the constant rate of $g(t) = 10$ patients per month. Approximately how many people will be receiving treatment at the clinic 15 months from now?

Solution

Since $f(15)$ is the fraction of patients whose treatment continues at least 15 months, it follows that of the current 300 patients, only $300f(15)$ will still be receiving treatment 15 months from now.

   To approximate the number of *new* patients who will be receiving treatment 15 months from now, divide the 15-month time interval $0 \le t \le 15$ into n equal subintervals of length Δt months and let t_j denote the beginning of the jth subinterval. Since new patients are accepted at the rate of 10 per month, the number of new patients accepted during the jth subinterval is $10\Delta t$. Fifteen months from now, approximately $15 - t_j$ months will have elapsed since these $10\Delta t$ new patients had their initial visits, and so approximately $(10\Delta t)f(15 - t_j)$ of them will still be receiving treatment at that time (see Figure 5.23). It follows that the total number of new patients still receiving

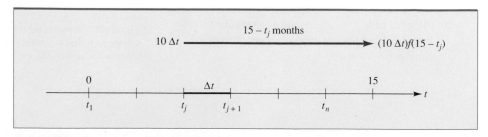

FIGURE 5.23   New members arriving during the jth subinterval.

treatment 15 months from now can be approximated by the sum

$$\sum_{j=1}^{n} 10f(15 - t_j)\Delta t$$

Adding this to the number of current patients who will still be receiving treatment in 15 months, you get

$$P \approx 300f(15) + \sum_{j=1}^{n} 10f(15 - t_j)\Delta t$$

where P is the total number of *all* patients (current and new) who will be receiving treatment 15 months from now.

As n increases without bound, the approximation improves and approaches the true value of P. It follows that

$$P = 300f(15) + \lim_{n \to +\infty} \sum_{j=1}^{n} 10f(15 - t_j)\Delta t$$

$$= 300f(15) + \int_0^{15} 10f(15 - t)\, dt$$

Since $f(t) = e^{-t/20}$, we have $f(15) = e^{-3/4}$ and $f(15 - t) = e^{-(15-t)/20} = e^{-3/4}e^{t/20}$. Hence,

$$P = 300e^{-3/4} + 10e^{-3/4}\int_0^{15} e^{t/20}\, dt$$

$$= 300e^{-3/4} + 10e^{-3/4}\left(\frac{e^{t/20}}{1/20}\right)\Big|_0^{15}$$

$$= 300e^{-3/4} + 200(1 - e^{-3/4})$$

$$\approx 247.24$$

That is, 15 months from now, the clinic will be treating approximately 247 patients.

In Example 5.6.1, we considered a variable survival function $f(t)$ and a constant renewal rate function $g(t)$. Essentially the same analysis applies when the renewal function also varies with time. Here is the result. Note that for definiteness, time is given in years, but the same basic formula would also apply for other units of time, for example, minutes, weeks, or months, as in Example 5.6.1.

Survival and Renewal ■ Suppose a population initially has P_0 members and that new members are added at the (renewal) rate of $R(t)$ individuals per year. Further suppose that the fraction of the population that remain for at least t years after arriving is given by the (survival) function $S(t)$. Then, at the end of a term of T years, the population will be

$$P(T) = P_0 S(T) + \int_0^T R(t)\, S(T - t)\, dt$$

In Example 5.6.1, each time period is 1 month, the initial "population" (membership) is $P_0 = 300$, the renewal rate is $R = 10$, the survival function is $f(t) = e^{-t/20}$, and the term is $T = 15$ months. Here is another example of survival/renewal, from biology.

EXAMPLE | 5.6.2

A mild toxin is introduced to a bacterial colony whose current population is 600,000. Observations indicate that $R(t) = 200\ e^{0.01t}$ bacteria per hour are born in the colony at time t and that the fraction of the population that survives for t hours after birth is $S(t) = e^{-0.015t}$. What is the population of the colony after 10 hours?

Solution

Substituting $P_0 = 600{,}000$, $R(t) = 200\ e^{0.01t}$, and $S(t) = e^{-0.015t}$ into the formula for survival and renewal, we find that the population at the end of the term of $T = 10$ hours is

$$P(10) = \underbrace{600{,}000}_{P_0}\underbrace{e^{-0.015(10)}}_{S(10)} + \int_0^{10} \underbrace{200\ e^{0.01t}}_{R(t)}\ \underbrace{e^{-0.015(10-t)}}_{S(T-t)}\ dt$$

$$\approx 516{,}425 + \int_0^{10} 200 e^{0.01t}[e^{-0.015(10)}\ e^{0.015t}]\ dt \qquad \text{since } e^{a-b} = e^a\ e^{-b}$$

$$\approx 516{,}425 + 200\ e^{-0.015(10)}\int_0^{10}[e^{0.01t}\ e^{0.015t}]\ dt \qquad \begin{array}{l}\text{factor } 200\ e^{-0.015(10)}\\ \text{outside the integral}\end{array}$$

$$\approx 516{,}425 + 172.14\int_0^{10} e^{0.025t}\ dt \qquad \begin{array}{l}\text{since } e^{a+b} = e^a\ e^b \text{ and}\\ 200\ e^{-0.015(10)} = 172.14\end{array}$$

$$\approx 516{,}425 + 172.14\left[\frac{e^{0.025t}}{0.025}\right]\Bigg|_0^{10} \qquad \begin{array}{l}\text{exponential rule for}\\ \text{integration}\end{array}$$

$$\approx 516{,}425 + \frac{172.14}{0.025}[e^{0.025(10)} - e^0]$$

$$\approx 518{,}381$$

Thus, the population of the colony declines from 600,000 to about 518,381 during the first 10 hours after the toxin is introduced.

The Flow of Blood through an Artery

Biologists have found that the speed of blood in an artery is a function of the distance of the blood from the artery's central axis. According to Poiseuille's law, the speed (in centimeters per second) of blood that is r centimeters from the central axis of the artery is $S(r) = k(R^2 - r^2)$, where R is the radius of the artery and k is a constant. In Example 5.6.3, you will see how to use this information to compute the rate at which blood flows through the artery.

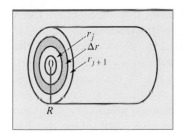

FIGURE 5.24 Subdividing a cross section of an artery into concentric rings.

EXAMPLE 5.6.3

Find an expression for the rate (in cubic centimeters per second) at which blood flows through an artery of radius R if the speed of the blood r centimeters from the central axis is $S(r) = k(R^2 - r^2)$, where k is a constant.

Solution

To approximate the volume of blood that flows through a cross section of the artery each second, divide the interval $0 \leq r \leq R$ into n equal subintervals of width Δr centimeters and let r_j denote the beginning of the jth subinterval. These subintervals determine n concentric rings as illustrated in Figure 5.24.

If Δr is small, the area of the jth ring is approximately the area of a rectangle whose length is the circumference of the (inner) boundary of the ring and whose width is Δr. That is,

$$\text{Area of } j\text{th ring} \approx 2\pi r_j \Delta r$$

If you multiply the area of the jth ring (square centimeters) by the speed (centimeters per second) of the blood flowing through this ring, you get the rate (cubic centimeters per second) at which blood flows through the jth ring. Since the speed of blood flowing through the jth ring is approximately $S(r_j)$ centimeters per second, it follows that

$$\begin{pmatrix} \text{Rate of flow} \\ \text{through } j\text{th ring} \end{pmatrix} \approx \begin{pmatrix} \text{area of} \\ j\text{th ring} \end{pmatrix}\begin{pmatrix} \text{speed of blood} \\ \text{through } j\text{th ring} \end{pmatrix}$$

$$\approx (2\pi r_j \Delta r) S(r_j)$$

$$\approx (2\pi r_j \Delta r)[k(R^2 - r_j^2)]$$

$$\approx 2\pi k(R^2 r_j - r_j^3)\Delta r$$

The rate of flow of blood through the entire cross section is the sum of n such terms, one for each of the n concentric rings. That is,

$$\text{Rate of flow} \approx \sum_{j=1}^{n} 2\pi k(R^2 r_j - r_j^3)\Delta r$$

As n increases without bound, this approximation approaches the true value of the rate of flow. In other words,

$$\text{Rate of flow} = \lim_{n \to +\infty} \sum_{j=1}^{n} 2\pi k(R^2 r_j - r_j^3)\Delta r$$

$$= \int_0^R 2\pi k(R^2 r - r^3)\, dr$$

$$= 2\pi k \left(\frac{R^2}{2} r^2 - \frac{1}{4} r^4 \right) \Big|_0^R$$

$$= \frac{\pi k R^4}{2}$$

Thus, the blood is flowing at the rate of $\dfrac{\pi k R^4}{2}$ cubic centimeters per second.

Cardiac Output   In studying the cardiovascular system, physicians and medical researchers are often interested in knowing the **cardiac output** of a person's heart, which is the volume of blood it pumps in unit time. Cardiac output is measured by a procedure called the **dye dilution method.*** A known amount of dye is injected into a vein near the heart. The dye then circulates with the blood through the right side of the heart, the lungs, and the left side of the heart before finally appearing in the arterial system. A monitoring probe is introduced into the aorta and blood samples are taken at regular time intervals to measure the concentration of dye leaving the heart until all the dye has passed the monitoring point. A typical concentration graph is shown in Figure 5.25.

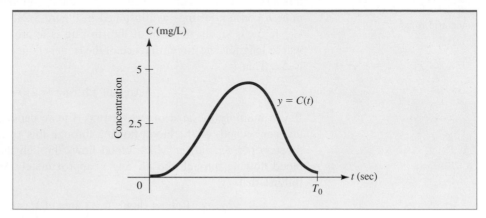

FIGURE 5.25   A typical graph showing concentration of dye in a patient's aorta.

Suppose the amount of dye injected is D mg, and that $C(t)$ (mg/L) is the concentration of dye at time t. Let T_0 denote the total time required for all the dye to pass the monitoring point, and divide the time interval $0 \le t \le T_0$ into n equal subintervals, each of length $\Delta t = (T_0 - 0)/n$. If R is the cardiac output (liters/min), then approximately $R\Delta t$ liters of blood flow past the monitoring probe during the kth time subinterval $t_{k-1} \le t \le t_k$, carrying $C(t_k)R\Delta t$ mg of dye. Adding up the amounts of dye over all n subintervals, we obtain the sum

$$\sum_{k=1}^{n} C(t_k)R\,\Delta t$$

as an approximation for the total amount of dye, and by taking the limit as $n \to +\infty$ we obtain the actual total amount of dye as a definite integral:

$$\lim_{n \to +\infty} \sum_{k=1}^{n} C(t_k)R\,\Delta t = \int_0^{T_0} C(t)R\,dt = R\int_0^{T_0} C(t)\,dt$$

Since D milligrams of dye were originally injected, we must have

$$D = R\int_0^{T_0} C(t)\,dt$$

so the cardiac output is given by

$$R = \frac{D}{\displaystyle\int_0^{T_0} C(t)\,dt}$$

---

*See the module, "Measuring Cardiac Output," by B. Horelick and S. Koont, *UMAP Modules 1977: Tools for Teaching,* Lexington, MA: Consortium for Mathematics and Its Applications, Inc., 1978. Another good source is *Calculus and Its Applications,* by S. Farlow and G. Haggard, Boston: McGraw-Hill, 1990, pp. 332–334.

EXAMPLE | 5.6.4

A physician injects 4 mg of dye into a vein near the heart of a patient, and a monitoring device records the concentration of dye in the blood at regular intervals over a 23-second period. It is determined that the concentration at time t $(0 \le t \le 23)$ is closely approximated by the function

$$C(t) = 0.09t^2 e^{-0.0007t^3} \quad \text{mg/L}$$

Based on this information, what is the patient's cardiac output?

EXPLORE!

Refer to Example 5.6.4. Graph the dye concentration function $C(t) = 0.09t^2 e^{-0.0007t^3}$ using the window [0, 23.5]5 by [−1, 6]1. Compute the patient's cardiac output assuming only a 20-second observation period, and compare with the results obtained in the example (for a 23-second period).

Solution

Integrating $C(t)$ over the time interval $0 \le t \le 23$, we find that

$$\int_0^{23} C(t)\, dt = \int_0^{23} 0.09t^2 e^{-0.0007t^3}\, dt$$

$$= 0.09 \int_0^{23} e^{-0.0007t^3} (t^2\, dt) \qquad \substack{\text{substitute } u = t^3 \\ du = 3t^2\, dt}$$

$$= 0.09 \int_0^{12,167} e^{-0.0007u} \left(\frac{1}{3}\, du\right) \qquad \substack{\text{when } x = 0,\ u = 0 \\ \text{when } x = 23,\ u = (23)^3 = 12{,}167}$$

$$= \frac{0.09}{3} \left.\left(\frac{e^{-0.0007u}}{-0.0007}\right)\right|_0^{12,167}$$

$$\approx -42.86 \left[e^{-0.0007(12,167)} - e^0\right]$$

$$\approx 42.85$$

Thus, the cardiac output is

$$R = \frac{4}{\displaystyle\int_0^{23} C(t)\, dt}$$

$$\approx \frac{4}{42.85} \approx 0.093 \quad \text{liters/sec}$$

or equivalently,

$$R \approx (0.093 \text{ liters/sec})(60 \text{ sec/min}) \approx 5.6 \text{ liters/min}$$

Population Density

The **population density** of an urban area is the number of people $p(r)$ per square mile that are located a distance r miles from the city center. We can determine the total population P of that part of the city that lies within R miles of the city center by using integration.

Our approach to using population density to determine total population will be similar to the approach used earlier in this section to determine the flow of blood through an artery. In particular, divide the interval $0 \le r \le R$ into n subintervals, each of width $\Delta r = R/n$, and let r_k denote the beginning (left endpoint) of the kth subinterval, for $k = 1, 2, \ldots, n$. These subintervals determine n concentric rings, centered on the city center as shown in Figure 5.26.

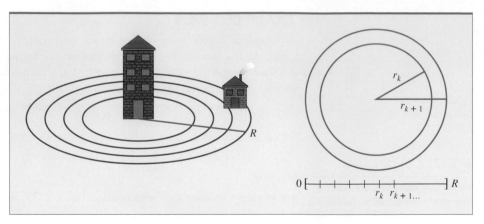

FIGURE 5.26　Subdividing an urban area into concentric rings.

The area of the kth ring is approximately the area of a rectangle whose length is the circumference of the inner boundary of the ring and whose width is Δr. That is,

$$\text{Area of } k\text{th ring} \approx 2\pi r_k \Delta r$$

and since the population density is $p(r)$ people per square mile, it follows that

$$\text{Population within } k\text{th ring } \approx \underbrace{p(r_k)}_{\substack{\text{population} \\ \text{per unit area}}} \cdot \underbrace{[2\pi r_k \Delta r]}_{\substack{\text{area} \\ \text{of ring}}} = 2\pi r_k\, p(r_k)\, \Delta r$$

We can estimate the total area with the bounding radius R by adding up the populations within the approximating rings; that is, by the Riemann sum

$$\begin{bmatrix} \text{Total population} \\ \text{within radius } R \end{bmatrix} = P(R) \approx \sum_{k=1}^{n} 2\pi r_k\, p(r_k)\, \Delta r$$

By taking the limit as $n \to \infty$, the estimate approaches the true value of the total population P, and since the limit of a Riemann sum is a definite integral, we have

$$P(R) = \lim_{n \to \infty} \sum_{k=1}^{n} 2\pi\, r_k\, p(r_k)\Delta r = \int_{0}^{R} 2\pi r p(r)\, dr$$

To summarize:

Total Population from Population Density ■ If a concentration of individuals has population density $p(r)$ individuals per square unit at a distance r from the center of concentration, then the total population $P(R)$ located within distance R from the center is given by

$$P(R) = \int_{0}^{R} 2\pi r p(r)\, dr$$

NOTE　We found it convenient to derive the population density formula by considering the population of a city. However, the formula also applies to more general population concentrations, such as bacterial colonies or even the "population" of water drops from a sprinkler system. ■

EXAMPLE | 5.6.5

A city has population density $p(r) = 3e^{-0.01r^2}$, where $p(r)$ is the number of people (in thousands) per square mile at a distance of r miles from the city center.

a. What population lives within a 5-mile radius of the city center?

b. The city limits are set at a radius R where the population density is 1,000 people per square mile. What is the total population within the city limits?

Solution

a. The population within a 5-mile radius is

$$P(5) = \int_0^5 2\pi r \, (3e^{-0.01r^2}) \, dr = 6\pi \int_0^5 e^{-0.01r^2} r \, dr$$

Using the substitution $u = -0.01 \, r^2$, we find that

$$du = -0.01(2r \, dr) \quad \text{or} \quad r \, dr = \frac{du}{-0.02} = -50 \, du$$

In addition, the limits of integration are transformed as follows:

$$\text{If } r = 5, \text{ then } u = -0.01(5)^2 = -0.25$$
$$\text{If } r = 0, \text{ then } u = -0.01(0)^2 = 0$$

Therefore, we have

$$P(5) = 6\pi \int_0^5 e^{-0.01r^2} r \, dr$$

$$= 6\pi \int_0^{-0.25} e^u (-50 \, du) \qquad \textit{since } r \, dr = -50 \, du$$

$$= 6\pi \, (-50)[e^u] \Big|_{u=0}^{u=-0.25}$$

$$= -300\pi[e^{-0.25} - e^0]$$

$$\approx 208.5$$

So roughly 208,500 people live within a 5-mile radius of the city center.

b. To find the radius R that corresponds to the city limits, we want the population density to be 1 (one thousand), so we solve

$$3e^{-0.01R^2} = 1$$

$$e^{-0.01R^2} = \frac{1}{3}$$

$$-0.01R^2 = \ln\left(\frac{1}{3}\right) \qquad \textit{take logarithms on both sides}$$

$$R^2 = \frac{\ln\left(\dfrac{1}{3}\right)}{-0.01} = 109.86$$

$$R \approx 10.48$$

Finally, using the substitution $u = -0.01r^2$ from part (a), we find that the population within the city limits is

$$P(10.48) = 6\pi \int_0^{10.48} e^{-0.01r^2} r\, dr$$

$$= 6\pi \int_0^{-1.1} e^u (-50\, du) \qquad \begin{array}{l} \textit{Limits of integration:} \\ \textit{when } r = 10.48, \textit{ then } u = -0.01(10.48)^2 \approx -1.1 \\ \textit{when } r = 0, \textit{ then } u = 0 \end{array}$$

$$\approx -300\pi\, [e^u] \Big|_{u=0}^{u=-1.1}$$

$$\approx -300\pi[e^{-1.1} - e^0]$$

$$\approx 628.75$$

Thus, approximately 628,750 people live within the city limits.

The Volume of a Solid of Revolution

The next application is a geometric one in which the characterization of the definite integral as the limit of a sum is used to find the volume of a **solid of revolution** formed by revolving a region R in the xy plane about the x axis.

The technique is to express the volume of the solid as the limit of a sum of the volumes of approximating disks. In particular, suppose that S is the solid formed by rotating about the x axis the region R under the curve $y = f(x)$ between $x = a$ and $x = b$, as shown in Figure 5.27a. Divide the interval $a \leq x \leq b$ into n equal subintervals of length Δx. Then approximate the region R by n rectangles and the solid S by the corresponding n cylindrical disks formed by rotating these rectangles about the x axis. The general approximation procedure is illustrated in Figure 5.27b for the case where $n = 3$.

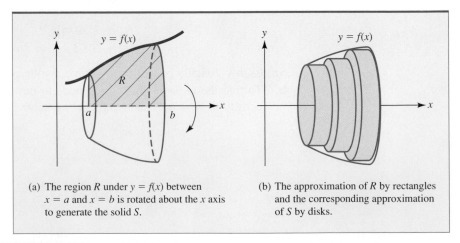

(a) The region R under $y = f(x)$ between $x = a$ and $x = b$ is rotated about the x axis to generate the solid S.

(b) The approximation of R by rectangles and the corresponding approximation of S by disks.

FIGURE 5.27 A solid S formed by rotating the region R about the x axis.

If x_j denotes the beginning (left endpoint) of the jth subinterval, then the jth rectangle has height $f(x_j)$ and width Δx as shown in Figure 5.28a. The jth approximating disk formed by rotating this rectangle about the x axis is as shown in Figure 5.28b.

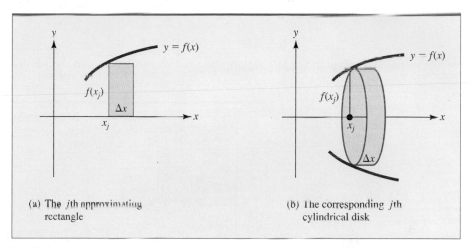

(a) The jth approximating rectangle

(b) The corresponding jth cylindrical disk

FIGURE 5.28 The volume of the solid S is approximated by adding volumes of approximating disks.

Since the jth approximating cylindrical disk has radius $r_j = f(x_j)$ and thickness Δx, its volume is

$$\text{Volume of } j\text{th disk} = (\text{area of circular cross section})(\text{width})$$
$$= \pi r_j^2(\text{width}) = \pi[f(x_j)]^2 \Delta x$$

The total volume of S is approximately the sum of the volumes of the n disks; that is,

$$\text{Volume of } S \approx \sum_{j=1}^{n} \pi[f(x_j)]^2 \Delta x$$

The approximation improves as n increases without bound and

$$\text{Volume of } S = \lim_{n \to \infty} \sum_{j=1}^{n} \pi[f(x_j)]^2 \Delta x = \pi \int_a^b [f(x)]^2 \, dx$$

To summarize:

Volume Formula

Suppose $f(x)$ is continuous and $f(x) \geq 0$ on $a \leq x \leq b$ and let R be the region under the curve $y = f(x)$ between $x = a$ and $x = b$. Then the solid S formed by revolving R about the x axis has volume

$$\text{Volume of } S = \pi \int_a^b [f(x)]^2 \, dx$$

Here are two examples.

EXAMPLE | 5.6.6

Find the volume of the solid S formed by revolving the region under the curve $y = x^2 + 1$ from $x = 0$ to $x = 2$ about the x axis.

Solution

The region, the solid of revolution, and the jth disk are shown in Figure 5.29.

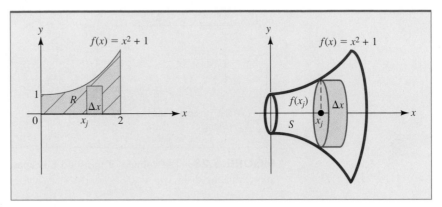

FIGURE 5.29 The solid formed by rotating about the x axis the region under the curve $y = x^2 + 1$ between $x = 0$ and $x = 2$.

The radius of the jth disk is $f(x_j) = x_j^2 + 1$. Hence,

$$\text{Volume of } j\text{th disk} = \pi[f(x_j)]^2 \, \Delta x = \pi(x_j^2 + 1)^2 \, \Delta x$$

and

$$\text{Volume of } S = \lim_{n \to \infty} \sum_{j=1}^{n} \pi(x_j^2 + 1)^2 \, \Delta x$$

$$= \pi \int_0^2 (x^2 + 1)^2 \, dx$$

$$= \pi \int_0^2 (x^4 + 2x^2 + 1) \, dx$$

$$= \pi \left(\frac{1}{5}x^5 + \frac{2}{3}x^3 + x \right) \Big|_0^2 = \frac{206}{15} \pi \approx 43.14$$

EXAMPLE | 5.6.7

A tumor has approximately the same shape as the solid formed by rotating the region under the curve $y = \dfrac{1}{3} \sqrt{16 - 4x^2}$ about the x axis, where x and y are measured in centimeters. Find the volume of the tumor.

Solution

The curve intersects the x axis when $y = 0$; that is, when

$$\frac{1}{3}\sqrt{16 - 4x^2} = 0$$

$$16 = 4x^2 \quad \text{since } \sqrt{a - b} = 0 \text{ only if } a = b$$

$$x^2 = 4 \quad \text{divide both sides by } 4$$

$$x = \pm 2$$

The curve (called an *ellipse*) is shown in Figure 5.30.

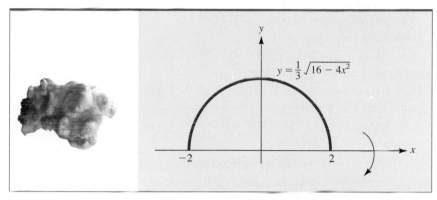

FIGURE 5.30   Tumor with the shape of the solid formed by rotating the curve $y = \frac{1}{3}\sqrt{16 - 4x^2}$ about the x axis.

Let $f(x) = \frac{1}{3}\sqrt{16 - 4x^2}$. Then the volume of the solid of revolution is given by

$$V = \int_{-2}^{2} \pi[f(x)]^2\, dx = \int_{2}^{2} \pi\left[\frac{1}{3}\sqrt{16 - 4x^2}\right]^2 dx$$

$$= \int_{2}^{2} \frac{\pi}{9}(16 - 4x^2)\, dx$$

$$= \frac{\pi}{9}\left[16x - \frac{4}{3}x^3\right]\Bigg|_{-2}^{2}$$

$$= \frac{\pi}{9}\left[16(2) - \frac{4}{3}(2)^3\right] - \frac{\pi}{9}\left[16(-2) \quad \frac{4}{3}(-2)^3\right]$$

$$\approx 14.89$$

Thus, the volume of the tumor is approximately 15 cm³.

PROBLEMS 5.6

SURVIVAL AND RENEWAL *In Problems 1 through 6, an initial population P_0 is given along with a renewal rate R, and a survival function S(t). In each case, use the given information to find the population at the end of the indicated term T.*

1. $P_0 = 50,000$; $R(t) = 40$; $S(t) = e^{-0.1t}$, t in months; term $T = 5$ months

2. $P_0 = 100,000$; $R(t) = 300$; $S(t) = e^{-0.02t}$, t in days; term $T = 10$ days

3. $P_0 = 500,000$; $R(t) = 800$; $S(t) = e^{-0.011t}$, t in years; term $T = 3$ years

4. $P_0 = 800,000$; $R(t) = 500$; $S(t) = e^{-0.005t}$, t in months; term $T = 5$ months

5. $P_0 = 500,000$; $R(t) = 100\, e^{0.01t}$; $S(t) = e^{-0.013t}$, t in years; term $T = 8$ years

6. $P_0 = 300,000$; $R(t) = 150\, e^{0.012t}$; $S(t) = e^{-0.02t}$, t in months; term $T = 20$ months

VOLUME OF A SOLID OF REVOLUTION *In Problems 7 through 14, find the volume of the solid of revolution formed by rotating the region R about the x axis.*

7. R is the region under the line $y = 3x + 1$ from $x = 0$ to $x = 1$.

8. R is the region under the curve $y = \sqrt{x}$ from $x = 1$ to $x = 4$.

9. R is the region under the curve $y = x^2 + 2$ from $x = -1$ to $x = 3$.

10. R is the region under the curve $y = 4 - x^2$ from $x = -2$ to $x = 2$.

11. R is the region under the curve $y = \sqrt{4 - x^2}$ from $x = -2$ to $x = 2$.

12. R is the region under the curve $y = \dfrac{1}{x}$ from $x = 1$ to $x = 10$.

13. R is the region under the curve $y = \dfrac{1}{\sqrt{x}}$ from $x = 1$ to $x = e^2$.

14. R is the region under the curve $y = e^{-0.1x}$ from $x = 0$ to $x = 10$.

15. **NET POPULATION GROWTH** It is projected that t years from now the population of a certain country will be changing at the rate of $e^{0.02t}$ million per year. If the current population is 50 million, what will be the population 10 years from now?

16. **NET POPULATION GROWTH** A study indicates that x months from now, the population of a certain town will be increasing at the rate of $10 + 2\sqrt{x}$ people per month. By how much will the population increase over the next 9 months?

17. **GROUP MEMBERSHIP** A national consumers' association has compiled statistics suggesting that the fraction of its members who are still active t months after joining is given by $f(t) = e^{-0.2t}$. A new local chapter has 200 charter members and expects to attract new members at the rate of 10 per month. How many members can the chapter expect to have at the end of 8 months?

18. **POLITICAL TRENDS** Sarah Greene is running for mayor. Polls indicate that the fraction of those who support her t weeks after first learning of her candidacy is given by $f(t) = e^{-0.03t}$. At the time she declared her candidacy, 25,000 people supported her, and new converts are being added at the constant rate of 100 people per week. Approximately how many people are likely to vote for her if the election is held 20 weeks from the day she entered the race?

19. **SPREAD OF DISEASE** A new strain of influenza has just been declared an epidemic by health officials. Currently, 5,000 people have the disease and 60 more victims are added each day. If the fraction of infected people who still have the disease t days after contracting it is given by $f(t) = e^{-0.02t}$, how many people will have the flu 30 days from now?

20. **NUCLEAR WASTE** A certain nuclear power plant produces radioactive waste in the form of strontium-90 at the constant rate of 500 pounds per year. The waste decays exponentially with a half-life of 28 years. How much of the radioactive waste from the nuclear plant will be present after 140 years? [*Hint:* Think of this as a survival and renewal problem.]

21. **ENERGY CONSUMPTION** It is estimated that the demand for oil is increasing exponentially at the rate of 10% per year. If the demand for oil is currently 30 billion barrels per year, how much oil will be consumed during the next 10 years?

22. **POPULATION GROWTH** The administrators of a town estimate that the fraction of people who will still be residing in the town t years after they arrive is given by the function $f(t) = e^{-0.04t}$. If the

current population is 20,000 people and new towns-people arrive at the rate of 500 per year, what will be the population 10 years from now?

23. **COMPUTER DATING** The operators of a new computer dating service estimate that the fraction of people who will retain their membership in the service for at least t months is given by the function $f(t) = e^{-t/10}$. There are 8,000 charter members, and the operators expect to attract 200 new members per month. How many members will the service have 10 months from now?

24. **FLOW OF BLOOD** Calculate the rate (in cubic centimeters per second) at which blood flows through an artery of radius 0.1 centimeter if the speed of the blood r centimeters from the central axis is $8 - 800r^2$ centimeters per second.

25. **CARDIAC OUTPUT** A physician injects 5 mg of dye into a vein near the heart of a patient and by monitoring the concentration of dye in the blood over a 24-second period, determines that the concentration of dye leaving the heart after t seconds ($0 \le t \le 24$) is given by the function
$$C(t) = -0.028t^2 + 0.672t \quad \text{mg/L}$$
   a. Use this information to find the patient's cardiac output.
   b. Sketch the graph of $C(t)$, and compare it to the graph in Figure 5.25. How are the two graphs alike? How are they different?

26. **CARDIAC OUTPUT** Answer the questions in Problem 25 for the dye concentration function
$$C(t) = \begin{cases} 0 & \text{for } 0 \le t \le 2 \\ -0.034(t^2 - 26t + 48) & \text{for } 2 \le t \le 24 \end{cases}$$

27. **CARDIAC OUTPUT** Answer the questions in Problem 25 for the dye concentration function
$$C(t) = \frac{1}{12,312}(t^4 - 48t^3 + 378t^2 + 4,752t)$$

28. **POPULATION DENSITY** The population density r miles from the center of a certain city is $D(r) = 5,000(1 + 0.5r^2)^{-1}$ people per square mile.
   a. How many people live within 5 miles of the city center?
   b. The city limits are set at a radius L where the population density is 1,000 people per square mile. What is L and what is the total population within the city limits?

29. **POPULATION DENSITY** The population density r miles from the center of a certain city is

$D(r) = 25,000e^{-0.05r^2}$ people per square mile. How many people live between 1 and 2 miles from the city center?

30. **POISEUILLE'S LAW** Blood flows through an artery of radius R. At a distance r centimeters from the central axis of the artery, the speed of the blood is given by $S(r) = k(R^2 - r^2)$. Show that the average velocity of the blood is one-half the maximum speed.

31. **CHOLESTEROL REGULATION** Fat travels through the bloodstream attached to protein in a combination called a *lipoprotein*. Low-density lipoprotein (LDL) picks up cholesterol from the liver and delivers it to the cells, dropping off any excess cholesterol on the artery walls. Too much LDL in the bloodstream increases the risk of heart disease and stroke. A patient with a high level of LDL receives a drug that is found to reduce the level at a rate given by
$$L'(t) = -0.3t(49 - t^2)^{0.4} \quad \text{units/day}$$
where t is the number of days after the drug is administered, for $0 \le t \le 7$.
   a. By how much does the patient's LDL level change during the first 3 days after the drug is administered?
   b. Suppose the patient's LDL level is 150 at the time the drug is administered. Find $L(t)$.
    c. The recommended "safe" LDL level is 130. How many days does it take for the patient's LDL level to be "safe"?

32. **CHOLESTEROL REGULATION** During his annual medical checkup, a man is advised by his doctor to adopt a regimen of exercise, diet, and medication to lower his blood cholesterol level to 220 milligrams per deciliter (mg/dL). Suppose the man finds that his cholesterol level t days after beginning the regimen is
$$L(t) = 190 + 65e^{-0.003t}$$
   a. What is the man's cholesterol level when he begins the regimen?
   b. How many days N must the man remain on the regimen to lower his cholesterol level to 220 mg/dL?
   c. What was the man's average cholesterol level during the first 30 days of the regimen? What was the average level over the entire period $0 \le t \le N$ of the regimen?

33. **BACTERIAL GROWTH**   An experiment is conducted with two bacterial colonies, each of which initially has a population of 100,000. In the first colony, a mild toxin is introduced that restricts growth so that only 50 new individuals are added per day and the fraction of individuals that survive at least t days is given by $f(t) = e^{-0.011t}$. The growth of the second colony is restricted indirectly, by limiting food supply and space for expansion, and after t days, it is found that this colony contains

$$P(t) = \frac{5,000}{1 + 49e^{0.009t}}$$

thousand individuals. Which colony is larger after 50 days? After 100 days? After 300 days?

34. **GROUP MEMBERSHIP**   A group has just been formed with an initial membership of 10,000. Suppose that the fraction of the membership of the group that remain members for at least t years after joining is $S(t) = e^{-0.03t}$, and that at time t, new members are being added at the rate of $R(t) = 10e^{0.017t}$ members per year. How many members will the group have 5 years from now?

35. **GROWTH OF AN ENDANGERED SPECIES**   Environmentalists estimate that the population of a certain endangered species is currently 3,000. The population is expected to be growing at the rate $R(t) = 10e^{0.01t}$ individuals per year t years from now, and the fraction that survive t years is given by $S(t) = e^{-0.07t}$. What will the population of the species be in 10 years?

36. **POPULATION TRENDS**   The population of a small town is currently 85,000. A study commissioned by the mayor's office finds that people are settling in the town at the rate of $R(t) = 1,200e^{0.01t}$ per year and that the fraction of the population who continue to live in the town t years after arriving is given by $S(t) = e^{-0.02t}$. How many people will live in the town in 10 years?

37. **POPULATION TRENDS**   Answer the question in Problem 36 for a constant renewal rate $R = 1,000$ and the survival function

$$S(t) = \frac{1}{t + 1}$$

38. **EVALUATING DRUG EFFECTIVENESS**   A pharmaceutical firm has been granted permission by the FDA to test the effectiveness of a new drug in combating a virus. The firm administers the drug to

a test group of uninfected but susceptible individuals, and using statistical methods, determines that t months after the test begins, people in the group are becoming infected at the rate of $D'(t)$ hundred individuals per month, where

$$D'(t) = 0.2 - 0.04t^{1/4}$$

Government figures indicate that without the drug, the infection rate would have been $W'(t)$ hundred individuals per month, where

$$W'(t) = \frac{0.8e^{0.13t}}{(1 + e^{0.13t})^2}$$

If the test is evaluated 1 year after it begins, how many people does the drug protect from infection? What percentage of the people who would have been infected if the drug had not been used were protected from infection by the drug?

39. **EVALUATING DRUG EFFECTIVENESS**   Repeat Problem 38 for another drug for which the infection rate is

$$D'(t) = 0.12 + \frac{0.08}{t + 1}$$

Assume the government comparison rate stays the same; that is,

$$W'(t) = \frac{0.8e^{0.13t}}{(1 + e^{0.13t})^2}$$

40. **LIFE EXPECTANCY**   In a certain undeveloped country, the life expectancy of a person t years old is $L(t)$ years, where
$$L(t) = 41.6(1 + 1.07t)^{0.13}$$

   a. What is the life expectancy of a person in this country at birth? At age 50?
   b. What is the average life expectancy of all people in this country between the ages of 10 and 70?
   c. Find the age T such that $L(T) = T$. Call T the *life limit* for this country. What can be said about the life expectancy of a person older than T years?
   d. Find the average life expectancy L_e over the age interval $0 \le t \le T$. Why is it reasonable to call L_e the *expected length of life* for people in this country?

41. **LIFE EXPECTANCY**   Answer the questions in Problem 40 for a country whose life expectancy function is

$$L(t) = \frac{110e^{0.015t}}{1 + e^{0.015t}}$$

42. ENERGY EXPENDED IN FLIGHT In an investigation by V. A. Tucker and K. Schmidt-Koenig,[*] it was determined that the energy E expended by a bird in flight varies with the speed v (km/hr) of the bird. For a particular kind of parakeet, the energy expenditure changes at a rate given by

$$\frac{dE}{dv} = \frac{0.31v^2 - 471.75}{v^2} \quad \text{for } v > 0$$

where E is given in joules per gram mass per kilometer. Observations indicate that the parakeet tends to fly at the speed v_{min} that minimizes E.

a. What is the most economical speed v_{min}?

b. Suppose that when the parakeet flies at the most economical speed v_{min} its energy expenditure is E_{min}. Use this information to find $E(v)$ for $v > 0$ in terms of E_{min}.

43. MEASURING RESPIRATION A pneumotachograph is a device used by physicians to graph the rate of air flow into and out of the lungs as a patient breathes. The graph in the accompanying figure shows the rate of inspiration (breathing in) for a particular patient. The area under the graph measures the total volume of air inhaled by the patient during the inspiration phase of one breathing cycle. Assume the inspiration rate is given by

$$R(t) = -0.41t^2 + 0.97t \quad \text{liters/sec}$$

a. How long is the inspiration phase?

b. Find the volume of air taken into the patient's lungs during the inspiration phase.

c. What is the average flow rate of air into the lungs during the inspiration phase?

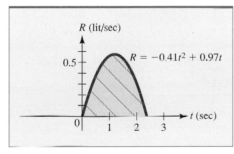

R (lit/sec)

0.5

$R = -0.41t^2 + 0.97t$

0   1   2   3   t (sec)

PROBLEM 43

44. MEASURING RESPIRATION Repeat Problem 43 with the inspiration rate function

$$R(t) = -1.2t^3 + 5.72t \quad \text{liters/sec}$$

and sketch the graph of $R(t)$.

45. WATER POLLUTION A ruptured pipe in an offshore oil rig produces a circular oil slick that is T feet thick at a distance r feet from the rupture, where

$$T(r) = \frac{3}{2 + r}$$

At the time the spill is contained, the radius of the slick is 7 feet. We wish to find the volume of oil that has been spilled.

a. Sketch the graph of $T(r)$. Notice that the volume we want is obtained by rotating the curve $T(r)$ about the T axis (vertical axis) rather than the r axis (horizontal axis).

b. Solve the equation $T = 3/(2 + r)$ for r in terms of T. Sketch the graph of $r(T)$, with T on the horizontal axis.

c. Find the required volume by rotating the graph of $r(T)$ found in part (b) about the T axis.

46. WATER POLLUTION Rework Problem 45 for a situation with spill thickness

$$T(r) = \frac{2}{1 + r^2}$$

(T and r in feet) and radius of containment 9 feet.

47. AIR POLLUTION Particulate matter emitted from a smokestack is distributed in such a way that r miles from the stack, the pollution density is $p(r)$ units per square mile, where

$$p(r) = \frac{200}{5 + 2r^2}$$

a. What is the total amount of pollution within a 3-mile radius of the smokestack?

b. Suppose a health agency determines that it is unsafe to live within a radius L of the smokestack where the pollution density is at least four units per square mile. What is L, and what is the total amount of pollution in the unsafe zone?

[*]E. Batschelet, *Introduction to Mathematics for Life Scientists,* 3rd ed., New York, Springer-Verlag, 1979, p. 299.

48. VOLUME OF A SPHERE Use integration to show that a sphere of radius r has volume

$$V = \frac{4}{3}\pi r^3$$

[*Hint:* Think of the sphere as the solid formed by rotating the region under the semicircle shown in the accompanying figure about the x axis.]

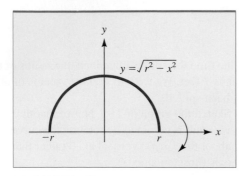

PROBLEM 48

49. VOLUME OF A CONE Use integration to show that a right circular cone of height h and top radius r has volume

$$V = \frac{1}{3}\pi r^2 h$$

[*Hint:* Think of the cone as a solid formed by rotating the triangle shown in the accompanying figure about the x axis.]

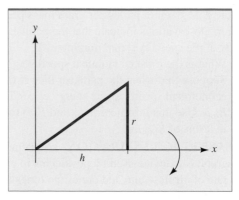

PROBLEM 49

Important Terms, Symbols, and Formulas

CHAPTER SUMMARY

Antiderivative; indefinite integral: (362, 364)

$$\int f(x)\,dx = F(x) + C \text{ if and only if } F'(x) = f(x)$$

Power rule: (365)

$$\int x^n\,dx = \frac{x^{n+1}}{n+1} + C \quad \text{for } n \ne -1$$

Logarithmic rule: $\int \frac{1}{x}\,dx = \ln |x| + C$ (365)

Exponential rule: $\int e^{kx}\,dx = \frac{1}{k}e^{kx} + C$ (365)

Constant multiple rule: $\int kf(x)\,dx = k\int f(x)\,dx$ (366)

Sum rule: (366)

$$\int [f(x) + g(x)]\,dx = \int f(x)\,dx + \int g(x)\,dx$$

Initial value problem (368)

Integration by substitution: (375)

$$\int g(u(x))u'(x)\,dx = \int g(u)\,du \quad \begin{array}{l} where\ u = u(x) \\ du = u'(x)\,dx \end{array}$$

Definite integral: (390)

$$\int_a^b f(x)\,dx = \lim_{n\to+\infty} [f(x_1) + \cdots + f(x_n)]\Delta x$$

Area under a curve: (388)

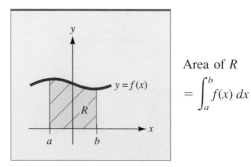

Area of R

$$= \int_a^b f(x)\,dx$$

Rules for definite integrals: (393)

$$\int_a^a f(x)\,dx = 0$$

$$\int_b^a f(x)\,dx = -\int_a^b f(x)\,dx$$

Constant multiple rule: (393)

$$\int_a^b kf(x)\,dx = k\int_a^b f(x)\,dx \quad \text{for constant } k$$

Sum rule: (393)

$$\int_a^b [f(x) + g(x)] \, dx = \int_a^b f(x) \, dx + \int_a^b g(x) \, dx$$

Difference rule: (393)

$$\int_a^b [f(x) - g(x)] \, dx = \int_a^b f(x) \, dx - \int_a^b g(x) \, dx$$

Subdivision rule: (393)

$$\int_a^b f(x) \, dx = \int_a^c f(x) \, dx + \int_c^b f(x) \, dx$$

Fundamental theorem of calculus: (391)

$$\int_a^b f(x) \, dx = F(b) - F(a) \quad \text{where } F'(x) = f(x)$$

Net change of $Q(x)$ over the interval $a \le x \le b$:   (396)

$$Q(b) - Q(a) = \int_a^b Q'(x) \, dx$$

Area between two curves: (405)

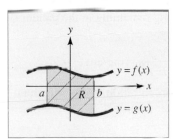

Area of R
$$= \int_a^b [f(x) - g(x)] \, dx$$

Average value of a function $f(x)$ over an interval
$a \le x \le b$:   (413)

$$V = \frac{1}{b-a} \int_a^b f(x) \, dx$$

Lorentz curve   (410)

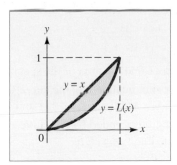

Gini index $= 2 \int_0^1 [x - L(x)] \, dx$

Net excess profit   (408)

Future value (amount) of an income stream   (424)

Present value of an income stream   (424)

Consumers' willingness to spend   (426)

Consumers' surplus: (428)

$$\text{CS} = \int_0^{q_0} D(q) \, dq - p_0 q_0, \text{ where } p = D(q) \text{ is demand}$$

Producers' surplus: (429)

$$\text{PS} = p_0 q_0 - \int_0^{q_0} S(q) \, dq, \text{ where } p = S(q) \text{ is supply}$$

Survival and renewal   (435)

Flow of blood through an artery   (437)

Cardiac output   (438)

Population from population density   (440)

Volume of a solid of revolution   (443)

Checkup for Chapter 5

1. Find these indefinite integrals (antiderivatives).

a. $\displaystyle\int x^3 - \sqrt{3x} + 5e^{-2x} \, dx$

b. $\displaystyle\int \frac{x^2 - 2x + 4}{x} \, dx$

c. $\displaystyle\int \sqrt{x} \left(x^2 - \frac{1}{x} \right) dx$

d. $\displaystyle\int \frac{x}{(3 + 2x^2)^{3/2}} \, dx$

e. $\displaystyle\int \frac{\ln \sqrt{x}}{x} \, dx$

f. $\displaystyle\int x e^{1+x^2} \, dx$

2. Evaluate each of these definite integrals.

a. $\displaystyle\int_1^4 x^{3/2} + \frac{2}{x} \, dx$

b. $\displaystyle\int_0^3 e^{3-x} \, dx$

c. $\displaystyle\int_0^1 \frac{x}{x+1}\,dx$

d. $\displaystyle\int_0^3 \frac{(x+3)}{\sqrt{x^2+6x+4}}\,dx$

3. In each case, find the area of the specified region.

   a. The region bounded by the curve $y = x + \sqrt{x}$, the x axis, and the lines $x = 1$ and $x = 4$.

   b. The region bounded by the curve $y = x^2 - 3x$ and the line $y = x + 5$.

4. Find the average value of the function $f(x) = \dfrac{x-2}{x}$ over the interval $1 \le x \le 2$.

5. **NET CHANGE IN REVENUE** The marginal revenue of producing q units of a certain commodity is $R'(q) = q(10 - q)$ hundred dollars per unit. How much additional revenue is generated as the level of production is increased from 4 to 9 units?

6. **BALANCE OF TRADE** The government of a certain country estimates that t years from now, imports will be increasing at the rate $I'(t)$ and exports at the rate $E'(t)$, both in billions of dollars per year, where
$$I'(t) = 12.5e^{0.2t} \quad \text{and} \quad E'(t) = 1.7t + 3$$
The trade deficit is $D(t) = I(t) - E(t)$. By how much will the trade deficit for this country change over the next 5 years? Will it increase or decrease during this time period?

7. **CONSUMERS' SURPLUS** Suppose q hundred units of a certain commodity are demanded by consumers when the price is $p = 25 - q^2$ dollars per unit. What is the consumers' surplus for the commodity when the level of production is $q_0 = 4$ (400 units)?

8. **AMOUNT OF AN INCOME STREAM** Money is transferred continuously into an account at the constant rate of \$5,000 per year. The account earns interest at the annual rate of 5% compounded continuously. How much will be in the account at the end of 3 years?

9. **POPULATION GROWTH** Demographers estimate that the fraction of people who will still be residing in a particular town t years after they arrive is given by the function $f(t) = e^{-0.02t}$. If the current population is 50,000 and new townspeople arrive at the rate of 700 per year, what will be the population 20 years from now?

10. **AVERAGE DRUG CONCENTRATION** A patient is injected with a drug, and t hours later, the concentration of the drug remaining in the patient's bloodstream is given by
$$C(t) = \frac{0.3t}{(t^2 + 16)^{1/2}} \quad \text{mg/cm}^3$$
What is the average concentration of the drug during the first 3 hours after the injection?

Review Problems

In Problems 1 through 20, find the indicated indefinite integral.

1. $\displaystyle\int (x^3 + \sqrt{x} - 9)\,dx$

2. $\displaystyle\int \left(x^{2/3} - \frac{1}{x} + 5 + \sqrt{x}\right) dx$

3. $\displaystyle\int (x^4 - 5e^{-2x})\,dx$

4. $\displaystyle\int \left(2\sqrt[3]{s} + \frac{5}{s}\right) ds$

5. $\displaystyle\int \left(\frac{5x^3 - 3}{x}\right) dx$

6. $\displaystyle\int \left(\frac{3e^{-x} + 2e^{3x}}{e^{2x}}\right) dx$

7. $\displaystyle\int \left(t^5 - 3t^2 + \frac{1}{t^2}\right) dt$

8. $\displaystyle\int (x+1)(2x^2 + \sqrt{x})\,dx$

9. $\displaystyle\int \sqrt{3x+1}\,dx$

10. $\displaystyle\int (3x+1)\sqrt{3x^2 + 2x + 5}\,dx$

11. $\displaystyle\int (x+2)(x^2 + 4x + 2)^5\,dx$

12. $\displaystyle\int \frac{x+2}{x^2 + 4x + 2}\,dx$

13. $\displaystyle\int \frac{3x + 6}{(2x^2 + 8x + 3)^2}\, dx$

14. $\displaystyle\int (t - 5)^{12}\, dt$

15. $\displaystyle\int v(v - 5)^{12}\, dv$

16. $\displaystyle\int \frac{\ln(3x)}{x}\, dx$

17. $\displaystyle\int 5xe^{-x^2}\, dx$

18. $\displaystyle\int \left(\frac{x}{x - 4}\right) dx$

19. $\displaystyle\int \left(\frac{\sqrt{\ln x}}{x}\right) dx$

20. $\displaystyle\int \left(\frac{e^x}{e^x + 5}\right) dx$

In Problems 21 through 30, evaluate the indicated definite integral.

21. $\displaystyle\int_0^1 (5x^4 - 8x^3 + 1)\, dx$

22. $\displaystyle\int_1^4 (\sqrt{t} + t^{-3/2})\, dt$

23. $\displaystyle\int_0^1 (e^{2x} + 4\sqrt[3]{x})\, dx$

24. $\displaystyle\int_1^9 \frac{x^2 + \sqrt{x} - 5}{x}\, dx$

25. $\displaystyle\int_{-1}^2 30(5x - 2)^2\, dx$

26. $\displaystyle\int_{-1}^1 \frac{(3x + 6)}{(x^2 + 4x + 5)^2}\, dx$

27. $\displaystyle\int_0^1 2te^{t^2 - 1}\, dt$

28. $\displaystyle\int_0^1 e^{-x}(e^{-x} + 1)^{1/2}\, dx$

29. $\displaystyle\int_0^{e-1} \left(\frac{x}{x + 1}\right) dx$

30. $\displaystyle\int_e^{e^2} \frac{1}{x(\ln x)^2}\, dx$

AREA BETWEEN CURVES *In Problems 31 through 38, sketch the indicated region R and find its area by integration.*

31. *R* is the region under the curve $y = x + 2\sqrt{x}$ over the interval $1 \le x \le 4$.

32. *R* is the region under the curve $y = e^x + e^{-x}$ over the interval $-1 \le x \le 1$.

33. *R* is the region under the curve $y = \dfrac{1}{x} + x^2$ over the interval $1 \le x \le 2$.

34. *R* is the region under the curve $y = \sqrt{9 - 5x^2}$ over the interval $0 \le x \le 1$.

35. *R* is the region bounded by the curve $y = \dfrac{4}{x}$ and the line $x + y = 5$.

36. *R* is the region bounded by the curves $y = \dfrac{8}{x}$ and $y = \sqrt{x}$ and the line $x = 8$.

37. *R* is the region bounded by the curve $y = 2 + x - x^2$ and the *x* axis.

38. *R* is the triangular region with vertices $(0, 0)$, $(2, 4)$, and $(0, 6)$.

AVERAGE VALUE OF A FUNCTION *In Problems 39 through 42, find the average value of the given function over the indicated interval.*

39. $f(x) = x^3 - 3x + \sqrt{2x}$; over $1 \le x \le 8$

40. $f(t) = t\sqrt[3]{8 - 7t^2}$; over $0 \le t \le 1$

41. $g(v) = ve^{-v^2}$; over $0 \le v \le 2$

42. $h(x) = \dfrac{e^x}{1 + 2e^x}$; over $0 \le x \le 1$

CONSUMERS' SURPLUS *In Problems 43 through 46, $p = D(q)$ is the demand curve for a particular commodity; that is, q units of the commodity will be demanded when the price is $p = D(q)$ dollars per unit. In each case, for the given level of production q_0, find $p_0 = D(q_0)$ and compute the corresponding consumers' surplus.*

43. $D(q) = 4(36 - q^2)$; $q_0 = 2$ units

44. $D(q) = 100 - 4q - 3q^2$; $q_0 = 5$ units

45. $D(q) = 10e^{-0.1q}$; $q_0 = 4$ units

46. $D(q) = 5 + 3e^{-0.2q}$; $q_0 = 10$ units

LORENTZ CURVES *In Problems 47 through 50, sketch the Lorentz curve $y = L(x)$ and find the corresponding Gini index.*

47. $L(x) = x^{3/2}$

48. $L(x) = x^{1.2}$

49. $L(x) = 0.3x^2 + 0.7x$

50. $L(x) = 0.75x^2 + 0.25x$

SURVIVAL AND RENEWAL *In Problems 51 through 54, an initial population P_0 is given along with a renewal rate $R(t)$ and a survival function $S(t)$. In each case, use the given information to find the population at the end of the indicated term T.*

51. $P_0 = 75,000$; $R(t) = 60$; $S(t) = e^{-0.09t}$; t in months; term $T = 6$ months

52. $P_0 = 125,000$; $R(t) = 250$; $S(t) = e^{-0.015t}$; t in years; term $T = 5$ years

53. $P_0 = 100,000$; $R(t) = 90\,e^{0.1t}$; $S(t) = e^{-0.2t}$; t in years; term $T = 10$ years

54. $P_0 = 200,000$; $R(t) = 50\,e^{0.12t}$; $S(t) = e^{-0.017t}$; t in hours; term $T = 20$ hours

VOLUME OF SOLID OF REVOLUTION *In Problems 55 through 58, find the volume of the solid of revolution formed by rotating the specified region R about the x axis.*

55. R is the region under the curve $y = x^2 + 1$ from $x = -1$ to $x = 2$.

56. R is the region under the curve $y = e^{-x/10}$ from $x = 0$ to $x = 10$.

57. R is the region under the curve $y = \dfrac{1}{\sqrt{x}}$ from $x = 1$ to $x = 3$.

58. R is the region under the curve $y = \dfrac{x+1}{\sqrt{x}}$ from $x = 1$ to $x = 4$.

In Problems 59 through 62, solve the given initial value problem.

59. $\dfrac{dy}{dx} = 2$, where $y = 4$ when $x = -3$

60. $\dfrac{dy}{dx} = x(x - 1)$, where $y = 1$ when $x = 1$

61. $\dfrac{dx}{dt} = e^{-2t}$, where $x = 4$ when $t = 0$

62. $\dfrac{dy}{dt} = \dfrac{t + 1}{t}$, where $y = 3$ when $t = 1$

63. Find the function whose tangent line has slope $x(x^2 + 1)^{-1}$ for each x and whose graph passes through the point $(1, 5)$.

64. Find the function whose tangent line has slope xe^{-2x^2} for each x and whose graph passes through the point $(0, -3)$.

65. **NET ASSET VALUE** It is estimated that t days from now a farmer's crop will be increasing at the rate of $0.5t^2 + 4(t + 1)^{-1}$ bushels per day. By how much will the value of the crop increase during the next 6 days if the market price remains fixed at $2 per bushel?

66. **DEPRECIATION** The resale value of a certain industrial machine decreases at a rate that changes with time. When the machine is t years old, the rate at which its value is changing is $200(t - 6)$ dollars per year. If the machine was bought new for $12,000, how much will it be worth 10 years later?

67. **TICKET SALES** The promoters of a county fair estimate that t hours after the gates open at 9:00 A.M. visitors will be entering the fair at the rate of $-4(t + 2)^3 + 54(t + 2)^2$ people per hour. How many people will enter the fair between 10:00 A.M. and noon?

68. **MARGINAL COST** At a certain factory, the marginal cost is $6(q - 5)^2$ dollars per unit when the level of production is q units. By how much will the total manufacturing cost increase if the level of production is raised from 10 to 13 units?

69. **PUBLIC TRANSPORTATION** It is estimated that x weeks from now, the number of commuters using a new subway line will be increasing at the rate of $18x^2 + 500$ per week. Currently, 8,000 commuters use the subway. How many will be using it 5 weeks from now?

70. **NET CHANGE IN BIOMASS** A protein with mass m (grams) disintegrates into amino acids at a rate given by

$$\frac{dm}{dt} = \frac{-15t}{t^2 + 5}$$

What is the net change in mass of the protein during the first 4 hours?

71. CONSUMPTION OF OIL   It is estimated that t years from the beginning of the year 2005, the demand for oil in a certain country will be changing at the rate of $D'(t) = (1 + 2t)^{-1}$ billion barrels per year. Will more oil be consumed (demanded) during 2006 or during 2009? How much more?

72. FUTURE VALUE OF AN INCOME STREAM
Money is transferred continuously into an account at the constant rate of $5,000 per year. The account earns interest at the annual rate of 5% compounded continuously. How much will be in the account at the end of 3 years?

73. FUTURE VALUE OF AN INCOME STREAM
Money is transferred continuously into an account at the constant rate of $1,200 per year. The account earns interest at the annual rate of 8% compounded continuously. How much will be in the account at the end of 5 years?

74. PRESENT VALUE OF AN INCOME STREAM
What is the present value of an investment that will generate income continuously at a constant rate of $1,000 per year for 10 years if the prevailing annual interest rate remains fixed at 7% compounded continuously?

75. REAL ESTATE INVENTORY   In a certain community the fraction of the homes placed on the market that remain unsold for at least t weeks is approximately $f(t) = e^{-0.2t}$. If 200 homes are currently on the market and if additional homes are placed on the market at the rate of 8 per week, approximately how many homes will be on the market 10 weeks from now?

76. AVERAGE REVENUE   A bicycle manufacturer expects that x months from now consumers will be buying 5,000 bicycles per month at the price of $P(x) = 80 + 3\sqrt{x}$ dollars per bicycle. What is the average revenue the manufacturer can expect from the sale of the bicycles over the next 16 months?

77. NUCLEAR WASTE   A nuclear power plant produces radioactive waste at a constant rate of 300 pounds per year. The waste decays exponentially with a half-life of 35 years. How much of the radioactive waste from the plant will remain after 200 years?

78. GROWTH OF A TREE   A tree has been transplanted and after x years is growing at the rate of

$$h'(x) = 0.5 + \frac{1}{(x + 1)^2}$$

meters per year. By how much does the tree grow during the second year?

79. FUTURE REVENUE   A certain oil well that yields 900 barrels of crude oil per month will run dry in 3 years. The price of crude oil is currently $16 per barrel and is expected to rise at the constant rate of 8 cents per barrel per month. If the oil is sold as soon as it is extracted from the ground, what will be the total future revenue from the well?

80. CONSUMERS' SURPLUS   Suppose that the consumers' demand function for a certain commodity is $D(q) = 50 - 3q - q^2$ dollars per unit.

  a. Find the number of units that will be bought if the market price is $32 per unit.

  b. Compute the consumers' willingness to spend to get the number of units in part (a).

  c. Compute the consumers' surplus when the market price is $32 per unit.

   d. Use the graphing utility of your calculator to graph the demand curve. Interpret the consumers' willingness to spend and the consumers' surplus as areas in relation to this curve.

81. AVERAGE PRICE   Records indicate that t months after the beginning of the year, the price of bacon in local supermarkets was $P(t) = 0.06t^2 - 0.2t + 1.2$ dollars per pound. What was the average price of bacon during the first 6 months of the year?

82. SURFACE AREA OF A HUMAN BODY   The surface area S of the body of an average person 4 feet tall who weighs w lb changes at the rate

$$S'(w) = 110w^{-0.575} \quad \text{in}^2/\text{lb}$$

The body of a particular child who is 4 feet tall and weighs 50 lb has surface area 1,365 in². If the child gains 3 lb while remaining the same height, by how much will the surface area of the child's body increase?

83. TEMPERATURE CHANGE   At t hours past midnight, the temperature T (°C) in a certain northern city is found to be changing at a rate given by

$$T'(t) = -0.02(t - 7)(t - 14) \quad °\text{C/hour}$$

By how much does the temperature change between 8 A.M. and 8 P.M.?

84. EFFECT OF A TOXIN   A toxin is introduced to a bacterial colony, and t hours later, the population $P(t)$ of the colony is changing at the rate

$$\frac{dP}{dt} = -(\ln 3)3^{4-t}$$

If there were 1 million bacteria in the colony when the toxin was introduced, what is $P(t)$? [*Hint:* Note that $3^x = e^{x \ln 3}$.]

85. **MARGINAL ANALYSIS** In a certain section of the country, the price of large Grade A eggs is currently $2.50 per dozen. Studies indicate that x weeks from now, the price $p(x)$ will be changing at the rate of $p'(x) = 0.2 + 0.003x^2$ cents per week.

a. Use integration to find $p(x)$ and then use the graphing utility of your calculator to sketch the graph of $p(x)$. How much will the eggs cost 10 weeks from now?

b. Suppose the rate of change of the price were $p'(x) = 0.3 + 0.003x^2$. How does this affect $p(x)$? Check your conjecture by sketching the new price function on the same screen as the original. Now how much will the eggs cost in 10 weeks?

86. **INVESTING IN A DOWN MARKET PERIOD** Jan opens a stock account with $5,000 at the beginning of January and subsequently, deposits $200 a month. Unfortunately, the market is depressed, and she finds that t months after depositing a dollar, only $100f(t)$ cents remain, where $f(t) = e^{-0.01t}$. If this pattern continues, what will her account be worth after 2 years? [*Hint:* Think of this as a survival and renewal problem.]

87. **DISTANCE AND VELOCITY** After t minutes, an object moving along a line has velocity $v(t) = 1 + 4t + 3t^2$ meters per minute. How far does the object travel during the third minute?

88. **AVERAGE POPULATION** The population (in thousands) of a certain city t years after January 1, 1995, is given by the function

$$P(t) = \frac{150e^{0.03t}}{1 + e^{0.03t}}$$

What is the average population of the city during the decade 1995–2005?

89. **DISTRIBUTION OF INCOME** A study suggests that the distribution of incomes for social workers and physical therapists may be represented by the Lorentz curves $y = L_1(x)$ and $y = L_2(x)$, respectively, where

$L_1(x) = x^{1.6}$ and $L_2(x) = 0.65x^2 + 0.35x$

For which profession is the distribution of income more equitably distributed?

90. **DISTRIBUTION OF INCOME** A study conducted by a certain state determines that the Lorentz curves for high school teachers and real estate brokers are given by the functions

$$L_1(x) = 0.67x^4 + 0.33x^3$$
$$L_2(x) = 0.72x^2 + 0.28x$$

respectively. For which profession is the distribution of income more fairly distributed?

91. **CONSERVATION** A lake has roughly the same shape as the bottom half of the solid formed by rotating the curve $2x^2 + 3y^2 = 6$ about the x axis, for x and y measured in miles. Conservationists wish the lake to contain 1,000 trout per cubic mile. If the lake currently contains 5,000 trout, how many more must be added to meet this requirement?

92. **HORTICULTURE** A sprinkler system sprays water onto a garden in such a way that $11e^{-r^2/10}$ inches of water per hour are delivered at a distance of r feet from the sprinkler. What is the total amount of water laid down by the sprinkler within a 5-foot radius during a 20-minute watering period?

93. **SPEED AND DISTANCE** A car is driven so that after t hours its speed is $S(t)$ miles per hour.

a. Write down a definite integral that gives the average speed of the car during the first N hours.

b. Write down a definite integral that gives the total distance the car travels during the first N hours.

c. Discuss the relationship between the integrals in parts (a) and (b).

94. Use the graphing utility of your calculator to draw the graphs of the curves $y = -x^3 - 2x^2 + 5x - 2$ and $y = x \ln x$ on the same screen. Use **ZOOM** and **TRACE** or some other feature of your calculator to find where the curves intersect, and then compute the area of the region bounded by the curves.

95. Repeat Problem 94 for the curves

$$y = \frac{x - 2}{x + 1} \quad \text{and} \quad y = \sqrt{25 - x^2}$$

Complete solutions for all EXPLORE! boxes throughout the text can be accessed at the book-specific website, www.mhhe.com/hoffmann.

Solution for Explore!
on Page 363

Store the constants $\{-4, -2, 2, 4\}$ into L1 and write Y1 = X^3 and Y2 = Y1 + L1. Graph Y1 in bold, using the modified decimal window $[-4.7, 4.7]1$ by $[-6, 6]1$. At $x = 1$ (where we have drawn a vertical line), the slopes for each curve appear equal.

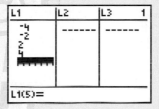  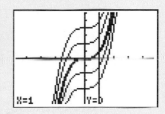

Using the tangent line feature of your graphing calculator, draw tangent lines at $x = 1$ for several of these curves. Every tangent line at $x = 1$ has a slope of 3, although each line has a different y intercept.

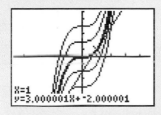 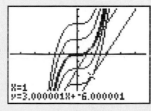 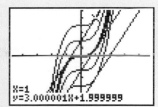

Solution for Explore!
on Page 364

The numerical integral, **fnInt**(expression, variable, lower limit, upper limit) can be found via the **MATH** key, **9:fnInt(**, which we use to write Y1 below. We obtain a family of graphs that appear to be parabolas with vertices on the y axis at $y = 0, -1$, and -4. The antiderivative of $f(x) = 2x$ is $F(x) = x^2 + C$, where $C = 0, -1$, and -4, in our case.

 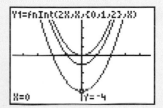

Solution for Explore! on Page 365

Place $y = F(x) = \ln |x|$ into Y1 as ln(abs(x)), using a bold graphing style, and store $f(x) = 1/x$ into Y2; then graph using a decimal window. Choose $x = 1$, and compare the derivative $F'(1)$, which is displayed in the graph on the left as $dy/dx = 1.0000003$, with the value $y = 1$ of $f(1)$ displayed on the right. The negligible difference in value in this case can be attributed to the use of numerical differentiation. In general, choosing any other nonzero x value, we can verify that $F'(x) = f(x)$. For instance, when $x = -2$, we have $F'(-2) = -0.5 = f(-2)$.

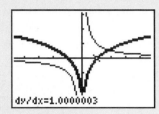
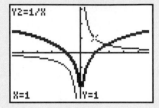

Solution for Explore! on Page 367

Place the integers from -5 to 5 into L1 (**STAT EDIT 1**). Set up the functions in the equation editor as shown here. Now graph with the designated window and notice that the antiderivative curves are generated sequentially from the lower to the upper levels. **TRACE** to the point $(2, 6)$ and observe that the antiderivative that passes through this point is the second on the listing of L1. This curve is $F(x) = x^3 + x - 4$, which can also be calculated analytically, as in Example 5.1.4. For $f(x) = 3x^2 - 2$, the family of antiderivatives is $F(x) = x^3 - 2x + L1$ and the same window dimensions can be used to produce the screen on the right. The desired antiderivative is the eighth in L1, corresponding to $F(x) = x^3 - 2x + 2$, whose constant term can also be confirmed algebraically.

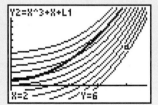
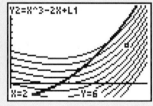

Solution for Explore! on Page 392

Following Example 5.3.3, set Y1 $= x^3 + 1$ and graph using a window $[-1, 3]1$ by $[-1, 2]1$. Access the numerical integration feature through **CALC, 7:∫f(x) dx**, specifying the lower limit as X $= 0$ and the upper limit as X $= 1$ to obtain $\int_0^1 (x^3 + 1) \, dx = 1.25$. Numerical integration can also be performed from the home screen via **MATH, 9: fnInt(,** as shown in the screen on the right.

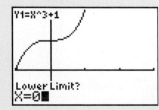
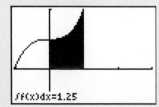
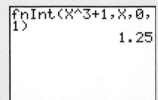

Solution for Explore! on Page 406

Following Example 5.4.2, set Y1 = 4X and Y2 = X^3 + 3X^2 in the equation editor of your graphing calculator. Graph using the window $[-6, 2]1$ by $[-25, 10]5$. The points of intersection are at $x = -4$, 0, and 1. Considering $y = 4x$ as a horizontal baseline, the area between Y1 and Y2 can be viewed as that of the difference curve Y3 = Y2 − Y1. Deselect (turn off) Y1 and Y2 and graph Y3 using the window $[-4.5, 1.5]0.5$ by $[-5, 15]5$. Numerical integration applied to this curve between $x = -4$ and 0 yields an area of 32 square units for the first sector enclosed by the two curves. The area of the second sector, between $x = 0$ and 1, has area -0.75. The total area enclosed by the two curves is $32 + |-0.75| = 32.75$.

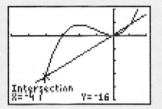
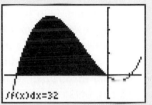
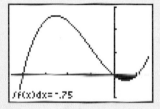

Solution for Explore! on Page 112

Set Y1 = $x^3 - 6x^2 + 10x - 1$ and use the **CALC, 7:∫f(x) dx,** feature to determine that the area under the curve from $x = 1$ to $x = 4$ is 9.75 square units, which equals the rectangular portion under Y2 = 9.75/(4 − 1) = 3.25 of length 3. It is as though the area under $f(x)$ over [1, 4] turned to water and became a level surface of height 3.25, the average $f(x)$ value. This value is attained at $x \approx 1.874$ (shown on the right) and also at $x = 3.473$. Note that you must clear the previous shading, using **DRAW, 1:ClrDraw,** before constructing the next drawing.

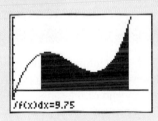
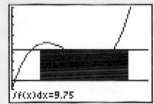
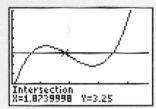

Solution for Explore! on Page 426

We graph $D_{new}(q)$ in bold as Y2 = 4(23 − X^2) with $D(q)$ in Y1 = 4(25 − X^2), using the viewing window $[0, 5]1$ by $[0, 150]10$. Visually, $D_{new}(q)$ is less than $D(q)$ for the observable range of values, supporting the conjecture that the area under the curve of $D_{new}(q)$ will be less than that of $D(q)$ over the range of values [0, 3]. This area is calculated to be $240, less than the $264 shown for $D(q)$ in Figure 5.20.

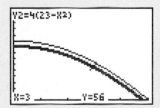
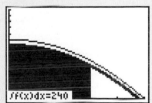

THINK ABOUT IT

JUST NOTICEABLE DIFFERENCES IN PERCEPTION

Calculus can help us answer questions about human perception, including questions relating to the number of different frequencies of sound or the number of different hues of light people can distinguish (see the accompanying figure). Our present goal is to show how integral calculus can be used to estimate the number of steps a person can distinguish as the frequency of sound increases from the lowest audible frequency of 15 hertz (Hz) to the highest audible frequency of 18,000 Hz. (Here hertz, abbreviated Hz, equals cycles per second.)

A mathematical model* for human auditory perception uses the formula $y = 0.767x^{0.439}$, where y Hz is the smallest change in frequency that is detectable at frequency x Hz. Thus, at the low end of the range of human hearing, 15 Hz, the smallest change of frequency a person can detect is $y = 0.767 \times 15^{0.439} \approx 2.5$ Hz, while at the upper end of human hearing, near 18,000 Hz, the least noticeable difference is approximately $y = 0.767 \times 18,000^{0.439} \approx 57$ Hz. If the smallest noticeable change of frequency were the same for all frequencies that people can hear, we could find the number of noticeable steps in human hearing by simply dividing the total frequency range by the size of this smallest noticeable change. Unfortunately, we have just seen that the smallest noticeable change of frequency increases as frequency increases, so the simple approach will not work. However, we can estimate the number of distinguishable steps using integration.

Toward this end, let $y = f(x)$ represent the just noticeable difference of frequency people can distinguish at frequency x. Next, choose numbers $x_0, x_1, \ldots, x_n$ beginning at $x_0 = 15$ Hz and working up through higher frequencies to $x_n = 18,000$ Hz in such a way that for $j = 0, 2, \ldots, n - 1$,

$$x_j + f(x_j) = x_{j+1}$$

*Part of this essay is based on *Applications of Calculus: Selected Topics from the Environmental and Life Sciences,* by Anthony Barcellos, New York: McGraw-Hill, 1994, pp. 21–24.

In other words, x_{j+1} is the number we get by adding the just noticeable difference at x_j to x_j itself. Thus, the jth step has length

$$\Delta x_j = x_{j+1} - x_j = f(x_j)$$

Dividing by $f(x_j)$, we get

$$\frac{\Delta x_j}{f(x_j)} = \frac{x_{j+1} - x_j}{f(x_j)} = 1$$

and it follows that

$$\sum_{j=0}^{n-1} \frac{\Delta x_j}{f(x_j)} = \sum_{j=0}^{n-1} \frac{x_{j+1} - x_j}{f(x_j)} = \frac{x_1 - x_0}{f(x_0)} + \frac{x_2 - x_1}{f(x_1)} + \cdots + \frac{x_n - x_{n-1}}{f(x_n)}$$

$$= \underbrace{1 + 1 + \cdots + 1}_{n \text{ terms}} = n$$

The sum on the left side of this equation is a Riemann sum, and since the step sizes $\Delta x_j = x_{j+1} - x_j$ are very small, the sum is approximately equal to a definite integral. Specifically, we have

$$\int_{x_0}^{x_n} \frac{dx}{f(x)} \approx \sum_{j=0}^{n-1} \frac{\Delta x_j}{f(x_j)} = n$$

Finally, using the modeling formula $f(x) = 0.767x^{0.439}$ along with $x_0 = 15$ and $x_n = 18,000$, we find that

$$\int_{x_0}^{x_n} \frac{dx}{f(x)} = \int_{15}^{18,000} \frac{dx}{0.767x^{0.439}}$$

$$= \frac{1}{0.767}\left(\frac{x^{0.561}}{0.561}\right)\Bigg|_{15}^{18,000}$$

$$= 2.324(18,000^{0.561} - 15^{0.561})$$

$$= 556.2$$

Thus, there are approximately 556 just noticeable steps in the audible range from 15 Hz to 18,000 Hz.

Here are some questions in which you are asked to apply these principles to issues involving both auditory and visual perception.

Questions

1.   The 88 keys of a piano range from 15 Hz to 4,186 Hz. If the number of keys were based on the number of just noticeable differences, how many keys would a piano have?

2.   An 8-bit gray-scale monitor can display 256 shades of gray. Let x represent the darkness of a shade of gray, where $x = 0$ for white and $x = 1$ for totally black. One model for gray-scale perception uses the formula $y = Ax^{0.3}$, where A is a positive constant and y is the smallest change detectable by the human eye at gray-level x. Experiments show that the human eye is incapable of distinguishing as many as 256 different shades of gray, so the number n of just noticeable

shading differences from $x = 0$ to $x = 1$ must be less than 256. Using the assumption that $n < 256$, find a lower bound for the constant A in the modeling formula $y = Ax^{0.3}$.

3. One model of the ability of human vision to distinguish colors of different hue uses the formula $y = 2.9 \times 10^{-24} x^{8.52}$, where y is the just noticeable difference for a color of wavelength x, with both x and y measured in nanometers (nm).

    a. Blue-green light has a wavelength of 580 nm. What is the least noticeable difference at this wavelength?

    b. Red light has a wavelength of 760 nm. What is the least noticeable difference at this wavelength?

    c. How many just noticeable steps are there in hue from blue-green light to red light?

4. Find a model of the form $y = ax^k$ for just noticeable differences in hue for the color spectrum from blue-green light at 580 nm to violet light at 400 nm. Use the fact that the minimum noticeable difference at the wavelength of blue-green light is 1 nm, while at the wavelength of violet light, the minimum noticeable difference is 0.043 nm.

CHAPTER 6

The concentration of pollutant in a lake can be determined using differential equations.

ADDITIONAL TOPICS IN INTEGRATION

SECTION 6.1 | Integration by Parts; Integral Tables

Integration by parts is a technique of integration based on the product rule for differentiation. In particular, if $u(x)$ and $v(x)$ are both differentiable functions of x, then

$$\frac{d}{dx}[u(x)v(x)] = u(x)\frac{dv}{dx} + v(x)\frac{du}{dx}$$

so that

$$u(x)\frac{dv}{dx} = \frac{d}{dx}[u(x)v(x)] - v(x)\frac{du}{dx}$$

Integrating both sides of this equation with respect to x, we obtain

$$\int\left[u(x)\frac{dv}{dx}\right]dx = \int\frac{d}{dx}[u(x)v(x)]\,dx - \int\left[v(x)\frac{du}{dx}\right]dx$$

$$= u(x)v(x) - \int\left[v(x)\frac{du}{dx}\right]dx$$

since $u(x)v(x)$ is an antiderivative of $\dfrac{d}{dx}[u(x)v(x)]$. Moreover, we can write this integral formula in the more compact form

$$\int u\,dv = uv - \int v\,du$$

since

$$dv = \frac{dv}{dx}dx \qquad \text{and} \qquad du = \frac{du}{dx}dx$$

The equation $\int u\,dv = uv - \int v\,du$ is called the **integration by parts formula.** The great value of this formula is that if we can find functions u and v so that a given integral $\int f(x)\,dx$ can be expressed in the form $\int f(x)\,dx = \int u\,dv$, then we have

$$\int f(x)\,dx = \int u\,dv = uv - \int v\,du$$

and the given integral is effectively exchanged for the integral $\int v\,du$. If the integral $\int v\,du$ is easier to compute than $\int u\,dv$, the exchange facilitates finding $\int f(x)\,dx$. Here is an example.

EXAMPLE | 6.1.1

Find $\int x^2 \ln x\,dx$.

Solution

Our strategy is to express $\int x^2 \ln x \, dx$ as $\int u \, dv$ by choosing u and v so that $\int v \, du$ is easier to evaluate than $\int u \, dv$. This strategy suggests that we choose

$$u = \ln x \quad \text{and} \quad dv = x^2 \, dx$$

since

$$du = \frac{1}{x} \, dx$$

is a simpler expression than $\ln x$, while v can be obtained by the relatively easy integration

$$v = \int x^2 \, dx = \frac{1}{3} x^3$$

(For simplicity, we leave the "$+ C$" out of the calculation until the final step.) Substituting this choice for u and v into the integration by parts formula, we obtain

$$\int x^2 \ln x \, dx = \int \overbrace{(\ln x)}^{u} \overbrace{(x^2 \, dx)}^{dv} = \overbrace{(\ln x)}^{u} \overbrace{\left(\frac{1}{3} x^3\right)}^{v} - \int \overbrace{\left(\frac{1}{3} x^3\right)}^{v} \overbrace{\left(\frac{1}{x} \, dx\right)}^{du}$$

$$= \frac{1}{3} x^3 \ln x - \frac{1}{3} \int x^2 \, dx = \frac{1}{3} x^3 \ln x - \frac{1}{3}\left(\frac{1}{3} x^3\right) + C$$

$$= \frac{1}{3} x^3 \ln x - \frac{1}{9} x^3 + C$$

Here is a summary of the procedure we have just illustrated.

Integration by Parts

To find an integral $\int f(x) \, dx$ using the integration by parts formula:

Step 1. Choose functions u and v so that $f(x) \, dx = u \, dv$. Try to pick u so that du is simpler than u and a dv that is easy to integrate.

Step 2. Organize the computation of du and v as

$$\begin{array}{cc} u & dv \\ du & v = \int dv \end{array}$$

Step 3. Complete the integration by finding $\int v \, du$. Then

$$\int f(x) \, dx = \int u \, dv = uv - \int v \, du$$

Add "$+ C$" only at the end of the computation.

Choosing a suitable u and dv for integration by parts requires insight and experience. For instance, in Example 6.1.1, things would not have gone so smoothly if we had chosen $u = x^2$ and $dv = \ln x\, dx$. Certainly $du = 2x\, dx$ is simpler than $u = x^2$, but what is $v = \int \ln x\, dx$? In fact, finding this integral is just as hard as finding the original integral $\int x^2 \ln x\, dx$ (see Example 6.1.4). Examples 6.1.2, 6.1.3, and 6.1.4 illustrate several ways of choosing u and dv in integrals that can be handled using integration by parts.

EXAMPLE 6.1.2

Find $\int x e^{2x}\, dx$.

Solution

Although both factors x and e^{2x} are easy to integrate, only x becomes simpler when differentiated. Therefore, we choose $u = x$ and $dv = e^{2x}\, dx$ and find

$$u = x \qquad dv = e^{2x}\, dx$$
$$du = dx \qquad v = \frac{1}{2}e^{2x}$$

Substituting into the integration by parts formula, we obtain

$$\int x(e^{2x}dx) = x\left(\frac{1}{2}e^{2x}\right) - \int\left(\frac{1}{2}e^{2x}\right)dx$$
$$= \frac{1}{2}xe^{2x} - \frac{1}{2}\left(\frac{1}{2}e^{2x}\right) + C$$
$$= \frac{1}{2}\left(x - \frac{1}{2}\right)e^{2x} + C$$

EXAMPLE 6.1.3

Find $\int x\sqrt{x+5}\, dx$.

Solution

Again, both factors x and $\sqrt{x+5}$ are easy to differentiate and to integrate, but x is simplified by differentiation, while the derivative of $\sqrt{x+5}$ is even more complicated than $\sqrt{x+5}$ itself. This observation suggests that you choose

$$u = x \qquad dv = \sqrt{x+5}\,dx = (x+5)^{1/2}dx$$

so that

$$du = dx \qquad v = \frac{2}{3}(x + 5)^{3/2}$$

Substituting into the integration by parts formula, you obtain

$$\int x(\sqrt{x + 5}\,dx) = x\left[\frac{2}{3}(x + 5)^{3/2}\right] - \int\left[\frac{2}{3}(x + 5)^{3/2}\right]dx$$

$$= \frac{2}{3}x(x + 5)^{3/2} - \frac{2}{3}\left[\frac{2}{5}(x + 5)^{5/2}\right] + C$$

$$= \frac{2}{3}x(x + 5)^{3/2} - \frac{4}{15}(x + 5)^{5/2} + C$$

NOTE   Some integrals can be evaluated by either substitution or integration by parts. For instance, the integral in Example 6.1.3 can be found by substituting as follows:

Let $u = x + 5$. Then $du = dx$ and $x = u - 5$, and

$$\int x\sqrt{x + 5}\,dx = \int (u - 5)\sqrt{u}\,du = \int (u^{3/2} - 5u^{1/2})\,du$$

$$- \frac{u^{5/2}}{5/2} - \frac{5u^{3/2}}{3/2} + C$$

$$= \frac{2}{5}(x + 5)^{5/2} - \frac{10}{3}(x + 5)^{3/2} + C$$

This form of the integral is not the same as that found in Example 6.1.3. To show that the two forms are equivalent, note that the antiderivative in Example 6.1.3 can be expressed as

$$\frac{2x}{3}(x + 5)^{3/2} - \frac{4}{15}(x + 5)^{5/2} = (x + 5)^{3/2}\left[\frac{2x}{3} - \frac{4}{15}(x + 5)\right]$$

$$= (x + 5)^{3/2}\left(\frac{2x}{5} - \frac{4}{3}\right) - (x + 5)^{3/2}\left[\frac{2}{5}(x + 5) - \frac{10}{3}\right]$$

$$= \frac{2}{5}(x + 5)^{5/2} - \frac{10}{3}(x + 5)^{3/2}$$

which is the form of the antiderivative obtained by substitution. This example shows that it is quite possible for you to do everything right and still not get the answer given at the back of the book.

Definite Integration by Parts

The integration by parts formula can be applied to definite integrals by noting that

$$\int_a^b u\,dv = uv\Big|_a^b - \int_a^b v\,du$$

Definite integration by parts is used in Example 6.1.4 to find an area.

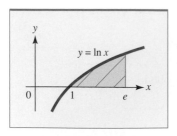

FIGURE 6.1 The region under $y = \ln x$ over $1 \le x \le e$.

EXAMPLE 6.1.4

Find the area of the region bounded by the curve $y = \ln x$, the x axis, and the lines $x = 1$ and $x = e$.

Solution

The region is shown in Figure 6.1. Since $\ln x \ge 0$ for $1 \le x \le e$, the area is given by the definite integral

$$A = \int_1^e \ln x \, dx$$

To evaluate this integral using integration by parts, think of $\ln x \, dx$ as $(\ln x)(1 \; dx)$ and use

$$u = \ln x \qquad dv = 1 \, dx$$
$$du = \frac{1}{x} \, dx \qquad v = \int 1 \, dx = x$$

Thus, the required area is

$$A = \int_1^e \ln x \, dx = x \ln x \Big|_1^e - \int_1^e x\left(\frac{1}{x} \, dx\right)$$
$$= x \ln x \Big|_1^e - \int_1^e 1 \, dx = (x \ln x - x)\Big|_1^e$$
$$= [e \ln e - e] - [1 \ln 1 - 1]$$
$$= [e(1) - e] - [1(0) - 1] \qquad \ln e = 1 \text{ and } \ln 1 = 0$$
$$= 1$$

As another illustration of the role played by integration by parts in applications, we use it in Example 6.1.5 to compute the future value of a continuous income stream (see Section 5.5).

EXAMPLE 6.1.5

Joyce is considering a 5-year investment, and estimates that t years from now it will be generating a continuous income stream of $3{,}000 + 50t$ dollars per year. If the prevailing annual interest rate remains fixed at 4% compounded continuously during the entire 5-year term, what should the investment be worth in 5 years?

Solution

We measure the "worth" of Joyce's investment by the future value of the income flow over the 5-year term. Recall (from Section 5.5) that an income stream deposited continuously at the rate $f(t)$ into an account that earns interest at an annual rate r compounded continuously for a term of T years has future value FV given by the integral

$$FV = \int_0^T f(t) e^{r(T-t)} \, dt$$

For this investment, we have $f(t) = 3{,}000 + 50t$, $r = 0.04$, and $T = 5$, so the future value is given by the integral

$$FV = \int_0^5 (3{,}000 + 50t)e^{0.04(5-t)}\,dt$$

Integrating by parts with

$$u = 3{,}000 + 50t \qquad dv = e^{0.04(5-t)}\,dt$$

$$du = 50\,dt \qquad v = \frac{e^{0.04(5-t)}}{-0.04} = -25e^{0.04(5-t)}$$

we get

$$FV = \left[(3{,}000 + 50t)(-25)e^{0.04(5-t)} \right]\Big|_0^5 - \int_0^5 (50)(-25)e^{0.04(5-t)}\,dt$$

$$= \left[(-75{,}000 - 1{,}250t)e^{0.04(5-t)} \right]\Big|_0^5 - 1{,}250\left[\frac{e^{0.04(5-t)}}{-0.04} \right]\Big|_0^5$$

$$= \left[(-106{,}250 - 1{,}250t)e^{0.04(5-t)} \right]\Big|_0^5 \qquad\qquad \textit{combine terms}$$

$$= \left[-106{,}250 - 1{,}250(5) \right]e^0 - \left[-106{,}250 - 1{,}250(0) \right]e^{0.04(5)}$$

$$\approx 17{,}274.04$$

Thus, in 5 years, Joyce's investment will be worth roughly $17,274.

Repeated Application of Integration by Parts

Sometimes integration by parts leads to a new integral that also must be integrated by parts. This situation is illustrated in Example 6.1.6.

EXAMPLE │ 6.1.6

Find $\displaystyle\int x^2 e^{2x}\,dx$.

Solution

Since the factor e^{2x} is easy to integrate and x^2 is simplified by differentiation, we choose

$$u = x^2 \qquad dv = e^{2x}\,dx$$

so that

$$du = 2x\,dx \qquad v = \int e^{2x}\,dx = \frac{1}{2}e^{2x}$$

Integrating by parts, we get

$$\int x^2 e^{2x}\,dx = x^2\left(\frac{1}{2}e^{2x}\right) - \int \left(\frac{1}{2}e^{2x}\right)(2x\,dx)$$

$$= \frac{1}{2}x^2 e^{2x} - \int x e^{2x}\,dx$$

The integral $\int xe^{2x}\,dx$ that remains can also be obtained using integration by parts. Indeed, in Example 6.1.2, we found that

$$\int xe^{2x}\,dx = \frac{1}{2}\left(x - \frac{1}{2}\right)e^{2x} + C$$

Thus,

$$\int x^2 e^{2x}\,dx = \frac{1}{2}x^2 e^{2x} - \int xe^{2x}\,dx$$

$$= \frac{1}{2}x^2 e^{2x} - \left[\frac{1}{2}\left(x - \frac{1}{2}\right)e^{2x}\right] + C$$

$$= \frac{1}{4}(2x^2 - 2x + 1)e^{2x} + C$$

Using Integral Tables

Most integrals you will encounter in the social, managerial, and life sciences can be evaluated using the basic formulas given in Section 5.1 along with substitution and integration by parts. However, occasionally you may encounter an integral that cannot be handled by these methods. Some integrals such as $\int \dfrac{e^x}{x}\,dx$ cannot be evaluated by any method, but others can be found by using a **table of integrals.**

A short table of integrals* is listed on pages 472 and 473. Note that the table is divided into sections such as "integrals involving $\sqrt{u^2 - a^2}$," and that formulas are given in terms of constants denoted a, b, and n. The use of this table is demonstrated in Examples 6.1.7 through 6.1.11.

EXAMPLE ┃ 6.1.7

Find $\displaystyle\int \frac{1}{x(3x - 6)}\,dx$.

Solution

Apply formula 6 with $a = -6$ and $b = 3$ to obtain

$$\int \frac{1}{x(3x - 6)}\,dx = \frac{-1}{6}\ln\left|\frac{x}{3x - 6}\right| + C$$

EXAMPLE ┃ 6.1.8

Find $\displaystyle\int \frac{1}{6 - 3x^2}\,dx$.

Solution

If the coefficient of x^2 were 1 instead of 3, you could use Formula 16. This suggests that you first rewrite the integrand as

$$\frac{1}{6 - 3x^2} = \frac{1}{3}\left(\frac{1}{2 - x^2}\right)$$

---

*A longer table of integrals can be found in reference books, such as Murray R. Spiegel, *Mathematical Handbook of Formulas and Tables*, Schaum Outline Series, New York: McGraw-Hill, 1968.

and then apply Formula 16 with $a = \sqrt{2}$:

$$\int \frac{1}{6 - 3x^2}\, dx = \frac{1}{3} \int \frac{1}{2 - x^2}\, dx$$

$$= \frac{1}{3}\left(\frac{1}{2\sqrt{2}}\right) \ln \left|\frac{\sqrt{2} + x}{\sqrt{2} - x}\right| + C$$

EXAMPLE | 6.1.9

Find $\displaystyle\int \frac{1}{3x^2 + 6}\, dx$.

Solution

It is natural to try to match this integral to the one in Formula 16 by writing

$$\int \frac{1}{3x^2 + 6}\, dx = -\frac{1}{3} \int \frac{1}{-2 - x^2}\, dx$$

However, since -2 is negative, it cannot be written as the square a^2 of any real number a, so the formula does not apply. There is a formula for this integral, but it involves inverse trigonometric functions, which are not covered in this text.

EXAMPLE | 6.1.10

Find $\displaystyle\int \frac{1}{\sqrt{4x^2 - 9}}\, dx$.

Solution

To put this integral in the form of Formula 20, rewrite the integrand as

$$\frac{1}{\sqrt{4x^2 - 9}} = \frac{1}{\sqrt{4(x^2 - 9/4)}} = \frac{1}{2\sqrt{x^2 - 9/4}}$$

Then apply the formula with $a^2 = 9/4$ to get

$$\int \frac{1}{\sqrt{4x^2 - 9}}\, dx = \frac{1}{2} \int \frac{1}{\sqrt{x^2 - 9/4}}\, dx = \frac{1}{2} \ln |x + \sqrt{x^2 - 9/4}| + C$$

Our final example involves the use of Formula 26. This is called a **reduction formula** because it allows us to express a given integral in terms of a simpler integral of the same form.

EXAMPLE | 6.1.11

Find $\displaystyle\int x^3 e^{5x}\, dx$.

Solution

Apply Formula 26 with $n = 3$ and $a = 5$ to get

$$\int x^3 e^{5x}\, dx = \frac{1}{5} x^3 e^{5x} - \frac{3}{5} \int x^2 e^{5x}\, dx$$

A SHORT TABLE OF INTEGRALS

Forms Involving $a + bu$

1. $\displaystyle\int \frac{u\,du}{a + bu} = \frac{1}{b^2}[a + bu - a\ln|a + bu|] + C$

2. $\displaystyle\int \frac{u^2\,du}{a + bu} = \frac{1}{2b^3}[(a + bu)^2 - 4a(a + bu) + 2a^2\ln|a + bu|] + C$

3. $\displaystyle\int \frac{u\,du}{(a + bu)^2} = \frac{1}{b^2}\left[\frac{a}{a + bu} + \ln|a + bu|\right] + C$

4. $\displaystyle\int \frac{u\,du}{\sqrt{a + bu}} = \frac{2}{3b^2}(bu - 2a)\sqrt{a + bu} + C$

5. $\displaystyle\int \frac{du}{u\sqrt{a + bu}} = \frac{1}{\sqrt{a}}\ln\left|\frac{\sqrt{a + bu} - \sqrt{a}}{\sqrt{a + bu} + \sqrt{a}}\right| + C \quad a > 0$

6. $\displaystyle\int \frac{du}{u(a + bu)} = \frac{1}{a}\ln\left|\frac{u}{a + bu}\right| + C$

7. $\displaystyle\int \frac{du}{u^2(a + bu)} = \frac{-1}{a}\left[\frac{1}{u} + \frac{b}{a}\ln\left|\frac{u}{a + bu}\right|\right] + C$

8. $\displaystyle\int \frac{du}{u^2(a + bu)^2} = \frac{-1}{a^2}\left[\frac{a + 2bu}{u(a + bu)} + \frac{2b}{a}\ln\left|\frac{u}{a + bu}\right|\right] + C$

Forms Involving $\sqrt{a^2 + u^2}$

9. $\displaystyle\int \sqrt{a^2 + u^2}\,du = \frac{u}{2}\sqrt{a^2 + u^2} + \frac{a^2}{2}\ln|u + \sqrt{a^2 + u^2}| + C$

10. $\displaystyle\int \frac{du}{\sqrt{a^2 + u^2}} = \ln|u + \sqrt{a^2 + u^2}| + C$

11. $\displaystyle\int \frac{du}{u\sqrt{a^2 + u^2}} = \frac{-1}{a}\ln\left|\frac{\sqrt{a^2 + u^2} + a}{u}\right| + C$

12. $\displaystyle\int \frac{du}{(a^2 + u^2)^{3/2}} = \frac{u}{a^2\sqrt{a^2 + u^2}} + C$

13. $\displaystyle\int u^2\sqrt{a^2 + u^2}\,du = \frac{u}{8}(a^2 + 2u^2)\sqrt{a^2 + u^2} - \frac{a^4}{8}\ln|u + \sqrt{a^2 + u^2}| + C$

Forms Involving $\sqrt{a^2 - u^2}$

14. $\displaystyle\int \frac{du}{u\sqrt{a^2 - u^2}} = \frac{-1}{a} \ln\left|\frac{a + \sqrt{a^2 - u^2}}{u}\right| + C$

15. $\displaystyle\int \frac{du}{u^2\sqrt{a^2 - u^2}} = \frac{\sqrt{a^2 - u^2}}{a^2 u} + C$

16. $\displaystyle\int \frac{du}{a^2 - u^2} = \frac{1}{2a} \ln\left|\frac{a + u}{a - u}\right| + C$

17. $\displaystyle\int \frac{\sqrt{a^2 - u^2}}{u}\,du = \sqrt{a^2 - u^2} - a\ln\left|\frac{a + \sqrt{a^2 - u^2}}{u}\right| + C$

Forms Involving $\sqrt{u^2 - a^2}$

18. $\displaystyle\int \sqrt{u^2 - a^2}\,du = \frac{u}{2}\sqrt{u^2 - a^2} - \frac{a^2}{2}\ln|u + \sqrt{u^2 - a^2}| + C$

19. $\displaystyle\int \frac{\sqrt{u^2 - a^2}}{u^2}\,du = \frac{-\sqrt{u^2 - a^2}}{u} + \ln|u + \sqrt{u^2 - a^2}| + C$

20. $\displaystyle\int \frac{du}{\sqrt{u^2 - a^2}} = \ln|u + \sqrt{u^2 - a^2}| + C$

21. $\displaystyle\int \frac{du}{u^2\sqrt{u^2 - a^2}} = \frac{\sqrt{u^2 - a^2}}{a^2 u} + C$

Forms Involving e^{au} and $\ln u$

22. $\displaystyle\int ue^{au}\,du = \frac{1}{a^2}(au - 1)e^{au} + C$

23. $\displaystyle\int \ln u\,du = u\ln|u| - u + C$

24. $\displaystyle\int \frac{du}{u\ln u} = \ln|\ln u| + C$

25. $\displaystyle\int u^m \ln u\,du = \frac{u^{m+1}}{m+1}\left(\ln u - \frac{1}{m+1}\right) \quad m \neq -1$

Reduction Formulas

26. $\displaystyle\int u^n e^{au}\,du = \frac{1}{a}u^n e^{au} - \frac{n}{a}\int u^{n-1}e^{au}\,du$

27. $\displaystyle\int (\ln u)^n\,du = u(\ln u)^n - n\int (\ln u)^{n-1}\,du$

28. $\displaystyle\int u^n\sqrt{a + bu}\,du = \frac{2}{b(2n + 3)}[u^n(a + bu)^{3/2} - na\int u^{n-1}\sqrt{a + bu}\,du] \quad \text{for } n \neq -\frac{3}{2}$

Now apply Formula 26 again, this time with $n = 2$ and $a = 5$ to get

$$\int x^2 e^{5x}\,dx = \frac{1}{5}x^2 e^{5x} - \frac{2}{5}\int x e^{5x}\,dx$$

Then apply the formula one more time (with $n = 1$, $a = 5$) to evaluate the integral on the right:

$$\int x e^{5x}\,dx = \frac{1}{5}x e^{5x} - \frac{1}{5}\int (1)e^{5x}\,dx$$

$$= \frac{1}{5}x e^{5x} - \frac{1}{5}\left(\frac{1}{5}e^{5x}\right) + C$$

Finally, we evaluate the original integral by combining these results:

$$\int x^3 e^{5x}\,dx = \frac{1}{5}x^3 e^{5x} - \frac{3}{5}\left[\frac{1}{5}x^2 e^{5x} - \frac{2}{5}\int x e^{5x}\,dx\right]$$

$$= \frac{1}{5}x^3 e^{5x} - \frac{3}{25}x^2 e^{5x} + \frac{6}{25}\left[\frac{1}{5}x e^{5x} - \frac{1}{5}\left(\frac{1}{5}e^{5x}\right)\right] + C$$

$$= \left[\frac{1}{5}x^3 - \frac{3}{25}x^2 + \frac{6}{125}x - \frac{6}{625}\right]e^{5x} + C$$

COMPUTER ALGEBRA SYSTEMS   Certain scientific calculators as well as computational software programs such as MAPLE, MATHCAD, and MATHE-MATICA contain a computer algebra system (CAS) that will compare a given integrand to integrands in a stored table. While electronic integration using a CAS is certainly helpful, it is by no means necessary since most integrals you encounter can be handled quite easily by hand or by using the short table on pages 472 and 473. ∎

P R O B L E M S   6.1

In Problems 1 through 26, use integration by parts to find the given integral.

1. $\displaystyle\int x e^{-x}\,dx$

2. $\displaystyle\int x e^{x/2}\,dx$

3. $\displaystyle\int (1 - x)e^x\,dx$

4. $\displaystyle\int (3 - 2x)e^{-x}\,dx$

5. $\displaystyle\int t \ln 2t\,dt$

6. $\displaystyle\int t \ln t^2\,dt$

7. $\displaystyle\int v e^{-v/5}\,dv$

8. $\displaystyle\int w e^{0.1w}\,dw$

9. $\displaystyle\int x\sqrt{x - 6}\,dx$

10. $\displaystyle\int x\sqrt{1 - x}\,dx$

11. $\int x(x + 1)^8\, dx$

12. $\int (x + 1)(x + 2)^6\, dx$

13. $\int \dfrac{x}{\sqrt{x + 2}}\, dx$

14. $\int \dfrac{x}{\sqrt{2x + 1}}\, dx$

15. $\int_{-1}^{4} \dfrac{x}{\sqrt{x + 5}}\, dx$

16. $\int_{0}^{2} \dfrac{x}{\sqrt{4x + 1}}\, dx$

17. $\int_{0}^{1} \dfrac{x}{e^{2x}}\, dx$

18. $\int_{1}^{e} \dfrac{\ln x}{x^2}\, dx$

19. $\int_{1}^{e^2} x \ln \sqrt[3]{x}\, dx$

20. $\int_{0}^{1} x(e^{-2x} + e^{-x})\, dx$

21. $\int_{1/2}^{e/2} t \ln 2t\, dt$

22. $\int_{1}^{2} (t - 1)e^{1-t}\, dt$

23. $\int \dfrac{\ln x}{x^2}\, dx$

24. $\int x(\ln x)^2\, dx$

25. $\int x^3 e^{x^2}\, dx$

26. $\int \dfrac{x^3}{\sqrt{x^2 + 1}}\, dx$

[*Hint:* Use $dv = xe^{x^2}\, dx$.]

$\left[\textit{Hint:} \text{ Use } dv = \dfrac{x}{\sqrt{x^2 + 1}}\, dx. \right]$

Use the table of integrals on pages 472–473 to find the integrals in Problems 27 through 38.

27. $\int \dfrac{x\, dx}{3 - 5x}$

28. $\int \sqrt{x^2 - 9}\, dx$

29. $\int \dfrac{\sqrt{4x^2 - 9}}{x^2}\, dx$

30. $\int \dfrac{dx}{(9 + 2x^2)^{3/2}}$

31. $\int \dfrac{dx}{x(2 + 3x)}$

32. $\int \dfrac{t\, dt}{\sqrt{4 - 5t}}$

33. $\int \dfrac{du}{16 - 3u^2}$

34. $\int we^{-3w}\, dw$

35. $\int (\ln x)^3\, dx$

36. $\int x^2 \sqrt{2 + 5x}\, dx$

37. $\int \dfrac{dx}{x^2(5 + 2x)^2}$

38. $\int \dfrac{\sqrt{9 - x^2}}{x}\, dx$

39. Find the function whose tangent line has slope $(x + 1)e^{-x}$ for each value of x and whose graph passes through the point (1, 5).

40. Find the function whose tangent line has slope $x \ln \sqrt{x}$ for each value of $x > 0$ and whose graph passes through the point (2, −3).

41. DISTANCE   After t seconds, an object is moving with velocity $te^{-t/2}$ meters per second. Express the position of the object as a function of time.

42. EFFICIENCY   After t hours on the job, a factory worker can produce $100te^{-0.5t}$ units per hour. How many units does the worker produce during the first 3 hours?

43. **FUND-RAISING** After t weeks, contributions in response to a local fund-raising campaign were coming in at the rate of $2,000te^{-0.2t}$ dollars per week. How much money was raised during the first 5 weeks?

44. **MARGINAL COST** A manufacturer has found that marginal cost is $(0.1q + 1)e^{0.03q}$ dollars per unit when q units have been produced. The total cost of producing 10 units is $200. What is the total cost of producing the first 20 units?

45. **POPULATION GROWTH** It is projected that t years from now the population of a certain city will be changing at the rate of $t \ln \sqrt{t + 1}$ thousand people per year. If the current population is 2 million, what will the population be 5 years from now?

46. **POPULATION GROWTH** The population $P(t)$ (thousands) of a bacterial colony t hours after the introduction of a toxin is changing at the rate

$$P'(t) = (1 - 0.5t)e^{0.5t} \text{ thousand bacteria per hour}$$

By how much does the population change during the fourth hour?

47. **AVERAGE DRUG CONCENTRATION** The concentration of a drug t hours after being injected into a patient's bloodstream is $C(t) = 4\, te^{(2 - 0.3t)}$ mg/mL. What is the average concentration of drug in the patient's bloodstream over the first 6 hours after the injection?

Problems 48 through 59 involve applications developed in Sections 5.4, 5.5, and 5.6 of Chapter 5.

48. **AVERAGE DEMAND** A manufacturer determines that when x hundred units of a particular commodity are produced, the profit generated is $P(x)$ thousand dollars, where

$$P(x) = \frac{500 \ln(x + 1)}{(x + 1)^2}$$

What is the average profit over the production range $0 \le x \le 10$?

49. **FUTURE VALUE OF AN INVESTMENT** Money is transferred into an account at the rate of $R(t) = 3,000 + 5t$ dollars per year for 10 years, where t is the number of years after 2000. If the account pays 5% interest compounded continuously, how much will be in the account at the end of the 10-year investment period (in 2010)?

50. **FUTURE VALUE OF AN INVESTMENT** Money is transferred into an account at the rate of $R(t) = 1,000te^{-0.3t}$ dollars per year for 5 years. If the account pays 4% interest compounded continuously, how much will accumulate in the account over a 5-year period?

51. **PRESENT VALUE OF AN INVESTMENT** An investment will generate income continuously at the rate of $R(t) = 20 + 3t$ hundred dollars per year for 5 years. If the prevailing interest rate is 7% compounded continuously, what is the present value of the investment?

52. **INVESTMENT EVALUATION** The management of a national chain of pizza parlors is selling a 6-year franchise to operate its newest outlet in Orlando, Florida. Experience in similar localities suggests that t years from now, the franchise will be generating profit continuously at the rate of $R(t) = 300 + 5t$ thousand dollars per year. If the prevailing rate of interest remains fixed during the next 6 years at 6% compounded continuously, what would be a fair price to charge for the franchise? [*Hint:* Use present value to measure what the franchise is "worth."]

53. **GROUP MEMBERSHIP** A book club has compiled statistics suggesting that the fraction of its members who are still active t months after joining is given by the function $S(t) = e^{-0.02t}$. The club currently has 5,000 members and expects to attract new members at the rate $R(t) = 5t$ per month. How many members can the club expect to have at the end of 9 months? [*Hint:* Think of this as a survival/renewal problem.]

54. **SPREAD OF DISEASE** A new virus has just been declared an epidemic by health officials. Currently, 10,000 people have the disease and it is estimated that t days from now, new cases will be reported at the rate of $R(t) = 10te^{-0.1t}$ people per day. If the fraction of victims who still have the virus t days after first contracting it is given by $S(t) = e^{-0.015t}$, how many people will be infected by the virus 90 days from now? One year (365 days) from now? [*Hint:* Think of this as a survival/renewal problem.]

55. CONSUMERS' SURPLUS A manufacturer has determined that when q thousand units of a particular commodity are produced, the price at which all the units can be sold is $p = D(q)$ dollars per unit, where D is the demand function

$$D(q) = 10 - qe^{0.02q}$$

　a. At what price are 5,000 ($q_0 = 5$) units demanded?
　b. Find the consumers' surplus when 5,000 units are demanded.

56. CONSUMERS' SURPLUS Answer the questions in Problem 55 for the demand function

$$D(q) = \ln\left(\frac{52}{q+1}\right)$$

when 12,000 ($q_0 = 12$) units are demanded.

57. LORENTZ CURVE Find the Gini index for an income distribution whose Lorentz curve is the graph of the function $L(x) = xe^{x-1}$ for $0 \le x \le 1$.

58. COMPARING INCOME DISTRIBUTIONS In a certain state, the Lorentz curves for the distributions of income for lawyers and engineers are $y = L_1(x)$ and $y = L_2(x)$, respectively, where

$$L_1(x) = 0.6x^2 + 0.4x \quad \text{and} \quad L_2(x) = x^2 e^{x-1}$$

Find the Gini index for each curve. Which profession has the more equitable distribution of income?

59. CARDIAC OUTPUT A physician injects 5 mg of dye into a vein near the heart of a patient and determines that the concentration of dye leaving the heart after t seconds is given by the function

$$C(t) = 1.54te^{-0.12t} - 0.007t^2$$

Assuming it takes 20 seconds for all the dye to clear the heart, what is the patient's cardiac output?

60. Use integration by parts to verify reduction formula 27:

$$\int (\ln u)^n \, du = u(\ln u)^n - n \int (\ln u)^{n-1} \, du$$

61. Use integration by parts to verify reduction formula 26:

$$\int u^n e^{au} \, du = \frac{1}{a} u^n e^{au} - \frac{n}{a} \int u^{n-1} e^{au} \, du$$

62. Find a reduction formula for

$$\int u^n (\ln u)^m \, du$$

CENTER OF A REGION Let $f(x)$ be a continuous function with $f(x) \ge 0$ for $a \le x \le b$. Let R be the region bounded by the curve $y = f(x)$, the x axis, and the lines $x = a$ and $x = b$. Then the centroid (or center) of R is the point $(\bar{x}, \bar{y})$, where

$$\bar{x} = \frac{1}{A} \int_a^b xf(x) \, dx \quad \text{and} \quad \bar{y} = \frac{1}{2A} \int_a^b [f(x)]^2 \, dx$$

and A is the area of R. In Problems 63 and 64, find the centroid of the region shown.

63.

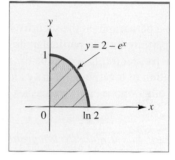

PROBLEM 63

64.

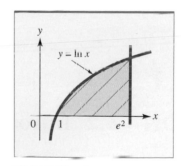

PROBLEM 64

[*Hint:* You may need to use a formula from the integral table on pages 472 and 473.]

65. SHOPPING MALL SECURITY The shaded region shown in the accompanying figure is a parking lot for a shopping mall. The dimensions are in hundreds of feet. To improve parking security, the mall manager plans to place a surveillance kiosk in the center of the lot (see the definition preceding Problems 63 and 64).

　a. Where would you place the kiosk? Describe the location you choose in terms of the coordinate system indicated in the figure. [*Hint:* You may require a formula from the integral table on pages 472 and 473.]

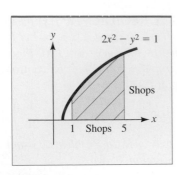

$2x^2 - y^2 = 1$

Shops

1 Shops 5

PROBLEM 65

 b. Write a paragraph on the security of shopping mall parking lots. In particular, do you think geometric centrality would be the prime consideration in locating a security kiosk or are there other, more important issues?

66. Use the graphing utility of your calculator to draw 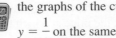 the graphs of the curves $y = x^2 e^{-x}$ and $y = \dfrac{1}{x}$ on the same screen. Use **ZOOM** and **TRACE** or some other feature of your calculator to find where the curves intersect, and then compute the area of the region bounded by the curves.

67. Repeat Problem 66 for the curves $\dfrac{x^2}{5} - \dfrac{y^2}{2} = 1$ and $y = x^3 - 3.5x^2 + 2x$

68. Repeat Problem 66 for the curves $y = \ln x$ and $y = x^2 - 5x + 4$

69. Repeat Problem 66 for the curves  $y = e^{2x} + 4$ and $y = 5e^x$

If your calculator has a numeric integration feature, use it to evaluate each of the integrals in Problems 70 through 73. In each case, verify your result by applying an appropriate integration formula.

70. $\displaystyle\int_1^2 x^2 \ln \sqrt{x}\, dx$

71. $\displaystyle\int_2^3 \sqrt{4x^2 - 7}\, dx$

72. $\displaystyle\int_0^1 x^3 \sqrt{4 + 5x}\, dx$

73. $\displaystyle\int_0^1 \dfrac{\sqrt{x^2 + 2x}}{(x + 1)^2}\, dx$

SECTION 6.2 Introduction to Differential Equations

In Section 1.4, we introduced mathematical modeling as a dynamic process involving three stages:

1. A real-world problem is given a mathematical formulation, called a *mathematical model*. This often involves making simplifying assumptions about the problem to make it more accessible.

2. The mathematical model is analyzed or solved by using tools from areas such as algebra, calculus, or statistics, among others, or by numerical methods involving graphing calculators or computers.

3. The solution of the mathematical model is interpreted in terms of the original real-world problem. This often leads to adjustments in the assumptions of the model.

The process may then be repeated, each time with a more refined model, until a satisfactory understanding of the real-world problem is attained.

We have used calculus to analyze mathematical models throughout this text, and many of these models have involved rates of change. Sometimes the mathematical formulation of a problem involves an equation in which a quantity and the rate of change of that quantity are related by an equation. Since rates of change are expressed in terms of derivatives or differentials, such an equation is appropriately called a **differential equation.** For example,

$$\frac{dy}{dx} = 3x^2 + 5 \quad \text{and} \quad \frac{dP}{dt} = kP \quad \text{and} \quad \left(\frac{dy}{dx}\right)^2 + 3\frac{dy}{dx} + 2y = e^x$$

are all differential equations. Differential equations are among the most useful tools for modeling continuous phenomena. Population dynamics, chemical kinetics, spread of disease, dynamic economic behavior, ecology, and the transmission of information are a few of the many areas in the managerial, physical, and social and life sciences that can be studied using differential equations. In this section, we introduce techniques for solving basic differential equations and examine a few practical applications.

The simplest type of differential equation has the form

$$\frac{dy}{dx} = g(x)$$

in which the derivative of the quantity y is given explicitly as a function of the independent variable x. Such an equation can be solved by simply finding the indefinite integral of $g(x)$. A complete characterization of all possible solutions of the equation is called a **general solution.** A differential equation coupled with a side condition is referred to as an **initial value problem,** and a solution that satisfies both the differential equation and the side condition is called a **particular solution** of the initial value problem. In Examples 6.2.1 through 6.2.3, we illustrate this terminology and demonstrate how differential equations can be used to build mathematical models.

EXAMPLE | 6.2.1

Find the general solution of the differential equation

$$\frac{dy}{dx} = x^2 + 3x$$

and the particular solution that satisfies $y = 2$ when $x = 1$.

Solution

Integrating, you get

$$y = \int \left(\frac{dy}{dx}\right) dx = \int (x^2 + 3x)\, dx$$

$$= \frac{1}{3}x^3 + \frac{3}{2}x^2 + C$$

This is the general solution since all solutions can be expressed in this form. For the particular solution, substitute $x = 1$ and $y = 2$ into the general solution:

$$2 = \frac{1}{3}(1)^3 + \frac{3}{2}(1)^2 + C$$

$$C = 2 - \frac{1}{3} - \frac{3}{2} = \frac{1}{6}$$

Thus, the required particular solution is $y = \frac{1}{3}x^3 + \frac{3}{2}x^2 + \frac{1}{6}$.

EXAMPLE | 6.2.2

The resale value of a certain industrial machine decreases over a 10-year period at a rate that depends on the age of the machine. When the machine is x years old, the rate at which its value is changing is $220(x - 10)$ dollars per year. Express the value of the machine as a function of its age and initial value. If the machine was originally worth $12,000, how much will it be worth when it is 10 years old?

Solution

Let $V(x)$ denote the value of the machine when it is x years old. The derivative $\dfrac{dV}{dx}$ is equal to the rate $220(x - 10)$ at which the value of the machine is changing. Hence, to model this problem, we can use the differential equation

$$\frac{dV}{dx} = 220(x - 10) = 220x - 2,200$$

To find V, solve this differential equation by integration:

$$V(x) = \int (220x - 2,200)\, dx = 110x^2 - 2,200x + C$$

Notice that C is equal to $V(0)$, the initial value of the machine. A more descriptive symbol for this constant is V_0. Using this notation, you can write the general solution as

$$V(x) = 110x^2 - 2,200x + V_0$$

If $V_0 = 12,000$, the corresponding particular solution is

$$V(x) = 110x^2 - 2,200x + 12,000$$

Thus, when the machine is $x = 10$ years old, its value is

$$V(10) = 110(10)^2 - 2,200(10) + 12,000 = \$1,000$$

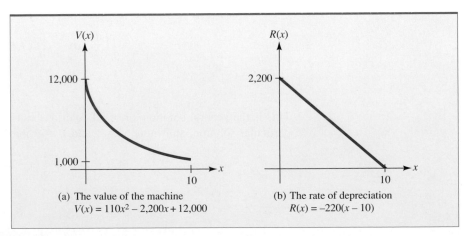

(a) The value of the machine
$V(x) = 110x^2 - 2,200x + 12,000$

(b) The rate of depreciation
$R(x) = -220(x - 10)$

FIGURE 6.2   The value of the machine and its rate of depreciation.

The negative of the rate of change of resale value of the machine

$$R = -\frac{dV}{dx} = -220(x - 10)$$

is called the **rate of depreciation.** Graphs of the resale value V of the machine and the rate of depreciation R are shown in Figure 6.2.

EXAMPLE | 6.2.3

An oil well that has just been opened is expected to yield 300 barrels of crude oil per month and, at that rate, is expected to run dry in 3 years. It is estimated that t months from now, the price of crude oil will be $p(t) = 28 + 0.3\sqrt{t}$ dollars per barrel. If the oil is sold as soon as it is extracted from the ground, what is the total revenue generated by the well during its operation?

Solution

Let $R(t)$ be the revenue generated during the first t months after the well is opened, so that $R(0) = 0$. To construct the relevant differential equation, use the rate relationship

$$\begin{pmatrix} \text{Rate of change of revenue with} \\ \text{respect to time (dollars/month)} \end{pmatrix} = \begin{pmatrix} \text{dollars per} \\ \text{barrel} \end{pmatrix}\begin{pmatrix} \text{barrels per} \\ \text{month} \end{pmatrix}$$

It follows that we can model this problem using the differential equation

$$\underbrace{\frac{dR}{dt}}_{\substack{\text{rate of change} \\ \text{of revenue}}} = \underbrace{(28 + 0.3\sqrt{t})}_{\substack{\text{dollars} \\ \text{per barrel}}}\ \underbrace{(300)}_{\substack{\text{barrels} \\ \text{per month}}}$$

and it follows that

$$\frac{dR}{dt} = (28 + 0.3\sqrt{t})(300)$$
$$= 8,400 + 90t^{1/2}$$

Integrating, we find that the general solution of this differential equation is

$$R(t) = \int(8,400 + 90t^{1/2})\,dt = 8,400t + 90\left(\frac{t^{3/2}}{3/2}\right) + C$$
$$= 8,400t + 60t^{3/2} + C$$

Since $R(0) = 0$, we find that

$$R(0) = 0 = 8,400(0) + 60(0)^{3/2} + C$$

so $C = 0$, and the appropriate particular solution is

$$R(t) = 8,400t + 60t^{3/2}$$

Therefore, since the well will run dry in 36 months, the total revenue generated by the well during its operation is

$$R(36) = 8,400(36) + 60(36)^{3/2}$$
$$= \$315,360$$

Separable Equations   Many useful differential equations can be formally rewritten so that all the terms containing the independent variable appear on one side of the equation and all the terms containing the dependent variable appear on the other. Differential equations with this special property are said to be **separable** and can be solved by the following procedure involving two integrations.

Separable Differential Equations   ■   A differential equation that can be written in the form

$$\frac{dy}{dx} = \frac{h(x)}{g(y)}$$

is said to be **separable.** The general solution of such an equation can be obtained by separating the variables and integrating both sides; that is,

$$\int g(y)\,dy = \int h(x)\,dx$$

EXAMPLE 6.2.4

Find the general solution of the differential equation $\dfrac{dy}{dx} = \dfrac{2x}{y^2}$.

Solution

To separate the variables, pretend that the derivative $\dfrac{dy}{dx}$ is actually a quotient and write

$$y^2\,dy = 2x\,dx$$

Now integrate both sides of this equation to get

$$\int y^2\,dy = \int 2x\,dx$$

$$\frac{1}{3}y^3 + C_1 = x^2 + C_2$$

where C_1 and C_2 are arbitrary constants. Solving for y, we get

$$\frac{1}{3}y^3 = x^2 + (C_2 - C_1) = x^2 + C_3 \quad \text{where } C_3 = C_2 - C_1$$

$$y^3 = 3x^2 + 3C_3 = 3x^2 + C \quad\quad \text{where } C = 3C_3$$

$$y = (3x^2 + C)^{1/3}$$

EXAMPLE 6.2.5

An object moves along the x axis in such a way that at each time t, its velocity is given by the differential equation

$$\frac{dx}{dt} = x^2 \ln t$$

If the object is at $x = -2$ when $t = 1$, where is it when $t = 3$?

Solution

Separating the variables in the given differential equation and integrating, we get

$$\int \frac{1}{x^2}\, dx = \int \ln t\, dt$$

$$\frac{-1}{x} + C_1 = t \ln t - \int \frac{1}{t}(t)\, dt \qquad \begin{array}{l} \textit{integration by parts with} \\ u = \ln t \quad dv = dt \end{array}$$

$$= t \ln t - t + C_2 \qquad du = \frac{1}{t}\, dt \quad v = t$$

so the general solution is

$$\frac{-1}{x} = t \ln t - t + C \qquad \textit{where } C = C_2 - C_1$$

Since $x = -2$ when $t = 1$, we have

$$\frac{-1}{(-2)} = (1)\ln(1) - (1) + C \qquad \textit{since } \ln 1 = 0$$

$$\frac{1}{2} = 0 - 1 + C$$

$$C = \frac{1}{2} + 1 = \frac{3}{2}$$

so that

$$\frac{-1}{x} = t \ln t - t + \frac{3}{2}$$

In particular, when $t = 3$, we have

$$\frac{-1}{x} = (3)\ln(3) - (3) + \frac{3}{2} \approx 1.80$$

Solving this equation for x, we get

$$x \approx \frac{-1}{1.8} \approx -0.56$$

which means that the object is at $x = -0.56$ when $t = 3$.

Exponential Growth and Decay

In Section 4.1, we obtained the formulas $Q = Q_0 e^{kt}$ for exponential growth and $Q = Q_0 e^{-kt}$ for exponential decay, and then in Section 4.3, we showed that the rate of change of a quantity undergoing either kind of exponential change (growth or decay) is proportional to its size; that is, $\frac{dQ}{dt} = mQ$ for some constant m. Example 6.2.6 shows that the converse of this result is also true.

EXAMPLE | 6.2.6

Show that a quantity Q that changes at a rate proportional to its size satisfies $Q(t) = Q_0 e^{mt}$, where $Q_0 = Q(0)$.

Solution

Since Q changes at a rate proportional to its size, we have

$$\frac{dQ}{dt} = mQ$$

for constant m. Separating the variables and integrating, we get

$$\int \frac{1}{Q} \, dQ = \int m \, dt$$
$$\ln Q = mt + C_1$$

and by taking exponentials on both sides

$$Q(t) = e^{\ln Q} = e^{mt + C_1} = e^{C_1} e^{mt}$$
$$= Ce^{mt} \qquad\qquad \textit{where } C = e^{C_1}$$

Since $Q(0) = Q_0$, it follows that

$$Q_0 = Q(0) = Ce^0 = C \qquad\qquad \textit{since } e^0 = 1$$

Therefore, $Q(t) = Q_0 e^{mt}$, as required.

Learning Models   In Chapter 4, we referred to the graphs of functions of the form $Q(t) = B - Ae^{-kt}$ as *learning curves* because functions of this form often describe the relationship between the efficiency with which an individual performs a given task and the amount of training or experience the "learner" has had. In general, any quantity that grows at a rate proportional to the difference between its size and a fixed upper limit is represented by a function of this form. Here is an example involving such a learning model.

EXAMPLE 6.2.7

The rate at which people hear about a new increase in postal rates is proportional to the number of people in the country who have not yet heard about it. Express the number of people who have heard already about the increase as a function of time.

Solution

Let $Q(t)$ denote the number of people who have already heard about the rate increase at time t and let B be the total population of the country. Then

$$\text{Rate at which people} \atop \text{hear about the increase} = \frac{dQ}{dt}$$

and

$$\text{Number of people who have} \atop \text{not yet heard about the increase} = B - Q$$

Because the rate at which people hear about the increase is proportional to the number of people who have not yet heard about the increase, if follows that

$$\frac{dQ}{dt} = k(B - Q)$$

where k is the constant of proportionality. Notice that the constant k must be positive because $\dfrac{dQ}{dt} > 0$ (since Q is an increasing function of t) and $B - Q > 0$ (since $B > Q$).

Separate the variables by writing

$$\frac{1}{B - Q} \, dQ = k \, dt$$

and integrate to get

$$\int \frac{1}{B - Q}\, dQ = \int k\, dt$$

$$-\ln |B - Q| = kt + C$$

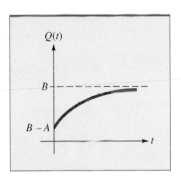

(Be sure you see where the minus sign came from.) This time you can drop the absolute value sign immediately since $B - Q > 0$. Hence,

$$-\ln (B - Q) = kt + C$$
$$\ln (B - Q) = -kt - C$$
$$B - Q = e^{-kt - C} = e^{-kt} e^{-C}$$
$$Q = B - e^{-C} e^{-kt}$$

Denoting the constant e^{-C} by A and using functional notation, you can conclude that

$$Q(t) = B - Ae^{-kt}$$

FIGURE 6.3 A learning curve: $Q(t) = B - Ae^{-kt}$.

which is precisely the general equation of a learning curve. For reference, the graph of Q is sketched in Figure 6.3.

Logistic Growth

Recall from Chapter 3 that the relative rate of growth of a quantity $Q(t)$ is given by the ratio $Q'(t)/Q(t)$. In exponential growth and decay, where $Q'(t) = kQ(t)$, the relative rate of growth is constant; namely,

$$\frac{Q'(t)}{Q(t)} = k$$

A population with no restrictions on environment or resources can often be modeled as having exponential growth. However, when such restrictions exist, it is reasonable to assume there is a largest population B supportable by the environment (called the **carrying capacity**) and that at any time t, the relative rate of population growth is proportional to the *potential* population $B - Q(t)$; that is,

$$\frac{Q'(t)}{Q(t)} = k(B - Q(t))$$

where k is a positive constant. This rate relationship leads to the differential equation

$$\frac{dQ}{dt} = kQ(B - Q)$$

This is called the **logistic equation,** and the corresponding model of restricted population growth is known as the **logistic model.**

The logistic differential equation is separable, but before solving it analytically, let us see what we can deduce using qualitative methods. First, note that the logistic equation can be written as $\dfrac{dQ}{dt} = R(Q)$, where the population rate $R(Q) = kQ(B - Q)$ is a quadratic function in Q. Since $k > 0$, the graph of $R(Q)$ is a downward opening parabola (see Figure 6.4), and since $R(0) = R(B) = 0$, the graph intersects the Q axis at 0 and at B. The high point (vertex) of the graph is where $Q = B/2$.

If the initial population $Q(0)$ is 0, then $Q(t) = 0$ for all $t > 0$, and likewise, if the initial population is $Q(0) = B$, then the population stays at the level of the carrying capacity for all $t > 0$. The two constant solutions $Q = 0$ and $Q = B$, called *equilibrium solutions* of the logistic equation, are shown in Figure 6.5a.

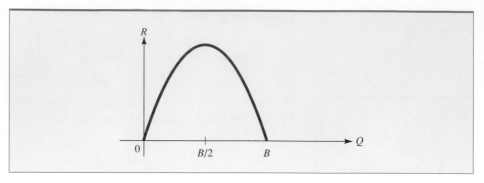

FIGURE 6.4 The graph of the population rate function $R(Q) = kQ(B - Q)$.

If the initial population $Q(0)$ satisfies $0 < Q(0) < B$, then

$$R(Q) = kQ(B - Q) > 0$$

since k, Q, and $B - Q$ are all positive. Since the rate of growth is positive, the population $Q(t)$ itself is increasing. As $Q(t)$ approaches B, the population growth rate $\dfrac{dQ}{dt}$ approaches 0, which suggests that the graph of the population function $Q(t)$ "flattens out," approaching the carrying capacity asymptotically. If the initial population satisfies $Q(0) > B$, we have

$$R(Q) = kQ(B - Q) < 0$$

since k and Q are positive and $B - Q$ is negative. In this case, the population *decreases* from its initial value, once again approaching the carrying capacity asymptotically as Q approaches B. Some nonconstant solutions of the logistic equation are shown in Figure 6.5b.

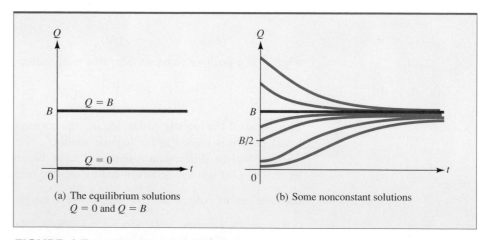

(a) The equilibrium solutions
$Q = 0$ and $Q = B$

(b) Some nonconstant solutions

FIGURE 6.5 Typical solutions of the logistic equation $dQ/dt = kQ(B - Q)$.

In addition to its role in modeling restricted population growth, the logistic equation can also be used to describe economic behavior as well as interactions within a social group, such as the spread of rumors (see Problem 40). In Example 6.2.8, we use a logistic model to describe the spread of an epidemic.

EXAMPLE 6.2.8

The rate at which an epidemic spreads through a community is jointly proportional to the number of residents who have been infected and the number of susceptible residents who have not. Express the number of residents who have been infected as a function of time.

Solution

Let $Q(t)$ denote the number of susceptible residents who have been infected by time t, and let B denote the total number of susceptible residents. Then

$$\frac{\text{Rate susceptible residents}}{\text{are being infected at time } t} = \frac{dQ}{dt}$$

and

$$\frac{\text{Number of susceptible}}{\text{residents not yet infected}} = B - Q$$

"Jointly proportional" means "proportional to the product," so the differential equation that describes the spread of the epidemic is

$$\frac{dQ}{dt} = kQ(B - Q)$$

where $k > 0$ is the constant of proportionality. (Do you see why k must be positive?) This is a separable differential equation whose solution is

$$\int \frac{dQ}{Q(B - Q)} = \int k\, dt$$

$$\frac{1}{B} \ln \left| \frac{Q}{B - Q} \right| = kt + C$$

where the integration on the left was performed using integral formula 6 in the integral table listed on page 472 in Section 6.1; specifically,

$$\int \frac{du}{u(a + bu)} = \frac{1}{a} \ln \left| \frac{u}{a + bu} \right| + C$$

with $u = Q$, $a = B$, and $b = -1$.

Since $Q > 0$ and $B > Q$, we can remove the absolute value bars from the solution and write

$$\frac{1}{B} \ln \left(\frac{Q}{B - Q} \right) = kt + C$$

$$\ln \left(\frac{Q}{B - Q} \right) = Bkt + BC$$

$$\frac{Q}{B - Q} = e^{Bkt + BC} = e^{Bkt} e^{BC}$$

$$= A_1 e^{Bkt} \quad \text{where } A_1 = e^{BC}$$

Multiplying both sides of the last equation by $B - Q$ and solving for Q, we get

$$Q = (B - Q)A_1 e^{Bkt} = (BA_1 - QA_1) e^{Bkt}$$

$$Qe^{-Bkt} = BA_1 - QA_1 \qquad \textit{multiply both sides by } e^{-Bkt}$$

EXPLORE!

Suppose in a small community of 1,000 residents the number contracting the flu over a 7-week period is given in the table:

Weeks	Number infected
1	45
2	75
3	200
4	450
5	595
6	700
7	760

Use the statistical modeling feature of your graphing calculator to fit a logistic curve to the data and determine the projected level of susceptible residents that will ultimately be infected, based on the logistic model.

$$Q(A_1 + e^{-Bkt}) = BA_1 \qquad \text{\textit{add } } QA_1 \text{ \textit{to both sides}}$$

$$Q = \frac{BA_1}{A_1 + e^{-Bkt}}$$

$$= \frac{B}{1 + \frac{1}{A_1} e^{-Bkt}} \qquad \begin{array}{l} \textit{divide all terms on} \\ \textit{the right by } A_1 \end{array}$$

Finally, setting $A = \dfrac{1}{A_1}$, we see that Q has the logistic form

$$Q(t) = \frac{B}{1 + Ae^{-Bkt}}$$

The graph of $Q(t)$, the number of infected residents, is shown in Figure 6.6. Note that this graph has the characteristic "S shape" of a logistic curve with carrying capacity B. At the beginning of the epidemic, only $Q(0) = \dfrac{B}{1 + A}$ residents are infected.

The number of infected residents increases rapidly at first and is spreading most rapidly at the inflection point on the graph. It is not difficult to show that this occurs when half the susceptible population is infected (see Problem 63). The infection rate then begins to decrease as the total number of infected residents asymptotically approaches the level B where all susceptible residents would be infected.

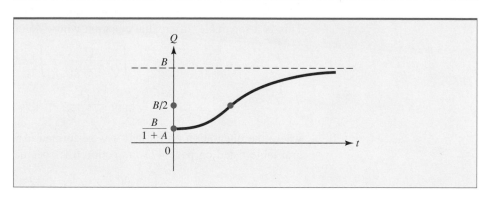

FIGURE 6.6   The logistic curve $y = \dfrac{B}{1 + Ae^{-Bkt}}$ showing the number of residents $Q(t)$ infected at time t during an epidemic.

Dilution Models   An important application of separable differential equations is to model situations in which a quantity is "diluted." Such models can be used in a variety of situations in areas such as finance, ecology, medicine, and chemistry. We illustrate the general dilution modeling procedure in a public health example.

EXAMPLE | 6.2.9

The residents of a certain community have voted to discontinue the fluoridation of their water supply. The local reservoir currently holds 200 million gallons of fluoridated water that contains 1,600 pounds of fluoride. The fluoridated water is flowing out of the reservoir at the rate of 4 million gallons per day and is being replaced at the same rate by unfluoridated water. At all times, the remaining fluoride is evenly distributed in the reservoir. Express the amount of fluoride in the reservoir as a function of time.

Solution

Let $Q(t)$ be the amount of fluoride (in pounds) in the reservoir t days after the fluoridation ends. Begin with the rate relationship

$$\begin{bmatrix} \text{Net rate of change of fluoride} \\ \text{with respect to time} \end{bmatrix} = \begin{bmatrix} \text{daily rate of fluoride} \\ \text{flowing in} \end{bmatrix} - \begin{bmatrix} \text{daily rate of fluoride} \\ \text{flowing out} \end{bmatrix}$$

The net rate of change of the fluoride with respect to time is $\dfrac{dQ}{dt}$, and since the fluoridation has been terminated, the rate of fluoride flowing in is 0. Since the volume of fluoridated water in the reservoir stays fixed at 200 million gallons and the fluoride is always evenly distributed in the reservoir, the concentration of fluoride in the reservoir at time t is given by the ratio

$$\frac{Q(t) \text{ pounds of fluoride}}{200 \text{ million gallons of fluoridated water}}$$

Therefore, since 4 million gallons of fluoridated water are being removed each day, the daily outflow rate of fluoride is given by the product

$$\begin{array}{l} \text{Daily rate of fluoride} \\ \text{flowing out} \end{array} = \left(\frac{Q(t) \text{ lb}}{200 \text{ million gal}} \right) \left(\frac{4 \text{ million gal}}{\text{day}} \right) = \frac{4Q}{200} \text{ lb/day}$$

The flow rate relationships are summarized in this diagram:

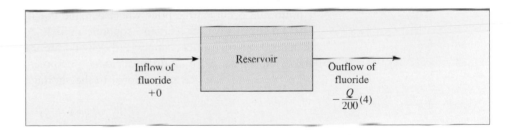

Since the rate of change of the amount of fluoride in the reservoir equals the inflow rate minus the outflow rate, it follows that

$$\frac{dQ}{dt} = \underbrace{0}_{\substack{\text{inflow} \\ \text{rate}}} - \underbrace{\frac{4Q}{200}}_{\substack{\text{outflow} \\ \text{rate}}} = -\frac{Q}{50}$$

Solving this differential equation by separation of variables, you get

$$\int \frac{1}{Q} \, dQ = -\int \frac{1}{50} \, dt$$

$$\ln Q = -\frac{t}{50} + C$$

$$Q = e^{C - t/50} = e^C e^{-t/50}$$

$$= Q_0 e^{-t/50} \qquad \text{where } Q_0 = e^C$$

Initially, the reservoir contained 1,600 pounds of fluoride, so that

$$1{,}600 = Q(0) = Q_0 e^0 = Q_0$$

Thus,

$$Q(t) = 1{,}600 e^{-t/50}$$

and the amount of fluoride decreases exponentially, as illustrated in Figure 6.7.

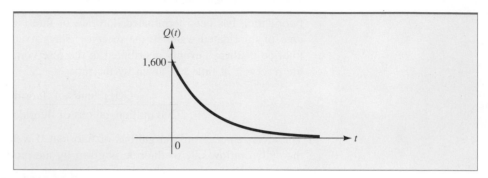

FIGURE 6.7 The amount of fluoride $Q(t) = 1{,}600 e^{-t/50}$.

A Price Adjustment Model

Let $S(p)$ denote the number of units of a particular commodity supplied to the market at a price of p dollars per unit, and let $D(p)$ denote the corresponding number of units demanded by the market at the same price. In static circumstances, market equilibrium occurs at the price where demand equals supply (recall the discussion in Section 1.4). However, certain economic models consider a more dynamic kind of economy in which price, supply, and demand are assumed to vary with time. One of these, the *Evans price adjustment model,** assumes that the rate of change of price with respect to time t is proportional to the shortage $D - S$, so that

$$\frac{dp}{dt} = k(D - S)$$

where k is a positive constant. Here is an example involving this model.

EXAMPLE 6.2.10

Suppose the price $p(t)$ of a particular commodity varies in such a way that its rate of change with respect to time is proportional to the shortage $D - S$, where $D(p)$ and $S(p)$ are the linear demand and supply functions $D = 8 - 2p$ and $S = 2 + p$.

a. If the price is \$5 when $t = 0$ and \$3 when $t = 2$, find $p(t)$.

b. Determine what happens to $p(t)$ in the "long run" (as $t \to +\infty$).

Solution

a. The rate of change of $p(t)$ is given by the separable differential equation

$$\frac{dp}{dt} = k(D - S) = k[(8 - 2p) - (2 + p)]$$

$$= k(6 - 3p)$$

---

*The Evans price adjustment model and several other dynamic economic models are examined in the text by J. E. Draper and J. S. Klingman, *Mathematical Analysis with Business and Economic Applications*, New York: Harper and Row, 1967, pp. 430–434.

Separating the variables, integrating, and solving for p, you get

$$\int \frac{dp}{6 - 3p} = \int k \, dt$$

$$\frac{-1}{3} \ln |6 - 3p| = kt + C_1$$

$$\ln |6 - 3p| = -3kt - 3C_1$$

$$6 - 3p = e^{-3kt - 3C_1}$$

$$-e^{-3kt} e^{3C_1} = Ce^{-3kt} \quad \text{where } C = e^{-3C_1}$$

$$p(t) = 2 - \frac{1}{3} Ce^{-3kt}$$

To evaluate the constant C, use the fact that $p(0) = 5$, so that

$$5 = p(0) = 2 - \frac{1}{3} Ce^0 = 2 - \frac{1}{3} C$$

$$C = -9$$

Thus, $$p(t) = 2 + 3e^{-3kt}$$

You still need to find k. Since $p = 3$ when $t = 2$, it follows that

$$3 = p(2) = 2 + 3e^{-3k(2)} = 2 + 3e^{-6k}$$

and by solving this equation for e^{-6k} and then taking logarithms, you get

$$e^{-6k} = \frac{3 - 2}{3} = \frac{1}{3}$$

$$-6k = \ln\left(\frac{1}{3}\right) = -1.0986$$

$$k = \frac{-1.0986}{-6} = 0.1831$$

You conclude that the price at time t is

$$p(t) = 2 + 3e^{-6(0.1831t)} = 2 + 3e^{-1.0986t}$$

b. As t increases without bound, $e^{-1.0986t}$ approaches 0 and $p(t)$ approaches 2, which is the price at which supply equals demand. That is, in the "long run," $p(t)$ approaches the equilibrium price of the commodity (see Figure 6.8).

EXPLORE!

Refer to Example 6.2.10. Graph $p(x) = 2 + 3e^{-1.0986x}$, using the window [0, 9.4]1 by [0, 7]1. Use the **TRACE** feature of your graphing calculator to find the smallest x value for which $p(x) < 2.01$. How about for $p(x) < 2.000001$?

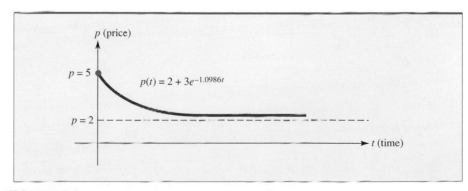

FIGURE 6.8 The price $p(t)$ approaches the equilibrium price $p = 2$ as $t \to +\infty$.

Why the Method of Separation of Variables Works

Consider the separable differential equation

$$\frac{dy}{dx} = \frac{h(x)}{g(y)}$$

or, equivalently,

$$g(y)\frac{dy}{dx} - h(x) = 0$$

The left-hand side of this equation can be rewritten in terms of the antiderivatives of g and h. In particular, if G is an antiderivative of g and H an antiderivative of h, it follows from the chain rule that

$$\frac{d}{dx}[G(y) - H(x)] = G'(y)\frac{dy}{dx} - H'(x) = g(y)\frac{dy}{dx} - h(x)$$

Hence, the differential equation $g(y)\dfrac{dy}{dx} - h(x) = 0$ says that

$$\frac{d}{dx}[G(y) - H(x)] = 0$$

But constants are the only functions whose derivatives are identically zero, and so

$$G(y) - H(x) = C$$

for some constant C. That is,

$$G(y) = H(x) + C$$

or, equivalently,

$$\int g(y)\,dy = \int h(x)\,dx + C$$

and the proof is complete.

PROBLEMS 6.2

In Problems 1 through 16, find the general solution of the given differential equation. You will need integration by parts in Problems 17 through 20.

1. $\dfrac{dy}{dx} = 3x^2 + 5x - 6$

2. $\dfrac{dP}{dt} = \sqrt{t} + e^{-t}$

3. $\dfrac{dy}{dx} = 3y$

4. $\dfrac{dy}{dx} = y^2$

5. $\dfrac{dy}{dx} = e^y$

6. $\dfrac{dy}{dx} = e^{x+y}$

7. $\dfrac{dy}{dx} = \dfrac{x}{y}$

8. $\dfrac{dy}{dx} = \dfrac{y}{x}$

9. $\dfrac{dy}{dx} = \sqrt{xy}$

10. $\dfrac{dy}{dx} = \dfrac{y^2 + 4}{xy}$

11. $\dfrac{dy}{dx} = \dfrac{y}{x - 1}$

12. $\dfrac{dy}{dx} = e^y\sqrt{x + 1}$

13. $\dfrac{dy}{dx} = \dfrac{y+3}{(2x-5)^6}$

14. $\dfrac{dy}{dx} = (e^y + 1)(x-2)^9$

15. $\dfrac{dx}{dt} = \dfrac{xt}{2t+1}$

16. $\dfrac{dy}{dt} = \dfrac{te^y}{2t-1}$

17. $\dfrac{dy}{dx} = xe^{x-y}$

18. $\dfrac{dw}{ds} = \dfrac{se^{2w}}{w}$

19. $\dfrac{dy}{dt} = y \ln \sqrt{t}$

20. $\dfrac{dx}{dt} = \dfrac{\ln t}{\ln x}$

In Problems 21 through 28, find the particular solution of the differential equation satisfying the indicated condition.

21. $\dfrac{dy}{dx} = e^{5x}$; $y = 1$ when $x = 0$

22. $\dfrac{dy}{dx} = 5x^4 - 3x^2 - 2$; $y = 4$ when $x = 1$

23. $\dfrac{dy}{dx} = \dfrac{x}{y^2}$; $y = 3$ when $x = 2$

24. $\dfrac{dy}{dx} = 4x^3y^2$; $y = 2$ when $x = 1$

25. $\dfrac{dy}{dx} = y\sqrt{4-x}$; $y = 2$ when $x = 4$

26. $\dfrac{dy}{dx} = xe^{y-x^2}$; $y = 0$ when $x = 1$

27. $\dfrac{dy}{dt} = \dfrac{y+1}{t(y-1)}$; $y = 2$ when $t = 1$

$$\left[\text{Hint:} \ \frac{y-1}{y+1} = 1 - \frac{2}{y+1} \right]$$

28. $\dfrac{dx}{dt} = xt\sqrt{t+1}$; $x = 1$ when $t = 0$

In Problems 29 through 40, write a differential equation describing the given situation. Define all variables you introduce. (Do not try to solve the differential equation at this time.)

29. GROWTH OF BACTERIA The number of bacteria in a culture grows at a rate that is proportional to the number present.

30. RADIOACTIVE DECAY A sample of radium decays at a rate that is proportional to its size.

31. INVESTMENT GROWTH An investment grows at a rate equal to 7% of its size.

32. CONCENTRATION OF DRUGS The rate at which the concentration of a drug in the bloodstream decreases is proportional to the concentration.

33. POPULATION GROWTH The population of a certain town increases at the constant rate of 500 people per year.

34. MARGINAL COST A manufacturer's marginal cost is $60 per unit.

35. TEMPERATURE CHANGE The rate at which the temperature of an object changes is proportional to the difference between its own temperature and the temperature of the surrounding medium.

36. DISSOLUTION OF SUGAR After being placed in a container of water, sugar dissolves at a rate proportional to the amount of undissolved sugar remaining in the container.

37. RECALL FROM MEMORY When a person is asked to recall a set of facts, the rate at which the facts are recalled is proportional to the number of relevant facts in the person's memory that have not yet been recalled.

38. THE SPREAD OF AN EPIDEMIC The rate at which an epidemic spreads through a community is jointly proportional to the number of people who have caught the disease and the number who have not.

39. CORRUPTION IN GOVERNMENT The rate at which people are implicated in a government scandal is jointly proportional to the number of people already implicated and the number of people involved who have not yet been implicated.

40. THE SPREAD OF A RUMOR The rate at which a rumor spreads through a community is jointly proportional to the number of people in the community who have heard the rumor and the number who have not.

41. Verify that the function $y = Ce^{kx}$ is a solution of the differential equation $\dfrac{dy}{dx} = ky$.

42. Verify that the function $Q = B - Ce^{-kt}$ is a solution of the differential equation $\dfrac{dQ}{dt} = k(B - Q)$.

43. Verify that $y = C_1 e^x + C_2 x e^x$ is a solution of the differential equation $\dfrac{d^2y}{dx^2} - 2\dfrac{dy}{dx} + y = 0$.

44. Verify that the function $y = \dfrac{1}{20}x^4 - \dfrac{C_1}{x} + C_2$ is a solution of the differential equation $x\dfrac{d^2y}{dx^2} + 2\dfrac{dy}{dx} = x^3$.

45. OIL PRODUCTION A certain oil well that yields 400 barrels of crude oil per month will run dry in 2 years. The price of crude oil is currently $56 per barrel and is expected to rise at the constant rate of 4 cents per barrel per month. If the oil is sold as soon as it is extracted from the ground, what will the total future revenue from the well be?

46. RECALL FROM MEMORY Some psychologists believe that when a person is asked to recall a set of facts, the rate at which the facts are recalled is proportional to the number of relevant facts in the subject's memory that have not yet been recalled. Express the number of facts that have been recalled as a function of time and draw the graph.

47. DISSOLUTION OF SUGAR After being placed in a container of water, sugar dissolves at a rate proportional to the amount of undissolved sugar remaining in the container. Express the amount of sugar that has been dissolved as a function of time and draw the graph.

48. NEWTON'S LAW OF COOLING The rate at which the temperature of an object changes is proportional to the difference between its own temperature and that of the surrounding medium. A cold drink is removed from a refrigerator on a hot summer day and placed in a room where the temperature is 80°F. Express the temperature of the drink as a function of time (minutes) if the temperature of the drink was 40°F when it left the refrigerator and was 50°F after 20 minutes in the room.

49. NEWTON'S LAW OF COOLING The rate at which the temperature of an object changes is proportional to the difference between its own temperature and that of the surrounding medium. Express the temperature of the object as a function of time and draw the graph if the initial temperature of the object is greater than that of its surroundings.

50. AGRICULTURAL PRODUCTION The **Mitscherlich model** for agricultural production specifies that the size $Q(t)$ of a crop changes at a rate proportional to the difference between the maximum possible crop size B and Q; that is,

$$\frac{dQ}{dt} = k(B - Q)$$

   a. Solve this equation for $Q(t)$. Express your answer in terms of k and the initial crop size $Q_0 = Q(0)$.

   b. A particular crop has a maximum size of 200 bushels per acre. At the start of the growing season ($t = 0$), the crop size is 50 bushels and 1 month later, it is 60 bushels. How large is the crop 3 months later ($t = 3$)?

   c. Note that this model is similar to the learning model discussed in Example 6.2.6. Is this just a coincidence or is there some meaningful analogy linking the two situations? Explain.

51. DILUTION A tank holds 200 gallons of brine containing 2 pounds of salt per gallon. Clear water flows into the tank at the rate of 5 gallons per minute, and the mixture, kept uniform by stirring, runs out at the same rate.

   a. If $S(t)$ is the amount of salt in solution at time t, then the amount of salt in a typical gallon of solution is given by

$$\frac{\text{Amount of salt}}{\text{Amount of fluid}} = \frac{S(t)}{200}$$

   At what rate is salt flowing out of the tank at time t?

   b. Write a differential equation for the time rate of change of $S(t)$ using the fact that

$$\frac{dS}{dt} = \begin{bmatrix} \text{rate at which} \\ \text{salt enters tank} \end{bmatrix} - \begin{bmatrix} \text{rate at which} \\ \text{salt leaves tank} \end{bmatrix}$$

   c. Solve the differential equation in part (b) to obtain $S(t)$. [*Hint:* What is $S(0)$?]

52. INVESTMENT PLAN An investor makes regular deposits totaling D dollars each year into an account that earns interest at the annual rate r compounded continuously.

   a. Explain why the account grows at the rate

$$\frac{dV}{dt} = rV + D$$

   where $V(t)$ is the value of the account t years after the initial deposit. Solve this differential equation to express $V(t)$ in terms of r and D.

b. Amanda wants to retire in 20 years. To build up a retirement fund, she makes regular annual deposits of $8,000. If the prevailing interest rate stays constant at 4% compounded continuously, how much will she have in her account at the end of the 20-year period?

c. Ray estimates he will need $800,000 to retire. If the prevailing rate of interest is 5% compounded continuously, how large should his regular annual deposits be so that he can retire in 30 years?

53. RETIREMENT INCOME A retiree deposits S dollars into an account that earns interest at an annual rate r compounded continuously, and annually withdraws W dollars.

a. Explain why the account changes at the rate

$$\frac{dV}{dt} = rV - W$$

where $V(t)$ is the value of the account t years after the account is initiated. Solve this differential equation to express $V(t)$ in terms of r, W, and S.

b. Frank and Jessie Jones deposit $500,000 in an account that pays 5% interest compounded continuously. If they withdraw $50,000 annually, what is their account worth at the end of 10 years?

c. What annual amount W can the couple in part (b) withdraw if their goal is to keep their account unchanged at $500,000?

d. If the couple in part (b) decide to withdraw $80,000 annually, how long does it take to exhaust their account?

54. THE SPREAD OF AN EPIDEMIC The rate at which an epidemic spreads through a community with 2,000 susceptible residents is jointly proportional to the number of residents who have been infected and the number of susceptible residents who have not. Express the number of residents who have been infected as a function of time (in weeks), if 500 residents had the disease initially and 855 residents had been infected by the end of the first week.

55. WORKER EFFICIENCY For $0 \leq p \leq 1$, let $p(t)$ be the likelihood that an assembly line worker will make a mistake t hours into the worker's 8-hour shift. A particular worker, Tom, never makes a mistake at the beginning of his shift and is only 5% likely to make a mistake at the end. Set up and solve a differential equation assuming that at each time t, the likelihood of Tom making an error increases at a rate proportional to the likelihood $1 - p(t)$ that a mistake has not already been made.

56. WORKER EFFICIENCY Sue, a coworker of Tom in Problem 55, has a 1% likelihood of making an error at the beginning of her shift but has a 3% likelihood of making an error at the end. Set up and solve a differential equation assuming that at each time t, the likelihood $p(t)$ of Sue making an error increases at a rate jointly proportional to $p(t)$ and the likelihood $1 - p(t)$ that a mistake has not already been made.

57. AIR PURIFICATION A 2,400-cubic-foot room contains an activated charcoal air filter through which air passes at the rate of 400 cubic feet per minute. The ozone in the air is absorbed by the charcoal as the air flows through the filter, and the purified air is recirculated in the room. Assuming that the remaining ozone is evenly distributed throughout the room at all times, determine how long it takes the filter to remove 50% of the ozone from the room.

58. CORRUPTION IN GOVERNMENT The number of people implicated in a certain major government scandal increases at a rate jointly proportional to the number of people already implicated and the number involved who have not yet been implicated. Suppose that 7 people were implicated when a Washington newspaper first made the scandal public, that 9 more were implicated over the next 3 months, and that another 12 were implicated during the following 3 months. Approximately how many people are involved in the scandal? [*Warning:* This problem will test your algebraic ingenuity!]

59. PRICE ADJUSTMENT Suppose the price $p(t)$ of a particular commodity varies in such a way that its rate of change $\frac{dp}{dt}$ is proportional to the shortage $D - S$, where $D = 7 - p$ and $S = 1 + p$ are the demand and supply functions for the commodity.

a. If the price is $6 when $t = 0$ and $4 when $t = 4$, find $p(t)$.

b. Show that as t increases without bound, $p(t)$ approaches the price where supply equals demand.

60. EVANS PRICE ADJUSTMENT MODEL Suppose that a particular commodity has linear demand and supply functions, $D(p) = a - bp$ and $S(p) = r + sp$, for price p and positive constants a, b, r, and s. Further assume that price is a function of time t and that the time rate of change of price is proportional to the shortage $D - S$, so that

$$\frac{dp}{dt} = k(D - S)$$

Solve this differential equation and sketch the graph of $p(t)$. What happens to $p(t)$ "in the long run" (as $t \to +\infty$)?

61. DOMAR DEBT MODEL Let D and I denote the national debt and national income, and assume that both are functions of time t. One of several **Domar debt models** assumes that the time rates of change of D and I are both proportional to I, so that

$$\frac{dD}{dt} = aI \quad \text{and} \quad \frac{dI}{dt} = bI$$

Suppose $I(0) = I_0$ and $D(0) = D_0$.

a. Solve both of these differential equations and express $D(t)$ and $I(t)$ in terms of a, b, I_0, and D_0.

b. The economist, Evsey Domar, who first studied this model, was interested in the ratio of national debt to national income. What happens to the ratio $\dfrac{D(t)}{I(t)}$ as $t \to +\infty$?

62. ALLOMETRY The different members or organs of an individual often grow at different rates, and an important part of **allometry** involves the study of relationships among these growth rates (recall the Think About It essay at the end of Chapter 1). Suppose $x(t)$ is the size (length, volume, or weight) at time t of one organ or member of an individual organism and $y(t)$ is the size of another organ or member of the same individual. Then the **law of allometry** states that the relative growth rates of x and y are proportional; that is,

$$\frac{y'(t)}{y(t)} = k\frac{x'(t)}{x(t)} \quad \text{for some constant } k > 0$$

First show that the allometry law can be written as

$$\frac{dy}{dx} = k\frac{y}{x}$$

then solve this equation for y in terms of x.

63. THE SPREAD OF AN EPIDEMIC The rate at which an epidemic spreads through a community is jointly proportional to the number of residents who have been infected and the number of susceptible resi-

dents who have not. Show that the epidemic is spreading most rapidly when one-half of the susceptible residents have been infected. [*Hint:* You do not have to solve a differential equation to do this. Just start with a formula for the *rate* at which the epidemic is spreading and use calculus to maximize this rate.]

64. LOGISTIC CURVES Show that if a quantity Q satisfies the differential equation $\dfrac{dQ}{dt} = kQ(B - Q)$, where k and B are positive constants, then the rate of change $\dfrac{dQ}{dt}$ is greatest when $Q(t) = \dfrac{B}{2}$. What does this result tell you about the inflection point of a logistic curve? Explain. [*Hint:* See the hint for Problem 63.]

65. CONCENTRATION OF BLOOD GLUCOSE Glucose is infused into the bloodstream of a patient at a constant rate R and, at the same time, is converted and excreted at a rate proportional to the present concentration of glucose $C(t)$; that is,

$$\frac{dC}{dt} = R - kC \quad \text{for } k > 0$$

Solve this equation for $C(t)$. Express your answer in terms of R, k, and the initial concentration of glucose $C_0 = C(0)$.

66. RESPONSE TO STIMULUS The Weber-Fechner law in experimental psychology specifies that the rate of change of a response R with respect to the level of stimulus S is inversely proportional to the stimulus; that is,

$$\frac{dR}{dS} = \frac{k}{S}$$

Let S_0 be the threshold stimulus; that is, the highest level of stimulus for which there is no response, so that $R = 0$ when $S = S_0$.

a. Solve this differential equation for $R(S)$. Express your answer in terms of k and S_0.

b. Sketch the graph of the response function $R(S)$ found in part (a).

SECTION 6.3 Improper Integrals; Continuous Probability

The definition of the definite integral $\displaystyle\int_a^b f(x)\,dx$, given in Section 5.3, requires the interval of integration $a \le x \le b$ to be bounded, but in certain applications, it is useful to consider integrals over *unbounded* intervals such as $x \ge a$. We will define such **improper integrals** in this section and examine a few properties and applications.

Then we will show how integration, including improper integration, can be used to compute probability.

The Improper Integral $\int_a^{+\infty} f(x)\,dx$

We denote the improper integral of $f(x)$ over the unbounded interval $x \geq a$ by $\int_a^{+\infty} f(x)\,dx$. If $f(x) \geq 0$ for $x \geq a$, this integral can be interpreted as the area of the region under the curve $y = f(x)$ to the right of $x = a$, as shown in Figure 6.9a. Although this region has infinite extent, its area may be finite or infinite, depending on how quickly $f(x)$ approaches zero as x increases indefinitely.

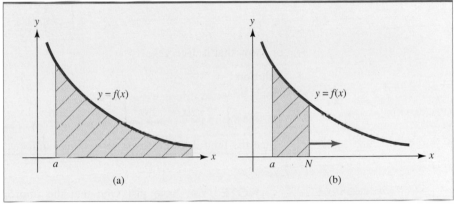

FIGURE 6.9 Area $= \int_0^{+\infty} f(x)\,dx = \lim_{N \to +\infty} \int_0^N f(x)\,dx.$

A reasonable strategy for finding the area of such a region is to first use a definite integral to compute the area from $x = a$ to some finite number $x = N$ and then to let N approach infinity in the resulting expression. That is,

$$\text{Total area} = \lim_{N \to +\infty} (\text{area from } a \text{ to } N) = \lim_{N \to +\infty} \int_a^N f(x)\,dx$$

This strategy is illustrated in Figure 6.9b and motivates the following definition of the improper integral.

The Improper Integral ■ If $f(x)$ is continuous for $x \geq a$, then

$$\int_a^{+\infty} f(x)\,dx = \lim_{N \to +\infty} \int_a^N f(x)\,dx$$

If the limit exists, the improper integral is said to **converge** to the value of the limit. If the limit does not exist, the improper integral **diverges.**

EXAMPLE 6.3.1

Either evaluate the improper integral

$$\int_1^{+\infty} \frac{1}{x^2}\,dx$$

or show that it diverges.

Solution

First compute the integral from 1 to N and then let N approach infinity. Arrange your work as follows:

$$\int_1^{+\infty} \frac{1}{x^2}\, dx = \lim_{N \to +\infty} \int_1^N \frac{1}{x^2}\, dx = \lim_{N \to +\infty} \left(-\frac{1}{x}\Big|_1^N \right) = \lim_{N \to +\infty} \left(-\frac{1}{N} + 1 \right) = 1$$

EXAMPLE | 6.3.2

Either evaluate the improper integral

$$\int_1^{+\infty} \frac{1}{x}\, dx$$

or show that it diverges.

Solution

$$\int_1^{+\infty} \frac{1}{x}\, dx = \lim_{N \to +\infty} \int_1^N \frac{1}{x}\, dx = \lim_{N \to +\infty} \left(\ln |x|\Big|_1^N \right) = \lim_{N \to +\infty} \ln N = +\infty$$

Since the limit does not exist (as a finite number), the improper integral diverges.

NOTE You have just seen that the improper integral $\int_1^{+\infty} \frac{1}{x^2}\, dx$ converges (Example 6.3.1) while $\int_1^{+\infty} \frac{1}{x}\, dx$ diverges (Example 6.3.2). In geometric terms, this says that the area under the curve $y = \frac{1}{x^2}$ to the right of $x = 1$ is *finite*, while the corresponding area under the curve $y = \frac{1}{x}$ to the right of $x = 1$ is *infinite*. The difference is due to the fact that as x increases without bound, $\frac{1}{x^2}$ approaches zero more quickly than does $\frac{1}{x}$. These observations are demonstrated in Figure 6.10. ■

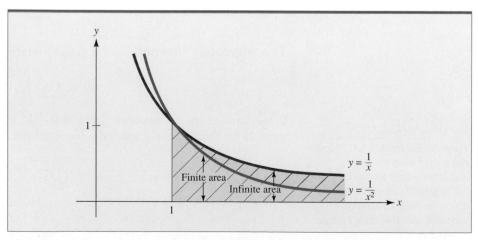

FIGURE 6.10 Comparison of the area under $y = 1/x$ with that under $y = 1/x^2$.

The evaluation of improper integrals arising from practical problems often involves limits of the form

$$\lim_{N \to \infty} \frac{N^p}{e^{kN}} = \lim_{N \to \infty} N^p e^{-kN} \quad (\text{for } k > 0)$$

In general, an exponential term e^{-kN} will grow faster than *any* power term N^p, so

$$N^p e^{-kN} = \frac{N^p}{e^{kN}}$$

will become very small "in the long run." To summarize:

A Useful Limit for Improper Integrals ■ For any power p and positive number k,

$$\lim_{N \to \infty} N^p e^{-kN} = 0$$

EXAMPLE | 6.3.3

Either evaluate the improper integral

$$\int_0^{+\infty} xe^{-2x} dx$$

or show that it diverges.

Solution

$$\int_0^{+\infty} xe^{-2x} dx = \lim_{N \to +\infty} \int_0^N xe^{-2x} dx$$

$$= \lim_{N \to +\infty} \left(-\frac{1}{2}xe^{-2x} \Big|_0^N + \frac{1}{2}\int_0^N e^{-2x} dx \right) \qquad \text{integration by parts}$$

$$= \lim_{N \to +\infty} \left(-\frac{1}{2}xe^{-2x} - \frac{1}{4}e^{-2x} \right) \Big|_0^N$$

$$= \lim_{N \to +\infty} \left(-\frac{1}{2}Ne^{-2N} - \frac{1}{4}e^{-2N} + 0 + \frac{1}{4} \right)$$

$$= \frac{1}{4}$$

since $e^{-2N} \to 0$ and $Ne^{-2N} \to 0$ as $N \to +\infty$.

Applications of the Improper Integral

Next, we examine two applications of the improper integral that generalize applications developed in Chapter 5. In each, the strategy is to express a quantity as a definite integral with a variable upper limit of integration that is then allowed to increase without bound. As you read through these applications, you may find it helpful to refer back to the analogous examples in Chapter 5.

Present Value of a Perpetual Income Flow

In Example 5.5.3 of Section 5.5, we showed how the present value of an investment that generates income continuously over a finite time period can be computed by a definite integral. If the generation of income continues in perpetuity, then an improper integral is required for computing its present value, as illustrated in Example 6.3.4.

EXAMPLE | 6.3.4

A donor wishes to endow a scholarship at a local college with a gift that provides a continuous income stream at the rate of $25,000 + 1,200t$ dollars per year in perpetuity. Assuming the prevailing annual interest rate stays fixed at 5% compounded continuously, what donation is required to finance the endowment?

Solution

The donor's gift should equal the present value of the income stream in perpetuity. Recall from Section 5.5 that an income stream $f(t)$ deposited continuously for a term of T years into an account that earns interest at an annual rate r compounded continuously has present value given by the integral

$$PV = \int_0^T f(t)e^{-rt}\,dt$$

For the donor's gift, we have $f(t) = 25,000 + 1,200t$ and $r = 0.05$, so for a term of T years, the present value is

Present value of
the gift for T years
$$PV = \int_0^T (25,000 + 1,200t)e^{-0.05t}\,dt$$

Integrating by parts, with

$$u = 25,000 + 1,200t \qquad dv = e^{-0.05t}\,dt$$

$$du = 1,200\,dt \qquad v = \frac{e^{-0.05t}}{-0.05} = -20e^{-0.05t}$$

we evaluate the present value as follows:

$$PV = \int_0^T (25,000 + 1,200t)e^{-0.05t}\,dt$$

$$= \left[(25,000 + 1,200t)(-20e^{-0.05t})\right]\Big|_0^T - \int_0^T 1,200(-20e^{-0.05t})\,dt$$

$$= \left[(-500,000 - 24,000t)e^{-0.05t}\right]\Big|_0^T + 24,000\left(\frac{e^{-0.05t}}{-0.05}\right)\Big|_0^T$$

$$= \left[(-980,000 - 24,000t)e^{-0.05t}\right]\Big|_0^T$$

$$= \left[(-980,000 - 24,000T)e^{-0.05T}\right] - \left[(-980,000 - 24,000(0))e^0\right]$$

$$= (-980,000 - 24,000T)e^{-0.05T} + 980,000$$

To find the present value in perpetuity, we take the limit as $T \to +\infty$; that is, we compute the improper integral

Present value of
the gift in perpetuity
$$\int_0^{+\infty} (25,000 + 1,200t)e^{-0.05t}\,dt$$

$$= \lim_{T \to +\infty} [(-980,000 - 24,000T)e^{-0.05T} + 980,000]$$

$$= 0 + 980,000 \qquad \begin{array}{l} \textit{since } e^{-0.05T} \to 0 \textit{ and} \\ Te^{-0.05T} \to 0 \textit{ as } T \to +\infty \end{array}$$

$$= 980,000$$

Therefore, the endowment is established with a gift of $980,000.

Nuclear Waste   In Example 5.6.1 of Section 5.6, we examined a survival and renewal problem over a term of finite length in which renewal occurred at a constant rate. In Example 6.3.5, we consider the more general situation in which survival and renewal both vary with time and the survival/renewal process occurs in perpetuity.

EXAMPLE | 6.3.5

It is estimated that t years from now, a certain nuclear power plant will be producing radioactive waste at the rate of $f(t) = 400t$ pounds per year. The waste decays exponentially at the rate of 2% per year. What will happen to the accumulation of radioactive waste from the plant in the long run?

Solution

To find the amount of radioactive waste present after N years, divide the N-year interval $0 \le t \le N$ into n equal subintervals of length Δt years and let t_j denote the beginning of the jth subinterval (Figure 6.11). Then,

$$\text{Amount of waste produced during } j\text{th subinterval} \approx 400t_j\,\Delta t$$

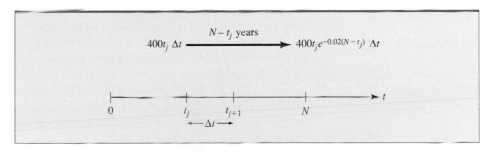

FIGURE 6.11   Radioactive waste generated during the jth subinterval.

Since the waste decays exponentially at the rate of 2% per year, and since there are $N - t_j$ years between times $t = t_j$ and $t = N$, it follows that

$$\begin{array}{c}\text{Amount of waste produced} \\ \text{during } j\text{th subinterval} \\ \text{still present at } t = N\end{array} \approx 400t_j e^{-0.02(N-t_j)}\Delta t$$

Thus,

$$\begin{array}{c}\text{Amount of waste present} \\ \text{after } N \text{ years}\end{array} = \lim_{n \to +\infty} \sum_{j=1}^{n} 400t_j e^{-0.02(N-t_j)}\Delta t$$

$$= \int_0^N 400t e^{-0.02(N-t)}\,dt$$

$$= 400e^{-0.02N}\int_0^N te^{0.02t}\,dt$$

The amount of radioactive waste present in the long run is the limit of this expression as N approaches infinity. That is,

$$\begin{array}{c}\text{Amount of waste present} \\ \text{in the long run}\end{array} = \lim_{N \to +\infty} 400e^{-0.02N}\int_0^N te^{0.02t}\,dt$$

$$= \lim_{N \to +\infty} 400e^{-0.02N}\left(50te^{0.02t} - 2,500e^{0.02t}\right)\Big|_0^N$$

$$= \lim_{N \to +\infty} 400e^{-0.02N}\left(50Ne^{0.02N} - 2,500e^{0.02N} + 2,500\right)$$

$$= \lim_{N \to +\infty} 400\left(50N - 2,500 + 2,500e^{-0.02N}\right)$$

$$= +\infty$$

That is, in the long run, the accumulation of radioactive waste from the plant will increase without bound.

Continuous Probability

Improper integrals also appear in the study of probability and statistics, which play an important role in certain areas of the social, managerial, and life sciences. We close this section with a brief introduction to this application.

For example, the life span of a lightbulb selected at random from a manufacturer's stock is a quantity that cannot be predicted with certainty. The process of selecting a bulb is called a **random experiment,** and the life span X of the bulb is a **continuous random variable.** Other examples of continuous random variables include the time a randomly selected motorist spends waiting at a traffic light, the weight of a randomly selected person, or the time it takes a randomly selected person to learn a particular task.

The **probability** of an event that can result from a random experiment is a number between 0 and 1 that specifies the likelihood of the event. For instance, in our lightbulb example, one possible event is that a bulb selected randomly from the manufacturer's stock has a life span between 20 and 35 hours. If X is the random variable denoting the life span of a randomly selected bulb, then this event can be described by the inequality $20 \leq X \leq 35$ and its probability denoted by $P(20 \leq X \leq 35)$. Similarly, the probability that the bulb will burn for at least 50 hours is denoted by $P(X \geq 50)$ or $P(50 \leq X < +\infty)$.

A **probability density function** for a continuous random variable X is a function f with the property that $f(x) \geq 0$ for all real x and that the area under the graph of f from $x = a$ to $x = b$ gives the probability $P(a \leq X \leq b)$. The graph of a possible probability density function for the life span of a lightbulb is sketched in Figure 6.12. Its shape reflects the fact that most bulbs burn out relatively quickly. For example, the probability that a bulb will fail within the first 40 hours is represented by the area under the curve between $x = 0$ and $x = 40$. This is much greater than the area under the curve between $x = 80$ and $x = 120$, which represents the probability that the bulb will fail between its 80th and 120th hours of use.

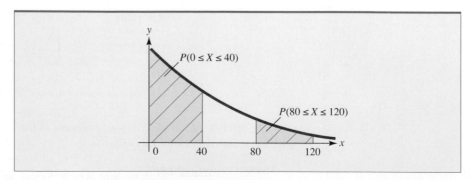

FIGURE 6.12 A possible probability density function for the life span of a lightbulb.

The basic property of probability density functions can be restated in terms of the integrals you would use to compute the appropriate areas.

Probability Density Functions ▪ A probability density function for the continuous random variable X is a function $f(x)$ that satisfies the following three conditions:

1. $f(x) \geq 0$ for all real x

2. The total area under the graph of $f(x)$ is 1

3. The probability that X lies in the interval $a \leq X \leq b$ is given by the integral

$$P(a \leq X \leq b) = \int_a^b f(x)\,dx$$

The values of a and b in this formula need not be finite, and if either is infinite, the corresponding probability is given by an improper integral. For example, the probability that $X \geq a$ is

$$P(X \geq a) = P(a \leq X < +\infty) = \int_a^{+\infty} f(x)\,dx$$

The second condition in the definition of a probability density function follows from the fact that the event $-\infty < X < +\infty$ is certain to occur. This condition can also be expressed as

$$\int_{-\infty}^{+\infty} f(x)\,dx = 1$$

where the improper integral on the left is defined as

$$\int_{-\infty}^{+\infty} f(x)\,dx = \lim_{N \to +\infty} \int_{-N}^0 f(x)\,dx + \lim_{N \to +\infty} \int_0^N f(x)\,dx$$

and converges if and only if both limits exist.

How to determine the appropriate probability density function for a particular random variable is a central problem in probability theory that is beyond the scope of this book. We restrict our attention to examining two examples illustrating particular probability density functions that prove useful in a variety of applications.

Uniform Density Functions   A **uniform probability density function** (Figure 6.13) is constant over a bounded interval $A \leq x \leq B$ and zero outside that interval. A random variable that has a uniform density function is said to be **uniformly distributed.** Roughly speaking, for a uniformly distributed random variable, all values in some bounded interval are "equally likely." More precisely, a continuous random variable is uniformly distributed if the probability that its value will be in a particular subinterval of the bounded interval is equal to the probability that it will be in any other subinterval that has the same length. An example of a uniformly distributed random variable is the waiting time of a motorist at a traffic light that remains red for, say, 40 seconds at a time. This random variable has a uniform distribution because all waiting times between 0 and 40 seconds are equally likely.

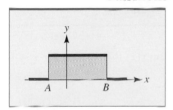

FIGURE 6.13 A uniform density function.

If k is the constant value of a uniform density function $f(x)$ on the interval $A \leq x \leq B$, the value of k is determined by the requirement that the total area under the graph of f be equal to 1. In particular,

$$1 = \int_{-\infty}^{+\infty} f(x)\,dx = \int_{A}^{B} f(x)\,dx \quad [\text{since } f(x) = 0 \text{ outside the interval } A \leq x \leq B]$$

$$= \int_{A}^{B} k\,dx = kx\Big|_{A}^{B} = k(B - A)$$

and so

$$k = \frac{1}{B - A}$$

This observation leads to the following formula for a uniform density function.

Uniform Density Function

$$f(x) = \begin{cases} \dfrac{1}{B - A} & \text{if } A \leq x \leq B \\ 0 & \text{otherwise} \end{cases}$$

Here is a typical application involving a uniform density function.

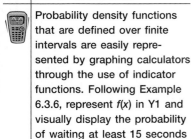

EXAMPLE 6.3.6

A certain traffic light remains red for 40 seconds at a time. You arrive (at random) at the light and find it red. Use an appropriate uniform density function to find the probability that you will have to wait at least 15 seconds for the light to turn green.

Solution

Let X denote the random variable that measures the time (in seconds) that you must wait. Since all waiting times between 0 and 40 are "equally likely," X is uniformly distributed over the interval $0 \leq x \leq 40$. The corresponding uniform density function is

$$f(x) = \begin{cases} \dfrac{1}{40} & \text{if } 0 \leq x \leq 40 \\ 0 & \text{otherwise} \end{cases}$$

and the desired probability is

$$P(15 \leq X \leq 40) = \int_{15}^{40} \frac{1}{40}\,dx = \frac{x}{40}\Big|_{15}^{40} = \frac{40 - 15}{40} = \frac{5}{8}$$

Exponential Density Functions   An **exponential probability density function** is a function $f(x)$ that is zero for $x < 0$ and decreases exponentially for $x \geq 0$. That is,

$$f(x) = \begin{cases} Ae^{-kx} & \text{for } x \geq 0 \\ 0 & \text{for } x < 0 \end{cases}$$

where A and k are positive constants.

The value of A is determined by the requirement that the total area under the graph of f be equal to 1. Thus,

$$1 = \int_{-\infty}^{+\infty} f(x)\,dx = \int_{0}^{+\infty} Ae^{-kx}\,dx = \lim_{N \to +\infty} \int_{0}^{N} Ae^{-kx}\,dx$$

$$= \lim_{N \to +\infty}\left(-\frac{A}{k}e^{-kx}\,\Big|_{0}^{N}\right) = \lim_{N \to +\infty}\left(-\frac{A}{k}e^{-kN} + \frac{A}{k}\right) = \frac{A}{k}$$

and so $A = k$.

This calculation leads to the following general formula for an exponential density function. The corresponding graph is shown in Figure 6.14.

Exponential Density Function

$$f(x) = \begin{cases} ke^{-kx} & \text{if } x \geq 0 \\ 0 & \text{if } x < 0 \end{cases}$$

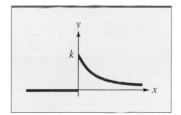

FIGURE 6.14 An exponential density function.

A random variable that has an exponential density function is said to be **exponentially distributed.** As you can see from the graph in Figure 6.14, the value of an exponentially distributed random variable is much more likely to be small than large. Such random variables include the life span of electronic components, the duration of telephone calls, and the interval between the arrivals of successive planes at an airport. Here is an example.

EXPLORE!

Refer to Example 6.3.7. Store the function $f(x)$ into Y1 as 0.5e^(−0.5X)∗(X ≥ 0), using the window [−2, 10]1 by [−0.1, 0.6]0.1. Confirm numerically that $f(x)$ is a probability density function and compute $P(2 \leq X \leq 3)$.

EXAMPLE 6.3.7

Let X be a random variable that measures the duration of telephone calls in a certain city and suppose that a probability density function for X is

$$f(x) = \begin{cases} 0.5e^{-0.5x} & \text{if } x \geq 0 \\ 0 & \text{if } x < 0 \end{cases}$$

where x denotes the duration (in minutes) of a randomly selected call.

a. Find the probability that a randomly selected call will last between 2 and 3 minutes.

b. Find the probability that a randomly selected call will last at least 2 minutes.

Solution

a. $$P(2 \leq X \leq 3) = \int_{2}^{3} 0.5e^{-0.5x}\,dx = -e^{-0.5x}\,\Big|_{2}^{3}$$

$$= -e^{-1.5} + e^{-1} = 0.1447$$

b. There are two ways to compute this probability. The first method is to evaluate an improper integral.

$$P(X \geq 2) = P(2 \leq X < +\infty) = \int_{2}^{+\infty} 0.5e^{-0.5x}\,dx$$

$$= \lim_{N \to +\infty} \int_{2}^{N} 0.5e^{-0.5x}\,dx = \lim_{N \to +\infty}\left(-e^{-0.5x}\,\Big|_{2}^{N}\right)$$

$$= \lim_{N \to +\infty}(-e^{-0.5N} + e^{-1}) = e^{-1} \approx 0.3679$$

The second method is to compute 1 minus the probability that X is less than 2. That is,

$$P(X \geq 2) = 1 - P(X < 2)$$

$$= 1 - \int_0^2 0.5e^{-0.5x}\,dx = 1 - \left(-e^{-0.5x} \Big|_0^2 \right)$$

$$= 1 - (-e^{-1} + 1) = e^{-1} \approx 0.3679$$

Normal Density Functions

The most widely used probability density functions are the **normal density functions.** You are probably already familiar with their famous "bell-shaped" graphs like the one in Figure 6.15. Detailed discussion of this important topic involves specialized ideas that are outside the scope of this text but may be found in any thorough statistics text.

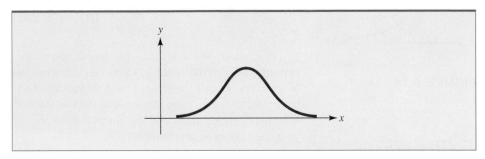

FIGURE 6.15   Graph of a normal density function.

The Expected Value of a Random Variable

A useful characteristic of a random variable X is its **expected value,** denoted by $E(X)$. If a random experiment is performed repeatedly and the results are recorded, then the arithmetic average of the recorded results will approach the expected value, so in this sense, $E(X)$ can be thought of as the "average" of the random variable X. Familiar expected values of random variables include the average highway mileage for a particular car model, the average waiting time to clear security at a certain airport, and the average life span of a particular appliance. Here is a formula for the expected value of a continuous random variable X in terms of an integral involving its probability density function.

Expected Value ▪ If X is a continuous random variable with probability density function f, the expected value (or mean) of X is

$$E(X) = \int_{-\infty}^{+\infty} xf(x)\,dx$$

NOTE **(Geometric Interpretation of Expected Value)** It may help to think of the probability density function for the random variable X as describing the distribution of mass on a beam lying along the x axis. Then the expected value of X is the point at which the beam will balance. ▪

Examples 6.3.8 and 6.3.9 illustrate how the integral formula for the expected value of a continuous random variable may be used.

EXAMPLE | 6.3.8

Find the expected value of the uniformly distributed random variable from Example 6.3.6 with density function

$$f(x) = \begin{cases} \dfrac{1}{40} & \text{if } 0 \le x \le 40 \\[2mm] 0 & \text{otherwise} \end{cases}$$

Solution

$$E(X) = \int_{-\infty}^{+\infty} xf(x)\,dx = \int_0^{40} \frac{x}{40}\,dx = \frac{x^2}{80}\Big|_0^{40} = \frac{1{,}600}{80} = 20$$

In the context of Example 6.3.6, this says that the average waiting time at the red light is 20 seconds, a conclusion that should come as no surprise since the random variable is uniformly distributed between 0 and 40.

EXAMPLE | 6.3.9

Find the expected value of the exponentially distributed random variable from Example 6.3.7 with density function

$$f(x) = \begin{cases} 0.5e^{-0.5x} & \text{if } x \ge 0 \\ 0 & \text{if } x < 0 \end{cases}$$

Solution

$$E(X) = \int_{-\infty}^{+\infty} xf(x)\,dx = \int_0^{+\infty} 0.5xe^{-0.5x}\,dx$$

$$= \lim_{N \to +\infty} \int_0^N 0.5xe^{-0.5x}\,dx$$

$$= \lim_{N \to +\infty} \left(-xe^{-0.5x}\Big|_0^N + \int_0^N e^{-0.5x}\,dx \right)$$

$$= \lim_{N \to +\infty} \left(-xe^{-0.5x} - 2e^{-0.5x} \right)\Big|_0^N$$

$$= \lim_{N \to +\infty} \left(-Ne^{-0.5N} - 2e^{-0.5N} + 2 \right)$$

$$= 2$$

That is, the average duration of telephone calls in the city in Example 6.3.7 is 2 minutes.

PROBLEMS | 6.3

In Problems 1 through 24, either evaluate the given improper integral or show that it diverges.

1. $\displaystyle\int_1^{+\infty} \frac{1}{x^3}\,dx$

2. $\displaystyle\int_1^{+\infty} x^{-3/2}\,dx$

3. $\displaystyle\int_1^{+\infty} \frac{1}{\sqrt{x}}\,dx$

4. $\displaystyle\int_1^{+\infty} x^{-2/3}\,dx$

5. $\displaystyle\int_3^{+\infty} \frac{1}{2x - 1}\, dx$

6. $\displaystyle\int_3^{+\infty} \frac{1}{\sqrt[3]{2x - 1}}\, dx$

7. $\displaystyle\int_3^{+\infty} \frac{1}{(2x - 1)^2}\, dx$

8. $\displaystyle\int_0^{+\infty} e^{-x}\, dx$

9. $\displaystyle\int_0^{+\infty} 5e^{-2x}\, dx$

10. $\displaystyle\int_1^{+\infty} e^{1-x}\, dx$

11. $\displaystyle\int_1^{+\infty} \frac{x^2}{(x^3 + 2)^2}\, dx$

12. $\displaystyle\int_1^{+\infty} \frac{x^2}{x^3 + 2}\, dx$

13. $\displaystyle\int_1^{+\infty} \frac{x^2}{\sqrt{x^3 + 2}}\, dx$

14. $\displaystyle\int_0^{+\infty} xe^{-x^2}\, dx$

15. $\displaystyle\int_1^{+\infty} \frac{e^{-\sqrt{x}}}{\sqrt{x}}\, dx$

16. $\displaystyle\int_0^{+\infty} xe^{-x}\, dx$

17. $\displaystyle\int_0^{+\infty} 2xe^{-3x}\, dx$

18. $\displaystyle\int_0^{+\infty} xe^{1-x}\, dx$

19. $\displaystyle\int_1^{+\infty} \frac{\ln x}{x}\, dx$

20. $\displaystyle\int_1^{+\infty} \frac{\ln x}{x^2}\, dx$

21. $\displaystyle\int_2^{+\infty} \frac{1}{x \ln x}\, dx$

22. $\displaystyle\int_2^{+\infty} \frac{1}{x\sqrt{\ln x}}\, dx$

23. $\displaystyle\int_0^{+\infty} x^2 e^{-x}\, dx$

24. $\displaystyle\int_1^{+\infty} \frac{e^{1/x}}{x^2}\, dx$

In Problems 25 through 30, determine whether the given function is a probability density function.

25. $f(x) = \begin{cases} \dfrac{10}{(x + 10)^2} & \text{for } x \geq 0 \\ 0 & \text{for } x < 0 \end{cases}$

26. $f(x) = \begin{cases} \dfrac{1}{2}e^{-2x} & \text{for } x \geq 0 \\ 0 & \text{for } x < 0 \end{cases}$

27. $f(x) = \begin{cases} xe^{-x} & \text{for } x \geq 0 \\ 0 & \text{for } x < 0 \end{cases}$

28. $f(x) = \begin{cases} \dfrac{1}{9}\sqrt{x} & \text{for } 0 \leq x \leq 9 \\ 0 & \text{otherwise} \end{cases}$

29. $f(x) = \begin{cases} \dfrac{3}{2}x^2 + 2x & \text{for } -1 \leq x \leq 1 \\ 0 & \text{otherwise} \end{cases}$

30. $f(x) = \begin{cases} \dfrac{1}{2}x^2 + \dfrac{5}{3}x & \text{for } 0 \leq x \leq 1 \\ 0 & \text{otherwise} \end{cases}$

In Problems 31 through 38, f(x) is a probability density function for a particular random variable X. Use integration to find the indicated probabilities.

31. $f(x) = \begin{cases} \dfrac{1}{3} & \text{if } 2 \leq x \leq 5 \\ 0 & \text{otherwise} \end{cases}$
  a. $P(2 \leq X \leq 5)$
  b. $P(3 \leq X \leq 4)$
  c. $P(X \geq 4)$

32. $f(x) = \begin{cases} \dfrac{x}{2} & \text{if } 0 \leq x \leq 2 \\ 0 & \text{otherwise} \end{cases}$
  a. $P(0 \leq X \leq 2)$
  b. $P(1 \leq X \leq 2)$
  c. $P(X \leq 1)$

33. $f(x) = \begin{cases} \frac{1}{8}(4-x) & \text{if } 0 \leq x \leq 4 \\ 0 & \text{otherwise} \end{cases}$

a. $P(0 \leq X \leq 4)$
b. $P(2 \leq X \leq 3)$
c. $P(X \geq 1)$

34. $f(x) = \begin{cases} \frac{3}{32}(4x - x^2) & \text{if } 0 \leq x \leq 4 \\ 0 & \text{otherwise} \end{cases}$

a. $P(0 \leq X \leq 4)$
b. $P(1 \leq X \leq 2)$
c. $P(X \leq 1)$

35. $f(x) = \begin{cases} \frac{3}{x^4} & \text{if } x \geq 1 \\ 0 & \text{if } x < 1 \end{cases}$

a. $P(1 \leq X < +\infty)$
b. $P(1 \leq X \leq 2)$
c. $P(X \geq 2)$

36. $f(x) = \begin{cases} \frac{1}{10}e^{-x/10} & \text{if } x \geq 0 \\ 0 & \text{if } x < 0 \end{cases}$

a. $P(0 \leq X < +\infty)$
b. $P(X \leq 2)$
c. $P(X \geq 5)$

37. $f(x) = \begin{cases} 2xe^{-x^2} & \text{if } x \geq 0 \\ 0 & \text{if } x < 0 \end{cases}$

a. $P(X \geq 0)$
b. $P(1 \leq X \leq 2)$
c. $P(X \leq 2)$

38. $f(x) = \begin{cases} \frac{1}{4}xe^{-x/2} & \text{if } x \geq 0 \\ 0 & \text{if } x < 0 \end{cases}$

a. $P(0 \leq X < +\infty)$
b. $P(2 \leq X \leq 4)$
c. $P(X \geq 6)$

In Problems 39 through 44, find the expected value for the random variable with the density function given in the indicated problem.

39. Problem 31
41. Problem 33
43. Problem 35

40. Problem 32
42. Problem 34
44. Problem 36

45. PRESENT VALUE OF AN INVESTMENT An investment will generate $2,400 per year in perpetuity. If the money is dispensed continuously throughout the year and if the prevailing annual interest rate remains fixed at 4% compounded continuously, what is the present value of the investment?

46. PRESENT VALUE OF RENTAL PROPERTY It is estimated that t years from now an apartment complex will be generating profit for its owner at the rate of $f(t) = 10,000 + 500t$ dollars per year. If the profit is generated in perpetuity and the prevailing annual interest rate remains fixed at 5% compounded continuously, what is the present value of the apartment complex?

47. PRESENT VALUE OF A FRANCHISE The management of a national chain of fast-food outlets is selling a permanent franchise in Seattle, Washington. Past experience in similar localities suggests that t years from now, the franchise will be generating profit at the rate of $f(t) = 12,000 + 900t$ dollars per year. If the prevailing interest rate remains fixed

at 5% compounded continuously, what is the present value of the franchise?

48. NUCLEAR WASTE A certain nuclear power plant produces radioactive waste at the rate of 600 pounds per year. The waste decays exponentially at the rate of 2% per year. How much radioactive waste from the plant will be present in the long run?

49. HEALTH CARE The fraction of patients who will still be receiving treatment at a certain health clinic t months after their initial visit is $f(t) = e^{-t/20}$. If the clinic accepts new patients at the rate of 10 per month, approximately how many patients will be receiving treatment at the clinic in the long run?

50. POPULATION GROWTH Demographic studies conducted in a certain city indicate that the fraction of the residents that will remain in the city for at least t years is $f(t) = e^{-t/20}$. The current population of the city is 200,000, and it is estimated that new residents will be arriving at the rate of 100 people per year. If this estimate is correct, what will happen to the population of the city in the long run?

51. MEDICINE A hospital patient receives intra-venously 5 units of a certain drug per hour. The drug is eliminated exponentially, so that the fraction that remains in the patient's body for t hours is $f(t) = e^{-t/10}$. If the treatment is continued indefi-nitely, approximately how many units of the drug will be in the patient's body in the long run?

52. USEFUL LIFE OF A MACHINE The useful life X of a particular kind of machine is a random variable with density function

$$f(x) = \begin{cases} \dfrac{3}{28} + \dfrac{3}{x^2} & \text{if } 3 \leq x \leq 7 \\ 0 & \text{otherwise} \end{cases}$$

where x is the number of years a randomly selected machine stays in use.

   a. Find the probability that a randomly selected machine will be useful for more than 4 years.

   b. Find the probability that a randomly selected machine will be useful for less than 5 years.

   c. Find the probability that a randomly selected machine will be useful between 4 and 6 years.

   d. What is the expected useful life of the machine?

53. TRAFFIC FLOW A certain traffic light remains red for 45 seconds at a time. You arrive (at random) at the light and find it red. Use an appropriate uni-form density function to find:

   a. The probability that the light will turn green within 15 seconds.

   b. The probability that the light will turn green within 5 and 10 seconds.

   c. The expected waiting time for cars arriving on red at the traffic light.

54. COMMUTING During the morning rush hour, commuter trains run every 20 minutes from the sta-tion near your home into the city center. You arrive (at random) at the station and find no train at the platform. Assuming that the trains are running on schedule, use an appropriate uniform density func-tion to find:

   a. The probability you will have to wait at least 8 minutes for a train.

   b. The probability you will have to wait between 2 and 5 minutes for a train.

   c. The expected wait for rush hour commuters arriv-ing at the station when no train is at the platform.

55. EXPERIMENTAL PSYCHOLOGY Suppose the length of time that it takes a laboratory rat to traverse a certain maze is measured by a random variable X that is exponentially distributed with density function

$$f(x) = \begin{cases} \dfrac{1}{3}e^{-x/3} & \text{if } x \geq 0 \\ 0 & \text{if } x < 0 \end{cases}$$

where x is the number of minutes a randomly se-lected rat spends in the maze.

   a. Find the probability that a randomly selected rat will take more than 3 minutes to traverse the maze.

   b. Find the probability that a randomly selected rat will take between 2 and 5 minutes to traverse the maze.

   c. Find the expected time required for a randomly selected laboratory rat to traverse the maze.

56. PRODUCT RELIABILITY The life span of the lightbulbs manufactured by a certain company is measured by a random variable X with probability density function

$$f(x) = \begin{cases} 0.01e^{-0.01x} & \text{if } x \geq 0 \\ 0 & \text{if } x < 0 \end{cases}$$

where x denotes the life span (in hours) of a ran-domly selected bulb.

   a. What is the probability that the life span of a ran-domly selected bulb is between 50 and 60 hours?

   b. What is the probability that the life span of a ran-domly selected bulb is no greater than 60 hours?

   c. What is the probability that the life span of a randomly selected bulb is greater than 60 hours?

   d. What is the expected life span of a randomly selected bulb?

57. PRODUCT RELIABILITY The useful life of a particular type of printer is measured by a random variable X with probability density function

$$f(x) = \begin{cases} 0.02e^{-0.02x} & \text{if } x \geq 0 \\ 0 & \text{if } x < 0 \end{cases}$$

where x denotes the number of months a randomly selected printer has been in use.

   a. What is the probability that a randomly selected printer will last between 10 and 15 months?

   b. What is the probability that a randomly selected printer will last less than 8 months?

   c. What is the probability that a randomly selected printer will last longer than 1 year?

   d. What is the expected life of a randomly selected printer?

58. CUSTOMER SERVICE Suppose the time X a customer must spend waiting in line at a certain bank is a random variable that is exponentially distributed with density function

$$f(x) = \begin{cases} \dfrac{1}{4}e^{-x/4} & \text{if } x \geq 0 \\ 0 & \text{if } x < 0 \end{cases}$$

where x is the number of minutes a randomly selected customer spends waiting in line.

a. Find the probability that a customer will have to stand in line at least 8 minutes.

b. Find the probability that a customer will have to stand in line between 1 and 5 minutes.

c. Find the expected waiting time for customers at the bank.

59. **MEDICAL RESEARCH** A group of patients with a potentially fatal disease has been treated with an experimental drug. Assume the survival time X for a patient receiving the drug is a random variable exponentially distributed with density function

$$f(x) = \begin{cases} ke^{-kx} & \text{if } x \geq 0 \\ 0 & \text{if } x < 0 \end{cases}$$

where x is the number of years a patient survives after first receiving the drug.

a. Research indicates that the expected survival time for a patient receiving the drug is 5 years. Based on this information, what is k?

b. Using the density function determined in part (a), what is the probability that a randomly selected patient survives for less than 2 years?

c. What is the probability that a randomly selected patient survives for more than 7 years?

60. **EXPERIMENTAL PSYCHOLOGY** Suppose the length of time that it takes a laboratory rat to traverse a certain maze is measured by a random variable X that is exponentially distributed with density function

$$f(x) = \begin{cases} \dfrac{1}{16}xe^{-x/4} & \text{if } x \geq 0 \\ 0 & \text{if } x < 0 \end{cases}$$

where x is the number of minutes a randomly selected rat spends in the maze.

a. Find the probability that a randomly selected rat will take no more than 5 minutes to traverse the maze.

b. Find the probability that a randomly selected rat will take at least 10 minutes to traverse the maze.

c. Find the expected time required for laboratory rats to traverse the maze.

61. **AIRPLANE ARRIVALS** The time interval between the arrivals of successive planes at a

certain airport is measured by a random variable X with probability density function

$$f(x) = \begin{cases} 0.2e^{-0.2x} & \text{for } x \geq 0 \\ 0 & \text{for } x < 0 \end{cases}$$

where x is the time (in minutes) between the arrivals of a randomly selected pair of successive planes.

a. What is the probability that two successive planes selected at random will arrive within 5 minutes of one another?

b. What is the probability that two successive planes selected at random will arrive more than 6 minutes apart?

c. If you arrive at the airport just in time to see a plane landing, how long would you expect to wait for the next arriving flight?

62. **SPY STORY** Still groggy from his encounter with the camel in Problem 56 of Section 5.1, our spy stumbles into a trap set by Scélérat and his men. He fights back, but eventually runs out of bullets and is forced to use his last weapon, a special stun grenade. Pulling the pin from the grenade, he recalls fondly the time exactly 1 year ago that he received it along with a pair of brass knuckles from his superior, "N", for Valentine's Day. As the grenade arcs through the air toward the enemy, he also recalls, somewhat less fondly, that this particular kind of grenade has a 1-year warranty and that its life span X is a random variable exponentially distributed with density function

$$f(t) = \begin{cases} 0.08e^{-0.08t} & \text{for } t \geq 0 \\ 0 & \text{for } t < 0 \end{cases}$$

where t is the number of months since the grenade left the munitions factory. Assuming the grenade was new when the spy received it and that it had been chosen randomly from the factory stock, what is the probability that the spy will expire before the warranty?

63. **ENDOWMENT** A wealthy patron of a small private college wishes to endow a chair in mathematics with a gift of G thousand dollars. Suppose the mathematician who occupies the chair is to receive $70,000 per year in salary and benefits. If money costs 8% per year compounded continuously, what is the smallest possible value for G?

64. **CAPITALIZED COST OF AN ASSET** The **capitalized cost** of an asset is the sum of the original cost of the asset and the present value of maintaining the asset. Suppose a company is considering the purchase of two different machines. Machine 1

costs $10,000 and t years from now will cost $M_1(t) = 1,000(1 + 0.06t)$ dollars to maintain. Machine 2 costs only $8,000, but its maintenance cost at time t is $M_2(t) = 1,100$ dollars.

   a. If the cost of money is 9% per year compounded continuously, what is the capitalized cost of each machine? Which one should the company purchase?

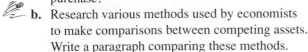 **b.** Research various methods used by economists to make comparisons between competing assets. Write a paragraph comparing these methods.

65. **LEGISLATIVE TURNOVER** A mathematical model in political science* asserts that the length of time served (continuously) by a legislator is a random variable X that is exponentially distributed with density function

$$N(t) = \begin{cases} ce^{-ct} & \text{for } t \geq 0 \\ 0 & \text{for } t < 0 \end{cases}$$

where t is the number of years of continuous service and c is a positive constant that depends on the nature and character of the legislative body.

   a. For the U.S. House of Representatives, it was found that $c = 0.0866$. Find the probability that a randomly selected House member will serve at least 6 years.

   b. For the British House of Commons, it was found that $c = 0.135$. What is the probability that a member of this body will serve at least 6 years?

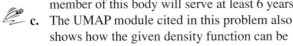 **c.** The UMAP module cited in this problem also shows how the given density function can be used to estimate how many members of the

*Thomas W. Casstevens, "Exponential Models for Legislative Turnover," *UMAP Modules 1978: Tools for Teaching*, Lexington, MA: Consortium for Mathematics and Its Applications, Inc., 1979.

Soviet Central Committee may have been purged by Nikita Khrushchev in the period 1956–1961. Read this module and write a paragraph on whether you think the method of analysis used by the author is valid.

66. **PRESENT VALUE** An investment will generate income continuously at the constant rate of Q dollars per year in perpetuity. Assuming a fixed annual interest rate of r (expressed as a decimal) compounded continuously, use an improper integral to show that the present value of the investment is Q/r dollars.

67. **PRESENT VALUE** In t years, an investment will be generating $f(t) = A + Bt$ dollars per year, where A and B are constants. If the income is generated in perpetuity and the prevailing annual interest rate of r (expressed as a decimal) compounded continuously does not change, show that the present value of this investment is $\dfrac{A}{r} + \dfrac{B}{r^2}$ dollars.

68. **EXPECTED VALUE** Show that a uniformly distributed random variable X with probability density function

$$f(x) = \begin{cases} \dfrac{1}{B - A} & \text{if } A \leq x \leq B \\ 0 & \text{otherwise} \end{cases}$$

has expected value $E(X) = \dfrac{A + B}{2}$.

69. **EXPECTED VALUE** Show that an exponentially distributed random variable X with probability density function

$$f(x) = \begin{cases} ke^{-kx} & \text{if } x \geq 0 \\ 0 & \text{if } x < 0 \end{cases}$$

has expected value $E(X) = 1/k$.

SECTION 6.4 | Numerical Integration

In this section, you will see some techniques you can use to approximate definite integrals. Numerical methods such as these are needed when the function to be integrated does not have an elementary antiderivative. For instance, neither $\sqrt{x^3 + 1}$ nor e^x/x has an elementary antiderivative.

Approximation by Rectangles

If $f(x)$ is positive on the interval $a \leq x \leq b$, the definite integral $\displaystyle\int_a^b f(x)\,dx$ is equal to the area under the graph of f between $x = a$ and $x = b$. As you saw in Section 5.3, one way to approximate this area is to use n rectangles, as shown in Figure 6.16. In particular, you divide the interval $a \le x \le b$ into n equal subintervals of width $\Delta x = \dfrac{b-a}{n}$ and let x_j denote the beginning of the jth subinterval. The base of the jth rectangle is the jth subinterval, and its height is $f(x_j)$. Hence, the area of the jth rectangle is $f(x_j)\Delta x$. The sum of the areas of all n rectangles is an approximation to the area under the curve, so an approximation to the corresponding definite integral is

$$\int_a^b f(x)\,dx \approx f(x_1)\Delta x + \cdots + f(x_n)\Delta x$$

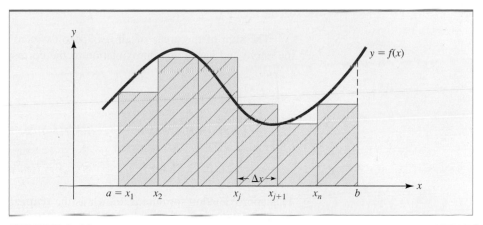

FIGURE 6.16 Approximation by rectangles.

This approximation improves as the number of rectangles increases, and you can estimate the integral to any desired degree of accuracy by making n large enough. However, since fairly large values of n are usually required to achieve reasonable accuracy, approximation by rectangles is rarely used in practice.

Approximation by Trapezoids

The accuracy of the approximation improves significantly if trapezoids are used instead of rectangles. Figure 6.17a shows the area from Figure 6.16 approximated by n trapezoids. Notice how much better the approximation is in this case.

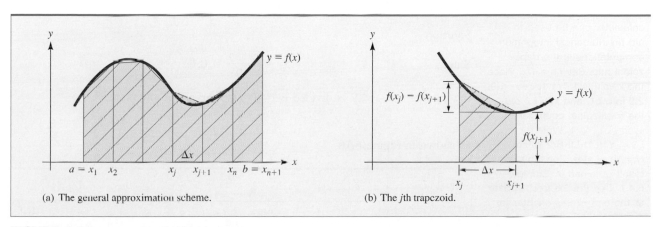

(a) The general approximation scheme.

(b) The jth trapezoid.

FIGURE 6.17 Approximation by trapezoids.

The jth trapezoid is shown in greater detail in Figure 6.17b. Notice that it consists of a rectangle with a right triangle on top of it. Since

$$\text{Area of rectangle} = f(x_{j+1})\Delta x$$

and

$$\text{Area of triangle} = \frac{1}{2}[f(x_j) - f(x_{j+1})]\Delta x$$

it follows that

$$\text{Area of } j\text{th trapezoid} = f(x_{j+1})\Delta x + \frac{1}{2}[f(x_j) - f(x_{j+1})]\Delta x$$

$$= \frac{1}{2}[f(x_j) + f(x_{j+1})]\Delta x$$

The sum of the areas of all n trapezoids is an approximation to the area under the curve and hence an approximation to the corresponding definite integral. Thus,

$$\int_a^b f(x)\,dx$$

$$\approx \frac{1}{2}[f(x_1) + f(x_2)]\Delta x + \frac{1}{2}[f(x_2) + f(x_3)]\Delta x + \cdots + \frac{1}{2}[f(x_n) + f(x_{n+1})]\Delta x$$

$$= \frac{\Delta x}{2}[f(x_1) + 2f(x_2) + \cdots + 2f(x_n) + f(x_{n+1})]$$

This approximation formula is known as the **trapezoidal rule** and applies even if the function f is not positive.

The Trapezoidal Rule

$$\int_a^b f(x)\,dx \approx \frac{\Delta x}{2}[f(x_1) + 2f(x_2) + \cdots + 2f(x_n) + f(x_{n+1})]$$

The use of the trapezoidal rule is shown in Example 6.4.1.

EXPLORE!

Refer to Example 6.4.1, where $a = 1$, $b = 2$, and $n = 10$. The list features of the graphing calculator can be used to aid the numerical integration computations in the trapezoidal rule. Set Y1 = 1/x. Place the x values 1.0, 1.1, . . . , 1.9, 2.0 into L1, and into L2 put the trapezoidal coefficients 1, 2, . . . , 2, 1. Write L3 = Y1(L1)*L2*H/2 where $H = (b - a)/n$. Confirm the result obtained in Example 6.4.1. See the Explore! Update at the end of this chapter for more details.

EXAMPLE | 6.4.1

Use the trapezoidal rule with $n = 10$ to approximate $\int_1^2 \frac{1}{x}\,dx$.

Solution

Since

$$\Delta x = \frac{2 - 1}{10} = 0.1$$

the interval $1 \leq x \leq 2$ is divided into 10 subintervals by the points

$$x_1 = 1, \, x_2 = 1.1, \, x_3 = 1.2, \, \ldots, \, x_{10} = 1.9, \, x_{11} = 2$$

as shown in Figure 6.18.

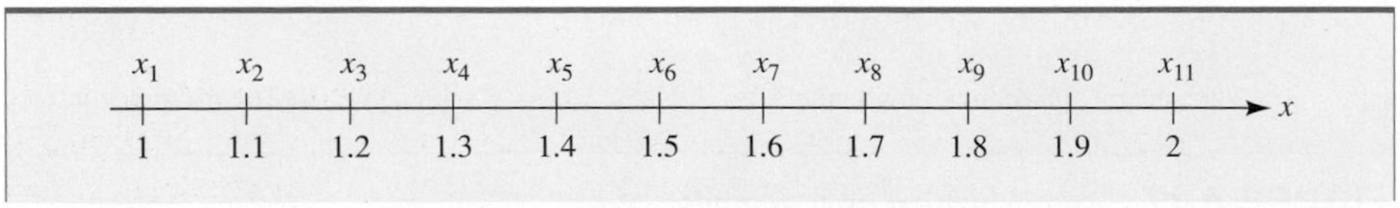

x_1	x_2	x_3	x_4	x_5	x_6	x_7	x_8	x_9	x_{10}	x_{11}	
1	1.1	1.2	1.3	1.4	1.5	1.6	1.7	1.8	1.9	2	x

FIGURE 6.18 Division of the interval $1 \leq x \leq 2$ into 10 subintervals.

Then, by the trapezoidal rule,

$$\int_1^2 \frac{1}{x}\,dx \approx \frac{0.1}{2}\left(\frac{1}{1} + \frac{2}{1.1} + \frac{2}{1.2} + \frac{2}{1.3} + \frac{2}{1.4} + \frac{2}{1.5} + \frac{2}{1.6} + \frac{2}{1.7} + \frac{2}{1.8} + \frac{2}{1.9} + \frac{1}{2}\right)$$

$$\approx 0.693771$$

The definite integral in Example 6.4.1 can be evaluated directly. In particular,

$$\int_1^2 \frac{1}{x}\,dx = \ln|x|\Big|_1^2 = \ln 2 \approx 0.693147$$

Thus, the approximation of this particular integral by the trapezoidal rule with $n = 10$ is accurate (after round off) to two decimal places.

Accuracy of the Trapezoidal Rule

The difference between the true value of the integral $\int_a^b f(x)\,dx$ and the approximation generated by the trapezoidal rule when n subintervals are used is denoted by E_n. Here is an estimate for the absolute value of E_n that is proved in more advanced courses.

Error Estimate for the Trapezoidal Rule ▪ If M is the maximum value of $|f''(x)|$ on the interval $a \le x \le b$, then

$$|E_n| \le \frac{M(b-a)^3}{12n^2}$$

The use of this formula is illustrated in Example 6.4.2.

EXAMPLE | 6.4.2

Estimate the accuracy of the approximation of $\int_1^2 \frac{1}{x}\,dx$ by the trapezoidal rule with $n = 10$.

Solution

Starting with $f(x) = \dfrac{1}{x}$, compute the derivatives

$$f'(x) = -\frac{1}{x^2} \qquad \text{and} \qquad f''(x) = \frac{2}{x^3}$$

and observe that the largest value of $|f''(x)|$ for $1 \le x \le 2$ is $|f''(1)| = 2$.

Apply the error formula with

$$M = 2 \qquad a = 1 \qquad b = 2 \qquad \text{and} \qquad n = 10$$

to get

$$|E_{10}| \le \frac{2(2-1)^3}{12(10)^2} \approx 0.00167$$

That is, the error in the approximation in Example 6.4.1 is guaranteed to be no greater than 0.00167. (In fact, to five decimal places, the error is 0.00062, as you can see by comparing the approximation obtained in Example 6.4.1 with the decimal representation of ln 2.)

With the aid of the error estimate you can decide in advance how many subintervals to use to achieve a desired degree of accuracy. Here is an example.

EXAMPLE 6.4.3

How many subintervals are required to guarantee that the error will be less than 0.00005 in the approximation of $\int_1^2 \frac{1}{x}\,dx$ using the trapezoidal rule?

Solution

From Example 6.4.2 you know that $M = 2$, $a = 1$, and $b = 2$, so that

$$|E_n| \leq \frac{2(2-1)^3}{12n^2} = \frac{1}{6n^2}$$

The goal is to find the smallest positive integer n for which

$$\frac{1}{6n^2} < 0.00005$$

or, equivalently,

$$n^2 > \frac{1}{6(0.00005)}$$

or

$$n > \sqrt{\frac{1}{6(0.00005)}} \approx 57.74$$

The smallest such integer is $n = 58$, and so 58 subintervals are required to ensure the desired accuracy.

Approximation Using Parabolas: Simpson's Rule

The relatively large number of subintervals required in Example 6.4.3 to ensure accuracy to within 0.00005 suggests that approximation by trapezoids may not be efficient enough for some applications. There is another approximation formula, called **Simpson's rule,** which is no harder to use than the trapezoidal rule but often requires substantially fewer calculations to achieve a specified degree of accuracy. Like the trapezoidal rule, it is based on the approximation of the area under a curve by columns, but unlike the trapezoidal rule, it uses parabolic arcs rather than line segments at the top of the columns.

To be more specific, the approximation of a definite integral using parabolas is based on the following construction (illustrated in Figure 6.19 for $n = 6$). Divide the interval $a \leq x \leq b$ into an **even** number of subintervals so that adjacent subintervals can be paired with none left over. Approximate the portion of the graph that lies above the first pair of subintervals by the (unique) parabola that passes through the three points $(x_1, f(x_1))$, $(x_2, f(x_2))$, and $(x_3, f(x_3))$ and use the area under this parabola between x_1 and x_3 to approximate the corresponding area under the curve. Do the same for the remaining pairs of subintervals and use the sum of the resulting areas to approximate the total area under the graph. Here is the approximation formula that results from this construction.

Simpson's Rule For an even integer n,

$$\int_a^b f(x)\,dx \approx \frac{\Delta x}{3}[f(x_1) + 4f(x_2) + 2f(x_3) + 4f(x_4) + \cdots + 2f(x_{n-1}) + 4f(x_n) + f(x_{n+1})]$$

Notice that the first and last function values in the approximating sum in Simpson's rule are multiplied by 1, while the others are multiplied alternately by 4 and 2.

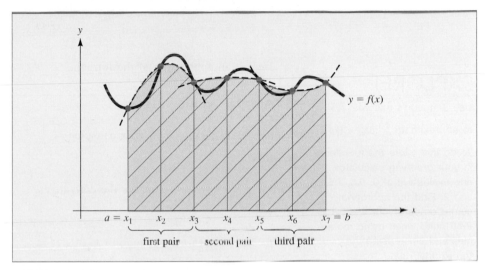

FIGURE 6.19   Approximation using parabolas.

The proof of Simpson's rule is based on the fact that the equation of a parabola is a polynomial of the form $y = Ax^2 + Bx + C$. For each pair of subintervals, the three given points are used to find the coefficients A, B, and C, and the resulting polynomial is then integrated to get the corresponding area. The details of the proof are straightforward but tedious and will be omitted.

EXPLORE!

Refer to Example 6.4.4, where $a = 1$, $b = 2$, and $n = 10$. The list feature of the graphing calculator can be used to facilitate the computations in Simpson's rule. Set Y1 = 1/x. Place the x values, x = 1.0, 1.1, . . . , 1.9, 2.0, into list L1, and the Simpson's rule coefficients 1, 4, 2, . . . , 4, 1 into L2. With H = (b − a)/n, write L3 = Y1(L1)*L2*H/3. Confirm the result obtained in Example 6.4.4. Refer to the EXPLORE! box on page 514 for trapezoidal rule details.

EXAMPLE | 6.4.4

Use Simpson's rule with $n = 10$ to approximate $\int_1^2 \frac{1}{x}\, dx$.

Solution

As in Example 6.4.1, $\Delta x = 0.1$, and hence the interval $1 \le x \le 2$ is divided into the 10 subintervals by

$$x_1 = 1,\ x_2 = 1.1,\ x_3 = 1.2,\ \ldots,\ x_{10} = 1.9,\ x_{11} = 2$$

Then, by Simpson's rule,

$$\int_1^2 \frac{1}{x}\, dx \approx \frac{0.1}{3}\left(\frac{1}{1} + \frac{4}{1.1} + \frac{2}{1.2} + \frac{4}{1.3} + \frac{2}{1.4} + \frac{4}{1.5} + \frac{2}{1.6} + \frac{4}{1.7} + \frac{2}{1.8} + \frac{4}{1.9} + \frac{1}{2} \right)$$

$$\approx 0.693150$$

Notice that this is an excellent approximation to the true value to 6 decimal places; namely, $\ln 2 = 0.693147$.

Accuracy of   The error estimate for Simpson's rule uses the fourth derivative $f^{(4)}(x)$ in much the
Simpson's Rule   same way the second derivative $f''(x)$ was used in the error estimate for the trapezoidal rule. Here is the estimation formula.

> **Error Estimate for Simpson's Rule** ■ If M is the maximum value of $|f^{(4)}(x)|$ on the interval $a \le x \le b$, then
>
> $$|E_n| \le \frac{M(b-a)^5}{180n^4}$$

EXPLORE!

Use Simpson's rule with $n = 4$ to approximate $\int_0^2 (3x^2 + 1)\,dx$. To do this, store the function in your graphing calculator and evaluate it at 0, 0.5, 1.5, and 2. Find the appropriate sum. Compare your answer to that found when using rectangles and when using the trapezoidal rule.

Here is an application of the formula.

EXAMPLE | 6.4.5

Estimate the accuracy of the approximation of $\int_1^2 \frac{1}{x}\,dx$ by Simpson's rule with $n = 10$.

Solution

Starting with $f(x) = \frac{1}{x}$, compute the derivatives

$$f'(x) = -\frac{1}{x^2} \quad f''(x) = \frac{2}{x^3} \quad f^{(3)}(x) = -\frac{6}{x^4} \quad f^{(4)}(x) = \frac{24}{x^5}$$

and observe that the largest value of $|f^{(4)}(x)|$ on the interval $1 \le x \le 2$ is $|f^{(4)}(1)| = 24$. Now apply the error formula with $M = 24$, $a = 1$, $b = 2$, and $n = 10$ to get

$$|E_{10}| \le \frac{24(2-1)^5}{180(10)^4} \approx 0.000013$$

That is, the error in the approximation in Example 6.4.4 is guaranteed to be no greater than 0.000013.

In Example 6.4.6, the error estimate is used to determine the number of subintervals that are required to ensure a specified degree of accuracy.

EXAMPLE | 6.4.6

How many subintervals are required to ensure accuracy to within 0.00005 in the approximation of $\int_1^2 \frac{1}{x}\,dx$ by Simpson's rule?

Solution

From Example 6.4.5 you know that $M = 24$, $a = 1$, and $b = 2$. Hence,

$$|E_n| \le \frac{24(2-1)^5}{180n^4} = \frac{2}{15n^4}$$

The goal is to find the smallest positive (even) integer for which

$$\frac{2}{15n^4} < 0.00005$$

or, equivalently,

$$n^4 > \frac{2}{15(0.00005)}$$

or

$$n > \left[\frac{2}{15(0.00005)}\right]^{1/4} \approx 7.19$$

The smallest such (even) integer is $n = 8$, and so eight subintervals are required to ensure the desired accuracy. Compare with the result of Example 6.4.3, where we found that 58 subintervals are required to ensure the same degree of accuracy using the trapezoidal rule.

Interpreting Data with Numerical Integration

Numerical integration can often be used to estimate a quantity $\int_a^b f(x)\,dx$ when all that is known about $f(x)$ is a set of experimentally determined data. Here are two examples.

EXAMPLE | 6.4.7

Jack needs to know the area of his swimming pool in order to buy a pool cover, but this is difficult because of the pool's irregular shape. Suppose Jack makes the measurements shown in Figure 6.20 at 4-ft intervals along the base of the pool (all measurements are in feet). How can he use the trapezoidal rule to estimate the area?

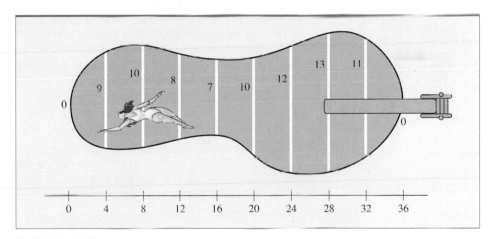

FIGURE 6.20 Measurements across a pool.

Solution

If Jack could find functions $f(x)$ for the top rim of the pool and $g(x)$ for the bottom rim, then the area would be given by the definite integral $A = \int_0^{36} [f(x) - g(x)]\,dx$.

The irregular shape makes it impossible or at least impractical to find formulas for f and g, but Jack's measurements tell him that

$$f(0) - g(0) = 0 \quad f(4) - g(4) = 9 \quad f(8) - g(8) = 10 \ldots f(36) - g(36) = 0$$

Substituting this information into the trapezoidal rule approximation and using $\Delta x = (36 - 0)/9 = 4$, Jack obtains

$$A = \int_0^{36} [f(x) - g(x)]\,dx$$

$$\approx \frac{4}{2}[0 + 2(9) + 2(10) + 2(8) + 2(7) + 2(10) + 2(12) + 2(13) + 2(11) + 0]$$

$$= \frac{4}{2}(160) = 320$$

Thus, Jack estimates the area of the pool to be approximately 320 ft^2.

EXAMPLE | 6.4.8

The management of a chain of pet supply stores is selling a 10-year franchise. Past records in similar localities suggest that t years from now, the franchise will be generating income at the rate of $f(t)$ thousand dollars per year, where $f(t)$ is as indicated in the following table for a typical decade.

Year t	0	1	2	3	4	5	6	7	8	9	10
Rate of income flow $f(t)$	510	580	610	625	654	670	642	610	590	573	550

If the prevailing rate of interest remains at 5% per year compounded continuously over the 10-year term, what is a fair price for the franchise, based on the given information?

Solution

If the rate of income flow $f(t)$ were a continuous function, a fair price for the franchise might be determined by computing the present value of the income flow over the 10-year term. According to the formula developed in Section 5.5, this present value would be given by the definite integral

$$PV = \int_0^{10} f(t)\, e^{-0.05t}\, dt$$

since the prevailing rate of interest is 5% ($r = 0.05$). Since we don't have such a continuous function $f(t)$, we will use Simpson's rule with $n = 10$ and $\Delta t = 1$ to *estimate* the present-value integral. We find that

$$\begin{aligned}
PV &= \int_0^{10} f(t)\, e^{-0.05t}\, dt \\
&\approx \frac{\Delta t}{3}\left[f(0)\,e^{-0.05(0)} + 4f(1)\,e^{-0.05(1)} + 2f(2)\,e^{-0.05(2)} + \cdots + 4f(9)\,e^{-0.05(9)} \right. \\
&\quad \left. + f(10)\,e^{-0.05(10)}\right] \\
&\approx \frac{1}{3}\left[(510)e^{-0.05(0)} + 4(580)e^{-0.05(1)} + 2(610)e^{-0.05(2)} + 4(625)e^{-0.05(3)} \right. \\
&\quad + 2(654)e^{-0.05(4)} + 4(670)e^{-0.05(5)} + 2(642)e^{-0.05(6)} + 4(610)e^{-0.05(7)} \\
&\quad \left. + 2(590)e^{-0.05(8)} + 4(573)e^{-0.05(9)} + (550)e^{-0.05(10)}\right] \\
&\approx \frac{1}{3}(14{,}387) \approx 4{,}796
\end{aligned}$$

Thus, the present value of the income stream over the 10-year term is approximately $4,796 thousand dollars ($4,796,000). The company may use this estimate as a fair asking price for the franchise.

PROBLEMS 6.4

In Problems 1 through 14, approximate the given integral using (a) the trapezoidal rule and (b) Simpson's rule with the specified number of subintervals.

1. $\int_1^2 x^2\, dx;\ n = 4$

2. $\int_4^6 \frac{1}{\sqrt{x}}\, dx;\ n = 10$

3. $\int_0^1 \frac{1}{1 + x^2}\, dx;\ n = 4$

4. $\int_2^3 \frac{1}{x^2 - 1}\, dx;\ n = 4$

5. $\int_{-1}^0 \sqrt{1 + x^2}\, dx;\ n = 4$

6. $\int_0^3 \sqrt{9 - x^2}\, dx;\ n = 6$

7. $\int_0^1 e^{-x^2}\, dx;\ n = 4$

8. $\int_0^2 e^{x^2}\, dx;\ n = 10$

9. $\int_2^4 \frac{dx}{\ln x};\ n = 6$

10. $\int_1^2 \frac{\ln x}{x + 2}\, dx;\ n = 4$

11. $\int_0^1 \sqrt[3]{1 + x^2}\, dx;\ n = 4$

12. $\int_0^1 \frac{dx}{\sqrt{1 + x^3}};\ n = 6$

13. $\int_0^2 e^{-\sqrt{x}}\, dx;\ n = 8$

14. $\int_1^2 \frac{e^x}{x}\, dx;\ n = 4$

In Problems 15 through 20, approximate the given integral and estimate the error $|E_n|$ using (a) the trapezoidal rule and (b) Simpson's rule with the specified number of subintervals.

15. $\int_1^2 \frac{1}{x^2}\, dx;\ n = 4$

16. $\int_0^2 x^3\, dx;\ n = 8$

17. $\int_1^3 \sqrt{x}\, dx;\ n = 10$

18. $\int_1^2 \ln x\, dx;\ n = 4$

19. $\int_0^1 e^{x^2}\, dx;\ n = 4$

20. $\int_0^{0.6} e^{x^3}\, dx;\ n = 6$

In Problems 21 through 26, determine how many subintervals are required to guarantee accuracy to within 0.00005 in the approximation of the given integral by (a) the trapezoidal rule and (b) Simpson's rule.

21. $\int_1^3 \frac{1}{x}\, dx$

22. $\int_0^4 (x^4 + 2x^2 + 1)\, dx$

23. $\int_1^2 \frac{1}{\sqrt{x}}\, dx$

24. $\int_1^2 \ln(1 + x)\, dx$

25. $\int_{1.2}^{2.4} e^x\, dx$

26. $\int_0^2 e^{x^2}\, dx$

27. A quarter circle of radius 1 has the equation $y = \sqrt{1 - x^2}$ for $0 \le x \le 1$ and has area $\pi/4$.

Thus, $\int_0^1 \sqrt{1 - x^2}\, dx = \pi/4$. Use this formula to estimate π by applying:
  a. The trapezoidal rule
  b. Simpson's rule
In each case, use $n = 8$ subintervals.

28. Use the trapezoidal rule to estimate the area bounded by the curve $y = \sqrt{x^3 + 1}$, the x axis, and the lines $x = 0$ and $x = 1$.

29. Use the trapezoidal rule with $n = 10$ to estimate the average value of the function $f(x) = \dfrac{e^{-0.4x}}{x}$ over the interval $1 \le x \le 6$.

30. Use the trapezoidal rule with $n = 6$ to estimate the average value of the function $y = \sqrt{\ln x}$ over the interval $1 \le x \le 4$.

31. Use the trapezoidal rule with $n = 7$ to estimate the volume of the solid generated by rotating the region under the curve $y = x/(1 + x)$ between $x = 0$ and $x = 1$ about the x axis.

32. Use Simpson's rule with $n = 6$ to estimate the volume of the solid generated by rotating the region under the curve $y = \ln x$ between $x = 1$ and $x = 2$ about the x axis.

33. **FUTURE VALUE OF AN INVESTMENT**   An investment generates income continuously at the rate of $f(t) = \sqrt{t}$ thousand dollars per year at time t (years). If the prevailing rate of interest is 6% per year compounded continuously, use the trapezoidal rule with $n = 5$ to estimate the future value of the investment over a 10-year term. (See Example 5.5.2 in Section 5.5.)

34. **PRESENT VALUE OF A FRANCHISE**   The management of a national chain of fast-food restaurants is selling a 5-year franchise to operate its newest outlet in Tulare, California. Past experience in similar localities suggests that t years from now, the franchise will be generating profit at the rate of $f(t) = 12{,}000\sqrt{t}$ dollars per year. Suppose the prevailing annual interest rate remains fixed during the next 5 years at 5% compounded continuously. Use Simpson's rule with $n = 10$ to estimate the present value of the franchise.

35. **SPREAD OF A DISEASE**   A new strain of influenza has just been declared an epidemic by health officials. Currently, 3,000 people have the disease and new victims are being added at the rate of $R(t) = 50\sqrt{t}$ people per week. Moreover, the fraction of infected people who still have the disease t weeks after contracting it is given by $S(t) = e^{-0.01t}$. Use Simpson's rule with $n = 8$ to estimate the number of people who have the flu 8 weeks from now. (Think of this as a survival/renewal problem, like Example 5.6.2 in Section 5.6.)

36. **CONSUMERS' SURPLUS**   An economist models the demand for a particular commodity by the function

$$p = D(q) = \frac{100}{q^2 + q + 1}$$

where q hundred units are sold when the price is p dollars per unit. Use Simpson's rule with $n = 6$ to estimate the consumers' surplus when the level of production is $q_0 = 5$. (See Example 5.5.5 in Section 5.5.)

37. **DISTANCE AND VELOCITY**   Sue and Tom are traveling in a car with a broken odometer. In order to determine the distance they travel between 2 and 3 P.M., Tom (the passenger) takes speedometer readings every 5 minutes:

Minutes after 2:00 P.M.	0	5	10	15	20	25	30	35	40	45	50	55	60
Speedometer reading	45	48	37	39	55	60	60	55	50	67	58	45	49

Use the trapezoidal rule to estimate the total distance traveled by the pair during the hour in question.

38. **MENTAL HEALTH CARE**   A county mental health clinic has just opened. The clinic initially accepts 300 people for treatment and plans to accept new patients at the rate of 10 per month. Let $f(t)$ denote the fraction of people receiving treatment continuously for at least t days. For the first 60 days, records are kept and these values of $f(t)$ are obtained:

t (days)	0	5	10	15	20	25	30	35	40	45	50	55	60
$f(t)$	1	$\frac{3}{4}$	$\frac{3}{5}$	$\frac{1}{2}$	$\frac{1}{3}$	$\frac{3}{10}$	$\frac{1}{5}$	$\frac{1}{6}$	$\frac{1}{7}$	$\frac{1}{9}$	$\frac{1}{12}$	$\frac{1}{15}$	$\frac{1}{20}$

Use this information together with the trapezoidal rule to estimate the number of patients in the clinic at the end of the 60-day period. [*Hint:* This is a survival/renewal problem. Recall Example 5.6.1 of Section 5.6.]

39. **FUTURE VALUE OF AN INVESTMENT** Marc has a small investment providing a variable income stream that is deposited continuously into an account earning interest at an annual rate of 4% compounded continuously. He spot-checks the monthly flow rate of the investment on the first day of every other month for a 1-year period, obtaining the results in this table:

Month	Jan.	Mar.	May	Jul.	Sep.	Nov.	Jan.
Rate of income flow	$437	$357	$615	$510	$415	$550	$593

For instance, income is entering the account at the rate of $615 per month on the first of May, but 2 months later, the rate of income flow is only $510 per month. Use this information together with Simpson's rule to estimate the future value of the income flow during this 1-year period. [*Hint:* Recall Example 5.5.2 of Section 5.5.]

40. **CARDIAC OUTPUT FROM DATA** When measuring cardiac output (Example 5.6.4 of Section 5.6), physicians generally do not have a specific formula $C(t)$ for the concentration of dye passing through the patient's heart. Instead, data analysis methods are used. Suppose 5 mg of dye are injected into a vein near a particular patient's heart and measurements are made every 5 seconds over a 30-second period, yielding these data:

Time t (seconds)	0	5	10	15	20	25	30
Concentration $C(t)$ (mg/L)	0	10	36	35	15	12	8

a. Use Simpson's rule to estimate the integral $\int_0^{30} C(t)\,dt$. Based on this result, what is the patient's approximate cardiac output?

b. Read an article on blood flow and write a paragraph on mathematical methods of measuring cardiac output under various circumstances.*

41. **POLLUTION CONTROL** An industrial plant spills pollutant into a river. The pollutant spreads out as it is carried downstream by the current of the river, and 3 hours later, the spill forms the pattern shown in the accompanying figure. Measurements (in feet) across the spill are made at 5-foot intervals, as indicated in the figure. Use this information together with the trapezoidal rule to estimate the area of the spill.

---

*A good place to start is the article, "Measuring Cardiac Output," *UMAP Modules 1977: Tools for Teaching,* Lexington, MA: Consortium for Mathematics and Its Applications, Inc., 1978.

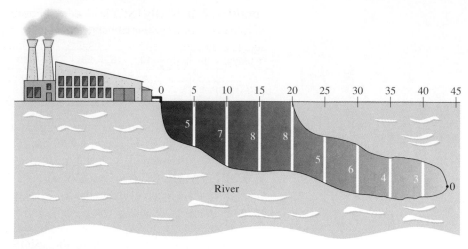

PROBLEM 41

42. **NET PROFIT FROM DATA** On the first day of each month, the manager of a small company estimates the rate at which profit is expected to increase during that month. The results are listed in the accompanying table for the first 6 months of the year, where $P'(t)$ is the rate of profit growth in thousands of dollars per month expected during the tth month ($t = 1$ for January, $t = 6$ for June). Use this information together with the trapezoidal rule to estimate the total profit earned by the company during this 6-month period (January through June).

t (month)	1	2	3	4	5	6
Rate of profit $P'(t)$	0.65	0.43	0.72	0.81	1.02	0.97

43. **PRODUCERS' SURPLUS FROM DATA** An economist studying the supply for a particular commodity gathers the data in the accompanying table, which lists the number of units q (in thousands) of the commodity that will be supplied to the market by producers at a price of p dollars per unit. Use this information together with the trapezoidal rule to estimate the producers' surplus when 7,000 units are supplied ($q_0 = 7$).

q (1,000 units)	0	1	2	3	4	5	6	7
p (dollars per unit)	1.21	3.19	3.97	5.31	6.72	8.16	9.54	11.03

44. **CONSUMERS' SURPLUS FROM DATA** An economist studying the demand for a particular commodity gathers the data in the accompanying table, which lists the number of units q (in thousands) of the commodity that will be demanded (sold) at a price of p dollars per unit. Use this information together with Simpson's rule to estimate the consumers' surplus when 24,000 units are produced; that is, when $q_0 = 24$.

q (1,000 units)	0	4	8	12	16	20	24
p (dollars per unit)	49.12	42.90	31.32	19.83	13.87	10.58	7.25

45. POPULATION DENSITY A demographic study determines that the population density of a certain city at a distance of r miles from the city center is $D(r)$ people per square mile (mi^2), where D is as indicated in the following table for $0 \le r \le 10$ at 2-mile intervals.

Distance r (miles) from city center	0	2	4	6	8	10
Population density $D(r)$ (people/mi^2)	3,120	2,844	2,087	1,752	1,109	879

Use the trapezoidal rule to estimate the total population of the city that is located within a 10-mile radius of the city center. (See Example 5.6.5 of Section 5.6.)

46. DISTRIBUTION OF INCOME A sociologist studying the distribution of income in an industrial society compiles the data displayed in the accompanying table, where $L(x)$ denotes the fraction of the society's total wealth earned by the lowest-paid $100x\%$ of the wage-earners in the society. Use this information together with the trapezoidal rule to estimate the Gini index (GI) for this society; namely,

$$GI = 2\int_0^1 [x - L(x)]\,dx$$

x	0	0.125	0.25	0.375	0.5	0.625	0.75	0.875	1
$L(x)$	0	0.0063	0.0631	0.1418	0.2305	0.3342	0.4713	0.6758	1

47. MORTALITY RATE FROM AIDS The accompanying table, which you first saw in the Think About It essay at the end of Chapter 3, gives the number of reported deaths due to AIDS, during the tth year after 1990, for the period 1990 to 2001. (*Source: Centers for Disease Control and Prevention, National Center for HIV, STD, and TB Prevention.*)

Year	t	Reported AIDS Deaths	Year	t	Reported AIDS Deaths
1990	0	31,120	1996	6	38,296
1991	1	36,175	1997	7	22,245
1992	2	40,587	1998	8	18,823
1993	3	45,850	1999	9	18,249
1994	4	50,842	2000	10	16,672
1995	5	51,670	2001	11	15,603

However, the table does not tell the complete story because many deaths actually due to AIDS go unreported or are attributed to other causes. Let $D(t)$ be the function that gives the cumulative number of AIDS deaths at time t. Then the data in the table can be thought of as rates of change of $D(t)$ at various times; that is, as mortality rates. For instance, the table tells us that in 1992 ($t = 2$), AIDS deaths were occurring at the rate of 40,587 per year.

a. Assuming that $D(t)$ is differentiable, explain why the total number of AIDS deaths N during the period 1990–2001 is given by the integral

$$N = \int_0^{11} D'(t)\,dt$$

b. Estimate N by using the data in the table together with the trapezoidal rule to approximate the integral in part (a).

 c. Why is the mortality function $D(t)$ probably *not* differentiable? Does this invalidate the estimate of N made in part (b)? Write a paragraph either defending or challenging the procedure used to estimate N in this problem.

48. PRODUCT RELIABILITY   The life span of a particular brand of food processor is measured by a random variable X with probability density function

$$f(x) = \begin{cases} 0.002xe^{-0.001x^2} & \text{for } x \geq 0 \\ 0 & \text{for } x < 0 \end{cases}$$

where x denotes the life span (in months) of a randomly selected processor. (Recall the discussion of continuous probability in Section 6.3.)

a. Verify that $f(x)$ is a probability density function by showing that

$$\int_0^{+\infty} f(x)\,dx = 1$$

b. The expected value of X is given by the improper integral

$$E(X) = \int_0^{+\infty} 0.002x^2 e^{-0.001x^2}\,dx$$

which can be evaluated only by numerical methods. Estimate $E(X)$ by applying Simpson's rule with $n = 10$ to the integral

$$\int_0^{100} 0.002x^2 e^{-0.001x^2}\,dx$$

Based on your result, how long would you expect a randomly chosen processor to last?

 c. The actual expected value $E(X)$ is roughly 28 (months). Discuss what (if anything) you could do to improve the result you obtained in part (b).

Important Terms, Symbols, and Formulas

CHAPTER SUMMARY

Improper integrals: (497)

$$\int_a^{+\infty} f(x)\,dx = \lim_{N \to +\infty} \int_a^N f(x)\,dx$$

$$\int_{-\infty}^{+\infty} f(x)\,dx = \lim_{N \to +\infty} \int_{-N}^0 f(x)\,dx + \lim_{N \to +\infty} \int_0^N f(x)\,dx$$

Continuous random variable   (502)

Probability density function: (503)

$$P(a \leq X \leq b) = \int_a^b f(x)\,dx$$

Uniform density function: (504)

$$f(x) = \begin{cases} \dfrac{1}{B - A} & \text{if } A \leq x \leq B \\ 0 & \text{otherwise} \end{cases}$$

Exponential density function: (505)

$$f(x) = \begin{cases} ke^{-kx} & \text{if } x \geq 0 \\ 0 & \text{if } x < 0 \end{cases}$$

Expected value (mean)·   (506)

$$E(X) = \int_{-\infty}^{\infty} xf(x)\,dx$$

Trapezoidal rule: (514)

$$\int_a^b f(x)\,dx$$

$$\approx \frac{\Delta x}{2}[f(x_1) + 2f(x_2) + \cdots + 2f(x_n) + f(x_{n+1})]$$

Error estimate: (515)

$$|E_n| \leq \frac{M(b - a)^3}{12n^2}$$

where M is the maximum value of $|f''(x)|$ on $[a, b]$

Simpson's rule: for n even,   (516)

$$\int_a^b f(x)\,dx \approx \frac{\Delta x}{3}[f(x_1) + 4f(x_2) + 2f(x_3) + \cdots$$

$$+ 2f(x_{n-1}) + 4f(x_n) + f(x_{n+1})]$$

Error estimate: (518)

$$|E_n| \leq \frac{M(b - a)^5}{180n^4}$$

where M is the maximum value of $|f^{(4)}(x)|$ on $[a, b]$

Checkup for Chapter 6

1. Use integration by parts to find each of these indefinite and definite integrals.

   a. $\displaystyle\int \sqrt{2x} \ln x^2\,dx$ **b.** $\displaystyle\int_0^1 xe^{0.2x}\,dx$

   c. $\displaystyle\int_{-4}^0 x\sqrt{1 - 2x}\,dx$ **d.** $\displaystyle\int \frac{x - 1}{e^x}\,dx$

2. In each case, either evaluate the given improper integral or show that it diverges.

   a. $\displaystyle\int_1^{+\infty} \frac{1}{x^{1.1}}\,dx$ **b.** $\displaystyle\int_1^{+\infty} xe^{-2x}\,dx$

   c. $\displaystyle\int_1^{+\infty} \frac{x}{(x + 1)^2}\,dx$ **d.** $\displaystyle\int_{-\infty}^{+\infty} xe^{-x^2}\,dx$

3. Use the integral table on pp. 472–473 to find these integrals.

   a. $\displaystyle\int \left(\ln \sqrt{3x}\right)^2\,dx$ **b.** $\displaystyle\int \frac{dx}{x\sqrt{4 + x^2}}$

   c. $\displaystyle\int \frac{dx}{x^2\sqrt{x^2 - 9}}$ **d.** $\displaystyle\int \frac{dx}{3x^2 - 4x}$

4. In each case, find the particular solution of the given differential equation that satisfies the specified condition,

   a. $\dfrac{dy}{dx} = \dfrac{-2}{x^2 y}$ where $y = 1$ when $x = -1$

   b. $\dfrac{dy}{dx} = \dfrac{xy}{x^2 + 1}$ where $y = -3$ when $x = 0$

   c. $\dfrac{dy}{dx} = xe^{y - x}$ where $y = 0$ when $x = 0$

5. **INVESTMENT GROWTH**   An investment of $10,000 is projected to grow at a rate equal to 5% of its size at any time t. What will the investment be worth in 10 years?

6. **PRESENT VALUE OF AN ASSET**   It is estimated that t years from now an office building will be generating profit for its owner at the rate of $R(t) = 50 + 3t$ thousand dollars per year. If the profit is generated in perpetuity and the prevailing annual interest rate remains fixed at 6% compounded continuously, what is the present value of the office building?

7. **PRODUCT RELIABILITY** The useful life of a brand of microwave oven is measured by a random variable X with probability density function

$$f(x) = \begin{cases} 0.03e^{-0.03x} & \text{if } x \geq 0 \\ 0 & \text{if } x < 0 \end{cases}$$

where x denotes the time (in months) that a randomly selected oven has been in use.

a. What is the probability that an oven selected randomly from a production line will last longer than 1 year?

b. What is the probability that a randomly selected oven will last between 3 and 6 months?

c. What is the expected life of a randomly selected oven?

8. **DRUG CONCENTRATION** A patient in a hospital receives 0.7 mg of a certain drug intravenously every hour. The drug is eliminated exponentially in such a way that the fraction of drug that remains in the patient's body after t hours is $f(t) = e^{-0.2t}$. If the treatment is continued indefinitely, approximately how many units will remain in the patient's bloodstream in the long run (as $t \to +\infty$)?

9. **DECAY OF A BIOMASS** A researcher determines that a certain protein with mass m(t) (grams) at time t (hours) disintegrates into amino acids at a rate jointly proportional to m(t) and t. Experiments indicate that half of any given sample will disintegrate in 12 hours.

a. Set up and solve a differential equation for m(t).

b. What fraction of a given sample remains after 9 hours?

10. Use the trapezoidal rule with $n = 8$ to estimate the value of the integral

$$\int_3^4 \frac{\sqrt{25 - x^2}}{x}\, dx$$

Then use the integral table on pp. 472–473 to compute the exact value of the integral and compare with your approximation.

Review Problems

In Problems 1 through 10, use integration by parts to find the given integral.

1. $\int te^{1-t}\, dt$

2. $\int (5 + 3x)e^{-x/2}\, dx$

3. $\int x\sqrt{2x + 3}\, dx$

4. $\int_{-9}^{-1} \frac{y\, dy}{\sqrt{4 - 5y}}$

5. $\int_1^4 \frac{\ln\sqrt{s}}{\sqrt{s}}\, ds$

6. $\int (\ln x)^2\, dx$

7. $\int_{-2}^1 (2x + 1)(x + 3)^{3/2}\, dx$

8. $\int \frac{w^3}{\sqrt{1 + w^2}}\, dw$

9. $\int x^3\sqrt{3x^2 + 2}\, dx$

10. $\int_0^1 \frac{x + 2}{e^{3x}}\, dx$

In Problems 11 through 16, use the integral table listed on pp. 472–473 to find the given integral.

11. $\int \frac{5\, dx}{8 - 2x^2}$

12. $\int \frac{2\, dt}{\sqrt{9t^2 + 16}}$

13. $\int w^2 e^{-w/3}\, dw$

14. $\int \frac{4\, dx}{x(9 + 5x)}$

15. $\int (\ln 2x)^3\, dx$

16. $\int \frac{dx}{x\sqrt{4 - x^2}}$

In Problems 17 through 26, either evaluate the given improper integral or show that it diverges.

17. $\displaystyle\int_0^{+\infty} \frac{1}{\sqrt[3]{1+2x}}\,dx$

18. $\displaystyle\int_0^{+\infty} (1+2x)^{-3/2}\,dx$

19. $\displaystyle\int_0^{+\infty} \frac{3t}{t^2+1}\,dt$

20. $\displaystyle\int_0^{+\infty} 3e^{-5x}\,dx$

21. $\displaystyle\int_0^{+\infty} xe^{-2x}\,dx$

22. $\displaystyle\int_0^{+\infty} 2x^2 e^{-x^3}\,dx$

23. $\displaystyle\int_0^{+\infty} x^2 e^{-2x}\,dx$

24. $\displaystyle\int_2^{+\infty} \frac{1}{t(\ln t)^2}\,dt$

25. $\displaystyle\int_1^{+\infty} \frac{\ln x}{\sqrt{x}}\,dx$

26. $\displaystyle\int_0^{+\infty} \frac{x-1}{x+2}\,dx$

In Problems 27 through 30 find the general solution of the given differential equation.

27. $\dfrac{dy}{dx} = x^3 - 3x^2 + 5$

28. $\dfrac{dy}{dx} = 0.02xy$

29. $\dfrac{dy}{dx} = k(80 - y)$

30. $\dfrac{dy}{dx} = e^{2x-y}$ [*Hint:* $e^{2x-y} = e^{2x}/e^y$.]

In Problems 31 through 34 find the particular solution of the given differential equation that satisfies the given condition.

31. $\dfrac{dy}{dx} = 5x^4 - 3x^2 - 2$; $y = 4$ when $x = 1$

32. $\dfrac{dy}{dx} = \dfrac{\ln x}{y}$; $y = 100$ when $x = 1$

33. $\dfrac{dy}{dx} = \dfrac{xy}{\sqrt{1-x^2}}$; $y = 2$ when $x = 0$

34. $\dfrac{d^2y}{dx^2} = 2$; $y = 5$ and $\dfrac{dy}{dx} = 3$ when $x = 0$

In Problems 35 through 38, integrate the given probability density functions to find the indicated probabilities.

35. $f(x) = \begin{cases} \dfrac{1}{3} & \text{if } 1 \le x \le 4 \\ 0 & \text{otherwise} \end{cases}$

   a. $P(1 \le X \le 4)$

   b. $P(2 \le X \le 3)$

   c. $P(X \le 2)$

36. $f(x) = \begin{cases} \dfrac{2}{9}(3-x) & \text{if } 0 \le x \le 3 \\ 0 & \text{otherwise} \end{cases}$

   a. $P(0 \le X \le 3)$

   b. $P(1 \le X \le 2)$

37. $f(x) = \begin{cases} 0.2e^{-0.2x} & \text{if } x \ge 0 \\ 0 & \text{if } x < 0 \end{cases}$

   a. $P(X > 0)$

   b. $P(1 \le X \le 4)$

   c. $P(X \ge 5)$

38. $f(x) = \begin{cases} \dfrac{5}{(x+5)^2} & \text{if } x \ge 0 \\ 0 & \text{if } x < 0 \end{cases}$

   a. $P(X > 0)$

   b. $P(1 \le X \le 9)$

   c. $P(X \ge 3)$

39. NET ASSET VALUE The resale value of a certain industrial machine decreases at a rate proportional to the difference between its current value and its scrap value of $5,000. The machine was bought new for $40,000 and was worth $30,000 after 4 years. How much will it be worth when it is 8 years old?

40. PRODUCTION After t hours on the job, a factory worker can produce $100te^{-0.5t}$ units per hour. How many units does a worker who arrives on the job at 8:00 A.M. produce between 10:00 A.M. and noon?

41. DILUTION A tank currently holds 200 gallons of brine that contains 3 pounds of salt per gallon. Clear water flows into the tank at the rate of 4 gallons per minute, while the mixture, which is kept uniform, runs out of the tank at the same rate. How much salt is in the tank at the end of 100 minutes?

42. POPULATION GROWTH The rate at which the population of a certain country is growing is jointly proportional to the upper bound of 10 million imposed by environmental factors and the difference between the upper bound and the size of the population. Express the population (in millions) of the country as a function of time (in years measured from 1995) if the population in 1995 was 4 million and the population in 2000 was 4.74 million.

43. CURRENCY RENEWAL A nation has 5 billion dollars in currency. Each day, about 18 million dollars comes into the banks and the same amount is paid out. Suppose the country decides that whenever a dollar bill comes into a bank, it is destroyed and replaced by a new style of currency. How long will it take for 90% of the currency in circulation to be the new style? [*Hint:* Think of this situation as if it were a dilution problem like Example 6.2.9 in Section 6.2.]

44. PARETO'S LAW **Pareto's law** in economics says that the rate of change (decrease) of the number of people P in a stable economy who have an income of at least x dollars is directly proportional to the number of such people and inversely proportional to their income. Express this law as a differential equation and solve for P in terms of x.

45. DEMOGRAPHICS (SURVIVAL/RENEWAL) Demographic studies conducted in a certain city indicate that the fraction of the residents that will remain in the city for at least t years is $f(t) = e^{-t/20}$. The current population of the city is 100,000, and it is estimated that t years from now, new people will be arriving at the rate of $100t$ people per year. If this estimate is correct, what will happen to the population of the city in the long run?

46. DURATION OF TELEPHONE CALLS The duration of telephone calls in a certain city is measured by a random variable X with probability density function

$$f(x) = \begin{cases} 0.5e^{-0.5x} & \text{for } x \geq 0 \\ 0 & \text{for } x < 0 \end{cases}$$

where x denotes the duration (in minutes) of a randomly selected call.

a. What is the probability that a randomly selected call lasts between 2 and 3 minutes?

b. What percentage of the calls can be expected to last 2 minutes or less? [*Hint:* Think of this as a probability.]

c. What percentage of the calls can be expected to last more than 2 minutes?

d. How long would you expect a randomly selected call to last?

47. MOVIE SHOW TIMES A 2-hour movie runs continuously at a local theater. You leave for the theater without first checking the show times. Use an appropriate uniform density function to find the probability that you will arrive at the theater within 10 minutes of (before or after) the start of the film.

48. SUBSCRIPTION GROWTH The publishers of a national magazine have found that the fraction of subscribers who remain subscribers for at least t years is $f(t) = e^{-t/10}$. Currently, the magazine has 20,000 subscribers and it is estimated that new subscriptions will be sold at the rate of 1,000 per year. Approximately how many subscribers will the magazine have in the long run?

49. PRESENT VALUE OF AN INVESTMENT It is estimated that t years from now, a certain investment will be generating income at the rate of $f(t) = 8,000 + 400t$ dollars per year. If the income is generated in perpetuity and the prevailing annual interest rate remains fixed at 5% compounded continuously, find the present value of the investment.

50. WARRANTY PROTECTION The life span X of a certain electrical appliance is exponentially distributed with density function

$$f(x) = \begin{cases} 0.08\,e^{-0.08x} & \text{if } x \geq 0 \\ 0 & \text{if } x < 0 \end{cases}$$

where x measures time (in months). The appliance carries a 1-year warranty from the manufacturer. Suppose you purchase one of these appliances, selected at random from the manufacturer's stock. Find the probability that the warranty will expire before your appliance becomes unusable.

51. SCHEDULING A SNACK A bakery turns out a fresh batch of chocolate chip cookies every 45 minutes. You arrive (at random) at the bakery, hoping to buy a fresh cookie. Use an appropriate uniform density function to find the probability that you arrive within 5 minutes of (before or after) the time that the cookies come out of the oven.

52. **TRAFFIC CONTROL** Suppose the time between the arrivals of successive cars at a toll booth is measured by a random variable X that is exponentially distributed with the density function

$$f(x) = \begin{cases} 0.5e^{-0.5x} & \text{if } x \geq 0 \\ 0 & \text{if } x < 0 \end{cases}$$

where x measures time (in minutes). Find the probability that a randomly selected pair of successive cars will arrive at the toll booth at least 6 minutes apart.

53. **PSYCHOLOGICAL TESTING** In a psychological experiment, it is found that the proportion of participants who require more than t minutes to finish a particular task is given by

$$\int_t^{+\infty} 0.07e^{-0.07u}\, du$$

   a. Find the proportion of participants who require more than 5 minutes to finish the task.

   b. What proportion requires between 10 and 15 minutes to finish?

54. **PRICE ADJUSTMENT OVER TIME** Suppose the price $p(t)$ of a particular commodity varies in such a way that its rate of change with respect to time is proportional to the shortage $D - S$, where $D(p)$ and $S(p)$ are the demand and supply functions for the commodity, respectively. That is,

$$\frac{dp}{dt} = k(D - S)$$

Find $p(t)$ for the case where $D(p) = 40 - 3p$ and $S(p) = 5 + 4p$, if the price is \$4 when $t = 0$ and is \$3 when $t = 5$.

55. **TIME ADJUSTMENT OF SUPPLY AND DEMAND** The supply $S(t)$ and demand $D(t)$ of a certain commodity vary with time t (months) in such a way that

$$\frac{dD}{dt} = -kD \quad \text{and} \quad \frac{dS}{dt} = 2kS$$

for some constant $k > 0$. It is known that $D(0) = 50$ units and $S(0) = 5$ units and that equilibrium occurs when $t = 10$ months; that is, $D(10) = S(10)$.

   a. Use this information to find k.

   b. Find $D(t)$ and $S(t)$.

   c. How many units are supplied and demanded at equilibrium?

56. **FICK'S LAW** When a cell is placed in a liquid containing a solute, the solute passes through the cell wall by diffusion. As a result, the concentration of the solute inside the cell changes, increasing if the concentration of the solute outside the cell is greater than the concentration inside and decreasing if the opposite is true. In biology, Fick's law asserts that the concentration of the solute inside the cell changes at a rate that is jointly proportional to the area of the cell wall and the difference between the concentrations of the solute inside and outside the cell. Assuming that the concentration of the solute outside the cell is constant and greater than the concentration inside, derive a formula for the concentration of the solute inside the cell.

57. **BIRTH AND DEATH RATES OF A POPULATION** Let $P(t)$ be the number of individuals in a population at time t, and for the same population, let $B(t)$ and $D(t)$ denote the number of births and deaths at time t, respectively. Then the population grows at the rate $P'(t) = B'(t) - D'(t)$, where $B'(t)$ and $D'(t)$ are, respectively, the birthrate and death rate.

   a. Suppose $B'(t) = bP(t)$ and $D'(t) = aP(t)$, where b and a are positive constants. Set up and solve a differential equation for $P(t)$. Express your solution in terms of b, a, and the initial population $P_0 = P(0)$.

   b. A population model used for emerging nations assumes that $B'(t) - D'(t) = kP^{1+1/c}$ for positive constants k and c. Set up and solve a differential equation for $P(t)$ with this assumption. Express your solution in terms of k, c, and $P_0 = P(0)$.

   c. Suppose $k = 0.02$ and $c = 3$ for a particular population modeled as indicated in part (b). If $P_0 = 1,000$ and t is in years, what is the population after 5 years?

58. **DEMOGRAPHY** A **Gompertz function** is used by demographers to predict populations. Let P_0 be the initial population of a region (a country, the world, etc.) and let β be the growth rate of the population. Then the Gompertz function $P(t)$ represents the population at time t and satisfies the differential equation

$$\frac{dP}{dt} = P(\ln P_0)(\ln \beta)\beta^t$$

Solve this equation to find P as a function of t. Assume that P_0 and β are positive constants.

59. **NUCLEAR WASTE** After t years of operation, a certain nuclear power plant produces radioactive waste at the rate of $R(t)$ pounds per year, where
$$R(t) = 300 - 200e^{-0.03t}$$
The waste decays exponentially at the rate of 2% per year. How much radioactive waste from the plant will be present in the long run?

NUMERICAL INTEGRATION *In Problems 60 through 63, approximate the given integral and estimate the error with the specified number of subintervals using:*

(a) *The trapezoidal rule*

(b) *Simpson's rule*

60. $\displaystyle\int_1^3 \frac{1}{x}\, dx; \ n = 10$

61. $\displaystyle\int_0^2 e^{x^2}\, dx; \ n = 8$

62. $\displaystyle\int_1^2 \frac{e^x}{x}\, dx; \ n = 10$

63. $\displaystyle\int_1^2 xe^{1/x}\, dx; \ n = 8$

NUMERICAL INTEGRATION *In Problems 64 and 65, determine how many subintervals are required to guarantee accuracy to within 0.00005 of the actual value of the given integral using:*

(a) *The trapezoidal rule*

(b) *Simpson's rule*

64. $\displaystyle\int_1^3 \sqrt{x}\, dx$

65. $\displaystyle\int_{0.5}^1 e^{-1.1x}\, dx$

66. **TOTAL COST FROM MARGINAL COST** A manufacturer determines that the marginal cost of producing q units of a particular commodity is $C'(q) = \sqrt{q}e^{0.01q}$ dollars per unit.

a. Express the total cost of producing the first 8 units as a definite integral.

b. Estimate the value of the total cost integral in part (a) using the trapezoidal rule with $n = 8$ subintervals.

67. **REVENUE FROM DEMAND DATA** An economist studying the demand for a particular commodity gathers the data in the accompanying table, which lists the number of units q (in thousands) of the commodity that will be demanded (sold) at a price of p dollars per unit.

q (1,000 units)	0	4	8	12	16	20	24
p ($/unit)		49.12	42.90	31.32	19.83	13.87	10.58	7.25

Use this information together with Simpson's rule to estimate the total revenue
$$R = \int_0^{24} qp(q)\, dq \quad \text{thousand dollars}$$
obtained as the level of production is increased from 0 to 24,000 units ($q = 0$ to $q = 24$).

68. A child, standing at the origin of a coordinate plane, holds a 5-foot length of rope that is attached to a sled. She walks along the x axis keeping the rope taut, as shown in the figure (units are in feet).

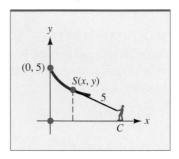

If the sled begins at $(0, 5)$ and is at $S(x, y)$ when the child is at C, it can be shown that
$$\frac{dy}{dx} = \frac{-y}{\sqrt{25 - y^2}}$$

a. Find the equation of the path followed by the sled. (The path is called a **tractrix.**) [*Hint:* Use an integral formula from the table on pp. 472–473.]

b. Use the graphing utility of your calculator to draw the path. Then use **ZOOM** and **TRACE** or other features of your calculator to find the point on the path that is exactly 3 feet from the y axis.

69. Use the graphing utility of your calculator to draw the graphs of the curves $y = -x^3 - 2x^2 + 5x - 2$ and $y = x \ln x$ on the same screen. Use **ZOOM** and **TRACE** or some other feature of your calculator to find where the curves intersect, and then compute the area of the region bounded by the curves.

70. Repeat Problem 69 for the curves
$$y = \frac{x - 2}{x + 1} \quad \text{and} \quad y = \sqrt{25 - x^2}$$

If your calculator has a numeric integration feature, use it to evaluate the integrals in Problems 71 and 72. In each case, verify your result by applying an appropriate integration formula from the table on pp. 472–473.

71. $\displaystyle\int_{-1}^{1} \frac{2 + 3x}{9 - x^2}\, dx$

72. $\displaystyle\int_{0}^{1} x^2\sqrt{9 + 4x^2}\, dx$

73. Use the numeric integration feature of your calculator to compute

$$I(N) = \int_{0}^{N} \frac{1}{\sqrt{\pi}} e^{-x^2}\, dx$$

for $N = 1, 10, 50$. Based on your results, do you think the improper integral

$$\int_{0}^{+\infty} \frac{1}{\sqrt{\pi}} e^{-x^2}\, dx$$

converges? If so, to what value?

74. Use the numeric integration feature of your calculator to compute

$$I(N) = \int_{1}^{N} \frac{\ln(x + 1)}{x}\, dx$$

for $N = 10, 100, 1{,}000, 10{,}000$. Based on your results, do you think the improper integral

$$\int_{1}^{+\infty} \frac{\ln(x + 1)}{x}\, dx$$

converges? If so, to what value?

75. ELIMINATION OF HAZARDOUS WASTE
To study the degradation of certain hazardous wastes with a high toxic content, biological researchers sometimes use the Haldane equation

$$\frac{dS}{dt} = \frac{aS}{b + cS + S^2}$$

where a, b, and c are positive constants and $S(t)$ is the concentration of substrate (the substance acted on by bacteria in the waste material).[*] Find the general solution of the Haldane equation. Express your answer in implicit form (as an equation involving S and t).

---

[*]Michael D. LaGrega, Philip L. Buckingham, and Jeffery C. Evans, *Hazardous Waste Management*, New York: McGraw-Hill, 1994, p. 578.

EXPLORE! UPDATE

Complete solutions for all EXPLORE! boxes throughout the text can be accessed at the book-specific website, www.mhhe.com/hoffmann.

**Solution for Explore!
on Page 488**

Place the data into the **STAT** editor, as shown in the left screen. In the **CALC** menu, select option **B:Logistic** and write the command **Logistic L1, L2, Y1** to obtain the parameters of the logistic equation, displayed in the middle screen. The data points and logistic graph can be presented using the **STAT PLOT** command (as in the right screen). After a long period (as x approaches a relatively large number), the denominator of the logistic formula becomes arbitrarily close to 1, indicating that the population of susceptible residents will level off at about $y = 766$. At approximately when did half of the susceptible community become infected?

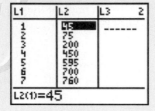 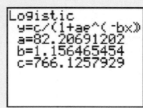 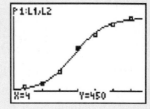

**Solution for Explore!
on Page 497**

Refer to Example 6.3.1. Place $f(x) = 1/x^2$ into Y1 of the equation editor and write Y2 = **fnInt**(Y1, X, 1, X, 0.001), using the numerical integration function, found in the **MATH** key menu, **9:fnInt(.** We deselect Y1 to only display the values of Y2. Set the table feature to start at X = 500 in increments of 500, as shown in the middle screen. The table on the right displays the area under $f(x) = 1/x^2$ from $x = 1$ to the designated X value. It appears that the integral values converge slowly to the value of 1. You may find your graphing calculator taking some seconds to compute the Y2 values.

Solution for Explore! on Page 504

Following Example 6.3.6, store $f(x)$ into Y1 as shown in the following, using the window $[0, 47]5$ by $[-0.01, 0.05]0.01$. The area below this uniform density function from $x = 15$ to 40 is $0.625 = 5/8$, as graphically displayed in the right screen.

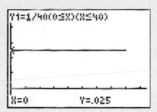
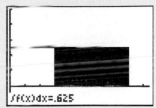

Solution for Explore! on Page 505

Following Example 6.3.7, set Y1 = $0.5e\wedge(-0.5X)*(X \geq 0)$, using the window $[-2, 10]1$ by $[-0.1, 0.6]0.1$. The graph of Y1 is shown in the following left screen. Now write Y2 = **fnInt(Y1, X, 0, X)** and observe the table of values for Y2 for X values, starting at $X = 20$ with unit increments. Notice that the cumulative density function Y2 has a value of 1 when $X = 25$, within the display's tolerance level of 10^{-5}. The numerical integration feature can be used to confirm that $P(2 \leq X \leq 3) \approx 0.1447$.

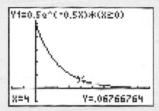
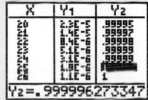
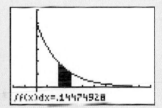

Solution for Explore! on Page 514

The list features of the graphing calculator can be used to facilitate the numerical integration computations required by the trapezoidal rule. Following Example 6.4.1, we have $f(x) = 1/x$, $a = 1$, $b = 2$, and $n = 10$. Set Y1 = $1/x$. On a cleared home screen, store values 1 into A, 2 into B, and 10 into N and define $(B - A)/N$ as H. A quick way to generate the numbers $1.0, 1.1, \ldots, 1.9, 2.0$ in L1 (accessed through **STAT, EDIT, 1:Edit**) is to write **L1 = seq(A + HX, X, 0, N)**, where the sequence command can be found in **LIST (2nd STAT), OPS, 5: seq(** .

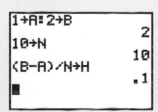
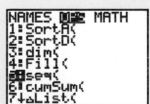
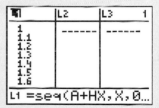

EXPLORE! UPDATE

Now place the trapezoidal rule coefficients 1, 2, . . . , 2, 1 into L2. Write **L3 = Y1(L1)*L2*H/2,** recalling that you must place the cursor at the top heading line of L3 to write the desired command. The following screens in the middle and on the right show all the data.

 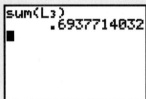 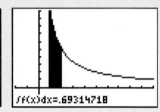

Finally, the sum of list L3 yields the desired trapezoidal rule approximation for $f(x) = 1/x$ over [1, 2] with $n = 10$. The sum command can be found in **LIST (2nd STAT), MATH, 5:sum(,** as shown here. The answer coincides with that in Example 6.4.1 and is a good approximation of the answer obtained by direct evaluation of the integral.

The graphing calculator steps shown here provide the key commands for writing a program to compute the trapezoidal rule of a function $f(x)$, over the interval [a, b] with n subdivisions. As an exercise, run through the same steps but use $n = 20$.

THINK ABOUT IT

MODELING EPIDEMICS

An outbreak of a disease affecting a large number of people is called an *epidemic* (from the Greek roots *epi* "upon" and *demos* "the people"). Mathematics plays an important role in the study of epidemics, a discipline known as *epidemiology*. Mathematical models have been developed for epidemics of influenza, bubonic plague, HIV, AIDS, smallpox, gonorrhea, and various other diseases. By making some simplifying assumptions, we can construct a model based on differential equations, called the **S-I-R model,** that is useful for making predictions about the course of an epidemic. Our model will be fairly straightforward, with enough simplifying assumptions to make the mathematics accessible. Later, we will examine the assumptions, with a view toward improving the model by making it more realistic.

To build our model, we consider an epidemic affecting a population of N people, each of whom falls into exactly one of the following groups at time t:

Susceptibles, $S(t)$, are people who have not yet become ill, but who could become sick later. When a susceptible person gets sick, it is assumed to happen with no time lag.

Infectives, $I(t)$, are people who have the disease and can infect others. Infectives are not quarantined and continue a normal pattern of interacting with other people.

Removeds, $R(t)$, are people who can no longer infect others. In particular, a recovered person cannot catch the disease again or infect others.

We assume that no people enter or leave the population (including births and deaths), and that at time $t = 0$, there are some susceptibles and infectives but no removeds; that is,

$$N = S(t) + I(t) + R(t) \qquad \text{for all } t \geq 0$$

where

$$S(0) > 0 \qquad I(0) > 0 \qquad R(0) = 0$$

so that $S(0) + I(0) = N$.

The key to modeling an epidemic or any other dynamic situation lies in the assumptions made about rates of change. In our model, we assume that the rate at which susceptible people are being infected at any time is proportional to the number of contacts between susceptibles and infectives; that is, to the product SI. Thus, the susceptible population $S(t)$ is *decreasing* (becoming infected) at a rate given by

$$(1) \qquad \frac{dS}{dt} = -aSI$$

We also assume that the rate people are being removed from the infected population is proportional to the number of people who are infective, so that

$$(2) \qquad \frac{dR}{dt} = bI$$

for some number $b > 0$. Finally, we assume that the rate at which the number of infectives changes is the rate at which susceptibles become infective minus the rate at which infective people are removed from the population of infectives. This means that

$$(3) \qquad \frac{dI}{dt} = aSI - bI = aI(S - b/a) = aI(S - c)$$

with $c = b/a$. [In Question 1, you are asked to obtain this last rate by differentiating both sides of the equation $I(t) = N - S(t) - R(t)$ with respect to t and substituting the rates in equations (1) and (2).]

The S-I-R model consists of the three simultaneous differential equations we have constructed; namely,

$$\frac{dS}{dt} = -aSI \qquad \frac{dR}{dt} = bI \qquad \frac{dI}{dt} = aI(S - c)$$

where $S(t) + I(t) + R(t) = N$ for all t, with $R(0) = 0$ and $S(0) > 0$, $I(0) > 0$. Although we cannot explicitly solve this system to obtain simple formulas for the functions $S(t)$, $I(t)$, and $R(t)$, we can use these differential equations to help us understand the progress of epidemics. We illustrate this by applying our S-I-R model to a historic case, an epidemic of bubonic plague in Eyam, a village near Sheffield, England, during 1665–1666. This analysis is possible because the village quarantined itself during the epidemic and kept sufficiently detailed records. We will use these records to estimate the constants a and b in equations (1) and (2).

First, note that the constant b in equation (2) can be interpreted as the rate at which people are removed from the population of infectives. The infective period of the bubonic plague in Eyam was 11 days or 0.367 months. Therefore, if we measure time t in months and assume, for simplicity, that infected people are removed from this population at a constant rate, we can estimate b by the ratio

$$b = \frac{1}{0.367} \approx 2.72$$

Next, records kept during the epidemic show that on July 19, 1666, there were 201 susceptibles and 22 infectives, while on August 19, 1666, there were 121 susceptibles and 21 infectives. Thus, the number of susceptibles changed by $121 - 201 = -80$

during this month-long period, and this change can be used to estimate $\dfrac{dS}{dt}$ on the day the period began. In other words, on July 19, 1666,

$$S = 201 \qquad I = 22 \qquad \text{and} \qquad \dfrac{dS}{dt} \approx -80$$

Substituting into equation (1) and solving for the constant a, we find that

$$a = \dfrac{-dS/dt}{SI} \approx \dfrac{-(-80)}{(201)(22)} \approx 0.018$$

It turns out that the resulting model, with the estimates we have just obtained for the values of a and b, does a good job of estimating the actual data. Figure 1 displays the values predicted by the model and the actual data. [This information is adapted from Raggett (1982) and from the discussion in Brauer and Castillo-Chavez (2001).]

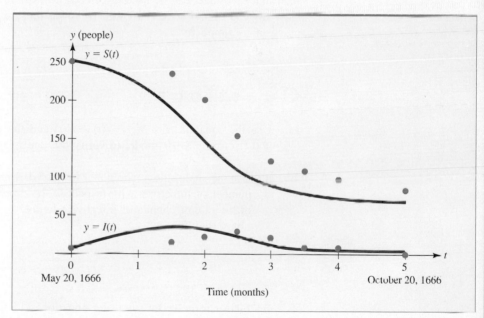

FIGURE 1 Comparison of Eyam plague data with values predicted by the S-I-R model.

The S-I-R model can be used to make some general observations about the course of epidemics. For example, unless the number of infectives increases initially, the epidemic will never get started (since the number of infectives does not grow beyond the initial number). In the language of calculus, an epidemic begins only if $\dfrac{dI}{dt} > 0$ at $t = 0$. Examining equation (3) of the S-I-R model, we see that this happens only if $S(0) > c$. For this reason, c is called the *threshold number of susceptibles*. When the initial number of susceptibles exceeds c, the epidemic spreads; otherwise, it dies out. For example, the records of the 1666 bubonic plague in Eyam indicate there were initially $S(0) = 254$ susceptibles. Using our estimates of $a \approx 0.018$ and $b \approx 2.72$, we estimate the threshold number for this model to be

$$c = \dfrac{b}{a} = \dfrac{2.72}{0.018} \approx 151$$

Since $S(0) = 254 > 151$, the model predicts that the epidemic should spread, which is exactly what happened historically.

You probably have noticed that in constructing our S-I-R model and applying it to the Eyam epidemic, we have made several fairly strong assumptions, not the least of which is that $S(t) + I(t) + R(t)$ is a constant. Is it realistic to assume that the size of a population subjected to an epidemic would never change? Is it any more realistic to assume that people become infective immediately with no time lag? Certainly not, since diseases usually have incubation periods, but inserting a time lag to account for the incubation period would make the resulting model too complicated to analyze without more sophisticated mathematical tools. This illustrates the central issue facing anyone who uses mathematical modeling: *a completely realistic model is often difficult if not impossible to analyze, while making assumptions to simplify the analysis can lead to results that are not entirely realistic.* The S-I-R model is indeed based on several simplifying assumptions that are somewhat unrealistic, but we have also seen that the model can be used effectively to explore the dynamics of an epidemic. How do you think the model could be made more realistic without sacrificing too much of its simplicity?

Questions

1. Use the equation $N = S(t) + I(t) + R(t)$ and the rate relations in equations (1) and (2) of the S-I-R model to verify the rate relation in equation (3).

2. **a.** Use the chain rule to show that in the S-I-R model, the rate of change of the number of infectives with respect to the number of susceptibles is the ratio of the rate of change of infectives divided by the rate of change of susceptibles, that is

$$\frac{dI}{dS} = \frac{dI/dt}{dS/dt}$$

   b. Find a formula for $\dfrac{dI}{dS}$ using equations (1) and (3) of the model. Use this to find $\dfrac{dI}{dS}$ for the 1666 Eyam bubonic plague epidemic.

   c. Use your answer in part (b) to find the number of susceptibles when the number of infected is a maximum.

3. An influenza epidemic broke out at a British boarding school in 1978. At the start of the epidemic there were 762 susceptible boys and 1 infective boy, and 1 day later, two more boys became ill. Assume that anyone sick with influenza is also infective.

   a. Use the data supplied to estimate the constant a in equation (1) of the S-I-R model for this epidemic with t representing time in days.

   b. Suppose that all infected boys are removed from the population the day after they become ill. Use this information to estimate the constant b in equation (2) of the S-I-R model for this epidemic.

c. Use the values of a and b you found in parts (a) and (b) to find the threshold number of susceptibles needed for this epidemic to get started.

d. Use the results of Question 2 to find the number of susceptibles when the number of students infected with influenza was the maximum.

4. Modify the S-I-R model for an epidemic to take into account vaccinations of people at a constant rate d, assuming that someone vaccinated immediately is no longer susceptible. Find a formula for $\dfrac{dI}{dS}$ for the resulting model.

5. An S-I model can be used to model an epidemic where everyone who gets the disease remains infective. (An S-I model is really the special case of the S-I-R model in which $b = 0$.) Answer these questions about such an epidemic.

a. Show that $\dfrac{dS}{dt} = -aS(N - S)$ where N is the population size (which we assume is constant).

b. Verify that $S(t) = \dfrac{N(N - 1)}{(N - 1) + e^{aNt}}$ is a solution of this differential equation and conclude that $I(t) = \dfrac{Ne^{aNt}}{(N - 1) + e^{aNt}}$.

c. Show that $\lim\limits_{t \to +\infty} S(t) = 0$ and $\lim\limits_{t \to +\infty} I(t) = N$. Draw a graph showing the susceptible and infective curves, $y = S(t)$ and $y - I(t)$, assuming that at time $t = 0$ there are I_0 infected people, where $0 < I_0 < N$.

References

W. O. Kermack and A. G. McKendrick, "A Contribution to the Mathematical Theory of Epidemics," *Proc. Royal Soc. London,* Vol. 115, 1927, pp. 700–721.

G. F. Raggett, "Modeling the Eyam Plague," *IMA Journal,* Vol. 18, 1982, pp. 221–226.

Fred Brauer and Carlos Castillo-Chavez, *Mathematical Models in Population Biology and Epidemiology,* New York: Springer-Verlag, 2001.

Leah Edelstein-Keshet, *Mathematical Models in Biology,* Birkhauser Mathematics Series, Boston: McGraw-Hill, 1988, pp. 243–256.

CHAPTER **7**

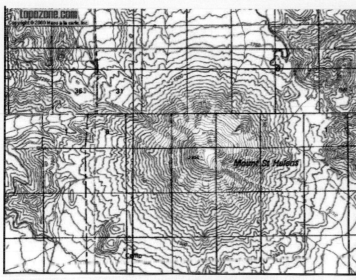

The shape of a surface can be visualized by mapping its level curves, which for a mountain are curves of constant elevation.

CALCULUS OF SEVERAL VARIABLES

SECTION   7.1 | Functions of Several Variables

In business, if a manufacturer determines that x units of a particular commodity can be sold domestically for $90 per unit, and y units can be sold to foreign markets for $110 per unit, then the total revenue obtained from all sales is given by

$$R = 90x + 110y$$

In psychology, a person's intelligence quotient (IQ) is measured by the ratio

$$IQ = \frac{100m}{a}$$

where a and m are the person's actual age and mental age, respectively. A carpenter constructing a storage box x feet long, y feet wide, and z feet deep knows that the box will have volume V and surface area S, where

$$V = xyz \qquad \text{and} \qquad S = 2xy + 2xz + 2yz$$

These are typical of practical situations in which a quantity of interest depends on the values of two or more variables. Other examples include the volume of water in a community's reservoir, which may depend on the amount of rainfall as well as the population of the community, and the output of a factory, which may depend on the amount of capital invested in the plant, the size of the labor force, and the cost of raw materials.

In this chapter, we will extend the methods of calculus to include functions of two or more independent variables. Most of our work will be with functions of two variables, which you will find can be represented geometrically by surfaces in three-dimensional space instead of curves in the plane. We begin with a definition and some terminology.

> **Function of Two Variables** ■ A function f of the two independent variables x and y is a rule that assigns to each ordered pair (x, y) in a given set D (the **domain** of f) exactly one real number, denoted by $f(x, y)$.

> **NOTE** **Domain Convention:** Unless otherwise stated, we assume that the domain of f is the set of all (x, y) for which the expression $f(x, y)$ is defined. ■

As in the case of a function of one variable, a function of two variables $f(x, y)$ can be thought of as a "machine" in which there is a unique "output" $f(x, y)$ for each "input" (x, y), as illustrated in Figure 7.1. The domain of f is the set of all possible

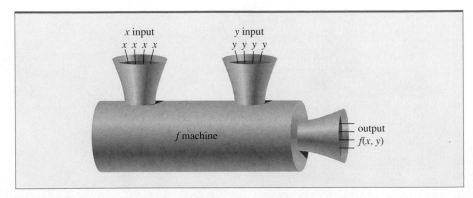

FIGURE 7.1   A function of two variables as a "machine."

inputs, and the set of all possible corresponding outputs is the **range** of f. Functions of three independent variables $f(x, y, z)$ or four independent variables $f(x, y, z, t)$, and so on can be defined in a similar fashion.

EXPLORE!

Graphing calculators can naturally represent functions of a single independent variable but require additional capabilities to portray multivariate functions, such as three-dimensional graphing features. A simple method to deal with multivariate functions is to graph one variable at specific values of the other variables. For example, store $f(x, y) = x^3 - x^2y^2 - xy^3 - y^4$ into Y1 as X^3 − X^2*L1^2 − X*L1^3 − L1^4, where L1 is the list of values {0, 1.5, 2.0, 2.25, 2.5}. Graph using the modified decimal window [−9.4, 9.4]1 by [−150, 100]20. How does varying the y values in L1 affect the shape of the graph?

EXAMPLE | 7.1.1

Suppose $f(x, y) = \dfrac{3x^2 + 5y}{x - y}$.

a. Find the domain of f.

b. Compute $f(1, -2)$.

Solution

a. Since division by any real number except zero is possible, the expression $f(x, y)$ can be evaluated for all ordered pairs (x, y) with $x - y \neq 0$ or $x \neq y$. Geometrically, this is the set of all points in the xy plane except for those on the line $y = x$.

b. $f(1, -2) = \dfrac{3(1)^2 + 5(-2)}{1 - (-2)} = \dfrac{3 - 10}{1 + 2} = -\dfrac{7}{3}$.

EXAMPLE | 7.1.2

Suppose $f(x, y) = xe^y + \ln x$.

a. Find the domain of f.

b. Compute $f(e^2, \ln 2)$.

Solution

a. Since xe^y is defined for all real numbers x and y and since $\ln x$ is defined only for $x > 0$, the domain of f consists of all ordered pairs (x, y) of real numbers for which $x > 0$.

b. $f(e^2, \ln 2) = e^2 e^{\ln 2} + \ln e^2 = 2e^2 + 2 = 2(e^2 + 1) \approx 16.78$

EXAMPLE | 7.1.3

Given the function of three variables $f(x, y, z) = xy + xz + yz$, evaluate $f(-1, 2, 5)$.

Solution

Substituting $x = -1$, $y = 2$, $z = 5$ into the formula for $f(x, y, z)$, we get

$$f(-1, 2, 5) = (-1)(2) + (-1)(5) + (2)(5) = 3$$

Applications   Here are four applications involving functions of several variables to business, economics, finance, and life science.

EXAMPLE | 7.1.4

A sports store in St. Louis carries two kinds of tennis rackets, the Serena Williams and the Jennifer Capriati autograph brands. The consumer demand for each brand depends not only on its own price, but also on the price of the competing brand. Sales figures indicate that if the Williams brand sells for x dollars per racket and the Capriati brand for y dollars per racket, the demand for Williams rackets will be $D_1 = 300 - 20x + 30y$ rackets per year and the demand for Capriati rackets will be $D_2 = 200 + 40x - 10y$ rackets per year. Express the store's total annual revenue from the sale of these rackets as a function of the prices x and y.

Solution

Let R denote the total monthly revenue. Then

$$R = \text{(number of Williams rackets sold)(price per Williams racket)}$$
$$+ \text{(number of Capriati rackets sold)(price per Capriati racket)}$$

Hence,

$$R(x, y) = (300 - 20x + 30y)(x) + (200 + 40x - 10y)(y)$$
$$= 300x + 200y + 70xy - 20x^2 - 10y^2$$

Output Q at a factory is often regarded as a function of the amount K of capital investment and the size L of the labor force. Output functions of the form

$$Q(K, L) = AK^\alpha L^\beta$$

where A, α, and β are positive constants with $\alpha + \beta = 1$, have proved to be especially useful in economic analysis and are known as **Cobb-Douglas production functions.*** Here is an example involving such a function.

EXPLORE!

Refer to Example 7.1.5. Store the equation

$0 = Q - 60K\verb|^|(1/3)*L\verb|^|(2/3)$

in the equation solver of your graphing calculator. Calculate the value of Q for $K = 512$ and $L = 1,000$. What happens to Q if the labor force L is doubled?

EXAMPLE 7.1.5

Suppose that at a certain factory, output is given by the Cobb-Douglas production function $Q(K, L) = 60K^{1/3}L^{2/3}$ units, where K is the capital investment measured in units of $\$1,000$ and L the size of the labor force measured in worker-hours.

a. Compute the output if the capital investment is $\$512,000$ and $1,000$ worker-hours of labor are used.

b. Show that the output in part (a) will double if both the capital investment and the size of the labor force are doubled.

Solution

a. Evaluate $Q(K, L)$ with $K = 512$ (thousand) and $L = 1,000$ to get

$$Q(512, 1,000) = 60(512)^{1/3}(1,000)^{2/3}$$
$$= 60(8)(100) = 48,000 \text{ units}$$

b. Evaluate $Q(K, L)$ with $K = 2(512)$ and $L = 2(1,000)$ as follows to get

$$Q[2(512), 2(1,000)] = 60[2(512)]^{1/3}[2(1,000)]^{2/3}$$
$$= 60(2)^{1/3}(512)^{1/3}(2)^{2/3}(1,000)^{2/3} = 96,000 \text{ units}$$

which is twice the output when $K = 512$ and $L = 1,000$.

---

*For instance, see Dominick Salvatore, *Managerial Economics,* New York: McGraw-Hill, 1989, pp. 332–336.

EXAMPLE | 7.1.6

Recall (from Section 4.1) that the present value of B dollars in t years invested at the annual rate r compounded k times per year is given by

$$P(B, r, k, t) = B\left(1 + \frac{r}{k}\right)^{-kt}$$

Find the present value of $10,000 in 5 years invested at 6% per year compounded quarterly.

Solution

We have $B = 10,000$, $r = 0.06$ (6% per year), $k = 4$ (compounded 4 times per year), and $t = 5$, so the present value is

$$P(10,000, 0.06, 4, 5) = 10,000\left(1 + \frac{0.06}{4}\right)^{-4(5)}$$

$$\approx 7,424.7$$

or approximately $7,425.

EXAMPLE | 7.1.7

In Section 6.2, we showed that a population that grows exponentially satisfies

$$P(A, k, t) = Ae^{kt}$$

where P is the population at time t, A is the initial population (when $t = 0$), and k is the relative (per capita) growth rate. The population of a certain country is currently 5 million people and is growing at the rate of 3% per year. What will the population be in 7 years?

Solution

Let P be measured in millions of people. Substituting $A = 5$, $k = 0.03$ (3% annual growth), and $t = 7$ into the population function, we find that

$$P(5, 0.03, 7) = 5e^{0.03(7)} \approx 6.16839$$

Therefore, in 7 years, the population will be approximately 6,168,400 people.

Graphs of Functions of Two Variables

The **graph** of a function of two variables $f(x, y)$ is the set of all triples (x, y, z) such that (x, y) is in the domain of f and $z = f(x, y)$. To "picture" such graphs, we need to construct a **three-dimensional** coordinate system. The first step in this construction is to add a third coordinate axis (the z axis) perpendicular to the familiar xy coordinate plane, as shown in Figure 7.2. Note that the xy plane is taken to be horizontal, and the positive z axis is "up."

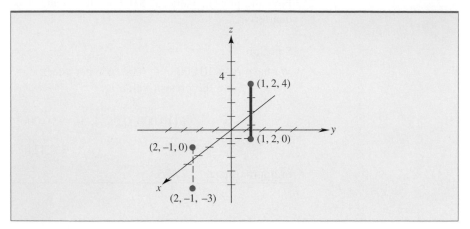

FIGURE 7.2   A three-dimensional coordinate system.

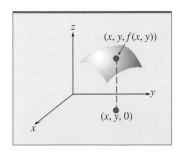

FIGURE 7.3   The graph of $z = f(x, y)$.

You can describe the location of a point in three-dimensional space by specifying three coordinates. For example, the point that is 4 units above the xy plane and lies directly above the point with xy coordinates $(x, y) = (1, 2)$ is represented by the ordered triple $(x, y, z) = (1, 2, 4)$. Similarly, the ordered triple $(2, -1, -3)$ represents the point that is 3 units directly below the point $(2, -1)$ in the xy plane. These points are shown in Figure 7.2.

To graph a function $f(x, y)$ of the two independent variables x and y, it is customary to introduce the letter z to stand for the dependent variable and to write $z = f(x, y)$ (Figure 7.3). The ordered pairs (x, y) in the domain of f are thought of as points in the xy plane, and the function f assigns a "height" z to each such point ("depth" if z is negative). Thus, if $f(1, 2) = 4$, you would express this fact geometrically by plotting the point $(1, 2, 4)$ in a three-dimensional coordinate space. The function may assign different heights to different points in its domain, and in general, its graph will be a surface in three-dimensional space.

Four such surfaces are shown in Figure 7.4. The surface in Figure 7.4a is a **cone,** Figure 7.4b shows a **paraboloid,** Figure 7.4c shows an **ellipsoid,** and Figure 7.4d shows what is commonly called a **saddle surface.** Surfaces such as these play an important role in examples and exercises in this chapter.

Level Curves

It is usually not easy to sketch the graph of a function of two variables. One way to visualize a surface is shown in Figure 7.5. Notice that when the plane $z = C$ intersects the surface $z = f(x, y)$, the result is a curve in space. The corresponding set of points (x, y) in the xy plane that satisfy $f(x, y) = C$ is called the **level curve** of f at C, and an entire family of level curves is generated as C varies over a set of numbers. By sketching members of this family in the xy plane, you can obtain a useful representation of the surface $z = f(x, y)$.

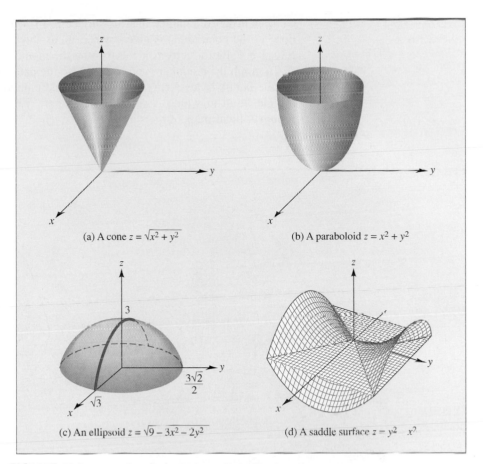

(a) A cone $z = \sqrt{x^2 + y^2}$ (b) A paraboloid $z = x^2 + y^2$

(c) An ellipsoid $z = \sqrt{9 - 3x^2 - 2y^2}$ (d) A saddle surface $z = y^2 - x^2$

FIGURE 7.4 Several surfaces in three-dimensional space.

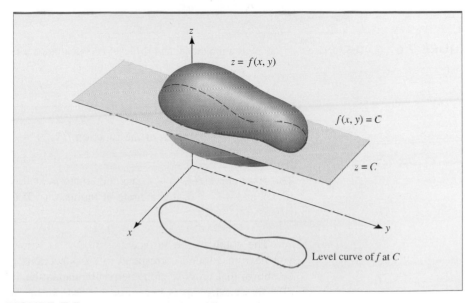

FIGURE 7.5 A level curve of the surface $z = f(x, y)$.

For instance, imagine that the surface $z = f(x, y)$ is a "mountain" whose "elevation" at the point (x, y) is given by $f(x, y)$, as shown in Figure 7.6a. The level curve $f(x, y) = C$ lies directly below a path on the mountain where the elevation is always C. To graph the mountain, you can indicate the paths of constant elevation by sketching the family of level curves in the plane and pinning a "flag" to each curve to show the elevation to which it corresponds (Figure 7.6b). This "flat" figure is called a **topographical map** of the surface $z = f(x, y)$.

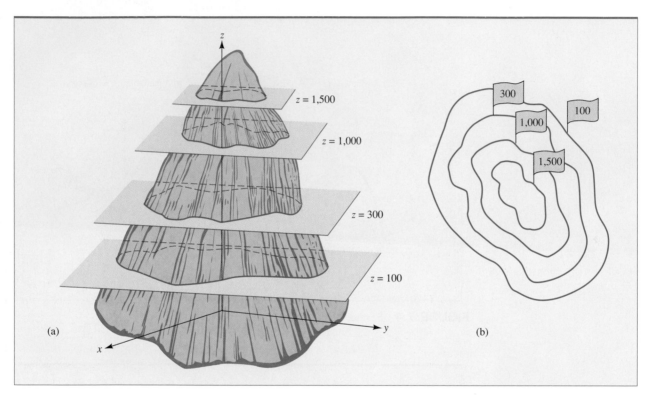

FIGURE 7.6 (a) The surface $z = f(x, y)$ as a mountain, and (b) level curves provide a topographical map of $z = f(x, y)$.

EXAMPLE | 7.1.8

Discuss the level curves of the function $f(x, y) = x^2 + y^2$.

Solution

The level curve $f(x, y) = C$ has the equation $x^2 + y^2 = C$. If $C = 0$, this is the point $(0, 0)$, and if $C > 0$, it is a circle of radius $\sqrt{C}$. If $C < 0$, there are no points that satisfy $x^2 + y^2 = C$.

The graph of the surface $z = x^2 + y^2$ is shown in Figure 7.7. The level curves you have just found correspond to cross sections perpendicular to the z axis. It can be shown that cross sections perpendicular to the x axis and the y axis are parabolas. (Try to see why this is true.) For this reason, the surface is shaped like a bowl. It is called a **circular paraboloid** or a **paraboloid of revolution.**

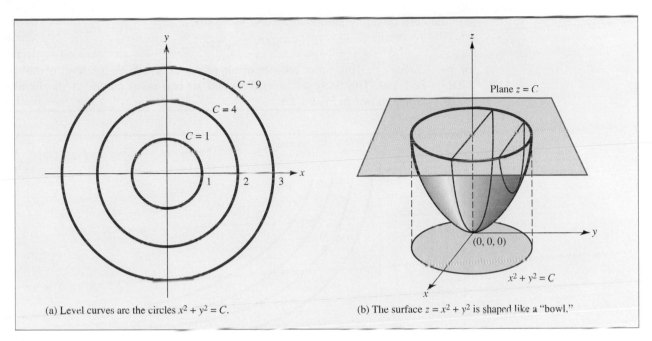

(a) Level curves are the circles $x^2 + y^2 = C$.   (b) The surface $z = x^2 + y^2$ is shaped like a "bowl."

FIGURE 7.7 Level curves help to visualize the shape of a surface.

Level Curves in Economics: Isoquants and Indifference Curves

Level curves appear in many different applications. For instance, in economics, if the output $Q(x, y)$ of a production process is determined by two inputs x and y (say, hours of labor and capital investment), then the level curve $Q(x, y) = C$ is called the **curve of constant product** C or, more briefly, an **isoquant** ("iso" means "equal").

Another application of level curves in economics involves the concept of indifference curves. A consumer who is considering the purchase of a number of units of each of two commodities is associated with a **utility function** $U(x, y)$, which measures the total satisfaction (or **utility**) the consumer derives from having x units of the first commodity and y units of the second. A level curve $U(x, y) = C$ of the utility function is called an **indifference curve** and gives all of the combinations of x and y that lead to the same level of consumer satisfaction. These terms are illustrated in Example 7.1.9.

EXAMPLE | 7.1.9

Suppose the utility derived by a consumer from x units of one commodity and y units of a second commodity is given by the utility function $U(x, y) = x^{3/2}y$. If the consumer currently owns $x = 16$ units of the first commodity and $y = 20$ units of the second, find the consumer's current level of utility and sketch the corresponding indifference curve.

Solution

The current level of utility is

$$U(16, 20) = (16)^{3/2}(20) = 1,280$$

and the corresponding indifference curve is

$$x^{3/2}y = 1{,}280$$

or $y = 1{,}280x^{-3/2}$. This curve consists of all points (x, y) where the level of utility $U(x, y)$ is 1,280. The curve $x^{3/2}y = 1{,}280$ and several other curves of the family $x^{3/2}y = C$ are shown in Figure 7.8.

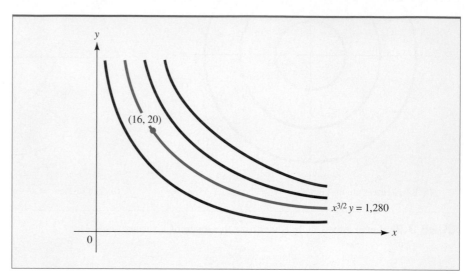

FIGURE 7.8   Indifference curves for the utility function $U(x, y) = x^{3/2}y$.

Computer Graphics

In practical work in the social, managerial, or life sciences, you will rarely, if ever, have to graph a function of two variables. Hence, we will spend no more time developing graphing procedures for such functions.

Computer software is now available for graphing functions of two variables. Such software often allows you to choose different scales along each coordinate axis and may also enable you to visualize a given surface from different viewpoints. These features permit you to obtain a detailed picture of the graph. A variety of computer-generated graphs are displayed in Figure 7.9 on page 553.

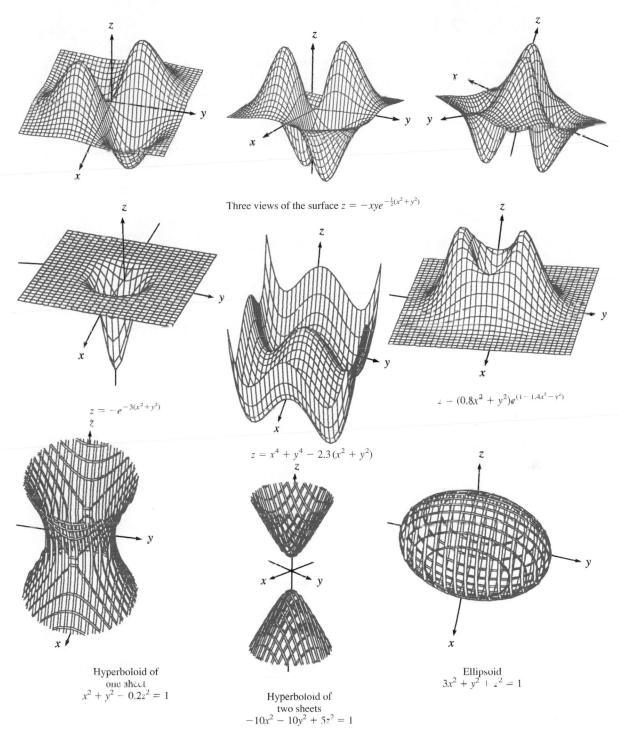

Three views of the surface $z = -xye^{-\frac{1}{2}(x^2 + y^2)}$

$z = -e^{-3(x^2 + y^2)}$

$z = x^4 + y^4 - 2.3(x^2 + y^2)$

$z = (0.8x^2 + y^2)e^{(1 - 1.4x^2 - y^2)}$

Hyperboloid of
one sheet
$x^2 + y^2 - 0.2z^2 = 1$

Hyperboloid of
two sheets
$-10x^2 - 10y^2 + 5z^2 = 1$

Ellipsoid
$3x^2 + y^2 + z^2 = 1$

FIGURE 7.9   Some computer-generated surfaces.

PROBLEMS 7.1

In Problems 1 through 12, compute the indicated function value.

1. $f(x, y) = (x - 1)^2 + 2xy^3; f(2, -1), f(1, 2)$

2. $f(x, y) = \dfrac{3x + 2y}{2x + 3y}; f(1, 2), f(-4, 6)$

3. $g(x, y) = \sqrt{y^2 - x^2}; g(4, 5), g(-1, 2)$

4. $g(u, v) = 10u^{1/2}v^{2/3}; g(16, 27), g(4, -1331)$

5. $f(r, s) = \dfrac{s}{\ln r}; f(e^2, 3), f(\ln 9, e^3)$

6. $f(x, y) = xye^{xy}; f(1, \ln 2), f(\ln 3, \ln 4)$

7. $g(x, y) = \dfrac{y}{x} + \dfrac{x}{y}; g(1, 2), g(2, -3)$

8. $f(s, t) = \dfrac{e^{st}}{2 - e^{st}}; f(1, 0), f(\ln 2, 2)$

9. $f(x, y, z) = xyz; f(1, 2, 3), f(3, 2, 1)$

10. $g(x, y, z) = (x + y)e^{yz}; g(1, 0, -1), g(1, 1, 2)$

11. $F(r, s, t) = \dfrac{\ln(r + t)}{r + s + t}; F(1, 1, 1), F(0, e^2, 3e^2)$

12. $f(x, y, z) = xye^z + xze^y + yze^x;$ $f(1, 1, 1), f(\ln 2, \ln 3, \ln 4)$

In Problems 13 through 18, describe the domain of the given function.

13. $f(x, y) = \dfrac{5x + 2y}{4x + 3y}$

14. $f(x, y) = \sqrt{9 - x^2 - y^2}$

15. $f(x, y) = \sqrt{x^2 - y}$

16. $f(x, y) = \dfrac{x}{\ln(x + y)}$

17. $f(x, y) = \ln(x + y - 4)$

18. $f(x, y) = \dfrac{e^{xy}}{\sqrt{x - 2y}}$

In Problems 19 through 26, sketch the indicated level curve $f(x, y) = C$ for each choice of constant C.

19. $f(x, y) = x + 2y; C = 1, C = 2, C = -3$

20. $f(x, y) = x^2 + y; C = 0, C = 4, C = 9$

21. $f(x, y) = x^2 - 4x - y; C = -4, C = 5$

22. $f(x, y) = \dfrac{x}{y}; C = -2, C = 2$

23. $f(x, y) = xy; C = 1, C = -1, C = 2, C = -2$

24. $f(x, y) = ye^x; C = 0, C = 1$

25. $f(x, y) = xe^y; C = 1, C = e$

26. $f(x, y) = \ln(x^2 + y^2); C = 4, C = \ln 4$

27. PRODUCTION   Using x skilled workers and y unskilled workers, a manufacturer can produce $Q(x, y) = 10x^2y$ units per day. Currently there are 20 skilled workers and 40 unskilled workers on the job.
  a. How many units are currently being produced each day?
  b. By how much will the daily production level change if 1 more skilled worker is added to the current workforce?
  c. By how much will the daily production level change if 1 more unskilled worker is added to the current workforce?
  d. By how much will the daily production level change if 1 more skilled worker *and* 1 more unskilled worker are added to the current workforce?

28. PRODUCTION COST   A manufacturer can produce scientific graphing calculators at a cost of $40 apiece and business calculators for $20 apiece.
  a. Express the manufacturer's total monthly production cost as a function of the number of graphing calculators and the number of business calculators produced.
  b. Compute the total monthly cost if 500 scientific and 800 business calculators are produced.
  c. The manufacturer wants to increase the output of scientific calculators by 50 a month from the level in part (b). What corresponding change should be made in the monthly output of business calculators so the total monthly cost will not change?

29. RETAIL SALES   A paint store carries two brands of latex paint. Sales figures indicate that if the first brand is sold for x_1 dollars per gallon and the second for x_2 dollars per gallon, the demand for the first brand will be $D_1(x_1, x_2) = 200 - 10x_1 + 20x_2$ gallons per month and the demand for the second brand will be $D_2(x_1, x_2) = 100 + 5x_1 - 10x_2$ gallons per month.

   a. Express the paint store's total monthly revenue from the sale of the paint as a function of the prices x_1 and x_2.

   b. Compute the revenue in part (a) if the first brand is sold for $6 per gallon and the second for $5 per gallon.

30. PRODUCTION   The output at a certain factory is $Q(K, L) = 120K^{2/3}L^{1/3}$ units, where K is the capital investment measured in units of $1,000 and L the size of the labor force measured in worker-hours.

   a. Compute the output if the capital investment is $125,000 and the size of the labor force is 1,331 worker-hours.

   b. What will happen to the output in part (a) if both the level of capital investment and the size of the labor force are cut in half?

31. PRODUCTION   The Easy Gro agricultural company estimates that when $100x$ worker-hours of labor are employed on y acres of land, the number of bushels of wheat produced is $f(x, y) = Ax^a y^b$, where A, a, and b are positive constants. Suppose the company decides to double the production factors x and y. Determine how this decision affects the production of wheat in each of these cases:

   a. $a + b > 1$

   b. $a + b < 1$

   c. $a + b = 1$

32. PRODUCTION   Suppose that when x machines and y worker-hours are used each day, a certain factory will produce $Q(x, y) = 10xy$ cell phones. Describe the relationship between the inputs x and y that results in an output of 1,000 phones each day. (Note that you are finding a level curve of Q.)

33. RETAIL SALES   A manufacturer with exclusive rights to a sophisticated new industrial machine is planning to sell a limited number of the machines to both foreign and domestic firms. The price the manufacturer can expect to receive for the machines will depend on the number of machines made available. It is estimated that if the manufacturer supplies x machines to the domestic market and y machines to the foreign market, the machines will sell for

$60 - \dfrac{x}{5} + \dfrac{y}{20}$ thousand dollars apiece domestically

and $50 - \dfrac{y}{10} + \dfrac{x}{20}$ thousand dollars apiece abroad. Express the revenue R as a function of x and y.

34. RETAIL SALES   A manufacturer is planning to sell a new product at the price of A dollars per unit and estimates that if x thousand dollars is spent on development and y thousand dollars on promotion, consumers will buy approximately $\dfrac{320y}{y + 2} + \dfrac{160x}{x + 4}$ units of the product. If manufacturing costs are $50 per unit, express profit in terms of x and y.

35. SURFACE AREA OF THE HUMAN BODY   Pediatricians and medical researchers sometimes use the following empirical formula* relating the surface area S (m^2) of a person to the person's weight W (kg) and height H (cm):

$$S(W, H) = 0.0072W^{0.425}H^{0.725}$$

   a. Find $S(15.83, 87.11)$. Sketch the level curve of $S(W, H)$ that passes through $(15.83, 87.11)$. Sketch several additional level curves of $S(W, H)$. What do these level curves represent?

   b. If Marc weighs 18.37 kg and has surface area 0.648 m^2, approximately how tall would you expect him to be?

   c. Suppose at some time in her life, Jenny weighs six times as much and is twice as tall as she was at birth. What is the corresponding percentage change in the surface area of her body?

   d. Ask your parents what your birth weight and length were (they will know!). Then obtain a doll that is approximately the same length you were at birth, and measure its surface area. Does the empirical formula accurately predict the result you obtained? Write a paragraph on any conclusions you draw from this "experiment."

36. PSYCHOLOGY   A person's intelligence quotient (IQ) is measured by the function

$$I(m, a) = \frac{100m}{a}$$

where a is the person's actual age and m is his or her mental age.

   a. Find $I(12, 11)$ and $I(16, 17)$.

   b. Sketch the graphs of several level curves of $I(m, a)$. How would you describe these curves?

---

*J. Routh, *Mathematical Preparation for Laboratory Technicians*, Philadelphia: Saunders Co., 1971, p. 92.

37. CONSTANT PRODUCTION CURVES Using x skilled and y unskilled workers, a manufacturer can produce $Q(x, y) = 3x + 2y$ units per day. Currently the workforce consists of 10 skilled workers and 20 unskilled workers.

   a. Compute the current daily output.

   b. Find an equation relating the levels of skilled and unskilled labor if the daily output is to remain at its current level.

   c. On a two-dimensional coordinate system, draw the isoquant (constant production curve) that corresponds to the current level of output.

   d. What change should be made in the level of unskilled labor y to offset an increase in skilled labor x of two workers so that the output will remain at its current level?

38. INDIFFERENCE CURVES Suppose the utility derived by a consumer from x units of one commodity and y units of a second commodity is given by the utility function $U(x, y) = 2x^3y^2$. The consumer currently owns $x = 5$ units of the first commodity and $y = 4$ units of the second. Find the consumer's current level of utility and sketch the corresponding indifference curve.

39. INDIFFERENCE CURVES The utility derived by a consumer from x units of one commodity and y units of a second is given by the utility function $U(x, y) = (x + 1)(y + 2)$. The consumer currently owns $x = 25$ units of the first commodity and $y = 8$ units of the second. Find the current level of utility and sketch the corresponding indifference curve.

40. AIR POLLUTION Dumping and other material-handling operations near a landfill may result in contaminated particles being emitted into the surrounding air. To estimate such particulate emission, the following empirical formula* can be used:

$$E(V, M) = k(0.0032)\left(\frac{V}{5}\right)^{1.3}\left(\frac{M}{2}\right)^{-1.4}$$

where E is the emission factor (pounds of particles released into the air per ton of soil moved), V is the mean speed of the wind (mph), M is the moisture content of the material (given as a percentage), and k is a constant that depends on the size of the particles.

  a. For a small particle (diameter 5 mm), it turns out that $k = 0.2$. Find $E(10, 13)$.

   b. The emission factor E can be multiplied by the number of tons of material handled to achieve a measure of total emissions. Suppose 19 tons of

the material in part (a) is handled. How many tons of a second kind of material with $k = 0.48$ (diameter 15 mm) and moisture content 27% must be handled to achieve the same level of total emissions if the wind velocity stays the same?

   c. Sketch several level curves of $E(V, M)$, assuming the size of the particle stays fixed. What is represented by these curves?

41. FLOW OF BLOOD One of Poiseuille's laws[†] says that the speed of blood V (cm/sec) flowing at a distance r (cm) from the axis of a blood vessel of radius R (cm) and length L (cm) is given by

$$V(P, L, R, r) = \frac{9.3P}{L}(R^2 - r^2)$$

where P (dynes/cm²) is the pressure in the vessel. Suppose a particular vessel has radius 0.0075 cm and is 1.675 cm long.

   a. How fast is the blood flowing at a distance 0.004 cm from the axis of this vessel if the pressure in the vessel is 3875 dynes/cm²?

   b. Since R and L are fixed for this vessel, V is a function of P and r alone. Sketch several level curves of $V(P, r)$. Explain what they represent.

42. CONSTANT RETURNS TO SCALE Suppose output Q is given by the Cobb-Douglas production function $Q(K, L) = AK^\alpha L^{1-\alpha}$, where A and α are positive constants and $0 < \alpha < 1$. Show that if K and L are both multiplied by the same positive number m, then the output Q will also be multiplied by m; that is, show that $Q(mK, mL) = mQ(K, L)$.

43. CHEMISTRY **Van der Waal's equation** of state says that 1 mole of a confined gas satisfies the equation

$$T(P, V) = 0.0122\left(P + \frac{a}{V^2}\right)(V - b) - 273.15$$

where T (°C) is the temperature of the gas, V (cm³) is its volume, P (atmospheres) is the pressure of the gas on the walls of its container, and a and b are constants that depend on the nature of the gas.

   a. Sketch the graphs of several level curves of T. These curves are called **curves of constant temperature** or **isotherms.**

   b. If the confined gas is chlorine, experiments show that $a = 6.49 \times 10^6$ and $b = 56.2$. Find $T(1.13, 31.275 \times 10^3)$, that is, the temperature that corresponds to 31,275 cm³ of chlorine under 1.13 atmospheres of pressure.

---

*M. D. LaGrega, P. L. Buckingham, and J. C. Evans, *Hazardous Waste Management,* New York: McGraw-Hill, 1994, p. 140.

[†]E. Batschelet, *Introduction to Mathematics for Life Scientists,* 2nd ed., New York: Springer-Verlag, 1979, pp. 102–103.

44. ICE AGE PATTERNS OF ICE AND TEMPERATURE The level curves in the land areas of the accompanying figure indicate ice elevations above sea level (in meters) during the last major ice age (approximately 18,000 years ago). The level curves in the sea areas indicate sea surface temperature. For instance, the ice was 1,000 meters thick above New York City, and the sea temperature near the Hawaiian Islands was about 24°C. Where on earth was the ice pack the thickest? What ice-bound land area was adjacent to the warmest sea?

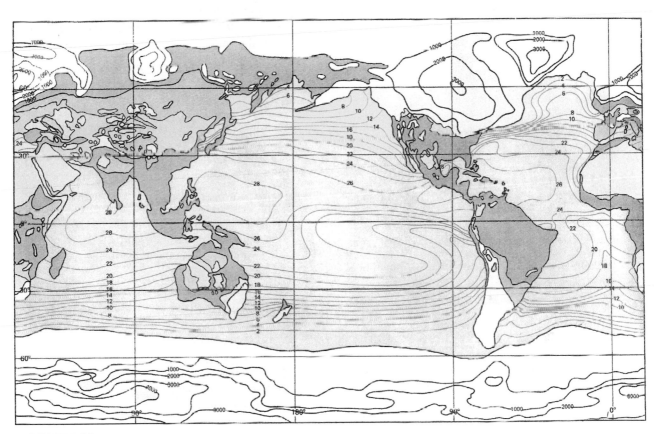

PROBLEM 44

SOURCE: *The Cambridge Encyclopedia of Earth Sciences*, New York: Crown/Cambridge Press, 1981, p. 302.

45. DAILY ENERGY EXPENDITURE Suppose a person of age A years has weight w in kilograms (kg), and height h in centimeters (cm). Then, the *Harris-Benedict equations* say that the daily basal energy expenditure in kilocalories is

$$B_m(w, h, A) = 66.47 + 13.75w + 5.00h - 6.77A$$

for a male

$$B_f(w, h, A) = 655.10 + 9.60w + 1.85h - 4.68A$$

for a female.

a. Find the basal energy expenditure of a man who weighs 90 kg, is 190 cm tall, and is 22 years old.

b. Find the basal energy expenditure of a woman who weighs 61 kg, is 170 cm tall, and is 27 years old.

c. A man maintains a weight of 85 kg and a height of 193 cm throughout his adult life. At what age will his daily basal energy expenditure be 2,018 kilocalories?

d. A woman maintains a weight of 67 kg and a height of 173 cm throughout her adult life. At what age will her daily basal energy expenditure be 1,504 kilocalories?

46. INVENTORY Suppose a company requires N units per year of a certain commodity. Suppose further that it costs D dollars to order a shipment of the commodity and that the storage cost per unit is S dollars per year. The units are used (or sold) at a constant rate throughout the year, and each shipment is used up just as a new shipment arrives.

a. Show that the total cost $C(x)$ of maintaining inventory when x units are ordered in each shipment is minimized when $x = Q$ where

$$Q(N, D, S) = \sqrt{\frac{2DN}{S}}$$

(See Example 3.5.7 in Section 3.5.)

b. The optimum order size Q found in part (a) is called the *economic order quantity* (EOQ). What is the EOQ when 9,720 units per year are ordered with an ordering cost of \$35 per shipment and a storage cost of 84 cents per unit per year?

47. AMORTIZATION OF DEBT Suppose a loan of A dollars is amortized over n years at an annual interest rate r compounded monthly. Let $i = \dfrac{r}{12}$ be the equivalent monthly rate of interest. Then the monthly payments will be M dollars, where

$$M(A, n, i) = \frac{Ai}{1 - (1 + i)^{-12n}}$$

(See Problems 58 through 61 in Section 4.1.)

a. Allison has a home mortgage of \$250,000 at the fixed rate of 5.2% per year for 15 years. What are her monthly payments? How much total interest does she pay for this loan?

b. Jorge also has a mortgage of \$250,000 but at the fixed rate of 5.6% per year for 30 years. What are his monthly payments? How much total interest does he pay?

48. WIND POWER A machine based on wind energy generally converts the kinetic energy of moving air to mechanical energy by means of a device such as a rotating shaft, as in a windmill. Suppose we have wind of velocity v traveling through a wind-collecting machine with cross-sectional area A. Then, in physics* it is shown that the total power generated by the wind is given by a formula of the form

$$P(v, A) = abAv^3$$

where $b = 1.2$ kg/m^3 is the density of the air, and a is a positive constant.

a. If a wind machine were perfectly efficient, then $a = \dfrac{1}{2}$ in the formula for $P(v, A)$. How much power would be produced by such an ideal windmill with blade radius 15 meters if the wind speed is 22 m/sec?

b. No wind machine is perfectly efficient. In fact, it has been shown that the best we can hope for is about 59% of ideal efficiency, and a good empirical formula for power is obtained by taking $a = \dfrac{8}{27}$. Compute $P(v, A)$ using this value of a if the blade radius of the windmill in part (a) is doubled and the wind velocity is halved.

 c. Mankind has been trying to harness the wind for at least 4,000 years, sometimes with interesting or bizarre consequences. For example, in the source quoted in the footnote, the author notes that during World War II, a windmill was built in Vermont with a blade radius of 175 feet! Read an article on windmills and other devices using wind power. Do you feel these devices have any place in the modern technological world? Explain.

49. REVERSE OSMOSIS In manufacturing semiconductors, it is necessary to use water with an extremely low mineral content, and to separate water from contaminants, it is common practice to use a membrane process called **reverse osmosis.** A key to the effectiveness of such a process is the **osmotic pressure,** which may be determined by the **van't Hoff equation:**[†]

$$P(N, C, T) = 0.075NC(273.15 + T)$$

where P is the osmotic pressure (in atmospheres), N is the number of ions in each molecule of solute, C is the concentration of solute (gram-mole/liter), and T is the temperature of the solute (°C). Find the osmotic pressure for a sodium chloride brine solution with concentration 0.53 gram-mole/liter at a temperature of 23°C. (You will need to know that each molecule of sodium chloride contains two ions: NaCl = Na$^+$ + Cl$^-$.)

50. Output in a certain factory is given by the Cobb-Douglas production function $Q(K, L) = 57K^{1/4}L^{3/4}$, where K is the capital in \$1,000 and L is the size of the labor force, measured in worker-hours.

---

*Raymond A. Serway, *Physics,* 3rd ed., Philadelphia: PA, Saunders, 1992, pp. 408–410.

[†]M. D. LaGrega, P. L. Buckingham, and J. C. Evans, *Hazardous Waste Management,* New York: McGraw-Hill, 1994, pp. 530–543.

a. Use your calculator to obtain $Q(K, L)$ for the values of K and L in this table:

$K(\$1,000)$	277	311	493	554	718
L	743	823	1,221	1,486	3,197
$Q(K, L)$					

b. Note from the next to last column that the output $Q(277, 743)$ is doubled when K is doubled from 277 to 554 and L is doubled from 743 to 1,486. In a similar manner, verify that output is tripled when K and L are both tripled, and that output is halved when K and L are both halved. Does anything interesting happen if K is doubled and L is halved? Verify your response with your calculator.

51. CONSTANT RETURNS TO SCALE   Generalize the results of Problem 50 by showing that for any number s, the Cobb-Douglas production function

$$Q(K, L) = AK^\alpha L^{1-\alpha}$$

satisfies

$$Q(sK, sL) = sQ(K, L)$$

A production function with this property is said to have *constant returns to scale.*

SECTION 7.2   Partial Derivatives

In many problems involving functions of two variables, the goal is to find the rate of change of the function with respect to one of its variables when the other is held constant. That is, the goal is to differentiate the function with respect to the particular variable in question while keeping the other variable fixed. This process is known as **partial differentiation,** and the resulting derivative is said to be a **partial derivative** of the function.

For example, suppose a manufacturer finds that

$$Q(x, y) = 5x^2 + 7xy$$

units of a certain commodity will be produced when x skilled workers and y unskilled workers are employed. Then if the number of unskilled workers remains fixed, the production rate with respect to the number of skilled workers is found by differentiating $Q(x, y)$ with respect to x while holding y constant. We call this the **partial derivative of Q with respect to x** and denote it by $Q_x(x, y)$; thus,

$$Q_x(x, y) = 5(2x) + 7(1)y = 10x + 7y$$

Similarly, if the number of skilled workers remains fixed, the production rate with respect to the number of unskilled workers is given by the **partial derivative of Q with respect to y,** which is obtained by differentiating $Q(x, y)$ with respect to y, holding x constant; that is, by

$$Q_y(x, y) = (0) + 7x(1) = 7x$$

Here is a general definition of partial derivatives and some alternative notation.

Partial Derivatives ■ Suppose $z = f(x, y)$. The partial derivative of f with respect to x is denoted by

$$\frac{\partial z}{\partial x} \qquad \text{or} \qquad f_x(x, y)$$

and is the function obtained by differentiating f with respect to x, treating y as a constant. The partial derivative of f with respect to y is denoted by

$$\frac{\partial z}{\partial y} \qquad \text{or} \qquad f_y(x, y)$$

and is the function obtained by differentiating f with respect to y, treating x as a constant.

NOTE Recall from Chapter 2 that the derivative of a function of one variable $f(x)$ is defined by the limit of a difference quotient; namely,

$$f'(x) = \lim_{h \to 0} \frac{f(x + h) - f(x)}{h}$$

With this definition in mind, the partial derivative $f_x(x, y)$ is given by

$$f_x(x, y) = \lim_{h \to 0} \frac{f(x + h, y) - f(x, y)}{h}$$

and the partial derivative $f_y(x, y)$ by

$$f_y(x, y) = \lim_{h \to 0} \frac{f(x, y + h) - f(x, y)}{h}$$ ∎

Computation of Partial Derivatives

No new rules are needed for the computation of partial derivatives. To compute f_x, simply differentiate f with respect to the single variable x, pretending that y is a constant. To compute f_y, differentiate f with respect to y, pretending that x is a constant. Here are some examples.

EXAMPLE 7.2.1

Find the partial derivatives f_x and f_y if $f(x, y) = x^2 + 2xy^2 + \dfrac{2y}{3x}$.

Solution

To simplify the computation, begin by rewriting the function as

$$f(x, y) = x^2 + 2xy^2 + \frac{2}{3}yx^{-1}$$

To compute f_x, think of f as a function of x and differentiate the sum term by term, treating y as a constant to get

$$f_x(x, y) = 2x + 2(1)y^2 + \frac{2}{3}y(-x^{-2}) = 2x + 2y^2 - \frac{2y}{3x^2}$$

To compute f_y, think of f as a function of y and differentiate term by term, treating x as a constant to get

$$f_y(x, y) = 0 + 2x(2y) + \frac{2}{3}(1)x^{-1} = 4xy + \frac{2}{3x}$$

EXAMPLE 7.2.2

Find the partial derivatives $\dfrac{\partial z}{\partial x}$ and $\dfrac{\partial z}{\partial y}$ if $z = (x^2 + xy + y)^5$.

Solution

Holding y fixed and using the chain rule to differentiate z with respect to x, you get

$$\frac{\partial z}{\partial x} = 5(x^2 + xy + y)^4 \frac{\partial}{\partial x}(x^2 + xy + y)$$

$$= 5(x^2 + xy + y)^4(2x + y)$$

Holding x fixed and using the chain rule to differentiate z with respect to y, you get

$$\frac{\partial z}{\partial y} = 5(x^2 + xy + y)^4 \frac{\partial}{\partial y}(x^2 + xy + y)$$

$$= 5(x^2 + xy + y)^4(x + 1)$$

EXAMPLE 7.2.3

Find the partial derivatives f_x and f_y if $f(x, y) = xe^{-2xy}$.

Solution

From the product rule,

$$f_x(x, y) = x(-2ye^{-2xy}) + e^{-2xy} = (-2xy + 1)e^{-2xy}$$

and from the constant multiple rule,

$$f_y(x, y) = x(-2xe^{-2xy}) = -2x^2e^{-2xy}$$

Geometric Interpretation of Partial Derivatives

Recall from Section 7.1 that functions of two variables can be represented graphically as surfaces drawn on three-dimensional coordinate systems. In particular, if $z = f(x, y)$, an ordered pair (x, y) in the domain of f can be identified with a point in the xy plane and the corresponding function value $z = f(x, y)$ can be thought of as assigning a "height" to this point. The graph of f is the surface consisting of all points (x, y, z) in three-dimensional space whose height z is equal to $f(x, y)$.

The partial derivatives of a function of two variables can be interpreted geometrically as follows. For each fixed number y_0, the points (x, y_0, z) form a vertical plane whose equation is $y = y_0$. If $z = f(x, y)$ and if y is kept fixed at $y = y_0$, then the corresponding points $(x, y_0, f(x, y_0))$ form a curve in a three-dimensional space that is the intersection of the surface $z = f(x, y)$ with the plane $y = y_0$. At each point on this curve, the partial derivative $\dfrac{\partial z}{\partial x}$ is simply the slope of the line in the plane $y = y_0$ that is tangent to the curve at the point in question. That is, $\dfrac{\partial z}{\partial x}$ is the slope of the tangent line "in the x direction." The situation is illustrated in Figure 7.10a.

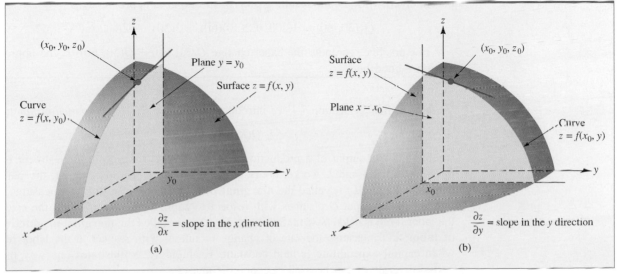

FIGURE 7.10 Geometric interpretation of partial derivatives.

Similarly, if x is kept fixed at $x = x_0$, the corresponding points $(x_0, y, f(x_0, y))$ form a curve that is the intersection of the surface $z = f(x, y)$ with the vertical plane $x = x_0$. At each point on this curve, the partial derivative $\dfrac{\partial z}{\partial y}$ is the slope of the tangent line in the plane $x = x_0$. That is, $\dfrac{\partial z}{\partial y}$ is the slope of the tangent line "in the y direction." The situation is illustrated in Figure 7.10b.

Marginal Analysis

In economics, the term **marginal analysis** refers to the practice of using a derivative to estimate the change in the value of a function resulting from a 1-unit increase in one of its variables. In Section 2.5, you saw some examples of marginal analysis involving ordinary derivatives of functions of one variable. Here is an example of how partial derivatives can be used in a similar fashion.

EXAMPLE 7.2.4

It is estimated that the weekly output of a certain plant is given by the function $Q(x, y) = 1{,}200x + 500y + x^2y - x^3 - y^2$ units, where x is the number of skilled workers and y the number of unskilled workers employed at the plant. Currently the workforce consists of 30 skilled workers and 60 unskilled workers. Use marginal analysis to estimate the change in the weekly output that will result from the addition of 1 more skilled worker if the number of unskilled workers is not changed.

Solution

The partial derivative

$$Q_x(x, y) = 1{,}200 + 2xy - 3x^2$$

is the rate of change of output with respect to the number of skilled workers. For any values of x and y, this is an approximation of the number of additional units that will be produced each week if the number of skilled workers is increased from x to $x + 1$ while the number of unskilled workers is kept fixed at y. In particular, if the workforce is increased from 30 skilled and 60 unskilled workers to 31 skilled and 60 unskilled workers, the resulting change in output is approximately

$$Q_x(30, 60) = 1{,}200 + 2(30)(60) - 3(30)^2 = 2{,}100 \text{ units}$$

For practice, compute the exact change $Q(31, 60) - Q(30, 60)$. Is the approximation a good one?

Recall that in Section 7.1 we introduced the *Cobb-Douglas production function*

$$Q(K, L) = AK^{\alpha}L^{\beta}$$

where Q is the output of a production process in which the capital investment is K thousand dollars and L worker-hours of labor are used. In this context, the partial derivative $Q_K(K, L)$ is called the **marginal productivity of capital** and measures the rate at which output Q changes with respect to capital expenditure when the size of the labor force is held constant. Similarly, $Q_L(K, L)$ is called the **marginal productivity of labor** and measures the rate of change of output with respect to the labor level when capital expenditure is held constant. Example 7.2.5 illustrates one way these partial derivatives can be used in economic analysis.

EXAMPLE 7.2.5

A manufacturer estimates that the monthly output at a certain factory is given by the Cobb-Douglas function

$$Q(K, L) = 50K^{0.4}L^{0.6}$$

where K is the capital expenditure in units of $1,000 and L is the size of the labor force, measured in worker-hours.

a. Find the marginal productivity of capital Q_K and the marginal productivity of labor Q_L when the capital expenditure is $750,000, and the level of labor is 991 worker-hours.

b. Should the manufacturer consider adding capital or increasing the labor level in order to increase output?

Solution

a.
$$Q_K(K, L) = 50(0.4K^{-0.6})L^{0.6} = 20K^{-0.6}L^{0.6}$$

and

$$Q_L(K, L) = 50K^{0.4}(0.6L^{-0.4}) = 30K^{0.4}L^{-0.4}$$

so with $K = 750$ ($750,000) and $L = 991$

$$Q_K(750, 991) = 20(750)^{-0.6}(991)^{0.6} \approx 23.64$$

and

$$Q_L(750, 991) = 30(750)^{0.4}(991)^{-0.4} \approx 26.84$$

b. From part (a), you see that an increase in 1 unit of capital (that is, $1,000) results in an increase in output of 23.64 units, which is less than the 26.84 unit increase in output that results from a unit increase in the labor level. Therefore, the manufacturer should increase the labor level by 1 worker-hour (from 991 worker-hours to 992) to increase output as quickly as possible from the current level.

Substitute and Complementary Commodities

Two commodities are said to be **substitute commodities** if an increase in the demand for either results in a decrease in demand for the other. Substitute commodities are competitive, like butter and margarine.

On the other hand, two commodities are said to be **complementary commodities** if a decrease in the demand of either results in a decrease in the demand of the other. An example is provided by cameras and film. If consumers buy fewer cameras, they likely buy less film, too.

We can use partial derivatives to obtain criteria for determining whether two commodities are substitute or complementary. Suppose $D_1(p_1, p_2)$ units of the first commodity and $D_2(p_1, p_2)$ of the second are demanded when the unit prices of the commodities are p_1 and p_2, respectively. It is reasonable to expect demand to decrease with increasing price, so

$$\frac{\partial D_1}{\partial p_1} < 0 \qquad \text{and} \qquad \frac{\partial D_2}{\partial p_2} < 0$$

For substitute commodities, the demand for each commodity increases with respect to the price of the other, so

$$\frac{\partial D_1}{\partial p_2} > 0 \qquad \text{and} \qquad \frac{\partial D_2}{\partial p_1} > 0$$

However, for complementary commodities, the demand for each decreases with respect to the price of the other, and

$$\frac{\partial D_1}{\partial p_2} < 0 \qquad \text{and} \qquad \frac{\partial D_2}{\partial p_1} < 0$$

Example 7.2.6 illustrates how these criteria can be used to determine whether a given pair of commodities are complementary, substitute, or neither.

EXAMPLE | 7.2.6

Suppose the demand function for flour in a certain community is given by

$$D_1(p_1, p_2) = 500 + \frac{10}{p_1 + 2} - 5p_2$$

while the corresponding demand for bread is given by

$$D_2(p_1, p_2) = 400 - 2p_1 + \frac{7}{p_2 + 3}$$

where p_1 is the dollar price of a pound of flour and p_2 is the price of a loaf of bread. Determine whether flour and bread are substitute or complementary commodities or neither.

Solution

You find that

$$\frac{\partial D_1}{\partial p_2} = -5 < 0 \qquad \text{and} \qquad \frac{\partial D_2}{\partial p_1} = -2 < 0$$

Since both partial derivatives are negative for all p_1 and p_2, it follows that flour and bread are complementary commodities.

Second-Order Partial Derivatives

Partial derivatives can themselves be differentiated. The resulting functions are called **second-order partial derivatives.** Here is a summary of the definition and notation for the four possible second-order partial derivatives of a function of two variables.

> **Second-Order Partial Derivatives**   ■   If $z = f(x, y)$, the partial derivative of f_x with respect to x is
>
> $$f_{xx} = (f_x)_x \qquad \text{or} \qquad \frac{\partial^2 z}{\partial x^2} = \frac{\partial}{\partial x}\left(\frac{\partial z}{\partial x}\right)$$
>
> The partial derivative of f_x with respect to y is
>
> $$f_{xy} = (f_x)_y \qquad \text{or} \qquad \frac{\partial^2 z}{\partial y\, \partial x} = \frac{\partial}{\partial y}\left(\frac{\partial z}{\partial x}\right)$$
>
> The partial derivative of f_y with respect to x is
>
> $$f_{yx} = (f_y)_x \qquad \text{or} \qquad \frac{\partial^2 z}{\partial x\, \partial y} = \frac{\partial}{\partial x}\left(\frac{\partial z}{\partial y}\right)$$
>
> The partial derivative of f_y with respect to y is
>
> $$f_{yy} = (f_y)_y \qquad \text{or} \qquad \frac{\partial^2 z}{\partial y^2} = \frac{\partial}{\partial y}\left(\frac{\partial z}{\partial y}\right)$$

The computation of second-order partial derivatives is illustrated in Example 7.2.7.

EXAMPLE 7.2.7

Compute the four second-order partial derivatives of the function

$$f(x, y) = xy^3 + 5xy^2 + 2x + 1$$

Solution

Since

$$f_x = y^3 + 5y^2 + 2$$

it follows that

$$f_{xx} = 0 \quad \text{and} \quad f_{xy} = 3y^2 + 10y$$

Since

$$f_y = 3xy^2 + 10xy$$

we have

$$f_{yy} = 6xy + 10x \quad \text{and} \quad f_{yx} = 3y^2 + 10y$$

> **NOTE** The two partial derivatives f_{xy} and f_{yx} are sometimes called the **mixed second-order partial derivatives** of f. Notice that the mixed partial derivatives in Example 7.2.7 are equal. This is not an accident. It turns out that for virtually all functions $f(x, y)$ you will encounter in practical work, the mixed partials will be equal; that is,
>
> $$f_{xy} = f_{yx}$$
>
> This means you will get the same answer if you first differentiate $f(x, y)$ with respect to x and then differentiate the resulting function with respect to y as you would if you performed the differentiation in the reverse order. ∎

Example 7.2.8 illustrates how a second-order partial derivative can convey useful information in a practical situation.

EXAMPLE 7.2.8

Suppose the output Q at a factory depends on the amount K of capital invested in the plant and equipment and also on the size L of the labor force, measured in worker-hours. Give an economic interpretation of the sign of the second-order partial derivative $\dfrac{\partial^2 Q}{\partial L^2}$.

Solution

If $\dfrac{\partial^2 Q}{\partial L^2}$ is negative, the marginal product of labor $\dfrac{\partial Q}{\partial L}$ decreases as L increases. This implies that for a fixed level of capital investment, the effect on output of one additional worker-hour of labor is greater when the workforce is small than when the workforce is large.

Similarly, if $\dfrac{\partial^2 Q}{\partial L^2}$ is positive, it follows that for a fixed level of capital investment, the effect on output of one additional worker-hour of labor is greater when the workforce is large than when it is small.

Typically, for a factory operating with an adequate workforce, the derivative $\dfrac{\partial^2 Q}{\partial L^2}$ will be negative. Can you give an economic explanation for this fact?

The Chain Rule for Partial Derivatives

In many practical situations, a particular quantity is given as a function of two or more variables, each of which can be thought of as a function of yet another variable, and the goal is to find the rate of change of the quantity with respect to this other variable. For example, the demand for a certain commodity may depend on the price of the commodity itself and on the price of a competing commodity, both of which are increasing with time, and the goal may be to find the rate of change of the demand with respect to time. You can solve problems of this type by using the following generalization of the chain rule

$$\frac{dz}{dt} = \frac{dz}{dx}\frac{dx}{dt}$$

obtained in Section 2.4.

> **Chain Rule for Partial Derivatives** ▪ Suppose z is a function of x and y, each of which is a function of t. Then z can be regarded as a function of t and
>
> $$\frac{dz}{dt} = \frac{\partial z}{\partial x}\frac{dx}{dt} + \frac{\partial z}{\partial y}\frac{dy}{dt}$$

Observe that the expression for $\dfrac{dz}{dt}$ is the sum of two terms, each of which can be interpreted using the chain rule for a function of one variable. In particular,

$$\frac{\partial z}{\partial x}\frac{dx}{dt} = \text{rate of change of } z \text{ with respect to } t \text{ for fixed } y$$

and $\qquad \dfrac{\partial z}{\partial y}\dfrac{dy}{dt} = \text{rate of change of } z \text{ with respect to } t \text{ for fixed } x$

The chain rule for partial derivatives says that the total rate of change of z with respect to t is the sum of these two "partial" rates of change. Here is a practical example illustrating the use of the chain rule for partial derivatives.

EXAMPLE | 7.2.9

A health store carries two kinds of multiple vitamins, brand A and brand B. Sales figures indicate that if brand A is sold for x dollars per bottle and brand B for y dollars per bottle, the demand for brand A will be

$$Q(x, y) = 300 - 20x^2 + 30y \quad \text{bottles per month}$$

It is estimated that t months from now the price of brand A will be

$$x = 2 + 0.05t \quad \text{dollars per bottle}$$

and the price of brand B will be

$$y = 2 + 0.1\sqrt{t} \quad \text{dollars per bottle}$$

At what rate will the demand for brand A be changing with respect to time 4 months from now?

Solution

Your goal is to find $\dfrac{dQ}{dt}$ when $t = 4$. Using the chain rule, you get

$$\frac{dQ}{dt} = \frac{\partial Q}{\partial x}\frac{dx}{dt} + \frac{\partial Q}{\partial y}\frac{dy}{dt} = -40x(0.05) + 30(0.05t^{-1/2})$$

When $t = 4$, $\qquad\qquad x = 2 + 0.05(4) = 2.2$

and hence,

$$\frac{dQ}{dt} = -40(2.2)(0.05) + 30(0.05)(0.5) = -3.65$$

That is, 4 months from now the monthly demand for brand A will be decreasing at the rate of 3.65 bottles per month.

In Section 2.5, you learned how to use increments to approximate the change in a function resulting from a small change in its independent variable. In particular, if y is a function of x, then

$$\Delta y \approx \frac{dy}{dx}\Delta x$$

where Δx is a small change in the variable x and Δy is the corresponding change in y. Here is the analogous approximation formula for functions of two variables, based on the chain rule for partial derivatives.

Incremental Approximation Formula for Functions of Two Variables ■ Suppose z is a function of x and y. If Δx denotes a small change in x and Δy a small change in y, the corresponding change in z is

$$\Delta z \approx \frac{\partial z}{\partial x}\Delta x + \frac{\partial z}{\partial y}\Delta y$$

The marginal analysis approximations made earlier in Examples 7.2.4 and 7.2.5 involved unit increments. However, the incremental approximation formula allows a more flexible range of marginal analysis computations, as illustrated in Example 7.2.10.

EXAMPLE | 7.2.10

At a certain factory, the daily output is $Q = 60K^{1/2}L^{1/3}$ units, where K denotes the capital investment measured in units of \$1,000 and L the size of the labor force measured in worker-hours. The current capital investment is \$900,000, and 1,000 worker-hours of labor are used each day. Estimate the change in output that will result if capital investment is increased by $1,000 and labor is increased by 2 worker-hours.

Solution

Apply the approximation formula with $K = 900$, $L = 1,000$, $\Delta K = 1$, and $\Delta L = 2$ to get

$$\Delta Q \approx \frac{\partial Q}{\partial K} \Delta K + \frac{\partial Q}{\partial L} \Delta L$$

$$= 30K^{-1/2}L^{1/3} \Delta K + 20K^{1/2}L^{-2/3} \Delta L$$

$$= 30\left(\frac{1}{30}\right)(10)(1) + 20(30)\left(\frac{1}{100}\right)(2)$$

$$= 22 \text{ units}$$

That is, output will increase by approximately 22 units.

PROBLEMS 7.2

In Problems 1 through 16, compute all first-order partial derivatives of the given function.

1. $f(x, y) = 2xy^5 + 3x^2y + x^2$

2. $z = 5x^2y + 2xy^3 + 3y^2$

3. $z = (3x + 2y)^5$

4. $f(x, y) = (x + xy + y)^3$

5. $f(s, t) = \dfrac{3t}{2s}$

6. $z = \dfrac{t^2}{s^3}$

7. $z = xe^{xy}$

8. $f(x, y) = xye^x$

9. $f(x, y) = \dfrac{e^{2-x}}{y^2}$

10. $f(x, y) = xe^{x+2y}$

11. $f(x, y) = \dfrac{2x + 3y}{y - x}$

12. $z = \dfrac{xy^2}{x^2y^3 + 1}$

13. $z = u \ln v$

14. $f(u, v) = u \ln uv$

15. $f(x, y) = \dfrac{\ln (x + 2y)}{y^2}$

16. $z = \ln \left(\dfrac{x}{y} + \dfrac{y}{x}\right)$

In Problems 17 through 20, evaluate the partial derivatives $f_x(x, y)$ and $f_y(x, y)$ at the given point $P_0(x_0, y_0)$.

17. $f(x, y) = 3x^2 - 7xy + 5y^3 - 3(x + y) - 1$; at $P_0(-2, 1)$

18. $f(x, y) = (x - 2y)^2 + (y - 3x)^2 + 5$; at $P_0(0, -1)$

19. $f(x, y) = xe^{-2y} + ye^{-x} + xy^2$; at $P_0(0, 0)$

20. $f(x, y) = xy \ln \left(\dfrac{y}{x}\right) + \ln (2x - 3y)^2$; at $P_0(1, 1)$

In Problems 21 through 26, find the second partials (including the mixed partials).

21. $f(x, y) = 5x^4y^3 + 2xy$

22. $f(x, y) = \dfrac{x + 1}{y - 1}$

23. $f(x, y) = e^{x^2y}$

24. $f(u, v) = \ln (u^2 + v^2)$

25. $f(s, t) = \sqrt{s^2 + t^2}$

26. $f(x, y) = x^2ye^x$

SUBSTITUTE AND COMPLEMENTARY COMMODITIES *In Problems 27 through 32, the demand functions for a pair of commodities are given. Use partial derivatives to determine whether the commodities are substitute, complementary, or neither.*

27. $D_1 = 500 - 6p_1 + 5p_2$;
$D_2 = 200 + 2p_1 - 5p_2$

28. $D_1 = 1,000 - 0.02p_1^2 - 0.05p_2^2$;
$D_2 = 800 - 0.001p_1^2 - p_1p_2$

29. $D_1 = 3,000 + \dfrac{400}{p_1 + 3} + 50p_2$;
$D_2 = 2,000 - 100p_1 + \dfrac{500}{p_2 + 4}$

30. $D_1 = 2,000 + \dfrac{100}{p_1 + 2} - 25p_2$;
$D_2 = 1,500 - \dfrac{p_2}{p_1 + 7}$

31. $D_1 = \dfrac{7p_2}{1 + p_1^2}$; $D_2 = \dfrac{p_1}{1 + p_2^2}$

32. $D_1 = 200p_1^{-1/2}p_2^{-1/2}$; $D_2 = 300p_1^{-1/2}p_2^{-3/2}$

LAPLACE'S EQUATION *The function $z = f(x, y)$ is said to satisfy **Laplace's equation** if $z_{xx} + z_{yy} = 0$. Functions that satisfy such an equation play an important role in a variety of applications in the physical sciences, especially in the theory of electricity and magnetism In Problems 33 through 36, determine whether the given function satisfies Laplace's equation.*

33. $z = x^2 - y^2$

34. $z = xy$

35. $z = xe^y - ye^x$

36. $z = [(x - 1)^2 + (y + 3)^2]^{-1/2}$

37. MARGINAL ANALYSIS At a certain factory, the daily output is $Q(K, L) = 60K^{1/2}L^{1/3}$ units, where K denotes the capital investment measured in units of \$1,000 and L the size of the labor force measured in worker-hours. Suppose that the current capital investment is \$900,000 and that 1,000 worker-hours of labor are used each day. Use marginal analysis to estimate the effect of an additional capital investment of \$1,000 on the daily output if the size of the labor force is not changed.

38. MARGINAL PRODUCTIVITY A manufacturer estimates that the annual output at a certain factory is given by

$$Q(K, L) = 30K^{0.3}L^{0.7}$$

units, where K is the capital expenditure in units of \$1,000 and L is the size of the labor force in worker-hours.

a. Find the marginal productivity of capital Q_K and the marginal productivity of labor Q_L when the capital expenditure is \$630,000 and the labor level is 830 worker-hours.

b. Should the manufacturer consider adding a unit of capital or a unit of labor in order to increase output more rapidly?

39. MARGINAL PRODUCTIVITY The productivity of a certain country is given by

$$Q(K, L) = 90K^{1/3}L^{2/3}$$

units, where K is the capital expenditure in units of \$1 million and L is the size of the labor force in thousands of worker-hours.

a. Find the marginal productivity of capital Q_K and the marginal productivity of labor Q_L when the capital expenditure is 5.495 billion dollars ($K = 5,495$) and the labor level is 4,587,000 worker-hours ($L = 4,587$).

b. Should the government of the country encourage capital investment or additional labor employment to increase productivity as rapidly as possible?

40. MARGINAL ANALYSIS A grocer's daily profit from the sale of two brands of apple juice is

$$P(x, y) = (x - 30)(70 - 5x + 4y) + (y - 40)(80 + 6x - 7y)$$

cents, where x is the price per can of the first brand and y is the price per can of the second. Currently the first brand sells for 50 cents per can and the second for 52 cents per can. Use marginal analysis to estimate the change in the daily profit that will result if the grocer raises the price of the second brand by one cent per can but keeps the price of the first brand unchanged.

41. FLOW OF BLOOD The smaller the resistance to flow in a blood vessel, the less energy is expended by the pumping heart. One of Poiseuille's laws* says that the resistance to the flow of blood in a blood vessel satisfies

$$F(L, r) = \frac{kL}{r^4}$$

where L is the length of the vessel, r is its radius, and k is a constant that depends on the viscosity of blood.

a. Find F, $\frac{\partial F}{\partial L}$, and $\frac{\partial F}{\partial r}$ in the case where $L = 3.17$ cm and $r = 0.085$ cm. Leave your answer in terms of k.

b. Suppose the vessel in part (a) is constricted and lengthened so that its new radius is 20% smaller than before and its new length is 20% greater. How do these changes affect the flow $F(L, r)$? How do they affect the values of $\frac{\partial F}{\partial L}$ and $\frac{\partial F}{\partial r}$?

42. CONSUMER DEMAND The monthly demand for a certain brand of toasters is given by a function $f(x, y)$, where x is the amount of money (measured in units of \$1,000) spent on advertising and y is the selling price (in dollars) of the toasters. Give economic interpretations of the partial derivatives f_x and f_y. Under normal economic conditions, what will be the sign of each of these derivatives?

43. CONSUMER DEMAND A bicycle dealer has found that if 10-speed bicycles are sold for x dollars apiece and the price of gasoline is y cents per gallon, approximately $F(x, y)$ bicycles will be sold each month, where

$$F(x, y) = 200 - 24\sqrt{x} + 4(0.1y + 3)^{3/2}$$

Currently, the bicycles sell for \$324 apiece and gasoline sells for \$2.20 per gallon. Use marginal analysis to estimate the change in the demand for bicycles that results when the price of bicycles is kept fixed but the price of gasoline decreases by one cent per gallon.

44. SURFACE AREA OF THE HUMAN BODY Recall from Problem 35 of Section 7.1 that the surface area of a person's body may be measured by the empirical formula

$$S(W, H) = 0.0072W^{0.425}H^{0.725}$$

where W (kg) and H (cm) are the person's weight and height, respectively. Currently, a child weighs 34 kg and is 120 cm tall.

a. Compute the partial derivatives $S_W (34, 120)$ and $S_H (34, 120)$, and interpret each as a rate of change.

b. Estimate the change in surface area that results if the child's height stays constant but her weight increases by one kilogram.

45. PACKAGING A soft drink can is a cylinder H cm tall with radius R cm. Its volume is given by the formula $V = \pi R^2 H$. A particular can is 12 cm tall with radius 3 cm. Use calculus to estimate the change in volume that results if the radius is increased by 1 cm while the height remains at 12 cm.

46. PACKAGING For the soft drink can in Problem 45, the surface area is given by $S = 2\pi R^2 + 2\pi RH$. Use calculus to estimate the change in surface that results if:

a. The radius is increased from 3 to 4 cm while the height stays at 12 cm.

b. The height is decreased from 12 to 11 cm while the radius stays at 3 cm.

47. CONSUMER DEMAND Two competing brands of power lawnmowers are sold in the same town. The price of the first brand is x dollars per mower, and the price of the second brand is y dollars per mower. The local demand for the first brand of mower is given by a function $D(x, y)$.

a. How would you expect the demand for the first brand of mower to be affected by an increase in x? By an increase in y?

b. Translate your answers in part (a) into conditions on the signs of the partial derivatives of D.

c. If $D(x, y) = a + bx + cy$, what can you say about the signs of the coefficients b and c if your conclusions in parts (a) and (b) are to hold?

48. CHEMISTRY The **ideal gas law** says that for n moles of an ideal gas, $PV = nRT$, where P is the pressure exerted by the gas, V is the volume of the gas, T is the temperature of the gas, and R is a constant (the **gas constant**). Compute the product

$$\frac{\partial V}{\partial T} \frac{\partial T}{\partial P} \frac{\partial P}{\partial V}$$

49. CARDIOLOGY To estimate the amount of blood that flows through a patient's lung, cardiologists use the empirical formula

$$P(x, y, u, v) = \frac{100xy}{xy + uv}$$

where P is a percentage of the total blood flow, x is the carbon dioxide output of the lung, y is the

*E. Batschelet, *Introduction to Mathematics for Life Scientists*, 2nd ed., New York: Springer-Verlag, 1979, p. 279.

arteriovenous carbon dioxide difference in the lung, u is the carbon dioxide output of the lung, and v is the arteriovenous carbon dioxide difference in the other lung.

It is known that blood flows into the lungs to pick up oxygen and dump carbon dioxide, so the arteriovenous carbon dioxide difference measures the extent to which this exchange is accomplished. (The actual measurement is accomplished by a device called a **cardiac shunt.**) The carbon dioxide is then exhaled from the lungs so that oxygen-bearing air can be inhaled.

Compute the partial derivatives P_x, P_y, P_u, and P_v, and give a physiological interpretation of each derivative.

50. **ELECTRICAL CIRCUIT** In an electrical circuit with two resistors of resistance R_1 and R_2 connected in parallel, the total resistance R is given by the formula

$$\frac{1}{R} = \frac{1}{R_1} + \frac{1}{R_2}$$

Show that

$$R_1 \frac{\partial R}{\partial R_1} + R_2 \frac{\partial R}{\partial R_2} = R$$

51. **BLOOD CIRCULATION** The flow of blood from an artery into a small capillary is given by the formula

$$F(x, y, z) = \frac{c\pi x^2}{4}\sqrt{y - z} \quad cm^3/sec$$

where c is a positive constant, x is the diameter of the capillary, y is the pressure in the artery, and z is the pressure in the capillary. What function gives the rate of change of blood flow with respect to capillary pressure, assuming fixed arterial pressure and capillary diameter? Is this rate increasing or decreasing?

52. **MARGINAL PRODUCTIVITY** Suppose the output Q of a factory depends on the amount K of capital investment measured in units of $1,000 and on the size L of the labor force measured in worker-hours. Give an economic interpretation of the second-order partial derivative $\frac{\partial^2 Q}{\partial K^2}$.

53. **MARGINAL PRODUCTIVITY** At a certain factory, the output is $Q = 120K^{1/2}L^{1/3}$ units, where K denotes the capital investment measured in units of $1,000 and L the size of the labor force measured in worker-hours.

a. Determine the sign of the second-order partial derivative $\frac{\partial^2 Q}{\partial L^2}$ and give an economic interpretation.
b. Determine the sign of the second-order partial derivative $\frac{\partial^2 Q}{\partial K^2}$ and give an economic interpretation.

54. **LAW OF DIMINISHING RETURNS** Suppose the daily output Q of a factory depends on the amount K of capital investment and on the size L of the labor force. A **law of diminishing returns** states that in certain circumstances, there is a value L_0 such that the marginal product of labor will be increasing for $L < L_0$ and decreasing for $L > L_0$.
a. Translate this law of diminishing returns into statements about the sign of a certain second-order partial derivative.
 b. Read about the principle of diminishing returns in an economics text. Then write a paragraph discussing the economic factors that might account for this phenomenon.

55. It is estimated that the weekly output at a certain plant is given by
$$Q(x, y) = 1{,}175x + 483y + 3.1x^2y - 1.2x^3 - 2.7y^2$$
units, where x is the number of skilled workers and y is the number of unskilled workers employed at the plant. Currently the workforce consists of 37 skilled and 71 unskilled workers.
a. Store the output function as
$$1{,}175X + 483Y + 3.1(X^2)*Y$$
$$- 1.2(X^3) - 2.7(Y^2)$$
Store 37 as X and 71 as Y and evaluate to obtain $Q(37, 71)$. Repeat for $Q(38, 71)$ and $Q(37, 72)$.
b. Store the partial derivative $Q_x(x, y)$ in your calculator and evaluate $Q_x(37, 71)$. Use the result to estimate the change in output resulting when the workforce is increased from 37 skilled workers to 38 and the unskilled workforce stays fixed at 71. Then compare with the actual change in output, given by the difference $Q(38, 71) - Q(37, 71)$.
c. Use the partial derivative $Q_y(x, y)$ to estimate the change in output that results when the number of unskilled workers is increased from 71 to 72 while the number of skilled workers stays at 37. Compare with the actual change $Q(37, 72) - Q(37, 71)$.

56. Repeat Problem 55 with the output function
$$Q(x, y) = 1{,}731x + 925y + x^2y - 2.7x^2 - 1.3y^{3/2}$$
and initial employment levels of $x = 43$ and $y = 85$.

Each of Problems 57 through 71 involves either the chain rule for partial derivatives or the incremental approximation formula for functions of two variables.

In Problems 57 through 62, use the chain rule to find $\dfrac{dz}{dt}$. Express your answer in terms of x, y, and t.

57. $z = 2x + 3y; x = t^2, y = 5t$

58. $z = x^2y; x = 3t + 1, y = t^2 - 1$

59. $z = \dfrac{3x}{y}; x = t, y = t^2$

60. $z = x^{1/2} y^{1/3}; x = 2t, y = 2t^2$

61. $z = xy; x = e^{2t}, y = e^{-3t}$

62. $z = \dfrac{x + y}{x - y}; x = t^3 + 1, y = 1 - t^2$

63. ALLOCATION OF LABOR   Using x hours of skilled labor and y hours of unskilled labor, a manufacturer can produce $Q(x, y) = 10xy^{1/2}$ units. Currently 30 hours of skilled labor and 36 hours of unskilled labor are being used. Suppose the manufacturer reduces the skilled labor level by 3 hours and increases the unskilled labor level by 5 hours. Use calculus to determine the approximate effect of these changes on production.

64. CONSUMER DEMAND   A car dealer determines that if gasoline-electric hybrid automobiles are sold for x dollars apiece and the price of gasoline is y cents per gallon, then approximately H hybrid cars will be sold each month, where

$$H(x, y) = 2,500 - 19x^{1/2} + 6(0.1y + 11)^{3/2}$$

She further determines that t months from now, the hybrid cars will be selling for

$$x(t) = 34,000 + 700t$$

dollars apiece and that gasoline will cost

$$y(t) = 220 + 10(3t)^{1/2}$$

cents per gallon. At what rate will the monthly demand for the hybrid cars be changing with respect to time 3 months from now? Will the demand be increasing or decreasing?

65. CONSUMER DEMAND   The demand for a certain product is

$$Q(x, y) = 200 - 10x^2 + 20xy$$

units per month, where x is the price of the product and y is the price of a competing product. It is esti-

mated that t months from now, the price of the product will be

$$x(t) = 10 + 0.5t$$

dollars per unit while the price of the competing product will be

$$y(t) = 12.8 + 0.2t^2$$

dollars per unit.
   a. At what rate will the demand for the product be changing with respect to time 4 months from now?
   b. At what percentage rate $100Q'(t)/Q(t)$ will the demand for the product be changing with respect to time 4 months from now?

66. ALLOCATION OF RESOURCES   At a certain factory, when the capital expenditure is K thousand dollars and L worker-hours of labor are employed, the daily output will be $Q = 120K^{1/2} L^{1/3}$ units. Currently capital expenditure is $400,000 ($K = 400$) and is increasing at the rate of $9,000 per day, while 1,000 worker-hours are being employed and labor is being decreased at the rate of 4 worker-hours per day. At what rate is production currently changing? Is it increasing or decreasing?

67. ALLOCATION OF LABOR   The output at a certain plant is

$$Q(x, y) = 0.08x^2 + 0.12xy + 0.03y^2$$

units per day, where x is the number of hours of skilled labor used and y is the number of hours of unskilled labor used. Currently, 80 hours of skilled labor and 200 hours of unskilled labor are used each day. Use calculus to estimate the change in output that will result if an additional $\frac{1}{2}$ hour of skilled labor is used each day, along with an additional 2 hours of unskilled labor.

68. **RETAIL SALES** A grocer's daily profit from the sale of two brands of orange juice is

$$P(x, y) = (x - 30)(70 - 3x + 4y) + (y - 40)(80 + 6x - 7y)$$

cents, where x is the price per can of the first brand and y is the price of the second brand, both in cents. Currently, the first brand sells for 50 cents per can and the second brand, for 52 cents per can. Use calculus to estimate the change in daily profit that will result if the grocer raises the price of the first brand by 1 cent per can and the second brand by 2 cents per can.

69. **PUBLISHING SALES** An editor estimates that if x thousand dollars are spent on development and y thousand dollars are spent on promotion, approximately $Q(x, y) = 20x^{3/2} y$ copies of a new book will be sold. Current plans call for the expenditure of $36,000 on development and $25,000 on promotion. Use calculus to estimate how sales will be affected if the amount spent on development is increased by $500 and the amount on promotion is decreased by $1,000.

70. **LANDSCAPING** A rectangular garden that is 30 yards long and 40 yards wide is bordered by a concrete path that is 0.8 yard wide. Use calculus to estimate the area of the concrete path.

71. **PACKAGING** A soft drink can is H centimeters (cm) tall and has a radius of R cm. The cost of material in the can is 0.0005 cents per cm^2 and the soda itself costs 0.001 cents per cm^3.
   a. Find a function $C(R, H)$ for the cost of the materials and contents of a can of soda. (You will need the formulas for volume and surface area given in Problems 45 and 46.)
   b. The cans are currently 12 cm tall and have a radius of 3 cm. Use calculus to estimate the effect on cost of increasing the radius by 0.3 cm and decreasing the height by 0.2 cm.

72. **SLOPE OF A LEVEL CURVE** Suppose $y = h(x)$ is a differentiable function of x and that $f(x, y) = C$ for some constant C. Use the chain rule (with x taking the role of t) to show that

$$\frac{\partial f}{\partial x} + \frac{\partial f}{\partial y}\frac{dy}{dx} = 0$$

Conclude that the slope at each point (x, y) on the level curve $F(x, y) = C$ is given by

$$\frac{dy}{dx} = -\frac{f_x}{f_y}$$

73. Use the formula obtained in Problem 72 to find the slope of the level curve

$$x^2 + xy + y^3 = 1$$

at the point $(-1, 1)$. What is the equation of the tangent line to the level curve at this point?

74. Use the formula obtained in Problem 72 to find the slope of the level curve

$$x^2y + 2y^3 - 2e^{-x} = 14$$

at the point $(0, 2)$. What is the equation of the tangent line to the level curve at this point?

75. **UTILITY** Suppose the utility derived by a consumer from x units of one commodity and y units of a second commodity is given by the utility function $U(x, y) = (x + 1)(y + 2)$. The consumer currently owns $x = 25$ units of the first commodity and $y = 8$ units of the second. Use calculus to estimate how many units of the first commodity the consumer could substitute for 1 unit of the second commodity without affecting the total utility.

SECTION 7.3 | Optimizing Functions of Two Variables

Suppose a manufacturer produces two DVD player models, the deluxe and the standard, and that the total cost of producing x units of the deluxe and y units of the standard is given by the function $C(x, y)$. How would you find the level of production $x = a$ and $y = b$ that results in minimal cost? Or perhaps the output of a certain production process is given by $Q(K, L)$, where K and L measure capital and labor expenditure, respectively. What levels of expenditure K_0 and L_0 result in maximum output?

In Section 3.4, you learned how to use the derivative $f'(x)$ to find the largest and smallest values of a function of a single variable $f(x)$, and the goal of this section is to extend those methods to functions of two variables $f(x, y)$. We begin with a definition.

> **Relative Extrema** ■ The function $f(x, y)$ is said to have a **relative maximum** at the point $P(a, b)$ in the domain of f if $f(a, b) \geq f(x, y)$ for all points (x, y) in a circular disk centered at P. Similarly, if $f(c, d) \leq f(x, y)$ for all points (x, y) in a circular disk centered at Q, then $f(x, y)$ has a **relative minimum** at $Q(c, d)$.

In geometric terms, there is a relative maximum of $f(x, y)$ at $P(a, b)$ if the surface $z = f(x, y)$ has a "peak" at the point $(a, b, f(a, b))$; that is, if $(a, b, f(a, b))$ is at least as high as any nearby point on the surface. Similarly, a relative minimum of $f(x, y)$ occurs at $Q(c, d)$ if the point $(c, d, f(c, d))$ is at the bottom of a "valley," so $(c, d, f(c, d))$ is at least as low as any nearby point on the surface. For example, in Figure 7.11, the function $f(x, y)$ has a relative maximum at $P(a, b)$ and a relative minimum at $Q(c, d)$.

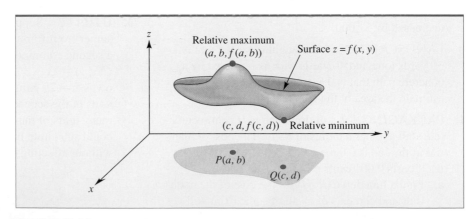

FIGURE 7.11 Relative extrema of the function $f(x, y)$.

Critical Points   The points (a, b) in the domain of $f(x, y)$ for which both $f_x(a, b) = 0$ and $f_y(a, b) = 0$ are said to be **critical points** of f. Like the critical numbers for functions of one variable, these critical points play an important role in the study of relative maxima and minima.

To see the connection between critical points and relative extrema, suppose $f(x, y)$ has a relative maximum at (a, b). Then the curve formed by intersecting the surface $z = f(x, y)$ with the vertical plane $y = b$ has a relative maximum and hence a horizontal tangent line when $x = a$ (see Figure 7.12a). Since the partial derivative $f_x(a, b)$ is the slope of this tangent line, it follows that $f_x(a, b) = 0$. Similarly, the curve formed by intersecting the surface $z = f(x, y)$ with the plane $x = a$ has a relative maximum when $y = b$ (see Figure 7.12b), and so $f_y(a, b) = 0$. This shows that a point at which a function of two variables has a relative maximum must be a critical point. A similar argument shows that a point at which a function of two variables has a relative minimum must also be a critical point.

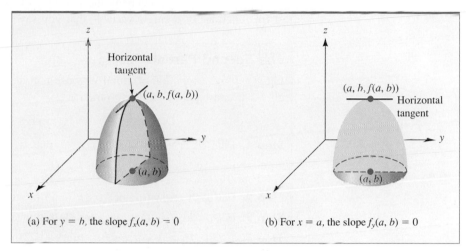

(a) For $y = b$, the slope $f_x(a, b) = 0$

(b) For $x = a$, the slope $f_y(a, b) = 0$

FIGURE 7.12   The partial derivatives are zero at a relative extremum.

Here is a more precise statement of the situation.

Critical Points and Relative Extrema ■ A point (a, b) in the domain of $f(x, y)$ for which the partial derivatives f_x and f_y both exist is called a *critical point* of f if both

$$f_x(a, b) = 0 \qquad \text{and} \qquad f_y(a, b) = 0$$

If the first-order partial derivatives of f exist at all points in some region R in the xy plane, then the relative extrema of f in R can occur only at critical points.

Saddle Points

Although all the relative extrema of a function must occur at critical points, not every critical point of a function corresponds to a relative extremum. For example, if $f(x, y) = y^2 - x^2$, then

$$f_x(x, y) = -2x \qquad \text{and} \qquad f_y(x, y) = 2y$$

so $f_x(0, 0) = f_y(0, 0) = 0$. Thus, the origin $(0, 0)$ is a critical point for $f(x, y)$, and the surface $z = y^2 - x^2$ has horizontal tangents at the origin along both the x axis and the y axis. However, in the xz plane (where $y = 0$) the surface has the equation $z = -x^2$, which is a downward opening parabola, while in the yz plane (where $x = 0$), we have the upward opening parabola $z = y^2$. This means that at the origin, the surface $z = y^2 - x^2$ has a *relative maximum* when viewed in the "x direction" and a *relative minimum* when viewed in the "y direction."

Instead of having a "peak" or a "valley" above the critical point $(0, 0)$, the surface $z = y^2 - x^2$ is shaped like a "saddle," as shown in Figure 7.13, and for this reason is

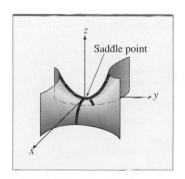

FIGURE 7.13   The saddle surface $z = y^2 - x^2$.

called a **saddle surface.** For a critical point to correspond to a relative extremum, the same extreme behavior (maximum or minimum) must occur in *all directions.* Any critical point (like the origin in this example) where there is a relative maximum in one direction and a relative minimum in another direction is called a **saddle point.**

The Second Partials Test

Here is a procedure involving second-order partial derivatives that you can use to decide whether a given critical point is a relative maximum, a relative minimum, or a saddle point. This procedure is the two-variable version of the second derivative test for functions of a single variable that you saw in Section 3.2.

The Second Partials Test

Let $f(x, y)$ be a function of x and y whose partial derivatives f_x, f_y, f_{xx}, f_{yy}, and f_{xy} all exist, and let $D(x, y)$ be the function

$$D(x, y) = f_{xx}(x, y)\, f_{yy}(x, y) - [f_{xy}(x, y)]^2$$

Step 1. Find all critical points of $f(x, y)$; that is, all points (a, b) so that

$$f_x(a, b) = 0 \quad \text{and} \quad f_y(a, b) = 0$$

Step 2. For each critical point (a, b) found in step 1, evaluate $D(a, b)$.

Step 3. If $D(a, b) < 0$, there is a **saddle point** at (a, b).

Step 4. If $D(a, b) > 0$, compute $f_{xx}(a, b)$:

If $f_{xx}(a, b) > 0$, there is a **relative minimum** at (a, b).

If $f_{xx}(a, b) < 0$, there is a **relative maximum** at (a, b).

If $D = 0$, the test is inconclusive and f may have either a relative extremum or a saddle point at (a, b).

Notice that there is a saddle point at the critical point (a, b) only when the quantity D in the second partials test is negative. If D is positive, there is either a relative maximum or a relative minimum *in all directions.* To decide which, you can restrict your attention to any one direction (say, the x direction) and use the sign of the second partial derivative f_{xx} in exactly the same way as the single variable second derivative was used in the second derivative test given in Chapter 3; namely,

$$\text{a relative minimum if } f_{xx}(a, b) > 0$$
$$\text{a relative maximum if } f_{xx}(a, b) < 0$$

You may find the following tabular summary a convenient way of remembering the conclusions of the second partials test:

Sign of D	Sign of f_{xx}	Behavior at (a, b)
+	+	Relative minimum
+	−	Relative maximum
−		Saddle point

The proof of the second partials test involves ideas beyond the scope of this text and is omitted. Examples 7.3.1 through 7.3.3 illustrate how the test can be used.

EXAMPLE | 7.3.1

Find all critical points for the function $f(x, y) = x^2 + y^2$ and classify each as a relative maximum, a relative minimum, or a saddle point.

Solution

Since

$$f_x = 2x \quad \text{and} \quad f_y = 2y$$

the only critical point of f is $(0, 0)$. To test this point, use the second-order partial derivatives

$$f_{xx} = 2 \quad f_{yy} = 2 \quad \text{and} \quad f_{xy} = 0$$

to get

$$D(x, y) = f_{xx}f_{yy} - (f_{xy})^2 = (2)(2) - 0^2 = 4$$

That is, $D(x, y) = 4$ for *all* points (x, y) and, in particular,

$$D(0, 0) = 4 > 0$$

Hence, f has a relative extremum at $(0, 0)$. Moreover, since

$$f_{xx}(0, 0) = 2 > 0$$

it follows that the relative extremum at $(0, 0)$ is a relative minimum. For reference, the graph of f is sketched in Figure 7.14.

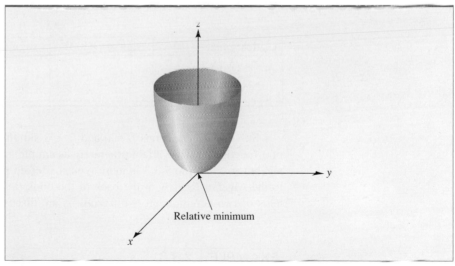

FIGURE 7.14   The surface $z = x^2 + y^2$ with a relative minimum at $(0, 0)$.

EXAMPLE | 7.3.2

Find all critical points for the function $f(x, y) = 12x - x^3 - 4y^2$ and classify each as a relative maximum, a relative minimum, or a saddle point.

Solution

Since

$$f_x = 12 - 3x^2 \quad \text{and} \quad f_y = -8y$$

you find the critical points by solving simultaneously the two equations

$$12 - 3x^2 = 0$$
$$-8y = 0$$

From the second equation, you get $y = 0$ and from the first,

$$3x^2 = 12$$
$$x = 2 \quad \text{or} \quad -2$$

Thus, there are two critical points, $(2, 0)$ and $(-2, 0)$.

To determine the nature of these points, you first compute

$$f_{xx} = -6x \qquad f_{yy} = -8 \qquad \text{and} \qquad f_{xy} = 0$$

and then form the function

$$D = f_{xx}f_{yy} - (f_{xy})^2 = (-6x)(-8) - 0 = 48x$$

Applying the second partials test to the two critical points, you find

$$D(2, 0) = 48(2) = 96 > 0 \qquad \text{and} \qquad f_{xx}(2, 0) = -6(2) = -12 < 0$$

and

$$D(-2, 0) = 48(-2) = -96 < 0$$

so a relative maximum occurs at $(2, 0)$ and a saddle point at $(-2, 0)$. These results are summarized in this table.

Critical point (a, b)	Sign of $D(a, b)$	Sign of $f_{xx}(a, b)$	Behavior at (a, b)
$(2, 0)$	+	−	Relative maximum
$(-2, 0)$	−		Saddle point

Solving the equations $f_x = 0$ and $f_y = 0$ simultaneously to find the critical points of a function of two variables is rarely as simple as in Examples 7.3.1 and 7.3.2. The algebra in Example 7.3.3 is more typical. Before proceeding, you may wish to refer to the Algebra Review at the back of the book, in which techniques for solving systems of two equations in two unknowns are discussed.

EXAMPLE | 7.3.3

Find all critical points for the function $f(x, y) = x^3 - y^3 + 6xy$ and classify each as a relative maximum, a relative minimum, or a saddle point.

Solution

Since

$$f_x = 3x^2 + 6y \qquad \text{and} \qquad f_y = -3y^2 + 6x$$

you find the critical points of f by solving simultaneously the two equations

$$3x^2 + 6y = 0 \qquad \text{and} \qquad -3y^2 + 6x = 0$$

From the first equation, you get $y = -\dfrac{x^2}{2}$ which you can substitute into the second equation to find

$$-3\left(\dfrac{-x^2}{2}\right)^2 + 6x = 0$$

$$-\dfrac{3x^4}{4} + 6x = 0$$

$$-x(x^3 - 8) = 0$$

The solutions of this equation are $x = 0$ and $x = 2$. These are the x coordinates of the critical points of f. To get the corresponding y coordinates, substitute these values of x into the equation $y = -\dfrac{x^2}{2}$ (or into either one of the two original equations). You will find that $y = 0$ when $x = 0$ and $y = -2$ when $x = 2$. It follows that the critical points of f are $(0, 0)$ and $(2, -2)$.

The second order partial derivatives of f are

$$f_{xx} = 6x \qquad f_{yy} = -6y \qquad \text{and} \qquad f_{xy} = 6$$

Hence,

$$D(x, y) = f_{xx}f_{yy} - (f_{xy})^2 = -36xy - 36 = -36(xy + 1)$$

Since

$$D(0, 0) = -36[(0)(0) + 1] = -36 < 0$$

it follows that f has a saddle point at $(0, 0)$. Since

$$D(2, -2) = -36[2(-2) + 1] = 108 > 0$$

and

$$f_{xx}(2, -2) = 6(2) - 12 > 0$$

you see that f has a relative minimum at $(2, -2)$. To summarize:

Critical point (a, b)	$D(a, b)$	$f_{xx}(a, b)$	Behavior at (a, b)
$(0, 0)$	$-$		Saddle point
$(2, -2)$	$+$	$+$	Relative minimum

Practical Optimization Problems

In Example 7.3.4, you will use the theory of relative extrema to solve an optimization problem from economics. Actually, you will be trying to find the *absolute* maximum of a certain function, which turns out to coincide with the *relative* maximum of the function. This is typical of two-variable optimization problems in the social and life sciences, and *in this text, you can assume that a relative extremum you find as the solution to any practical optimization problem is actually the absolute extremum.*

EXAMPLE | 7.3.4

The only grocery store in a small rural community carries two brands of frozen apple juice, a local brand that it obtains at the cost of 30 cents per can and a well-known national brand that it obtains at the cost of 40 cents per can. The grocer estimates that if the local brand is sold for x cents per can and the national brand for y cents per can, then approximately $70 - 5x + 4y$ cans of the local brand and $80 + 6x - 7y$ cans of the national brand will be sold each day. How should the grocer price each brand to maximize the profit from the sale of the juice?

Solution

Since

$$\begin{pmatrix} \text{Total} \\ \text{profit} \end{pmatrix} = \begin{pmatrix} \text{profit from the sale} \\ \text{of the local brand} \end{pmatrix} + \begin{pmatrix} \text{profit from the sale} \\ \text{of the national brand} \end{pmatrix}$$

it follows that the total daily profit from the sale of the juice is given by the function

$$f(x, y) = \underbrace{(70 - 5x + 4y)}_{\substack{\text{items sold}}} \cdot \underbrace{(x - 30)}_{\substack{\text{profit per item}}} + \underbrace{(80 + 6x - 7y)}_{\substack{\text{items sold}}} \cdot \underbrace{(y - 40)}_{\substack{\text{profit per item}}}$$

$$\substack{\textbf{\textit{local brand}} \qquad\qquad\qquad \textbf{\textit{national brand}}}$$

$$= -5x^2 + 10xy - 20x - 7y^2 + 240y - 5{,}300$$

Compute the partial derivatives

$$f_x = -10x + 10y - 20 \quad \text{and} \quad f_y = 10x - 14y + 240$$

and set them equal to zero to get

$$-10x + 10y - 20 = 0 \quad \text{and} \quad 10x - 14y + 240 = 0$$

or

$$-x + y = 2 \quad \text{and} \quad 5x - 7y = -120$$

Then solve these equations simultaneously to get

$$x = 53 \quad \text{and} \quad y = 55$$

It follows that $(53, 55)$ is the only critical point of f.

Next apply the second partials test. Since

$$f_{xx} = -10 \quad f_{yy} = -14 \quad \text{and} \quad f_{xy} = 10$$

you get

$$D(x, y) = f_{xx}f_{yy} - (f_{xy})^2 = (-10)(-14) - (10)^2 = 40$$

Because you have

$$D(53, 55) = 40 > 0 \quad \text{and} \quad f_{xx}(53, 55) = -10 < 0$$

it follows that f has a (relative) maximum when $x = 53$ and $y = 55$. That is, the grocer can maximize profit by selling the local brand of juice for 53 cents per can and the national brand for 55 cents per can.

EXAMPLE 7.3.5

The business manager for Acme Corporation plots a grid on a map of the region Acme serves and determines that the company's three most important customers are located at points $A(1, 5)$, $B(0, 0)$, and $C(8, 0)$, where units are in miles. At what point $W(x, y)$ should a warehouse be located in order to minimize the sum of the squares of the distances from W to A, B, and C (see Figure 7.15)?

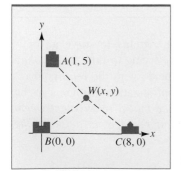

FIGURE 7.15 Locations of businesses A, B, and C and warehouse W.

Solution

The sum of the squares of the distances from W to A, B, and C is given by the function

$$\underbrace{S(x, y)}_{\substack{\text{sum of squares} \\ \text{of distances}}} = \underbrace{[(x - 1)^2 + (y - 5)^2]}_{\substack{\text{square of distance} \\ \text{from } W \text{ to } A}} + \underbrace{(x^2 + y^2)}_{\substack{\text{square of distance} \\ \text{from } W \text{ to } B}} + \underbrace{[(x - 8)^2 + y^2]}_{\substack{\text{square of distance} \\ \text{from } W \text{ to } C}}$$

To minimize $S(x, y)$, you begin by computing the partial derivatives

$$S_x = 2(x - 1) + 2x + 2(x - 8) = 6x - 18$$
$$S_y = 2(y - 5) + 2y + 2y = 6y - 10$$

Then $S_x = 0$ and $S_y = 0$ when

$$6x - 18 = 0$$
$$6y - 10 = 0$$

or $x = 3$ and $y = \dfrac{5}{3}$. Since $S_{xx} = 6$, $S_{xy} = 0$, and $S_{yy} = 6$, you get

$$D = S_{xx}S_{yy} - S_{xy}^2 = (6)(6) - 0^2 = 36 > 0$$

and

$$S_{xx}\left(3, \frac{5}{3}\right) = 6 > 0$$

Thus, the sum of squares is minimized at the map point $W\left(3, \dfrac{5}{3}\right)$.

PROBLEMS 7.3

In Problems 1 through 20, find the critical points of the given function and classify each as a relative maximum, a relative minimum, or a saddle point.

1. $f(x, y) = 5 - x^2 - y^2$

2. $f(x, y) = 2x^2 - 3y^2$

3. $f(x, y) = xy$

4. $f(x, y) = x^2 + 2y^2 - xy + 14y$

5. $f(x, y) = \dfrac{16}{x} + \dfrac{6}{y} + x^2 - 3y^2$

6. $f(x, y) = xy + \dfrac{8}{x} + \dfrac{8}{y}$

7. $f(x, y) = 2x^3 + y^3 + 3x^2 - 3y - 12x - 4$

8. $f(x, y) = (x - 1)^2 + y^3 - 3y^2 - 9y + 5$

9. $f(x, y) = x^3 + y^2 - 6xy + 9x + 5y + 2$

10. $f(x, y) = -x^4 - 32x + y^3 - 12y + 7$

11. $f(x, y) = (x^2 + 2y^2)e^{1 - x^2 - y^2}$

12. $f(x, y) = (x - 4)\ln(xy)$

13. $f(x, y) = x^3 - 4xy + y^3$

14. $f(x, y) = \dfrac{x}{x^2 + y^2 + 4}$

15. $f(x, y) = e^{-(x^2 + y^2 - 6y)}$

16. $f(x, y) = 2x^4 + x^2 + 2xy + 3x + y^2 + 2y + 5$

17. $f(x, y) = \dfrac{1}{x^2 + y^2 + 3x - 2y + 1}$

18. $f(x, y) = xye^{\left(\frac{16x^2 + 9y^2}{288}\right)}$

19. $f(x, y) = x\ln\left(\dfrac{y^2}{x}\right) + 3x - xy^2$

20. $f(x, y) = 4xy - 2x^4 - y^2 + 4x - 2y$

21. **RETAIL SALES** A T-shirt shop carries two competing shirts, one endorsed by Tim Duncan and the other by Shaq O'Neal. The owner of the store can obtain both types at a cost of $2 per shirt and estimates that if Duncan shirts are sold for x dollars apiece and O'Neal shirts for y dollars apiece, consumers will buy approximately $40 - 50x + 40y$ Duncan shirts and $20 + 60x - 70y$ O'Neal shirts each day. How should the owner price the shirts in order to generate the largest possible profit?

22. **PRICING** The telephone company is planning to introduce two new types of executive communications systems that it hopes to sell to its largest commercial customers. It is estimated that if the first type of system is priced at x hundred dollars per system and the second type at y hundred dollars per system, approximately $40 - 8x + 5y$ consumers will buy the first type and $50 + 9x - 7y$ will buy the second type. If the cost of manufacturing the first type is $1,000 per system and the cost of manufacturing the second type is $3,000 per system, how should the telephone company price the systems to generate the largest possible profit?

23. **CONSTRUCTION** Suppose you wish to construct a rectangular box with a volume of 32 ft³. Three different materials will be used in the construction. The material for the sides costs $1 per square foot, the material for the bottom costs $3 per square foot, and the material for the top costs $5 per square foot. What are the dimensions of the least expensive such box?

24. **CONSTRUCTION** A farmer wishes to fence off a rectangular pasture along the bank of a river. The area of the pasture is to be 6,400 yd², and no fencing is needed along the river bank. Find the dimensions of the pasture that will require the least amount of fencing.

25. **RETAIL SALES** A company produces x units of commodity A and y units of commodity B. All the units can be sold for $p = 100 - x$ dollars per unit of A and $q = 100 - y$ dollars per unit of B. The cost (in dollars) of producing these units is given by the joint-cost function $C(x, y) = x^2 + xy + y^2$. What should x and y be to maximize profit?

26. **RETAIL SALES** Repeat Problem 25 for the case where $p = 20 - 5x$, $q = 4 - 2y$, and $C = 2xy + 4$.

27. **RESPONSE TO STIMULI** Consider an experiment in which a subject performs a task while being exposed to two different stimuli (for example, sound and light). For low levels of the stimuli, the subject's performance might actually improve, but as the stimuli increase, they eventually become a distraction and the performance begins to deteriorate. Suppose in a certain experiment in which x units of stimulus A and y units of stimulus B are applied, the performance of a subject is measured by the function

$$f(x, y) = C + xye^{1 - x^2 - y^2}$$

where C is a positive constant. How many units of each stimulus result in maximum performance?

28. **SOCIAL CHOICES** The social desirability of an enterprise often involves making a choice between the commercial advantage of the enterprise and the social or ecological loss that may result. For instance, the lumber industry provides paper products to society and income to many workers and entrepreneurs, but the gain may be offset by the destruction of habitable territory for spotted owls and other endangered species. Suppose the social desirability of a particular enterprise is measured by the function

$$D(x, y) = (16 - 6x)x - (y^2 - 4xy + 40)$$

where x measures commercial advantage (profit and jobs) and y measures ecological disadvantage (species displacement, as a percentage) with $x \geq 0$ and $y \geq 0$. The enterprise is deemed desirable if $D \geq 0$ and undesirable if $D < 0$.

a. What values of x and y will maximize social desirability? Interpret your result. Is it possible for this enterprise to be desirable?

 b. The function given in part (a) is artificial, but the ideas are not. Research the topic of ethics in industry and write a paragraph on how you feel these choices should be made.*

29. **PARTICLE PHYSICS** A particle of mass m in a rectangular box with dimensions x, y, and z has ground state energy

$$E(x, y, z) = \frac{k^2}{8m}\left(\frac{1}{x^2} + \frac{1}{y^2} + \frac{1}{z^2}\right)$$

where k is a physical constant. If the volume of the box satisfies $xyz = V_0$ for constant V_0, find the values of x, y, and z that minimize the ground state energy.

---

*Start with the article by K. R. Stollery, "Environmental Controls in Extractive Industries," *Land Economics*, Vol. 61, 1985, p. 169.

30. ALLOCATION OF FUNDS   A manufacturer is planning to sell a new product at the price of $150 per unit and estimates that if x thousand dollars is spent on development and y thousand dollars is spent on promotion, consumers will buy approximately $\dfrac{320y}{y+2} + \dfrac{160x}{x+4}$ units of the product. If manufacturing costs for this product are $50 per unit, how much should the manufacturer spend on development and how much on promotion to generate the largest possible profit from the sale of this product? [*Hint:* Profit = (number of units)(price per unit − cost per unit) − total amount spent on development and promotion.]

31. PROFIT UNDER MONOPOLY   A manufacturer with exclusive rights to a sophisticated new industrial machine is planning to sell a limited number of the machines to both foreign and domestic firms. The price the manufacturer can expect to receive for the machines will depend on the number of machines made available. (For example, if only a few of the machines are placed on the market, competitive bidding among prospective purchasers will tend to drive the price up.) It is estimated that if the manufacturer supplies x machines to the domestic market and y machines to the foreign market, the machines will sell for $60 - \dfrac{x}{5} + \dfrac{y}{20}$ thousand dollars apiece domestically and for $50 - \dfrac{y}{10} + \dfrac{x}{20}$ thousand dollars apiece abroad. If the manufacturer can produce the machines at the cost of $10,000 apiece, how many should be supplied to each market to generate the largest possible profit?

32. PROFIT UNDER MONOPOLY   A manufacturer with exclusive rights to a new industrial machine is planning to sell a limited number of them and estimates that if x machines are supplied to the domestic market and y to the foreign market, the machines will sell for $150 - \dfrac{x}{6}$ thousand dollars apiece domestically and for $100 - \dfrac{y}{20}$ thousand dollars apiece abroad.

   a. How many machines should the manufacturer supply to the domestic market to generate the largest possible profit at home?

   b. How many machines should the manufacturer supply to the foreign market to generate the largest possible profit abroad?

   c. How many machines should the manufacturer supply to each market to generate the largest possible *total* profit?

    d. Is the relationship between the answers in parts (a), (b), and (c) accidental? Explain. Does a similar relationship hold in Problem 31? What accounts for the difference between these two problems in this respect?

33. CITY PLANNING   Four small towns in a rural area wish to pool their resources to build a television station. If the towns are located at the points $(-5, 0)$, $(1, 7)$, $(9, 0)$, and $(0, -8)$ on a rectangular map grid, where units are in miles, at what point $S(a, b)$ should the station be located to minimize the sum of squares of the distances from the towns?

34. MAINTENANCE   In relation to a rectangular map grid, four oil rigs are located at the points $(-300, 0)$, $(-100, 500)$, $(0, 0)$, and $(400, 300)$ where units are in feet. Where should a maintenance shed $M(a, b)$ be located to minimize the sum of squares of the distances from the rigs?

35. GENETICS   Alternative forms of a gene are called *alleles*. Three alleles, designated A, B, and O, determine the four human blood types A, B, O, and AB. Suppose that p, q, and r are the proportions of A, B, and O in a particular population, so that $p + q + r = 1$. Then according to the Hardy-Weinberg law in genetics, the proportion of individuals in the population who carry two different alleles is given by $P = 2pq + 2pr + 2rq$. What is the largest value of P?

36. LEARNING   In a learning experiment, a subject is first given x minutes to examine a list of facts. The fact sheet is then taken away and the subject is allowed y minutes to prepare mentally for an exam based on the fact sheet. Suppose it is found that the score achieved by a particular subject is related to x and y by the formula
$$S(x, y) = -x^2 + xy + 10x - y^2 + y + 15$$

   a. What score does the subject achieve if he takes the test "cold" (with no study or contemplation)?

   b. How much time should the subject spend in study and contemplation to maximize his score? What is the maximum score?

37. Tom, Dick, and Mary are participating in a cross-country relay race. Tom will trudge as fast as he can through thick woods to the edge of a river; then Dick will take over and row to the opposite shore. Finally, Mary will take the baton and run along the river road to the finish line. The course is shown in the accompanying figure. Teams must start at point S and finish at point F, but they may position one member anywhere along the shore of the river and another anywhere along the river road.

   a. Suppose Tom can trudge at 2 mph, Dick can row at 4 mph, and Mary can run at 6 mph. Where should Dick and Mary wait to receive the baton in order for the team to finish the course as quickly as possible?

 b. The main competition for Tom, Dick, and Mary is the team of Ann, Jan, and Phineas. If Ann can trudge at 1.7 mph, Jan can row at 3.5 mph, and Phineas can run at 6.3 mph, which team should win? By how much?

 c. Does this problem remind you of the spy story in Problem 18 of Section 3.5? Create your own spy story problem based on the mathematical ideas in this problem.

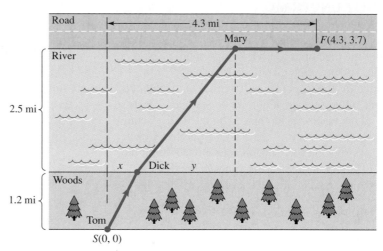

PROBLEM 37

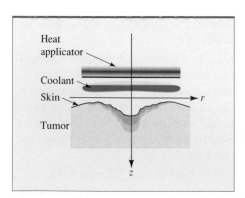

PROBLEM 38

38. CANCER THERAPY   Certain malignant tumors that do not respond to conventional methods of treatment such as surgery or chemotherapy may be treated by **hyperthermia,** which involves applying extreme heat to tumors using microwave transmissions.* One particular kind of microwave applicator used in this kind of therapy produces an absorbed energy density that falls off exponentially. Specifically, the temperature at each point located r units from the central axis of a tumor and z units inside the tumor is given by a formula of the form

$$T(r, z) = Ae^{-pr^2}(e^{-qz} - e^{-sz})$$

where A, p, q, and r are positive constants that depend on properties of both blood and the heating appliance. At what depth inside the tumor does maximum temperature occur? Express your answer in terms of A, p, q, and r.

---

*The ideas in this essay are based on the article by Leah Edelstein-Keshet, "Heat Therapy for Tumors," *UMAP Modules 1991: Tools for Teaching,* Lexington, MA: Consortium for Mathematics and Its Applications, Inc., 1992, pp. 73–101.

39. LIVABLE SPACE Define the *livable space* of a building to be the volume of space in the building where a person 6 feet tall can walk (upright). An A-frame cabin is y feet long and has equilateral triangular ends x feet on a side, as shown in the accompanying figure. If the surface area of the cabin (roof and two ends) is to be 500 ft^2, what dimensions x and y will maximize the livable space?

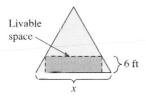

PROBLEM 39

40. BUTTERFLY WING PATTERNS The beautiful patterns on the wings of butterflies have long been a subject of curiosity and scientific study. Mathematical models used to study these patterns often focus on determining the level of morphogen (a chemical that effects change). In a model dealing with eyespot patterns,* a quantity of morphogen is released from an eyespot and the morphogen concentration t days later is given by

$$S(r, t) = \frac{1}{\sqrt{4\pi t}} e^{-\left(\gamma k t + \frac{r^2}{4t}\right)} \qquad t > 0$$

where r measures the radius of the region on the wing affected by the morphogen, and k and γ are positive constants.

a. Find t_m so that $\dfrac{\partial S}{\partial t} = 0$. Show that the function $S_m(t)$ formed from $S(r, t)$ by fixing r has a relative maximum at t_m. Is this the same as saying that the function of two variables $S(r, t)$ has a relative maximum?

b. Let $M(r)$ denote the maximum found in part (a); that is, $M(r) = S(r, t_m)$. Find an expression for M in terms of $z = (1 + 4\gamma k r^2)^{1/2}$.

 c. It turns out that $M(z)$ is what is really needed to analyze the eyespot wing pattern. Read pages 461–468 in the text cited with this problem, and write a paragraph on how biology and mathematics are blended in the study of butterfly wing patterns.

41. Let $f(x, y) = x^2 + y^2 - 4xy$. Show that f does *not* have a relative minimum at its critical point $(0, 0)$, even though it does have a relative minimum at $(0, 0)$ in both the x and y directions. [*Hint:* Consider the direction defined by the line $y = x$. That is, substitute x for y in the formula for f and analyze the resulting function of x.]

---

*J. D. Murray, *Mathematical Biology*, 2nd ed., New York: Springer-Verlag, 1993, pp. 461–468.

In Problems 42 through 45 find the partial derivatives f_x and f_y, and then use your graphing utility to determine the critical points of each function.

42. $f(x, y) = (x^2 + 3y - 5)e^{-x^2 - 2y^2}$

43. $f(x, y) = \dfrac{x^2 + xy + 7y^2}{x \ln y}$

44. $f(x, y) = 6x^2 + 12xy + y^4 + x - 16y - 3$

45. $f(x, y) = 2x^4 + y^4 - x^2(11y - 18)$

46. LEVEL CURVES Sometimes you can classify the critical points of a function by inspecting its level curves. In each case shown in the accompanying figure, determine the nature of the critical point of f at $(0, 0)$.

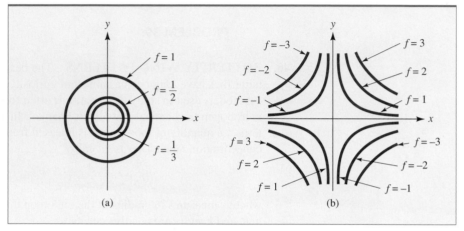

PROBLEM 46

SECTION 7.4 | The Method of Least-Squares

Throughout this text, you have seen applied functions, many of which are derived from published research, and you may have wondered how researchers come up with such functions. A common procedure for associating a function with an observed physical phenomenon is to gather data, plot it on a graph, and then find a function whose graph "best fits" the data in some mathematically meaningful way. We will now develop such a procedure, called the **method of least-squares** or **regression analysis,** which was first mentioned in Example 1.3.7 of Section 1.3, in connection with fitting a line to unemployment data.

The Least-Squares Procedure

Suppose you wish to find a function $y = f(x)$ that fits a particular data set reasonably well. The first step is to decide what type of function to try. Sometimes this can be done by a theoretical analysis of the observed phenomenon and sometimes by inspecting the plotted data. Two sets of data are plotted in Figure 7.16. These are called **scatter diagrams.** In Figure 7.16a, the points lie roughly along a straight line, suggesting that a linear function $y = mx + b$ be used. However, in Figure 7.16b, the points appear to follow an exponential curve, and a function of the form $y = Ae^{-kx}$ would be more appropriate.

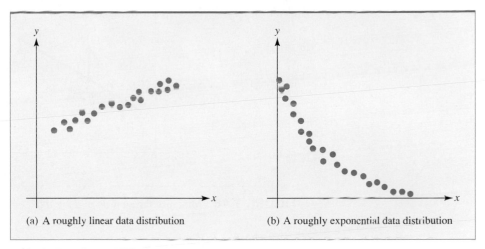

(a) A roughly linear data distribution

(b) A roughly exponential data distribution

FIGURE 7.16 Two scatter diagrams.

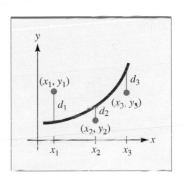

FIGURE 7.17 Sum of squares of the vertical distances $d_1^2 + d_2^2 + d_3^2$.

Once the type of function has been chosen, the next step is to determine the particular function of this type whose graph is "closest" to the given set of points. A convenient way to measure how close a curve is to a set of points is to com pute the sum of the squares of the vertical distances from the points to the curve. In Figure 7.17, for example, this is the sum $d_1^2 + d_2^2 + d_3^2$. The closer the curve is to the points, the smaller this sum will be, and the curve for which this sum is smallest is said to best fit the data according to the **least-squares criterion.**

The use of the least-squares criterion to fit a linear function to a set of points is illustrated in Example 7.4.1. The computation involves the technique from Section 7.3 for minimizing a function of two variables.

EXAMPLE 7.4.1

Use the least-squares criterion to find the equation of the line that is closest to the three points (1, 1), (2, 3), and (4, 3).

Solution

As indicated in Figure 7.18, the sum of the squares of the vertical distances from the three given points to the line $y = mx + b$ is

$$d_1^2 + d_2^2 + d_3^2 = (m + b - 1)^2 + (2m + b - 3)^2 + (4m + b - 3)^2$$

This sum depends on the coefficients m and b that define the line, and so the sum can be thought of as a function $S(m, b)$ of the two variables m and b. The goal, therefore, is to find the values of m and b that minimize the function

$$S(m, b) = (m + b - 1)^2 + (2m + b - 3)^2 + (4m + b - 3)^2$$

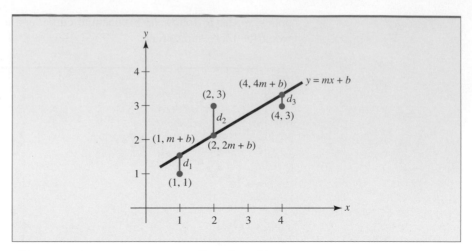

FIGURE 7.18   Minimize the sum $d_1^2 + d_2^2 + d_3^2$.

You do this by setting the partial derivatives $\dfrac{\partial S}{\partial m}$ and $\dfrac{\partial S}{\partial b}$ equal to zero to get

$$\frac{\partial S}{\partial m} = 2(m + b - 1) + 4(2m + b - 3) + 8(4m + b - 3)$$

$$= 42m + 14b - 38 = 0$$

and

$$\frac{\partial S}{\partial b} = 2(m + b - 1) + 2(2m + b - 3) + 2(4m + b - 3)$$

$$= 14m + 6b - 14 = 0$$

Solving the resulting equations

$$42m + 14b = 38$$
$$14m + 6b = 14$$

simultaneously for m and b you conclude that

$$m = \frac{4}{7} \qquad \text{and} \qquad b = 1$$

It can be shown that the critical point $(m, b) = \left(\dfrac{4}{7}, 1\right)$ does indeed minimize the function $S(m, b)$, and so it follows that

$$y = \frac{4}{7}x + 1$$

is the equation of the line that is closest to the three given points.

The Least-Squares Line

The line that is closest to a set of points according to the least-squares criterion is called the **least-squares line** for the points. (The term **regression line** is also used, especially in statistical work.) The procedure used in Example 7.4.1 can be generalized to give formulas for the slope m and the y intercept b of the least-squares line for an arbitrary set of n points $(x_1, y_1), (x_2, y_2), \ldots, (x_n, y_n)$. The formulas involve sums of the x and y values. All the sums run from $j - 1$ to $j = n$, and to simplify the notation, the indices are omitted. For example, Σx is used instead of $\displaystyle\sum_{j=1}^{n} x_j$.

The Least-Squares Line ■ The equation of the least-squares line for the n points $(x_1, y_1), (x_2, y_2), \ldots, (x_n, y_n)$ is $y = mx + b$, where

$$m = \frac{n\Sigma xy - \Sigma x \Sigma y}{n\Sigma x^2 - (\Sigma x)^2} \quad \text{and} \quad b = \frac{\Sigma x^2 \Sigma y - \Sigma x \Sigma xy}{n\Sigma x^2 - (\Sigma x)^2}$$

EXPLORE!

Your graphing calculator can easily assist you in producing and displaying the lists and summations needed to calculate the coefficients of the least-squares equation. Using the data in Example 7.4.2, place the x values into L1 and y values into L2, and write L3 = L1*L2 and L4 = L1². Use the summation features of your calculator to obtain the column totals needed to calculate the slope and y-intercept formulas displayed on this page above Example 7.4.2.

EXAMPLE | 7.4.2

Use the formulas to find the least-squares line for the points $(1, 1)$, $(2, 3)$, and $(4, 3)$ from Example 7.4.1.

Solution

Arrange your calculations as follows:

x	y	xy	x^2
1	1	1	1
2	3	6	4
4	3	12	16
$\Sigma x = 7$	$\Sigma y = 7$	$\Sigma xy = 19$	$\Sigma x^2 = 21$

Then use the formulas with $n = 3$ to get

$$m = \frac{3(19) - 7(7)}{3(21) - (7)^2} = \frac{4}{7} \quad \text{and} \quad b = \frac{21(7) - 7(19)}{3(21) - (7)^2} = 1$$

from which it follows that the equation of the least-squares line is

$$y = \frac{4}{7}x + 1$$

Least-Squares Prediction

The least-squares line (or curve) that best fits the data collected in the past can be used to make rough predictions about the future. This is illustrated in Example 7.4.3.

EXAMPLE | 7.4.3

A college admissions officer has compiled these data relating students' high school and college grade-point averages:

High school GPA	2.0	2.5	3.0	3.0	3.5	3.5	4.0	4.0
College GPA	1.5	2.0	2.5	3.5	2.5	3.0	3.0	3.5

Find the equation of the least-squares line for these data and use it to predict the college GPA of a student whose high school GPA is 3.7.

Solution

Let x denote the high school GPA and y the college GPA and arrange the calculations as follows:

x	y	xy	x^2
2.0	1.5	3.0	4.0
2.5	2.0	5.0	6.25
3.0	2.5	7.5	9.0
3.0	3.5	10.5	9.0
3.5	2.5	8.75	12.25
3.5	3.0	10.5	12.25
4.0	3.0	12.0	16.0
4.0	3.5	14.0	16.0
$\Sigma x = 25.5$	$\Sigma y = 21.5$	$\Sigma xy = 71.25$	$\Sigma x^2 = 84.75$

Use the least-squares formula with $n = 8$ to get

$$m = \frac{8(71.25) - 25.5(21.5)}{8(84.75) - (25.5)^2} \approx 0.78$$

and

$$b = \frac{84.75(21.5) - 25.5(71.25)}{8(84.75) - (25.5)^2} \approx 0.19$$

The equation of the least-squares line is therefore

$$y = 0.78x + 0.19$$

To predict the college GPA y of a student whose high school GPA x is 3.7, substitute $x = 3.7$ into the equation of the least-squares line. This gives

$$y = 0.78(3.7) + 0.19 \approx 3.08$$

which suggests that the student's college GPA might be about 3.1.

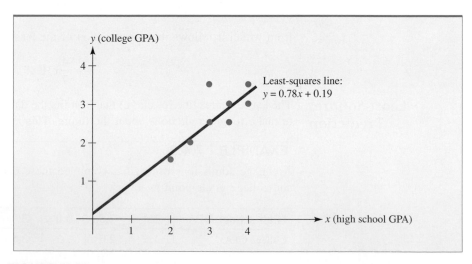

FIGURE 7.19   The least-squares line for high school and college GPAs.

The original data are plotted in Figure 7.19, together with the least-squares line $y = 0.78x + 0.19$. Actually, in practice, it is a good idea to plot the data before proceeding with the calculations. By looking at the graph you will usually be able to tell whether approximation by a straight line is appropriate or whether a better fit might be possible with a curve of some sort.

Nonlinear Curve-Fitting

In each of Examples 7.4.1, 7.4.2, and 7.4.3, the least-squares criterion was used to fit a linear function to a set of data. With appropriate modifications, the procedure can also be used to fit nonlinear functions to data. One kind of modified curve-fitting procedure is illustrated in Example 7.4.4.

EXAMPLE | 7.4.4

A manufacturer gathers these data relating the level of production x (hundred units) of a particular commodity to the demand price p (dollars per unit) at which all x units will be sold.

k	Production x (hundred units)	Demand Price p (dollars per unit)
1	6	743
2	10	539
3	17	308
4	22	207
5	28	128
6	35	73

a. Plot a scatter diagram for the data on a graph with production level on the x axis and demand price on the y axis.

b. Notice that the scatter diagram in part (a) suggests that the demand function is exponential. Modify the least squares procedure to find a curve of the form $p = Ae^{mx}$ that best fits the data in the table.

c. Use the exponential demand function you found in part (b) to predict the revenue the manufacturer should expect if 4,000 ($x = 40$) units are produced.

Solution

a. The scatter diagram is plotted in Figure 7.20.

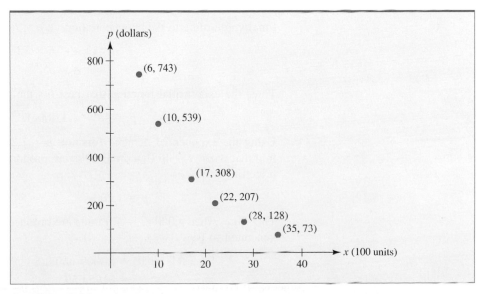

FIGURE 7.20 A scatter diagram for the demand data in Example 7.4.4.

Just-In-Time Review

Recall that

$$\ln(ab) = \ln a + \ln b$$

b. Taking logarithms on both sides of the equation $p = Ae^{mx}$, we find that

$$
\begin{aligned}
\ln p &= \ln(Ae^{mx}) \\
&= \ln A + \ln(e^{mx}) \quad \textit{product rule for logarithms} \\
&= \ln A + mx \qquad\quad \ln e^{u} = u
\end{aligned}
$$

or equivalently, $y = mx + b$, where $y = \ln p$ and $b = \ln A$. Thus, to find the curve of the form $p = Ae^{mx}$ that best fits the given data points (x_k, p_k) for $k = 1, \ldots, 6$, we first find the least-squares line $y = mx + b$ for the data points $(x_k, \ln p_k)$. Arrange the calculations as follows:

k	x_k	p_k	$y_k = \ln p_k$	$x_k y_k$	x_k^2
1	6	743	6.61	39.66	36
2	10	539	6.29	62.90	100
3	17	308	5.73	97.41	289
4	22	207	5.33	117.26	484
5	28	128	4.85	135.80	784
6	35	73	4.29	150.15	1,225
	$\Sigma x = 118$		$\Sigma y = 33.10$	$\Sigma xy = 603.18$	$\Sigma x^2 = 2{,}918$

Use the least-squares formula with $n = 6$ to get

$$m = \frac{6(603.18) - (118)(33.10)}{6(2{,}918) - (118)^2} = -0.08$$

and

$$b = \frac{2{,}918(33.10) - (118)(603.18)}{6(2{,}918) - (118)^2} = 7.09$$

which means the least-squares line has the equation

$$y = -0.08x + 7.09$$

Finally, returning to the exponential curve $p = Ae^{mx}$, recall that $\ln A = b$, so that

$$
\begin{aligned}
\ln A &= b = 7.09 \\
A &= e^{7.09} = 1{,}200
\end{aligned}
$$

Thus, the exponential function that best fits the given demand data is

$$p = Ae^{mx} = 1{,}200e^{-0.08x}$$

c. Using the exponential demand function $p = 1{,}200e^{-0.08x}$ found in part (b), we find that when $x = 40$ (hundred) units are produced, they can all be sold at a unit price of

$$p = 1{,}200e^{-0.08(40)} = \$48.91$$

Therefore, when 4,000 ($x = 40$) units are produced, we would expect the revenue generated to be

$$
\begin{aligned}
R = xp(x) &= (4{,}000 \text{ units})(\$48.91 \text{ per unit}) \\
&= \$195{,}659
\end{aligned}
$$

The procedure illustrated in Example 7.4.4 is sometimes called **log-linear regression.** A curve of the form $y = Ax^k$ that best fits given data can also be found by log-linear regression. The specific procedure is outlined in Problem 32, where it is used to verify an allometric formula first mentioned in the Think About It essay at the end of Chapter 1.

The least-squares procedure can also be used to fit other nonlinear functions to data. For instance, to find the quadratic function $y = Ax^2 + Bx + C$ whose graph (a parabola) best fits a particular set of data, you would proceed as in Example 7.4.1, minimizing the sum of squares of vertical distances from the given points to the graph. Such computations are algebraically complicated and usually require the use of a computer or graphing calculator.

PROBLEMS 7.4

In Problems 1 through 4 plot the given points and use the method of Example 7.4.1 to find the corresponding least-squares line.

1. $(0, 1), (2, 3), (4, 2)$

2. $(1, 1), (2, 2), (6, 0)$

3. $(1, 2), (2, 4), (4, 4), (5, 2)$

4. $(1, 5), (2, 4), (3, 2), (6, 0)$

In Problems 5 through 12, plot the given points and use the formula to find the corresponding least-squares line.

5. $(1, 2), (2, 2), (2, 3), (5, 5)$

6. $(-4, -1), (-3, 0), (-1, 0), (0, 1), (1, 2)$

7. $(-2, 5), (0, 4), (2, 3), (4, 2), (6, 1)$

8. $(-6, 2), (-3, 1), (0, 0), (0, -3), (1, -1), (3, -2)$

9. $(0, 1), (1, 1.6), (2.2, 3), (3.1, 3.9), (4, 5)$

10. $(3, 5.72), (4, 5.31), (6.2, 5.12), (7.52, 5.32), (8.03, 5.67)$

11. $(-2.1, 3.5), (-1.3, 2.7), (1.5, 1.3), (2.7, -1.5)$

12. $(-1.73, -4.33), (0.03, -2.19), (0.93, 0.15), (3.82, 1.61)$

In Problems 13 through 16, modify the least-squares procedure as illustrated in Example 7.4.4 to find a curve of the form $y = Ae^{mx}$ that best fits the given data.

13. $(1, 15.6), (3, 17), (5, 18.3), (7, 20), (10, 22.4)$

14. $(5, 9.3), (10, 10.8), (15, 12.5), (20, 14.6), (25, 17)$

15. $(2, 13.4), (4, 9), (6, 6), (8, 4), (10, 2.7)$

16. $(5, 33.5), (10, 22.5), (15, 15), (20, 10), (25, 6.8), (30, 4.5)$

17. COLLEGE ADMISSIONS Over the past 4 years, a college admissions officer has compiled the following data (measured in units of 1,000) relating the number of college catalogs requested by high school students by December 1 to the number of completed applications received by March 1:

Catalogs requested	4.5	3.5	4.0	5.0
Applications received	1.0	0.8	1.0	1.5

a. Plot these data on a graph.

b. Find the equation of the least-squares line.

c. Use the least-squares line to predict how many completed applications will be received by March 1 if 4,800 catalogs are requested by December 1.

18. **SALES**   A company's annual sales (in units of 1 billion dollars) for its first 5 years of operation are shown in this table:

Year	1	2	3	4	5
Sales	0.9	1.5	1.9	2.4	3.0

    a. Plot these data on a graph.
    b. Find the equation of the least-squares line.
    c. Use the least-squares line to predict the company's sixth-year sales.

19. **DEMAND AND REVENUE**   A manufacturer gathers the data listed in the accompanying table relating the level of production x (hundred units) of a particular commodity to the demand price p (dollars per unit) at which all the units will be sold:

Production x (hundreds of units)	5	10	15	20	25	30	35
Demand price p (dollars per unit)	44	38	32	25	18	12	6

    a. Plot these data on a graph.
    b. Find the equation of the least-squares line for the data.
    c. Use the linear demand equation you found in part (b) to predict the revenue the manufacturer should expect if 4,000 units ($x = 40$) are produced.

20. **DRUG ABUSE**   For each of five different years, the accompanying table gives the percentage of high school students who had used cocaine at least once in their lives up to that year:

Year	1991	1993	1995	1997	1999
Percentage who had used cocaine at least once	6.0	4.9	7.0	8.2	9.5

SOURCE: The White House Office of National Drug Control Policy, "2002 National Drug Control Strategy" (http://www.whitehousedrugpolicy.gov).

    a. Plot these data on a graph, with the number of years after 1991 on the x axis and the percentage of cocaine users on the y axis.
    b. Find the equation of the least-squares line for the data.
    c. Use the least-squares line to predict the percentage of high school students who used cocaine at least once by the year 2005.

21. **VOTER TURNOUT**   On election day, the polls in a certain state open at 8:00 A.M. Every 2 hours after that, an election official determines what percentage of the registered voters have already cast their ballots. The data through 6:00 P.M. are shown here:

Time	10:00	12:00	2:00	4:00	6:00
Percentage turnout	12	19	24	30	37

    a. Plot these data on a graph.
    b. Find the equation of the least-squares line. (Let x denote the number of hours after 8:00 A.M.)
    c. Use the least-squares line to predict what percentage of the registered voters will have cast their ballots by the time the polls close at 8:00 P.M.

22. **POPULATION PREDICTION** The accompanying table gives the U.S. decennial census figures (in millions) for the period 1950–2000:

Year	1950	1960	1970	1980	1990	2000
Population	150.7	179.3	203.2	226.5	248.7	291.4

SOURCE: U.S. Census Bureau (http://www.census.gov).

     a. Find the least-squares line $y = mt + b$ for these data, where y is the U.S. population t decades after 1950.

     b. Use the least-squares line found in part (a) to predict the U.S. population for the year 2010.

23. **POPULATION PREDICTION** Modify the least-squares procedure, as illustrated in Example 7.4.4, to find a function of the form $P(t) = Ae^{rt}$ whose graph best fits the population data in Problem 22, where $P(t)$ is the U.S. population t decades after 1950.

     a. Roughly at what percentage rate is the U.S. population growing?

     b. Based on your population function, what would you expect the U.S. population to be in the year 2005?

24. **PUBLIC HEALTH** In a study of five industrial areas, a researcher obtained these data relating the average number of units of a certain pollutant in the air and the incidence (per 100,000 people) of a certain disease:

Units of pollutant	3.4	4.6	5.2	8.0	10.7
Incidence of disease	48	52	58	76	96

     a. Plot these data on a graph.

     b. Find the equation of the least-squares line.

     c. Use the least-squares line to estimate the incidence of the disease in an area with an average pollution level of 7.3 units.

25. **INVESTMENT ANALYSIS** Jennifer has several different kinds of investments, whose total value $V(t)$ (in thousands of dollars) at the beginning of the tth year after she began investing is given in this table, for $1 \le t \le 10$:

Year t	1	2	3	4	5	6	7	8	9	10
Value $V(t)$ of all investments	57	60	62	65	62	65	70	75	79	85

     a. Modify the least-squares procedure, as illustrated in Example 7.4.4, to find a function of the form $V(t) = Ae^{rt}$ whose graph best fits these data. Roughly at what annual rate, compounded continuously, is her account growing?

     b. Use the function you found in part (a) to predict the total value of her investments at the beginning of the 20th year after she began investing.

     c. Jennifer estimates she will need $300,000 for retirement. Use the function from part (a) to determine how long it will take her to attain this goal.

     d. Jennifer's friend, Frank Kornerkutter, looks over her investment analysis and snorts, "What a waste of time! You can find the A and r in your function $V(t) = Ae^{rt}$ by just using $V(1) = 57$ and $V(10) = 85$ and a little algebra." Find A and r using Frank's method, and comment on the relative merits of the two approaches.

26. DISPOSABLE INCOME AND CONSUMPTION The accompanying table gives the personal consumption expenditure and the corresponding disposable income (in billions of dollars) for the United States in the period 1996–2001:

Year	1996	1997	1998	1999	2000	2001
Disposable income	5,677.7	5,968.2	6,355.6	6,627.4	7,120.0	7,393.2
Personal consumption	5,237.5	5,529.3	5,856.0	6,246.5	6,683.7	6,967.0

SOURCE: U.S. Department of Commerce, Bureau of Economic Analysis, "Personal Consumption Expenditures by Major Type of Product" (http://www.dea.doc.gov).

a. Plot these data on a graph, with disposable income on the *x* axis and consumption expenditure on the *y* axis.

b. Find the equation of the least-squares line for the data.

c. Use the least-squares line to predict the consumption that would correspond to $8,000 billion of disposable income.

 d. Write a paragraph on the relationship between disposable income and consumption.

27. STOCK MARKET AVERAGE The accompanying table gives the Dow Jones Industrial Average (DJIA) at the close of the first trading day of the year shown:

Year	1990	1992	1996	1998	2001	2002
DJIA	2,810	3,172	5,177	7,965	10,646	10,073

SOURCE: Dow Jones (http://www.djindexes.com).

a. Plot these data on a graph, with the number of years after 1990 on the *x* axis and the DJIA on the *y* axis.

b. Find the equation of the least-squares line for the data.

c. What does the least-squares line predict for the DJIA on the first day of trading in 2003? Use the Internet to find where the DJIA actually closed on that day (January 2, 2003), and compare with the predicted value.

d. Write a paragraph on whether you think it is possible to find a curve that fits the DJIA well enough to usefully predict future market behavior.

28. GASOLINE PRICES The average retail price per gallon (in cents) of regular unleaded gasoline at 3-year intervals from 1988 to 2003 is given in this table:

Year	1988	1991	1994	1997	2000	2003
Price per gallon (cents)	95	114	111	123	151	159

SOURCE: U.S. Department of Energy (http://www.eia.doe.gov).

a. Plot these data on a graph, with the number of years after 1988 on the *x* axis and the average price of gasoline on the *y* axis.

b. Find the equation of the least-squares line for the data.

c. What price per gallon does the least-squares line predict you will pay for a gallon of regular unleaded gasoline in 2006?

29. GROSS DOMESTIC PRODUCT This table lists the Gross Domestic Product (GNP) figures for China (billions of yuan) for the period 1996–2001:

Year	1996	1997	1998	1999	2000	2001
GDP	6,788	7,446	7,835	8,191	8,940	9,593

SOURCE: Chinese government website (http://www.china.org.cn).

a. Find the least-squares line $y = mt + b$ for these data, where y is the GDP of China t years after 1996.

b. Use the least-squares line found in part (a) to predict the GDP of China for the year 2005.

30. BACTERIAL GROWTH   A biologist studying a bacterial colony measures its population each hour and records these data:

Time t (hours)	1	2	3	4	5	6	7	8
Population $P(t)$ (thousands)	280	286	292	297	304	310	316	323

a. Plot these data on a graph. Does the scatter diagram suggest that the population growth is linear or exponential?

b. If you think the scatter diagram in part (a) suggests linear growth, find a population function of the form $P(t) = mt + b$ that best fits the data. However, if you think the scatter diagram suggests exponential growth, modify the least squares procedure, as illustrated in Example 7.4.4, to obtain a best-fitting population function of the form $P(t) = Ae^{kt}$.

c. Use the population function you obtained in part (b) to predict how long it will take for the population to reach 400,000. How long will it take for the population to double from 280,000?

31. SPREAD OF AIDS   The number of reported cases of AIDS in the United States by year of reporting at 4-year intervals since 1980 is given in this table:

Year	1980	1984	1988	1992	1996	2000
Reported cases of AIDS	99	6,360	36,064	79,477	61,109	42,156

SOURCE: World Health Organization and the United Nations (http://www.unaids.org).

a. Plot these data on a graph with time t (years after 1980) on the x axis.

b. Find the equation of the least-squares line for the given data.

c. How many cases of AIDS does the least-squares line in part (b) predict will be reported in 2005?

 d. Do you think the least-squares line fits the given data well? If not, write a paragraph explaining which (if any) of the following four curve types would fit the data better:

   1. (quadratic) $y = At^2 + Bt + C$
   2. (cubic) $y = At^3 + Bt^2 + Ct + D$
   3. (exponential) $y = Ae^{kt}$
   4. (power-exponential) $y = Ate^{kt}$

(You may find it helpful to review the Think About It essay at the end of Chapter 3, which involves a similar analysis of the number of deaths due to AIDS.)

32. ALLOMETRY   The determination of relationships between measurements of various parts of a particular organism is a topic of interest in the branch of biology called *allometry*.* (Recall the Think About It essay at the end of Chapter 1.) Suppose a biologist observes that the shoulder height h and antler size w of an elk, both in centimeters (cm), are related as indicated in this table:

---

*Roger V. Jean, "Differential Growth, Huxley's Allometric Formula, and Sigmoid Growth," *UMAP Modules 1983: Tools for Teaching*, Lexington, MA: Consortium for Mathematics and Its Applications, Inc., 1984.

Shoulder Height h (cm)	Antler Size w (cm)
87.9	52.4
95.3	60.3
106.7	73.1
115.4	83.7
127.2	98.0
135.8	110.2

a. For each data point (h, w) in the table, plot the point $(\ln h, \ln w)$ on a graph. Note that the scatter diagram suggests that $y = \ln w$ and $x = \ln h$ are linearly related.

b. Find the least-squares line $y = mx + b$ for the data $(\ln h, \ln w)$ you obtained in part (a).

c. Find numbers a and c so that $w = ah^c$. [*Hint:* Substitute $y = \ln w$ and $x = \ln h$ from part (a) into the least-squares equation found in part (b).]

33. ALLOMETRY The accompanying table relates the weight C of the large claw of a fiddler crab to the weight W of the rest of the crab's body, both measured in milligrams (mg).

Weight W (mg) of the body	57.6	109.2	199.7	300.2	355.2	420.1	535.7	743.3
Weight C (mg) of the claw	5.3	13.7	38.3	78.1	104.5	135.0	195.6	319.2

a. For each data point (W, C) in the table, plot the point $(\ln W, \ln C)$ on a graph. Note that the scatter diagram suggests that $y = \ln C$ and $x = \ln W$ are linearly related.

b. Find the least-squares line $y = mx + b$ for the data $(\ln W, \ln C)$ you obtained in part (a).

c. Find positive numbers a and k so that $C = aW^k$. [*Hint:* Substitute $y = \ln C$ and $x = \ln W$ from part (a) into the least-squares equation found in part (b).]

SECTION 7.5 Constrained Optimization: The Method of Lagrange Multipliers

In many applied problems, a function of two variables is to be optimized subject to a restriction or **constraint** on the variables. For example, an editor, constrained to stay within a fixed budget of $60,000, may wish to decide how to divide this money between development and promotion in order to maximize the future sales of a new book. If x denotes the amount of money allocated to development, y the amount allocated to promotion, and $f(x, y)$ the corresponding number of books that will be sold, the editor would like to maximize the sales function $f(x, y)$ subject to the budgetary constraint that $x + y = 60,000$.

For a geometric interpretation of the process of optimizing a function of two variables subject to a constraint, think of the function itself as a surface in three-dimensional space and of the constraint (which is an equation involving x and y) as a curve in the *xy* plane. When you find the maximum or minimum of the function subject to the given constraint, you are restricting your attention to the portion of the surface that lies directly above the constraint curve. The highest point on this portion of the surface is the constrained maximum, and the lowest point is the constrained minimum. The situation is illustrated in Figure 7.21.

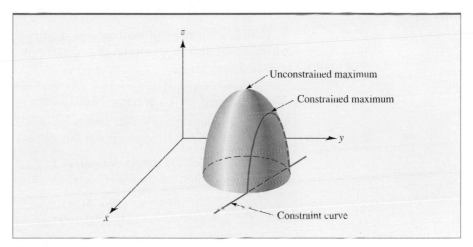

FIGURE 7.21   Constrained and unconstrained extrema.

You have already seen some constrained optimization problems in Chapter 3. (For instance, recall Example 3.5.1 of Section 3.5.) The technique you used in Chapter 3 to solve such a problem involved reducing it to a problem of a single variable by solving the constraint equation for one of the variables and then substituting the resulting expression into the function to be optimized. The success of this technique depended on solving the constraint equation for one of the variables, which is often difficult or even impossible to do in practice. In this section, you will see a more versatile technique called the **method of Lagrange multipliers,** in which the introduction of a *third* variable (the multiplier) enables you to solve constrained optimization problems without first solving the constraint equation for one of the variables.

More specifically, the method of Lagrange multipliers uses the fact that any relative extremum of the function $f(x, y)$ subject to the constraint $g(x, y) = k$ must occur at a critical point (a, b) of the function

$$F(x, y) = f(x, y) - \lambda[g(x, y) - k]$$

where λ is a new variable (the **Lagrange multiplier**). To find the critical points of F, compute its partial derivatives

$$F_x = f_x - \lambda g_x \qquad F_y = f_y - \lambda g_y \qquad F_\lambda = -(g - k)$$

and solve the equations $F_x = 0$, $F_y = 0$, and $F_\lambda = 0$ simultaneously, as follows:

$$F_x = f_x - \lambda g_x = 0 \qquad \text{or} \qquad f_x = \lambda g_x$$
$$F_y = f_y - \lambda g_y = 0 \qquad \text{or} \qquad f_y = \lambda g_y$$
$$F_\lambda = -(g - k) = 0 \qquad \text{or} \qquad g = k$$

Finally, evaluate $f(a, b)$ at each critical point (a, b) of F.

> **NOTE** The method of Lagrange multipliers tells you only that any constrained extrema must occur at critical points of the function $F(x, y)$. The method cannot be used to show that constrained extrema exist or to determine whether any particular critical point (a, b) corresponds to a constrained maximum, a minimum, or neither. However, *for the functions considered in this text, you can assume that if f has a constrained maximum (minimum) value, it will be given by the largest (smallest) of the critical values $f(a, b)$.* ∎

Here is a summary of the Lagrange multiplier procedure for finding the largest and smallest values of a function of two variables subject to a constraint.

The Method of Lagrange Multipliers

Step 1. (*Formulation*) Find the largest (or smallest) value of $f(x, y)$ subject to the constraint $g(x, y) = k$, assuming that this extreme value exists.

Step 2. Compute the partial derivatives f_x, f_y, g_x, and g_y, and find all numbers $x = a$, $y = b$, and λ that satisfy the system of equations

$$f_x(a, b) = \lambda g_x(a, b)$$
$$f_y(a, b) = \lambda g_y(a, b)$$
$$g(a, b) = k$$

These are the *Lagrange equations*.

Step 3. Evaluate f at each point (a, b) that satisfies the system of equations in step 2.

Step 4. (*Interpretation*) If $f(x, y)$ has a largest (smallest) value subject to the constraint $g(x, y) = k$, it will be the largest (smallest) of the values found in step 3.

A geometric justification of the multiplier method is given at the end of this section. In Example 7.5.1, the method is used to solve the problem from Example 3.5.1 in Section 3.5.

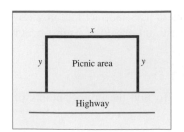

FIGURE 7.22 Rectangular picnic area.

EXAMPLE | 7.5.1

The highway department is planning to build a picnic area for motorists along a major highway. It is to be rectangular with an area of 5,000 square yards and is to be fenced off on the three sides not adjacent to the highway. What is the least amount of fencing that will be needed to complete the job?

Solution

Label the sides of the picnic area as indicated in Figure 7.22 and let f denote the amount of fencing required. Then,

$$f(x, y) = x + 2y$$

The goal is to minimize f given the requirement that the area must be 5,000 square yards; that is, subject to the constraint

$$g(x, y) = xy = 5,000$$

Find the partial derivatives

$$f_x = 1 \qquad f_y = 2 \qquad g_x = y \qquad \text{and} \qquad g_y = x$$

and obtain the three Lagrange equations

$$1 = \lambda y \qquad 2 = \lambda x \qquad \text{and} \qquad xy = 5{,}000$$

From the first and second equations you get

$$\lambda = \frac{1}{y} \qquad \text{and} \qquad \lambda = \frac{2}{x}$$

(since $y \neq 0$ and $x \neq 0$), which implies that

$$\frac{1}{y} = \frac{2}{x} \qquad \text{or} \qquad x = 2y$$

Now substitute $x = 2y$ into the third Lagrange equation to get

$$2y^2 = 5{,}000 \qquad \text{or} \qquad y = \pm 50$$

and use $y = 50$ in the equation $x = 2y$ to get $x = 100$. Thus, $x = 100$ and $y = 50$ are the values that minimize the function $f(x, y) = x + 2y$ subject to the constraint $xy = 5{,}000$. The optimal picnic area is 100 yards wide (along the highway), extends 50 yards back from the road, and requires $100 + 50 + 50 = 200$ yards of fencing.

EXAMPLE 7.5.2

Find the maximum and minimum values of the function $f(x, y) = xy$ subject to the constraint $x^2 + y^2 = 8$.

Solution

Let $g(x, y) = x^2 + y^2$ and use the partial derivatives

$$f_x = y \qquad f_y = x \qquad g_x = 2x \qquad \text{and} \qquad g_y = 2y$$

to get the three Lagrange equations

$$y = 2\lambda x \qquad x = 2\lambda y \qquad \text{and} \qquad x^2 + y^2 = 8$$

Neither x nor y can be zero if all three of these equations are to hold (do you see why?), and so you can rewrite the first two equations as

$$2\lambda = \frac{y}{x} \qquad \text{and} \qquad 2\lambda = \frac{x}{y}$$

which implies that

$$\frac{y}{x} = \frac{x}{y} \qquad \text{or} \qquad x^2 = y^2$$

Now substitute $x^2 = y^2$ into the third equation to get

$$2x^2 = 8 \qquad \text{or} \qquad x = \pm 2$$

If $x = 2$, it follows from the equation $x^2 = y^2$ that $y = 2$ or $y = -2$. Similarly, if $x = -2$, it follows that $y = 2$ or $y = -2$. Hence, the four points at which the constrained extrema can occur are $(2, 2)$, $(2, -2)$, $(-2, 2)$, and $(-2, -2)$. Since

$$f(2, 2) = f(-2, -2) = 4 \qquad \text{and} \qquad f(2, -2) = f(-2, 2) = -4$$

it follows that when $x^2 + y^2 = 8$, the maximum value of $f(x, y)$ is 4, which occurs at the points $(2, 2)$ and $(-2, -2)$, and the minimum value is -4, which occurs at $(2, -2)$ and $(-2, 2)$.

For practice, check these answers by solving the optimization problem using the methods of Chapter 3.

> **NOTE**   In Examples 7.5.1 and 7.5.2, the first two Lagrange equations were used to eliminate the new variable λ, and then the resulting expression relating x and y was substituted into the constraint equation. For most constrained optimization problems you encounter, this particular sequence of steps will often lead quickly to the desired solution.   ■

Lagrange Multipliers for Functions of Three Variables

The method of Lagrange multipliers can be extended to constrained optimization problems involving functions of more than two variables and more than one constraint. For instance, to optimize $f(x, y, z)$ subject to the constraint $g(x, y, z) = k$, you solve

$$f_x = \lambda g_x \qquad f_y = \lambda g_y \qquad f_z = \lambda g_z \qquad \text{and} \qquad g = k$$

Here is an example of a problem involving this kind of constrained optimization.

EXAMPLE | 7.5.3

A jewelry box is to be constructed of material that costs \$1 per square inch for the bottom, \$2 per square inch for the sides, and \$5 per square inch for the top. If the total volume is to be 96 in.^{3}, what dimensions will minimize the total cost of construction?

Solution

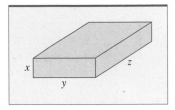

Let the box be x inches deep, y inches long, and z inches wide where x, y, and z are all positive, as indicated in the accompanying figure. Then the volume of the box is $V = xyz$ and the total cost of construction is given by

$$C = \underbrace{1yz}_{bottom} + \underbrace{2(2xy + 2xz)}_{sides} + \underbrace{5yz}_{top} = 6yz + 4xy + 4xz$$

You wish to minimize $C = 6yz + 4xy + 4xz$ subject to $V = xyz = 96$. The Lagrange equations are

$$C_x = \lambda V_x \qquad \text{or} \qquad 4y + 4z = \lambda(yz)$$
$$C_y = \lambda V_y \qquad \text{or} \qquad 6z + 4x = \lambda(xz)$$
$$C_z = \lambda V_z \qquad \text{or} \qquad 6y + 4x = \lambda(xy)$$

and $xyz = 96$. Solving each of the first three equations for λ, you get

$$\frac{4y + 4z}{yz} = \frac{6z + 4x}{xz} = \frac{6y + 4x}{xy} = \lambda$$

By multiplying each expression by xyz, you obtain

$$4xy + 4xz = 6yz + 4yx$$
$$4xy + 4xz = 6yz + 4xz$$
$$6yz + 4yx = 6yz + 4xz$$

which can be further simplified by canceling common terms on both sides of each equation to get

$$4xz = 6yz$$
$$4xy = 6yz$$
$$4yx = 4xz$$

By dividing z from both sides of the first equation, y from the second, and x from the third, you obtain

$$4x = 6y \quad \text{and} \quad 4x = 6z \quad \text{and} \quad 4y = 4z$$

so that $y = \dfrac{2}{3}x$ and $z = \dfrac{2}{3}x$. Substituting these values into the constraint equation $xyz = 96$, you first find that

$$x\left(\frac{2}{3}x\right)\left(\frac{2}{3}x\right) = 96$$

$$\frac{4}{9}x^3 = 96$$

$$x^3 = 216 \quad \text{so} \quad x = 6$$

and then

$$y = z = \frac{2}{3}(6) = 4$$

Thus, the minimal cost occurs when the jewelry box is 6 inches deep with a square base, 4 inches on a side.

Maximization of Utility

A *utility function* $U(x, y)$ measures the total satisfaction or *utility* a consumer receives from having x units of one particular commodity and y units of another. Example 7.5.4 illustrates how the method of Lagrange multipliers can be used to determine how many units of each commodity the consumer should purchase to maximize utility while staying within a fixed budget.

EXAMPLE | 7.5.4

A consumer has $600 to spend on two commodities, the first of which costs $20 per unit and the second $30 per unit. Suppose that the utility derived by the consumer from x units of the first commodity and y units of the second commodity is given by the **Cobb-Douglas utility function** $U(x, y) = 10x^{0.6}y^{0.4}$. How many units of each commodity should the consumer buy to maximize utility?

Solution

The total cost of buying x units of the first commodity at $20 per unit and y units of the second commodity at $30 per unit is $20x + 30y$. Since the consumer has only $600 to spend, the goal is to maximize utility $U(x, y)$ subject to the budgetary constraint $20x + 30y = 600$.

The three Lagrange equations are

$$6x^{-0.4}y^{0.4} = 20\lambda \qquad 4x^{0.6}y^{-0.6} = 30\lambda \qquad \text{and} \qquad 20x + 30y = 600$$

From the first two equations you get

$$\frac{6x^{-0.4}y^{0.4}}{20} = \frac{4x^{0.6}y^{-0.6}}{30} = \lambda$$

$$9x^{-0.4}y^{0.4} = 4x^{0.6}y^{-0.6}$$

$$9y = 4x \qquad \text{or} \qquad y = \frac{4}{9}x$$

Substituting $y = \frac{4}{9}x$ into the third Lagrange equation, you get

$$20x + 30\left(\frac{4}{9}x\right) = 600$$

$$\left(\frac{300}{9}\right)x = 600$$

so that

$$x = 18 \qquad \text{and} \qquad y = \frac{4}{9}(18) = 8$$

That is, to maximize utility, the consumer should buy 18 units of the first commodity and 8 units of the second.

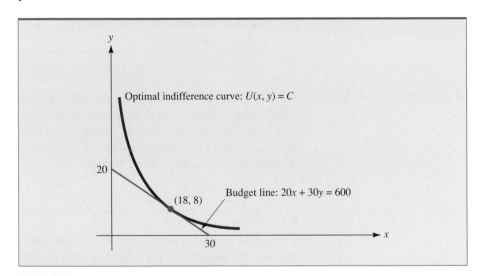

FIGURE 7.23 Budgetary constraint and optimal indifference curve.

Recall from Section 7.1 that the level curves of a utility function are known as *indifference curves*. A graph showing the relationship between the optimal indifference curve $U(x, y) = C$, where $C = U(18, 8)$ and the budgetary constraint $20x + 30y = 600$, is sketched in Figure 7.23.

Allocation of Resources   An important class of problems in business and economics involves determining an optimal allocation of resources subject to a constraint on those resources. Here is an example in which sales are maximized subject to a budgetary constraint.

EXAMPLE | 7.5.5

An editor has been allotted $60,000 to spend on the development and promotion of a new book. It is estimated that if x thousand dollars is spent on development and y thousand on promotion, approximately $f(x, y) = 20x^{3/2}y$ copies of the book will be sold. How much money should the editor allocate to development and how much to promotion in order to maximize sales?

Solution

The goal is to maximize the function $f(x, y) = 20x^{3/2}y$ subject to the constraint $g(x, y) = 60$, where $g(x, y) = x + y$. The corresponding Lagrange equations are

$$30x^{1/2}y = \lambda \qquad 20x^{3/2} = \lambda \qquad \text{and} \qquad x + y = 60$$

From the first two equations you get

$$30x^{1/2}y = 20x^{3/2}$$

Since the maximum value of f clearly does not occur when $x = 0$, you may assume that $x \neq 0$ and divide both sides of this equation by $30x^{1/2}$ to obtain

$$y = \frac{2}{3}x$$

Substituting this expression into the third Lagrange equation, you get

$$x + \frac{2}{3}x = 60 \qquad \text{or} \qquad \frac{5}{3}x = 60$$

from which it follows that

$$x = 36 \qquad \text{and} \qquad y = \frac{2}{3}(36) = 24$$

That is, to maximize sales, the editor should spend $36,000 on development and $24,000 on promotion. If this is done, approximately $f(36, 24) = 103,680$ copies of the book will be sold.

A graph showing the relationship between the budgetary constraint and the level curve for optimal sales is shown in Figure 7.24.

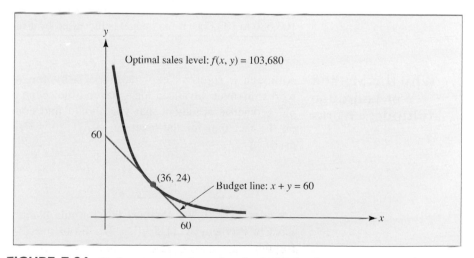

FIGURE 7.24   Budgetary constraint and optimal sales level.

The Significance of the Lagrange Multiplier λ

Usually, solving a constrained optimization problem by the method of Lagrange multipliers does not require actually finding a numerical value for the *Lagrange multiplier λ*. In some problems, however, you may want to compute λ, which has this useful interpretation.

The Lagrange Multiplier ▪ Suppose M is the maximum (or minimum) value of $f(x, y)$, subject to the constraint $g(x, y) = k$. The Lagrange multiplier λ is the rate of change of M with respect to k. That is,

$$\lambda = \frac{dM}{dk}$$

Hence,

$\lambda \approx$ change in M resulting from a 1-unit increase in k

---

EXAMPLE | 7.5.6

Suppose the editor in Example 7.5.5 is allotted $61,000 instead of $60,000 to spend on the development and promotion of the new book. Estimate how the additional $1,000 will affect the maximum sales level.

Solution

In Example 7.5.5, in order to find the maximum value M of $f(x, y) = 20x^{3/2}y$ subject to the constraint $x + y = 60$, you solved the three Lagrange equations

$$30x^{1/2}y = \lambda \qquad 20x^{3/2} = \lambda \qquad \text{and} \qquad x + y = 60$$

to obtain $x = 36$ and $y = 24$. The multiplier λ can then be found by substituting these values of x and y into the first or second Lagrange equation. Using the second equation you get

$$\lambda = 20(36)^{3/2} = 4,320$$

which means that maximal sales will increase by approximately 4,320 copies (from 103,680 to 108,000) if the budget is increased by 1 unit, from 60 to 61 thousand dollars.

Why the Method of Lagrange Multipliers Works

Although a rigorous explanation of why the method of Lagrange multipliers works involves advanced ideas beyond the scope of this text, there is a rather simple geometric argument that you should find convincing. This argument depends on the fact that for the level curve $F(x, y) = C$, the slope at each point (x, y) is given by

$$\frac{dy}{dx} = -\frac{F_x}{F_y}$$

provided $F_y \neq 0$. We obtained this formula using the chain rule for partial derivatives in Problem 72 in Section 7.2. Problems 52 and 53 illustrate the use of this formula.

Now, consider the constrained optimization problem:

Maximize $f(x, y)$ subject to $g(x, y) = k$

Geometrically, this means you must find the highest level curve of f that intersects the constraint curve $g(x, y) = k$. As Figure 7.25 suggests, the critical intersection will occur at a point where the constraint curve is tangent to a level curve; that is, where the slope of the constraint curve $g(x, y) = k$ is equal to the slope of a level curve $f(x, y) = C$.

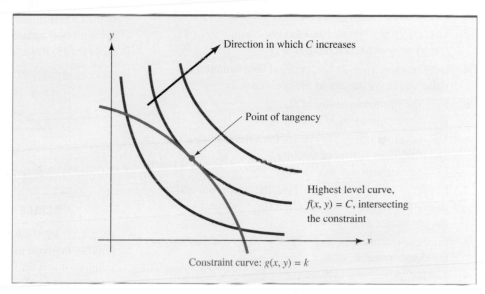

FIGURE 7.25 Increasing level curves and the constraint curve.

According to the formula stated at the beginning of this discussion, you have

Slope of constraint curve = Slope of level curve

$$-\frac{g_x}{g_y} = -\frac{f_x}{f_y}$$

or, equivalently,

$$\frac{f_x}{g_x} = \frac{f_y}{g_y}$$

If you let λ denote this common ratio, then

$$\frac{f_x}{g_x} = \lambda \quad \text{and} \quad \frac{f_y}{g_y} = \lambda$$

from which you get the first two Lagrange equations

$$f_x = \lambda g_x \quad \text{and} \quad f_y = \lambda g_y$$

The third Lagrange equation

$$g(x, y) = k$$

is simply a statement of the fact that the point of tangency actually lies on the constraint curve.

PROBLEMS 7.5

In Problems 1 through 16, use the method of Lagrange multipliers to find the indicated extremum. You may assume the extremum exists.

1. Find the maximum value of $f(x, y) = xy$ subject to the constraint $x + y = 1$.

2. Find the maximum and minimum values of the function $f(x, y) = xy$ subject to the constraint $x^2 + y^2 = 1$.

3. Let $f(x, y) = x^2 + y^2$. Find the minimum value of $f(x, y)$ subject to the constraint $xy = 1$.

4. Let $f(x, y) = x^2 + 2y^2 - xy$. Find the minimum value of $f(x, y)$ subject to the constraint $2x + y = 22$.

5. Find the minimum value of $f(x, y) = x^2 - y^2$ subject to the constraint $x^2 + y^2 = 4$.

6. Let $f(x, y) = 8x^2 - 24xy + y^2$. Find the maximum and minimum values of the function $f(x, y)$ subject to the constraint $8x^2 + y^2 = 1$.

7. Let $f(x, y) = x^2 - y^2 - 2y$. Find the maximum and minimum values of the function $f(x, y)$ subject to the constraint $x^2 + y^2 = 1$.

8. Find the maximum value of $f(x, y) = xy^2$ subject to the constraint $x + y^2 = 1$.

9. Let $f(x, y) = 2x^2 + 4y^2 - 3xy - 2x - 23y + 3$. Find the minimum value of the function $f(x, y)$ subject to the constraint $x + y = 15$.

10. Let $f(x, y) = 2x^2 + y^2 + 2xy + 4x + 2y + 7$. Find the minimum value of the function $f(x, y)$ subject to the constraint $4x^2 + 4xy = 1$.

11. Find the maximum and minimum values of $f(x, y) = e^{xy}$ subject to $x^2 + y^2 = 4$.

12. Find the maximum value of $f(x, y) = \ln(xy^2)$ subject to $2x^2 + 3y^2 = 8$ for $x > 0$ and $y > 0$.

13. Find the maximum value of $f(x, y, z) = xyz$ subject to $x + 2y + 3z = 24$.

14. Find the maximum and minimum values of $f(x, y, z) = x + 3y - z$ subject to $z = 2x^2 + y^2$.

15. Let $f(x, y, z) = x + 2y + 3z$. Find the maximum and minimum values of $f(x, y, z)$ subject to the constraint $x^2 + y^2 + z^2 = 16$.

16. Find the minimum value of $f(x, y, z) = x^2 + y^2 + z^2$ subject to $4x^2 + 2y^2 + z^2 = 4$.

17. **CONSTRUCTION** A farmer wishes to fence off a rectangular pasture along the bank of a river. The area of the pasture is to be 3,200 square meters, and no fencing is needed along the river bank. Find the dimensions of the pasture that will require the least amount of fencing.

18. **CONSTRUCTION** There are 320 meters of fencing available to enclose a rectangular field. How should the fencing be used so that the enclosed area is as large as possible?

19. **POSTAL PACKAGING** According to postal regulations, the girth plus length of parcels sent by fourth-class mail may not exceed 108 inches. What is the largest possible volume of a rectangular parcel with two square sides that can be sent by fourth-class mail? (Refer to the accompanying figure.)

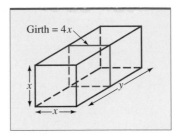

Girth = $4x$

PROBLEM 19

20. **POSTAL PACKAGING** According to the postal regulation given in Problem 19, what is the largest volume of a cylindrical can that can be sent by fourth-class mail? (A cylinder of radius R and length H has volume $\pi R^2 H$.)

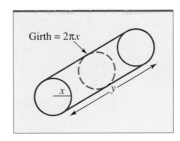

Girth = $2\pi x$

PROBLEM 20

21. **PACKAGING** Use the fact that 12 fluid ounces is approximately 6.89π cubic inches to find the dimensions of the 12-ounce soda can that can be constructed using the least amount of metal. (Recall that the volume of a cylinder of radius r and height h is $\pi r^2 h$, and that a circle of radius r has area πr^2 and circumference $2\pi r$.)

22. **PACKAGING** A cylindrical can is to hold 4π cubic inches of frozen orange juice. The cost per square inch of constructing the metal top and bottom is twice the cost per square inch of constructing the cardboard side. What are the dimensions of the least expensive can? (See the measurement information in Problem 21.)

23. **ALLOCATION OF FUNDS**   A manufacturer has $8,000 to spend on the development and promotion of a new product. It is estimated that if x thousand dollars is spent on development and y thousand is spent on promotion, sales will be approximately $f(x, y) = 50x^{1/2}y^{3/2}$ units. How much money should the manufacturer allocate to development and how much to promotion to maximize sales?

24. **ALLOCATION OF FUNDS**   If x thousand dollars is spent on labor and y thousand dollars is spent on equipment, the output at a certain factory will be $Q(x, y) = 60x^{1/3}y^{2/3}$ units. If $120,000 is available, how should this be allocated between labor and equipment to generate the largest possible output?

25. **MARGINAL ANALYSIS**   Use the Lagrange multiplier λ to estimate the change in the maximum output of the factory in Problem 24 that will result if the money available for labor and equipment is increased by $1,000.

26. **SURFACE AREA OF THE HUMAN BODY**   Recall from Problem 35 of Section 7.1 that an empirical formula for the surface area of a person's body is

$$S(W, H) = 0.0072W^{0.425}H^{0.725}$$

where W (kg) is the person's weight and H (cm) is his or her height. Suppose for a short period of time, Maria's weight adjusts as she grows taller so that $W + H = 160$. With this constraint, what height and weight will maximize the surface area of Maria's body?

In Problems 27 and 28, you will need to know that a closed cylinder of radius R and length L has volume $V = \pi R^2 L$ and surface area $S = 2\pi RL + 2\pi R^2$. The volume of a hemisphere of radius R is $V = \dfrac{2}{3}\pi R^3$ and its surface area is $S = 2\pi R^2$.

27. **MICROBIOLOGY**   A bacterium is shaped like a cylindrical rod. If the volume of the bacterium is fixed, what relationship between the radius R and length H of the bacterium will result in minimum surface area?

PROBLEM 27   **PROBLEM 28**

28. **MICROBIOLOGY**   A bacterium is shaped like a cylindrical rod with two hemispherical "caps" on the ends. If the volume of the bacterium is fixed, what must be true about its radius R and length L to achieve minimum surface area?

29. **OPTICS**   The thin lens formula in optics says that the focal length L of a thin lens is related to the object distance d_o and image distance d_i by the equation

$$\frac{1}{d_o} + \frac{1}{d_i} = \frac{1}{L}$$

If L remains constant while d_o and d_i are allowed to vary, what is the maximum distance $s = d_o + d_i$ between the object and the image?

30. **CONSTRUCTION**   A jewelry box is constructed by partitioning a box with a square base as shown in the accompanying figure. If the box is designed to have volume 800 cm³, what dimensions should it have to minimize its total surface area (top, bottom, sides, and interior partitions)? Notice that we have said nothing about where the partitions are located. Does it matter?

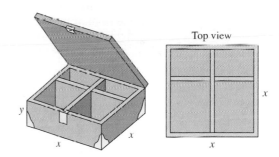

Top view

PROBLEM 30

31. **CONSTRUCTION**   Suppose the jewelry box in Problem 30 is designed so that the material in the top costs twice as much as the material in the bottom and sides and three times as much as the material in the interior partitions. What dimensions minimize the total cost of constructing the box.

32. **SPY STORY**   Having disposed of Scélérat's gunmen (Problem 62 in Section 6.3), the spy goes looking for his enemy. He enters a room and the door slams behind him. Immediately, he begins to feel warm, and too late, he realizes he is trapped inside Scélérat's dreaded broiler room. Searching desperately for a way to survive, he notices that the room is shaped like the circle $x^2 + y^2 = 60$ and that he is standing at the center $(0, 0)$. He presses the stem on

his special heat-detecting wristwatch and sees that the temperature at each point (x, y) in the room is given by

$$T(x, y) = x^2 + y^2 + 3xy + 5x + 15y + 130$$

From an informant's report, he knows that somewhere in the walls of this room there is a trap door leading outside the castle, and he reasons that it must be located at the coolest point. Where is it? Just how cool will the spy be when he gets there?

33. **PARTICLE PHYSICS**   A particle of mass m in a rectangular box with dimensions x, y, and z has ground state energy

$$E(x, y, z) = \frac{k^2}{8m}\left(\frac{1}{x^2} + \frac{1}{y^2} + \frac{1}{z^2}\right)$$

where k is a physical constant. In Problem 29, Section 7.3, you were asked to minimize the ground state energy subject to the fixed volume constraint $V_0 = xyz$ using substitution. Solve the same constrained optimization problem using the method of Lagrange multipliers.

34. **CONSTRUCTION**   A rectangular building is to be constructed of material that costs $31 per square foot for the roof, $27 per square foot for the sides and the back, and $55 per square foot for the facing and glass used in constructing the front. If the building is to have a volume of $16,000 \text{ ft}^3$, what dimensions will minimize the total cost of construction?

35. **CONSTRUCTION**   A storage shed is to be constructed of material that costs $15 per square foot for the roof, $12 per square foot for the two sides and back, and $20 per square foot for the front. What are the dimensions of the largest shed (in volume) that can be constructed for $8,000?

36. **ALLOCATION OF FUNDS**   A manufacturer is planning to sell a new product at the price of $150 per unit and estimates that if x thousand dollars is spent on development and y thousand dollars is spent on promotion, approximately $\dfrac{320y}{y + 2} + \dfrac{160x}{x + 4}$ units of the product will be sold. The cost of manufacturing the product is $50 per unit. If the manufacturer has a total of $8,000 to spend on development and promotion, how should this money be allocated to generate the largest possible profit? [*Hint:* Profit = (number of units)(price per unit − cost per unit) − amount spent on development and production.]

37. **MARGINAL ANALYSIS**   Suppose the manufacturer in Problem 36 decides to spend $8,100 instead

of $8,000 on the development and promotion of the new product. Use the Lagrange multiplier λ to estimate how this change will affect the maximum possible profit.

38. **ALLOCATION OF UNRESTRICTED FUNDS**
    a. If unlimited funds are available, how much should the manufacturer in Problem 36 spend on development and how much on promotion in order to generate the largest possible profit? [*Hint:* Use the methods of Section 7.3.]
    b. Suppose the allocation problem in part (a) is solved by the method of Lagrange multipliers. What is the value of λ that corresponds to the optimal budget? Interpret your answer in terms of dM/dk.
    c. Your answer to part (b) should suggest another method for solving the problem in part (a). Solve the problem using this new method.

39. **UTILITY**   A consumer has $280 to spend on two commodities, the first of which costs $2 per unit and the second $5 per unit. Suppose that the utility derived by the consumer from x units of the first commodity and y units of the second is given by $U(x, y) = 100x^{0.25}y^{0.75}$.
    a. How many units of each commodity should the consumer buy to maximize utility?
    b. Compute the Lagrange multiplier λ and interpret in economic terms. (In the context of maximizing utility, λ is called the **marginal utility of money.**)

40. **UTILITY**   A consumer has k dollars to spend on two commodities, the first of which costs a dollars per unit and the second b dollars per unit. Suppose that the utility derived by the consumer from x units of the first commodity and y units of the second commodity is given by the Cobb-Douglas utility function $U(x, y) = x^\alpha y^\beta$, where $0 < \alpha < 1$ and $\alpha + \beta = 1$. Show that utility is maximized when

$$x = \frac{k\alpha}{a} \text{ and } y = \frac{k\beta}{b}.$$

41. In Problem 40, how does the maximum utility change if k is increased by 1 dollar?

In Problems 42 through 44, let $Q(x, y)$ be a production function, where x and y represent units of labor and capital, respectively. If unit costs of labor and capital are given by p and q, respectively, then $px + qy$ represents the total cost of production.

42. MINIMUM COST Use Lagrange multipliers to show that subject to a fixed production level c, the total cost is minimized when

$$\frac{Q_x}{p} = \frac{Q_y}{q} \quad \text{and} \quad Q(x, y) = c$$

provided Q_x and Q_y are not both 0 and $p \neq 0$ and $q \neq 0$. (This is often referred to as the **minimum cost problem,** and its solution is called the **least-cost combination of inputs.**)

43. FIXED BUDGET Show that the inputs x and y that maximize the production level $Q(x, y)$ subject to a fixed cost k satisfy

$$\frac{Q_x}{p} = \frac{Q_y}{q} \quad \text{with } px + qy = k$$

(Assume that neither p nor q is 0.) This is called a **fixed budget problem.**

44. MINIMUM COST Show that subject to the fixed production level $Ax^\alpha y^\beta = k$, where k is a constant and α and β are positive with $\alpha + \beta = 1$, the joint cost function $C(x, y) = px + qy$ is minimized when

$$x = \frac{k}{A}\left(\frac{\alpha q}{\beta p}\right)^\beta, y = \frac{k}{A}\left(\frac{\beta p}{\alpha q}\right)^\alpha$$

CES PRODUCTION *A constant elasticity of substitution (CES) production function is one with the general form*

$$Q(K, L) = A[\alpha K^{-\beta} + (1 - \alpha)L^{-\beta}]^{-1/\beta}$$

where K is capital expenditure; L is the level of labor; and A, α, and β are constants that satisfy $A > 0, 0 < \alpha < 1$, and $\beta > -1$. Problems 45 through 47 are concerned with such production functions.

---

45. Use the method of Lagrange multipliers to maximize the CES production function

$$Q = 55[0.6K^{-1/4} + 0.4L^{-1/4}]^{-4}$$

subject to the constraint

$$2K + 5L = 150$$

46. Use the method of Lagrange multipliers to maximize the CES production function

$$Q = 50[0.3K^{-1/5} + 0.7L^{-1/5}]^{-5}$$

subject to the constraint

$$5K + 2L = 140$$

47. Show that a CES production function

$$Q(K, L) = A[\alpha K^{-\beta} + (1 - \alpha)L^{-\beta}]^{-1/\beta}$$

satisfies $Q(sK, sL) = sQ(K, L)$ for any constant s. In other words, such a function has *constant returns to scale.* (Recall that you showed that Cobb-Douglas production functions have this property in Problem 51 of Section 7.1.)

48. SATELLITE CONSTRUCTION A space probe has the shape of the surface

$$4x^2 + y^2 + 4z^2 = 16$$

where x, y, and z are in feet. When it reenters the earth's atmosphere, the probe begins to heat up in such a way that the temperature at each point $P(x, y, z)$ on the probe's surface is given by

$$T(x, y, z) = 8x^2 + 4yz - 16z + 600$$

where T is in degrees Fahrenheit. Use the method of Lagrange multipliers to find the hottest and coolest points on the probe's surface. What are the extreme temperatures?

49. Use Lagrange multipliers to find the possible maximum or minimum points on that part of the surface $z - x \quad y$ for which $y = x^5 + x - 2$. Then use your calculator to sketch the curve $y = x^5 + x - 2$ and the level curves to the surface $f(x, y) = x - y$ and show that the points you have just found do not represent relative maxima or minima. What do you conclude from this observation?

50. HAZARDOUS WASTE MANAGEMENT A study conducted at a waste disposal site reveals soil contamination over a region that may be described roughly as the interior of the ellipse

$$\frac{x^2}{4} + \frac{y^2}{9} = 1$$

where x and y are in miles. The manager of the site plans to build a circular enclosure to contain all polluted territory.

 a. If the office at the site is at the point $S(1, 1)$, what is the radius of the smallest circle centered at S that contains the entire contaminated region? [*Hint:* The function

$$f(x, y) = (x - 1)^2 + (y - 1)^2$$

measures the square of the distance from $S(1, 1)$ to the point $P(x, y)$. The required radius can be found by maximizing $f(x, y)$ subject to a certain constraint.]

b. Read an article on waste management, and write a paragraph on how management decisions are made regarding landfills and other disposal sites.*

51. **MARGINAL ANALYSIS**   Let $P(K, L)$ be a production function, where K and L represent the capital and labor required for a certain manufacturing procedure. Suppose we wish to maximize $P(K, L)$ subject to a cost constraint, $C(K, L) = A$, for constant A. Use the method of Lagrange multipliers to show that optimal production is attained when

$$\frac{\frac{\partial P}{\partial K}}{\frac{\partial C}{\partial K}} = \frac{\frac{\partial P}{\partial L}}{\frac{\partial C}{\partial L}}$$

that is, when the ratio of marginal production from capital to the marginal cost of capital equals the ratio of marginal production of labor to the marginal cost of labor.

52. Let $F(x, y) = x^2 + 2xy - y^2$.
   a. If $F(x, y) = k$ for constant k, use the method of implicit differentiation developed in Chapter 2 to find $\dfrac{dy}{dx}$.

---

*An excellent case study may be found in M. D. LaGrega, P. L. Buckingham, and J. C. Evans, *Hazardous Waste Management,* New York: McGraw-Hill, 1994, pp. 946–955.

   b. Find the partial derivatives F_x and F_y and verify that

$$\frac{dy}{dx} = -\frac{F_x}{F_y}$$

53. Repeat Problem 52 for the function

$$F(x, y) = xe^{xy^2} + \frac{y}{x} + x \ln (x + y)$$

 In Problems 54 through 57, use the method of Lagrange multipliers to find the indicated maximum or minimum. You will need to use the graphing utility or the solve application on your calculator.

54. Maximize $f(x, y) = e^{x+y} - x \ln \left(\dfrac{y}{x}\right)$ subject to $x + y = 4$.

55. Minimize $f(x, y) = \ln (x + 2y)$ subject to $xy + y = 5$.

56. Minimize $f(x, y) = \dfrac{1}{x^2} + \dfrac{3}{xy} + \dfrac{1}{y^2}$ subject to $x + 2y = 7$.

57. Maximize $f(x, y) = xe^{x^2-y}$ subject to $x^2 + 2y^2 = 1$.

SECTION 7.6 Double Integrals

In Chapters 5 and 6, you integrated a function of one variable $f(x)$ by reversing the process of differentiation, and a similar procedure can be used to integrate a function of two variables $f(x, y)$. However, since two variables are involved, we shall integrate $f(x, y)$ by holding one variable fixed and integrating with respect to the other.

For instance, to evaluate the partial integral $\displaystyle\int_1^2 xy^2 dx$ you would integrate with respect to x, using the fundamental theorem of calculus with y held constant:

$$\int_1^2 xy^2 dx = \frac{1}{2}x^2y^2 \Big|_{x=1}^{x=2}$$

$$= \left[\frac{1}{2}(2)^2y^2\right] - \left[\frac{1}{2}(1)^2y^2\right] = \frac{3}{2}y^2$$

Similarly, to evaluate $\int_{-1}^{1} xy^2\,dy$, you integrate with respect to y, holding x constant:

$$\int_{-1}^{1} xy^2\,dy = x\left(\frac{1}{3}y^3\right)\Big|_{y=-1}^{y=1}$$

$$= \left[x\left(\frac{1}{3}(1)^3\right)\right] - \left[x\left(\frac{1}{3}(-1)^3\right)\right] = \frac{2}{3}x$$

In general, partially integrating a function $f(x, y)$ with respect to x results in a function of y alone, which can then be integrated as a function of a single variable, thus producing what we call an **iterated integral** $\int\left[\int f(x, y)\,dx\right]dy$. Similarly, the iterated integral $\int\left[\int f(x, y)\,dy\right]dx$ is obtained by first integrating with respect to y, holding x constant, and then with respect to x. Returning to our example, you find,

$$\int_{-1}^{1}\left(\int_{1}^{2} xy^2\,dx\right)dy = \int_{-1}^{1}\frac{3}{2}y^2\,dy = \frac{1}{2}y^3\Big|_{y=-1}^{y=1} = 1$$

and

$$\int_{1}^{2}\left(\int_{-1}^{1} xy^2\,dy\right)dx = \int_{1}^{2}\frac{2}{3}x\,dx = \frac{1}{3}x^2\Big|_{x=1}^{x=2} = 1$$

In our example, the two iterated integrals turned out to have the same value, and you can assume this will be true for all iterated integrals considered in this text. The double integral of $f(x, y)$ over a rectangular region in the xy plane has the following definition in terms of iterated integrals.

The Double Integral over a Rectangular Region   ■   The **double integral** $\iint_{R} f(x, y)\,dA$ over the rectangular region

$$R: a \leq x \leq b,\ c \leq y \leq d$$

is given by the common value of the two iterated integrals

$$\int_{a}^{b}\left[\int_{c}^{d} f(x, y)\,dy\right]dx \quad \text{and} \quad \int_{c}^{d}\left[\int_{a}^{b} f(x, y)\,dx\right]dy$$

that is,

$$\iint_{R} f(x, y)\,dA = \int_{a}^{b}\left[\int_{c}^{d} f(x, y)\,dy\right]dx = \int_{c}^{d}\left[\int_{a}^{b} f(x, y)\,dx\right]dy$$

Example 7.6.1 illustrates the computation of this kind of double integral.

EXAMPLE | 7.6.1

Evaluate the double integral

$$\iint_R xe^{-y}\, dA$$

where R is the rectangular region $-2 \le x \le 1$, $0 \le y \le 5$ using:

a. x integration first

b. y integration first

Solution

a. With x integration first:

$$\iint_R xe^{-y}\, dA = \int_0^5 \left(\int_{-2}^1 xe^{-y}\, dx \right) dy$$

$$= \int_0^5 \frac{1}{2} x^2 e^{-y} \Big|_{x=-2}^{x=1} dy$$

$$= \int_0^5 \frac{1}{2} e^{-y}[(1)^2 - (-2)^2]\, dy = \int_0^5 -\frac{3}{2} e^{-y} dy$$

$$= -\frac{3}{2}(-e^{-y}) \Big|_{y=0}^{y=5} = \frac{3}{2}(e^{-5} - e^0) = \frac{3}{2}(e^{-5} - 1)$$

b. Integrating with respect to y first, you get

$$\iint_R xe^{-y}\, dA = \int_{-2}^1 \left(\int_0^5 xe^{-y}\, dy \right) dx$$

$$= \int_{-2}^1 x(-e^{-y}) \Big|_{y=0}^{y=5} dx = \int_{-2}^1 [-x(e^{-5} - e^0)]\, dx$$

$$= \left[-(e^{-5} - 1)\left(\frac{1}{2} x^2 \right) \right] \Big|_{x=-2}^{x=1}$$

$$= -\frac{1}{2}(e^{-5} - 1)[(1)^2 - (-2)^2] = \frac{3}{2}(e^{-5} - 1)$$

In Example 7.6.1, the order of integration made no difference. Not only do the computations yield the same result, but the integrations are essentially of the same level of difficulty. However, sometimes the order does matter, as illustrated in Example 7.6.2.

EXAMPLE | 7.6.2

Evaluate the double integral

$$\iint_R xe^{xy}\, dA$$

where R is the rectangular region $0 \le x \le 2$, $0 \le y \le 1$.

Solution

If you evaluate the integral in the order

$$\int_0^1 \left(\int_0^2 xe^{xy}\,dx \right) dy$$

it will be necessary to use integration by parts for the inner integration:

$$u = x \qquad dv = e^{xy}\,dx$$

$$du = dx \qquad v = \frac{1}{y}e^{xy}$$

$$\int_0^2 xe^{xy}\,dx = \frac{x}{y}e^{xy}\Big|_{x=0}^{x=2} - \int_0^2 \frac{1}{y}e^{xy}\,dx$$

$$= \left(\frac{x}{y} - \frac{1}{y^2}\right)e^{xy}\Big|_{x=0}^{x=2} = \left(\frac{2}{y} - \frac{1}{y^2}\right)e^{2y} - \left(\frac{-1}{y^2}\right)$$

Then the outer integration becomes

$$\int_0^1 \left[\left(\frac{2}{y} - \frac{1}{y^2}\right)e^{2y} + \frac{1}{y^2} \right] dy$$

Now what? Any ideas?

On the other hand, if you use y integration first, both computations are easy:

$$\int_0^2 \left(\int_0^1 xe^{xy}\,dy \right) dx = \int_0^2 \frac{xe^{xy}}{x}\Big|_{y=0}^{y=1} dx$$

$$= \int_0^2 (e^x - 1)\,dx = (e^x - x)\Big|_{x=0}^{x=2}$$

$$= (e^2 - 2) - e^0 = e^2 - 3$$

Double Integrals over Nonrectangular Regions

In each of the preceding examples, the region of integration is a rectangle, but double integrals can also be defined over nonrectangular regions. Before doing so, however, we will introduce an efficient procedure for describing certain such regions in terms of inequalities.

Vertical Cross Sections

The region R shown in Figure 7.26 is bounded below by the curve $y = g_1(x)$, above by the curve $y = g_2(x)$, and on the sides by the vertical lines $x = a$ and $x = b$. This region can be described by the inequalities

$$R: a \le x \le b,\ g_1(x) \le y \le g_2(x)$$

The first inequality specifies the interval in which x must lie, while the second indicates the lower and upper bounds of the vertical cross section of R for each x in this interval. In words:

R is the region such that for each x between a and b,

y varies from $g_1(x)$ to $g_2(x)$.

This method for describing a region is illustrated in Example 7.6.3.

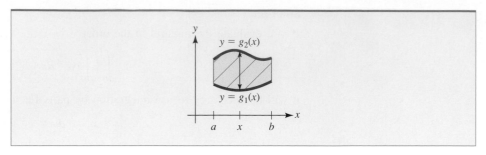

FIGURE 7.26 Vertical cross sections. The region

$$R: a \leq x \leq b, \, g_1(x) \leq y \leq g_2(x)$$

EXAMPLE | 7.6.3

Let R be the region bounded by the curve $y = x^2$ and the line $y = 2x$. Use inequalities to describe R in terms of its vertical cross sections.

Solution

Begin with a sketch of the curve and line as shown in Figure 7.27. Identify the region R, and, for reference, draw a vertical cross section. Solve the equations $y = x^2$ and $y = 2x$ simultaneously to find the points of intersection, $(0, 0)$ and $(2, 4)$. Observe that in the region R, the variable x takes on all values from $x = 0$ to $x = 2$ and that for each such value of x, the vertical cross section is bounded below by $y = x^2$ and above by $y = 2x$. Hence, R can be described by the inequalities

$$0 \leq x \leq 2 \qquad \text{and} \qquad x^2 \leq y \leq 2x$$

EXPLORE!

Graph the region over which the double integral is being evaluated in Example 7.6.3. Store $y = x^2$ into Y1 and $y = 2x$ into Y2 of the equation editor and graph using the window [−0.15, 2.2]1 by [−0.5, 4.5]1. Use the TRACE or the intersection-finding feature of your calculator to locate the intersection points of Y1 and Y2. Use the vertical line feature of your calculator to display the vertical cross sections over which the integration will be performed.

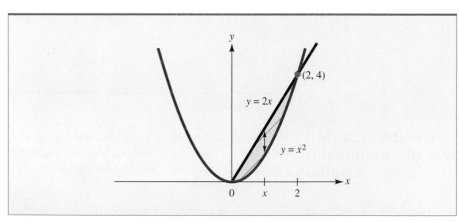

FIGURE 7.27 The region R between $y = x^2$ and $y = 2x$ described by vertical cross sections as $R: 0 \leq x \leq 2, x^2 \leq y \leq 2x$.

Horizontal Cross Sections

The region R in Figure 7.28 is bounded on the left by the curve $x = h_1(y)$, on the right by $x = h_2(y)$, below by the horizontal line $y = c$, and above by $y = d$. This region can be described by the pair of inequalities

$$R: c \leq y \leq d, \, h_1(y) \leq x \leq h_2(y)$$

The first inequality specifies the interval in which y must lie, and the second indicates the left-hand ("trailing") and right-hand ("leading") bounds of a horizontal cross section. In words:

R is the region such that for each y between c and d,

x varies from $h_1(y)$ to $h_2(y)$.

This method of description is illustrated in Example 7.6.4 for the same region described using vertical cross sections in Example 7.6.3.

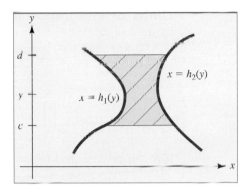

FIGURE 7.28 Horizontal cross sections: The region

$$R:\ c < y \le d,\ h_1(y) \le x \le h_2(y)$$

EXAMPLE 7.6.4

Describe the region R bounded by the curve $y = x^2$ and the line $y = 2x$ in terms of inequalities using horizontal cross sections.

Solution

As in Example 7.6.3, sketch the region and find the points of intersection of the line and curve, but this time draw a horizontal cross section (Figure 7.29).

In the region R, the variable y takes on all values from $y = 0$ to $y = 4$. For each such value of y, the horizontal cross section extends from the line $y = 2x$ on the left to the curve $y = x^2$ on the right. Since the equation of the line can be rewritten as $x = \frac{1}{2}y$ and the equation of the curve as $x = \sqrt{y}$, the inequalities describing R in terms of its horizontal cross sections are

$$0 \le y \le 4 \qquad \text{and} \qquad \frac{1}{2}y \le x \le \sqrt{y}$$

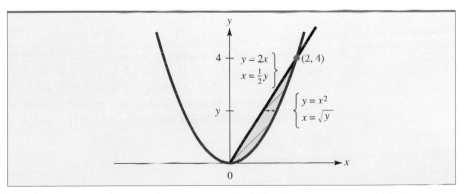

FIGURE 7.29 The region R between $y = x^2$ and $y = 2x$ described by horizontal cross sections as $R: 0 \le y \le 4,\ \frac{1}{2}y \le x \le \sqrt{y}$.

To evaluate a double integral over a region R described using either vertical or horizontal cross sections, you use an iterated integral whose limits of integration come from the inequalities describing the region. Here is a more precise description of how the limits of integration are determined.

Limits of Integration for Double Integrals ■ If R can be described by the inequalities

$$a \le x \le b \quad \text{and} \quad g_1(x) \le y \le g_2(x)$$

then

$$\iint\limits_R f(x, y)\, dA = \int_a^b \left[\int_{g_1(x)}^{g_2(x)} f(x, y)\, dy \right] dx$$

If R can be described by the inequalities

$$c \le y \le d \quad \text{and} \quad h_1(y) \le x \le h_2(y)$$

then

$$\iint\limits_R f(x, y)\, dA = \int_c^d \left[\int_{h_1(y)}^{h_2(y)} f(x, y)\, dx \right] dy$$

EXAMPLE 7.6.5

Let I be the double integral

$$I = \int_0^1 \int_0^y y^2 e^{xy}\, dx\, dy$$

a.   Sketch the region of integration and rewrite the integral with the order of integration reversed.

b.   Evaluate I using either order of integration.

Solution

a.   Comparing I with the general form for the order $dx\, dy$, we see that the region of integration is

$$R: \quad \underbrace{0 \le y \le 1}_{\substack{\text{outer limits} \\ \text{of integration}}}, \quad \underbrace{0 \le x \le y}_{\substack{\text{inner limits} \\ \text{of integration}}}$$

Thus, if y is a number in the interval $0 \le y \le 1$, the horizontal cross section of R at y extends from $x = 0$ on the left to $x = y$ on the right. The region is the triangle shown in Figure 7.30a. As shown in Figure 7.30b, the same region R can be described by taking vertical cross sections at each number x in the interval $0 \le x \le 1$ that are bounded below by $y = x$ and above by $y = 1$. Expressed in terms of inequalities, this means that

$$R: 0 \le x \le 1, \ x \le y \le 1$$

so the integral can also be written as

$$I = \int_0^1 \int_x^1 y^2 e^{xy}\, dy\, dx$$

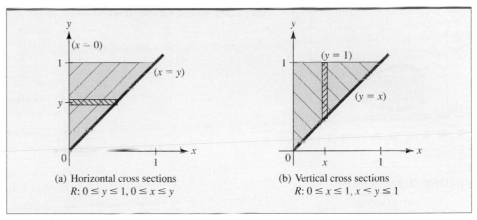

FIGURE 7.30 The region of integration for $I = \displaystyle\int_0^1 \int_0^y y^2 e^{xy} \, dx \, dy$.

b. Here is the evaluation of I using the given order of integration. For practice, you should compute I by reversing the order of integration.

$$\int_0^1 \int_0^y y^2 e^{xy} \, dx \, dy = \int_0^1 \left(y e^{xy} \Big|_{x=0}^{x=y} \right) dy \qquad since \int e^{xy} \, dx = \frac{1}{y} e^{xy}$$

$$= \int_0^1 \left(y e^{y^2} - y \right) dy$$

$$= \left(\frac{1}{2} e^{y^2} - \frac{1}{2} y^2 \right) \Big|_0^1$$

$$= \left(\frac{1}{2} e - \frac{1}{2} \right) - \left(\frac{1}{2} - 0 \right) = \frac{1}{2} e - 1$$

Applications of Double Integrals

Next, we will examine a few applications of double integrals, all of which are generalizations of familiar applications of definite integrals of functions of one variable. Specifically, we will see how double integration can be used to compute area, volume, and average value.

The Area of a Region in the Plane

The area of a region R in the xy plane can be computed as the double integral over R of the constant function $f(x, y) = 1$.

Area Formula ▪ The area of a region R in the xy plane is given by the formula

$$\text{Area of } R = \iint\limits_R 1 \, dA$$

To get a feel for why the area formula holds, consider the elementary region R shown in Figure 7.31, which is bounded above by the curve $y = g_2(x)$ and below by the curve $y = g_1(x)$, and which extends from $x = a$ to $x = b$.

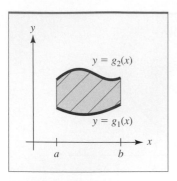

FIGURE 7.31 Area of

$$R = \iint\limits_R 1 \, dA.$$

According to the double-integral formula for area,

$$\text{Area of } R = \iint\limits_R 1 \, dA$$

$$= \int_a^b \int_{g_1(x)}^{g_2(x)} 1 \, dy \, dx$$

$$= \int_a^b \left[y \Big|_{y=g_1(x)}^{y=g_2(x)} \right] dx$$

$$= \int_a^b [g_2(x) - g_1(x)] \, dx$$

which is precisely the formula for the area between two curves that you saw in Section 5.4. Here is an example illustrating the use of the area formula.

EXAMPLE | 7.6.6

Find the area of the region R bounded by the curves $y = x^3$ and $y = x^2$.

Solution

The region is shown in Figure 7.32. Using the area formula, you get

$$\text{Area of } R = \iint\limits_R 1 \, dA = \int_0^1 \int_{x^3}^{x^2} 1 \, dy \, dx$$

$$= \int_0^1 \left(y \Big|_{y=x^3}^{y=x^2} \right) dx$$

$$= \int_0^1 (x^2 - x^3) \, dx$$

$$= \left[\frac{1}{3} x^3 - \frac{1}{4} x^4 \right] \Big|_0^1$$

$$= \frac{1}{12}$$

EXPLORE!

Find the area of the region *R* by using the numeric integration feature of your graphing utility to evaluate $\int_0^1 (x^2 - x^3) \, dx$. Explain why this gives the same answer as the method used in the solution to Example 7.6.6.

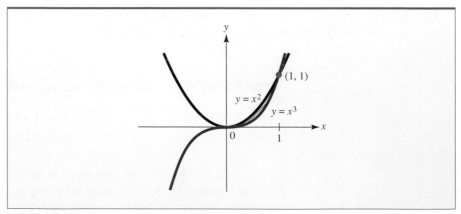

FIGURE 7.32 The region bounded by $y = x^2$ and $y = x^3$.

Volume as a
Double Integral

Recall from Section 5.3 that the region under the curve $y = f(x)$ over an interval $a \leq x < b$, where $f(x)$ is continuous and $f(x) \geq 0$, has area given by the definite integral $A = \int_a^b f(x)\,dx$. An analogous argument for a continuous, nonnegative function of two variables $f(x, y)$ yields this formula for volume as a double integral.

> **Volume as a Double Integral** ▪ If $f(x, y)$ is continuous and $f(x, y) > 0$ on the rectangular region R, then the solid region under the surface $z = f(x, y)$ over the region R has volume given by
>
> $$V = \int\int_R f(x, y)\,dA$$

EXAMPLE 7.6.7

A biomass covers the triangular bottom of a container with vertices $(0, 0)$, $(6, 0)$, and $(3, 3)$, to a depth $h(x, y) = \dfrac{x}{(y + 2)}$ at each point (x, y) in the region, where all dimensions are in centimeters. What is the total volume of the biomass?

Solution
The volume is given by the double integral $V = \int\int_R h(x, y)\,dA$, where R is the triangular region R shown in Figure 7.33.

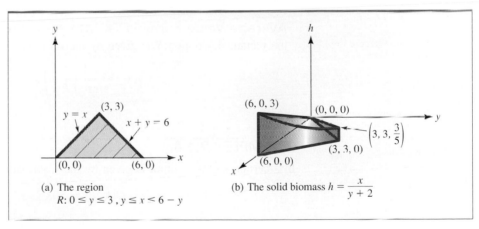

(a) The region
$R: 0 \leq y \leq 3, y \leq x < 6 - y$

(b) The solid biomass $h = \dfrac{x}{y + 2}$

FIGURE 7.33 The volume of a biomass.

Note that this region is bounded by the x axis ($y = 0$), and the lines $x = y$ and $x + y = 6$, so it can be described as

$$R: 0 \leq y \leq 3, y \leq x \leq 6 - y$$

Therefore, the volume of the biomass is given by

$$V = \int_0^3 \int_y^{6-y} \frac{x}{y + 2}\,dx\,dy$$

$$= \int_0^3 \frac{1}{y + 2}\left(\frac{x^2}{2}\right)\Bigg|_y^{6-y} dy$$

$$= \int_0^3 \frac{1}{2(y+2)} [(6-y)^2 - y^2] \, dy = \int_0^3 \frac{1}{2(y+2)} [36 - 12y] \, dy$$

$$= \int_0^3 \left[-6 + \left(\frac{60}{2y+4} \right) \right] dy \qquad \textit{divide } 2y + 4 \textit{ into } -12y + 36$$

$$= -6y + 30 \ln |2y+4| \Big|_0^3$$

$$= [-6(3) + 30 \ln(2(3) + 4)] - [-6(0) + 30 \ln(0 + 4)]$$

$$\approx 9.489$$

We conclude that the volume of the biomass is approximately 9.5 cm^3.

Average Value of a Function $f(x, y)$ In Section 5.4, you saw that the average value of a function $f(x)$ over an interval $a \le x \le b$ is given by the integral formula

$$\text{AV} = \frac{1}{b-a} \int_a^b f(x) \, dx$$

That is, to find the average value of a function of one variable over an interval, you integrate the function over the interval and divide by the length of the interval. The two-variable procedure is similar. In particular, to find the average value of a function of two variables $f(x, y)$ over a rectangular region R, you integrate the function over R and divide by the area of R.

Average Value Formula ■ The average value of the function $f(x, y)$ over the rectangular region R is given by the formula

$$\text{AV} = \frac{1}{\text{area of } R} \iint_R f(x, y) \, dA$$

EXAMPLE | 7.6.8

In a certain factory, output is given by the Cobb-Douglas production function

$$Q(K, L) = 50K^{3/5}L^{2/5}$$

where K is the capital investment in units of \$1,000 and L is the size of the labor force measured in worker-hours. Suppose that monthly capital investment varies between \$10,000 and \$12,000, while monthly use of labor varies between 2,800 and 3,200 worker-hours. Find the average monthly output for the factory.

Solution

It is reasonable to estimate the average monthly output by the average value of $Q(K, L)$ over the rectangular region R: $10 \le K \le 12$, $2,800 \le L \le 3,200$. The region has area

$$A = \text{area of } R = (12 - 10) \times (3,200 - 2,800)$$
$$= 800$$

so the average output is

$$AV = \frac{1}{800}\int\int_R 50K^{3/5}L^{2/5}dA$$

$$= \frac{1}{800}\int_{2,800}^{3,200}\left(\int_{10}^{12} 50K^{3/5}L^{2/5}dK\right) dL$$

$$= \frac{1}{800}\int_{2,800}^{3,200} 50L^{2/5}\left(\frac{5}{8}K^{8/5}\right)\Big|_{K=10}^{K=12} dL$$

$$= \frac{1}{800}(50)\left(\frac{5}{8}\right)\int_{2,800}^{3,200} L^{2/5}(12^{8/5} - 10^{8/5}) dL$$

$$= \frac{1}{800}(50)\left(\frac{5}{8}\right)(12^{8/5} - 10^{8/5})\left(\frac{5}{7}L^{7/5}\right)\Big|_{L=2,800}^{L=3,200}$$

$$= \frac{1}{800}(50)\left(\frac{5}{8}\right)\left(\frac{5}{7}\right)(12^{8/5} - 10^{8/5})[(3,200)^{7/5} - (2,800)^{7/5}]$$

$$\approx 5,181.23$$

Thus, the average monthly output is approximately 5,181 units.

PROBLEMS 7.6

Evaluate the double integrals in Problems 1 through 18.

1. $\int_0^1 \int_1^2 x^2 y \, dx \, dy$

2. $\int_1^2 \int_0^1 x^2 y \, dy \, dx$

3. $\int_0^{\ln 2} \int_{-1}^0 2xe^y \, dx \, dy$

4. $\int_2^3 \int_{-1}^1 (x + 2y) \, dy \, dx$

5. $\int_1^3 \int_0^1 \frac{2xy}{x^2 + 1} \, dx \, dy$

6. $\int_0^1 \int_0^1 x^2 e^{xy} \, dy \, dx$

7. $\int_0^4 \int_{-1}^1 x^2 y \, dy \, dx$

8. $\int_0^1 \int_1^5 y\sqrt{1 - y^2} \, dx \, dy$

9. $\int_2^3 \int_1^2 \frac{x + y}{xy} \, dy \, dx$

10. $\int_1^2 \int_2^3 \left(\frac{y}{x} + \frac{x}{y}\right) dy \, dx$

11. $\int_0^4 \int_0^{\sqrt{x}} x^2 y \, dy \, dx$

12. $\int_0^1 \int_1^5 xy\sqrt{1 - y^2} \, dx \, dy$

13. $\int_0^1 \int_{y-1}^{1-y} (2x + y) \, dx \, dy$

14. $\int_0^1 \int_{x^2}^x 2xy \, dy \, dx$

15. $\int_0^1 \int_0^4 \sqrt{xy} \, dy \, dx$

16. $\int_0^1 \int_x^{2x} e^{y-x} \, dy \, dx$

17. $\int_1^e \int_0^{\ln x} xy \, dy \, dx$

18. $\int_0^3 \int_{y^2/4}^{\sqrt{10-y^2}} xy \, dx \, dy$

In Problems 19 through 24, use inequalities to describe R in terms of its vertical and horizontal cross sections.

19. R is the region bounded by $y = x^2$ and $y = 3x$.

20. R is the region bounded by $y = \sqrt{x}$ and $y = x^2$.

21. R is the rectangle with vertices $(-1, 1)$, $(2, 1)$, $(2, 2)$, and $(-1, 2)$.

22. R is the triangle with vertices $(1, 0)$, $(1, 1)$, and $(2, 0)$.

23. R is the region bounded by $y = \ln x$, $y = 0$, and $x = e$.

24. R is the region bounded by $y = e^x$, $y = 2$, and $x = 0$.

In Problems 25 through 36, evaluate the given double integral for the specified region R.

25. $\displaystyle\iint_R 3xy^2 \, dA$, where R is the rectangle bounded by the lines $x = -1$, $x = 2$, $y = -1$, and $y = 0$.

26. $\displaystyle\iint_R (x + 2y) \, dA$, where R is the triangle with vertices $(0, 0)$, $(1, 0)$, and $(0, 2)$.

27. $\displaystyle\iint_R xe^y \, dA$, where R is the triangle with vertices $(0, 0)$, $(1, 0)$, and $(1, 1)$.

28. $\displaystyle\iint_R 48xy \, dA$, where R is the region bounded by $y = x^3$ and $y = \sqrt{x}$.

29. $\displaystyle\iint_R (2y - x) \, dA$, where R is the region bounded by $y = x^2$ and $y = 2x$.

30. $\displaystyle\iint_R 12x \, dA$, where R is the region bounded by $y = x^2$ and $y = 6 - x$.

31. $\displaystyle\iint_R (2x + 1) \, dA$, where R is the triangle with vertices $(-1, 0)$, $(1, 0)$, and $(0, 1)$.

32. $\displaystyle\iint_R 2x \, dA$, where R is the region bounded by $y = \dfrac{1}{x^2}$, $y = x$, and $x = 2$.

33. $\displaystyle\iint_R \dfrac{1}{y^2 + 1} \, dA$, where R is the triangle bounded by the lines $y = \dfrac{1}{2}x$, $y = -x$, and $y = 2$.

34. $\displaystyle\iint_R e^{y^3} \, dA$, where R is the region bounded by $y = \sqrt{x}$, $y = 1$, and $x = 0$.

35. $\displaystyle\iint_R 12x^2 e^{y^2} \, dA$, where R is the region in the first quadrant bounded by $y = x^3$ and $y = x$.

36. $\displaystyle\iint_R y \, dA$, where R is the region bounded by $y = \ln x$, $y = 0$, and $x = e$.

In Problems 37 through 44, sketch the region of integration for the given integral and set up an equivalent integral with the order of integration reversed.

37. $\displaystyle\int_0^2 \int_0^{4x^2} f(x, y)\, dy\, dx$

38. $\displaystyle\int_0^1 \int_0^{2y} f(x, y)\, dx\, dy$

39. $\displaystyle\int_0^1 \int_{x^3}^{\sqrt{x}} f(x, y)\, dy\, dx$

40. $\displaystyle\int_0^4 \int_{y/2}^{\sqrt{y}} f(x, y)\, dx\, dy$

41. $\displaystyle\int_1^{e^2} \int_{\ln x}^2 f(x, y)\, dy\, dx$

42. $\displaystyle\int_0^{\ln 3} \int_{e^x}^3 f(x, y)\, dy\, dx$

43. $\displaystyle\int_{-1}^1 \int_{x^2+1}^2 f(x, y)\, dy\, dx$

44. $\displaystyle\int_{-1}^1 \int_{-\sqrt{y+1}}^{\sqrt{y+1}} f(x, y)\, dx\, dy$

In Problems 45 through 54, use a double integral to find the area of R.

45. R is the triangle with vertices $(-4, 0)$, $(2, 0)$, and $(2, 6)$.

46. R is the triangle with vertices $(0, -1)$, $(-2, 1)$, and $(2, 1)$.

47. R is the region bounded by $y = \dfrac{1}{2}x^2$ and $y = 2x$.

48. R is the region bounded by $y = \sqrt{x}$ and $y = x^2$.

49. R is the region bounded by $y = x^2 - 4x + 3$ and the x axis.

50. R is the region bounded by $y = x^2 + 6x + 5$ and the x axis.

51. R is the region bounded by $y = \ln x$, $y = 0$, and $x = e$.

52. R is the region bounded by $y = x$, $y = \ln x$, $y = 0$, and $y = 1$.

53. R is the region in the first quadrant bounded by $y = 4 - x^2$, $y = 3x$, and $y = 0$.

54. R is the region bounded by $y = \dfrac{16}{x}$, $y = x$, and $x = 8$.

In Problems 55 through 64, find the volume of the solid under the surface $z = f(x, y)$ and over the given region R.

55. $f(x, y) = 6 - 2x - 2y$;
$R: 0 \le x \le 1, 0 \le y \le 2$

56. $f(x, y) = 9 - x^2 - y^2$,
$R: -1 \le x \le 1, -2 \le y \le 2$

57. $f(x, y) = \dfrac{1}{xy}$;
$R: 1 \le x \le 2, 1 \le y \le 3$

58. $f(x, y) = e^{x+y}$;
$R: 0 \le x \le 1, 0 \le y \le \ln 2$

59. $f(x, y) = xe^{-y}$;
$R: 0 \le x \le 1, 0 \le y \le 2$

60. $f(x, y) = (1 - x)(4 - y)$;
$R: 0 \le x \le 1, 0 \le y \le 4$

61. $f(x, y) = 2x + y$; R is bounded by $y = x$, $y = 2 - x$, and $y = 0$.

62. $f(x, y) = e^{y^2}$; R is bounded by $x = 2y$, $x = 0$, and $y = 1$.

63. $f(x, y) = x + 1$; R is bounded by $y = 8 - x^2$ and $y = x^2$.

64. $f(x, y) = 4xe^y$; R is bounded by $y = 2x$, $y = 2$, and $x = 0$.

In Problems 65 through 72, find the average value of the function f(x, y) over the given region R.

65. $f(x, y) = xy(x - 2y)$;
$R: -2 \le x \le 3, -1 \le y \le 2$

66. $f(x, y) = \dfrac{y}{x} + \dfrac{x}{y}$;
$R: 1 \le x \le 4, 1 \le y \le 3$

67. $f(x, y) = xye^{x^2y}$;
$R: 0 \le x \le 1, 0 \le y \le 2$

68. $f(x, y) = \dfrac{\ln x}{xy}$;
$R: 1 \le x \le 2, 2 \le y \le 3$

69. $f(x, y) = 6xy$; R is the triangle
with vertices $(0, 0), (0, 1), (3, 1)$.

70. $f(x, y) = e^{x^2}$; R is the triangle
with vertices $(0, 0), (1, 0), (1, 1)$.

71. $f(x, y) = x$; R is the region
bounded by $y = 4 - x^2$ and $y = 0$.

72. $f(x, y) = e^x y^{-1/2}$; R is the region
bounded by $x = \sqrt{y}, y = 0$, and $x = 1$.

In Problems 73 through 76, evaluate the double integral over the specified region R. Choose the order of integration carefully.

73. $\displaystyle\iint_R \dfrac{\ln (xy)}{y}\, dA$; $R: 1 \le x \le 3, 2 \le y \le 5$

74. $\displaystyle\iint_R ye^{xy}\, dA$; $R: -1 \le x \le 1, 1 \le y \le 2$

75. $\displaystyle\iint_R x^3 e^{x^2 y}\, dA$; $R: 0 \le x \le 1, 0 \le y \le 1$

76. $\displaystyle\iint_R e^{x^3}\, dA$; $R: \sqrt{y} \le x \le 1, 0 \le y \le 1$

77. PRODUCTION   At a certain factory, output Q is related to inputs x and y by the expression
$$Q(x, y) = 2x^3 + 3x^2 y + y^3$$
If $0 \le x \le 5$ and $0 \le y \le 7$, what is the average output of the factory?

78. PRODUCTION   A bicycle dealer has found that if 10-speed bicycles are sold for x dollars apiece and the price of gasoline is y cents per gallon, then approximately
$$Q(x, y) = 200 - 24\sqrt{x} + 4(0.1y + 3)^{3/2}$$
bicycles will be sold each month. If the price of bicycles varies between \$289 and \$324 during a typical month, and the price of gasoline varies between \$2.30 and \$2.50, approximately how many bicycles will be sold each month on average?

79. AVERAGE PROFIT   A manufacturer estimates that when x units of a particular commodity are sold domestically and y units are sold to foreign markets, the profit is given by
$$P(x, y) = (x - 30)(70 + 5x - 4y)$$
$$+ (y - 40)(80 - 6x + 7y)$$
hundred dollars. If monthly domestic sales vary between 100 and 125 units and foreign sales between 70 and 89 units, what is the average monthly profit?

80. AVERAGE RESPONSE TO STIMULI   In a psychological experiment, x units of stimulus A and y units of stimulus B are applied to a subject, whose performance on a certain task is then measured by the function
$$P(x, y) = 10 + xye^{1 - x^2 - y^2}$$
Suppose x varies between 0 and 1 while y varies between 0 and 3. What is the subject's average response to the stimuli?

81. AVERAGE ELEVATION   A map of a small regional park is a rectangular grid, bounded by the lines $x = 0, x = 4, y = 0$, and $y = 3$, where units are in miles. It is found that the elevation above sea level at each point (x, y) in the park is given by
$$E(x, y) = 90(2x + y^2) \text{ feet}$$
Find the average elevation in the park. (Remember, 1 mi = 5,280 feet.)

82. PROPERTY VALUE   A community is laid out as a rectangular grid in relation to two main streets that intersect at the city center. Each point in the community has coordinates (x, y) in this grid, for $-10 \le x \le 10, -8 \le y \le 8$ with x and y measured in miles. Suppose the value of the land located at the point (x, y) is V thousand dollars, where

$$V(x, y) = (250 + 17x)e^{-0.01x - 0.05y}$$

Estimate the value of the block of land occupying the rectangular region $1 \leq x \leq 3, 0 \leq y \leq 2$.

83. PROPERTY VALUE   Repeat Problem 82 for

$$V(x, y) = (300 + x + y)e^{-0.01x}$$

and the region $-1 \leq x \leq 1, -1 \leq y \leq 1$.

84. PROPERTY VALUE   Repeat Problem 82 for

$$V(x, y) = 400xe^{-y}$$

and the region $R: 0 \leq y \leq x, 0 \leq x \leq 1$.

85. AVERAGE SURFACE AREA OF THE HUMAN BODY   Recall from Problem 35, Section 7.1, that the surface area of a person's body may be estimated by the empirical formula

$$S(W, H) = 0.0072W^{0.425}H^{0.725}$$

where W is the person's weight in kilograms, H is the person's height in centimeters, and the surface area S is measured in square meters.

a. Find the average value of the function $S(W, H)$ over the region

$$R: 3.2 \leq W \leq 80, 38 \leq H \leq 180$$

b. A child weighs 3.2 kg and is 38 cm tall at birth and as an adult, has a stable weight of 80 kg and height of 180 cm. Can the average value in part (a) be interpreted as the average lifetime surface area of this person's body? Explain.

86. ARCHITECTURAL DESIGN   A building is to have a curved roof above a rectangular base. In relation to a rectangular grid, the base is the rectangular region $-30 \leq x \leq 30, -20 \leq y \leq 20$, where x and y are measured in meters. The height of the roof above each point (x, y) in the base is given by

$$h(x, y) = 12 - 0.003x^2 - 0.005y^2$$

a. Find the volume of the building.

b. Find the average height of the roof.

87. CONSTRUCTION   A storage bin is to be constructed in the shape of the solid bounded above by the surface

$$z = 20 - x^2 - y^2$$

below by the xy plane, and on the sides by the plane $y = 0$ and the parabolic cylinder $y = 4 - x^2$, where x, y, and z are in meters. Find the volume of the bin.

88. CONSTRUCTION   A fancy jewelry box has the shape of the solid bounded above by the plane

$$3x + 4y + 2z = 12$$

below by the xy plane, and on the sides by the planes $x = 0$ and $y = 0$, where x, y, and z are in inches. Find the volume of the box.

89. EXPOSURE TO DISEASE   The likelihood that a person with a contagious disease will infect others in a social situation may be assumed to be a function $f(s)$ of the distance s between individuals. Suppose contagious individuals are uniformly distributed throughout a rectangular region R in the xy plane. Then the likelihood of infection for someone at the origin $(0, 0)$ is proportional to the exposure index E, given by the double integral

$$E = \iint_R f(s) \, dA$$

where $s = \sqrt{x^2 + y^2}$ is the distance between $(0, 0)$ and (x, y). Find E for the case where

$$f(s) = 1 - \frac{s^2}{9}$$

and R is the square

$$R: -2 \leq x \leq 2, -2 \leq y \leq 2$$

In Problems 90 through 92, use double integration to find the required quantity. In some cases, you may need to use the numeric integration feature of your calculator.

90. Find the area of the region bounded above by the curve (ellipse) $4x^2 + 3y^2 = 7$ and below by the parabola $y = x^2$.

91. Find the volume of the solid bounded above by the graph of $f(x, y) = x^2e^{-xy}$ and below by the rectangular region $R: 0 \leq x \leq 2, 0 \leq y \leq 3$.

92. Find the average value of $f(x, y) = xy \ln\left(\frac{y}{x}\right)$ over the rectangular region bounded by the lines $x = 1$, $x = 2, y = 1$, and $y = 3$.

Important Terms, Symbols, and Formulas

Function of two variables: $z = f(x, y)$ (544)

Domain convention (544)

Cobb-Douglas production function (546)

Three-dimensional coordinate system (548)

Level curve: $f(x, y) = C$ (548)

Topographical map (550)

Utility (551)

Indifference curve (551)

Partial derivatives of $z = f(x, y)$: (559)

$$f_x = \frac{\partial z}{\partial x} \qquad f_y = \frac{\partial z}{\partial y}$$

Marginal productivity (562)

Complementary and substitute commodities (563)

Second-order partial derivatives: (564)

$$f_{xx} = \frac{\partial^2 z}{\partial x^2} \quad f_{xy} = \frac{\partial^2 z}{\partial y \partial x} \quad f_{yx} = \frac{\partial^2 z}{\partial x \partial y} \quad f_{yy} = \frac{\partial^2 z}{\partial y^2}$$

Equality of mixed second-order partial derivatives:

$$f_{xy} = f_{yx} \quad (565)$$

Chain rule for partial derivatives: (566)

$$\frac{dz}{dt} = \frac{\partial z}{\partial x} \frac{dx}{dt} + \frac{\partial z}{\partial y} \frac{dy}{dt}$$

Incremental approximation formula for a function of two variables $z = f(x, y)$: (567)

$$\Delta z \approx \frac{\partial z}{\partial x} \Delta x + \frac{\partial z}{\partial y} \Delta y$$

Relative maximum; relative minimum; saddle point (574) and (575)

Critical point: $f_x = f_y = 0$ (575)

Second partials test at a critical point (a, b): (576)
Let $D(a, b) = f_{xx} f_{yy} - (f_{xy})^2$.

If $D < 0$, f has a saddle point at (a, b).

If $D > 0$, and $f_{xx} < 0$, f has a relative maximum at (a, b).

If $D > 0$, and $f_{xx} > 0$, f has a relative minimum at (a, b).

If $D = 0$, the test is inconclusive.

Scatter diagram (586)

Least-squares criterion (587)

Least-squares line: $y = mx + b$, where

$$m = \frac{n\Sigma xy - \Sigma x \Sigma y}{n\Sigma x^2 - (\Sigma x)^2} \quad \text{and} \quad b = \frac{\Sigma x^2 \Sigma y - \Sigma x \Sigma xy}{n\Sigma x^2 - (\Sigma x)^2} \quad (589)$$

Log-linear regression (593)

Method of Lagrange multipliers: (600)

To find extreme values of $f(x, y)$ subject to $g(x, y) = k$, solve the equations

$$f_x = \lambda g_x \qquad f_y = \lambda g_y \qquad \text{and} \qquad g = k$$

The Lagrange multiplier: (606)

$$\lambda = \frac{dM}{dk}, \text{ where } M \text{ is the optimal value of } f(x, y) \text{ subject to } g(x, y) = k.$$

Double integral (618)
over the region R: $a \leq x \leq b$, $g_1(x) \leq y \leq g_2(x)$

$$\iint_R f(x, y) dA = \int_a^b \left[\int_{g_1(x)}^{g_2(x)} f(x, y) dy \right] dx$$

over the region R: $c \leq y \leq d$, $h_1(y) \leq x \leq h_2(y)$

$$\iint_R f(x, y) dA = \int_c^d \left[\int_{h_1(y)}^{h_2(y)} f(x, y) dx \right] dy$$

Area of the region R in the xy plane is

$$\text{Area of } R = \iint_R 1 \, dA \quad (619)$$

Volume under $z = f(x, y)$ over a rectangle R where $f(x, y) \geq 0$ is $V = \iint_R f(x, y) \, dA$ (621)

Average value of $f(x, y)$ over the rectangle R: (622)

$$\text{AV} = \frac{1}{\text{area of } R} \iint_R f(x, y) dA$$

Checkup for Chapter 7

1. In each case, first describe the domain of the given function and then find the partial derivatives $f_x, f_y,$ $f_{xx},$ and $f_{yx}.$

   a. $f(x, y) = x^3 + 2xy^2 - 3y^4$

   b. $f(x, y) = \dfrac{2x + y}{x - y}$

   c. $f(x, y) = e^{2x-y} + \ln(y^2 - 2x)$

2. Describe the level curves of each of these functions:

   a. $f(x, y) = x^2 + y^2$

   b. $f(x, y) = x + y^2$

3. In each case, find all critical points of the given function $f(x, y)$ and use the second partials test to classify each as a relative maximum, a relative minimum, or a saddle point.

   a. $f(x, y) = 4x^3 + y^3 - 6x^2 - 6y^2 + 5$

   b. $f(x, y) = x^2 - 4xy + 3y^2 + 2x - 4y$

   c. $f(x, y) = xy - \dfrac{1}{y} - \dfrac{1}{x}$

4. Use the method of Lagrange multipliers to find these constrained extrema:

   a. The smallest value of $f(x, y) = x^2 + y^2$ subject to $x + 2y = 4.$

   b. The largest and the smallest values of the function $f(x, y) = xy^2$ subject to $2x^2 + y^2 = 6.$

5. Evaluate each of these double integrals:

   a. $\displaystyle\int_{-1}^{3}\int_{0}^{2} x^3 y\, dx\, dy$

   b. $\displaystyle\int_{0}^{2}\int_{-1}^{1} x^2 e^{xy}\, dx\, dy$

   c. $\displaystyle\int_{1}^{2}\int_{1}^{y} \frac{y}{x}\, dx\, dy$

   d. $\displaystyle\int_{0}^{2}\int_{0}^{2-x} xe^{-y}\, dy\, dx$

6. **MARGINAL PRODUCTIVITY** A company will produce $Q(K, L) = 120K^{3/4}L^{1/4}$ hundred units of a particular commodity when the capital expenditure is K thousand dollars and the size of the workforce is L worker-hours. Find the marginal productivity of capital Q_K and the marginal productivity of labor Q_L when the capital expenditure is $1,296,000$ dollars and the labor level is 20,736 worker-hours.

7. **UTILITY** Everett has just received $300 as a birthday gift, and has decided to spend it on DVDs and computer video games. He has determined that the utility (satisfaction) derived from the purchase of x DVDs and y video games is

   $$U(x, y) = \ln(x^2\sqrt{y})$$

   If each DVD costs $20 and each video game costs $30, how many DVDs and video games should he purchase in order to maximize utility?

8. **MEDICINE** A certain disease can be treated by administering at least 70 units of drug C, but that level of medication sometimes results in serious side effects. Looking for a safer approach, a physician decides instead to use drugs A and B, which result in no side effects as long as their combined dosage is less than 60 units. Moreover, she determines that when x units of drug A and y units of drug B are administered to a patient, the effect is equivalent to administering E units of drug C, where

   $$E = 0.05(xy - 2x^2 - y^2 + 95x + 20y)$$

   What dosages of drugs A and B will maximize the equivalent level E of drug C? If the physician administers the optimum dosages of drugs A and B, will the combined effect be enough to help the patient without running the risk of side effects?

9. **AVERAGE TEMPERATURE** A flat metal plate lying in the xy plane is heated in such a way that the temperature at the point (x, y) is $T\,°C$, where

   $$T(x, y) = 10ye^{-xy}$$

   Find the average temperature over a rectangular portion of the plate for which $0 \le x \le 2$ and $0 \le y \le 1.$

10. **LEAST-SQUARES APPROXIMATION OF PROFIT DATA** A company's annual profit (in millions of dollars) for the first 5 years of operation is shown in this table:

Year	1	2	3	4	5
Profit (millions of dollars)	1.03	1.52	2.03	2.41	2.84

   a. Plot these data on a graph.

   b. Find the equation of the least-squares line through the data.

   c. Use the least-squares line to predict the company's sixth year profit.

Review Problems

In Problems 1 through 5, find the partial derivatives f_x and f_y.

1. $f(x, y) = 2x^3y + 3xy^2 + \dfrac{y}{x}$

2. $f(x, y) = (xy^2 + 1)^5$

3. $f(x, y) = xye^{xy}$

4. $f(x, y) = \dfrac{x^2 - y^2}{2x + y}$

5. $f(x, y) = \ln\left(\dfrac{xy}{x + 3y}\right)$

6. For each of these functions, compute the second-order partial derivatives f_{xx}, f_{yy}, f_{xy}, and f_{yx}.
   a. $f(x, y) = x^2 + y^3 - 2xy^2$
   b. $f(x, y) = e^{x^2+y^2}$
   c. $f(x, y) = x \ln y$

7. At a certain factory, the daily output is approximately $40K^{1/3}L^{1/2}$ units, where K denotes the capital investment measured in units of $1,000 and L denotes the size of the labor force measured in worker-hours. Suppose that the current capital investment is $125,000 and that 900 worker-hours of labor are used each day. Use marginal analysis to estimate the effect that an additional capital investment of $1,000 will have on the daily output if the size of the labor force is not changed.

8. In economics, the marginal product of labor is the rate at which output Q changes with respect to labor L for a fixed level of capital investment K. An economic law states that, under certain circumstances, the marginal product of labor increases as the level of capital investment increases. Translate this law into a mathematical statement involving a second-order partial derivative.

9. For each of these functions, sketch the indicated level curves:
   a. $f(x, y) = x^2 - y; f = 2, f = -2$
   b. $f(x, y) = 6x + 2y; f = 0, f = 1, f = 2$

10. For each of these functions, find the slope of the indicated level curve at the specified value of x:
    a. $f(x, y) = x^2 - y^3; f = 2; x = 1$
    b. $f(x, y) = xe^y; f = 2; x = 2$

11. **MARGINAL ANALYSIS** Using x skilled workers and y unskilled workers, a manufacturer can produce $Q(x, y) = 60x^{1/3}y^{2/3}$ units per day. Currently the manufacturer employs 10 skilled workers and 40 unskilled workers and is planning to hire 1 additional skilled worker. Use calculus to estimate the corresponding change that the manufacturer should make in the level of unskilled labor so that the total output will remain the same.

In Problems 12 through 15, find all critical points of the given function and use the second partials test to classify each as a relative maximum, a relative minimum, or a saddle point.

12. $f(x, y) = x^3 + y^3 + 3x^2 - 18y^2 + 81y + 5$

13. $f(x, y) = x^2 + y^3 + 6xy - 7x - 6y$

14. $f(x, y) = 3x^2y + 2xy^2 - 10xy - 8y^2$

15. $f(x, y) = xe^{2x^2+5xy+2y^2}$

16. Use the method of Lagrange multipliers to find the maximum and minimum values of the function $f(x, y) = x^2 + 2y^2 + 2x + 3$ subject to the constraint $x^2 + y^2 = 4$.

17. Use the method of Lagrange multipliers to prove that of all rectangles with a given perimeter, the square has the largest area.

18. **ALLOCATION OF FUNDS** A manufacturer is planning to sell a new product at the price of $350 per unit and estimates that if x thousand dollars is spent on development and y thousand dollars is spent on promotion, consumers will buy approximately $\dfrac{250y}{y + 2} + \dfrac{100x}{x + 5}$ units of the product. If manufacturing costs for the product are $150 per unit, how much should the manufacturer spend on development and how much on promotion to generate the largest possible profit if unlimited funds are available?

19. **ALLOCATION OF FUNDS** Suppose the manufacturer in Problem 18 has only $11,000 to spend on the development and promotion of the new product. How should this money be allocated to generate the largest possible profit?

20. **ALLOCATION OF FUNDS** Suppose the manufacturer in Problem 19 decides to spend $12,000

instead of \$11,000 on the development and promotion of the new product. Use the Lagrange multiplier λ to estimate how this change will affect the maximum possible profit.

21. Let $f(x, y) = \dfrac{12}{x} + \dfrac{18}{y} + xy$, where $x > 0$, $y > 0$.

    How do you know that f must have a minimum in the region $x > 0$, $y > 0$? Find the minimum.

In Problems 22 through 29, evaluate the double integral. You may need to exchange the order of integration.

22. $\displaystyle\int_0^1 \int_{-2}^0 (2x + 3y)\,dy\,dx$

23. $\displaystyle\int_0^1 \int_0^2 e^{-x-y}\,dy\,dx$

24. $\displaystyle\int_0^1 \int_0^2 x\sqrt{1-y}\,dx\,dy$

25. $\displaystyle\int_0^1 \int_{-1}^1 xe^{2y}\,dy\,dx$

26. $\displaystyle\int_0^2 \int_1^1 \dfrac{6xy^2}{x^2+1}\,dy\,dx$

27. $\displaystyle\int_1^e \int_1^e \ln(xy)\,dy\,dx$

28. $\displaystyle\int_0^1 \int_0^{1-x} x(y-1)^2\,dy\,dx$

29. $\displaystyle\int_1^2 \int_0^x e^{y/x}\,dy\,dx$

In Problems 30 and 31, evaluate the given double integral for the specified region R.

30. $\displaystyle\iint_R 6x^2y\,dA$, where R is the rectangle with vertices $(-1, 0)$, $(2, 0)$, $(2, 3)$, and $(-1, 3)$.

31. $\displaystyle\iint_R (x + 2y)\,dA$, where R is the rectangular region bounded by $x = 0$, $x = 1$, $y = -2$, and $y = 2$.

32. Find the volume under the surface $z = 2xy$ and above the rectangle with vertices $(0, 0)$, $(2, 0)$, $(0, 3)$, and $(2, 3)$.

33. Find the volume under the surface $z = xe^{-y}$ and above the rectangle bounded by the lines $x = 1$, $x = 2$, $y = 2$, and $y = 3$.

34. Find the average value of $f(x, y) = xy^2$ over the rectangular region with vertices $(-1, 3)$, $(1, 5)$, $(2, 3)$, and $(2, 5)$.

35. Find three positive numbers x, y, and z so that $x + y + z = 20$ and the product $P = xyz$ is a maximum. [*Hint:* Use the fact that $z = 20 - x - y$ to express P as a function of only two variables.]

36. Find three positive numbers x, y, and z so that $2x + 3y + z = 60$ and the sum $S = x^2 + y^2 + z^2$ is minimized. (See the hint to Problem 35.)

37. Find the shortest distance from the origin to the surface $y^2 - z^2 = 10$. [*Hint:* Express the distance $\sqrt{x^2 + y^2 + z^2}$ from the origin to a point (x, y, z) on the surface in terms of the two variables x and y, and minimize the *square* of the resulting distance function.]

38. Plot the points $(1, 1)$, $(1, 2)$, $(3, 2)$, and $(4, 3)$ and use partial derivatives to find the corresponding least-squares line.

39. **SALES** The marketing manager for a certain company has compiled these data relating monthly advertising expenditure and monthly sales (both measured in units of \$1,000):

Advertising	3	4	7	9	10
Sales	78	86	138	145	156

   a. Plot these data on a graph.

   b. Find the least-squares line.

   c. Use the least-squares line to predict monthly sales if the monthly advertising expenditure is \$5,000.

40. **UTILITY** Suppose the utility derived by a consumer from x units of one commodity and y units of a second commodity is given by the utility function $U(x, y) = x^3 y^2$. The consumer currently owns $x = 5$ units of the first commodity and $y = 4$ units of the second. Use calculus to estimate how many units of the second commodity the consumer could substitute for 1 unit of the first commodity without affecting total utility.

41. **CONSUMER DEMAND** A paint store carries two brands of latex paint. Sales figures indicate that if the first brand is sold for x dollars per quart and the second for y dollars per quart, the demand for the first brand will be Q quarts per month, where
$$Q(x, y) = 200 - 10x^2 + 20y$$
It is estimated that t months from now the price of the first brand will be $x(t) = 5 + 0.02t$ dollars per quart and the price of the second will be $y(t) = 6 + 0.4\sqrt{t}$ dollars per quart. At what rate will the demand for the first brand of paint be changing with respect to time 9 months from now?

42. **COOLING AN ANIMAL'S BODY** The difference between an animal's surface temperature and that of the surrounding air causes a transfer of energy by convection. The coefficient of convection h is given by

$$h = \frac{kV^{1/3}}{D^{2/3}}$$

where V (cm/sec) is wind velocity, D (cm) is the diameter of the animal's body, and k is a constant.

   a. Find the partial derivatives h_V and h_D. Interpret these derivatives as rates.

   b. Compute the ratio $\dfrac{h_V}{h_D}$.

43. **CONSUMER DEMAND** Suppose that when apples sell for x cents per pound and bakers earn y dollars per hour, the price of apple pies at a certain supermarket chain is

$$p(x, y) = \frac{1}{4}x^{1/3}y^{1/2}$$

dollars per pie. Suppose also that t months from now, the price of apples will be

$$x = 129 - \sqrt{8t}$$

cents per pound and bakers' wages will be

$$y = 15.60 + 0.2t$$

dollars per hour. If the supermarket chain can sell

$$Q = \frac{4,184}{p}$$ pies per week when the price is p dollars

per pie, at what rate will the weekly demand Q for pies be changing with respect to time 2 months from now?

44. Arnold, the heat-seeking mussel, is the world's smartest mollusk. Arnold likes to stay warm, and by using the crustacean coordinate system he learned from a passing crab, he has determined that at each nearby point (x, y) on the ocean floor, the temperature (°C) is

$$T(x, y) = 2x^2 - xy + y^2 - 2y + 1$$

Arnold's world consists of a rectangular portion of ocean bed with vertices $(-1, -1)$, $(-1, 1)$, $(1, -1)$, and $(1, 1)$, and since it is very hard for him to move, he plans to stay where he is as long as the average temperature of this region is at least 5°C. Does Arnold move or stay put?

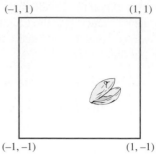

(−1, 1)   (1, 1)

(−1, −1)   (1, −1)

PROBLEM 44

45. **AIR POLLUTION** At a certain factory, the amount of air pollution generated each day is measured by the function $Q(E, T) = 125E^{2/3}T^{1/2}$, where E is the number of employees and T (°C) is the average temperature during the workday. Currently, there are $E = 151$ employees and the average temperature is $T = 10$ °C. If the average daily temperature is falling at the rate of 0.21 °C per day and the number of employees is increasing at the rate of 2 per month, use calculus to estimate the corresponding effect on the rate of pollution. Express your answer in units per day. You may assume that there are 22 workdays per month.

46. **POPULATION** A demographer sets up a grid to describe location within a suburb of a major metropolitan area. In relation to this grid, the population density at each point (x, y) is given by

$$f(x, y) = 1 + 3y^2$$

hundred people per square mile, where x and y are in miles. A housing project occupies the region R bounded by the curve $y^2 = 4 - x$, and the y axis ($x = 0$). What is the total population within the project region R?

47. **POLLUTION** There are two sources of air pollution that affect the health of a certain community. Health officials have determined that at a point located r miles from source A and s miles from source B, there will be

$$N(r, s) = 40e^{-r/2}\,e^{-s/3}$$

units of pollution. A housing project lies in a region R for which

$$2 \le r \le 3 \quad \text{and} \quad 1 \le s \le 2$$

What is the total pollution within the region R?

48. **NUCLEAR WASTE DISPOSAL** Nuclear waste is often disposed of by sealing it into containers that are then dumped into the ocean. It is important to dump the containers into water shallow enough to ensure that they do not break when they hit bottom. Suppose as the container falls through the water, there is a drag force that is proportional to the container's velocity. Then, it can be shown that the depth s (in meters) of a container of weight W newtons at time t seconds is given by the formula

$$s(W, t) = \left(\frac{W - B}{k}\right)t + \frac{W(W - B)}{k^2 g}\left[e^{-(kgt/W)} - 1\right]$$

where B is a (constant) buoyancy force, k is the drag constant, and $g = 9.8$ m/sec^2 is the constant acceleration due to gravity.

a. Find $\dfrac{\partial s}{\partial W}$ and $\dfrac{\partial s}{\partial t}$. Interpret these derivatives as rates. Do you think it is possible for either partial derivative to ever be zero?

b. For a fixed weight, the speed of the container is $\dfrac{\partial s}{\partial t}$. Suppose the container will break when its speed on impact with the ocean floor exceeds 10 m/sec. If $B = 1{,}983$ newtons and $k = 0.597$ kg/sec, what is the maximum depth for safely dumping a container of weight $W = 2{,}417$ newtons?

 c. Research the topic of nuclear waste disposal, and write a paragraph on whether you think it is best done on land or at sea.

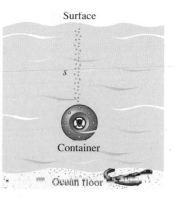

PROBLEM 48

49. **PRODUCTION** For the production function given by $Q = x^a y^b$, where $a > 0$ and $b > 0$, show that

$$x\frac{\partial Q}{\partial x} + y\frac{\partial Q}{\partial y} = (a + b)Q$$

In particular, if $b = 1 - a$ with $0 < a < 1$, then

$$x\frac{\partial Q}{\partial x} + y\frac{\partial Q}{\partial y} = Q$$

Complete solutions for all EXPLORE! boxes throughout the text can be accessed at the book-specific website, www.mhhe.com/hoffmann.

Solution for Explore! on Page 545

Store $f(x, y) = x^3 - x^2y^2 - xy^3 - y^4$ into Y1 as X^3 − X^2*L1^2 − X*L1^3 − L1^4, where L1 is the list of values {0, 1.5, 2.0, 2.25, 2.5}. Graph using the modified decimal window $[-9.4, 9.4]1$ by $[-150, 100]20$. Press the **TRACE** key and arrow right to $x = 2$ to observe the different $Y = f(x, L1)$ values that occur for varying L1 values. For larger L1 values the curves take on larger cubic dips in the positive x domain.

 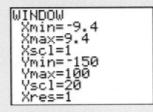 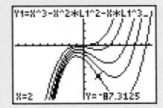

Solution for Explore! on Page 589

Using the data in Example 7.4.2, place the x values into L1 and the y values into L2. You can write L3 = L1*L2 and L4 = L1^2 if you wish to see the lists of values. To obtain all the sums needed to compute formulas for the slope m and y intercept b, press the **STAT** key, arrow right to **CALC,** select option **2:2-Var Stats,** and insert symbols for list L1 and L2, shown next in the middle screen. Pressing **ENTER** and arrowing up or down this screen gives all the desired sums, as shown in the far right screen.

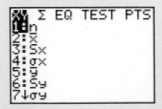 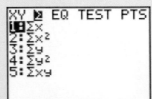 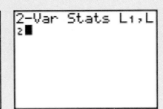

However, a more symbolic, presentable method on your calculator is to use the statistics symbolism available through the **VARS** key, **5:Statistics,** using both the **XY** and the **Σ** menus, shown next in the left and middle screens. The formulas for the slope m and y intercept b are computed in the screen next on the right, yielding $m = 0.5714 = 4/7$ and $b = 1$.

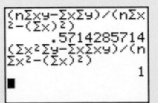

Solution for Explore!
on Page 591

Following Example 7.4.4, store the production and demand price data into lists L1 and L2, respectively. Using the **STAT PLOT** procedure explained in the Calculator Introduction, obtain the scatterplot of decreasing prices shown in the following middle screen, which suggests an exponential curve with a negative exponent. Press **STAT,** arrow right to **CALC,** and arrow down to **0:ExpReg,** making sure to indicate the desired lists and function location. Specifically, write **ExpReg L1, L2, Y1** before completing the final keystroke. Recall that the symbol **Y1** is found through the keystroke sequence, **VARS, Y-VARS, 1:Function, 1:Y1.**

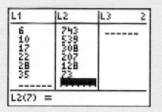 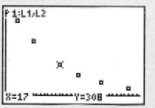 

The form of the exponential equation is $Y = a*b^x$, and we find that $a = 1,200$, with base $b = 0.92315$. If we write $b^x = e^{mx}$, we would find that $m = -0.079961$; that is, $0.92315 = e^{-0.079961}$. Pressing **ZOOM, 9:ZoomStat,** yields the following right screen, showing an almost perfect fit to the data for an exponential curve, with an equation $Y = 1,200(0.92315)^x = 1,200e^{(-0.079961)x}$. This equation coincides with the solution shown on page 592. We have computed a log-linear regression without having to take logarithms of the production and demand price variables, by directly selecting the Exponential Regression model

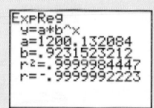 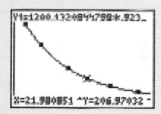

Solution for Explore!
on Page 616

Refer to Example 7.6.3. Store $y = x^2$ into Y1 and $y = 2x$ into Y2 of the equation editor and graph using the window $[-0.15, 2.2]1$ by $[-0.5, 4.5]1$. By tracing, the intersection points of Y1 and Y2 are easily located. The vertical line feature can be found using the **DRAW** key (**2nd PRGM**), **4: Vertical.** By arrowing left or right, cross sections of the area between Y1 and Y2, over which the integration will be performed, can be shown.

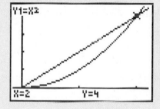  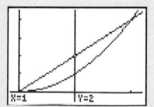

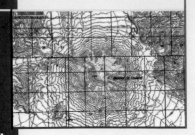

THINK ABOUT IT

MODELING POPULATION DIFFUSION

In 1905, five muskrats were accidentally released near Prague in the current Czech Republic. Subsequent to 1905, the range of the muskrat population expanded and the front (the outer limit of the population) moved as indicated in Figure 1. In the figure, the closed curves labeled with dates are equipopulation contours; that is, curves of constant, minimally detectable populations of muskrats. For instance, the curve labeled 1920 indicates that the muskrat population had expanded from Prague to the gates of Vienna in the 15 years after their release. A population dispersion such as this can be studied using mathematical models based on *partial differential equations*, that is, equations involving functions of two or more variables and their partial derivatives. We will examine such a model and then return to our illustration to see how well the model can be used to describe the dispersion of the muskrats.

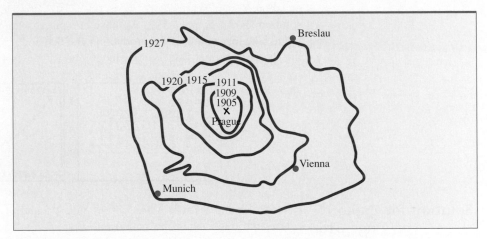

FIGURE 1 Equipopulation curves for a muskrat population in Europe.

SOURCE: Leah Edelstein-Keshet, *Mathematical Models in Biology,* Boston: McGraw-Hill, 1988, p. 439.

The model we will discuss is based on the **diffusion equation,** an extremely versatile partial differential equation with important applications in the physical and life sciences as well as economics. *Diffusion* is the name used for the process by which particles spread out as they continually collide and randomly change direction after being inserted at a source. Suppose the particles can only move in one spatial direction (say, along a thin rod or tube). Then $C(x, t)$, the concentration of particles at time t located x units from the source (the point of insertion), satisfies the one-dimensional diffusion equation

$$\frac{\partial C}{\partial t} = \alpha \frac{\partial^2 C}{\partial x^2}$$

where α is a positive constant called the *diffusion coefficient*. Similarly, the two-dimensional diffusion equation

$$\frac{\partial C}{\partial t} = \alpha\left(\frac{\partial^2 C}{\partial x^2} + \frac{\partial^2 C}{\partial y^2}\right)$$

is used to model the dispersion of particles moving randomly in a plane, where $C(x, y, t)$ is the concentration of particles at the point (x, y) at time t.

Mathematical biologists have adapted the diffusion equation to model the spread of living organisms, including both plants and animals. We will examine such a model, due to J. G. Skellam. First, suppose that at a particular time ($t = 0$), an organism is introduced at a point (called a "source"), where it had previously not been present. Skellam assumed that the population of the organism disperses from the source in two ways:

a. By growing exponentially at the continuous reproduction rate r.

b. By moving randomly in an xy coordinate plane, with the source at the origin.

Based on these assumptions, he then modeled the dispersion of the population by the modified two-dimensional diffusion equation

$$(1) \quad \frac{\partial N}{\partial t} = D\underbrace{\left(\frac{\partial^2 N}{\partial x^2} + \frac{\partial^2 N}{\partial y^2}\right)}_{\substack{\text{expansion} \\ \text{by random} \\ \text{movement}}} + \underbrace{rN}_{\substack{\text{growth by} \\ \text{exponential} \\ \text{reproduction}}}$$

where $N(x, y, t)$ is the population density at the point (x, y) at time t, and D is a positive constant, called the *dispersion coefficient*, that is analogous to the diffusion coefficient.

It can be shown that one solution of Skellam's equation is

$$(2) \quad N(x, y, t) = \frac{M}{4\pi Dt}\, e^{rt - (x^2 + y^2)/(4Dt)}$$

where M is the number of individuals initially introduced at the source (see Question 5). The *asymptotic rate of population expansion, V*, is the distance between locations with equal population densities in successive years, and Skellam's model can be used to show that

$$(3) \quad V = \sqrt{4rD}$$

(see Question 4). Likewise, the *intrinsic rate of growth*, r, can be estimated using data of the growth of existing populations, and the dispersion coefficient, D, can be estimated using the formula

$$(4) \quad D \approx \frac{2A^2(t)}{\pi t}$$

where $A(t)$ is average distance organisms have traveled at time t.

Skellam's model has been used to study the spread of a variety of organisms, including oak trees, cereal leaf beetles, cabbage butterflies, and others. As an illustration of how the model can be applied, we return to the population of Central European muskrats introduced in the opening paragraph and Figure 1. Population studies indicate that r, the intrinsic rate of muskrat population increase, was no greater than 1.1 per year, and that D, the dispersion coefficient, was no greater than 230 km²/year. Consequently, the solution to Skellam's model stated in Equation (2) predicts that the distribution of muskrats, under the best circumstances for the species, is given by

$$(5) \quad N(x, y, t) = \frac{5}{4\pi(230)t} e^{1.1t - (x^2 + y^2)/(920t)}$$

where (x, y) is the point x km east and y km north from the release point near Prague and t is the time in years (after 1905). Formula (3) predicts that the maximum rate of population expansion is

$$V = \sqrt{4rD} = \sqrt{4(1.1)(230)} \approx 31.8 \text{ km/yr}$$

which is a little greater than the observed rate of 25.4 km/yr.

The derivation of the diffusion equation may be found in many differential equations texts, or see *Introduction to Mathematics for Life Scientists,* 3rd ed., by Edward Batschelet, Springer-Verlag, New York, pages 392–395. Skellam's model and several variations are discussed in *Mathematical Models in Biology* by Leah Edelstein-Keshet, McGraw-Hill, Boston, 1988, pages 436–441. It is important to emphasize that Skellam's modified diffusion equation given in formula (1) has solutions other than formula (2). In general, solving partial differential equations is very difficult, and often the best that can be done is to focus on finding solutions with certain specified forms. Such solutions can then be used to analyze practical situations, as we did with the muskrat problem.

Questions

1. Verify that $C(x, t) = \dfrac{M}{2\sqrt{\pi Dt}} e^{-(x^2/4Dt)}$ satisfies the diffusion equation

$$\frac{\partial C}{\partial t} = \alpha \frac{\partial^2 C}{\partial x^2}$$

Do this by calculating the partial derivatives and inserting them into the equation.

2. What relationship must hold between the coefficients a and b for $C(x, t) = e^{ax + bt}$ to be a solution of the diffusion equation

$$\frac{\partial C}{\partial t} = \alpha \frac{\partial^2 C}{\partial x^2}$$

3. Suppose that a population of organisms spreads out along a one-dimensional line according to the partial differential equation

$$\frac{\partial N}{\partial t} = D\frac{\partial^2 N}{\partial x^2} + rN$$

Show that the function $N(x, t) = \dfrac{M}{2\sqrt{\pi Dt}} e^{rt - (x^2/4Dt)}$ is a solution to this partial differential equation where M is the initial population of organisms located at the point $x = 0$ when $t = 0$.

4. Show that on the contours of equal population density, that is, the curves of the form $N(x, t) = A$ where A is a constant, the ratio x/t equals

$$\frac{x}{t} = \pm\left[4rD - \frac{2D}{t}\ln t - \frac{4D}{t}\ln\left(\sqrt{2\pi D}\,\frac{A}{M}\right)\right]^{1/2}$$

Using this formula, it can be shown that the ratio $\frac{x}{t}$ can be approximated by $\frac{x}{t} \approx \pm 2\sqrt{rD}$, which gives us a formula for the rate of population expansion.

5. Verify that $N(x, y, t) = \dfrac{M}{4\pi Dt} e^{rt - (x^2 + y^2)/(4Dt)}$ is a solution of the partial differential equation

$$\frac{\partial N}{\partial t} = D\left(\frac{\partial^2 N}{\partial x^2} + \frac{\partial^2 N}{\partial y^2}\right) + rN$$

6. Use Equation (5), which we obtained using Skellam's model, to find the population density for muskrats in 1925 at the location 50 km north and 50 km west of the release point near Prague.

7. Use Skellam's model to construct a function that estimates the population density of the small cabbage white butterfly if the largest diffusion coefficient observed is 129 km²/year and the largest intrinsic rate of increase observed is 31.5/year. What is the predicted maximum rate of population expansion? How does this compare with the largest observed rate of population expansion of 170 km/year?

8. In this question, you are asked to explore an alternative approach to analyzing the muskrat problem using Skellam's model. Recall that the intrinsic rate of growth of the muskrat population was $r = 1.1$ and that the maximum rate of dispersion was observed to be $V = 25.4$.

   a. Use these values for r and V in formula (3) to estimate the dispersion coefficient D.

   b. By substituting $r = 1.1$ into formula (2) along with the value for D you obtained in part (a), find the population density in 1925 for the muskrats at the location 50 km north and 50 km west of the release point (source) near Prague. Compare your answer with the answer to Question 6.

   c. Use formula (4) to estimate the average distance A of the muskrat population in 1925 from its source near Prague.

References

D. A. Andow, P. M. Kareiva, Simon A. Levin, and Akira Okubo, "Spread of Invading Organisms," *Landscape Ecology,* Vol. 4, nos. 2/3, 1990, pp. 177–188.
Leah Edelstein-Keshet, *Mathematical Models in Biology,* Boston: McGraw-Hill, 1988.
J. G. Skellam, "The Formulation and Interpretation of Mathematical Models of Diffusionary Processes in Population Biology," in *The Mathematical Theory of the Dynamics of Biological Populations,* edited by M. S. Bartlett and R. W. Hiorns, New York: Academic Press, 1973, pp. 63–85.
J. G. Skellam, "Random Dispersal in Theoretical Populations," *Biometrika,* Vol. 28, 1951, pp. 196–218.

APPENDIX A

ALGEBRA REVIEW

SECTION A1 | A Brief Review of Algebra

There are many techniques from elementary algebra that are needed in calculus. This appendix contains a review of such topics, and we begin by examining numbering systems.

The Real Numbers

An **integer** is a "whole number," either positive or negative. For example, 1, 2, 875, -15, -83, and 0 are integers, while $\frac{2}{3}$, 8.71, and $\sqrt{2}$ are not.

A **rational number** is a number that can be expressed as the quotient $\frac{a}{b}$ of two integers, where $b \neq 0$. For example, $\frac{2}{3}$, $\frac{8}{5}$, and $\frac{-4}{7}$ are rational numbers, as are

$$-6\frac{1}{2} = \frac{-13}{2} \qquad \text{and} \qquad 0.25 = \frac{25}{100} = \frac{1}{4}$$

Every integer is a rational number since it can be expressed as itself divided by 1. When expressed in decimal form, rational numbers are either terminating or infinitely repeating decimals. For example,

$$\frac{5}{8} = 0.625 \qquad \frac{1}{3} = 0.33\ldots \qquad \text{and} \qquad \frac{13}{11} = 1.181818\ldots$$

A number that cannot be expressed as the quotient of two integers is called an **irrational number.** For example,

$$\sqrt{2} \approx 1.41421356 \qquad \text{and} \qquad \pi \approx 3.14159265$$

are irrational numbers.

The rational numbers and irrational numbers form the **real numbers** and can be visualized geometrically as points on a **number line** as illustrated in Figure A.1.

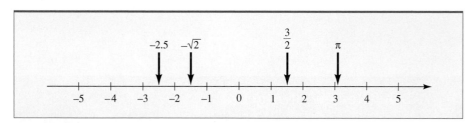

FIGURE A.1 The number line.

Inequalities

If a and b are real numbers and a is to the right of b on the number line, we say that ***a* is greater than *b*** and write $a > b$. If a is to the left of b, we say that ***a* is less than *b*** and write $a < b$ (Figure A.2). For example,

$$5 > 2 \qquad -12 < 0 \qquad \text{and} \qquad -8.2 < -2.4$$

FIGURE A.2 Inequalities.

Moreover,

$$\frac{6}{7} < \frac{7}{8}$$

as you can see by noting that

$$\frac{6}{7} = \frac{48}{56} \quad \text{and} \quad \frac{7}{8} = \frac{49}{56}$$

The symbol $\geq$ stands for **greater than or equal to,** and the symbol $\leq$ stands for **less than or equal to.** Thus, for example,

$$-3 \geq -4 \qquad -3 \geq -3 \qquad -4 \leq -3 \qquad \text{and} \qquad -4 \leq -4$$

Intervals   A set of real numbers that can be represented on the number line by a line segment is called an **interval.** Inequalities can be used to describe intervals. For example, the interval $a \leq x < b$ consists of all real numbers x that are between a and b, including a but excluding b. This interval is shown in Figure A.3. The numbers a and b are known as the **endpoints** of the interval. The square bracket at a indicates that a is included in the interval, while the rounded bracket at b indicates that b is excluded.

Intervals may be finite or infinite in extent and may or may not contain either endpoint. The possibilities (including customary notation and terminology) are illustrated in Figure A.4.

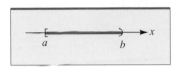

FIGURE A.3 The interval $a \leq x < b$.

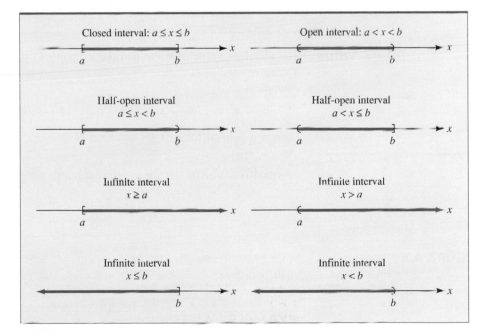

FIGURE A.4 Intervals of real numbers.

EXAMPLE | A1.1

Use inequalities to describe these intervals.

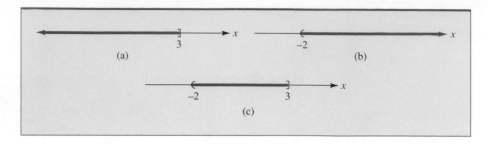

Solution

a. $x \le 3$ **b.** $x > -2$ **c.** $-2 < x \le 3$

EXAMPLE | A1.2

Represent each of these intervals as a line segment on a number line.

a. $x < -1$ **b.** $-1 \le x \le 2$ **c.** $x > 2$

Solution

a.

b.

c.

Absolute Value

The **absolute value** of a real number x, denoted by $|x|$, is the distance from x to 0 on a number line. Since distance is always nonnegative, it follows that $|x| \ge 0$. For example,

$$|4| = 4 \qquad |-4| = 4 \qquad |0| = 0 \qquad |5 - 9| = 4 \qquad |\sqrt{3} - 3| = 3 - \sqrt{3}$$

Here is a general formula for absolute value.

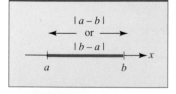

FIGURE A.5 The distance between a and $b = |a - b|$.

> **Absolute Value** ▪ For any real number x, the absolute value of x is
>
> $$|x| = \begin{cases} x & \text{if } x \ge 0 \\ -x & \text{if } x < 0 \end{cases}$$

More generally, the distance between any two numbers a and b is the absolute value of their difference taken in either order, as illustrated in Figure A.5.

EXAMPLE | A1.3

Find the distance on the number line between -2 and 3.

Solution

The distance between two numbers is the absolute value of their difference. Hence,

$$\text{Distance} = |-2 - 3| = |-5| = 5$$

The situation is illustrated in Figure A.6.

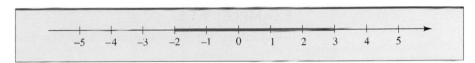

FIGURE A.6   Distance between -2 and 3.

Absolute Values and Intervals

The geometric interpretation of absolute value as distance can be used to simplify certain algebraic inequalities involving absolute values. Here is an example.

EXAMPLE | A1.4

Find the interval consisting of all real numbers x such that $|x - 1| \le 3$.

Solution

In geometric terms, the numbers x for which $|x - 1| \le 3$ are those whose distance from 1 is less than or equal to 3. As illustrated in Figure A.7, these are the numbers that satisfy $-2 \le x \le 4$.

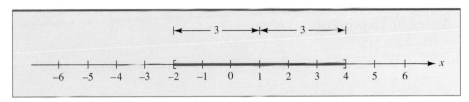

FIGURE A.7   The interval on which $|x - 1| \le 3$ is $-2 \le x \le 4$.

To find this interval algebraically, without relying on the geometry, rewrite the inequality $|x - 1| \le 3$ as

$$-3 \le x - 1 \le 3$$

and add 1 to each term to get

$$-3 + 1 \le x - 1 + 1 \le 3 + 1$$

or

$$-2 \le x \le 4$$

Exponential Notation

These rules define the expression a^x for $a > 0$ and all rational values of x.

> **Definition of a^x for $x \ge 0$**   ■   Integer powers: If n is a positive integer, then
>
> $$a^n = a \cdot a \cdots a$$
>
> where the product $a \cdot a \cdots a$ contains n factors.
> Fractional powers: If n and m are positive integers, then
>
> $$a^{n/m} = (\sqrt[m]{a})^n = \sqrt[m]{a^n}$$
>
> where $\sqrt[m]{}$ denotes the positive mth root.
>
> Negative powers: $a^{-n} = \dfrac{1}{a^n}$
>
> Zero power: $a^0 = 1$

EXAMPLE | A1.5

Evaluate these expressions (without using your calculator).

a. $9^{1/2}$ **b.** $27^{2/3}$ **c.** $8^{-1/3}$ **d.** $\left(\dfrac{1}{100}\right)^{-3/2}$ **e.** 5^0

Solution

a. $9^{1/2} = \sqrt{9} = 3$

b. $27^{2/3} = (\sqrt[3]{27})^2 = 3^2 = 9$
$$= \sqrt[3]{(27)^2} = \sqrt[3]{729} = 9$$

c. $8^{-1/3} = \dfrac{1}{8^{1/3}} = \dfrac{1}{\sqrt[3]{8}} = \dfrac{1}{2}$

d. $\left(\dfrac{1}{100}\right)^{-3/2} = 100^{3/2} = (\sqrt{100})^3 = 10^3 = 1{,}000$

e. $5^0 = 1$

Laws of Exponents

Exponents obey these useful laws.

> **Laws of Exponents** ■ For a real number a, we have
>
> The product law: $a^r a^s = a^{r+s}$
>
> The quotient law: $\dfrac{a^r}{a^s} = a^{r-s}$ if $a \neq 0$
>
> The power law: $(a^r)^s = a^{rs}$

The laws of exponents are illustrated in Examples A1.6 and A1.7.

EXAMPLE | A1.6

Evaluate these expressions (without using a calculator).

a. $(2^{-2})^3$ **b.** $\dfrac{3^3}{3^{1/3}(3^{2/3})}$ **c.** $2^{7/4}(8^{-1/4})$

Solution

a. $(2^{-2})^3 = 2^{-6} = \dfrac{1}{2^6} = \dfrac{1}{64}$

b. $\dfrac{3^3}{3^{1/3}(3^{2/3})} = \dfrac{3^3}{3^{1/3+2/3}} = \dfrac{3^3}{3^1} = 3^2 = 9$

c. $2^{7/4}(8^{-1/4}) = 2^{7/4}(2^3)^{-1/4} = 2^{7/4}(2^{-3/4}) = 2^{7/4-3/4} = 2^1 = 2$

EXAMPLE | A1.7

Solve each of these equations for n.

a. $\dfrac{a^5}{a^2} = a^n$ **b.** $(a^n)^5 = a^{20}$

Solution

a. Since $\dfrac{a^5}{a^2} = a^{5-2} = a^3$, it follows that $n = 3$.

b. Since $(a^n)^5 = a^{5n}$, it follows that $5n = 20$ or $n = 4$.

Factoring

To factor an expression is to write it as a product of two or more terms, called **factors**. Factoring is used to simplify complicated expressions and to solve equations and is based on the distributive law for addition and multiplication.

> **The Distributive Law** ■ For any real numbers a, b, and c,
>
> $$ab + ac = a(b + c)$$

The factoring techniques you will need in this book are illustrated in Examples A1.8 through A1.10.

EXAMPLE | A1.8

Factoring Out Common Terms

Factor the expression $3x^4 - 6x^3$.

Solution

Since $3x^3$ is a factor of each of the terms in this expression, you can use the distributive law to "factor out" $3x^3$ and write

$$3x^4 - 6x^3 = 3x^3(x - 2)$$

EXAMPLE | A1.9

Simplify the expression $10(1 - x)^4(x + 1)^4 + 8(x + 1)^5(1 - x)^3$.

Solution

The greatest common factor is $2(1 - x)^3(x + 1)^4$. Factor this out to get

$$10(1 - x)^4(x + 1)^4 + 8(x + 1)^5(1 - x)^3 = 2(1 - x)^3(x + 1)^4[5(1 - x) + 4(x + 1)]$$

Since no further factorization is possible, do the multiplication in the square brackets and combine the resulting terms to conclude that

$$10(1 - x)^4(x + 1)^4 + 8(x + 1)^5(1 - x)^3 = 2(1 - x)^3(x + 1)^4(5 - 5x + 4x + 4)$$
$$= 2(1 - x)^3(x + 1)^4(9 - x)$$

There are times in calculus when it is necessary to factor expressions involving radicals or fractional exponents. Here is an example of such a factorization.

EXAMPLE | A1.10

Factor the expression

$$\frac{2}{3}x^{-1/3}(x + 1) + x^{2/3}$$

Solution

We find that

$$\frac{2}{3}x^{-1/3}(x+1) + x^{2/3} = \frac{\frac{2}{3}(x+1)}{x^{1/3}} + x^{2/3}$$

$$= \frac{\frac{2}{3}(x+1) + x^{2/3}(x^{1/3})}{x^{1/3}} \qquad \text{\textit{put terms over the common denominator } }x^{1/3}\text{ \textit{and simplify}}$$

$$= \frac{\frac{2}{3}(x+1) + x}{x^{1/3}} = \frac{\frac{5}{3}x + \frac{2}{3}}{x^{1/3}} \qquad \text{\textit{since }}x^{2/3}x^{1/3} = x$$

$$= \frac{1}{3}(5x + 2)x^{-1/3}$$

Simplifying Quotients by Factoring and Canceling

Example A1.11 illustrates how you can combine factoring and canceling to simplify certain types of quotients that arise frequently in calculus.

EXAMPLE ┃ A1.11

Simplify the quotient

$$\frac{4(x+3)^4(x-2)^2 - 6(x+3)^3(x-2)^3}{(x+3)(x-2)^3}$$

Solution

First simplify the numerator to get

$$\frac{4(x+3)^4(x-2)^2 - 6(x+3)^3(x-2)^3}{(x+3)(x-2)^3} = \frac{2(x+3)^3(x-2)^2[2(x+3) - 3(x-2)]}{(x+3)(x-2)^3}$$

$$= \frac{2(x+3)^3(x-2)^2(2x+6 - 3x+6)}{(x+3)(x-2)^3}$$

$$= \frac{2(x+3)^3(x-2)^2(12 - x)}{(x+3)(x-2)^3}$$

and then "cancel" the common factor of $(x+3)(x-2)^2$ from the numerator and denominator to conclude that

$$\frac{4(x+3)^4(x-2)^2 - 6(x+3)^3(x-2)^3}{(x+3)(x-2)^3} = \frac{2(x+3)^2(12 - x)}{x - 2}$$

Summation Notation

Sums of the form

$$a_1 + a_2 + \cdots + a_n$$

appear in Chapters 5 and 6. To describe such a sum, it suffices to specify the general term a_j and to indicate that n terms of this form are to be added, starting with the term in which $j = 1$ and ending with the term in which $j = n$. It is customary to use the Greek letter Σ (sigma) to denote summation and to express the sum compactly as follows.

Summation Notation   ▪   The sum of the numbers $a_1 \cdots a_n$ is given by

$$a_1 + a_2 + \cdots + a_n = \sum_{j=1}^{n} a_j$$

The use of summation notation is illustrated in Examples A1.12 and A1.13.

EXAMPLE | A1.12

Use summation notation to represent these sums.

a. $1 + 4 + 9 + 16 + 25 + 36 + 49 + 64$

b. $(1 - x_1)^2 \Delta x + (1 - x_2)^2 \Delta x + \cdots + (1 - x_{15})^2 \Delta x$

Solution

a.   This is a sum of 8 terms of the form j^2, starting with $j = 1$ and ending with $j = 8$. Hence,

$$1 + 4 + 9 + 16 + 25 + 36 + 49 + 64 = \sum_{j=1}^{8} j^2$$

b.   The jth term of this sum is $(1 - x_j)^2 \Delta x$. Hence,

$$(1 - x_1)^2 \Delta x + (1 - x_2)^2 \Delta x + \cdots + (1 - x_{15})^2 \Delta x = \sum_{j=1}^{15} (1 - x_j)^2 \Delta x$$

EXAMPLE | A1.13

Evaluate these sums.

a. $\displaystyle\sum_{j=1}^{4} (j^2 + 1)$

b. $\displaystyle\sum_{j=1}^{3} (-2)^j$

Solution

a. $\displaystyle\sum_{j=1}^{4} (j^2 + 1) = (1^2 + 1) + (2^2 + 1) + (3^2 + 1) + (4^2 + 1)$

$$= 2 + 5 + 10 + 17 = 34$$

b. $\displaystyle\sum_{j=1}^{3} (-2)^j = (-2)^1 + (-2)^2 + (-2)^3 = -2 + 4 - 8 = -6$

PROBLEMS A1

INTERVALS   *In Problems 1 through 4, use inequalities to describe the indicated interval.*

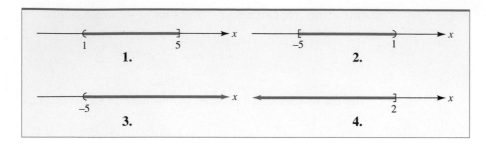

In Problems 5 through 8, represent the given interval as a line segment on a number line.

5. $x \geq 2$ 6. $-6 \leq x < 4$
7. $-2 < x \leq 0$ 8. $x > 3$

DISTANCE   *In Problems 9 through 12, find the distance on the number line between the given pair of real numbers.*

9. 0 and -4 10. 2 and 5
11. -2 and 3   12. -3 and -1

ABSOLUTE VALUE AND INTERVALS   *In Problems 13 through 18, find the interval or intervals consisting of all real numbers x that satisfy the given inequality.*

13. $|z| \leq 3$ 14. $|x - 2| \leq 5$
15. $|x + 4| \leq 2$ 16. $|1 - x| < 3$
17. $|x + 2| \geq 5$ 18. $|x - 1| > 3$

EXPONENTIAL NOTATION   *In Problems 19 through 26, evaluate the given expression without using a calculator.*

19. 5^3 20. 2^{-3}
21. $16^{1/2}$ 22. $36^{-1/2}$
23. $8^{2/3}$ 24. $27^{-4/3}$
25. $\left(\dfrac{1}{4}\right)^{1/2}$ 26. $\left(\dfrac{1}{4}\right)^{-3/2}$

In Problems 27 through 34, evaluate the given expression without using a calculator.

27. $\dfrac{2^5(2^2)}{2^8}$ 28. $\dfrac{3^4(3^3)}{(3^2)^3}$

29. $\dfrac{2^{4/3}(2^{5/3})}{2^5}$ 30. $\dfrac{5^{-3}(5^2)}{(5^{-2})^3}$

31. $\dfrac{2(16^{3/4})}{2^3}$ 32. $\dfrac{\sqrt{27}\,(\sqrt{3})^3}{9}$

33. $[\sqrt{8}\,(2^{5/2})]^{-1/2}$ 34. $[\sqrt{27}\,(3^{5/2})]^{1/2}$

In Problems 35 through 42, solve the given equation for n. (Assume a > 0 and a ≠ 1.)

35. $a^3 a^7 - a^n$

36. $\dfrac{a^5}{a^2} = a^n$

37. $a^4 a^{-3} = a^n$

38. $a^2 a^n = \dfrac{1}{a}$

39. $(a^3)^n = a^{12}$

40. $(a^n)^5 = \dfrac{1}{a^{10}}$

41. $a^{3/5} a^{\ n} = \dfrac{1}{a^2}$

42. $(a^n)^3 = \dfrac{1}{\sqrt{a}}$

SIMPLIFYING EXPRESSIONS   *In Problems 43 through 54, factor and simplify the given expression as much as possible.*

43. $x^5 - 4x^4$

44. $3x^3 - 12x^4$

45. $100 - 25(x - 3)$

46. $60 - 20(4 - x)$

47. $8(x + 1)^3(x - 2)^2 + 6(x + 1)^2(x - 2)^3$

48. $12(x + 3)^5(x - 1)^3 - 8(x + 3)^6(x - 1)^2$

49. $x^{-1/2}(2x + 1) + 4x^{1/2}$

50. $x^{-1/4}(3x + 5) + 4x^{3/4}$

51. $\dfrac{(x + 3)^3(x + 1) - (x + 3)^2(x + 1)^2}{(x + 3)(x + 1)}$

52. $\dfrac{3(x - 2)^2(x + 1)^2 - 2(x - 2)(x + 1)^3}{(x - 2)^4}$

53. $\dfrac{4(1 - x)^2(x + 3)^3 + 2(1 - x)(x + 3)^4}{(1 - x)^4}$

54. $\dfrac{6(x + 2)^5(1 - x)^4 - 4(x + 2)^6(1 - x)^3}{(x + 2)^8(1 - x)^2}$

SUMMATION NOTATION   *In Problems 55 through 58, evaluate the given sum.*

55. $\displaystyle\sum_{j=1}^{4}(3j + 1)$

56. $\displaystyle\sum_{j=1}^{5} j^2$

57. $\displaystyle\sum_{j=1}^{10}(-1)^j$

58. $\displaystyle\sum_{j=1}^{5} 2^j$

In Problems 59 through 64, use summation notation to represent the given sum.

59. $1 + \dfrac{1}{2} + \dfrac{1}{3} + \dfrac{1}{4} + \dfrac{1}{5} + \dfrac{1}{6}$

60. $3 + 6 + 9 + 12 + 15 + 18 + 21 + 24 + 27 + 30$

61. $2x_1 + 2x_2 + 2x_3 + 2x_4 + 2x_5 + 2x_6$

62. $1 - 1 + 1 - 1 + 1 - 1$

63. $1 - 2 + 3 - 4 + 5 - 6 + 7 - 8$

64. $x - x^2 + x^3 - x^4 + x^5$

65. ECOLOGY   The atmosphere above each square centimeter of the earth's surface weighs 1 kilogram (kg).
  a. Assuming the earth is a sphere of radius $R = 6,440$ km, use the formula $S = 4\pi R^2$ to calculate the surface area of the earth and then find the total mass of the atmosphere.
  b. Oxygen occupies approximately 22% of the total mass of the atmosphere, and it is estimated that plant life produces approximately 0.9×10^{13} kg of oxygen per year. If none of this oxygen were used up by plants or animals (or combustion), how long would it take to build up the total mass of oxygen in the atmosphere (part a)?*

*Adapted from a problem in E. Batschelet, *Introduction to Mathematics for Life Scientists*, 2nd ed., New York: Springer-Verlag, 1979, p. 31.

SECTION   A2   Factoring Polynomials and Solving Systems of Equations

A **polynomial** is an expression of the form

$$a_0 + a_1 x + a_2 x^2 + \cdots + a_n x^n$$

where n is a nonnegative integer and $a_0, a_1, a_2, \cdots, a_n$ are real numbers, known as the **coefficients** of the polynomial. If $a_n \neq 0$, n is said to be the **degree** of the polynomial. For example, $3x^5 - 7x^2 + 12$ is a polynomial of degree 5. Polynomial expressions appear throughout calculus, as examples and in practical applications. In this section, we shall study techniques for factoring polynomials and shall also see how to solve systems of equations involving polynomials.

Factoring Polynomials with Integer Coefficients

Many of the polynomials that arise in practice have integer coefficients (or are closely related to polynomials that do). Techniques for factoring polynomials with integer coefficients are illustrated in the following examples. In each, the goal is to rewrite the given polynomial as a product of polynomials of lower degree that also have integer coefficients.

EXAMPLE | A2.1

Factor the polynomial $x^2 - 2x - 3$ using integer coefficients.

Solution

The goal is to write the polynomial as a product of the form

$$x^2 - 2x - 3 = (x + a)(x + b)$$

where a and b are integers. The distributive law implies that

$$(x + a)(x + b) = x^2 + (a + b)x + ab$$

Hence, the goal is to find integers a and b such that

$$x^2 - 2x - 3 = x^2 + (a + b)x + ab$$

or, equivalently, such that

$$a + b = -2 \qquad \text{and} \qquad ab = -3$$

From the list

$$1, -3 \qquad \text{and} \qquad -1, 3$$

of pairs of integers whose product is -3, choose $a = -3$ and $b = 1$ as the only pair whose sum is -2. It follows that

$$x^2 - 2x - 3 = (x - 3)(x + 1)$$

which you should check by multiplying out the right-hand side.

EXAMPLE | A2.2

Factor the polynomial $2x^2 + x - 10$ using integer coefficients.

Solution

We wish to find integers a, b, c, and d so that

$$2x^2 + x - 10 = (ax + b)(cx + d) \qquad \textit{distributive law}$$
$$= (ac)x^2 + (bc + ad)x + (bd)$$

so we must have

$$ac = 2$$
$$bc + ad = 1$$
$$bd = -10$$

Because a, b, c, and d must all be integers, there are only a limited number of possibilities for the choices of a, c and b, d; namely,

4 possible choices for the pair a, c: 2, 1; 1, 2; -2, -1; and -1, -2
8 possible choices for the pair b, d: 2, -5; 5, -2; 1, -10; 10, -1; -2, 5;
-5, 2; -1, 10; and -10, 1

There are $(4)(8) = 32$ possible ways of forming the expression $bc + ad$. We find that the condition $bc + ad = 1$ is satisfied when $a = 2$, $b = 5$, $c = 1$, $d = -2$, so that

$$2x^2 + x - 10 = (2x + 5)(x - 2)$$

(Note that $bc + ad = 1$ is also satisfied by $a = -2$, $b = -5$, $c = -1$, and $d = 2$. Why is it not necessary to list this as a second factorization of the given polynomial?)

EXAMPLE | A2.3

Factor the polynomial $x^3 - 8$ using integer coefficients.

Solution

The fact that $2^3 = 8$ tells you that $x - 2$ must be a factor of this expression. That is, there are integers a and b for which

$$x^3 - 8 = (x - 2)(x^2 + ax + b)$$

Since

$$(x - 2)(x^2 + ax + b) = x^3 + (a - 2)x^2 + (b - 2a)x - 2b$$

(which you should check for yourself), the goal is to find integers a and b for which

$$a - 2 = 0 \qquad b - 2a = 0 \qquad \text{and} \qquad 2b = 8$$

Clearly, the only such integers are $a = 2$ and $b = 4$. Hence,

$$x^3 - 8 = (x - 2)(x^2 + 2x + 4)$$

Convince yourself by examining pairs of integers whose product is 4 that the polynomial $x^2 + 2x + 4$ cannot be factored further with integer coefficients.

The Difference of Two Squares   Here is a formula (which you can verify by multiplication) for the factorization of the difference between two perfect squares. It is particularly useful, and you should memorize it.

> **The Difference of Two Squares** ■ For any real numbers a and b,
>
> $$a^2 - b^2 = (a + b)(a - b)$$

The use of this formula is illustrated in Example A2.4.

EXAMPLE | A2.4

Factor the polynomial $x^5 - 4x^3$ using integer coefficients.

Solution

First factor out x^3 to get

$$x^5 - 4x^3 = x^3(x^2 - 4)$$

and then factor $x^2 - 4$ (which is the difference of two squares) to conclude that

$$x^5 - 4x^3 = x^3(x + 2)(x - 2)$$

The Solution of Equations by Factoring

The **solutions** of an equation are the values of the variable that make the equation true. For example, $x = 2$ is a solution of the equation

$$x^3 - 6x^2 + 12x - 8 = 0$$

because substitution of 2 for x gives

$$2^3 - 6(2^2) + 12(2) - 8 = 8 - 24 + 24 - 8 = 0$$

In Examples A2.5 and A2.6, you will see how factoring can be used to solve certain equations. The technique is based on the fact that if the product of two (or more) terms is equal to zero, then at least one of the terms must be equal to zero. For example, if $ab = 0$, then either $a = 0$ or $b = 0$ (or both).

EXAMPLE | A2.5

Solve the equation $x^2 - 3x = 10$.

Solution

First subtract 10 from both sides to get

$$x^2 - 3x - 10 = 0$$

and then factor the resulting polynomial on the left-hand side to get

$$(x - 5)(x + 2) = 0$$

Since the product $(x - 5)(x + 2)$ can be zero only if one (or both) of its factors is zero, it follows that the solutions are $x = 5$ (which makes the first factor zero) and $x = -2$ (which makes the second factor zero).

EXAMPLE | A2.6

Solve the equation $1 - \dfrac{1}{x} - \dfrac{2}{x^2} = 0$.

Solution

Put the fractions on the left-hand side over the common denominator x^2 and add to get

$$\frac{x^2}{x^2} - \frac{x}{x^2} - \frac{2}{x^2} = 0$$

or

$$\frac{x^2 - x - 2}{x^2} = 0$$

Now factor the polynomial in the numerator to get

$$\frac{(x + 1)(x - 2)}{x^2} = 0$$

A quotient is zero only if its numerator is zero and its denominator is *not* zero, so it follows that $x = -1$ and $x = 2$ are the required solutions.

The Solution of Quadratic Equations

An equation of the form

$$ax^2 + bx + c = 0 \qquad (\text{for } a \neq 0)$$

is said to be a **quadratic equation.** A quadratic equation can have at most two solutions. As you have seen, one way to find the solutions is to factor the equation. When the factors are not obvious or when the equation cannot be factored at all, you can use this special formula to solve quadratic equations.

The Quadratic Formula ■ The solutions of the quadratic equation

$$ax^2 + bx + c = 0 \qquad (\text{for } a \neq 0)$$

are given by the formula

$$x = \frac{-b \pm \sqrt{b^2 - 4ac}}{2a}$$

The term $b^2 - 4ac$ in the quadratic formula is called the **discriminant** of the quadratic equation. If the discriminant is positive, the equation has two solutions, one coming from the formula with the sign $\pm$ replaced by $+$ and the other with $\pm$ replaced by $-$. If the discriminant is zero, the equation has only one solution since the formula reduces to $x = \dfrac{-b}{2a}$. If the discriminant is negative, the equation has no real solutions since negative numbers do not have real square roots.

The use of the quadratic formula is illustrated in Examples A2.7 through A2.9.

EXAMPLE | A2.7

Solve the equation $x^2 + 3x + 1 = 0$.

Solution

This is a quadratic equation with $a = 1$, $b = 3$, and $c = 1$. Using the quadratic formula, you get

$$x = \frac{-3 + \sqrt{5}}{2} \approx -0.38 \qquad \text{and} \qquad x = \frac{-3 - \sqrt{5}}{2} \approx -2.62$$

EXAMPLE | A2.8

Solve the equation $x^2 + 18x + 81 = 0$.

Solution

This is a quadratic equation with $a = 1$, $b = 18$, and $c = 81$. Using the quadratic formula, you find that the discriminant is zero and that the formula for x gives

$$x = \frac{-18 \pm \sqrt{0}}{2} = -\frac{18}{2} = -9$$

EXAMPLE | A2.9

Solve the equation $x^2 + x + 1 = 0$.

Solution

This is a quadratic equation with $a = 1$, $b = 1$, and $c = 1$. Using the quadratic formula, you get

$$x = \frac{-1 \pm \sqrt{-3}}{2}$$

Since there is no real square root of -3, it follows that the equation has no real solution.

Systems of Equations

A collection of equations that are to be solved simultaneously is called a **system of equations.** Some of the calculus problems in Chapter 7 involve the solution of systems of two (or more) equations in two (or more) unknowns. A typical example is to find the real numbers x and y that satisfy the system

$$2x + 3y = 5$$
$$x + 2y = 4$$

The procedure for solving a system of two equations in two unknowns is to (temporarily) eliminate one of the variables, thereby reducing the problem to a single equation in one variable, which you then solve for its variable. Once you have found the value of one of the variables, you can substitute it into either of the original equations and solve to get the value of the other variable.

The most common techniques for the elimination of variables are illustrated in Examples A2.10 and A2.11.

EXAMPLE | A2.10

Elimination by Multiplication and Addition

Solve the system

$$4x + 3y = 13$$
$$3x + 2y = 7$$

Solution

To eliminate y, multiply both sides of the first equation by 2 and both sides of the second equation by -3 so that the system becomes

$$8x + 6y = 26$$
$$-9x - 6y = -21$$

Then add the equations to get

$$-x + 0 = 5 \quad \text{or} \quad x = -5$$

To find y, you can substitute $x = -5$ into either of the original equations. If you choose the second equation, you get

$$3(-5) + 2y = 7 \quad 2y = 22 \quad \text{or} \quad y = 11$$

That is, the solution of the system is $x = -5$ and $y = 11$.

To check this answer, substitute $x = -5$ and $y = 11$ into each of the original equations. From the first equation you get

$$4(-5) + 3(11) = -20 + 33 = 13$$

and from the second equation you get

$$3(-5) + 2(11) = -15 + 22 = 7$$

as required.

EXAMPLE | A2.11

Elimination by Substitution

Solve the system

$$2y^2 - x^2 = 14$$
$$x - y = 1$$

Solution

Solve the second equation for x to get

$$x = y + 1$$

and substitute this into the first equation to eliminate x. This gives

$$2y^2 - (y + 1)^2 = 14$$
$$2y^2 - (y^2 + 2y + 1) = 14$$
$$2y^2 - y^2 - 2y - 1 = 14$$
$$y^2 - 2y - 15 = 0$$

or

$$(y + 3)(y - 5) = 0$$

from which it follows that

$$y = -3 \quad \text{or} \quad y = 5$$

If $y = -3$, the second equation gives

$$x - (-3) = 1 \quad \text{or} \quad x = -2$$

and if $y = 5$, the second equation gives

$$x - 5 = 1 \quad \text{or} \quad x = 6$$

Hence the system has two solutions,

$$x = 6, y = 5 \quad \text{and} \quad x = -2, y = -3$$

To check these answers, substitute each pair x, y into the first equation. If $x = 6$ and $y = 5$, you get

$$2(5^2) - 6^2 = 50 - 36 = 14$$

and if $x = -2$ and $y = -3$, you get

$$2(-3)^2 - (-2)^2 = 18 - 4 = 14$$

as required.

PROBLEMS   A2

FACTORING POLYNOMIALS WITH INTEGER COEFFICIENTS   *In Problems 1 through 14 factor the given polynomial using integer coefficients.*

1. $x^2 + x - 2$
2. $x^2 + 3x - 10$
3. $x^2 - 7x + 12$
4. $x^2 + 8x + 12$
5. $x^2 - 2x + 1$
6. $x^2 + 6x + 9$
7. $16x^2 - 25$
8. $3x^2 - x - 14$
9. $x^3 - 1$
10. $x^3 - 27$
11. $x^7 - x^5$
12. $x^3 + 2x^2 + x$
13. $2x^3 - 8x^2 - 10x$
14. $x^4 + 5x^3 - 14x^2$

SOLUTION OF EQUATIONS BY FACTORING   *In Problems 15 through 30 solve the given equation by factoring.*

15. $x^2 - 2x - 8 = 0$
16. $x^2 - 4x + 3 = 0$
17. $x^2 + 10x + 25 = 0$
18. $x^2 + 8x + 16 = 0$
19. $x^2 - 16 = 0$
20. $x^2 - 25 = 0$
21. $2x^2 + 3x + 1 = 0$
22. $x^2 - 2x + 1 = 0$
23. $4x^2 + 12x + 9 = 0$
24. $6x^2 + 7x - 3 = 0$
25. $1 + \dfrac{4}{x} - \dfrac{5}{x^2} = 0$
26. $\dfrac{9}{x^2} - \dfrac{6}{x} + 1 = 0$
27. $2 + \dfrac{2}{x} - \dfrac{4}{x^2} = 0$
28. $\dfrac{3}{x^2} - \dfrac{5}{x} - 2 = 0$
29. $\dfrac{x}{x - 2} - \dfrac{4}{x + 3} - \dfrac{10}{x^2 + x - 6} = 0$
30. $\dfrac{x}{x + 1} + \dfrac{3}{2x + 3} - \dfrac{11x + 10}{2x^2 + 5x + 3} = 0$

QUADRATIC FORMULA   *In Problems 31 through 36 use the quadratic formula to solve the given equation.*

31. $2x^2 + 3x + 1 = 0$
32. $-x^2 + 3x - 1 = 0$
33. $x^2 - 2x + 3 = 0$
34. $x^2 - 2x + 1 = 0$
35. $4x^2 + 12x + 9 = 0$
36. $x^2 + 12 = 0$

SYSTEMS OF EQUATIONS   *In Problems 37 through 42 solve the given system of equations.*

37. $x + 5y = 13$
$3x - 10y = -11$

38. $2x - 3y = 4$
$3x - 5y = 2$

39. $5x - 4y = 12$
$2x - 3y = 2$

40. $3x^2 - 9y = 0$
$3y^2 - 9x = 0$

41. $2y^2 - x^2 - 1$
$x - 2y = 3$

42. $2x^2 - y^2 = -7$
$2x + y = 1$

SECTION   A3   Evaluating Limits with L'Hôpital's Rule

L'Hôpital's Rule:
$$\frac{0}{0} \text{ and } \frac{\infty}{\infty} \text{ Forms}$$

In curve sketching and other applications of calculus, it is often necessary to compute a limit of the form

$$\lim_{x \to c} \frac{f(x)}{g(x)}$$

where c is either a finite number or ∞. If $\lim_{x \to c} g(x) \neq 0$, then the quotient rule for limits may be used, but if both $f(x)$ and $g(x)$ approach 0 as x approaches c, practically anything can happen. For example,

$$\lim_{x \to \infty} \frac{(1/x^3) - (1/x^2)}{1/x} \qquad \lim_{x \to 0} \frac{2x^3 + 3x^2}{x^5 + x^2} \qquad \text{and} \qquad \lim_{x \to 1} \frac{x - 1}{x^3 - 1}$$

all have this property, but the limit on the left is 0, the one in the center is ∞, and the one on the right is $\frac{1}{3}$.

Limits such as these are called $\frac{0}{0}$ **indeterminate forms.** Similarly, limits of quotients in which both the numerator and denominator increase or decrease without bound as $x \to c$ are called $\frac{\infty}{\infty}$ **indeterminate forms.**

There is a powerful technique, known as **L'Hôpital's rule,** which you can use to analyze indeterminate forms. The rule says, in effect, that if your attempt to find the limit of a quotient leads to either a $\frac{0}{0}$ or $\frac{\infty}{\infty}$ indeterminate form, then take derivatives of the numerator and the denominator and try again. Here is a more symbolic statement of the procedure.

L'Hôpital's Rule

If $\lim_{x \to c} f(x) = 0$ and $\lim_{x \to c} g(x) = 0$, then

$$\lim_{x \to c} \frac{f(x)}{g(x)} = \lim_{x \to c} \frac{f'(x)}{g'(x)}$$

If $\lim_{x \to c} f(x) = \infty$ and $\lim_{x \to c} g(x) = \infty$, then

$$\lim_{x \to c} \frac{f(x)}{g(x)} = \lim_{x \to c} \frac{f'(x)}{g'(x)}$$

The use of L'Hôpital's rule is illustrated in Examples A3.1 through A3.3. As you read through these examples, pay particular attention to the following two points:

1. L'Hôpital's rule involves differentiation of the numerator and the denominator *separately*. A common mistake is to differentiate the entire quotient using the quotient rule.

2. L'Hôpital's rule applies only to quotients whose limits are indeterminate forms $\frac{0}{0}$ or $\frac{\infty}{\infty}$. Limits of the form $\frac{0}{\infty}$ or $\frac{\infty}{0}$ are *not* indeterminate (the first is 0, and the second is ∞).

EXAMPLE | A3.1

Use L'Hôpital's rule to compute the limit

$$\lim_{x \to \infty} \frac{x}{(x+1)^2}$$

Solution

This is a $\frac{\infty}{\infty}$ indeterminate form, so L'Hôpital's rule applies, and we get

$$\lim_{x \to \infty} \frac{x}{(x+1)^2} = \lim_{x \to \infty} \frac{1}{2(x+1)} = 0$$

EXAMPLE | A3.2

Use L'Hôpital's rule to compute the limit

$$\lim_{x \to 1} \frac{x^5 - 3x^4 + 5x - 3}{4x^5 + 2x^3 - 5x^2 - 1}$$

Solution

By substituting $x = 1$ into the numerator and denominator, we see that this is a $\frac{0}{0}$ indeterminate form. We could evaluate this limit by the factor method developed in Chapter 1, but notice how much easier it is to use L'Hôpital's rule:

$$\lim_{x \to 1} \frac{x^5 - 3x^4 + 5x - 3}{4x^5 + 2x^3 - 5x^2 - 1} = \lim_{x \to 1} \frac{5x^4 - 12x^3 + 5}{20x^4 + 6x^2 - 10x} = \frac{-2}{16} = \frac{-1}{8}$$

EXAMPLE | A3.3

Evaluate $\lim_{x \to 2} \dfrac{2x + 5}{x^2 + 3x - 10}$.

Solution

If you blindly apply L'Hôpital's rule, you get

$$\lim_{x \to 2} \frac{2x + 5}{x^2 + 3x - 10} = \lim_{x \to 2} \frac{2}{2x + 3} = \frac{2}{7}$$

However, if you use your calculator to evaluate the given quotient at a number very close to 2 (say, at 2.0001), you find that the number you get is much larger than $\frac{2}{7}$. Why? The answer you got with L'Hôpital's rule was wrong because the given limit is not indeterminate. In fact, by simply substituting $x = 2$, you get

$$\lim_{x \to 2} \frac{2x + 5}{x^2 + 3x - 10} = \frac{9}{0} = \infty$$

Limits Involving e^x and $\ln x$

In certain applications, you will need to compute limits involving e^x and $\ln x$. Sometimes these limits can be computed directly, using the basic formulas

$$\lim_{x \to \infty} e^x = \infty \qquad \text{and} \qquad \lim_{x \to \infty} \ln x = \infty$$

but other limits are indeterminate forms and require the use of L'Hôpital's rule. Here are two examples.

EXAMPLE | A3.4

Find $\lim_{x \to \infty} \dfrac{e^{-x}}{1 + e^{-x}}$.

Solution

We find that

$$\lim_{x \to \infty} \frac{e^{-x}}{1 + e^{-x}} = \frac{0}{1} = 0$$

Notice that the limit is *not* an indeterminate form, so L'Hôpital's rule does *not* apply. However, if we blindly apply L'Hôpital's rule, we get

$$\lim_{x \to \infty} \frac{e^{-x}}{1 + e^{-x}} = \lim_{x \to \infty} \frac{-e^{-x}}{-e^{-x}} = 1$$

which is *wrong!* This example reminds us that it is *very* important to make sure we are dealing with an indeterminate form before using L'Hôpital's rule.

EXAMPLE | A3.5

Find $\lim_{x \to \infty} \dfrac{3 - e^x}{x^2}$.

Solution

This time the limit is indeterminate of the form $\frac{\infty}{\infty}$. Applying L'Hôpital's rule, we get

$$\lim_{x \to \infty} \frac{3 - e^x}{x^2} = \lim_{x \to \infty} \frac{-e^x}{2x}$$

Since this new limit is also of the form $\dfrac{\infty}{\infty}$, we apply L'Hôpital's rule again to get

$$\lim_{x \to \infty} \frac{-e^x}{2x} = \lim_{x \to \infty} \frac{-e^x}{2} = -\infty$$

and we conclude that

$$\lim_{x \to \infty} \frac{3 - e^x}{x^2} = -\infty$$

Although L'Hôpital's rule only applies to $\dfrac{0}{0}$ and $\dfrac{\infty}{\infty}$ indeterminate forms, other kinds of indeterminate forms can often be computed by combining L'Hôpital's rule with a little algebra. This procedure is illustrated in Examples A3.6 and A3.7.

EXAMPLE | A3.6

Find $\lim\limits_{x \to \infty} e^{-x} \ln x$.

Solution

This limit is of the indeterminate form $0 \cdot \infty$ and can be rewritten as

$$\lim_{x \to \infty} \frac{e^{-x}}{1/\ln x} \quad \left(\text{of the form } \frac{0}{0}\right)$$

or as

$$\lim_{x \to \infty} \frac{\ln x}{e^x} \quad \left(\text{of the form } \frac{\infty}{\infty}\right)$$

Applying L'Hôpital's rule to the simpler second quotient, you get

$$\lim_{x \to \infty} e^{-x} \ln x = \lim_{x \to \infty} \frac{\ln x}{e^x} = \lim_{x \to \infty} \frac{1/x}{e^x} = 0$$

As a final illustration of this technique, here is the limit that was used in Section 4.1 to define the number e.

EXAMPLE | A3.7

Find $\lim\limits_{x \to \infty} \left(1 + \dfrac{1}{x}\right)^x$.

Solution

This limit is of the indeterminate form 1^∞. To simplify the problem let

$$y = \left(1 + \frac{1}{x}\right)^x$$

Then

$$\ln y = x \ln \left(1 + \frac{1}{x}\right)$$

and

$$\lim_{x \to \infty} \ln y = \lim_{x \to \infty} x \ln \left(1 + \frac{1}{x}\right) \quad (\infty \cdot 0)$$

$$\lim_{x \to \infty} \ln y = \lim_{x \to \infty} \frac{\ln (1 + 1/x)}{1/x} \quad \left(\frac{0}{0}\right)$$

$$= \lim_{x \to \infty} \frac{\dfrac{(-1/x^2)}{(1 + 1/x)}}{-1/x^2} \quad \text{L'Hôpital's rule}$$

$$= \lim_{x \to \infty} \frac{1}{1 + 1/x} \quad \text{algebraic simplification}$$

$$= 1$$

Since $\ln y \to 1$, it follows that $y \to e^1 = e$. That is,

$$\lim_{x \to \infty} \left(1 + \frac{1}{x}\right)^x = e$$

PROBLEMS | A3

In Problems 1 through 16, use L'Hôpital's rule to evaluate the given limit if the limit is an indeterminate form.

1. $\displaystyle \lim_{x \to 0} \frac{x^3 - 3x^2}{3x^4 + 2x}$

2. $\displaystyle \lim_{x \to 0} \frac{x^2(x - 1)}{3x^3 + 2x - 5}$

3. $\displaystyle \lim_{x \to \infty} \frac{x^2 - 2x + 3}{2x^2 + 5x + 1}$

4. $\displaystyle \lim_{x \to \infty} \frac{x^2 + x - 5}{1 - 2x - x^3}$

5. $\displaystyle \lim_{x \to \infty} \frac{(1/x) - (2/x^2)}{(1/x^3) + (2/x^2) - (3/x)}$

6. $\displaystyle \lim_{x \to 3} \frac{x^2 + 2x - 15}{x^3 - 19x + 3}$

7. $\displaystyle \lim_{x \to -1} \frac{x^3 + 3x^2 + 3x + 1}{2x^3 + 3x^2 - 1}$
   [*Hint:* Use L'Hôpital's rule twice.]

8. $\displaystyle \lim_{x \to 1/2} \frac{-8x^3 + 2x^2 + 3x - 1}{(2x - 1)^3}$

9. $\displaystyle \lim_{x \to \infty} \frac{e^{-x}}{1 + e^{-2x}}$

10. $\displaystyle \lim_{x \to \infty} x^2 e^{-x}$

11. $\displaystyle \lim_{t \to 0} \frac{\sqrt{t}}{e^t}$

12. $\displaystyle \lim_{t \to \infty} \frac{\ln \sqrt{t}}{t}$

13. $\displaystyle \lim_{x \to \infty} \frac{(\ln x)^2}{x}$

14. $\displaystyle \lim_{x \to \infty} x^{1/x}$

15. $\displaystyle \lim_{x \to 0} (1 + 2x)^{1/x}$

16. $\displaystyle \lim_{x \to \infty} \left(1 + \frac{1}{x}\right)^{x^2}$

Important Terms, Symbols, and Formulas

Integer   (642)

Rational number   (642)

Irrational number   (642)

Real numbers   (642)

Number line   (642)

Inequality   (642)

Interval   (643)

Absolute value: (644)

$$|x| = \begin{cases} x & \text{if } x \geq 0 \\ -x & \text{if } x < 0 \end{cases}$$

Distance on a number line   (644)

Exponential notation: (645)

$$a^n = \underbrace{a \cdot a \cdots a}_{n \text{ terms}}$$

$$a^{-n} = \frac{1}{a^n} \quad \text{(negative power)}$$

$$a^{n/m} = (\sqrt[m]{a})^n = \sqrt[m]{a^n} \quad \text{(fractional powers)}$$

Laws of exponents: (646)

$$a^r a^s = a^{r+s} \quad \text{(product law)}$$

$$\frac{a^r}{a^s} = a^{r-s} \quad \text{(quotient law)}$$

$$(a^r)^s = a^{rs} \quad \text{(power law)}$$

Factoring   (647)

The summation notation: (649)

$$\sum_{j=1}^{n} a_j = a_1 + a_2 + \cdots + a_n$$

Polynomial:

Coefficients of a polynomial   (652)

Degree of a polynomial   (652)

Laws of numbers used in factoring:

Distributive law $ab + ac = a(b + c)$ (647)

Difference of two squares   (654)

$$a^2 - b^2 = (a + b)(a - b)$$

Solving an equation by factoring   (654)

The quadratic formula: (655)

The solutions of $ax^2 + bx + c = 0$ for $a \neq 0$ are

$$x = \frac{-b \pm \sqrt{b^2 - 4ac}}{2a}$$

Discriminant of the quadratic equation $ax^2 + bx + c = 0$ is $b^2 - 4ac$ (655)

System of equations   (656)

Solving a system of equations by elimination   (656)

L'Hôpital's rule: (659)

$$\left(\frac{0}{0} \text{ form}\right) \text{ If } \lim_{x \to c} f(x) = 0 \quad \text{and} \quad \lim_{x \to c} g(x) = 0, \text{ then}$$

$$\lim_{x \to c} \frac{f(x)}{g(x)} = \lim_{x \to c} \frac{f'(x)}{g'(x)}$$

$$\left(\frac{\infty}{\infty} \text{ form}\right) \text{ If } \lim_{x \to c} f(x) = \infty \quad \text{and} \quad \lim_{x \to c} g(x) = \infty, \text{ then}$$

$$\lim_{x \to c} \frac{f(x)}{g(x)} = \lim_{x \to c} \frac{f'(x)}{g'(x)}$$

Review Problems

In Problems 1 and 2, use inequalities to describe the given interval.

In Problems 3 through 6, represent the given interval as a line segment on a number line.

3. $-3 \leq x < 2$

5. $x \geq 1$

4. $-1 < x < 5$

6. $2 \leq x < 7$

In Problems 7 and 8, find the distance on the number line between the given pair of real numbers.

7. 0 and 3

8. −5 and −2

In Problems 9 and 10, find the interval or intervals consisting of all real numbers x that satisfy the given inequality

9. $|x - 3| \leq 1$

10. $|2x + 1| > 3$

In Problems 11 through 20, evaluate the given expression without using a calculator.

11. 3^5

12. 4^{-2}

13. $8^{2/3}$

14. $49^{3/2}$

15. $\dfrac{4(32)^{3/4}}{(\sqrt{2})^3}$

16. $\left(\dfrac{1}{9}\right)^{-5/2}$

17. $16^{3/2} + 27^{2/3}$

18. $\dfrac{2^{3/2}(4^{5/2})}{8^{2/3}}$

19. $\dfrac{\sqrt[3]{54}\ \sqrt[6]{2}}{\sqrt{8}}$

20. $\dfrac{\sqrt[3]{81}\,(6^{2/3})}{2^{4/3}}$

In Problems 21 through 24, solve the given equation for n (assume $a > 0$, $a \neq 1$).

21. $a^{2/3}a^{1/2} = a^{3n}$

22. $\dfrac{a^3}{(\sqrt{a})^5} = a^{2n}$

23. $a^2 a^{-5} = (a^n)^3$

24. $a^{2n}a^3 = a^{-7}$

In Problems 25 through 28, evaluate the given sum.

25. $\displaystyle\sum_{k=1}^{3} 2k + 3$

26. $\displaystyle\sum_{k=1}^{4} (k + 1)^2$

27. $\displaystyle\sum_{k=1}^{5} (2k^2 - k)$

28. $\displaystyle\sum_{k=1}^{4}\left[\dfrac{k - 1}{k + 3}\right]^2$

In Problems 29 and 30, express the given sum in terms of the summation notation.

29. $1 + \dfrac{1}{2} - \dfrac{1}{3} + \dfrac{1}{4} - \dfrac{1}{5} + \dfrac{1}{6} - \dfrac{1}{7}$

30. $3 + 12 + 27 + 48 + 75$

In Problems 31 through 34, factor the given expression.

31. $x^4 - 9x^2$

32. $x^3 + 3(x - 12)$

33. $x^{16} - (2x)^4$

34. $2(x - 3)^2(x + 1) - 5(x - 3)^3(2x)$

In Problems 35 and 36, simplify the given quotient as much as possible.

35. $\dfrac{x^2(x - 1)^3 - 2x(x - 1)^2}{x^2 - x - 2}$

36. $\dfrac{x(x + 2)^4 - x^3(x + 2)^2}{x^2 + 3x + 2}$

In Problems 37 through 42, factor the given polynomial using integral coefficients.

37. $x^2 + 2x - 15$

38. $2x^2 + 5x - 3$

39. $4x^2 + 12x + 9$

40. $12x^2 + 5x - 3$

41. $x^3 + 3x^2 - x - 3$

42. $x^4 - 2x^3 - 3x^2 + 8x - 4$

In Problems 43 through 48, solve the given equation by factoring.

43. $x^2 + 3x - 4 = 0$

44. $2x^2 - 3x - 2 = 0$

45. $x^2 + 14x + 49 = 0$

46. $x^2 - 64 = 0$

47. $1 - \dfrac{1}{x} - \dfrac{2}{x^2} = 0$

48. $4 + \dfrac{9}{x^2} = \dfrac{12}{x}$

In Problems 49 through 54, use the quadratic formula to find all real numbers x that satisfy the given equation.

49. $14x^2 - x - 3 = 0$

50. $24x^2 + x - 10 = 0$

51. $x^2 - 3x + 5 = 0$

52. $7x^2 + 3x - 2 = 0$

53. $3x^2 + 5x - 2 = 0$

54. $2x^2 + 12x + 11 = 0$

In Problems 55 through 58, solve the given system of equations.

55. $3x + 5y = -1$
    $2x + 7y = 3$

56. $2x + y = 7$
    $-x + 4y = 1$

57. $3x^2 - y^2 = -1$
    $2x + y = 4$

58. $5x^2 - 2y^2 = 1$
    $5x - 2y = 4$

In Problems 59 through 64, use L'Hôpital's rule to evaluate the given limit if the limit is an indeterminate form.

59. $\displaystyle\lim_{x \to 1} \frac{x^2 - 1}{2x^3 + x - 3}$

60. $\displaystyle\lim_{x \to -2} \frac{x^3 + 8}{3x^3 - 7x + 10}$

61. $\displaystyle\lim_{x \to \infty} \frac{e^{-2x}}{3 + 2e^{-2x}}$

62. $\displaystyle\lim_{x \to \infty} x^3 e^{-\sqrt{x}}$

63. $\displaystyle\lim_{x \to \infty} x(e^{1/x} - 1)$

64. $\displaystyle\lim_{x \to \infty} \left(1 - \frac{3}{x}\right)^{2x}$

THINK ABOUT IT

PROBLEM

For comfortable modern living, it is estimated that each person needs roughly 60 m^2 for housing, 40 m^2 for his or her job, 50 m^2 for public buildings and recreation facilities, 90 m^2 for transportation (e.g., highways), and 4,000 m^2 for the production of food.

Questions

1. Switzerland has approximately 11,000 km^2 of livable space (arable and habitable land). How many people can comfortably live in Switzerland? Look up the actual population of Switzerland. Based on the figures given here, is Switzerland overcrowded or is there still room for comfortable growth?

2. Use an almanac, an encyclopedia, or some other source to obtain the population of India. How much livable space would there have to be in India to accommodate its current population without overcrowding? Now look up the total area of India. Even if all of India is comprised of livable space, is there enough room for the population to live comfortably?

 3. You probably knew that India is overcrowded. Pick another country where the answer is not so obvious (Bolivia? Zimbabwe? San Marino?). Describe the "comfort of living" in that country.

Source: Adapted from a problem in E. Batschelet, *Introduction to Mathematics for Life Scientists,* 2nd ed., New York: Springer-Verlag, 1979, p. 31. You may find it interesting to examine several of the applied life science problems on pages 31–33 of the Batschelet text.

TABLES

TABLE I   Powers of e

x	e^x	e^{-x}	x	e^x	e^{-x}	x	e^x	e^{-x}
0.00	1.0000	1.00000	0.50	1.6487	.60653	1.00	2.7183	.36788
0.01	1.0101	0.99005	0.51	1.6653	.60050	1.10	3.0042	.33287
0.02	1.0202	.98020	0.52	1.6820	.59452	1.20	3.3201	.30119
0.03	1.0305	.97045	0.53	1.6989	.58860	1.30	3.6693	.27253
0.04	1.0408	.96079	0.54	1.7160	.58275	1.40	4.0552	.24660
0.05	1.0513	.95123	0.55	1.7333	.57695	1.50	4.4817	.22313
0.06	1.0618	.94176	0.56	1.7507	.57121	1.60	4.9530	.20190
0.07	1.0725	.93239	0.57	1.7683	.56553	1.70	5.4739	.18268
0.08	1.0833	.92312	0.58	1.7860	.55990	1.80	6.0496	.16530
0.09	1.0942	.91393	0.59	1.8040	.55433	1.90	6.6859	.14957
0.10	1.1052	.90484	0.60	1.8221	.54881	2.00	7.3891	.13534
0.11	1.1163	.89583	0.61	1.8404	.54335	3.00	20.086	.04979
0.12	1.1275	.88692	0.62	1.8589	.53794	4.00	54.598	.01832
0.13	1.1388	.87809	0.63	1.8776	.53259	5.00	148.41	.00674
0.14	1.1503	.86936	0.64	1.8965	.52729	6.00	403.43	.00248
0.15	1.1618	.86071	0.65	1.9155	.52205	7.00	1096.6	.00091
0.16	1.1735	.85214	0.66	1.9348	.51685	8.00	2981.0	.00034
0.17	1.1853	.84366	0.67	1.9542	.51171	9.00	8103.1	.00012
0.18	1.1972	.83527	0.68	1.9739	.50662	10.00	22026.5	.00005
0.19	1.2092	.82696	0.69	1.9937	.50158			
0.20	1.2214	.81873	0.70	2.0138	.49659			
0.21	1.2337	.81058	0.71	2.0340	.49164			
0.22	1.2461	.80252	0.72	2.0544	.48675			
0.23	1.2586	.79453	0.73	2.0751	.48191			
0.24	1.2712	.78663	0.74	2.0959	.47711			
0.25	1.2840	.77880	0.75	2.1170	.47237			
0.26	1.2969	.77105	0.76	2.1383	.46767			
0.27	1.3100	.76338	0.77	2.1598	.46301			
0.28	1.3231	.75578	0.78	2.1815	.45841			
0.29	1.3364	.74826	0.79	2.2034	.45384			
0.30	1.3499	.74082	0.80	2.2255	.44933			
0.31	1.3634	.73345	0.81	2.2479	.44486			
0.32	1.3771	.72615	0.82	2.2705	.44043			
0.33	1.3910	.71892	0.83	2.2933	.43605			
0.34	1.4049	.71177	0.84	2.3164	.43171			
0.35	1.4191	.70469	0.85	2.3396	.42741			
0.36	1.4333	.69768	0.86	2.3632	.42316			
0.37	1.4477	.69073	0.87	2.3869	.41895			
0.38	1.4623	.68386	0.88	2.4109	.41478			
0.39	1.4770	.67706	0.89	2.4351	.41066			
0.40	1.4918	.67032	0.90	2.4596	.40657			
0.41	1.5068	.66365	0.91	2.4843	.40252			
0.42	1.5220	.65705	0.92	2.5093	.39852			
0.43	1.5373	.65051	0.93	2.5345	.39455			
0.44	1.5527	.64404	0.94	2.5600	.39063			
0.45	1.5683	.63763	0.95	2.5857	.38674			
0.46	1.5841	.63128	0.96	2.6117	.38289			
0.47	1.6000	.62500	0.97	2.6379	.37908			
0.48	1.6161	.61878	0.98	2.6645	.37531			
0.49	1.6323	.61263	0.99	2.6912	.37158			

TABLE II The Natural Logarithm (Base e)

x	ln x	x	ln x	x	ln x	x	ln x
.01	−4.60517	0.50	−0.69315	1.00	0.00000	1.5	0.40547
.02	−3.91202	.51	.67334	1.01	.00995	1.6	7000
.03	.50656	.52	.65393	1.02	.01980	1.7	0.53063
.04	.21888	.53	.63488	1.03	.02956	1.8	8779
		.54	.61619	1.04	.03922	1.9	0.64185
.05	−2.99573	.55	.59784	1.05	.04879	2.0	9315
.06	.81341	56	.57982	1.06	.05827	2.1	0.74194
.07	.65926	.57	.56212	1.07	.06766	2.2	8846
.08	.52573	.58	.54473	1.08	.07696	2.3	0.83291
.09	.40795	.59	.52763	1.09	.08618	2.4	7547
0.10	−2.30259	0.60	−0.51083	1.10	.09531	2.5	0.91629
.11	.20727	.61	.49430	1.11	.10436	2.6	5551
.12	.12026	.62	.47804	1.12	.11333	2.7	9325
.13	.04022	.63	.46204	1.13	.12222	2.8	1.02962
.14	−1.96611	.64	.44629	1.14	.13103	2.9	6471
.15	.89712	.65	.43078	1.15	.13976	3.0	9861
.16	.83258	.66	.41552	1.16	.14842	4.0	1.38629
.17	.77196	.67	.40048	1.17	.15700	5.0	1.60944
.18	.71480	.68	.38566	1.18	.16551	10.0	2.30258
.19	.66073	.69	.37106	1.19	.17395		
0.20	−1.60944	0.70	−0.35667	1.20	.18232		
.21	.56065	.71	.34249	1.21	.19062		
.22	.51413	.72	.32850	1.22	.19885		
.23	.46968	.73	.31471	1.23	.20701		
.24	.42712	.74	.30111	1.24	.21511		
.25	.38629	.75	.28768	1.25	.22314		
.26	.34707	.76	.27444	1.26	.23111		
.27	.30933	.77	.26136	1.27	.23902		
.28	.27297	.78	.24846	1.28	.24686		
.29	.23787	.79	.23572	1.29	.25464		
0.30	−1.20397	0.80	−0.22314	1.30	.26236		
.31	.17118	.81	.21072	1.31	.27003		
.32	.13943	.82	.19845	1.32	.27763		
.33	.10866	.83	.18633	1.33	.28518		
.34	.07881	.84	.17435	1.34	.29267		
.35	−1.04982	.85	−0.16252	1.35	.30010		
.36	.02165	.86	.15032	1.36	.30748		
.37	−0.99425	.87	.13926	1.37	.31481		
.38	.96758	.88	.12783	1.38	.32208		
.39	.94161	.89	.11653	1.39	.32930		
0.40	−0.91629	0.90	0.10536	1.40	.33647		
.41	.89160	.91	.09431	1.41	.34359		
.42	.86750	.92	.08338	1.42	.35066		
.43	.84397	.93	.07257	1.43	.35767		
.44	.82098	.94	.06188	1.44	.36464		
.45	.79851	.95	.05129	1.45	.37156		
.46	.77653	.96	.04082	1.46	.37844		
.47	.75502	.97	.03046	1.47	.38526		
.48	.73397	.98	.02020	1.48	.39204		
.49	.71335	.99	.01005	1.49	.39878		

From S. K. Stein, *Calculus and Analytic Geometry.* Copyright © 1973 by McGraw-Hill, Inc. Used with permission of McGraw-Hill Book Company.

Answers to Odd-Numbered Problems, Chapter Checkup Problems, and Odd-Numbered Chapter Review Problems

CHAPTER 1 | Section 1

1. $f(0) = -2; f(-2) = 0; f(1) = 6$

3. $g(-1) = 2; g(1) = 2; g(2) = \dfrac{5}{2}$

5. $h(2) = 2\sqrt{3}; h(0) = 2; h(-4) = 2\sqrt{3}$

7. $f(1) = 1; f(5) = \dfrac{1}{27}; f(13) = \dfrac{1}{125}$

9. $f(1) = 0; f(2) = 2; f(3) = 2$

11. $f(-6) = 3; f(-5) = -4; f(16) = 4$

13. Yes

15. No, $f(t)$ is not defined for $t > 1$

17. All real numbers x except $x = -2$

19. All real numbers x for which $x \geq -3$

21. All real numbers t for which $-3 < t < 3$

23. $f(g(x)) = 3x^2 + 14x + 10$

25. $f(g(x)) = x^3 + 2x^2 + 4x + 2$

27. $f(g(x)) = \dfrac{1}{(x-1)^2}$

29. $f(g(x)) = |x|$

31. $\dfrac{f(x+h) - f(x)}{h} = 2x + h$

33. $\dfrac{f(x+h) - f(x)}{h} = \dfrac{-1}{x(x+h)}$

35. $f(g(x)) = \sqrt{1 - 3x}; g(f(x)) = 1 - 3\sqrt{x};$
$f(g(x)) = g(f(x))$ if $x = 0$

37. $f(g(x)) = x; g(f(x)) = x; f(g(x)) = g(f(x))$ for all real numbers except $x = 1$ and $x = 2$

39. $f(x - 2) = 2x^2 - 11x + 15$

41. $f(x - 1) = x^5 - 3x^2 + 6x - 3$

43. $f(x^2 + 3x - 1) = \sqrt{x^2 + 3x - 1}$

45. $f(x + 1) = \dfrac{x}{x+1}$

Note: In answers 47 to 51, answers may vary.

47. $h(x) = x - 1; g(u) = u^2 + 2u + 3$

49. $h(x) = x^2 + 1; g(u) = \dfrac{1}{u}$

51. $h(x) = 2 - x; g(u) = \sqrt[3]{u} + \dfrac{4}{u}$

53. a. $f(2) = 46$
    b. $f(2) - f(1) = 26$

55. a. $P(9) = \dfrac{97}{5}$; 19,400 people
    b. $P(9) - P(8) = \dfrac{1}{15}$; 67 people
    c. $P(t)$ approaches 20 (20,000 people)

57. a. $S(0) = 25.344$ cm/sec
    b. $S(6 \times 10^{-3}) = 19.008$ cm/sec

59. a. $s(8) = 5.8 \approx 6$ species
    b. $s_2 = \sqrt[3]{2}\, s_1$
    c. Approximately 41,000 square miles

61. a. All real numbers x except $x = 200$
    b. All real numbers x for which $0 \leq x \leq 100$
    c. $C(50) = \$50$ million
    d. $C(100) - C(50) = \$100$ million
    e. $C(x) = 37.5$ million implies that $x = 40\%$

63. a. $c(p(t)) = 4.2 + 0.08t^2$
    b. $c(p(2)) = 4.52$ parts per million
    c. 5 years

65. a. $R(x) = -0.02x^2 + 29x;$
    $P(x) = -1.45x^2 + 10.7x - 15.6$
    b. $P(x) > 0$ if $2 < x < 5.38$

67. a. $R(x) = -0.5x^2 + 39x;$
    $P(x) = -2x^2 + 29.8x - 67$
    b. $P(x) > 0$ if $2.76 < x < 12.14$

69. a. $Q(p(t)) = \dfrac{4,374}{(0.04t^2 + 0.2t + 12)^2}$
    b. $Q(p(10)) = 13.5$ kilograms/week
    c. $t = 0$

71. All real numbers x except $x = 1$ and $x = -1.5$

73. $f(g(2.3)) = 6.31$

ANSWERS

75. a.

Level of Education \ Year	1991	1992	1993	1994	1995	1996	1997	1998	1999	2000
No high school diploma	1.00	1.00	1.00	1.00	1.00	1.00	1.00	1.00	1.00	1.00
High school diploma	1.45	1.41	1.52	1.48	1.53	1.48	1.42	1.47	1.52	1.45
Some college	1.63	1.63	1.68	1.62	1.70	1.68	1.63	1.72	1.76	1.67
Bachelor's degree	2.48	2.55	2.74	2.72	2.64	2.54	2.51	2.73	2.83	2.76
Advanced degree	3.65	3.80	4.35	4.10	4.04	4.08	3.92	3.95	4.20	3.85

b. 1.45, 1.67, 2.76, 3.85

CHAPTER 1 | Section 2

1. a. Power function
b. Polynomial
c. Polynomial
d. Rational function

3. $f(x) = x$

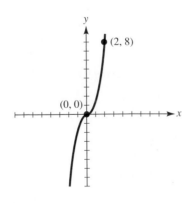

5. $f(x) = x^3$

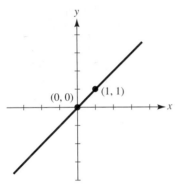

7. $f(x) = 2x - 1$

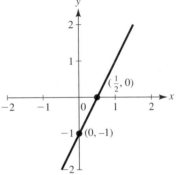

9. $f(x) = x(2x + 5)$

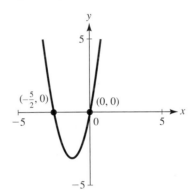

11. $f(x) = -x^2 - 2x + 15$

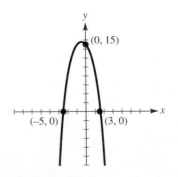

13. $f(x) = \sqrt{x}$

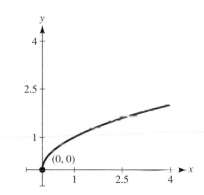

15. $f(x) = \begin{cases} x - 1 & \text{if } x \leq 0 \\ x + 1 & \text{if } x > 0 \end{cases}$

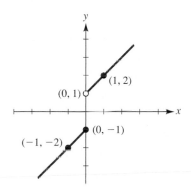

17. $f(x) = \begin{cases} x^2 + x - 3 & \text{if } x < 1 \\ 1 - 2x & \text{if } x \geq 1 \end{cases}$

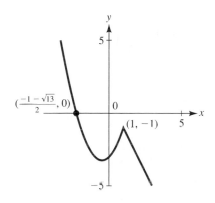

19. $y = 3x + 5$ and $y = -x + 3$; $\left(-\dfrac{1}{2}, \dfrac{7}{2}\right)$

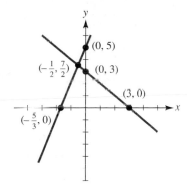

21. $y = x^2$ and $y = 3x - 2$; $(2, 4)$ and $(1, 1)$

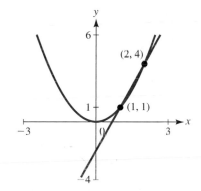

23. $3y - 2x = 5$ and $y + 3x = 9$; $(2, 3)$

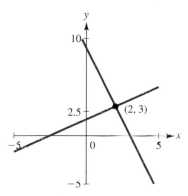

25. a. $(0, -1)$
   b. $(1, 0)$
   c. $f(x) = 3$ when $x = 4$
   d. $f(x) = -3$ when $x = -2$

27. a. $(0, 2)$
   b. $(-1, 0), (3.5, 0)$
   c. $f(x) = 3$ when $x = 2$
   d. $f(x) = -3$ when $x = 4$

29. $P(p) = (p - 40)(120 - p)$; optimal price is \$80 per recorder

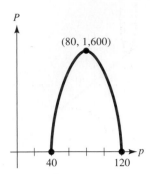

31. a. $E(p) = -200p^2 + 12{,}000p$

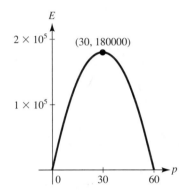

b. The p intercepts represent prices at which consumers spend no money on the commodity.

c. \$30 per unit

33. a. $P(x) = -0.07x^2 + 35x - 574.77$

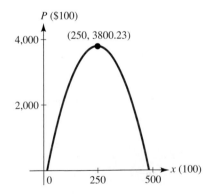

b. 25,000 units; \$25.50

35. $D(v) = 0.065v^2 + 0.148v$

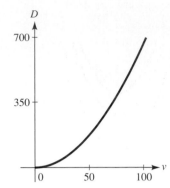

37. a. $R(p) = -0.05p^2 + 210p$

b.

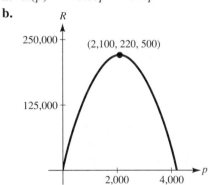

c. \$2,100 per month; \$220,500

39. a.

b. 3,967 thousand tons

c. 4.27 years after 1990, or in March 1994.

d. No, the formula predicts negative emissions after December 2004.

41. Function

43. Not a function

45. Answers will vary.

47. a. The graph of $y = x^2 + 3$ is the graph of $y = x^2$ shifted upward 3 units.

b.

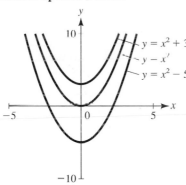

c. The graph of $g(x)$ is the graph of $f(x)$ shifted $|c|$ units upward if $c > 0$ or downward if $c < 0$.

49. a. The graph of $y = (x - 2)^2$ is the graph of $y = x^2$ shifted to the right 2 units.

b.

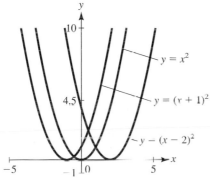

c. The graph of $g(x)$ is the graph of $f(x)$ shifted $|c|$ units to the right if $c > 0$ and to the left if $c < 0$.

51. a.

Days of Training	Mowers per Day
2	6
3	7.23
5	8.15
10	8.69
50	8.96

b. The number of mowers per day approaches 9.

c.

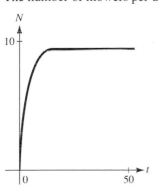

53.

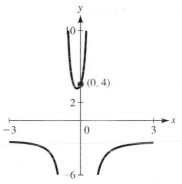

$f(x)$ is defined for $x \neq \dfrac{-1 \pm \sqrt{17}}{8}$.

55.

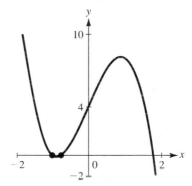

x intercepts: $x = -1, -0.76, 1.76$.

57. A point (x, y) is on the circle when the distance from (x, y) to (a, b) is R. That is,
$$\sqrt{(x - a)^2 + (y - b)^2} = R$$
or
$$(x - a)^2 + (y - b)^2 = R^2.$$

CHAPTER 1 Section 3

1. $m = -\dfrac{7}{2}$

3. $m = -1$

5. m is undefined.

7. Slope: 2; intercepts: $(0, 0)$; $y = 2x$

9. Slope: $\dfrac{5}{8}$; intercepts: $(-4, 0)$, $\left(0, \dfrac{5}{2}\right)$;

$y = \dfrac{5}{8}x + \dfrac{5}{2}$

11. Slope: undefined; intercepts: (3, 0)

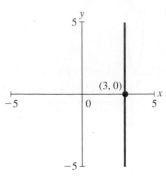

13. Slope: 3; intercepts: (0, 0)

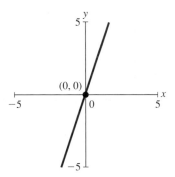

15. Slope: $-\dfrac{3}{2}$; intercepts: (2, 0), (0, 3)

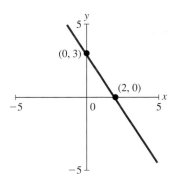

17. Slope: $-\dfrac{5}{2}$; intercepts: (2, 0), (0, 5)

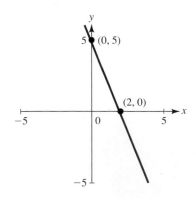

19. $y = x - 2$

21. $y = -\dfrac{1}{2}x + \dfrac{1}{2}$

23. $y = 5$

25. $y = -x + 1$

27. $y = \dfrac{-45}{52}x + \dfrac{43}{52}$

29. $y = 5$

31. $y = -2x + 9$

33. $y = x + 2$

35. $y = C(x) = 60x + 5{,}000$

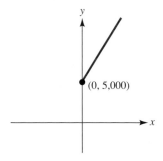

37. a. $y = f(t) = 35t + 220$ **b.** 325 **c.** 220

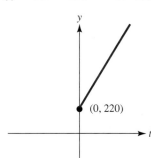

39. $f(t) = -150t + 1{,}500$

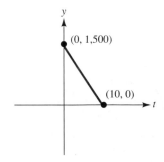

41. a. $y = f(t) = -4t + 248$
   b. $f(8) = 216$ million gallons

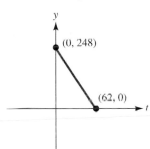
(0, 248)
(62, 0)

43.

a.

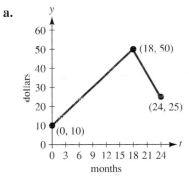
(18, 50)
(24, 25)
(0, 10)
months

b.

(24, 20)
(2, 15)
(0, 10) (11, 8)
months

c.

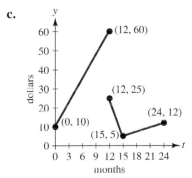
(12, 60)
(12, 25)
(24, 12)
(0, 10)
(15, 5)
months

45. a. 95.5 cm
   b. 15.4 years old
   c. 50 cm; yes
   d. 180 cm; yes

47. a. $F = \dfrac{9}{5}C + 32$

   b. 59°F
   c. 20°C
   d. $-40°$ Celsius $= -40°$ Fahrenheit

49. a. $v(1930) = \$800$; $v(1990) = \$51,200$;
   $v(2000) = \$102,400$
   b. No, it is not linear.

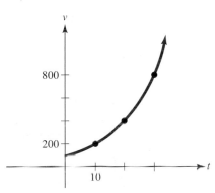
800
200
10

51. a. $y = -6(t - 1995) + 575$
   b. 515
   c. 2003

53. a. $B = \left(\dfrac{S - V}{N}\right)t + V$

   b. \$30,800

55. The two lines are not parallel; they do not have the same slope.

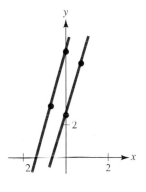

ANSWERS

57. **a.**

Hours rented	2	5	10	t
Total cost	$70.00	$85.00	$110.00	$60 + 5t$

b. $y = 60 + 5t, t \geq 0$

c.

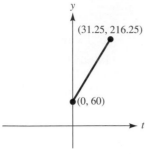

d. Solve $60 + 5t = 216.25$;
$t = 31.25$ hours (31 hours, 15 minutes)

59. The slope of -0.389 means the unemployment rate drops by approximately 0.389% from year to year.

61. Lines L_1 and L_2 are given as perpendicular, and the slope of L_1 is $m_1 = b/a$ while that of L_2 is $m_2 = c/a$. In right triangle OAB, the hypotenuse has length $|AB| = |b - c|$ while the two legs have lengths $|OA| = \sqrt{a^2 + b^2}$ and $|OB| = \sqrt{a^2 + c^2}$. Thus, by the Pythagorean theorem

$$(a^2 + b^2) + (a^2 + c^2) = (b - c)^2$$
$$a^2 + b^2 + a^2 + c^2 = b^2 - 2bc + c^2$$
$$2a^2 = -2bc$$
$$\frac{bc}{a^2} = -1$$
$$\left(\frac{b}{a}\right)\left(\frac{c}{a}\right) = -1$$
$$m_1 m_2 = -1$$
so $$m_2 = \frac{-1}{m_1}$$

CHAPTER 1 | Section 4

1. $A = 2w(500 - w)$

3. $P = x(18 - x)$

5. $R = x(35x + 15)$

7. $A = x(160 - x)$; 80 m by 80 m

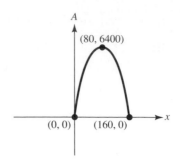

9. $V = x\left(1{,}000 - \dfrac{x^2}{2}\right)$

11. $V = \pi r(60 - r^2)$

13. $C = 0.08\pi\left(r^2 + \dfrac{2}{r}\right)$

15. $R = kP$; R = rate of population growth; P = size of population

17. $R = k(T_0 - T_e)$; R = rate of temperature change; T_0 = temperature of object; T_e = temperature of surrounding medium

19. $R = kP(T - P)$; R = rate of implication; P = number of people already implicated; T = total number of people involved

21. $C = \dfrac{k_1}{R} + k_2 R$; R = speed of truck

23.

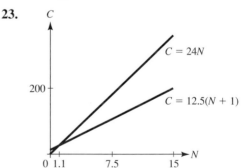

25. **a.** 95.2 mg

b. Since $0.0072(2W)^{0.425}(2H)^{0.725}$
$$= 0.0072(2)^{0.425}(W)^{0.425}(2)^{0.725}(H)^{0.725}$$
$$= (2)^{0.425}(2)^{0.725}0.0072(W)^{0.425}(H)^{0.725}$$
$$\approx 2.22[0.0072(W)^{0.425}(H)^{0.725}]$$

the larger child has 2.22 times the surface area of the smaller. If S is multiplied by 2.22 in the formula $C = \dfrac{SA}{1.7}$, then C grows by the same factor.

27. $R(x) = \begin{cases} 2{,}400 & \text{if } 1 \leq x \leq 40 \\ x\left(80 - \dfrac{1}{2}x\right) & \text{if } 40 < x < 80 \\ 40x & \text{if } x \geq 80 \end{cases}$

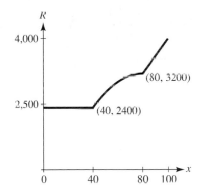

29. a. $C(x) = \begin{cases} 3.50x & \text{if } x < 50 \\ 3x & \text{if } x \geq 50 \end{cases}$

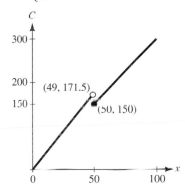

b. $21.50

31. $V(x) = 4x - \dfrac{x^3}{3}$

33. $A(x) = 8x + \dfrac{100}{x} + 57$

35. $P(p) = (33{,}000 - 1{,}000p)(p - 18)$

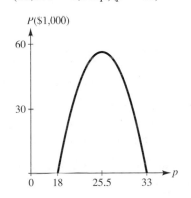

Optimal price is $25.50. The manufacturer will actually make more profit by *reducing* the price by $4.50 per unit.

37. $C(x) = 20x + \dfrac{5{,}120}{x}$

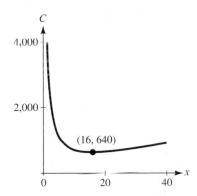

The number of machines is 16

39. $R(x) = (3 - 0.02x)(140 + x)$, where x = days from July 1

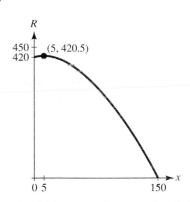

Farmer should harvest when $x = 5$ or July 6

41. a. $x_e = 25$; $p_e = 225$

b.

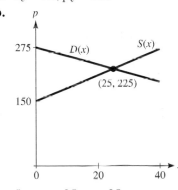

c. $0 < x < 25$; $x > 25$

ANSWERS

43. **a.** $x_e = 9; p_e = 25.43$

b.

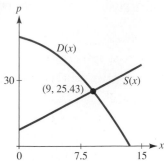

c. $0 < x < 9; x > 9$

45. **a.** $x_e = 10; p_e = 35$

b.

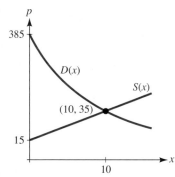

c. $S(0) = 15$; No units will be produced until the price is at least \$15.

47. 2 hours, 45 minutes after the second plane leaves

49. If $N \le 6,000$ or $N \ge 126,000$, choose publisher A, otherwise publisher B.

51. $I = k\pi r^2$

53. $y = \dfrac{K_m}{R_m}x + \dfrac{1}{R_m}$

55. **a.**

x	2,000	4,000	6,000	8,000
$C(x)$	85,200	96,200	107,200	118,200

b.

x	2,000	4,000	6,000	8,000
$R(x)$	39,000	78,000	117,000	156,000

c. $C(x) = 5.5x + 74,200$

d. $R(x) = 19.5x$

e.

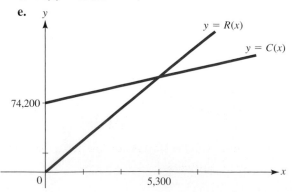

f. $C(x) = R(x) = \$103,350$ when $x = 5,300$ books

g. Approximately 4,359 books must be sold for a revenue of \$85,000. The "profit" is actually a loss of \$13,174.

CHAPTER 1 | Section 5

1. $\lim_{x \to a} f(x) = b$

3. $\lim_{x \to a} f(x) = b$

5. Limit does not exist.

7. 4

9. 7

11. 16

13. $\dfrac{4}{7}$

15. Limit does not exist.

17. 2

19. 7

21. $\dfrac{5}{3}$

23. 5

25. $\dfrac{1}{4}$

27. $+\infty; -\infty$

29. $-\infty; -\infty$

31. $\dfrac{1}{2}; \dfrac{1}{2}$

33. $0; 0$

35. $+\infty; -\infty$

37. $1; -1$

39.

x	1.9	1.99	1.999	2	2.001	2.01	2.1
$f(x)$	1.71	1.9701	1.997001		2.003001	2.0301	2.31

$\lim_{x \to 2} f(x) = 2$

41.

x	0.9	0.99	0.999	1	1.001	1.01	1.1
$f(x)$	-17.29	-197.0299	$-1,997.002999$		2,003.003001	203.0301	23.31

$\lim\limits_{x\to 1} f(x)$ does not exist.

43. $\lim\limits_{x\to c} [2f(x) - 3g(x)] = 2(5) - 3(-2) = 16$

45. $\lim\limits_{x\to c} \dfrac{f(x)}{g(x)} = \dfrac{5}{-2} = \dfrac{5}{2}$

47. $\lim\limits_{x\to\infty} \sqrt{g(x)} = \sqrt{4} = 2$

49. 1.8 in.

51. a. $P(t) = \dfrac{\sqrt{9t^2 + 0.5t + 179}}{0.2t + 1,500}$ thousand dollars per person

b. $\lim\limits_{t\to\infty} P(t) = \$15,000$ per person

53. a. $\lim\limits_{S\to +\infty} I(S) = a$. No matter how large the bite size, the required vigilance limits the amount of food intake.

b. Answers will vary.

55. $7.50; As the number of units produced increases, the contribution of fixed costs to the average cost decreases to 0.

57. $\lim\limits_{x\to 0} f(x)$ does not exist

59. a. 0

b. $\dfrac{a_n}{b_m}$

c. $+\infty$ if a_n and b_m have the same sign, $-\infty$ if a_n and b_m have different signs

CHAPTER 1 Section 6

1. -2; 1; does not exist

3. 2; 2; exists and equals 2

5. 39

7. 0

9. 0

11. $\dfrac{1}{4}$

13. 15; 0

15. Yes

17. Yes

19. No

21. No

23. No

25. Yes

27. $f(x)$ is continuous for all real numbers x.

29. $f(x)$ is continuous for all real numbers x except $x = 2$.

31. $f(x)$ is continuous for all $x \neq -1$.

33. $f(x)$ is continuous for all $x \neq -3, 6$.

35. $f(x)$ is continuous for all $x \neq 0, 1$.

37. $f(x)$ is continuous for all real numbers x.

39. $f(x)$ is continuous for all $x \neq 0$.

41. a. $W(20) \approx 3.8$
$W(50) = -7$

b. $v = 25$ mph [*Note:* $v = 99$ mph is out of range.]

c. W is continuous at both $v = 4$ and $v = 45$, assuming we round $W(4)$ and $W(45)$ to the nearest tenth. The slight discrepancy may be viewed as an inaccuracy in the model.

43. $p(x)$ is discontinuous at $x = 1, 2, 3, 4,$ and 5.

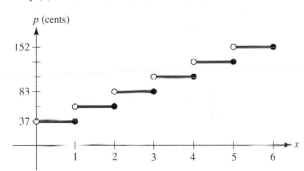

45. The graph is discontinuous at $t = 10$ and $t = 25$. Sue stops at a gas station and purchases some amount of gas at these times.

47. a. $4,000; $12,000

b.

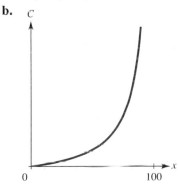

c. $\lim\limits_{x\to 100^-} C(x) = \infty$; it is not possible to remove all the pollution.

49. a. $C(0) \approx 0.333$; $C(100) \approx 7.179$

b. $C(x)$ is not continuous on the interval $0 \le x \le 100$ because $C(80)$ is not defined.

51. $A = 6$

53. $f(x)$ is continuous on the open interval $0 < x < 1$. $f(x)$ is continuous at every point of the closed interval $0 \le x \le 1$ except $x = 0$.

55. Let $f(x) = \sqrt[3]{x - 8} + 9x^{2/3} - 29$. Then $f(x)$ is continuous on $0 \le x \le 8$, $f(0) = -31 < 0$ and $f(8) = 7 > 0$. By the IVT, $f(x) = 0$ for some $0 < x < 8$.

57. a. $\lim\limits_{x \to 2} f(x)$ exists at $x = 2$ but $f(x)$ is not continuous there.

b. $\lim\limits_{x \to -2} f(x)$ does not exist, so $f(x)$ is not continuous at $x = -2$.

59. During each hour, the minute hand moves continuously from being behind the hour hand to being ahead. Therefore, at some time, they must coincide.

CHAPTER 1 | Checkup

1. All real numbers x such that $-2 < x < 2$

2. $g(h(x)) = \dfrac{2x + 1}{4x + 5}$; $x \ne -\dfrac{1}{2}$

3. a. $y = -\dfrac{1}{2}x + \dfrac{3}{2}$

b. $y = 2x - 3$

4. a.

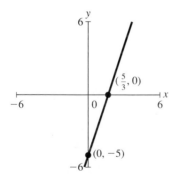

b.

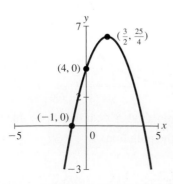

5. a. 2

b. 4

c. 1

d. $-\infty$

6. $f(x)$ is not continuous at $x = 1$.

7. a. $p(t) = 0.02t + 1.7$

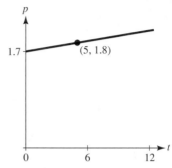

b. $1.70 per gallon

c. $1.88 per gallon

8. a. $A = 3$, $B = -1$; $52

b.

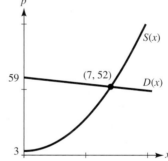

c. $-$26$; $54

9. a. $t = 9$

b. The function $g(t) = f(t) - 10$ is continuous and $g(1) = -2$ and $g(7) = 6$, so by the IVT $g(t) = 0$ or $f(t) = 10$ sometime between $t = 1$ and $t = 7$.

10. $M = 2.5D + 0.2$; 0.2%

CHAPTER 1 | Review Problems

1. a. All real numbers x

b. All real numbers x except $x = 1$ and $x = -2$

c. All real numbers x for which $|x| \ge 3$

3. a. $g(h(x)) = x^2 - 4x + 4$

b. $g(h(x)) = \dfrac{1}{2x + 5}$

c. $g(h(x)) = \sqrt{-2x - 3}$

5. *Note:* Answers will vary.

    a. $g(u) = u^5$; $h(x) = x^2 + 3x + 4$

    b. $g(u) = (u - 1)^2 + \dfrac{5}{2u^3}$; $h(x) = 3x + 2$

7. $c = -4$

9. **a.** $m = 3$, $b = 2$

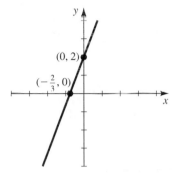

    b. $m = \dfrac{5}{4}$, $b = -5$

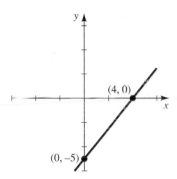

    c. $m = -\dfrac{3}{2}$, $b - 0$

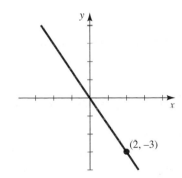

    d. $m = -\dfrac{2}{3}$, $b = 8$

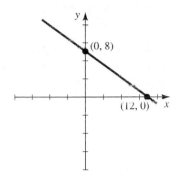

11. **a.**

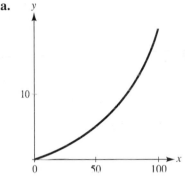

    b. 5 weeks

    c. 20 weeks

13. $V = \dfrac{4\pi}{3}\left(\sqrt{\dfrac{S}{4\pi}}\right)^3 = \dfrac{S^{3/2}}{6\sqrt{\pi}}$; V is multiplied by $2^{3/2}$.

15. **a.** $(3, -4)$

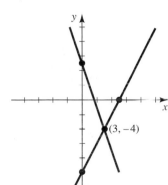

b. No intersection

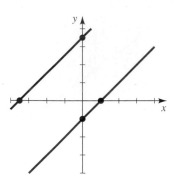

c. $(1, 0)$ and $(-1, 0)$

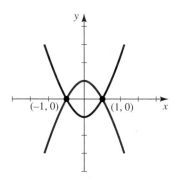

d. $(3, 9)$ and $(-5, 25)$

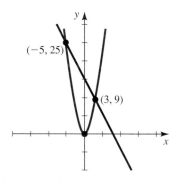

17. $P(p) = 2(360 - p)(p - 150)$; optimal price $p = \$255$

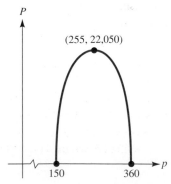

19. For x machines $C(x) = 80x + \dfrac{11,520}{x}$

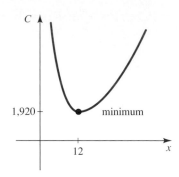

Minimum cost when $x = 12$

21. a. 150 units

b. \$1,500 profit

c. 180 units

23. $y = k(N - x)$, where y is the recall rate, x is the number of facts that have been recalled, and N is the total number of facts.

25. $C = 60x + (2\pi - 6)x^2$

27. $C(x) = 1,500 + 2x$ for $0 \le x \le 5,000$

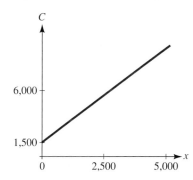

$C(x)$ is continuous for $0 \le x \le 5,000$.

29. Choose Proposition A if $V < 30,000$, otherwise choose Proposition B.

31. $\dfrac{3}{2}$

33. -12

35. Limit does not exist

37. 0

39. $-\infty$

41. 0

43. 0

45. Not continuous for $x = -3$

47. Continuous for all real numbers x

49. $A = \dfrac{B}{(4{,}000)^3}$

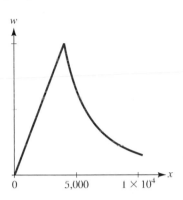

51.

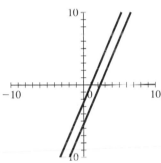

No, because $\dfrac{21}{9} \neq \dfrac{654}{279}$.

53. The function is discontinuous at $x = 1$.

55. The limit exists and is 0.

CHAPTER 2 | Section 1

1. $f'(x) = 5; m = 5$

3. $f'(x) = 4x - 3; m = -3$

5. $g'(t) = -\dfrac{2}{t^2}; m = -8$

7. $f'(x) = \dfrac{1}{2\sqrt{x}}; m = \dfrac{1}{6}$

9. $f'(x) = 2x; y = 2x - 1$

11. $f'(x) = -2; y = -2x + 7$

13. $f'(x) = \dfrac{2}{x^2}; y = 2x + 4$

15. $f'(x) = \dfrac{1}{\sqrt{x}}; y = \dfrac{1}{2}x + 2$

17. $\dfrac{dy}{dx} = 0$

19. $f'(x) = 1 - 2x; f'(-1) = 3$

21. $\dfrac{dy}{dx} = 2x - 2; \left.\dfrac{dy}{dx}\right|_{x=1} = 0$

23. **a.** $m_{\text{sec}} = 3.31$
   b. $m_{\text{tan}} = 3$

25. **a.** $m_{\text{sec}} = 1.5$
   b. $m_{\text{tan}} = 2$

27. **a.** $\text{rate}_{\text{ave}} = \dfrac{-13}{16}$
   b. $\text{rate}_{\text{ins}} = -1$

29. **a.** $\text{rate}_{\text{ave}} = 4$
   b. $\text{rate}_{\text{ins}} = 8$

31. **a.** The average rate of temperature change between t_0 and $t_0 + h$ hours after midnight. The instantaneous rate of temperature change t_0 hours after midnight.
   b. The average rate of change in blood alcohol level between t_0 and $t_0 + h$ hours after consumption. The instantaneous rate of change in blood alcohol level t_0 hours after consumption.
   c. The average rate of change of the 30-year fixed mortgage rate between t_0 and $t_0 + h$ years after 2000. The instantaneous rate of change of the 30-year fixed mortgage rate t_0 years after 2000.

33. $V'(30) \approx \dfrac{65 - 50}{50 - 30} = \dfrac{3}{4}$; decreases to 0

35. Approx. $-0.009°C/\text{meter}$; approx. $0°C/\text{meter}$

37. **a.** $P = (120 - q)q - 20q$
   b. $80 per unit
   c. $60 per unit; increasing

39. **a.** $P'(x) = 4{,}000(17 - 2x)$
   b. $P'(x) = 0$ when $x = \dfrac{17}{2}$ or 850 units. At this level of production, profits are neither increasing nor decreasing.

41. **a.** $5.94 per unit
   b. $5.90 per unit; increasing

45. **a.** 0.0211 mm per mm of mercury
   b. 0.022 mm per mm of mercury; increasing
   c. 72.22 mm of mercury. At this pressure, the aortic diameter is neither increasing nor decreasing.

47. **a.** The graph of $y = x^2 - 3$ is the same as the graph of $y = x^2$ shifted down by 3 units. Thus, both

ANSWERS

curves have the same slope for each x, and their derivatives are the same, both equal to $y' = 2x$.

b. $y' = 2x$

49. a. $\dfrac{dy}{dx} = 2x; \dfrac{dy}{dx} = 3x^2$

b. $\dfrac{dy}{dx} = 4x^3; \dfrac{dy}{dx} = 27x^{26}$

51. For $x > 0$, we have $f(x) = x$ and
$$f'(x) = \lim_{h \to 0} \frac{(x + h) - (x)}{h} = 1$$
and for $x < 0$, we have $f(x) = -x$ so that
$$f'(x) = \lim_{h \to 0} \frac{-(x + h) - (-x)}{h} = -1$$
However, the derivative for $x = 0$ would be
$$f'(0) = \lim_{h \to 0} \frac{|0 + h| - 0}{h} = \lim_{h \to 0} \frac{|h|}{h}$$
which does not exist since the two one-sided limits at $x = 0$ are not the same (limit from the left is -1 and from the right is $+1$).

53. $f(x)$ is not continuous at $x = 1$, so it cannot be differentiable there.

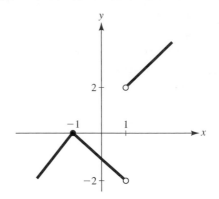

55.

h	-0.02	-0.01	-0.001	0	0.001	0.01	0.02
$x + h$	3.83	3.84	3.849	3.85	3.851	3.86	3.87
$f(x)$	4.37310	4.37310	4.37310	4.37310	4.37310	4.37310	4.37310
$f(x + h)$	4.35192	4.36251	4.37204	4.37310	4.37415	4.38368	4.39426
$\dfrac{f(x + h) - f(x)}{h}$	1.059	1.059	1.059	undefined	1.059	1.058	1.058

CHAPTER 2 | Section 2

1. $\dfrac{dy}{dx} = -4x^{-5}$

3. $\dfrac{dy}{dx} = 0$

5. $\dfrac{dy}{dr} = 2\pi r$

7. $\dfrac{dy}{dx} = \dfrac{\sqrt{2}}{2\sqrt{x}}$

9. $\dfrac{dy}{dt} = \dfrac{-9}{2\sqrt{t^3}}$

11. $\dfrac{dy}{dx} = 2x + 2$

13. $f'(x) = 9x^8 - 40x^7 + 1$

15. $f'(x) = -0.06x^2 + 0.3$

17. $\dfrac{dy}{dt} = \dfrac{-1}{t^2} - \dfrac{2}{t^3} + \dfrac{1}{2\sqrt{t^3}}$

19. $f'(x) = \dfrac{3}{2}\sqrt{x} - \dfrac{3}{2\sqrt{x^5}}$

21. $\dfrac{dy}{dx} = -\dfrac{x}{8} - \dfrac{2}{x^2} - \dfrac{3}{2}x^{1/2} - \dfrac{2}{3x^3} + \dfrac{1}{3}$

23. $\dfrac{dy}{dx} = \dfrac{4}{x^3} + \dfrac{2}{3x^{1/3}} - \dfrac{1}{4x^{3/2}} + \dfrac{x}{2} + \dfrac{1}{3}$

25. $\dfrac{dy}{dx} = 2x + \dfrac{4}{x^2}$

27. $y = 10x + 2$

29. $y = -\dfrac{1}{16}x + 2$

31. $y = x + 3$

33. $y = -4x - 1$

35. $y = 3x - 3$

37. $y = -3x + \dfrac{22}{3}$

39. $f'(-1) = -5$

41. $f'(1) = -\dfrac{3}{2}$

43. $f'(1) = \dfrac{1}{2}$

45. a. 0.2 parts per million per year
 b. 0.15 parts per million
 c. 0.4 parts per million

47. a. $f'(x) = -6$
 b. The change in the average SAT score is the same each year. The average SAT score decreases each year.

49. a. $T'(0) = \$40$ per year
 b. \$480

51. Approximately 2,435 people per day.

53. a. $C(x) = 1.9x + \dfrac{7,280}{x}$
 b. $C'(40) = -2.65$; decreasing

55. $\dfrac{f'(x)}{f(x)} = \dfrac{6x^2 - 10x}{2x^3 - 5x^2 + 4}$; $\dfrac{f'(1)}{f(1)} = -4$

57. a. \$10,800 per year
 b. 17.53%

59. a. $f(t) = \dfrac{100(2,000)}{45,000 + 2,000t} = \dfrac{200}{45 + 2t}$

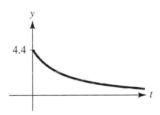

 b. 4.26%
 c. The percentage rate of change approaches zero.

61. $P'(x) = 2 + 6x^{1/2}$
 a. $P'(9) = 20$ persons per month
 b. 0.39%

63. a. $T'(t) = -204.21t^2 + 61.96t + 12.52$
 b. $T'(0) = 12.52$, increasing; $T'(0.713) = -47.12$, decreasing
 c. $t = 0.442$ days or 10.61 hours; $T(0.442) = 42.8°C$, which is the maximum temperature during the period.

65. a. $v(t) = 6t + 2; a(t) = 6$
 b. Not stationary

67. a. $v(t) = 4t^3 - 12t^2 + 8; a(t) = 12t^2 - 24t$
 b. $t = 1, 1+\sqrt{3}$

69. a. 32 ft/sec
 b. 128 ft
 c. -32 ft/sec
 d. -96 ft/sec

71. $a = -1, b = 5, c = 0$

73. $(f + g)'(x)$

$= \lim_{h \to 0} \dfrac{(f + g)(x + h) - (f + g)(x)}{h}$

$= \lim_{h \to 0} \dfrac{[f(x + h) + g(x + h)] - [f(x) + g(x)]}{h}$

$= \lim_{h \to 0} \dfrac{[f(x + h) - f(x)] + [g(x + h) - g(x)]}{h}$

$= \lim_{h \to 0} \left[\dfrac{f(x + h) - f(x)}{h} + \dfrac{g(x + h) - g(x)}{h} \right]$

$= \lim_{h \to 0} \dfrac{f(x + h) - f(x)}{h} + \lim_{h \to 0} \dfrac{g(x + h) - g(x)}{h}$

$= f'(x) + g'(x)$

75. a. \$150 per ton of carbon
 b. 0.2% per dollar

CHAPTER 2 | Section 3

1. $f'(x) = 12x - 1$

3. $\dfrac{dy}{du} = -300u - 20$

5. $f'(x) = \dfrac{1}{3}\left(6x^5 - 12x^3 + 4x + 1 + \dfrac{1}{x^2} \right)$

7. $\dfrac{dy}{dx} = \dfrac{-3}{(x - 2)^2}$

9. $f'(t) = \dfrac{-(t^2 + 2)}{(t^2 - 2)^2}$

11. $\dfrac{dy}{dx} = \dfrac{-3}{(x + 5)^2}$

13. $f'(x) = \dfrac{11x^2 - 10x - 7}{(2x^2 + 5x - 1)^2}$

15. $f'(x) = \dfrac{2(x^2 + 2x + 4)}{(x + 1)^2}$

17. $f'(x) = 10(2 + 5x)$

19. $g'(t) = \dfrac{4\sqrt{t^5} + 20\sqrt{t^3} - 2t + 5}{2\sqrt{t}(2t + 5)^2}$

21. $y = 17x - 4$

23. $y = 3x + 2$

25. $y = -\dfrac{11}{2}x + \dfrac{19}{2}$

27. $(1, -4), (-1, 0)$

29. $(0, 1), \left(-2, -\dfrac{1}{3}\right)$

31. $(5, 0), (3, 108), (0, 0)$

33. -18

35. 4

37. $y = \dfrac{2}{5}x + \dfrac{3}{5}$

39. $y = \dfrac{1}{31}(-x - 371)$

41. a.–d. $y' = \dfrac{9 - 4x}{x^4}$

43. $f''(x) = 8x^3 - 24x + 18$

45. $\dfrac{d^2y}{dx^2} = \dfrac{4}{3x^3} + \dfrac{\sqrt{2}}{4x^{3/2}} - \dfrac{1}{8x^{5/2}}$

47. $\dfrac{d^2y}{dx^2} = 36x^2 + 30x + 12$

49. a. $S'(2) = 378.07$
   b. Sales approach a limit of \$6,666.67.

51. a. $P'(5) = \dfrac{1,900}{441} \approx 4.3\%$ increase per week.

   b. $P(t)$ approaches 100% in the long run; the rate of change approaches 0.

53. a. $P'(16) \approx -0.631$; decreasing
   b. Increasing for $0 < x < 10$; decreasing for $x > 10$

55. a. $R(t) = -3t^2 + 16t + 15$
   b. 10 units per hour per hour

57. a. $v(t) = 15t^4 - 15t^2; a(t) = 60t^3 - 30t$
   b. $a(t) = 0$ when $t = 0, \dfrac{\sqrt{2}}{2}$

59. a. $v(t) = -3t^2 + 14t + 1$
   $a(t) = -6t + 14$
   b. $a(t) = 0$ when $t = \dfrac{7}{3}$

61. a. 9.8 meters per minute
   b. 9.83 meters

63. a. $S = \dfrac{2}{3}KM - M^2$

   b. $\dfrac{dS}{dM} = \dfrac{2}{3}K - 2M$ and represents the rate of change sensitivity with respect to the amount of medicine absorbed.

65. a. $a(t) = -32$
   b. $a(t)$ is constant.
   c. The object is accelerating downward.

67. $\dfrac{3}{8x^{5/2}} + \dfrac{3}{x^4}$

69. a. $\dfrac{d}{dx}\left(\dfrac{fg}{h}\right) = \dfrac{hfg' + hf'g - fgh'}{h^2}$

   b. $\dfrac{dy}{dx} = \dfrac{12x^3 + 51x^2 + 70x - 33}{(3x + 5)^2}$

73.

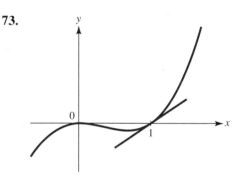

$f'(x) = 0$ where $x = 0$ and $x = \dfrac{2}{3}$.

75.

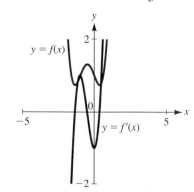

The x intercepts of $f'(x)$ occur at $x = -0.5$, $x = -\dfrac{1}{2} + \dfrac{\sqrt{3}}{2} \approx 0.366$, and $x = -\dfrac{1}{2} - \dfrac{\sqrt{3}}{2} \approx -1.366$. The x intercepts are those points where

$f'(x) = 0$, which are the points where the tangent to the graph of $f(x)$ is horizontal, that is, the maxima and minima of $f(x)$.

CHAPTER 2 | Section 4

1. $\dfrac{dy}{dx} = 6(3x - 2)$

3. $\dfrac{dy}{dx} = \dfrac{x + 1}{\sqrt{x^2 + 2x - 3}}$

5. $\dfrac{dy}{dx} = \dfrac{-4x}{(x^2 + 1)^3}$

7. $\dfrac{dy}{dx} = \dfrac{-x}{(x^2 - 9)^{3/2}}$

9. $\dfrac{dy}{dx} = \dfrac{-2x}{(x^2 - 1)^2}$

11. -160

13. $\dfrac{2}{3}$

15. -16

17. $f'(x) = 8(2x + 1)^3$

19. $f'(x) = 8x^2(x^5 - 4x^3 - 7)^7(5x^2 - 12)$

21. $f'(t) = \dfrac{-2(5t - 3)}{(5t^2 - 6t + 2)^2}$

23. $g'(x) = \dfrac{-4x}{(4x^2 + 1)^{3/2}}$

25. $f'(x) = \dfrac{24x}{(1 - x^2)^5}$

27. $h'(s) = \dfrac{15(1 + \sqrt{3s})^4}{2\sqrt{3s}}$

29. $f'(x) = (x + 2)^2(2x - 1)^4(16x + 17)$

31. $G'(x) = -\dfrac{5}{2}(3x + 1)^{-1/2}(2x - 1)^{-3/2}$

33. $f'(x) = \dfrac{(x + 1)^4(9 - x)}{(1 - x)^5}$

35. $f'(y) = \dfrac{5 - 6y}{(1 - 4y)^{3/2}}$

37. $y = -48x - 32$

39. $y = -12x + 13$

41. $y = \dfrac{2}{3}x - \dfrac{1}{3}$

43. $x = 0; x = -1; x = -\dfrac{1}{2}$

45. $x = -\dfrac{2}{3}$

47. $x = 2$

49. $f'(x) = 6(3x + 5)$

51. $f''(x) = 180(3x + 1)^3$

53. $\dfrac{d^2h}{dt^2} = 80(t^2 + 5)^6(3t^2 + 1)$

55. $f''(x) = (1 + x^2)^{-3/2}$

57. **a.** \$2,295 per year
   b. 10.4% per year

59. **a.** -12 pounds per dollar
   b. $(-12)(0.5) = -6$ pounds per week; decreasing

61. Demand will be decreasing by 12%.

63. **a.** $L'(w) = 0.65w^{1.6}; L'(60) \approx 455$ mm/kg
   b. A 100-day-old tiger weighs $w(100) = 24$ kg and is $L(24) \approx 969$ mm long. By the chain rule,
   $$L'(A) = L'(w)w'(A) = (0.65w^{1.6})(0.21)$$
   so that when $A = 100$, $w = 24$, we have
   $$L'(100) = (0.65)(0.21)(24)^{1.6} \approx 22.1$$
   That is, the tiger's length is increasing at the rate of about 22.1 mm per day.

65. **a.** \$64,000; 8,000 units
   b. 6,501 units per month; increasing

67. **a.** 6 hours per day; \$16,000
   b. $-$\$968 per month; decreasing

69. **a.** $V'(T) = 0.41(-0.02T + 0.4)$
   b. $m'(V) = \dfrac{0.39}{(1 + 0.09V)^2}$
   c. $V(10) = 2.6732$ cm^3; 0.02078 gm/°C

71. **a.** $\dfrac{dT}{dL} = \dfrac{a(3L - 2b)}{2\sqrt{L - b}}$

73. **a.** $v(t) = \dfrac{3}{2}(1 - 2t)(3 + t - t^2)^{1/2}$
   $$a(t) = \dfrac{24t^2 - 24t - 33}{4(3 + t - t^2)^{1/2}}$$
   b. $t = \dfrac{1}{2}, 2; s\left(\dfrac{1}{2}\right) = \dfrac{13\sqrt{13}}{8} \approx 5.86$
   $$a\left(\dfrac{1}{2}\right) = \dfrac{-3\sqrt{13}}{2} \approx -5.41$$

ANSWERS

c. $a(t) = 0$ when $t = \dfrac{2 + \sqrt{26}}{4} \approx 1.775$

$s(1.775) \approx 2.07; v(1.775) \approx -4.875$

d.

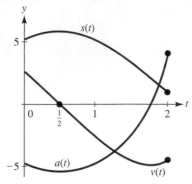

e. The object is slowing down for $0 \le t < 0.5$ and $1.775 < t \le 2$.

75. $\dfrac{dy}{dx} = \dfrac{d}{dx}[h(x)h(x)] = h(x)\dfrac{dh(x)}{dx} + \dfrac{dh(x)}{dx}h(x)$

$= 2h(x)\dfrac{dh(x)}{dx}$

77. $f'(1) \approx 0.2593, f'(-3) \approx -0.4740$; one horizontal tangent, $y = 2.687$, where $x = 0$.

CHAPTER 2 Section 5

1. a. $C'(x) = \dfrac{2}{5}x + 4; R'(x) = 9 - \dfrac{x}{2}$

 b. $C'(3) = \$5.20$

 c. $C(4) - C(3) = \$5.40$

 d. $R'(3) = \$7.50$

 e. $R(4) - R(3) = \$7.25$

3. a. $C'(x) = \dfrac{2}{3}x + 2; R'(x) = -3x^2 - 8x + 80$

 b. $C'(3) = 4$

 c. $C(4) - C(3) \approx 4.33$

 d. $R'(3) = 29$

 e. $R(4) - R(3) = 15$

5. a. $C'(x) = \dfrac{x}{2}; R'(x) = \dfrac{2x^2 + 4x + 3}{(1 + x)^2}$

 b. $C'(3) = \$1.50$

 c. $C(4) - C(3) = \$1.75$

 d. $R'(3) = \dfrac{33}{16} \approx \$2.06$

 e. $R(4) - R(3) = \$2.05$

7. 2.1

9. $\dfrac{100f'(4)}{f(4)}(0.3) \approx 0.20; 20\%$

11. a. $C'(3) = \$499.70$

 b. $C(4) - C(3) = \$500.20$

13. a. \$241

 b. \$244

15. 200

17. Revenue will decrease by approximately \$150.80.

19. Daily output will increase by approximately 8 units.

21. 5.12%

23. Daily output will increase by approximately 825 units.

25. 0.2 units

27. $\dfrac{3c}{|c - b|}$

29. 46.67%

31. 0.0385 ft

33. The tangent line at $x = x_0$ is $y = f'(x_0)(x - x_0) + f(x_0)$. Setting $y = 0$ and solving for $x = x_1$ gives

$0 = f'(x_0)(x_1 - x_0) + f(x_0) \Rightarrow x_1 - x_0 = \dfrac{-f(x_0)}{f'(x_0)} \Rightarrow$

$x_1 = x_0 - \dfrac{f(x_0)}{f'(x_0)}$. Repeating the process shows

$x_n = x_{n-1} - \dfrac{f(x_{n-1})}{f'(x_{n-1})}.$

35. 3.82070437, 1.61179338

37. a. $x_{n+1} = x_n - \dfrac{f(x_n)}{f'(x_n)} = x_n - \dfrac{x_n^{1/3}}{\frac{1}{3}x_n^{-2/3}} =$

$x_n - 3x_n = -2x_n$

 b. For any $x_0 \ne 0$ the sequence $x_0, -2x_0, 4x_0, \ldots$ grows larger in magnitude and alternates sign, so it cannot approach a limit.

CHAPTER 2 Section 6

1. $\dfrac{dy}{dx} = -\dfrac{2}{3}$

3. a. $\dfrac{dy}{dx} = \dfrac{-y}{x + 2} = \dfrac{-[3/(x + 2)]}{x + 2}$

 b. $\dfrac{dy}{dx} = \dfrac{-3}{(x + 2)^2}$

5. **a.** $\dfrac{dy}{dx} = \dfrac{2x - y}{x + 2} = \dfrac{x^2 + 4x}{(x + 2)^2}$

   b. $\dfrac{dy}{dx} = \dfrac{x(x + 4)}{(x + 2)^2}$

7. $\dfrac{dy}{dx} = -\dfrac{x}{y}$

9. $\dfrac{dy}{dx} = \dfrac{y - 3x^2}{3y^2 - x}$

11. $\dfrac{dy}{dx} = \dfrac{3 - 2y^2}{2y(1 + 2x)}$

13. $\dfrac{dy}{dx} = -\dfrac{\sqrt{y}}{\sqrt{x}}$

15. $\dfrac{dy}{dx} = \dfrac{y - 1}{1 - x}$

17. $\dfrac{dy}{dx} = \dfrac{1}{3(2x + y)^2} - 2$

19. $\dfrac{dy}{dx} = \dfrac{y - 5x(x^2 + 3y^2)^4}{15y(x^2 + 3y^2)^4 - x}$

21. $y = \dfrac{1}{3}x + \dfrac{4}{3}$

23. $y = \dfrac{1}{2}x + 2$

25. $y = \dfrac{13}{12}x + \dfrac{11}{12}$

27. **a.** None
   b. $(9, 0)$

29. **a.** None
   b. None

31. **a.** $(1, -2), (-1, 2)$
   b. $(-2, 1), (2, -1)$

33. $\dfrac{d^2y}{dx^2} = \dfrac{-3y^2 - x^2}{9y^3} = \dfrac{-5}{9y^3}$

35. $\Delta y \approx -1.704$

37. $\dfrac{dx}{dt} = 1.74$ or 174 units per month

39. 0.476 cm^3 per month

41. $\dfrac{dr}{dt} = 20$ mm/min

43. $\dfrac{dx}{xt} = 0.15419$ or 15.419 units per month

45. **a.** 14.04
   b. -9.87

47. $\dfrac{dK}{dt} = \dfrac{-2}{5}(\$1{,}000) = -\$400$ per week

49. 4 feet per second

51. $\Delta y \approx 0.566$ units

53. Since $v = \dfrac{KR^2}{L} = $ constant at the center of the vessel $(r - 0)$, we have

$$0 = v'(t) = K\left[\dfrac{2R}{L}R' - \dfrac{R^2}{L^2}L'\right]$$

Thus,

$$\dfrac{2R}{L}R' = \dfrac{R^2}{L^2}L'$$

$$\dfrac{L'}{L} = 2\dfrac{R'}{R}$$

so the relative rate of change of L is twice that of R.

55. $\dfrac{x^2}{a^2} + \dfrac{y^2}{b^2} = 1;\ \dfrac{2x}{a^2} + \dfrac{2yy'}{b^2} = 0;\ 2b^2x + 2a^2yy' = 0,$

$y' = -\dfrac{2b^2x}{2a^2y} = -\dfrac{b^2x}{a^2y}.$ At $P(x_0, y_0)$, $m = -\dfrac{b^2x_0}{a^2y_0}$

so the equation of the tangent line is

$$y - y_0 = -\dfrac{b^2x_0}{a^2y_0}(x - x_0)$$

$$a^2yy_0 - a^2y_0^2 = -b^2xx_0 + b^2x_0^2$$

$$b^2xx_0 + a^2yy_0 = b^2x_0^2 + a^2y_0^2$$

$$\dfrac{x_0x}{a^2} + \dfrac{y_0y}{b^2} = \dfrac{x_0^2}{a^2} + \dfrac{y_0^2}{b^2}$$

$$= 1$$

since $P(x_0, y_0)$ lies on the curve and thus satisfies the equation of the curve.

57. Let $y = x^{r/s}$, then $y^s = x^r$ and $sy^{s-1}\dfrac{dy}{dx} = rx^{r-1}$,

$\dfrac{dy}{dx} = \dfrac{rx^{r-1}}{sy^{s-1}}.$ But $y^{s-1} = \dfrac{y^s}{x^{r/s}} = \dfrac{x^r}{x^{r/s}}$, so

$$\dfrac{dy}{dx} = \dfrac{r}{s} \cdot x^{r-1} \cdot \dfrac{x^{r/s}}{x^r}$$

$$= \dfrac{r}{s} \cdot x^{r-1+r/s-r}$$

$$= \dfrac{r}{s} \cdot x^{r/s - 1}$$

ANSWERS

59. Horizontal tangent lines are $y = 1.24$ and $y = -1.24$.

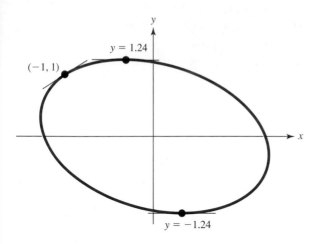

61. Horizontal tangent lines are $y = 1.23$ and $y = -1.23$.

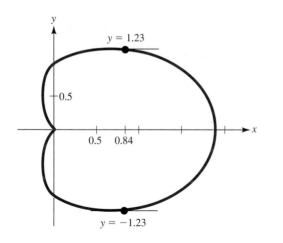

CHAPTER 2 | Checkup

1. a. $\dfrac{dy}{dx} = 12x^3 - \dfrac{2}{\sqrt{x}} - \dfrac{10}{x^3}$

b. $\dfrac{dy}{dx} = -15x^4 + 39x^2 - 2x - 4$

c. $\dfrac{dy}{dx} = \dfrac{-10x^2 + 10x + 1}{(1 - 2x)^2}$

d. $\dfrac{dy}{dx} = (9x - 6)(3 - 4x + 3x^2)^{1/2}$

2. $f''(t) = 24t + 8$

3. $y = -4x$

4. $\dfrac{3}{8}$

5. a. 58 dollars per year
   b. 2.98%

6. a. $v(t) = 6t^2 - 6t$; $a(t) = 12t - 6$
   b. $t = 0, 1$; retreating $0 < t < 1$; advancing $1 < t < 2$.
   c. 6

7. a. $C'(5) = 5.4(\$100) = \$540$
   b. $C(6) - C(5) = 5.44(\$100) = \$544$

8. Output increases by approximately $\dfrac{75{,}000}{7}$ units

9. 0.001586 m^2 per week

10. $\dfrac{8}{3}\%$

CHAPTER 2 | Review Problems

1. $f'(x) = 2x - 3$

3. $f'(x) = 24x^3 - 21x^2 + 2$

5. $\dfrac{dy}{dx} = \dfrac{-14x}{(3x^2 + 1)^2}$

7. $f'(x) = 10(20x^3 - 6x + 2)(5x^4 - 3x^2 + 2x + 1)^9$

9. $\dfrac{dy}{dx} = 2\left(x + \dfrac{1}{x}\right)\left(1 - \dfrac{1}{x^2}\right) + \dfrac{5}{2\sqrt{3x^3}}$

11. $f'(x) = 3\sqrt{6x + 5} + \dfrac{9x + 3}{\sqrt{6x + 5}}$

13. $\dfrac{dy}{dx} = \dfrac{-7}{2(3x + 2)^2}\sqrt{\dfrac{3x + 2}{1 - 2x}}$

15. $y = -x - 1$

17. $y = -\dfrac{2}{3}x + \dfrac{5}{3}$

19. a. 87.5%
   b. -400%
   c. -100%

21. a. 2
   b. $\dfrac{3}{2}$
   c. $\dfrac{27}{320}$

23. a. $\dfrac{dy}{dx} = -\dfrac{5}{3}$

   b. $\dfrac{dy}{dx} = -\dfrac{2y}{x}$

   c. $\dfrac{dy}{dx} = \dfrac{1 - 10(2x + 3y)^4}{15(2x + 3y)^4}$

   d. $\dfrac{dy}{dx} = -\left[\dfrac{1 + 10y^3(1 - 2xy^3)^4}{4 + 30xy^2(1 - 2xy^3)^4}\right]$

25. $\dfrac{d^2y}{dx^2} = \dfrac{6y^2 - 9x^2}{4y^3} = \dfrac{-9}{2y^3}$

27. a. 75,000 per year

   b. 0 per year

29. a. $v(t) = \dfrac{-2(t + 4)(t - 3)}{(t^2 + 12)^2}$,

   $a(t) = \dfrac{2(2t^3 + 3t^2 - 72t - 12)}{(t^2 + 12)^3}$. The object is advancing for $0 < t < 3$, retreating for $3 < t < 4$. It is always decelerating for $0 < t < 4$.

   b. $\dfrac{11}{42}$

31. a. Output will increase by approximately 12,000 units.

   b. Output will increase by 12,050 units.

33. Output will decrease by approximately 5,000 units per day.

35. Pollution will increase by approximately 10%.

37. a. 0.2837 individuals per square kilometer

   b. 2.61 million

   c. 55 years; 60.67 animals per year

39. 1.5%

41. a. $\dfrac{dF}{dC} = \dfrac{-kD^2}{2\sqrt{A - C}}$; F decreases as C increases

   b. $\dfrac{50}{A - C}\%$

43. $425.25 \le A \le 479.53$; accurate to 6%

45. $100\dfrac{\Delta Q}{Q} \approx 0.67\%$

47. $6.59 \le V \le 7.89$; accurate to 9%

49. 2 toasters per month

51. 10.7%

53. 5.5 seconds; 242 feet

55. a. \$195 per unit per month

   b. $-\$16$ per unit per month per month

   c. $-\$8$ per unit per month

   d. $-\$8.75$ per unit per month

57. $-\$99$ per month

59. 28.37 cubic inches

61. 3 ft/sec

63. 2.25 ft/sec

65. -3.29 ft/sec

67. $\dfrac{8}{5}$ ft/sec

69. The percentage rate of change approaches 0 since, if $y = mx + b$, $\dfrac{100y'}{y} = \dfrac{100m}{mx + b}$, which approaches 0 as x approaches ∞.

71.

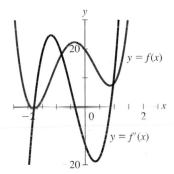

73. a.

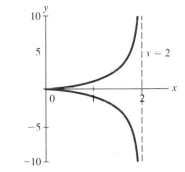

   b. At $(1, 1)$, the tangent line is $y = 2x - 1$ and at $(1, -1)$, it is $y = -2x + 1$.

   c. As $x \to 2^-$, the top branch of the curve rises indefinitely ($y \to +\infty$), while the bottom branch falls indefinitely ($y \to -\infty$).

   d. The x axis is a (double) tangent line to the curve at the origin.

CHAPTER 3 | Section 1

1. $f'(x) > 0$ for $-2 < x < 2$; $f'(x) < 0$ for $x < -2$ and $x > 2$

3. $f'(x) > 0$ for $x < -4$ and $0 < x < 2$; $f'(x) < 0$ for $-4 < x < -2$, $-2 < x < 0$, and $x > 2$

5. B

7. D

9. $f(x)$ is increasing for $x > 2$; $f(x)$ is decreasing for $x < 2$.

11. $f(x)$ is increasing for $x < -1$ and $x > 1$; $f(x)$ is decreasing for $-1 < x < 1$.

13. $g(t)$ is increasing for $t < 0$ and $t > 4$; $g(t)$ is decreasing for $0 < t < 4$.

15. $f(t)$ is increasing for $0 < t < 2$ and $t > 2$; $f(t)$ is decreasing for $t < -2$ and $-2 < t < 0$.

17. $h(u)$ is increasing for $-3 < u < 0$; $h(u)$ is decreasing for $0 < u < 3$.

19. $F(x)$ is increasing for $x < -3$ and $x > 3$; $F(x)$ is decreasing for $-3 < x < 0$ and $0 < x < 3$.

21. $f(x)$ is increasing for $x > 1$; $f(x)$ is decreasing for $0 < x < 1$.

23. $x = 0, 1$; $(0, 2)$ relative minimum; $(1, 3)$ neither

25. $x = -1$; $(-1, 3)$ neither

27. $x = 1$; $(1, 0)$ neither

29. $t = -\sqrt{3}, \sqrt{3}$; $\left(\sqrt{3}, \dfrac{\sqrt{3}}{6}\right)$ relative maximum; $\left(-\sqrt{3}, -\dfrac{\sqrt{3}}{6}\right)$ relative minimum

31. $t = -2, 0, 1, 4$; $(0, 0)$ relative maximum; $\left(4, \dfrac{8}{9}\right)$ relative minimum

33. $t = 0, -1, 1$; $(0, 1)$ relative maximum; $(-1, 0)$ and $(1, 0)$ relative minima

35.

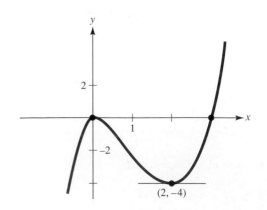

37.

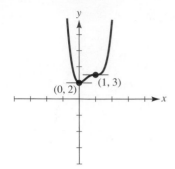

39.

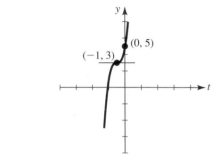

41.

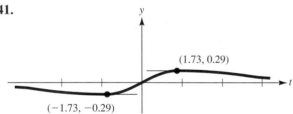

43.

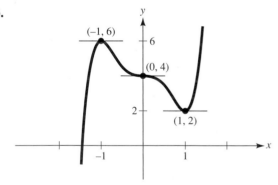

45.

Critical Numbers	Classification
-2	Relative minimum
0	Neither
2	Relative maximum

47.

Critical Numbers	Classification
-1	Neither
$\dfrac{4}{3}$	Relative maximum

49. One possibility:

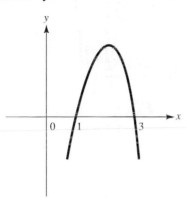

51. One possibility:

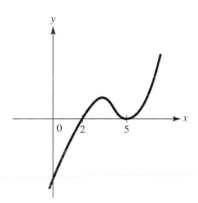

53. a. $A'(x) = 2x - 20 - \dfrac{242}{x^2}$

   b. Increasing: $x > 11$; decreasing: $0 \leq x < 11$

   c. Average cost is minimized when $x = 11$; Minimum average cost: \$102,000/unit.

55. $R(x) = x(10 - 3x)^2$; $\dfrac{dR}{dx} = (10 - 3x)(10 - 9x)$;

   Revenue is maximized when $x = \dfrac{10}{9}$.

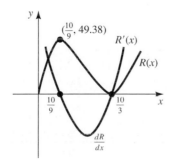

57. Maximum concentration occurs when $t = 0.9$ hours.

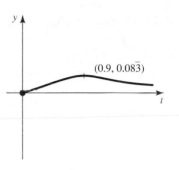

59. a.

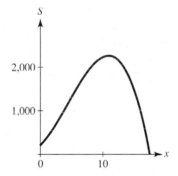

   b. 207

   c. \$11,000; 2,264 units

61. a. $1 \leq r \leq 5.495$

   b. 5.495%; 1,137 mortgages

63. a. 1971, 1976, 1980, 1983, 1988, 1994

   b. 1973, 1979, 1981, 1985, 1989

   c. Approximately $\dfrac{1}{2}$% per year

   d. Approximately $\dfrac{1}{2}$% per year

65. a. $Y(t) = \dfrac{9,300}{31 + t}(3 + t - 0.05t^2)$

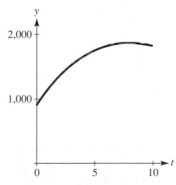

   b. 8 weeks; 1,860 pounds

67.

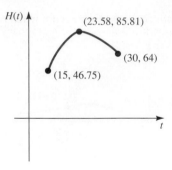

$H(t)$

(23.58, 85.81)

(30, 64)

(15, 46.75)

Maximum of 85.81% at 23.58°C.

69.

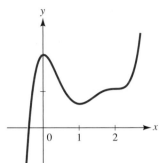

71.

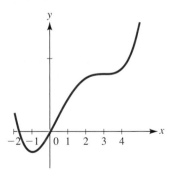

73. $a = 2; b = 3; c = -12; d = -12$

75.

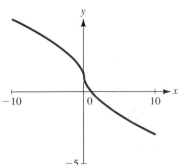

77. By the product rule, $\dfrac{dy}{dx} = (x - p)(1) + (1)(x - q)$

$= 2x - p - q$. Solving $\dfrac{dy}{dx} = 0$ yields $x = \dfrac{p + q}{2}$,

the point midway between the x intercepts p and q.

79. $f'(x) = 0$ at $x = -3; -2.529, -1.618, -0.346, 0.618$

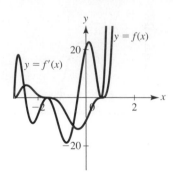

$y = f(x)$

$y = f'(x)$

81. $f'(x)$ is never 0.

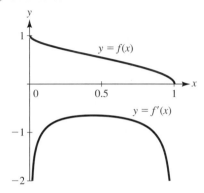

$y = f(x)$

$y = f'(x)$

83. The top half of a circle.

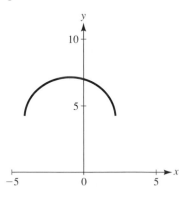

CHAPTER 3 | Section 2

1. $f''(x) > 0$ for $x > 2; f''(x) < 0$ for $x < 2$

3. $f''(x) > 0$ for $x < -1$ and $x > 1; f''(x) < 0$ for $-1 < x < 1$

5. Concave upward for $x > -1$; concave downward for $x < -1$; inflection at $(-1, 2)$

7. Concave upward for $x > -\dfrac{1}{3}$; concave downward for $x < -\dfrac{1}{3}$; inflection at $\left(-\dfrac{1}{3}, -\dfrac{1}{27}\right)$

9. Concave upward for $t < 0$ and $t > 1$; concave downward for $0 < t < 1$; inflection at $(1, 0)$

11. Concave upward for $x < 0$ and $x > 3$; concave downward for $0 < x < 3$; inflection at $(0, -5)$ and $(3, -65)$

13. Increasing for $x < -3$ and $x > 3$; decreasing for $-3 < x < 3$; concave upward for $x > 0$; concave downward for $x < 0$; maximum at $(-3, 20)$; minimum at $(3, -16)$; inflection at $(0, 2)$

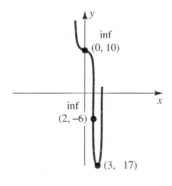

15. Increasing for $x > 3$; decreasing for $x < 3$; concave upward for $x < 0$ and $x > 2$; concave downward for $0 < x < 2$; minimum at $(3, -17)$; inflection at $(0, 10)$ and $(2, -6)$

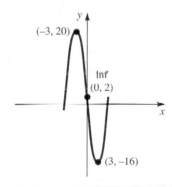

17. Increasing for all x; concave upward for $x > 2$; concave downward for $x < 2$; inflection at $(2, 0)$

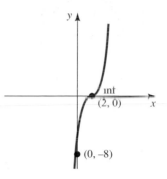

19. Increasing for $x > 0$; decreasing for $x < 0$; concave upward for $x < -\sqrt{5}$, $-1 < x < 1$, $x > \sqrt{5}$; concave downward for $-\sqrt{5} < x < -1$ and $1 < x < \sqrt{5}$; minimum at $(0, -125)$; inflection points at $(-\sqrt{5}, 0)$, $(\sqrt{5}, 0)$, $(-1, -64)$, and $(1, -64)$

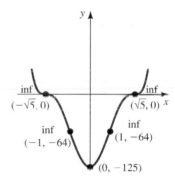

21. Increasing for $s > -1$; decreasing for $s < -1$; concave upward for $s < -4$ and $s > -2$; concave downward at $-4 < s < -2$; minimum at $(-1, -54)$; inflection at $(-4, 0)$ and $(-2, -32)$

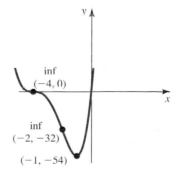

23. Increasing for $x > 0$; decreasing for $x < 0$; concave upward for all real x; minimum at $(0, 1)$

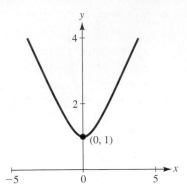

25. Increasing for $x < -\dfrac{1}{2}$; decreasing for $x > -\dfrac{1}{2}$; concave upward for $x < -1$ and $x > 0$; concave downward for $-1 < x < 0$; maximum at $\left(-\dfrac{1}{2}, \dfrac{4}{3}\right)$ inflection at $(-1, 1)$ and $(0, 1)$.

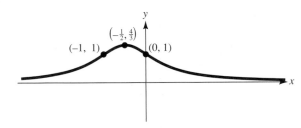

27. $f''(x) = 6(x + 1)$; maximum at $(-2, 5)$; minimum at $(0, 1)$

29. $f''(x) = 12(x^2 - 3)$; maximum at $(0, 81)$; minimum at $(3, 0)$ and $(-3, 0)$

31. $f''(x) = \dfrac{36}{x^3}$; maximum at $(-3, -11)$; minimum at $(3, 13)$

33. $f''(x) = 12x^2 - 60x + 50$; maximum at $\left(\dfrac{5}{2}, \dfrac{625}{16}\right)$; minimum at $(0, 0)$ and $(5, 0)$

35. $f''(t) = \dfrac{4(3t^2 - 1)}{(1 + t^2)^3}$; maximum at $(0, 2)$

37. $f''(x) = \dfrac{24(x - 2)}{x^4}$; maximum at $(-4, -13.5)$. Test fails for $x = 2$ (there is an inflection point at $(2, 0)$).

39. Concave upward for $x < 0$, for $0 < x < 1$, and for $x > 3$; concave downward for $1 < x < 3$; inflection points at $x = 1, 3$

41. Concave upward for $x > 1$; concave downward for $x < 1$; inflection point at $x = 1$

43. a. Increasing for $x < 0$ and $x > 4$; decreasing for $0 < x < 4$

   b. Concave upward for $x > 2$ and concave downward for $x < 2$

   c. Relative minimum at $x = 4$, relative maximum at $x = 0$; inflection point at $x = 2$.

   d.

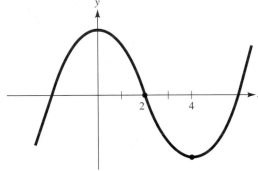

45. a. Increasing for $-\sqrt{5} < x < \sqrt{5}$; decreasing for $x > \sqrt{5}$ and $x < -\sqrt{5}$

   b. Concave upward for $x < 0$ and concave downward for $x > 0$

   c. Relative maximum at $x = \sqrt{5}$ and relative minimum at $x = -\sqrt{5}$; inflection point at $x = 0$

   d.

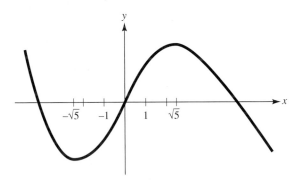

47. A typical graph is shown.

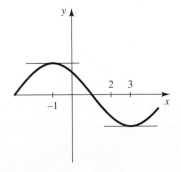

49. $f(x)$ is increasing for $x > 2$.
$f(x)$ is decreasing for $x < 2$.
$f(x)$ is concave upward for all real x.
$f(x)$ has a relative minimum at $x = 2$.
$f(x)$ has no inflection points.

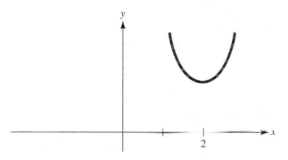

51. $f(x)$ is increasing for $x > 2$.
$f(x)$ is decreasing for $x < -3$ and $-3 < x < 2$.
$f(x)$ is concave upward for $x < -3$ and $x > -1$.
$f(x)$ is concave downward for $-3 < x < -1$.
$f(x)$ has a relative minimum at $x = 2$.
$f(x)$ has inflection points at $x = -3$ and $x = -1$.

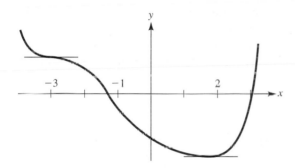

53. $C(x) = 0.3x^3 - 5x^2 + 28x + 200$
a. $C'(x) = 0.9x^2 - 10x + 28$

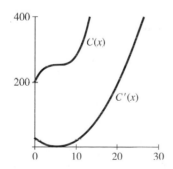

b. Only inflection number of $C(x)$ is $x = 5.56$. It corresponds to a minimum on the graph of $C'(x)$.

55. Output rate is $Q'(t) = -3t^2 + 9t + 15$.
a. Rate is maximized at $t = 1.5$ (9:30 A.M.).
b. Rate is minimized at $t = 4$ (noon).

57. Rate of growth is $P'(t) = -3t^2 + 18t + 48$.
a. Rate is largest when $t = 3$ years.
b. Rate is smallest when $t = 0$ years.
c. Rate of growth of $P'(t)$ is $P''(t) = -6t + 18$, which is largest when $t = 0$.

59. **a.** $N'(t) = \dfrac{60 - 5t^2}{(12 + t^2)^2};$

$N''(t) = \dfrac{10t^3 - 360t}{(12 + t^2)^3}$

b. 3.5 weeks after the outbreak; 722 new cases
c. 6 weeks after the outbreak; 62.5 new cases

61. **a.** $M'(r) = \dfrac{0.02 - 0.018r - 0.00018r^2}{(1 + 0.009r^2)^2}$

$M''(r) = \dfrac{-0.018 - 0.00108r + 0.000486r^2 + 0.00000324r^3}{(1 + 0.009r^2)^3}$

b.

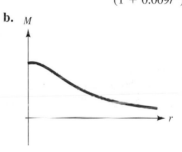

c. 7.10%

63. $\dfrac{100P'(t)}{P(t)} = A - BP(t);\ P'(t) = \dfrac{P(t)(A - BP(t))}{100};$

$P''(t) = \dfrac{1}{100^2}[P(t)(A - BP(t))(A - 2BP(t))];$

$P''(t) = 0$ when $P(t) = \dfrac{A}{B}$ or $P(t) = \dfrac{A}{2B}$. When

$P(t) = \dfrac{A}{B}$, the population is at the maximum

possible, and there is no change ($P'(t) = 0$). When

$P(t) = \dfrac{A}{2B}$, the population is changing most rapidly.

65. **a.** $R'(t) = A''(t) = \dfrac{d}{dt}[(k\sqrt{A(t)}\,(M - A(t))]$

$= k\left(\dfrac{1}{2}\right)\dfrac{A'(t)}{\sqrt{A(t)}}[M - A(t)] + k\sqrt{A(t)}(-A'(t))$

$= \dfrac{kA'(t)}{2\sqrt{A(t)}}[M - 3A(t)]$

$= 0$ when $A = \dfrac{M}{3}$.

b. Greatest

c. The graph of $A(t)$ has an inflection point where
$$A(t) = \frac{M}{3}.$$

67. $f'(x) = 4x^3 + 1; f''(x) = 12x^2.$ Although $f''(0) = 0$, neither $f'(x)$ nor $f''(x)$ change sign at $x = 0$, so the graph of f has neither a relative extremum nor an inflection point where $x = 0$.

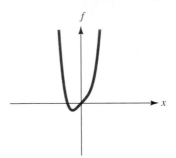

69. a.

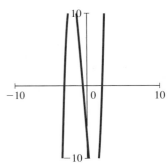

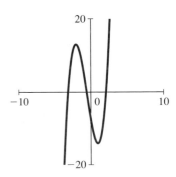

b.

x	-4	-2	-1	0	1	2
$f(x)$	-39	13	6	-7	-14	-3
$f'(x)$	60	0	-12	-12	0	24
$f''(x)$	-42	-18	-6	6	18	30

c. $(-3.08, 0), (-0.54, 0), (2.12, 0); (0, -7)$

d. Relative maximum at $(-2, 13)$; relative minimum at $(1, -14)$

e. $x < -2$ and $x > 1$

f. $-2 < x < 1$

g. $\left(-\dfrac{1}{2}, -\dfrac{1}{2}\right)$

h. $x > -\dfrac{1}{2}$

i. $x < -\dfrac{1}{2}$

j. Answers will vary.

k. $13; -39$

CHAPTER 3 | Section 3

1. Vertical asymptote, $x = 0$; horizontal asymptote, $y = 0$

3. No vertical asymptotes; horizontal asymptote at $y = 0$

5. Vertical asymptotes, $x = -2, x = 2$; horizontal asymptotes, $y = 2$ and $y = 0$ (x axis)

7. Vertical asymptote, $x = 2$; horizontal asymptote, $y = 0$

9. Vertical asymptote, $x = -2$; horizontal asymptote, $y = 3$

11. No vertical asymptotes; horizontal asymptote, $y = 1$

13. Vertical asymptotes, $t = 2, t = 3$; horizontal asymptote, $y = 1$

15. Vertical asymptotes, $x = 0, x = 1$; horizontal asymptote, $y = 0$

17.

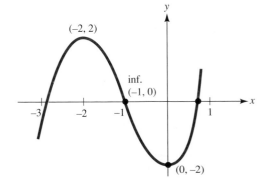

19.

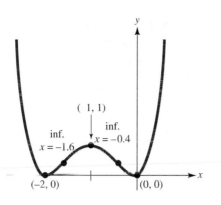

inf.
$x = -1.6$

$(1, 1)$

inf.
$x = -0.4$

$(-2, 0)$

$(0, 0)$

21.

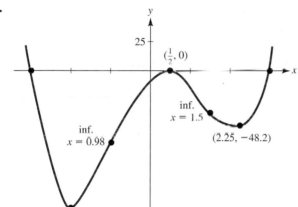

25

$(\frac{1}{2}, 0)$

inf.
$x = 1.5$

inf.
$x = 0.98$

$(2.25, -48.2)$

$(-2, -125)$

23.

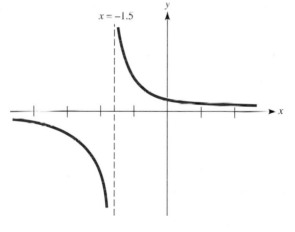

$x = -1.5$

25.

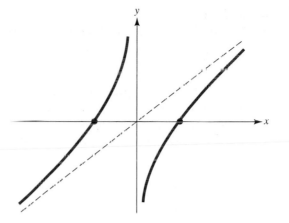

27.

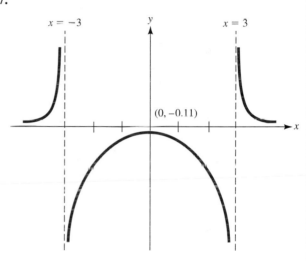

$x = -3$

$x = 3$

$(0, -0.11)$

29.

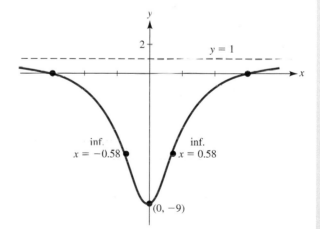

2

$y = 1$

inf.
$x = -0.58$

inf.
$x = 0.58$

$(0, -9)$

ANSWERS

31.

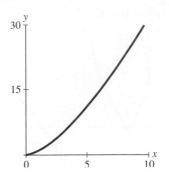

33. Answers may vary.

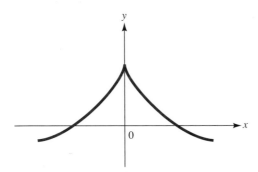

35. Answers may vary.

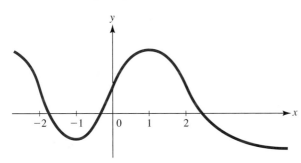

37. Answers may vary.

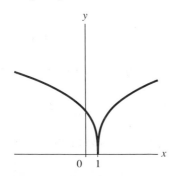

39. a. $f(x)$ is increasing ($f'(x) > 0$) for $0 < x < 2$ or $x > 2$; $f(x)$ is decreasing ($f'(x) < 0$) for $x < 0$.
   b. Relative minimum at $x = 0$
   c. $f''(x) = x^2(x - 2)(5x - 6)$; $f(x)$ is concave up for $x < 0$, $0 < x < \dfrac{6}{5}$, and $x > 2$; concave down for $\dfrac{6}{5} < x < 2$.
   d. $x = \dfrac{6}{5}$ and $x = 2$

41. a. $f(x)$ is increasing ($f'(x) > 0$) for $-3 < x < 2$ or $2 < x$; $f(x)$ is decreasing ($f'(x) < 0$) for $x < -3$.
   b. Relative minimum at $x = -3$
   c. $f''(x) = \dfrac{-x - 8}{(x - 2)^3}$; $f(x)$ is concave up for $-8 < x < 2$; concave down for $x < -8$ and $x > 2$.
   d. $x = -8$

43. $B = -\dfrac{5}{2}$; $A = -10$

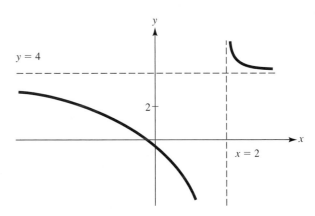

45. Answers may vary.

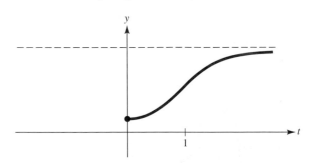

47.

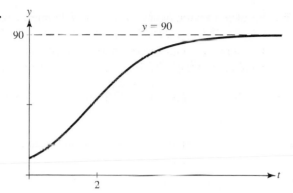

49. a. Vertical asymptote, $x = 0$; no horizontal asymptotes

b. The average cost curve $A(x)$ approaches the line $y = 3x + 1$ as x gets larger.

c.

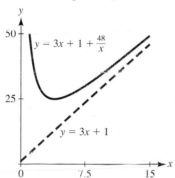

51. a.

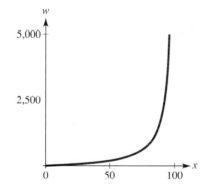

b. 11.8%

53. a.

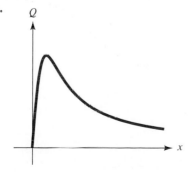

b. \$5,196; 674 units

c. \$9,000

55. a.

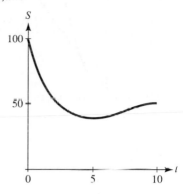

b. $t = 5$; 41.2%

c. Positive; decreasing

57. a. $C(x) = \dfrac{7{,}672}{x} + 2.95x$

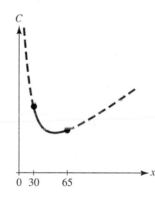

b. 51 mph; \$300.58

59. a. $f'(x) = \dfrac{10(x - 1)}{3x^{1/3}}$; $f(x)$ is increasing for $x < 0$ and $x > 1$; $f(x)$ is decreasing for $0 < x < 1$; relative minimum at $(1, -3)$; relative maximum at $(0, 0)$.

b. $f''(x) = \dfrac{10(2x + 1)}{9x^{4/3}}$; $f(x)$ is concave upward for $x > -\dfrac{1}{2}$; $f(x)$ is concave downward for $x < -\dfrac{1}{2}$; inflection point at $\left(-\dfrac{1}{2}, -3\sqrt[3]{2}\right)$.

c. $(0, 0)$, $\left(\dfrac{5}{2}, 0\right)$; no asymptotes

ANSWERS

d.

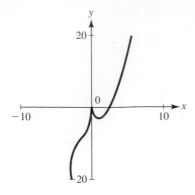

61. a.

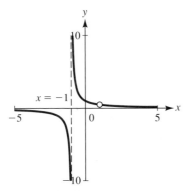

The graph has a hole at $x = 1$.

b.

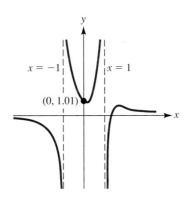

Vertical asymptotes at $x = 1$, $x = -1$; horizontal asymptote $y = 0$.

CHAPTER 3 | Section 4

1. Absolute maximum at $(1, 10)$; absolute minimum at $(-2, 1)$

3. Absolute maximum at $(0, 2)$; absolute minimum at $\left(2, -\dfrac{40}{3}\right)$

5. Absolute maximum at $(-1, 2)$; absolute minimum at $(-2, -56)$

7. Absolute maximum at $(-3, 3{,}125)$; absolute minimum at $(0, -1024)$

9. Absolute maximum at $\left(3, \dfrac{10}{3}\right)$; absolute minimum at $(1, 2)$

11. Absolute minimum at $(1, 2)$; no absolute maximum

13. $f(x)$ has no absolute maximum or minimum for $x > 0$.

15. Absolute maximum at $(0, 1)$; no absolute minimum

17. **a.** $R(q) = 49q - q^2$; $R'(q) = 49 - 2q$;

$C'(q) = \dfrac{1}{4}q + 4$;

$P(q) = -\dfrac{9}{8}q^2 + 45q - 200$;

maximum when $q = 20$

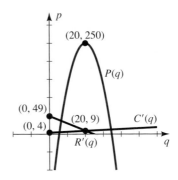

b. $A(q) = \dfrac{1}{8}q + 4 + \dfrac{200}{q}$; $A(q)$ is minimized at $q = 40$.

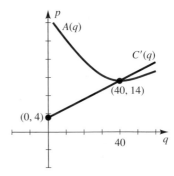

19. a. $R(q) = 180q - 2q^2$; $R'(q) = 180 - 4q$; $C'(q) = 3q^2 + 5$; $P(q) = -q^3 - 2q^2 + 175q - 162$; $P(q)$ is maximized at $q = 7$.

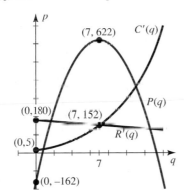

b. $A(q) = q^2 + 5 + \dfrac{162}{q}$; $A(q)$ is minimized at $q = 4.327$.

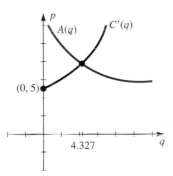

21. a. $R(q) = 1.0625q - 0.0025q^2$;
$R'(q) = 1.0625 - 0.005q$;
$C'(q) = \dfrac{q^2 + 6q - 1}{(q + 3)^2}$;
$P(q) = \dfrac{-0.0025q^3 + 0.055q^2 + 3.1875q - 1}{q + 3}$;
Profit $P(q)$ is maximized when $P'(q) = 0$; when $q = 17.3$ units.

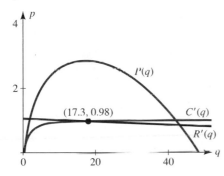

b. $A(q) = \dfrac{q^2 + 1}{q(q + 3)}$ is minimized when $q = \dfrac{1 + \sqrt{10}}{3} \approx 1.3874$.

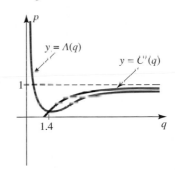

23. $E(p) = \dfrac{-1.3p}{-1.3p + 10}$; $E(4) = -\dfrac{13}{12}$, elastic

25. $F(p) = \dfrac{2p^2}{p^2 - 200}$; $E(10) = -2$, elastic

27. $E(p) = \dfrac{30}{p - 30}$; $E(10) = -\dfrac{3}{2}$, elastic

29. $R'(q) = -4q + 68$;
$A(q) = \dfrac{R(q)}{q} = -2q + 68 - \dfrac{128}{q}$

a. $R'(q) = A(q)$ when $q = 8$

b. A increasing ($A' > 0$) if $0 < q < 8$ and decreasing ($A' < 0$) if $q > 8$

c.

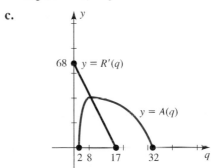

31. The slope $f'(x) = 4x - x^2$ has its largest absolute value when $x = -1$. The graph is steepest at $\left(-1, \dfrac{7}{3}\right)$. The slope of the tangent is -5.

33. a. $x = 4$ or in 1996. The membership in 1996 was 46,400.

b. 0 in 1992 (when $t = 0$)

35. $p = \dfrac{n}{m}$

ANSWERS

37. The speed of the blood is greatest when $r = 0$, that is, at the central axis.

39. a. $E = \dfrac{-3p}{2q + 3p}$

   b. When $p = 3$, $q = 2$, and $E = -\dfrac{9}{13}$; demand is inelastic.

41. a. $225 \le p \le 250$

   b. $E(p) = \dfrac{p}{p - 250}$

   Demand is elastic when $p > 125$, inelastic when $p < 125$, and of unit elasticity when $p = 125$.

   c. Total revenue is increasing for $p < 125$ and decreasing for $p > 125$.

   d. If any number of prints are available, then $p = 125$ maximizes total revenue. If only 50 prints are available, then $p = 225$ maximizes total revenue.

43. a. $v = 39$ km/hr

   b. Writing exercise; responses will vary.

45. a. $D = \dfrac{C}{2}$; $\dfrac{C^2}{4}$

   b. $\dfrac{C^3}{12}$

47. $R = r$

49. a. $R(x) = \dfrac{AB + A(1 - m)x^m}{(B + x^m)^2}$; $x = \left(\dfrac{B}{m - 1}\right)^{1/m}$

   b. $R'(x) = \dfrac{-Amx^{m-1}[(1 - m)x^m + (1 + m)B]}{(B + x^m)^3}$;

   $x = 0$ and $x = \left[\dfrac{B(m + 1)}{m - 1}\right]^{1/m}$

   c. Relative maximum; use the first derivative test

51. a. $F(r) = a\pi r^4(r_0 - r)$

   b. $r = \dfrac{4}{5}r_0$

53. a. $E(p) = \dfrac{ap}{ap - b}$

   b. $E(p) = -1 \Rightarrow ap = (-1)(ap - b)$

   $\Rightarrow 2ap = b \Rightarrow p = \dfrac{b}{2a}$

   c. Elastic for $\dfrac{b}{2a} < p \le \dfrac{b}{a}$, inelastic for $0 \le p < \dfrac{b}{2a}$

55. $E(p) = \dfrac{p}{q}\dfrac{dq}{dp} = \dfrac{p}{\left(\dfrac{a}{p^m}\right)}\left(\dfrac{-ma}{p^{m+1}}\right) =$

   $\left(\dfrac{p^{m+1}}{a}\right)\left(\dfrac{-ma}{p^{m+1}}\right) = -m$

   If $m = 1$, demand is of unit elasticity. If $m > 1$, demand is elastic, and if $0 < m < 1$, demand is inelastic.

CHAPTER 3 | Section 5

1. $\dfrac{1}{2}$

3. $x = 25$, $y = 25$

5. $\$40.83 \approx \$41.00$

7. 80 trees

9. Make the playground square with side $S = 60$ m.

11. Let x be the length of the field and y the width, and let p be the fixed value of the perimeter, so that

   $p = 2(x + y)$ and $y = \dfrac{1}{2}(p - 2x)$. The area is

   $$A = xy = x\left[\dfrac{1}{2}(p - 2x)\right] = -x^2 + \dfrac{1}{2}px$$

   Differentiating, we find that

   $$A' = -2x + \dfrac{1}{2}p = 0$$

   when $x = \dfrac{p}{4}$. Since $A'' = -2 < 0$, the maximum

   area occurs when $x = \dfrac{p}{4}$ and

   $$y = \dfrac{1}{2}\left[p - 2\left(\dfrac{p}{4}\right)\right] = \dfrac{p}{4}$$

   that is, when the figure is a square.

13. 6 by 2.5

15. 2 by 2 by $\dfrac{4}{3}$ meters

17. 2 hours after the initial time; minimizing the square of x will also minimize x when $x \ge 0$.

19. $r = 1.51$ inches; $h = 3.02$ inches

21. The most economical route is to run the cable 2,000 meters under water and 400 meters over land.

23. 27π cubic inches

25. $r = \dfrac{2}{3}h$

27. **a.** 10 machines
    b. $400
    c. $200

29. **a.** 200 bottles
    b. every three months

31. **a.** $C(N) = 8N + \dfrac{200(N + 1)}{N}$
    b. 5 withdrawals of $2,000 each

33. 5 years from now

35. $x = 18; y = 36; V = 11,664 \text{ in}^3$

37. $C(x) = 1,200 + 1.20x + \dfrac{100}{x^2}; x = 6$

39. The point P should be $\dfrac{5\sqrt{3}}{3} \approx 2.9$ miles from A.

41. $S = Kwh^3 = Kh^3\sqrt{225 - h^2};$

    $S'(h) = \dfrac{675h^2 - 4h^4}{\sqrt{225 - h^2}} = 0$ when $h \approx 13$ in.;

    $w \approx 7.5$ in.

43. 4.5 miles from plant A

45. Suppose the setup cost and operating cost are aN

    and $\dfrac{b}{N}$, respectively, for positive constants a and b.

    Total cost is then $C = aN + \dfrac{b}{N}$. The minimum cost

    occurs when

    $$C' = a - \dfrac{b}{N^2} = 0$$

    $$aN = \dfrac{b}{N}$$

    that is, when setup cost aN equals operating cost $\dfrac{b}{N}$.

47. Frank is right; in Example 3.5.5 replace 3,000 with any fixed distance $D \geq 1,200$. The result is the same since D drops out when $C'(x)$ is computed.

49. **a.** Let x be the number of machines and t the number of hours required to produce Q units. The setup cost is $C_s = xs$ and the operating cost (for all x machines) is $C_o = pt$. Since n units can be produced per machine per hour, $Q = nxt$ or

    $t = \dfrac{Q}{nx}$. The total cost is

    $$C = C_s + C_o = xs + \dfrac{pQ}{nx}$$

so that

$$C' = s - \dfrac{pQ}{nx^2} = 0$$

when $x = \sqrt{\dfrac{pQ}{ns}}$. Since $C'' > 0$, this corresponds to a minimum.

b. $C_s = xs = \sqrt{\dfrac{pQs}{n}}, C_o = \dfrac{pQ}{nx} = \sqrt{\dfrac{pQs}{n}}$

51. **a.** $P(x) = x\left(15 - \dfrac{3}{8}x\right) - \dfrac{7}{8}x^2 - 5x - 100 - tx;$

    Thus, $P'(x) = -\dfrac{5}{2}x + 10 - t = 0$

    when $x = \dfrac{2}{5}(10 - t)$

    b. $t = 5$
    c. The monopolist will absorb $4.25 of the $5 tax per unit. $0.75 will be passed on to the consumer.
    d. Writing exercise; responses will vary.

CHAPTER 3 Checkup

1. (a) is the graph of $f(x)$ and (b) is the graph of $f'(x)$ Answers will vary. One reason is the x intercepts of the graph in (b) correspond to the high and low points in (a).

2. **a.** Increasing for $x < 0$ and $0 < x < 3$; decreasing for $x > 3$

Critical Numbers	Classification
0	Neither
3	Relative maximum

   b. Increasing for $t < 1$ and $t > 2$; decreasing for $1 < t < 2$

Critical Numbers	Classification
1	Relative maximum
2	Relative minimum

   c. Increasing for $-3 < t < 3$; decreasing for $t < -3$ and $t > 3$

Critical Numbers	Classification
-3	Relative minimum
3	Relative maximum

d. Increasing for $x < -1$ and $x > 9$; decreasing for $-1 < x < 9$

Critical Numbers	Classification
-1	Relative maximum
9	Relative minimum

3. a. Concave upward for $x > 2$; concave downward for $x < 0$ and $0 < x < 2$; inflection at $x = 2$

b. Concave upward for $-5 < x < 0$ and $x > 1$; concave downward for $x < -5$ and $0 < x < 1$; inflection at $x = -5$, $x = 0$, and $x = 1$

c. Concave upward for $t > 1$; concave downward for $t < 1$; no inflection points

d. Concave upward for $-1 < t < 1$; concave downward for $t < -1$ and $t > 1$; inflection at $t = -1$ and $t = 1$

4. a. Vertical asymptote, $x = -3$; horizontal asymptote, $y = 2$

b. Vertical asymptotes, $x = -1$, $x = 1$; horizontal asymptote, $y = 0$

c. Vertical asymptotes, $x = -\frac{3}{2}$, $x = 1$; horizontal asymptote, $y = \frac{1}{2}$

d. Vertical asymptote, $x = 0$; horizontal asymptote, $y = 0$

5. a. No asymptotes; intercepts at $(0, 0)$ and $\left(\frac{4}{3}, 0\right)$; relative minimum at $(1, -1)$; inflection points at $(0, 0)$ and $\left(\frac{2}{3}, -\frac{16}{27}\right)$

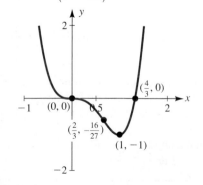

b. No asymptotes; y intercept at $(0, 1)$; relative minimum at $(0, 1)$; inflection points at $(1, 2)$ and $\left(\frac{1}{2}, \frac{23}{16}\right)$

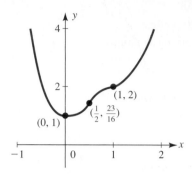

c. Vertical asymptote, $x = 0$; horizontal asymptote, $y = 1$; x intercept at $(-1, 0)$; relative minimum at $(-1, 0)$; inflection point at $\left(-\frac{3}{2}, \frac{1}{9}\right)$

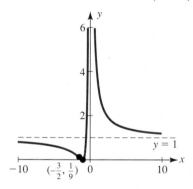

d. Vertical asymptote, $x = 1$; horizontal asymptote, $y = 0$; x intercept at $\left(\frac{1}{2}, 0\right)$; y intercept at $(0, 1)$; relative maximum at $(0, 1)$; inflection point at $\left(-\frac{1}{2}, \frac{8}{9}\right)$

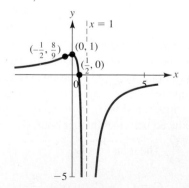

6. Answers will vary.

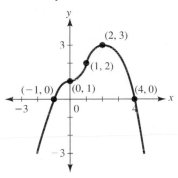

7. 8:20 A.M.

8. $70

9. a.

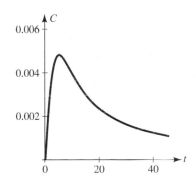

b. $t = 9$

c. The concentration tends to 0.

10. a. 1.667 million

b. $t = 2$ hours; 5 million

c.

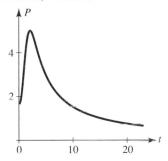

The population dies off.

CHAPTER 3 | Review

1. $f(x)$ is increasing for $-1 < x < 2$; decreasing for $x < -1$ and $x > 2$; concave up for $x < \frac{1}{2}$; concave

down for $x > \frac{1}{2}$; relative maximum $(2, 15)$; relative minimum $(-1, -12)$; inflection point $\left(\frac{1}{2}, \frac{3}{2}\right)$.

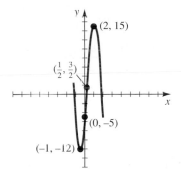

3. $f(x)$ is increasing for $x < -0.79$ and $1.68 < x$; $f(x)$ is decreasing for $-0.79 < x < 1.68$;

$f(x)$ is concave downward for $x < \frac{4}{9}$;

$f(x)$ is concave upward for $x > \frac{4}{9}$.

There is a relative maximum at $(-0.79, 22.51)$, a relative minimum at $(1.68, -0.23)$.
There is one inflection point, at $(0.44, 11.14)$.

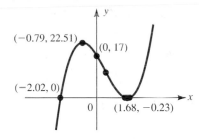

5. $f(t)$ is increasing for $t < -2$ and $t > 2$; decreasing for $-2 < t < 2$; concave up for $-\sqrt{2} < t < 0$ and for $t > \sqrt{2}$; concave down for $t < -\sqrt{2}$ and for $0 < t < \sqrt{2}$. Relative maximum at $(-2, 64)$; relative minimum at $(2, 64)$; inflection points $(-\sqrt{2}, 39.6)$ and $(\sqrt{2}, -39.6)$.

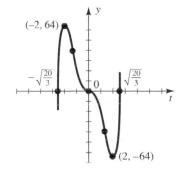

ANSWERS

7. $g(t)$ is increasing for $t < -2$ and for $t > 0$; decreasing for $-2 < t < -1$ and for $-1 < t < 0$; concave up for $t > -1$; concave down for $t < -1$. Relative maximum $(-2, -4)$; relative minimum $(0, 0)$; no inflection points.

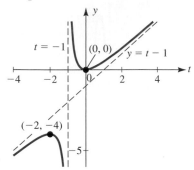

9. $F(t)$ is increasing for $x < -2$ and for $x > 2$; decreasing for $-2 < x < 0$ and for $0 < x < 2$; concave up for $x > 0$; concave down for $x < 0$. Relative maximum $(-2, -6)$; relative minimum $(2, 10)$; no inflection points.

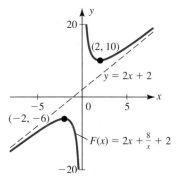

11. (b) is the graph of $f(x)$ and (a) is the graph $f'(x)$. Answers will vary. One reason is that the graph in (b) is always increasing and the graph in (a) is always positive.

13.

Critical Numbers	Classification
-1	Relative minimum
0	Relative maximum
3/2	Neither
7	Relative minimum

15.

Critical Numbers	Classification
0	Relative minimum
2	Neither

17. Here is one possible graph.

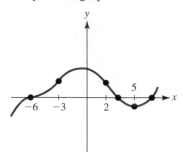

19. Here is one possible graph.

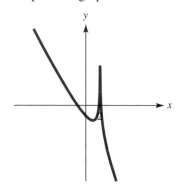

21. Relative maximum at $(2, 15)$; relative minimum at $(-1, -12)$

23. Relative maximum at $(-2, -4)$; relative minimum at $(0, 0)$

25. Absolute maximum value of 40 where $x = -3$; absolute minimum value of -12 where $x = -1$

27. Absolute maximum value of $\frac{1}{2}$ where $s = -\frac{1}{2}$ or $s = 1$; absolute minimum value of 0 where $s = 0$

29. a. $f(x)$ is increasing for $0 < x < 1$ and $x > 1$; it is decreasing for $x < 0$.

 b. $f(x)$ is concave upward for $x < \frac{1}{3}$ and $x > 1$; it is concave downward for $\frac{1}{3} < x < 1$.

 c. $f(x)$ has a relative minimum at $x = 0$ and inflection points at $x = \frac{1}{3}$ and $x = 1$.

d.

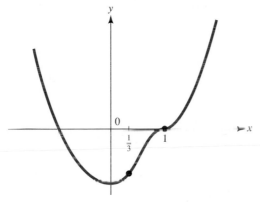

31. $12.50

33. $R = kN(P - N)$ so $R' = k(P - 2N)$ and $R' = 0$ when $N = \dfrac{P}{2}$. Since $R''\left(\dfrac{P}{2}\right) = -2k < 0$, there is a maximum at $N = \dfrac{P}{2}$.

35. $r = \dfrac{2}{3}h$

37. $p = \$8.12$ (or $8.13) per card

39. 17 floors

41. Row all the way to town.

43. 12 machines

45. a. $E = \dfrac{-2p^2}{100 - p^2}$

 b. At $p = 6$, $E = -\dfrac{9}{8}$, so $|E| = \left|\dfrac{9}{8}\right| = \dfrac{9}{8}$. Since $|E| > 1$, demand is elastic (i.e., as price increases, revenue decreases).

 c. $5.77

47. a. $E(p) = \dfrac{1.4p^2}{0.7p^2 - 300}$

 b. $E(8) = -0.351$; raise the price

49. Rectangle: 3.9 feet by 4.2 feet; side of triangle: 3.9 feet

51. 400 Floppsies and 700 Moppsies

53. 4,000 maps per batch

55. *Hint:* If x units are ordered, $C = k_1 x + \dfrac{k_2}{x}$.

57. a. Relative minimum at $x = \dfrac{1}{c}$

 b. Maximum of $\dfrac{\pi}{3}(5 - 3\sqrt{2})$; minimum of $\dfrac{\pi}{6}$

 c. Maximum of $\dfrac{\sqrt{3}\pi}{16}$; minimum of $\dfrac{\sqrt{3}\pi}{4(2 + \sqrt{2})^2}$

 d. $\lim_{r \to \infty} f(x) - Kc^r$, when r is much larger than R, the packing fraction depends only on the cell structure in the lattice.

59. a.

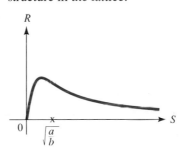

The graph appears to have a highest point $\left(\text{at } x = \sqrt{\dfrac{a}{b}}\right)$, a lowest point (at $x = 0$), and a point of inflection. The growth rate seems to level off (toward 0) as S grows larger and larger.

61. a. $E_1(p) = 0.287$; 0.574%; 1.435%

 b. $E_2(p) = 0.349$; 0.698%; 1.745%

 c. 0.2917; 1.49%

 d. $147,000

CHAPTER 4 Section 1

1. $e^2 \approx 7.389$, $e^{-2} \approx 0.135$, $e^{0.05} \approx 1.051$, $e^{-0.05} \approx 0.951$, $e^0 = 1$, $e \approx 2.718$, $\sqrt{e} \approx 1.649$, $\dfrac{1}{\sqrt{e}} \approx 0.607$

3.

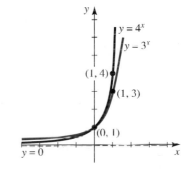

5. a. 9

   b. $\dfrac{1}{27}$

7. a. 12

   b. $\dfrac{189}{1,331}\sqrt{7}$

9. a. 3

   b. 4

11. a. 243

   b. $e^{14/3}$

13. a. $9x^4$

   b. $2x^{2/3}y$

15. a. $\dfrac{1}{x^{1/3}y^{1/2}}$

   b. $x^{1.1}y^2$

17. a. $\dfrac{1}{t}$

   b. t

19. $\dfrac{3}{2}$

21. 1

23. 1

25. $b = 2, C = 3$

27. a. \$1,967.15

   b. \$2,001.60

   c. \$2,009.66

   d. \$2,013.75

29. \$3,534.12

31. a. \$6,361.42

   b. \$6,342.19

33. a. 50,000,000

   b. 91,105,940

35. 400

37. 320

39. The original investment will be quadrupled.

41. 13,570

43. 20,480 bacteria

45. 4,000

47. a. 12,000 people per square mile

   b. 5,959 people per square mile

49. 204.8 grams

51. a. 0.5488

   b. 0.1812

   c. 0.1215

53. $\dfrac{1}{\sqrt[3]{10}}I_0 \approx 0.46I_0$

55. 0.05%

57. a. 310

   b. No. Prediction of 169 is close enough to actual 167 and is a prediction, not an absolute.

   c. Writing exercise; responses will vary.

59. \$1,206.93

61. a. No. A fair monthly payment is \$166.07.

   b. Writing exercise; responses will vary.

63.

x	-2.2	-1.5	0	1.5	2.3
$f(x)$	10.5561	4	0.5	0.0625	0.0206

65. As $n \to -\infty, \left(1 + \dfrac{1}{n}\right)^n \to e \approx 2.71828$

67. $\displaystyle\lim_{n \to +\infty}\left(2 - \dfrac{5}{2n}\right)^{n/3} = +\infty$

CHAPTER 4 | Section 2

1. $\ln 1 = 0, \ln 2 \approx 0.693, \ln e = 1, \ln 5 \approx 1.609,$

$\ln \dfrac{1}{5} \approx -1.609, \ln e^2 = 2, \ln 0$ and $\ln -2$ are undefined; e^x cannot be negative or equal to zero.

3. 3

5. 5

7. $\dfrac{8}{25}$

9. $3 + \log_3 2 + \log_3 5$

11. $2 \log_3 2 + 2 \log_3 5$

13. $4 \log_2 x + 3 \log_2 y$

15. $\dfrac{1}{3}[\ln x + \ln(x - 1)]$

17. $2 \ln x + \dfrac{2}{3}\ln(3 - x) - \dfrac{1}{2}\ln(x^2 + x + 1)$

19. $3 \ln x - x^2$

21. $\dfrac{\ln 53}{\ln 4} \approx 2.864$

23. 5

25. $\dfrac{\ln 2}{0.06} \approx 11.552$

27. $\dfrac{\ln 5}{4} \approx 0.402$

29. $e^{-C-t/50}$

31. 4

33. $\dfrac{2}{\ln 3} \approx 1.820$

35. $10 \ln 2$

37. $5 \ln 2 \approx 3.4657$

39. $7 \ln 5 - \ln 2 \approx 10.5729$

41. -5.5

43. $\dfrac{\ln 2}{0.06} \approx 11.55$ years

45. $\dfrac{\ln 2}{13} \approx 5.33\%$

47. $\dfrac{12 \ln 3}{\ln 2} \approx 19.02$ years

49. **a.** 0.765 g/cm^3; 0.784 g/cm^3

  b. $-50 \ln\left(\dfrac{0.125}{0.13}\right) \approx 1.96$/sec

51. $5,614$ years

53. $Q(t) = 6,000e^{0.0203t}$; $20,250$

55. $Q(t) = 500 - 200e^{-0.1331t}$; 459.5 units

57. $10,523$ years

59. 24.84 years ago; 95.8%

61. $f(t) = 70 + 142e^{-kt}$; ideal temperature is $72.04°$F

63. Scélérat; Wednesday morning at $1{:}27$ A.M.

65. **a.** 8.25

  b. 10^{14} joules

67. **a.** 2×10^8

  b. $10^{1.2} \approx 15.85$ times more intense

69. **a.** 45%

  b. 2.34%

71. The year 2095

73. **a.**

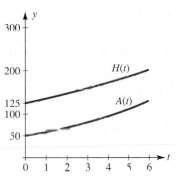

  b. $A = \dfrac{2H^2}{625}$

75. The line $y = x$ has slope 1, so it is perpendicular to the line through $A(a, b)$ and $B(b, a)$, which has slope $m = \dfrac{(a - b)}{(b - a)} = -1$. If O is the origin and M is the point where $y = x$ intersects the line segment AB, then right triangles OMA and OMB are similar since they share side OM and
$$|OA| = \sqrt{a^2 + b^2} = \sqrt{b^2 + a^2} = |OB|$$
It follows that $|AM| = |BM|$, which means that A and B are reflections of one another in $y = x$.

77. $Y = mX + b$ where $Y = \ln y$, $X = \ln x$, $m = k$, and $b = \ln c$.

79. $x \approx -17.4213$ (using a graphing utility)

81. $x \approx 1.1697$ (using a graphing utility)

83. **a.** $(\log b)(\log a) = \left(\dfrac{\ln b}{\ln a}\right)\left(\dfrac{\ln a}{\ln b}\right) = 1$

  b. $\log_a x = \dfrac{\ln x}{\ln a}$

  $= \dfrac{(\ln x)(\ln b)}{(\ln b)(\ln a)} = (\log_b x)(\log_a b)$

  $= \dfrac{\log_b x}{\log_a b}$ using part (a)

CHAPTER 4 | Section 3

1. $f'(x) = 5e^{5x}$

3. $f'(x) = xe^x + e^x$

5. $f'(x) = -0.5e^{-0.05x}$

7. $f'(x) = (6x^2 + 20x + 33)e^{6x}$

9. $f'(x) = -6e^x(1 - 3e^x)$

11. $f'(x) = \dfrac{3}{2\sqrt{3x}}\, e^{\sqrt{3x}}$

ANSWERS

13. $f'(x) = \dfrac{3}{x}$

15. $f'(x) = 2x \ln x + x$

17. $f'(x) = \dfrac{2}{3}e^{2x/3}$

19. $f'(x) = \dfrac{-2}{(x + 1)(x - 1)}$

21. $f'(x) = -2e^{-2x} + 3x^2$

23. $g'(s) = (e^s + 1)(2e^{-s} + s) +$
$\qquad (e^s + s + 1)(-2e^{-s} + 1)$
$\qquad = 1 + 2s + e^s + se^s - 2se^{-s}$

25. $h'(t) = \dfrac{te^t \ln t + t \ln t - e^t - t}{t(\ln t)^2}$

27. $f'(x) = \dfrac{e^x - e^{-x}}{2}$

29. $f'(t) = \dfrac{t + 1}{2t\sqrt{\ln t + t}}$

31. $f'(x) = \dfrac{1 - e^{-x}}{x + e^{-x}}$

33. $g'(u) = \dfrac{1}{\sqrt{u^2 + 1}}$

35. $\dfrac{1}{e}; 0$

37. $\dfrac{3\sqrt{3}}{8}e^{-3/2}; 0$

39. $y = x$

41. $y = e^2$

43. $y = \dfrac{1}{2}x - \dfrac{1}{2}$

45. $f''(x) = 4e^{2x} + 2e^{-x}$

47. $2 \ln t + 3$

49. $f'(x) = f(x)\left[\dfrac{4}{2x + 3} + \dfrac{1 - 10x}{2(x - 5x^2)}\right]$

51. $f'(x) = f(x)\left[\dfrac{5}{x + 2} - \dfrac{1}{2(3x - 5)}\right]$

53. $f'(x) = f(x)\left[\dfrac{3}{x + 1} - \dfrac{2}{6 - x} + \dfrac{2}{3(2x + 1)}\right]$

55. $f'(x) = (2 \ln 5)x5^{x^2}$

57. a. $E(p) = -0.04p$; elastic for $p > 25$, inelastic for
$p < 25$, of unit elasticity for $p = 25$
b. Demand will decrease by approximately 1.2%.
c. $R(p) = 3,000pe^{-0.04p}$; $p = 25$

59. a. $E(p) = \dfrac{-p^2 - p}{10(p + 11)}$; elastic when $p > 15.91$,
inelastic when $p < 15.91$, of unit elasticity
when $p = 15.91$
b. Demand will decrease by approximately 1.85%.
c. $R(p) = 5,000\,p(p + 11)\,e^{-0.1p}$;
$p = 15.91$

61. a. $C'(x) = 0.2e^{0.2x}$
b. 5 units

63. a. $C'(x) = \dfrac{6\,e^{x/10}}{\sqrt{x^7}}\left(1 + \dfrac{x}{5}\right)$
b. 5 units

65. a. Population is increasing at the rate of 1.22 million people per year.
b. Constant rate of 2% per year

67. a. Value is decreasing at the rate of $1,082.68 per year.
b. Constant rate of -40% per year

69. a. Approximately 406 copies
b. 368 copies

71. a. $F'(t) = -k(1 - B)e^{-kt}$ is the rate at which you are forgetting material.
b. $F'(t) = -k(F(t) - B)$, which says that the rate you forget is proportional to the fraction you have left to forget.
c.

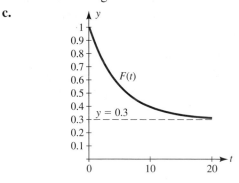

73. a. $E'(p) = 3,000e^{-0.01p}(1 - 0.01p)$
b. $p = 100$
c. $p = 200$

75. a. $N'(t) = \dfrac{36e^{-0.02t}}{(1 + 3e^{-0.02t})^2}$; the population is increasing at all times t.
b. Increasing for $t < 50 \ln 3$, decreasing for $t > 50 \ln 3$
c. $N(t)$ approaches 600.

77. a. $P_1'(10) \approx 1.556$ cm/day
$P_2''(10) = -0.257$ cm/day per day; decreasing

b. Plants have the same height, approximately 21 cm after 20.71 days; $P_1'(20.71) \approx 0.286$, $P_2'(20.71) \approx 0.001$, so the first plant is growing more rapidly.

79. $\dfrac{R'(t_0)}{R(t_0)} - \dfrac{0.09(11) - 0.02(8)}{19} \approx 0.0437; 4.37\%$

81. The population of the town x years from now will be

$$P(x) = 5,000\sqrt{x^2 + 4x + 19}$$

The percentage rate of change x years from now will be

$$\frac{100\,P'(x)}{P(x)} = 100\frac{d}{dx}[\ln 5,000\sqrt{x^2 + 4x + 19}]$$

$$= 100\frac{d}{dx}[\ln 5,000 + \frac{1}{2}\ln(x^2 + 4x + 19)]$$

$$= \frac{100(x + 2)}{x^2 + 4x + 19}$$

Hence the percentage rate of change 3 years from now will be

$$\frac{100(3 + 2)}{3^2 + 12 + 19} = 12.5\%\text{ per year}$$

83. $f'(x) = \dfrac{2^x(x \ln 2 - 1)}{x^2}$

85. $\dfrac{1 + \ln x}{\ln 10}$

87. $f(x) = \dfrac{1}{3}\ln(x + 1) - 4\ln(1 + 3x)$

$f'(x) = \dfrac{1}{3(x + 1)} - \dfrac{12}{1 + 3x}; f'(0.65) \approx -3.87$

tangent line at $(0.65, -4.16)$ is $y = -3.87x - 1.65$.

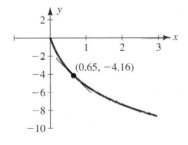

CHAPTER 4 | Section 4

1. $f_5(x)$

3. $f_3(x)$

5. $f(t)$ is increasing for all real t; concave upward for all real t. There is a horizontal asymptote at $y = 2$.

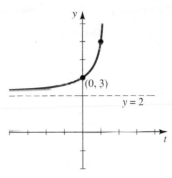

7. $g(x)$ is decreasing for all real x; concave downward for all real x; $y = 2$ is a horizontal asymptote.

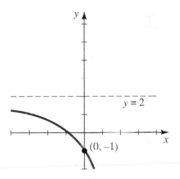

9. $f(x)$ is increasing for all real x; concave upward for $x < 0.549$; concave downward for $x > 0.549$. Inflection point is $(0.549, 1)$, and $y = 2$ and $y = 0$ are horizontal asymptotes.

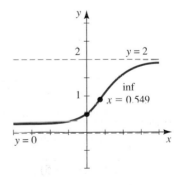

ANSWERS

11. $f(x)$ is increasing for $x > -1$; decreasing for $x < -1$; concave upward for $x > -2$; concave downward for $x < -2$. Relative minimum is $\left(-1, -\dfrac{1}{e}\right)$ and inflection point is $\left(-2, -\dfrac{2}{e^2}\right)$. The x axis ($y = 0$) is a horizontal asymptote.

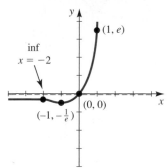

13. $f(x)$ is increasing for $x < 1$; decreasing for $x > 1$; concave upward for $x > 2$; concave downward for $x < 2$. Relative maximum is $(1, e)$, inflection point is $(2, 2)$. The x axis ($y = 0$) is a horizontal asymptote.

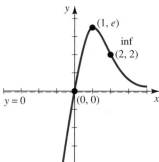

15. $f(x)$ is increasing for $0 < x < 2$; decreasing for $x < 0$ and $x > 2$; concave upward for $x < 0.6$ and $x > 3.4$; concave downward for $0.6 < x < 3.4$. Relative minimum is $(0, 0)$; relative maximum is $\left(2, \dfrac{4}{e^2}\right)$; inflection points are $(0.6, 0.2)$ and $(3.4, 0.4)$. The x axis is a horizontal asymptote.

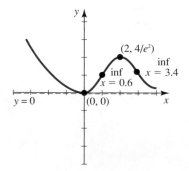

17. $f(x)$ is increasing for all real x; concave upward for $x < 0$; concave downward for $x > 0$. Inflection point is $(0, 3)$. The x axis ($y = 0$) and $y = 6$ are horizontal asymptotes.

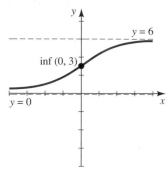

19. $f(x)$ is increasing for $x > 1$; decreasing for $x < 1$; concave upward for $x < e$; concave downward for $x > e$. Relative minimum is $(1, 0)$; inflection point is $(e, 1)$. The y axis ($x = 0$) is a vertical asymptote.

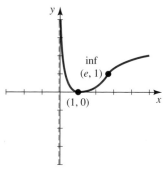

21. a. As $t \to \infty$, $f(t) \to 1$

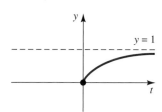

b. 0.741

c. 0.089

23. a. $A = 85, k = \dfrac{1}{20}\ln\dfrac{17}{6}$

b.

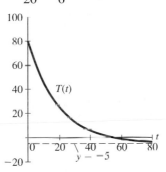

The temperature approaches $-5°C$.

c. $12.8°C$

d. 54.4 minutes

25. a.

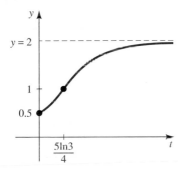

b. 500

c. 1,572

d. 2,000

27. 37.5 units per day

29. a. Approximately 403 copies

b. 348 copies

31. a. $e - 1 \approx 1.7$ years

b. The learning rate $L'(t)$ is largest when $t = 0$ (at birth).

33. $C = \dfrac{1}{199}; k = 0.1745$

Rate of change of proportion of day-trading is

$$p'(t) = \frac{199ke^{kt}}{(199 + e^{kt})^2}$$

which is maximized when

$$p''(t) = 0; t \approx 30.33 \text{ weeks}$$
$$p(30.33) \approx 0.5$$

35. a. $P(x) = 1,000e^{-0.02x}(x - 125)$

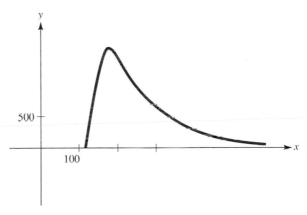

b. $175

37. 69.44 years from now

39. 0.45 years

41. a. $C = \dfrac{b}{aR}; E''\left(\dfrac{b}{aR}\right) = \dfrac{2a^3R^3}{b} > 0$

b. $k = \dfrac{b}{aR}; m = \dfrac{4}{e}a^2R^2$

43. a. $E(t) = 1,000w(t)p(t)$

b. $t = 82; 3,527$ pounds

c.

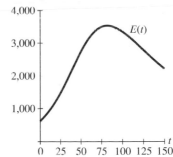

45. a. $C = 9, k = \dfrac{1}{2}\ln 3$

b. 4 hours

c. 4 hours

47. a. $N(0) = 15$ employees; $N(5) = 482$ employees; 2.10 years; 500 employees

b.

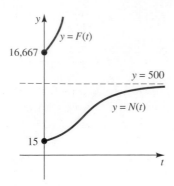

$s > 1$. The rate of blood cell production is maximized and minimized at the two inflection points, respectively.

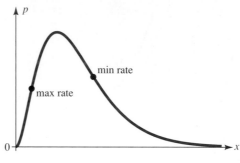

49. a. $C'(t) = Ae^{-kt}(1 - kt)$; $C(t)$ is increasing for $t < \dfrac{1}{k}$ and decreasing for $t > \dfrac{1}{k}$. Maximum concentration is $\dfrac{A}{ke}$ at $t = \dfrac{1}{k}$.

b. $C''(t) = kAe^{-kt}(kt - 2)$; the graph of $C(t)$ is concave upward for $t > \dfrac{2}{k}$ and for $t < \dfrac{2}{k}$ is concave downward. Then $\left(\dfrac{2}{k}, \dfrac{2A}{ke^2}\right)$ is the inflection point. The rate of change of drug concentration is a minimum at the inflection point.

c.

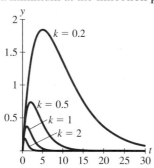

The point where the maximum occurs shifts left as k increases and the maximum height decreases.

c.

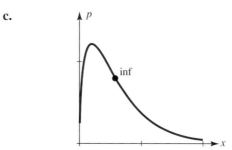

There is only one inflection point.

55. $f''(t) = 0$ when $t = \dfrac{\ln c}{k}$ and $f\left(\dfrac{\ln c}{k}\right) = \dfrac{A}{2}$ or half the number of susceptible residents.

57.

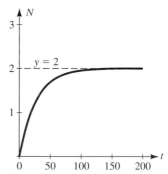

The value of N approaches the maximum of 2 million people.

51. a. $Q(t) = 1{,}139e^{0.06t}$, $Q(7) = 1{,}734$ staff members

b. $t \approx 11.5$ years

c. Writing exercise; responses will vary.

53. a. $x = r$; $p''(r) < 0$

b. The inflection points occur at $x = \dfrac{r(s - \sqrt{s})}{s}$ and $x = \dfrac{r(s + \sqrt{s})}{s}$. Both are positive since

59. a. $t = \dfrac{1}{b - a}\ln\left(\dfrac{b}{a}\right)$

"In the long run," the concentration approaches 0.

b.

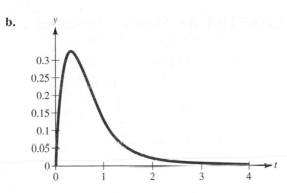

c. Writing exercise; responses will vary.

61. a. $V(5) = \$207.64$;

$$V'(t) = V_0\left(1 - \frac{2}{L}\right)^t \ln\left(1 - \frac{2}{L}\right)$$

$V'(5) \approx -\$60$/year

b. $100 \ln\left(1 - \frac{2}{L}\right)$

CHAPTER 4 | Checkup

1. a. 1

b. $\dfrac{10}{3}$

c. 0

d. $\dfrac{16}{81}$

2. a. $27x^6 y^3$

b. $\dfrac{1}{\sqrt[3]{3}xy^{2/3}}$

c. $\dfrac{y^{7/6}}{x^{1/6}}$

d. $\dfrac{1}{x^{6.5}y^8}$

3. a. $x = 3, x = -1$

b. $x = \dfrac{1}{\ln 4}$

c. $x = -4, x = 4$

d. $t = -2 \ln \dfrac{11}{3}$

4. a. $\dfrac{dy}{dx} = \dfrac{e^x(x^2 - 5x + 3)}{(x^2 - 3x)^2}$

b. $\dfrac{dy}{dx} = \dfrac{3x^2 + 4x - 3}{x^3 + 2x^2 - 3x}$

c. $\dfrac{dy}{dx} = x^2(1 + 3 \ln x)$

d. $\dfrac{dy}{dx} = y\left(-2 + \dfrac{6}{2x - 1} + \dfrac{2x}{1 - x^2}\right)$

5. a. $f(x)$ is increasing for $0 < x < 2$; decreasing for $x < 0$ and $x > 2$; concave upward for $x < 2 - \sqrt{2}$ and $x > 2 + \sqrt{2}$; concave downward for $2 - \sqrt{2} < x < 2 + \sqrt{2}$. There is a relative maximum at $x = 2$, a relative minimum at $x = 0$. There are two inflection points, at $x = 2 - \sqrt{2}$ and $x = 2 + \sqrt{2}$. The x axis $(y = 0)$ is a horizontal asymptote.

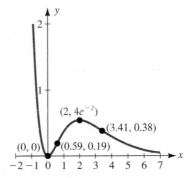

b. $f(x)$ is increasing for $0 < x < \sqrt{e}$; decreasing for $x > \sqrt{e}$; concave upward for $x > e^{5/6}$; concave downward for $x < e^{5/6}$. There is a relative maximum at $x = \sqrt{e}$. There is an inflection point at $(2.30, 0.08)$, where $x = e^{5/6}$. The y axis $(x = 0)$ is a vertical asymptote, and the x axis $(y = 0)$ is a horizontal asymptote.

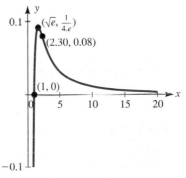

c. $f(x)$ is increasing for $0 < x < \dfrac{1}{4}$ and $x > 1$; decreasing for $\dfrac{1}{4} < x < 1$; concave downward for $x \neq 1$. There is a relative maximum at

$x = \dfrac{1}{4}$. There are no inflection points, and $x = 0$ and $x = 1$ are vertical asymptotes.

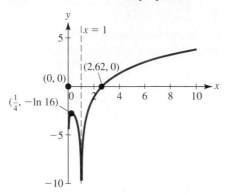

d. $f(x)$ is increasing for all x; concave upward for $x < 0$; concave downward for $x > 0$. There is an inflection point at $x = 0$, and $y = 0$ and $y = 4$ are horizontal asymptotes.

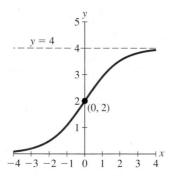

6. $2,323.67; 8.1 years

7. a. Increasing for $-1 < t < e - 1$; decreasing for $t > e - 1$
  b. $t = e^{3/2} - 1$
  c. The price approaches $500.

8. a. $q'(p) = -1000e^{-p}(p + 1) < 0$ for $p \geq 0$
  b. $p = \sqrt{2}$ or $141.42; $117,384.14

9. 6,601 years old

10. a. 80,000
  b. 2 hours; 81,873
  c. The population dies off completely.

CHAPTER 4 | Review Problems

1.

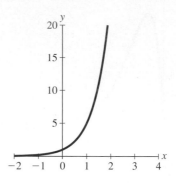

3.

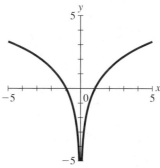

5. a. $f(4) = \dfrac{3,125}{8}$

  b. $f(3) = \dfrac{100}{3}$

  c. $f(9) = \dfrac{65}{2}$

  d. $f(10) = \dfrac{6}{5}$

7. $x = 25 \ln 4$

9. $x = e^2$

11. $x = \dfrac{41}{2}$

13. $x = 0$

15. $\dfrac{dy}{dx} = xe^{-x}(2 - x)$

17. $\dfrac{dy}{dx} = 2 \ln x + 2$

19. $\dfrac{dy}{dx} = \dfrac{2}{x \ln 3}$

21. $\dfrac{dy}{dx} = e^x$ (Note that $y = e^x$.)

23. $\dfrac{dy}{dx} = \dfrac{-(1 + 2e^{-x})}{1 + e^{-x}}$

25. $\dfrac{dy}{dx} = \dfrac{-e^{-x}(x^2 + x + 1 + x \ln x)}{x(x + \ln x)^2}$

27. $\dfrac{dy}{dx} = \dfrac{ye^{x-x^2}(2x - 1) + 1}{e^{x-x^2} - 1}$

29. $\dfrac{dy}{dx} = 2y\left[\dfrac{3x + 3e^{2x}}{x^2 + e^{2x}} - 1 - \dfrac{1 - 2x}{3(1 + x - x^2)}\right]$

31. $f(x)$ is increasing for all x; concave upward for $x > 0$; concave downward for $x < 0$. There is an inflection point at $x = 0$.

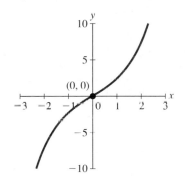

33. $f(t)$ is increasing for $t > 0$; decreasing for $t < 0$; concave upward for all t. There is a relative minimum at $t = 0$. There are no inflection points.

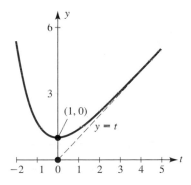

35. $F(u)$ is increasing for $-2 < u < -1$ and $u > -1$; concave upward for $u > -1$; concave downward

for $-2 < u < -1$. There is an inflection point at $u = -1$, and $u = -2$ is a vertical asymptote.

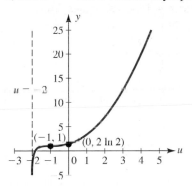

37. $G(x)$ is decreasing for all x. $G(x)$ is concave upward for all x.

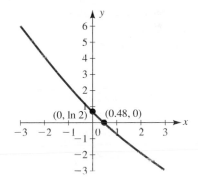

39. $\ln 4; \ln 3$

41. $\left(e + \dfrac{1}{e}\right)^5; 32$

43. $y = 2x - 2$

45. $y = 4x$

47. **a.**

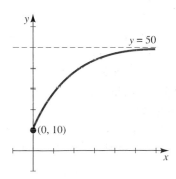

   b. 10,000

   c. 32,027

   d. 9.81 thousand dollars ($9,808.29)

   e. Just under 50,000 units

49. a.

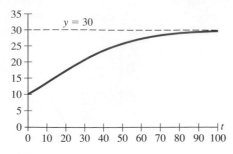

b. 10 million

c. 17.28 million (17,283,507)

d. The population will approach 30 million.

51. 8.20% per year compounded continuously

53. a. $4,323.25

b. $4,282.09

55. 5.83%

57. a. 0.13 parts per million per year

b. Constant rate of 3%

59. After 200 years

61. a. Since $\lambda = \dfrac{\ln 2}{k}$, we have $k = \dfrac{\ln 2}{\lambda}$ and $Q(t) = Q_0 e^{-(\ln 2/\lambda)t}$.

b. $Q_0(0.5)^{kt} = Q_0 e^{-(\ln 2/\lambda)t}$

$kt \ln 0.5 = -\left(\dfrac{\ln 2}{\lambda}\right)t$

So, $k = \dfrac{1}{\lambda}$

63. Bronze age began about 5,000 years ago (3,000 B.C.); maximum percentage is 55%.

65. 0.8110 minutes = 48.66 seconds; $-8.64°C$ per minute

67. a. $A = 5e, k = \dfrac{1}{2}$

b. $t = 9.78$ hours

69. a. 0.15% per year

b. 70.24 years, 0.15% per year

71. $10^{-1.6} \approx 0.0251$

73. a. $D(10) = 0.00195; D(25) = 0.000591$

b.

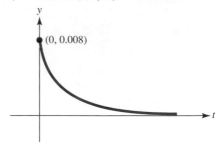

75. a. 2.31×10^{-200} percent left, too little for proper measurement.

b. Writing exercise.

77. a.

1790	3,867,087
1800	5,256,550
1830	12,956,719
1860	30,207,500
1880	50,071,364
1900	77,142,427
1920	108,425,601
1940	138,370,607
1960	162,289,823
1980	178,782,499
1990	184,566,653
2000	189,034,385

b. The model predicts that the population was increasing most rapidly around 1915.

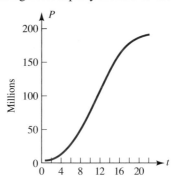

c. Writing exercise; responses will vary.

79.

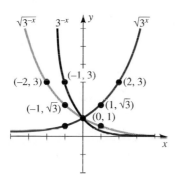

81. $x = 1.066$

83.

n	$(\sqrt{n})^{\sqrt{n+1}}$	$(\sqrt{n+1})^{\sqrt{n}}$
8	22.63	22.36
9	32.27	31.62
12	88.21	85.00
20	957.27	904.84
25	3,665	3,447
31	16,528	15,494
37	68,159	63,786
38	85,679	80,166
43	261,578	244,579
50	1,165,565	1,089,362
100	1.12×10^{10}	1.05×10^{10}
1,000	2.87×10^{47}	2.76×10^{47}

$$(\sqrt{n})^{\sqrt{n+1}} \; > \; (\sqrt{n+1})^{\sqrt{n}}$$

CHAPTER 5 | Section 1

1. $-3x + C$

3. $\dfrac{x^6}{6} + C$

5. $-\dfrac{1}{x} + C$

7. $4\sqrt{t} + C$

9. $\dfrac{5}{3}u^{3/5} + C$

11. $t^3 - \dfrac{2\sqrt{5}}{3}t^{3/2} + 2t + C$

13. $2y^{3/2} + y^{-2} + C$

15. $\dfrac{e^x}{2} + \dfrac{2}{5}x^{5/2} + C$

17. $\dfrac{u^{1.1}}{3.3} - \dfrac{u^{2.1}}{2.1} + C$

19. $x + \ln x^2 - \dfrac{1}{x} + C$

21. $-\dfrac{5}{4}x^4 + \dfrac{11}{3}x^3 - x^2 + C$

23. $\dfrac{2}{7}t^{7/2} - \dfrac{2}{3}t^{3/2} + C$

25. $\dfrac{1}{2}e^{2t} + 2e^t + t + C$

27. $\dfrac{1}{3}\ln|y| - 10\sqrt{y} - 2e^{-y/2} + C$

29. $\dfrac{2}{5}t^{5/2} - \dfrac{2}{3}t^{3/2} + 4t^{1/2} + C$

31. $y = \dfrac{3}{2}x^2 - 2x - \dfrac{3}{2}$

33. $y = \ln x^2 + \dfrac{1}{x} - 2$

35. $f(x) = 2x^2 + x - 1$

37. $f(x) = \dfrac{x^4}{4} + \dfrac{2}{x} + 2x - \dfrac{5}{4}$

39. $f(x) = -e^{-x} + \dfrac{x^3}{3} + 5$

41. 10,128 people

43. $\dfrac{5}{2}(e^{0.4} - e^{0.2})$

45. $22,360

47. a. $18\dfrac{1}{3}$ (18 items)

   b. $48\dfrac{1}{3}$ (48 items)

49. 3,253

51. a. $T(t) = 16 - 20e^{-0.35t}$
   b. 6.1°C
   c. 3.44 hours

53. a. $P(q) = 100q - q^2 - 200$
   b. $q = 50$; $2,300

55. $c(x) = 0.9x + 0.2x^{3/2} + 10$

57. $f'(x)$ is maximized when $x = 10$.
   a. Seven items per minute
   b. $f(x) = x + 0.6x^2 - 0.02x^3$
   c. 100 items

59. $2,265.80

61. $v(r) = \dfrac{1}{2}a(R^2 - r^2)$

63. 20 meters

65. $\int b^x dx = \int e^{x \ln b} dx = \dfrac{e^{x \ln b}}{\ln b} + C = \dfrac{b^x}{\ln b} + C$

67. a. $v(t) = -23t + 67;\ s(t) = -\dfrac{23}{2}t^2 + 67t$

b.

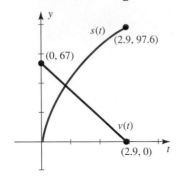

c. $v(t) = 0$ when $t = 2.9$ sec and $s(2.9) = 97.6$ ft; $s(t) = 45$ ft when $t \approx 0.77$ sec or 5.05 sec and $v(0.78) \approx 49.30$ ft/sec while $v(5.05) \approx -49.15$.

CHAPTER 5 | Section 2

1. a. $u = 3x + 4$
  b. $u = 3 - x$
  c. $u = 2 - t^2$
  d. $u = 2 + t^2$

3. $\dfrac{1}{12}(2x + 6)^6 + C$

5. $\dfrac{1}{6}(4x - 1)^{3/2} + C$

7. $-e^{1-x} + C$

9. $\dfrac{1}{2}e^{x^2} + C$

11. $\dfrac{1}{12}(t^2 + 1)^6 + C$

13. $\dfrac{4}{21}(x^3 + 1)^{7/4} + C$

15. $\dfrac{2}{5}\ln|y^5 + 1| + C$

17. $\dfrac{1}{26}(x^2 + 2x + 5)^{13} + C$

19. $\dfrac{3}{5}\ln|x^5 + 5x^4 + 10x + 12| + C$

21. $-\dfrac{3}{2}\left(\dfrac{1}{u^2 - 2u + 6}\right) + C$

23. $\dfrac{1}{2}(\ln 5x)^2 + C$

25. $\dfrac{-1}{\ln x} + C$

27. $\dfrac{1}{2}[\ln(x^2 + 1)]^2 + C$

29. $\ln|e^x - e^{-x}| + C$

31. $\dfrac{1}{2}x - \dfrac{1}{4}\ln|2x + 1| + C$

33. $\dfrac{1}{10}(2x + 1)^{5/2} - \dfrac{1}{6}(2x + 1)^{3/2} + C$

35. $2\ln(\sqrt{x} + 1) + C$

37. $y = \ln|x + 1| + 1$

39. $y = \dfrac{1}{2}\ln|x^2 + 4x + 5| - \dfrac{1}{2}\ln 2 + 3$

41. $f(x) = \dfrac{1}{5} - \dfrac{1}{5}(1 - 2x)^{5/2}$

43. $f(x) = \dfrac{3}{2} - \dfrac{1}{2}e^{4-x^2}$

45. a. $x(t) = -\dfrac{4}{9}(3t + 1)^{3/2} + \dfrac{40}{9}$
  b. $x(4) = -16.4$
  c. $t = 0.4$

47. a. $x(t) = \sqrt{2t + 1} - 1$
  b. $x(4) = 2$
  c. $t = \dfrac{15}{2}$

49. a. $C(q) = (q - 4)^3 + 64 + k$, where k is the overhead
  b. \$1,500

51. 2.3 meters

53. a. $R(x) = 50x - 175e^{-0.01x^2} + 175$
  b. \$50,175

55. a. $C(t) = \dfrac{1}{e^{0.01t} + 1}$
  b. 0.3543 mg/cm^3; 0.1419 mg/cm^3
  c. 294 minutes

57. a. $L(t) = 0.03\sqrt{-t^2 + 16t + 36} + 0.07$; at $t = 8$ (3:00 P.M.); 0.37 parts per million

b. The ozone level at 11:00 A.M. ($t = 4$) is $L(4) = 0.345$. The same level occurs at $t = 12$ (7:00 P.M.).

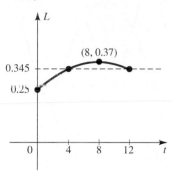

59. a. $p(x) = \dfrac{300}{\sqrt{x^2 + 9}} + 15$

b. \$66.45; \$115

c. 265

61. $R'(x) = \dfrac{11}{\sqrt{14 - x}} \cdot \dfrac{x}{}$, $C'(x) = 2 + x + x^2$

Let $14 - x = u$, $du = -dx$. Then,

$$R(x) = -\int \dfrac{u - 3}{\sqrt{u}} du$$

$$= -\int (u^{1/2} - 3u^{-1/2}) du$$

$$= -\left(\dfrac{2}{3}u^{3/2} - 6u^{1/2}\right) + K_1$$

$$= 6\sqrt{14 - x} - \dfrac{2}{3}(14 - x)^{3/2} + K_1$$

We also have

$$C(x) = \int (2 + x + x^2)dx = 2x + \dfrac{x^2}{2} + \dfrac{x^3}{3} + K_2$$

so that the profit $P(x) = R(x) - C(x)$ satisfies $P(9) - P(5) = R(9) - C(9) - [R(5) - C(5)] = -295.54 - (-64.167) = -231.373$

63. $\dfrac{3}{7}(x^{2/3} + 1)^{7/2} - \dfrac{3}{5}(x^{2/3} + 1)^{5/2} + C$

65. $e^x + 1 - \ln(e^x + 1) + C$

CHAPTER 5 | Section 3

1. 15

3. $\dfrac{95}{2}$

5. $\dfrac{6}{5}$

7. $-\dfrac{6}{5}$

9. $3 - \dfrac{4}{e}$

11. 1.95

13. 144

15. $\dfrac{8}{3} + \ln 3 \approx 3.7653$

17. $\dfrac{2}{9}$

19. 3.2

21. $\dfrac{4}{3}$

23. $\dfrac{7}{6}$

25. e

27. $\dfrac{8}{3}$

29. $e^3 - e^2$

31. -20

33. 0

35. 3

37. $\dfrac{33}{5}$

39. 4

41. $\dfrac{112}{9}$

43. $V(5) - V(0)$

45. \$480

47. 0.75 ppm

49. About 98 people

51. \$75

53. $1,500\left(\dfrac{3}{2} + \dfrac{5}{4}\ln\dfrac{11}{9}\right) \approx 2,626$ telephones

55. The concentration decreases by 0.8283 mg/cm^3

57. A decrease of \$1,870

59. $2\ln 2 \approx 1.386$ grams

61. $8\sqrt{11} - 8\sqrt{6}$ or about 7 facts

63. 96 ft

ANSWERS

65. a. $\dfrac{\pi}{4}$

b. $\dfrac{\pi}{4}$; part of the area under the circle
$(x - 1)^2 + y^2 = 1$

CHAPTER 5 | Section 4

1. $\dfrac{5}{12}$

3. $2 \ln 2 - \dfrac{1}{2}$

5. Area = 1

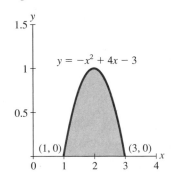

7. Area = $\dfrac{4}{3}$

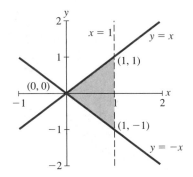

9. Area = $\dfrac{4}{3}$

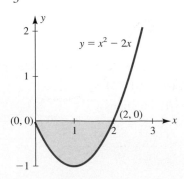

11. Area = 9

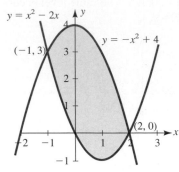

13. Area = $\dfrac{443}{6}$

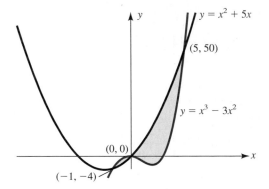

15. Area = 18

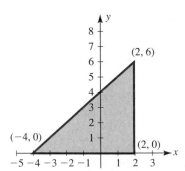

17. Area = 14

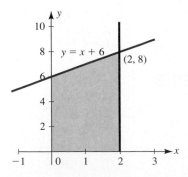

19. -2

21. $\dfrac{3}{2}\left(e - \dfrac{1}{e}\right)$

23. $\dfrac{\ln 5 - \ln 3}{\ln 3}$

25. Average value $= \dfrac{2}{3}$

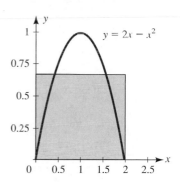

27. Average value $= \dfrac{\ln 2}{2}$

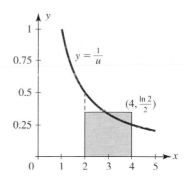

29. $\dfrac{1}{2}$

31. 0.1833

33. 0.383

35. \$1.57 per pound

37. 2,272.2

39. 30,000 kg

41. **a.** 16 years
   b. \$209,067

c.

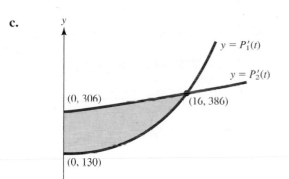

43. **a.** 14.7 years
   b. \$582,221
   c.

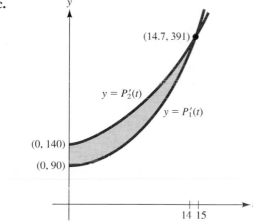

45. 0.412 million

47. $\dfrac{1}{40}$ mg/cm^3

49. **a.** 0°C
   b. 8 A.M. and 2 P.M.

51. **a.** 39.25 mph
   b. 3:30 P.M.

53. **a.** $M_0 + 20.833$
   b. $t = \sqrt{5}, M(\sqrt{5}) - M_0 + 50\sqrt{5}/e$

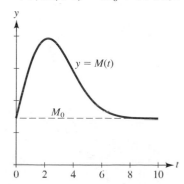

55. Baseball: $\frac{1}{3}$; football: $\frac{5}{18}$; basketball: $\frac{9}{25}$. Football is the most equitable sport, basketball the least.

57. 5,710 people

59. $241,223.76

61. a. $S' = F''(M) = \frac{1}{3}(2k - 6M) = 0$ for $M = k/3$.
Maximum since $S'' = -2 < 0$.

b. $\frac{k^3}{108}$

63.

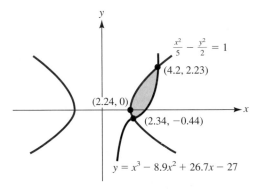

$$A = \int_{\sqrt{5}}^{2.34} \left[\sqrt{\frac{2x^2}{5} - 2} - \left(-\sqrt{\frac{2x^2}{5} - 2} \right) \right] dx$$
$$+ \int_{2.34}^{4.2} \left[\sqrt{\frac{2x^2}{5} - 2} - (x^3 - 8.9x^2 + 26.7x - 27) \right] dx$$
$$\approx 2.097$$

65. Writing exercise; responses will vary.

CHAPTER 5 | Section 5

1. a. $624

b.

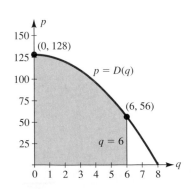

3. a. $1,600 \ln 2 \approx \$1,109.04$

b.

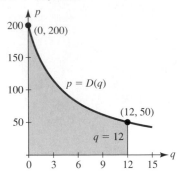

5. a. $800\left(1 - \frac{1}{\sqrt{e}}\right) \approx \$314.78$

b.

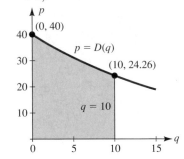

7. $p_0 = \$110$; CS = $36

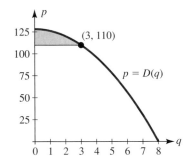

9. $p_0 = \$31.15$; CS = $21.21

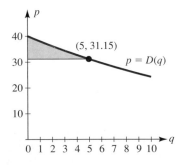

11. $p_0 = \$34.80$; PS = $12.80

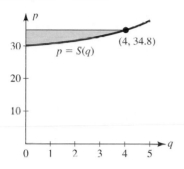

13. $p_0 = \$26.41$; PS = $2.14

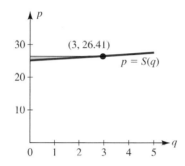

15. a. $104
   b. CS = $162, PS = $324

17. a. $40
   b. CS = $200, PS = $116.67

19. a. $1
   b. CS = $3.09, PS = $0.67

21. a. 11 years
   b. $26,620
   c.

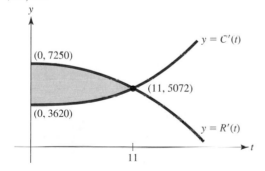

23. a. 8
   b. $15,069
   c. Net earnings are represented as the area between the curve $R'(t) = 6{,}537e^{-0.3t}$ and the horizontal line $y = 593$.

25. $17,182.82

27. $237,730; $319,453

29. $5,308.78

31. The $50,000 plan is better, producing net income of $37,465 versus $22,479 for the $30,000 plan over 5 years.

33. a. $P(q) = -q^3 + 24q^2 + 108q - 3{,}000$
   b. 18 units
   c. $162

35. a. $P(t) = 32.5e^{0.04t} - 32.5$; 4.14 billion barrels; 4.67 billion barrels
   b. 12 years
   c. 823.22 billion dollars
   d. Answers will vary.

37. a. $P(t) = 60e^{0.02t} - 60$; 3.71 billion barrels; 3.94 billion barrels
   b. 9.12 years
   c. 608.89 billion dollars
   d. Answers will vary.

39. $1,929,148

41. a. $137,334.29
   b. $44,585.04

43. a. $207,360
   b. Answers will vary.

CHAPTER 5 | Section 6

1. 30,484

3. 468,130

5. 451,404

7. $7\pi \approx 21.99$ cubic units

9. $\dfrac{1{,}532}{15}\pi \approx 320.86$ cubic units

11. $\dfrac{32}{3}\pi \approx 33.51$ cubic units

13. $2\pi \approx 6.28$ cubic units

15. 61,070,138

17. About 80 members

19. 4097.62 (4098 people)

21. 515.48 billion barrels

23. 4,207 members

ANSWERS

25. a. 0.0775 liters/sec

b.

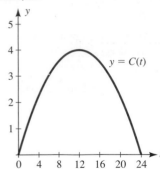

27. a. 0.0853 liters/sec

b.

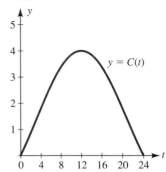

29. Approximately 208,128 people

31. a. The LDL level decreases 6.16 units.

b. $L(t) = \dfrac{3}{28}(49 - t^2)^{1.4} + 150 - \dfrac{21}{4}(49)^{0.4}$

c. 5.8 days

33. After T years, the first population is

$$P_1(T) = \left(100,000 - \frac{50}{0.011}\right)e^{-0.011T} + \frac{50}{0.011}$$

Then $P_1(T) < P(T)$ for $T = 50$ and $T = 100$, but $P_1(300) > P(300)$.

35. 1,565.83 (1,566 animals)

37. 10,125 people

39. $\displaystyle\int_0^{12} [W'(t) - D'(t)]dt = 0.363$;

about 36 people; 18.1%

41. a. 55 years

b. 70.78 years

c. 86.36 years; they have already exceeded their life expectancy.

d. 71.69 years

43. a. 2.37 sec

b. 0.905 L

c. 0.382 L/sec

45. a.

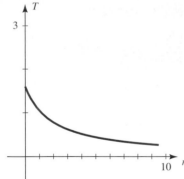

b. $r(T) = \dfrac{3}{T} - 2$

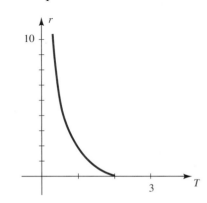

c. $\pi\left[12\left(\ln\dfrac{1}{3} - \ln\dfrac{3}{2}\right) + \dfrac{77}{3}\right] \approx 23.93 \text{ ft}^3$

47. a. $100\pi \ln \dfrac{23}{5} \approx 479.42$ units

b. $L = \dfrac{3\sqrt{10}}{2} \approx 4.74$ miles; $100\pi \ln 10 \approx 723.38$ units

49. The hypotenuse of the triangle has equation $y = \dfrac{r}{h}x$, and the volume is

$$V = \pi\int_0^h \left(\frac{r}{h}x\right)^2 dx = \pi\int_0^h \frac{r^2}{h^2}x^2\, dx$$

$$= \frac{\pi r^2}{h^2}\left(\frac{1}{3}x^3\right)\Big|_0^h = \frac{\pi r^2}{3h^2}(h^3 - 0)$$

$$= \frac{1}{3}\pi r^2 h$$

CHAPTER 5 | Checkup

1. a. $\dfrac{x^4}{4} - \dfrac{2\sqrt{3}}{3}x^{3/2} - \dfrac{5}{2}e^{-2x} + C$

b. $\dfrac{x^2}{2} - 2x + 4 \ln |x| + C$

c. $\dfrac{2}{7}x^{7/2} - 2x^{1/2} + C$

d. $\dfrac{-1}{2\sqrt{3 + 2x^2}} + C$

e. $\dfrac{1}{4}(\ln x)^2 + C$

f. $\dfrac{1}{2}e^{1+x^2} + C$

2. a. $\dfrac{62}{5} + 4 \ln 2$

b. $e^3 - 1$

c. $1 - \ln 2$

d. $\sqrt{31} - 2$

3. a. $\dfrac{73}{6}$

b. 36

4. $1 - 2 \ln 2$

5. $10,333.33

6. 71.14 billion dollars; increase

7. $4,266.67

8. $16,183.42

9. 45,055

10. 0.1 mg/cm^3

CHAPTER 5 | Review Problems

1. $\dfrac{1}{4}x^4 + \dfrac{2}{3}x^{3/2} - 9x + C$

3. $\dfrac{x^5}{5} + \dfrac{5}{2}e^{-2x} + C$

5. $\dfrac{5}{3}x^3 - 3 \ln |x| + C$

7. $\dfrac{1}{6}t^6 - t^3 - \dfrac{1}{t} + C$

9. $\dfrac{2}{9}(3x + 1)^{3/2} + C$

11. $\dfrac{1}{12}(x^2 + 4x + 2)^6 + C$

13. $\dfrac{-3}{4(2x^2 + 8x + 3)} + C$

15. $\dfrac{1}{14}(v - 5)^{14} + \dfrac{5}{13}(v - 5)^{13} + C$

17. $-\dfrac{5}{2}e^{-x^2} + C$

19. $\dfrac{2}{3}(\ln x)^{3/2} + C$

21. 0

23. $\dfrac{1}{2}(e^2 + 5)$

25. 1,710

27. $1 - \dfrac{1}{e}$

29. $e - 2$

31. Area $= \dfrac{101}{6}$

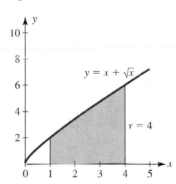

33. Area $= \ln 2 + \dfrac{7}{3}$

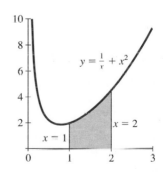

ANSWERS

35. Area $= \dfrac{15}{2} - 8 \ln 2$

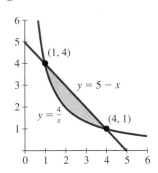

37. Area $= \dfrac{9}{2}$

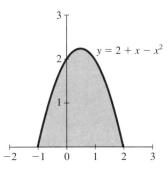

39. $\dfrac{11,407}{84} - \dfrac{2\sqrt{2}}{21} \approx 135.7$

41. $\dfrac{1}{4}\left(1 - \dfrac{1}{e^4}\right)$

43. \$128; \$21.33

45. \$6.70; \$6.16

47. $GI = \dfrac{1}{5}$

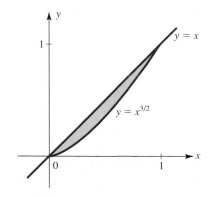

49. $GI = \dfrac{1}{10}$

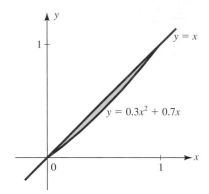

51. 43,984

53. 14,308

55. $\dfrac{78}{5}\pi \approx 49.01$ cubic units

57. $\pi \ln 3 \approx 3.45$ cubic units

59. $y = 2x + 10$

61. $x = \dfrac{9}{2} - \dfrac{1}{2}e^{-2t}$

63. $y = \dfrac{1}{2}\ln(x^2 + 1) + 5 - \dfrac{1}{2}\ln 2$

65. \$87.57

67. 1,220 people

69. 11,250 commuters

71. In 2006 (0.2554 billion barrels versus 0.1003 billion barrels in 2009)

73. \$7,377.37

75. 61.65 (about 62 homes)

77. 14,868 pounds

79. \$565,056

81. \$1.32 per pound

83. Temperature decreases by 2.88°C

85. **a.** $p_1(x) = 0.2x + 0.001x^3 + 250$; $p_1(10) = \$2.53$ per dozen

b. $p_2(x) = 0.3x + 0.001x^3 + 250$; $p_2(10) = \$2.54$ per dozen

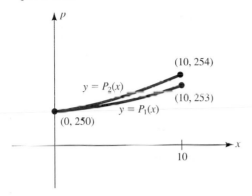

87. 30 meters

89. Physical therapists

91. 2,255 trout

93. a. $\dfrac{1}{N}\displaystyle\int_0^N S(t)\,dt$

b. $\displaystyle\int_0^N S(t)\,dt$

c. The average speed is equal to the total distance divided by the total number of hours.

95. The region bounded by the curves is between $x = -4.66$ and $x = -1.82$; the curves also intersect at $x = 4.98$. The area is approximately 3.

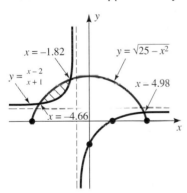

CHAPTER 6 | Section 1

1. $-(x + 1)e^{-x} + C$

3. $(2 - x)e^x + C$

5. $\dfrac{1}{2}t^2\left(\ln 2t - \dfrac{1}{2}\right) + C$

7. $-5(v + 5)e^{-v/5} + C$

9. $\dfrac{2}{3}x(x - 6)^{3/2} - \dfrac{4}{15}(x - 6)^{5/2} + C$

11. $\dfrac{1}{9}x(x + 1)^9 - \dfrac{1}{90}(x + 1)^{10} + C$

13. $2x\sqrt{x + 2} - \dfrac{4}{3}(x + 2)^{3/2} + C$

15. $\dfrac{8}{3}$

17. $\dfrac{1}{4}(1 - 3e^{-2})$

19. $\dfrac{1}{12}(3e^4 + 1)$

21. $\dfrac{1}{16}(e^2 + 1)$

23. $-\dfrac{1}{x}(\ln x + 1) + C$

25. $\dfrac{1}{2}e^{x^2}(x^2 - 1) + C$

27. $\dfrac{1}{25}(3 - 5x - 3\ln|3 - 5x|) + C$

29. $\dfrac{-\sqrt{4x^2 - 9}}{x} + 2\ln|2x + \sqrt{4x^2 - 9}| + C$

31. $\dfrac{1}{2}\ln\left|\dfrac{x}{2 + 3x}\right| + C$

33. $\dfrac{\sqrt{3}}{24}\ln\left|\dfrac{4 + \sqrt{3}u}{4 - \sqrt{3}u}\right| + C$

35. $x(\ln x)^3 - 3x(\ln x)^2 + 6x\ln x - 6x + C$

37. $-\dfrac{1}{25}\left[\dfrac{5 + 4x}{x(5 + 2x)} + \dfrac{4}{5}\ln\left|\dfrac{x}{5 + 2x}\right|\right] + C$

39. $f(x) = 5 + \dfrac{3}{e} - \dfrac{x + 2}{e^x}$

41. $s(t) = 4 - 2e^{-t/2}(t + 2)$

43. \$13,212

45. 2,008,876

47. $\dfrac{40}{27}(5e^2 - 14e^{1/5}) \approx 29.4$ mg/mL

49. $62{,}000e^{1/2} - 63{,}000 \approx \$39{,}220.72$

51. \$11,417

53. 4,367

55. a. \$4.47 per unit
   b. \$14,284.56

ANSWERS

57. $1 - \dfrac{2}{e} \approx 0.2642$

59. 0.09043 L/sec

61. Let
$$U = u^n \qquad dV = e^{au}\,du$$
$$dU = nu^{n-1}\,du \qquad V = \dfrac{1}{a}e^{au}$$

Then

$$\int u^n e^{au}\,du$$

$$= u^n\!\left(\dfrac{1}{a}e^{au}\right) - \int \dfrac{1}{a}e^{au}(nu^{n-1}\,du)$$

$$= \dfrac{1}{a}u^n e^{au} - \dfrac{n}{a}\int u^{n-1}e^{au}\,du$$

63. (0.244, 0.353)

65. a. Locate the kiosk at the center $(\bar{x}, \bar{y})$ of the parking lot. The curved boundary of the lot has the equation $2x^2 - y^2 = 1$ or, equivalently, $y = \sqrt{2x^2 - 1}$, so the lot has area

$$A = \int_1^5 \sqrt{2x^2 - 1}\,dx = 16.38$$

(Use integral formula 18 in the table on page 472.) Therefore, we have

$$\bar{x} = \dfrac{1}{A}\int_1^5 x\sqrt{2x^2 - 1}\,dx = \dfrac{57}{16.38} = 3.48$$

and

$$\bar{y} = \dfrac{1}{2A}\int_1^5 \left(\sqrt{2x^2 - 1}\right)^2 dx = \dfrac{236/3}{2(16.38)} = 2.41$$

The kiosk should be located at (3.48, 2.41).

b. Writing exercise.

67. Area ≈ 0.75834

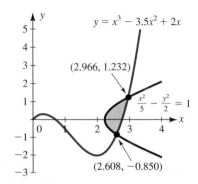

$y = x^3 - 3.5x^2 + 2x$

(2.966, 1.232)

$\dfrac{x^2}{5} - \dfrac{y^2}{2} = 1$

(2.608, −0.850)

69. Area ≈ 1.95482

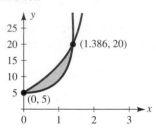

(1.386, 20)

(0, 5)

71. 4.2265

73. 0.4509

CHAPTER 6 Section 2

1. $y = x^3 + \dfrac{5}{2}x^2 - 6x + C$

3. $y = Ce^{3x}$

5. $e^{-y} = C - x$

7. $y^2 = x^2 + C$

9. $\sqrt{y} = \dfrac{1}{3}x^{3/2} + C$

11. $y = C|x - 1|$

13. $\ln(y + 3)^{10} = -(2x - 5)^{-5} + C$

15. $x = Ce^{t/2}(2t + 1)^{-1/4}$

17. $y = \ln(xe^x - e^x + C)$ or $e^y = (x - 1)e^x + C$

19. $y = Ce^{(\ln t - 1)t/2}$

21. $y = \dfrac{1}{5}e^{5x} + \dfrac{4}{5}$

23. $y^3 = \dfrac{3}{2}x^2 + 21$

25. $y = \dfrac{6}{4(4 - x)^{3/2} + 3}$

27. $y - 2\ln|y + 1| = \ln|t| + 2(1 - \ln 3)$

29. Let Q denote the number of bacteria.
$$\dfrac{dQ}{dt} = kQ$$

31. Let Q denote the value of the investment.
$$\dfrac{dQ}{dt} = 0.07Q$$

33. Let P denote the population.
$$\dfrac{dP}{dt} = 500$$

35. $\frac{dT(t)}{dt} = k[T_m - T(t)]$, where k is the proportionality constant, T_m = temperature of the surrounding medium, and $T(t)$ = object's temperature at time t

37. $\frac{dR(t)}{dt} = k[F - R(t)]$, where k is the proportionality constant, F is the total number of facts, and $R(t)$ = number of facts recalled at time t

39. $\frac{dP(t)}{dt} = kP(t)[N - P(t)]$, where k is the proportionality constant, N is the number of people involved, and $P(t)$ is the number of people implicated at time t

41. $y = Ce^{kx} \Rightarrow \frac{dy}{dx} = C(ke^{kx}) = k(Ce^{kx}) = ky$

43. $y = C_1e^x + C_2xe^x$

$\frac{dy}{dx} = C_1e^x + C_2(xe^x + e^x)$
$\qquad = (C_1 + C_2)e^x + C_2xe^x$

$\frac{d^2y}{dx^2} = (C_1 + C_2)e^x + C_2(xe^x + e^x)$
$\qquad = (C_1 + 2C_2)e^x + C_2xe^x$

$\Rightarrow \frac{d^2y}{dx^2} - 2\frac{dy}{dx} + y = (C_1 + 2C_2)e^x + C_2xe^x -$
$\qquad 2C_1e^x - 2C_2xe^x - 2C_2e^x + C_1e^x + C_2xe^x$
$\qquad = (C_1 + 2C_2 - 2C_1 - 2C_2 + C_1)e^x +$
$\qquad (C_2 - 2C_2 + C_2)xe^x$
$\qquad = 0 \cdot e^x + 0 \cdot xe^x = 0$

45. $542,208

47. If S_0 is the initial amount of sugar and $S(t)$ is the amount dissolved at time t, then $S_0 - S$ is the amount of undissolved sugar at time t. The rate relationship can be expressed as the differential equation

$$\frac{dS}{dt} = k(S_0 - S)$$

Separate variables and solve to get
$$S(t) = S_0(1 - e^{-kt})$$
The graph of $S(t)$ is shown in this figure:

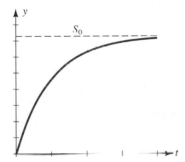

49. Let T_0 be the initial temperature of the object, let $T(t)$ be the temperature at time t, and let T be the temperature of the surrounding medium. The rate relationship can be expressed as the differential equation

$$\frac{dT}{dt} = k(T_m - T)$$

Separate variables and solve to get
$$T(t) = T_m + (T_0 - T_m)e^{-kt}$$
The graph of $T(t)$ is shown in this figure:

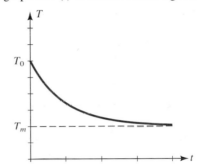

51. a. $\frac{S(t)}{40}$ gallons per minute

b. $\frac{dS}{dt} = -\frac{S}{40}$

c. $S(t) = 400e^{-t/40}$

53. a. The rate relationship can be expressed as
$$\frac{dV}{dt} = \left[\begin{array}{c}\text{rate money is}\\\text{added to account}\end{array}\right] - \left[\begin{array}{c}\text{rate money}\\\text{is withdrawn}\end{array}\right]$$
$$= rV - W$$

Separating the variables, we find that
$$\frac{dV}{rV - W} = dt$$
The particular solution with $V(0) = S$ is
$$V = \frac{W}{r} + \left(S - \frac{W}{r}\right)e^{rt}$$

b. $175,639
c. $25,000
d. 7.5 years

55. $\frac{dp}{dt} = k(1 - p), p(8) = 0.05$

$p(t) = 1 - (0.95)^{t/8}$

57. 4.16 minutes

59. a. $p(t) = 3 + 3e^{-(\ln 3/4)t}$

b. $S = D$ when $7 - p = 1 + p$ or $p = 3$, which is the same as $\lim\limits_{t\to\infty} p(t)$.

61. a. $D(t) = \dfrac{a}{b}I_0 e^{bt} + \left(D_0 - \dfrac{a}{b}I_0\right)$

$I(t) = I_0 e^{bt}$

b. The ratio approaches $\dfrac{a}{b}$.

63. Let $P(t)$ = number of infected residents and C = total number of susceptible residents.

$\dfrac{dP}{dt} = kP(C - P)$ is maximized when

$\dfrac{d^2P}{dt^2} = k(C - P) - kP = k(C - 2P) = 0$

that is, when $P = \dfrac{C}{2}$.

65. $C(t) = \dfrac{R}{k} + \left(C_0 - \dfrac{R}{k}\right)e^{-kt}$

CHAPTER 6 | Section 3

1. $\dfrac{1}{2}$

3. Diverges.

5. Diverges.

7. $\dfrac{1}{10}$

9. $\dfrac{5}{2}$

11. $\dfrac{1}{9}$

13. Diverges.

15. $\dfrac{2}{e} = 2e^{-1}$

17. $\dfrac{2}{9}$

19. Diverges.

21. Diverges.

23. 2

25. Yes

27. Yes

29. No, $f(x)$ is negative for $-1 \le x < 0$.

31. a. 1

b. $\dfrac{1}{3}$

c. $\dfrac{1}{3}$

33. a. 1

b. $\dfrac{3}{16}$

c. $\dfrac{9}{16}$

35. a. 1

b. $\dfrac{7}{8}$

c. $\dfrac{1}{8}$

37. a. 1

b. $\dfrac{1}{e} - \dfrac{1}{e^4}$

c. $1 - \dfrac{1}{e^4}$

39. $\dfrac{7}{2}$

41. $\dfrac{4}{3}$

43. $\dfrac{3}{2}$

45. $60,000

47. $600,000

49. 200 patients

51. 50 units

53. a. $\dfrac{1}{3}$

b. $\dfrac{1}{9}$

c. $\dfrac{45}{2}$ seconds

55. a. $\dfrac{1}{e} \approx 0.368$

b. $\dfrac{1}{e^{2/3}} - \dfrac{1}{e^{5/3}} \approx 0.325$

c. 3 minutes

57. a. $\dfrac{1}{e^{1/5}} - \dfrac{1}{e^{3/10}} \approx 0.078$

b. $1 - \dfrac{1}{e^{4/25}} \approx 0.148$

c. $\dfrac{1}{e^{6/25}} \approx 0.787$

d. 50 months

59. a. $\dfrac{1}{5}$

b. $1 - \dfrac{1}{e^{2/5}} \approx 0.33$

c. $e^{-7/5} \approx 0.247$

61. a. $1 - \dfrac{1}{e} \approx 0.632$

b. $e^{-6/5} \approx 0.301$

c. 5 minutes

63. $875,000

65. a. 0.595

b. 0.445

67. $\displaystyle \lim_{N \to +\infty} \int_0^N (A + Bt)e^{-rt}\,dt$

$\displaystyle = \lim_{N \to +\infty} -\frac{A}{r}e^{-rt} + B\left(-\frac{t}{r}e^{-rt} + \frac{1}{r}\int_0^N e^{-rt}\,dt\right)$

$\displaystyle = \lim_{N \to +\infty} \left(-\frac{A}{r}e^{-rt} - \frac{Bt}{r}e^{-rt} - \frac{B}{r^2}e^{-rt}\right)\Bigg|_0^N$

$\displaystyle = 0 - \left(-\frac{A}{r}e^0 - 0 - \frac{B}{r^2}e^0\right) = \frac{A}{r} + \frac{B}{r^2}$

69. $\displaystyle E(x) = \int_{-\infty}^{\infty} xf(x)\,dx = \int_0^{\infty} kxe^{-kx}\,dx$

$\displaystyle = \lim_{N \to \infty} \int_0^N kxe^{-kx}\,dx$

$\displaystyle = \lim_{N \to \infty} \left(-xe^{-kx}\Big|_0^N + \int_0^N e^{-kx}\,dx\right)$

$\displaystyle = \lim_{N \to \infty} \left(-xe^{-kx} - \frac{1}{k}e^{-kx}\right)\Bigg|_0^N$

$\displaystyle = \lim_{N \to \infty} \left(-Ne^{-kN} - \frac{1}{k}e^{-kN} + \frac{1}{k}\right) = \frac{1}{k}$

CHAPTER 6 | Section 4

1. a. 2.343750

b. $\dfrac{7}{3}$

3. a. 0.782794

b. 0.785392

5. a. 1.151479

b. 1.147782

7. a. 0.742984

b. 0.746855

9. a. 1.930756

b. 1.922752

11. a. 1.096997

b. 1.094800

13. a. 0.849195

b. 0.836203

15. a. $0.508994; |E_n| \le 0.031250$

b. $0.500418; |E_n| \le 0.002604$

17. a. $2.796731; |E_n| < 0.001667$

b. $2.797432; |E_n| \le 0.000017$

19. a. $1.490679; |E_n| \le 0.084946$

b. $1.463711; |E_n| \le 0.004483$

21. a. 164

b. 18

23. a. 36

b. 6

25. a. 179

b. 8

27. a. 3.0898

b. 3.1212

29. 0.138569

31. 0.358531 cubic units

33. $26,072.45

35. 3,496 people

37. 51.75 miles

39. $5,950

41. 235 ft^2

43. $34,200

45. 475,197

47. a. The integral gives the difference between the cumulative number of AIDS deaths by 2001 and the cumulative number by 1990. This is exactly the number of deaths in the period 1990–2001.

b. 362,771

c. Answers will vary.

ANSWERS

CHAPTER 6 | Checkup

1. a. $\dfrac{4\sqrt{2}}{9}x^{3/2}(3 \ln x - 2) + C$

   b. $25 - 20e^{0.2}$

   c. $-\dfrac{298}{15}$

   d. $-xe^{-x} + C$

2. a. 10

   b. $\dfrac{3}{4}e^{-2}$

   c. Diverges.

   d. 0

3. a. $\dfrac{x}{4}\left[(\ln 3x)^2 - 2 \ln 3x + 2\right]$

   b. $-\dfrac{1}{2} \ln \left|\dfrac{\sqrt{4 + x^2} + 2}{x}\right| + C$

   c. $\dfrac{\sqrt{x^2 - 9}}{9x} + C$

   d. $-\dfrac{1}{4} \ln \left|\dfrac{x}{3x - 4}\right| + C$

4. a. $y = \sqrt{\dfrac{4}{x} + 5}$

   b. $y = -3\sqrt{x^2 + 1}$

   c. $y = x - \ln(x + 1)$

5. $16,487.21$

6. $1,666,667$

7. a. 0.6977

   b. 0.0787

   c. 33.3 months

8. 3.5 mg

9. a. $\dfrac{dm}{dt} = kmt;\ m(t) = Ce^{-(\ln 2/144)t^2}$

   b. 67.7%

10. $1.027552;\ 5 \ln\left(\dfrac{3}{2}\right) - 1$

CHAPTER 6 | Review Problems

1. $-e^{1-t}(t + 1) + C$

3. $\dfrac{1}{3}x(2x + 3)^{3/2} - \dfrac{1}{15}(2x + 3)^{5/2} + C$

5. $4 \ln 2 - 2$

7. $\dfrac{74}{7}$

9. $\dfrac{1}{9}x^2(3x^2 + 2)^{3/2} - \dfrac{2}{135}(3x^2 + 2)^{5/2} + C$

11. $\dfrac{5}{8} \ln \left|\dfrac{2 + x}{2 - x}\right| + C$

13. $-3w^2e^{-w/3} - 18we^{-w/3} - 54e^{-w/3} + C$

15. $x\left[(\ln 2x)^3 - 3(\ln 2x)^2 + 6 \ln 2x - 6\right] + C$

17. Diverges.

19. Diverges.

21. $\dfrac{1}{4}$

23. $\dfrac{1}{4}$

25. Diverges.

27. $y = \dfrac{x^4}{4} - x^3 + 5x + C$

29. $y = 80 - Ce^{-kx}$

31. $y = x^5 - x^3 - 2x + 6$

33. $y = 2e^{1 - \sqrt{1 - x^2}}$

35. a. 1

   b. $\dfrac{1}{3}$

   c. $\dfrac{1}{3}$

37. a. 1

   b. 0.3694

   c. 0.3679

39. $22,857$

41. $\dfrac{dQ}{dt} = \dfrac{-Q}{50};\ 81.2$ pounds

43. $\dfrac{dQ}{dt} = 18 - \dfrac{18Q}{5,000};\ 640$ days

45. The population will increase without bound.

47. $\dfrac{1}{6}$

49. $320,000$

51. $\dfrac{2}{9}$

53. a. 0.7047
  b. 0.1466

55. a. $k = \dfrac{\ln 10}{30}$
  b. $D(t) = 50e^{-(\ln 10/30)t}, S(t) = 5e^{(\ln 10/15)t}$
  c. 23.2 units

57. a. $P'(t) = (b - d)P(t); P(t) = P_0 e^{(b-d)t}$
  b. $P'(t) = kP^{1+1/e}; P(t) = \left(P_0^{-1/e} - \dfrac{kt}{c}\right)^{-c}$
  c. 3,375

59. 15,000

61. a. 17.565086; error $|E_8| \leq 10.3$
  b. 16.538595; error $|E_8| \leq 1.1$

63. a. 2.94940, error $|E_8| \leq 0.0035$
  b. 2.94834; error $|E_8| \leq 0.00008$

65. a. 13
  b. 2

67. 4,804.8 thousand dollars

69. The graphs intersect at $x = 1$ and approximately $x = 0.41$; area ≈ 0.1692.

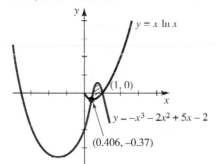

71. $\dfrac{2}{3} \ln 2$

73. Answers will vary; 0.5.

75. $\dfrac{b}{a} \ln S + \dfrac{c}{a} S + \dfrac{1}{2a} S^2 = t + C$

CHAPTER 7 | Section 1

1. $f(2, -1) = -3, f(1, 2) = 16$

3. $g(4, 5) = 3, g(-1, 2) = \sqrt{3} \approx 1.7321$

5. $f(e^2, 3) = \dfrac{3}{2}; f(\ln 9, e^3) = 25.515$

7. $g(1, 2) = 2.5; g(2, -3) = -2.167$

9. $f(1, 2, 3) = 6; f(3, 2, 1) = 6$

11. $F(1, 1, 1) \approx 0.2310; F(0, e^2, 3e^2) \approx 0.1048$

13. All ordered pairs (x, y) of real numbers for which $y \neq \dfrac{-4}{3}x$

15. All ordered pairs (x, y) of real numbers for which $y \leq x^2$

17. All ordered pairs (x, y) of real numbers for which $x > 4 - y$

19.

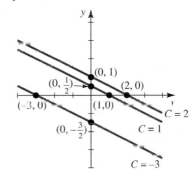

21.

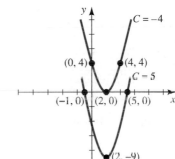

23.

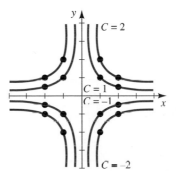

25.

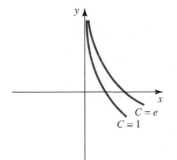

27. a. 160,000 units
   b. Production will increase by 16,400 units.
   c. Production will increase by 4,000 units.
   d. Production will increase by 20,810 units.

29. a. $R(x_1, x_2) = 200x_1 - 10x_1^2 + 25x_1x_2 + 100x_2 - 10x_2^2$
   b. $1,840

31. a. If $a + b > 1$, production is more than doubled.
   b. If $a + b < 1$, production is increased (but not doubled).
   c. If $a + b = 1$, production is doubled.

33. $R(x, y) = 60x - \dfrac{x^2}{5} + \dfrac{xy}{10} + 50y - \dfrac{y^2}{10}$

35. a. $S(15.83, 87.11) = 0.5938$

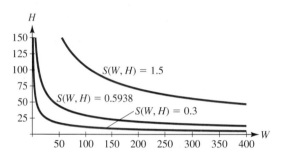

The curves represent different combinations of height and weight that result in the same surface area.

   b. Height = 90.05 cm
   c. 254%
   d. Writing exercise; responses will vary.

37. a. 70 units
   b. $y = -\dfrac{3}{2}x + 35$
   c.

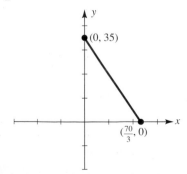

   d. Unskilled labor should be decreased by three workers.

39. 260

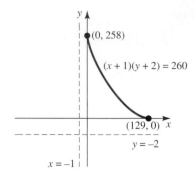

41. a. 0.866 cm/sec
   b.

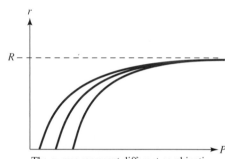

The curves represent different combinations of pressure and distance from the axis that result in the same speed.

43. a.

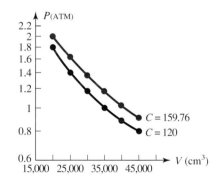

   b. 159.76°C

45. a. 2105.03 kilocalories
   b. 1428.84 kilocalories
   c. Approximately 27 years
   d. Approximately 24.4 years

47. a. $2,003.13; $110,563.40
   b. $1,435.20; $266,672

49. 23.54

51. For $Q(K, L) = AK^\alpha L^{1-\alpha}$,

$$Q(sK, sL) = A(sK)^\alpha (sL)^{1-\alpha}$$
$$= As^\alpha K^\alpha s^{1-\alpha} L^{1-\alpha}$$
$$= s^{\alpha+1-\alpha} AK^\alpha L^{1-\alpha}$$
$$= s AK^\alpha L^{1-\alpha}$$
$$= sQ(K, L)$$

CHAPTER 7 | Section 2

1. $f_x = 2y^5 + 6xy + 2x; f_y = 10xy^4 + 3x^2$

3. $\dfrac{\partial z}{\partial x} = 15(3x + 2y)^4; \dfrac{\partial z}{\partial y} = 10(3x + 2y)^4$

5. $f_s = -\dfrac{3t}{2s^2}; f_t = \dfrac{3}{2s}$

7. $\dfrac{\partial z}{\partial x} = (xy + 1)e^{xy}; \dfrac{\partial z}{\partial y} = x^2 e^{xy}$

9. $f_x = \dfrac{-e^{2-x}}{y^2}; f_y = \dfrac{-2e^{2-x}}{y^3}$

11. $f_x = \dfrac{5y}{(y - x)^2}; f_y = \dfrac{-5x}{(y - x)^2}$

13. $\dfrac{\partial z}{\partial u} = \ln v; \dfrac{\partial z}{\partial v} = \dfrac{u}{v}$

15. $f_x = \dfrac{1}{y^2(x + 2y)}; f_y = \dfrac{2[y - (x + 2y)\ln(x + 2y)]}{y^3(x + 2y)}$

17. $f_x(-2, 1) = -22; f_y(-2, 1) = 26$

19. $f_x(0, 0) = 1; f_y(0, 0) = 1$

21. $f_{xx} = 60x^2 y^3; f_{xy} = 2(30x^3 y^2 + 1);$
$f_{yx} = 2(30x^3 y^2 + 1); f_{yy} = 30x^4 y$

23. $f_{xx} = 2y(2x^2 y + 1)e^{x^2 y}; f_{xy} = 2x(x^2 y + 1)e^{x^2 y};$
$f_{yx} = 2x(x^2 y + 1)e^{x^2 y}; f_{yy} = x^4 e^{x^2 y}$

25. $f_{ss} = \dfrac{t^2}{\sqrt{(s^2 + t^2)^3}}; f_{st} = \dfrac{st}{\sqrt{(s^2 + t^2)^3}};$

$f_{ts} = \dfrac{-st}{\sqrt{(s^2 + t^2)^3}}; f_{tt} = \dfrac{s^2}{\sqrt{(s^2 + t^2)^3}}$

27. Substitute

29. Neither

31. Substitute

33. Yes

35. No

37. Daily output will increase by approximately 10 units.

39. a. The marginal productivity of capital is approximately 27 units. The marginal productivity of labor is approximately 64 units.

b. Additional labor employment

41. $F(L, r) = \dfrac{kL}{r^4}$

a. $F(3.17, 0.085) = 60,727.24k;$
$\dfrac{\partial F}{\partial L} = \dfrac{k}{r^4} = 19,156.86k;$
$\dfrac{\partial F}{\partial r} = -\dfrac{4kL}{r^5} = -2,857,752.58k$

b. $F(1.2L, 0.8r) = \dfrac{k(1.2L)}{(0.8r)^4} = 2.93F(L, r);$

$\dfrac{\partial F}{\partial r}(1.2L, 0.8r) = 3.66\dfrac{\partial F}{\partial r}(L, r);$

$\dfrac{\partial F}{\partial L}(1.2L, 0.8r) = 2.44\dfrac{\partial F}{\partial L}(L, r)$

43. The monthly demand for bicycles decreases by approximately 3.

45. The volume is increased by 72π cm^3.

47. a. An increase in x will decrease the demand $D(x, y)$ for the first brand of mower. An increase in y will increase the demand $D(x, y)$ for the first brand of mower.

b. $\dfrac{\partial D}{\partial x} < 0, \dfrac{\partial D}{\partial y} > 0$

c. $b < 0, c > 0$

49. $P(x, y, u, v) = \dfrac{100xy}{xy + uv};$

$P_x = \dfrac{(xy + uv)100y - 100xy^2}{(xy + uv)^2} = \dfrac{100uvy}{(xy + uv)^2};$

$P_y = \dfrac{(xy + uv)100x - 100x^2 y}{(xy + uv)^2} = \dfrac{100uvx}{(xy + uv)^2};$

$P_u = \dfrac{-100xyv}{(xy + uv)^2}; P_v = \dfrac{-100uxy}{(xy + uv)^2}$

All of these partials measure the rate of change of percentage of total blood flow with respect to the quantities x, y, u, and v, respectively.

51. $\dfrac{\partial F}{\partial z} = \dfrac{-c\pi x^2}{8\sqrt{y - z}};$ decreasing since $F_{zz} < 0$

53. a. $\dfrac{\partial^2 Q}{\partial L^2} < 0;$ for a fixed level of capital investment, the effect on output of the addition of one worker hour is greater when the workforce is small than when it is large.

b. $\dfrac{\partial^2 Q}{\partial K^2} < 0;$ for a fixed workforce, the effect on output of the addition of $1,000 in capital investment is greater when the capital investment is small than when it is large.

55. a. $Q(37, 71) = 304{,}691$; $Q(38, 71) = 317{,}310$; $Q(37, 72) = 309{,}031$
  b. $Q_x(37, 71) = 12{,}534$ units; $Q(38, 71) - Q(37, 71) = 12{,}619$ units
  c. $Q_y(37, 71) = 4{,}344$ units; $Q(37, 72) - Q(37, 71) = 4{,}340$ units

57. $\dfrac{dz}{dt} = 4t + 15$

59. $\dfrac{dz}{dt} = \dfrac{3}{y} - \dfrac{6xt}{y^2}$

61. $\dfrac{dz}{dt} = 2ye^{2t} - 3xe^{-3t}$

63. The number of units produced decreases by about 55.

65. a. 424 units/month
  b. 16.31

67. The number of units produced increases by about 61.6 units per day.

69. Sales will decrease by approximately 2,070 books.

71. a. $C(R, H) = 0.001\pi(R^2 + RH + R^2H)$
  b. The cost increases by about 0.08 cents per can.

73. $\dfrac{dy}{dx} = \dfrac{1}{2}$; $x - 2y = -1$

75. 2.6 units

CHAPTER 7 | Section 3

	Relative maximum	Relative minimum	Saddle point
1.	$(0, 0)$	None	None
3.	None	None	$(0, 0)$
5.	None	None	$(2, -1)$
7.	$(-2, -1)$	$(1, 1)$	$(-2, 1); (1, -1)$
9.	None	$\left(4, \dfrac{19}{2}\right)$	$\left(2, \dfrac{7}{2}\right)$
11.	$(0, 1); (0, -1)$	$(0, 0)$	$(1, 0); (-1, 0)$
13.	None	$\left(\dfrac{4}{3}, \dfrac{4}{3}\right)$	$(0, 0)$
15.	$(0, 3)$	None	None
17.	$\left(-\dfrac{3}{2}, 1\right)$	None	None
19.	$(e, 1); (e, -1)$	None	None

21. Duncan shirts $x = \$2.70$; O'Neal shirts $y = \$2.50$

23. The base of the box is a 2 ft by 2 ft square. The height is 8 ft.

25. Whole milk, $x = 20$ gallons; skim milk, $y = 20$ gallons

27. $x = \dfrac{\sqrt{2}}{2}$; $y = \dfrac{\sqrt{2}}{2}$

29. $x = y = z = \sqrt[3]{V_0}$

31. $x = 200$; $y = 300$

33. $S\left(\dfrac{5}{4}, -\dfrac{1}{4}\right)$

35. Maximum of $P = \dfrac{2}{3}$, when $p = q = r = \dfrac{1}{3}$.

37. a. Dick should wait at $x = 0.424$ miles; Mary should wait at $y = 2.236$ miles (1.6397 miles from F); optimum time is 1.748 hours
  b. Tom, Dick, and Mary will win by 0.2080 hours (12.5 minutes).
  c. Writing exercise; responses will vary.

39. Problem can be stated as:
maximize $V = 6\left[x - 2\left(\dfrac{6}{\sqrt{3}}\right)\right]y$
subject to $2xy + 2\left(\dfrac{1}{2}\dfrac{\sqrt{3}}{2}x^2\right) = 500$
Solution is $x \approx 13.87$ ft; $y \approx 12.02$ ft.

41. $\dfrac{\partial f}{\partial x} = 2x - 4y$; $\dfrac{\partial f}{\partial y} = 2y - 4x$. Thus, $(0, 0)$ is a critical point. Since $\dfrac{\partial^2 f}{\partial^2 x} = 2 > 0$, the second derivative test tells us there is a minimum in the x direction. Likewise, $\dfrac{\partial^2 f}{\partial^2 y} = 2 > 0$ implies a minimum in the y direction. However, along the curve determined by $y = x$, we have $f = -2x^2$, which has a relative maximum at $(0, 0)$.

43. $f_x = \dfrac{x^2 - 7y^2}{x^2 \ln y}$;
$f_y = \dfrac{y(x + 14y)\ln y - (x^2 + xy + 7y^2)}{xy(\ln y)^2}$
Critical points: $(\sqrt{7}e, e), (-\sqrt{7}e, e)$

45. $f_x = 8x^3 - 22xy + 36x$; $f_y = 4y^3 - 11x^2$
Critical point: $(0, 0)$

CHAPTER 7 | Section 4

1. $y = \dfrac{1}{4}x + \dfrac{3}{2}$

3. $y = 3$

5. $y = \dfrac{7}{9}x + \dfrac{19}{18}$

7. $y = -\dfrac{1}{2}x + 4$

9. $y = 1.018x + 0.802$

11. $y = -0.915x + 1.683$

13. $y = 15.018e^{0.04x}$

15. $y = 20.03e^{-0.201x}$

17. a.

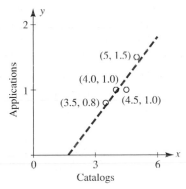

b. $y = 0.42x - 0.71$

c. When 4,800 catalogs ($x = 4.8$) are requested, $y = 0.42(4.8) - 0.71 = 1.306$ or 1,306 applications are predicted to be received.

19. a.

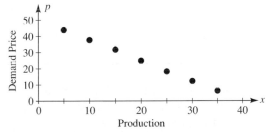

b. $p = -1.29x + 50.71$

c. The predicted price is negative. All 4,000 units cannot be sold at any price.

21. a. Let x denote the number of hours after the polls open and y the corresponding percentage of registered voters that have already cast their ballots. Then

x	2	4	6	8	10
y	12	19	24	30	37

x	y	xy	x^2
2	12	24	4
4	19	76	16
6	24	144	36
8	30	240	64
10	37	370	100
$\sum x$	$\sum y$	$\sum xy$	$\sum x^2$
$= 30$	$= 122$	$= 854$	$= 220$

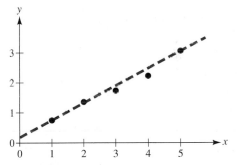

b. $y = 3.05x + 6.10$

c. When the polls close at 8:00 P.M., $x = 12$ and so $y = 3.05(12) + 6.1 = 42.7$, which means that approximately 42.7% of the registered voters can be expected to vote.

23. a. Approximately 12.5% per decade

b. 308.4 million

25. a. $V = 53.90e^{0.041t}$; 4.09%

b. $122,380

c. Approximately 42 years

d. Frank's formula is $V = 54.52e^{0.044t}$. It is easier to find but does not take into account the trend in account value.

27. a.

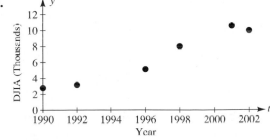

b. $y = 693.75t + 2131.1$

c. Predicted: 11,150; actual: 8,608

29. a. $y = 538.9t + 6{,}784.8$
   b. 11,634 billion yuan

31. a.

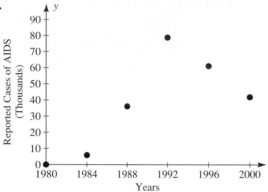

   b. $y = 2{,}985.3t + 7{,}691$
   c. 82,324 cases
   d. No. The line has positive slope, but the number of cases appears to be decreasing.

33. a.

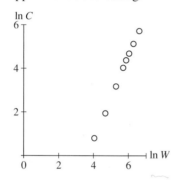

   b. $y = 1.631x - 4.975$
   c. $C = 0.0069W^{1.631}$

CHAPTER 7 | Section 5

1. $f\left(\dfrac{1}{2}, \dfrac{1}{2}\right) = \dfrac{1}{4}$

3. $f(1, 1) = f(-1, -1) = 2$

5. $f(0, 2) = f(0, -2) = -4$

7. $f\left(\dfrac{\sqrt{3}}{2}, -\dfrac{1}{2}\right) = f\left(-\dfrac{\sqrt{3}}{2}, -\dfrac{1}{2}\right) = \dfrac{3}{2}$ (max);
   $f(0, 1) = -3$ (min)

9. $f(8, 7) = -18$

11. $f(\sqrt{2}, \sqrt{2}) = f(-\sqrt{2}, -\sqrt{2}) = e^2$ (max);
   $f(\sqrt{2}, -\sqrt{2}) = f(-\sqrt{2}, \sqrt{2}) = e^{-2}$ (min)

13. $f\left(8, 4, \dfrac{8}{3}\right) = \dfrac{256}{3}$ (max)

15. $f\left(\dfrac{4}{\sqrt{14}}, \dfrac{8}{\sqrt{14}}, \dfrac{12}{\sqrt{14}}\right) = \dfrac{56}{\sqrt{14}}$ (max);
   $f\left(-\dfrac{4}{\sqrt{14}}, -\dfrac{8}{\sqrt{14}}, -\dfrac{12}{\sqrt{14}}\right) = \dfrac{-56}{\sqrt{14}}$ (min)

17. 40 meters by 80 meters

19. 11,664 cubic inches, when $x = 18$, $y = 36$

21. $r = 1.51$ inches; $h = 3.02$ inches

23. Development, $x = \$2{,}000$; promotion, $y = \$6{,}000$

25. Output will be increased by 31.75.

27. $H = 2R$

29. $s_{\max} = 4L$

31. $x = 8.93$ cm, $y = 10.04$ cm

33. $x = y = z = \sqrt[3]{V_0}$

35. Front length 11.5 ft; side length 15.4 ft; height 7.2 ft

37. $\lambda = 306.12$, which gives the approximate change per $1,000. Since the difference is only $100, the maximum profit is increased by approximately $0.1(\$306.12) = \$30.61$.

39. a. $x = 35$ units, $y = 42$ units
   b. $\lambda = 14.33$ is the approximate change in the maximum utility resulting from a one-unit increase in the budget.

41. Increases by $\lambda = \left(\dfrac{\alpha}{a}\right)^{\alpha}\left(\dfrac{\beta}{b}\right)^{\beta}$

43. Let $Q(x, y) =$ production; $C(x, y) = px + qy = k$.
   $C_x = p$, $C_y = q$. Therefore, $Q_x = \lambda p$; $Q_y = \lambda q$;
   $\dfrac{Q_x}{p} = \dfrac{Q_y}{q}$.

45. $Q(40, 14) \approx 1{,}398$

47. For $Q(K, L) = A[\alpha K^{-\beta} + (1 - \alpha) L^{-\beta}]^{-1/\beta}$,
   $$\begin{aligned}
   Q(sK, sL) &= A[\alpha(sK)^{-\beta} + (1 - \alpha)(sL)^{-\beta}]^{-1/\beta} \\
   &= A[\alpha s^{-\beta} K^{-\beta} + (1 - \alpha) s^{-\beta} L^{-\beta}]^{-1/\beta} \\
   &= A[s^{-\beta}\{\alpha K^{-\beta} + (1 - \alpha) L^{-\beta}\}]^{-1/\beta} \\
   &= A[s^{-\beta}]^{-1/\beta}[\alpha K^{-\beta} + (1 - \alpha) L^{-\beta}]^{-1/\beta} \\
   &= sA[\alpha K^{-\beta} + (1 - \alpha) L^{-\beta}]^{-1/\beta} \\
   &= sQ(K, L)
   \end{aligned}$$

49. $x = 0$, $y = -2$. The critical point $(0, -2)$ is an inflection point, not a relative extremum.

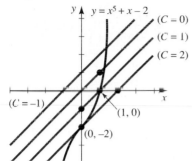

51. $\dfrac{\partial P}{\partial K} = \lambda \dfrac{\partial C}{\partial K}; \dfrac{\partial P}{\partial L} = \lambda \dfrac{\partial C}{\partial L}; \lambda = \dfrac{\dfrac{\partial P}{\partial K}}{\dfrac{\partial C}{\partial K}} = \dfrac{\dfrac{\partial P}{\partial L}}{\dfrac{\partial C}{\partial L}}$

53. $\dfrac{dy}{dx} = \dfrac{\dfrac{y}{x^2} - (1 + xy^2)e^{xy^2} - \ln(x + y) - \dfrac{x}{x + y}}{2x^2 y e^{xy^2} + \dfrac{1}{x} + \dfrac{x}{x + y}}$

55. $f(2.1623, 1.5811) = 1.6723$

57. $f(0.9729, -0.1635) = 2.9522$

CHAPTER 7 | Section 6

1. $\dfrac{7}{6}$

3. -1

5. $4 \ln 2 = \ln 16$

7. 0

9. $\ln 3$

11. 32

13. $\dfrac{1}{3}$

15. $\dfrac{32}{9}$

17. $\dfrac{e^2 - 1}{8}$

19. Vertical cross sections: $0 \le x \le 3$
$x^2 \le y \le 3x$
Horizontal cross sections: $0 \le y \le 9$
$\dfrac{y}{3} \le x \le \sqrt{y}$

21. Vertical cross sections: $-1 \le x \le 2$
$1 \le y \le 2$
Horizontal cross sections: $1 \le y \le 2$
$-1 \le x \le 2$

23. Vertical cross sections: $1 \le x \le e$
$0 \le y \le \ln x$
Horizontal cross sections: $0 \le y \le 1$
$e^y \le x \le e$

25. $\dfrac{3}{2}$

27. $\dfrac{1}{2}$

29. $\dfrac{44}{15}$

31. 1

33. $\dfrac{3}{2} \ln 5$

35. $2(e - 2)$

37.

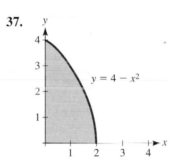

$$\int_{y=0}^{y=4} \int_{x=0}^{x=\sqrt{4-y}} f(x, y)\, dx\, dy$$

39.

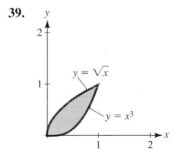

$$\int_{y=0}^{y=1} \int_{x=y^2}^{x=y^{1/3}} f(x, y)\, dx\, dy$$

ANSWERS

41.

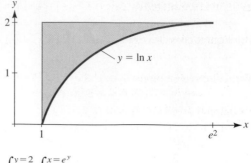

$$\int_{y=0}^{y=2} \int_{x=1}^{x=e^y} f(x, y)\, dx\, dy$$

43.

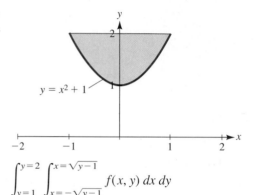

$$\int_{y=1}^{y=2} \int_{x=-\sqrt{y-1}}^{x=\sqrt{y-1}} f(x, y)\, dx\, dy$$

45. $\displaystyle\int_{x=-4}^{x=2} \int_{y=0}^{y=x+4} 1\, dy\, dx = 18$

47. $\displaystyle\int_{x=0}^{x=4} \int_{y=\frac{1}{2}x^2}^{y=2x} 1\, dy\, dx = \frac{16}{3}$

49. $\displaystyle\int_{x=1}^{x=3} \int_{y=x^2-4x+3}^{0} 1\, dy\, dx = \frac{4}{3}$

51. $\displaystyle\int_{x=1}^{x=e} \int_{y=0}^{y=\ln x} 1\, dy\, dx = 1$

53. $\displaystyle\int_{y=0}^{y=3} \int_{x=y/3}^{x=\sqrt{4-y}} 1\, dx\, dy = \frac{19}{6}$

55. $\displaystyle\int_{y=0}^{y=2} \int_{x=0}^{x=1} (6 - 2x - 2y)\, dx\, dy = 6$

57. $\displaystyle\int_{x=1}^{x=2} \int_{y=1}^{y=3} \frac{1}{xy}\, dy\, dx = (\ln 3)(\ln 2)$

59. $\displaystyle\int_{x=0}^{x=1} \int_{y=0}^{y=2} xe^{-y}\, dy\, dx = \frac{1}{2}\left(1 - \frac{1}{e^2}\right)$

61. $\displaystyle\int_{y=0}^{y=1} \int_{x=y}^{x=2-y} (2x + y)\, dx\, dy = \frac{7}{3}$

63. $\displaystyle\int_{x=-2}^{x=2} \int_{y=x^2}^{y=8-x^2} (x + 1)\, dy\, dx = \frac{64}{3}$

65. Area $= \displaystyle\int_{x=-2}^{x=3} \int_{y=-1}^{y=2} 1\, dy\, dx = 15$

Average $= \dfrac{1}{15} \displaystyle\int_{x=-2}^{x=3} \int_{y=-1}^{y=2} xy(x - 2y)\, dy\, dx = \frac{1}{6}$

67. Area $= \displaystyle\int_{x=0}^{x=1} \int_{y=0}^{y=2} 1\, dy\, dx = 2$

Average $= \dfrac{1}{2} \displaystyle\int_{x=0}^{x=1} \int_{y=0}^{y=2} xye^{x^2y}\, dy\, dx = \frac{e^2 - 3}{4}$

69. Area $= \displaystyle\int_{x=0}^{x=3} \int_{y=x/3}^{y=1} 1\, dy\, dx = \frac{3}{2}$

Average $= \dfrac{2}{3} \displaystyle\int_{x=0}^{x=3} \int_{y=x/3}^{y=1} 6xy\, dy\, dx = \frac{9}{2}$

71. Area $= \displaystyle\int_{x=-2}^{x=2} \int_{y=0}^{y=4-x^2} 1\, dy\, dx = \frac{32}{3}$

Average $= \dfrac{3}{32} \displaystyle\int_{x=-2}^{x=2} \int_{y=0}^{y=4-x^2} x\, dy\, dx = 0$

73. $\displaystyle\int_{x=1}^{x=3} \int_{y=2}^{y=5} \frac{\ln(xy)}{y}\, dy\, dx$

$= (3\ln 3 - 2)\ln\frac{5}{2} + (\ln 5)^2 - (\ln 2)^2$

75. $\displaystyle\int_{x=0}^{x=1} \int_{y=0}^{y=1} x^3 e^{x^2y}\, dy\, dx = \frac{e - 2}{2}$

77. Area $= \displaystyle\int_{x=0}^{x=5} \int_{y=0}^{y=7} 1\, dy\, dx = 35$

Average $= \dfrac{1}{35} \displaystyle\int_{x=0}^{x=5} \int_{y=0}^{y=7} (2x^3 + 3x^2y + y^3)\, dy\, dx$

$= \dfrac{943}{4}$

79. Average $= \dfrac{1}{25 \cdot 19} \displaystyle\int_{x=100}^{x=125} \int_{y=70}^{y=89} [(x - 30)(70 + 5x$

$- 4y) + (y - 40)(80 - 6x + 7y)]\, dy\, dx$

$= 24{,}896.5 \ (\$2{,}489{,}650)$

81. Average $= \dfrac{1}{12} \displaystyle\int_{x=0}^{x=4} \int_{y=0}^{y=3} 90(2x + y^2)\, dy\, dx$

$= 630$ feet

83. Value $= \displaystyle\int_{x=-1}^{x=1} \int_{y=-1}^{y=1} (300 + x + y)\, e^{-0.01x} dy\, dx$

$= 79{,}800\, e^{0.01} - 80{,}200\, e^{-0.01} \approx 1{,}200$

85. **a.** 0.991 square meters

b. No, it can only be considered the average surface area from birth until the time at which the person reached adulthood.

87. $\dfrac{17{,}408}{105} \approx 166 \text{ m}^3$

89. $\dfrac{304}{27} \approx 11.26$

91. $\dfrac{7e^{-6}}{9} + \dfrac{17}{9} \approx 1.891$ cubic units

CHAPTER 7 | Checkup

1. **a.** Domain: all ordered pairs (x, y) of real numbers
$$f_x = 3x^2 + 2y^2$$
$$f_y = 4xy - 12y^3$$
$$f_{xx} = 6x$$
$$f_{yx} = 4y$$

b. Domain: all ordered pairs (x, y) of real numbers for which $x \neq y$
$$f_x = \frac{-3y}{(x - y)^2}$$
$$f_y = \frac{3x}{(x - y)^2}$$
$$f_{xx} = \frac{6y}{(x - y)^3}$$
$$f_{yx} = \frac{-3(x + y)}{(x - y)^3}$$

c. Domain: all ordered pairs (x, y) of real numbers for which $y^2 > 2x$
$$f_x = 2e^{2x-y} - \frac{2}{y^2 - 2x}$$
$$f_y = -e^{2x-y} + \frac{2y}{y^2 - 2x}$$
$$f_{xx} = 4e^{2x-y} - \frac{4}{(y^2 - 2x)^2}$$
$$f_{yx} = -2e^{2x-y} + \frac{4y}{(y^2 - 2x)^2}$$

2. **a.** Circles centered at the origin and the single point $(0, 0)$

b. Parabolas with vertices on the x axis and opening to the left

3. **a.** Relative maximum: $(0, 0)$; relative minimum: $(1, 4)$; saddle points: $(1, 0)$, $(0, 4)$

b. Saddle point: $(-1, 0)$

c. Relative minimum: $(-1, -1)$

4. **a.** $\dfrac{16}{5}$ at $\left(\dfrac{4}{5}, \dfrac{8}{5}\right)$

b. Maximum value of 4 at $(1, 2)$ or $(1, -2)$; minimum value of -4 at $(-1, 2)$ or $(-1, -2)$

5. **a.** 16

b. $\dfrac{1}{4}(e^2 + 3e^{-2})$

c. $2 \ln 2 - \dfrac{3}{4}$

d. $1 - \dfrac{1}{e^2}$

6. $Q_K = 180;\ Q_L = 3.75$

7. 12 DVDs and 2 video games

8. 30 units of drug A and 25 units of drug B, which results in an equivalent dosage of $E(30, 25) = 83.75$ units. Since the total number of units is 55, which is less than 60, there is no risk of side effects, and since $E(30, 25) > 70$, the combination is effective.

9. $\dfrac{5}{2}(1 + e^{-2}) \approx 2.84\ ^{\circ}\text{C}$

10. **a.**

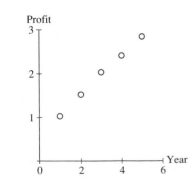

b. $y = 0.45x + 0.61$

c. 3.31 million dollars

ANSWERS

CHAPTER 7 | Review Problems

1. $f_x - 6x^2y + 3y^2 - \dfrac{y}{x^2}; f_y = 2x^3 + 6xy + \dfrac{1}{x}$

3. $f_x = y(xy + 1)e^{xy}; f_y = x(xy + 1)e^{xy}$

5. $f_x - \dfrac{3y}{x(x + 3y)} = \dfrac{1}{x} - \dfrac{1}{x + 3y};$

   $f_y = \dfrac{x}{y(x + 3y)} = \dfrac{1}{y} - \dfrac{3}{x + 3y}$

7. Daily output will increase by approximately 16 units.

9. a.

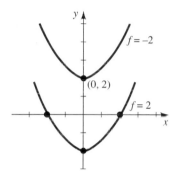

   b.

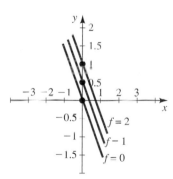

11. The level of unskilled labor should be decreased by approximately two workers.

13. Relative minimum at $\left(-\dfrac{23}{2}, 5\right)$; saddle point at $\left(\dfrac{1}{2}, 1\right)$

15. Saddle points at $\left(\dfrac{2}{3}, -\dfrac{5}{6}\right)$ and $\left(-\dfrac{2}{3}, \dfrac{5}{6}\right)$

17. Maximize area $A = xy$ subject to fixed perimeter $P = 2x + 2y$. Lagrange conditions are $y = \lambda(x)$, $x = \lambda(y)$, and $2x + 2y - C$. We must have $\lambda > 0$ since x and y are positive, so $x = y$ and the optimum rectangle is actually a square.

19. Development $x = \$4,000$; promotion $y = \$7,000$

21. We have $f_x - y - \dfrac{12}{x^2}$ and $f_y = x - \dfrac{18}{y^2}$,

    so $f_x = f_y = 0$ when $x = 2$ and $y = 3$. Since $f(x, y)$ is large when either x or y is large or small, a relative minimum is indicated at $(2, 3)$. To verify this claim, note that

    $$D = \left(\dfrac{24}{x^3}\right)\left(\dfrac{24}{y^3}\right) \quad 1 \quad \text{and} \quad f_{xx} = \dfrac{24}{x^3}$$

    so that $D(2, 3) > 0$ and $f_{xx}(2, 3) > 0$.

23. 0.5466

25. $\dfrac{1}{4}(e^2 - e^{-2})$

27. $2e - 2$

29. $\dfrac{3}{2}(e - 1)$

31. 2

33. $\dfrac{3}{2}(e^{-2} - e^{-3})$ cubic units

35. $x = y = z = \dfrac{20}{3}$

37. $\sqrt{10}$; at $(0, \pm\sqrt{10}, 0)$

39. a.

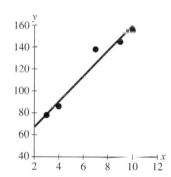

   b. $y = 11.54x + 44.45$

   c. Approximately $\$102,150$

41. -0.74; demand is decreasing at the rate of about $\dfrac{3}{4}$ quart per month.

43. -3; demand is decreasing at the rate of 3 pies per week.

45. The amount of pollution is decreasing by about 113 units per day.

47. About 7.056 units

49. $Q(x, y) = x^a y^b$

$Q_x = ax^{a-1}y^b$; $Q_y = bx^a y^{b-1}$

$xQ_x + yQ_y = x(ax^{a-1}y^b) + y(bx^a y^{b-1})$

$\qquad\qquad = (a + b)x^a y^b = (a + b)Q$

If $a + b = 1$, then $xQ_x + yQ_y = Q$.

APPENDIX | Algebra Review, Section A1

1. $1 < x \le 5$

3. $x > -5$

5.

7.

9. 4

11. 5

13. $-3 \le x \le 3$

15. $-6 \le x \le -2$

17. $x \le -7$ or $x \ge 3$

19. 125

21. 4

23. 4

25. $\dfrac{1}{2}$

27. $\dfrac{1}{2}$

29. $\dfrac{1}{4}$

31. 2

33. $\dfrac{1}{4}$

35. $n = 10$

37. $n = 1$

39. $n = 4$

41. $n = \dfrac{13}{5}$

43. $x^4(x - 4)$

45. $-25(x - 7) = 25(7 - x)$

47. $2(x + 1)^2(x - 2)^2(7x - 2)$

49. $x^{-1/2}(6x + 1)$

51. $2(x + 3)$

53. $\dfrac{2(x + 3)^3(5 - x)}{(1 - x)^3}$

55. 34

57. 0

59. $\displaystyle\sum_{j=1}^{6} \dfrac{1}{j}$

61. $\displaystyle\sum_{j=1}^{6} 2x_j$

63. $\displaystyle\sum_{j=1}^{8} (-1)^{j+1}j$

65. a. Surface area is approximately 5.212×10^8 km²; mass of the atmosphere is 5.212×10^{18} kg.

   b. 127,400 years

APPENDIX | Algebra Review, Section A2

1. $(x + 2)(x - 1)$

3. $(x - 3)(x - 4)$

5. $(x - 1)^2$

7. $(4x + 5)(4x - 5)$

9. $(x - 1)(x^2 + x + 1)$

11. $x^5(x + 1)(x - 1)$

13. $2x(x - 5)(x + 1)$

15. $x = 4$; $x = -2$

17. $x = -5$

19. $x = 4$; $x = -4$

21. $x = -\dfrac{1}{2}$; $x = -1$

23. $x = -\dfrac{3}{2}$

25. $x = 1$; $x = -5$

27. $x = 1$; $x = -2$

29. $x = -1$

31. $x = -\dfrac{1}{2}$; $x = -1$

33. No real solutions

35. $x = -\dfrac{3}{2}$

37. $x = 3; y = 2$

39. $x = 4, y = 2$

41. $x = -7; y = -5$ and $x = 1, y = -1$

APPENDIX | Algebra Review, Section A3

1. 0

3. $\dfrac{1}{2}$

5. $\dfrac{1}{3}$

7. 0

9. 0

11. 0

13. 0

15. e^2

APPENDIX | Algebra Review, Review Problems

1. $-2 \le x < 3$

3.

5.

7. 3

9. $2 \le x \le 4$

11. 243

13. 4

15. $16\sqrt[4]{2}$

17. 73

19. $\dfrac{3}{2}$

21. $n = \dfrac{7}{18}$

23. $n = -1$

25. 21

27. 95

29. $1 + \displaystyle\sum_{k=2}^{7} \dfrac{(-1)^k}{k}$

31. $x^2(x + 3)(x - 3)$

33. $x^4(x^6 + 4)(x^3 + 2)(x^3 - 2)$

35. $x(x - 1)^2$

37. $(x + 5)(x - 3)$

39. $(2x + 3)^2$

41. $(x + 1)(x - 1)(x + 3)$

43. $x = -4; x = 1$

45. $x = -7$

47. $x = -1; x = 2$

49. $x = -\dfrac{3}{7}; x = \dfrac{1}{2}$

51. No real solutions

53. $x = -2; x = \dfrac{1}{3}$

55. $x = -2; y = 1$

57. $x = 1, y = 2$ and $x = 15, y = -26$

59. $\dfrac{2}{7}$

61. 0

63. 1

INDEX